Student Solutions Manual

Beginning and Intermediate Algebra
An Integrated Approach

FIFTH EDITION

R. David Gustafson
Rock Valley College

Peter D. Frisk
Rock Valley College

Prepared by

Michael G. Welden
Rock Valley College

BROOKS/COLE
CENGAGE Learning™

Australia • Brazil • Japan • Korea • Mexico • Singapore • Spain • United Kingdom • United States

BROOKS/COLE
CENGAGE Learning™

For product information and technology assistance, contact us at
Cengage Learning Customer & Sales Support, 1-800-354-9706

For permission to use material from this text or product, submit all requests online at
www.cengage.com/permissions
Further permissions questions can be emailed to
permissionrequest@cengage.com

ISBN-13: 978-0-495-11815-2
ISBN-10: 0-495-11815-X

Brooks/Cole
10 Davis Drive
Belmont, CA 94002-3098
USA

Cengage Learning is a leading provider of customized learning solutions with office locations around the globe, including Singapore, the United Kingdom, Australia, Mexico, Brazil, and Japan. Locate your local office at: **international.cengage.com/region**

Cengage Learning products are represented in Canada by Nelson Education, Ltd.

For your course and learning solutions, visit **academic.cengage.com**

Purchase any of our products at your local college store or at our preferred online store **www.ichapters.com**

Printed in Canada
2 3 4 5 6 7 11 10 09 08

Preface

This manual contains detailed solutions to all of the odd exercises of the text *Beginning and Intermediate Algebra*, fifth edition, by R. David Gustafson and Peter D. Frisk. It also contains solutions to all chapter summary, chapter test, and cumulative review exercises found in the text.

Many of the exercises in the text may be solved using more than one method, but it is not feasible to list all possible solutions in this manual. Also, some of the exercises may have been solved in this manual using a method that differs slightly from that presented in the text. There are a few exercises in the text whose solutions may vary from person to person. Some of these solutions may not have been included in this manual. For the solution to an exercise like this, the notation "answers may vary" has been included.

Please remember that only reading a solution does not teach you how to solve a problem. To repeat a commonly used phrase, mathematics is not a spectator sport. You MUST make an honest attempt to solve each exercise in the text without using this manual first. This manual should be viewed more or less as a last resort. Above all, DO NOT simply copy the solution from this manual onto your own paper. Doing so will not help you learn how to do the exercise, nor will it help you to do better on quizzes or tests.

I would like to thank the members of the mathematics faculty at Rock Valley College and Laura Localio of Brooks/Cole Publishing Company for their help and support. This solutions manual was prepared using EXP 5.1.

This book is dedicated to my nieces Hannah and Mairin Welden.

May your study of this material be successful and rewarding.

Michael G. Welden

Contents

Exercise 1.1 (page 11)

1. set **3.** whole **5.** subset **7.** prime

9. is not equal to **11.** is greater than or equal to **13.** number **15.** natural: $1, 2, 6, 9$

17. positive integers: $1, 2, 6, 9$ **19.** integers: $-3, -1, 0, 1, 2, 6, 9$

21. real: $-3, -\frac{1}{2}, -1, 0, 1, 2, \frac{5}{3}, \sqrt{7}, 3.25, 6, 9$ **23.** odd integers: $-3, -1, 1, 9$

25. composite: $6, 9$

27. $4 + 5 = 9$
9: natural, odd, composite, whole

29. $15 - 15 = 0$
0: even, whole

31. $3 \cdot 8 = 24$
24: natural, even, composite, whole

33. $24 \div 8 = 3$
3: natural, odd, prime, whole

35. $3 \boxed{<} 5$

37. $5 \boxed{} 3 + 2$
$5 \boxed{=} 5$

39. $25 \boxed{<} 32$

41. $5 + 7 \boxed{} 10$
$12 \boxed{>} 10$

43. $3 + 9 \boxed{} 20 - 8$
$12 \boxed{=} 12$

45. $4 \cdot 2 \boxed{} 2 \cdot 4$
$8 \boxed{=} 8$

47. $8 \div 2 \boxed{} 4 + 2$
$4 \boxed{<} 6$

49. $45 \div 9 \boxed{} 36 \div 12$
$5 \boxed{>} 3$

51. $3 + 2 + 5 \boxed{} 5 + 2 + 3$
$10 \boxed{=} 10$

53. $7 > 3$ **55.** $8 \le 8$ **57.** $3 + 4 = 7$ **59.** $\sqrt{2} \approx 1.41$

61. $3 \le 7 \Rightarrow \boxed{7 \ge 3}$ **63.** $6 > 0 \Rightarrow \boxed{0 < 6}$

65. $3 + 8 > 8 \Rightarrow \boxed{8 < 3 + 8}$ **67.** $6 - 2 < 10 - 4 \Rightarrow \boxed{10 - 4 > 6 - 2}$

69. $2 \cdot 3 < 3 \cdot 4 \Rightarrow \boxed{3 \cdot 4 > 2 \cdot 3}$ **71.** $\frac{12}{4} < \frac{24}{6} \Rightarrow \boxed{\frac{24}{6} > \frac{12}{4}}$

73.

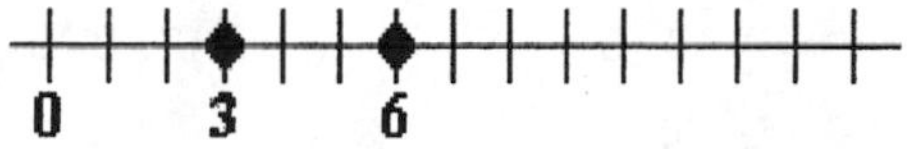

6 is greater than 3. 6 is to the right of 3.

75.

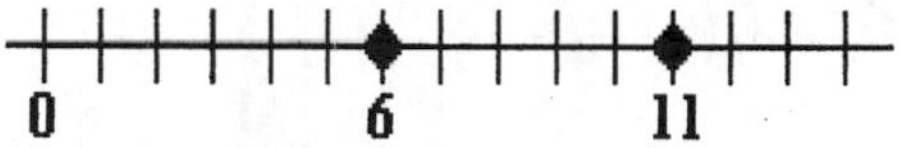

11 is greater than 6. 11 is to the right of 6.

77. 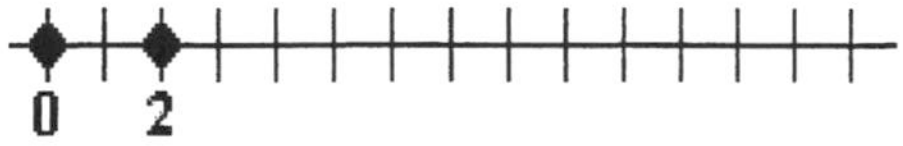

2 is greater than 0. 2 is to the right of 0.

79. 

8 is greater than 0. 8 is to the right of 0.

81. 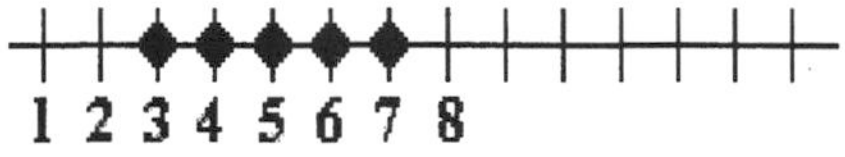

83. 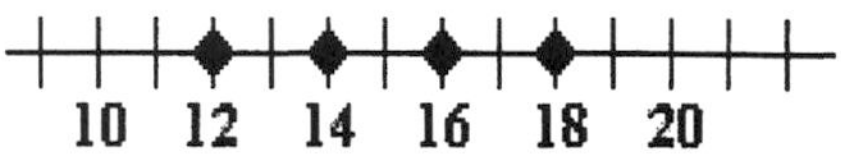

85. 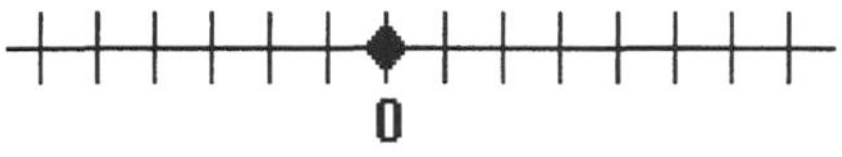

87. 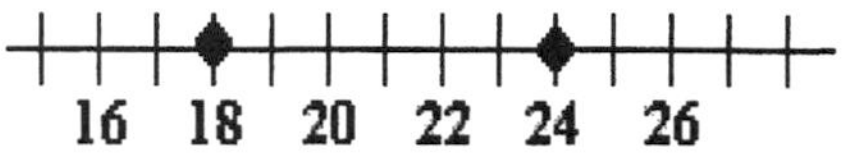

89. 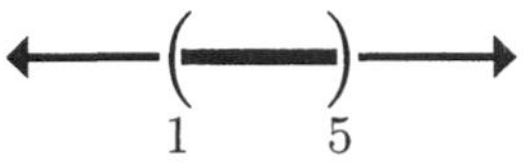

91. 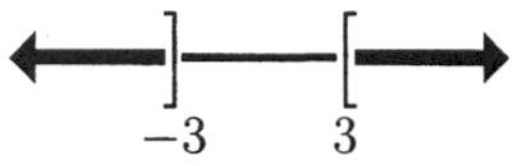

93. 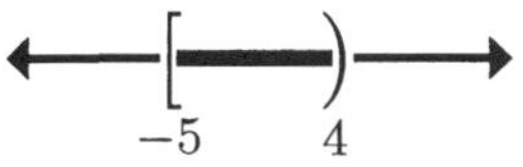

95. $|36| = 36$

97. $|0| = 0$

99. $|-230| = 230$

101. $|12 - 4| = |8| = 8$

103. $|21 - 19| = |2| = 2$

105-109. Answers may vary.

Exercise 1.2 (page 25)

1. true

3. false; 21 has factors of 3 and 7.

5. true; 8 is to the right of -2.

7. true; $|-9| = 9$, so $9 \leq |-9|$.

9. $3 + 7 \boxed{=} 10$

11. $|-2| = 2$, so $|-2| \boxed{=} 2$

13. numerator

15. simplify

17. proper

19. 1

21. multiply

23. numerators, denominator

25. plus (or $+$)

27. repeating

29. $30 = 6 \cdot 5 = 2 \cdot 3 \cdot 5$

31. $70 = 7 \cdot 10 = 7 \cdot 2 \cdot 5$

33. $\frac{6}{12} = \frac{1 \cdot \cancel{6}^{1}}{2 \cdot \cancel{6}_{1}} = \frac{1}{2}$

35. $\frac{15}{20} = \frac{3 \cdot \cancel{5}^{1}}{4 \cdot \cancel{5}_{1}} = \frac{3}{4}$

37. $\frac{24}{18} = \frac{4 \cdot \cancel{6}^{1}}{3 \cdot \cancel{6}_{1}} = \frac{4}{3}$

39. $\frac{72}{64} = \frac{9 \cdot \cancel{8}^{1}}{8 \cdot \cancel{8}_{1}} = \frac{9}{8}$

41. $\frac{1}{2}\cdot\frac{3}{5}=\frac{1\cdot 3}{2\cdot 5}=\frac{3}{10}$

43. $\frac{4}{3}\cdot\frac{6}{5}=\frac{4\cdot 6}{3\cdot 5}=\frac{4\cdot 2\cdot \overset{1}{\cancel{3}}}{\underset{1}{\cancel{3}}\cdot 5}=\frac{8}{5}$

45. $\frac{5}{12}\cdot\frac{18}{5}=\frac{5\cdot 18}{12\cdot 5}=\frac{\overset{1}{\cancel{5}}\cdot 3\cdot \overset{1}{\cancel{6}}}{2\cdot \underset{1}{\cancel{6}}\cdot \underset{1}{\cancel{5}}}=\frac{3}{2}$

47. $\frac{17}{34}\cdot\frac{3}{6}=\frac{17\cdot 3}{34\cdot 6}=\frac{\overset{1}{\cancel{17}}\cdot \overset{1}{\cancel{3}}}{2\cdot \underset{1}{\cancel{17}}\cdot 2\cdot \underset{1}{\cancel{3}}}=\frac{1}{4}$

49. $12\cdot\frac{5}{6}=\frac{12}{1}\cdot\frac{5}{6}=\frac{12\cdot 5}{1\cdot 6}=\frac{2\cdot \overset{1}{\cancel{6}}\cdot 5}{1\cdot \underset{1}{\cancel{6}}}=\frac{10}{1}=10$

51. $\frac{10}{21}\cdot 14=\frac{10}{21}\cdot\frac{14}{1}=\frac{10\cdot 14}{21\cdot 1}=\frac{10\cdot 2\cdot \overset{1}{\cancel{7}}}{3\cdot \underset{1}{\cancel{7}}}=\frac{20}{3}$

53. $\frac{3}{5}\div\frac{2}{3}=\frac{3}{5}\cdot\frac{3}{2}=\frac{3\cdot 3}{5\cdot 2}=\frac{9}{10}$

55. $\frac{3}{4}\div\frac{6}{5}=\frac{3}{4}\cdot\frac{5}{6}=\frac{3\cdot 5}{4\cdot 6}=\frac{\overset{1}{\cancel{3}}\cdot 5}{4\cdot 2\cdot \underset{1}{\cancel{3}}}=\frac{5}{8}$

57. $\frac{2}{13}\div\frac{8}{13}=\frac{2}{13}\cdot\frac{13}{8}=\frac{2\cdot 13}{13\cdot 8}=\frac{\overset{1}{\cancel{2}}\cdot \overset{1}{\cancel{13}}}{\underset{1}{\cancel{13}}\cdot 4\cdot \underset{1}{\cancel{2}}}=\frac{1}{4}$

59. $\frac{21}{35}\div\frac{3}{14}=\frac{21}{35}\cdot\frac{14}{3}=\frac{21\cdot 14}{35\cdot 3}=\frac{\overset{1}{\cancel{7}}\cdot \overset{1}{\cancel{3}}\cdot 14}{5\cdot \underset{1}{\cancel{7}}\cdot \underset{1}{\cancel{3}}}=\frac{14}{5}$

61. $6\div\frac{3}{14}=\frac{6}{1}\div\frac{3}{14}=\frac{6}{1}\cdot\frac{14}{3}=\frac{6\cdot 14}{1\cdot 3}=\frac{2\cdot \overset{1}{\cancel{3}}\cdot 14}{1\cdot \underset{1}{\cancel{3}}}=\frac{28}{1}=28$

63. $\frac{42}{30}\div 7=\frac{42}{30}\div\frac{7}{1}=\frac{42}{30}\cdot\frac{1}{7}=\frac{42\cdot 1}{30\cdot 7}=\frac{\overset{1}{\cancel{6}}\cdot \overset{1}{\cancel{7}}\cdot 1}{5\cdot \underset{1}{\cancel{6}}\cdot \underset{1}{\cancel{7}}}=\frac{1}{5}$

65. $\frac{3}{5}+\frac{3}{5}=\frac{3+3}{5}=\frac{6}{5}$

67. $\frac{4}{13}-\frac{3}{13}=\frac{4-3}{13}=\frac{1}{13}$

69. $\frac{1}{6}+\frac{1}{24}=\frac{1\cdot \mathbf{4}}{6\cdot \mathbf{4}}+\frac{1}{24}=\frac{4}{24}+\frac{1}{24}=\frac{4+1}{24}=\frac{5}{24}$

71. $\frac{3}{5}+\frac{2}{3}=\frac{3\cdot \mathbf{3}}{5\cdot \mathbf{3}}+\frac{2\cdot \mathbf{5}}{3\cdot \mathbf{5}}=\frac{9}{15}+\frac{10}{15}=\frac{9+10}{15}=\frac{19}{15}$

73. $\frac{9}{4}-\frac{5}{6}=\frac{9\cdot \mathbf{3}}{4\cdot \mathbf{3}}-\frac{5\cdot \mathbf{2}}{6\cdot \mathbf{2}}=\frac{27}{12}-\frac{10}{12}=\frac{27-10}{12}=\frac{17}{12}$

75. $\frac{7}{10}-\frac{1}{14}=\frac{7\cdot \mathbf{7}}{10\cdot \mathbf{7}}-\frac{1\cdot \mathbf{5}}{14\cdot \mathbf{5}}=\frac{49}{70}-\frac{5}{70}=\frac{49-5}{70}=\frac{44}{70}=\frac{22\cdot \overset{1}{\cancel{2}}}{35\cdot \underset{1}{\cancel{2}}}=\frac{22}{35}$

77. $3-\frac{3}{4}=\frac{3}{1}-\frac{3}{4}=\frac{3\cdot\mathbf{4}}{1\cdot\mathbf{4}}-\frac{3}{4}=\frac{12}{4}-\frac{3}{4}=\frac{12-3}{4}=\frac{9}{4}$

79. $\frac{17}{3}+4=\frac{17}{3}+\frac{4}{1}=\frac{17}{3}+\frac{4\cdot\mathbf{3}}{1\cdot\mathbf{3}}=\frac{17}{3}+\frac{12}{3}=\frac{17+12}{3}=\frac{29}{3}$

81. $4\frac{3}{5}+\frac{3}{5}=\left(4+\frac{3}{5}\right)+\frac{3}{5}=\left(\frac{20}{5}+\frac{3}{5}\right)+\frac{3}{5}=\frac{23}{5}+\frac{3}{5}=\frac{26}{5}=5\frac{1}{5}$

83. $3\frac{1}{3}-1\frac{2}{3}=\left(3+\frac{1}{3}\right)-\left(1+\frac{2}{3}\right)=\left(\frac{9}{3}+\frac{1}{3}\right)-\left(\frac{3}{3}+\frac{2}{3}\right)=\frac{10}{3}-\frac{5}{3}=\frac{5}{3}=1\frac{2}{3}$

85. $3\frac{3}{4}-2\frac{1}{2}=\left(3+\frac{3}{4}\right)-\left(2+\frac{1}{2}\right)=\left(\frac{12}{4}+\frac{3}{4}\right)-\left(\frac{8}{4}+\frac{2}{4}\right)=\frac{15}{4}-\frac{10}{4}=\frac{5}{4}=1\frac{1}{4}$

87. $8\frac{2}{9}-7\frac{2}{3}=\left(8+\frac{2}{9}\right)-\left(7+\frac{2}{3}\right)=\left(\frac{72}{9}+\frac{2}{9}\right)-\left(\frac{63}{9}+\frac{6}{9}\right)=\frac{74}{9}-\frac{69}{9}=\frac{5}{9}$

89.

$$\begin{array}{r} 0.\ 8 \\ 5\overline{\big)\,4.\ 0} \\ \underline{4\ \ 0} \\ 0 \end{array}$$

$\frac{4}{5}=0.8$, terminating

91.

$$\begin{array}{rl} & 0.\ 4\ 0\ 9\ 0\ 9 \\ 22 & \overline{\big)\,9.\ 0\ 0\ 0\ 0\ 0} \\ & \underline{8\ 8} \\ & 2\ 0 \\ & \underline{0} \\ & 2\ 0\ 0 \\ & \underline{1\ 9\ 8} \\ & 2\ 0 \\ & \underline{0} \\ & 2\ 0\ 0 \\ & \underline{1\ 9\ 8} \\ & 2 \end{array}$$

$\frac{9}{22}=0.4\overline{09}$; repeating

93.

$$\begin{array}{r} 23.45 \\ +\ 135.2 \\ \hline 158.65 \end{array}$$

95.

$$\begin{array}{r} \ 6\ \ \ \ ^{11}_{\not 1}\ \ 13 \\ 6\ \not 7\,.\ \not 2\ \not 3\ 5 \\ -\ 2\ 2\,.\ 4\ 5 \\ \hline 4\ 4\,.\ 7\ 8\ 5 \end{array}$$

97.

$$\begin{array}{r} 3.\ 4 \\ \times\quad 1\ 3.\ 2 \\ \hline 6\ 8 \\ 1\ 0\ 2 \\ 3\ 4 \\ \hline 4\ 4\ 8\ 8 \end{array}$$

Put two digits to the right of the decimal point. Answer: 44.88

99. $0.23 \overline{\smash{)}1.0465}$

Move decimal points 2 places right.

$$\begin{array}{r} 4.\ 5\ 5 \\ 23.\ \overline{\smash{)}\,1\ 0\ 4.\ 6\ 5} \\ 9\ 2 \\ \hline 1\ 2\ 6 \\ 1\ 1\ 5 \\ \hline 1\ 1\ 5 \\ 1\ 1\ 5 \\ \hline 0 \end{array}$$

Problems 101-107 are to be solved using a calculator. The keystrokes needed to solve each problem using a TI-83 graphing calculator appear in each solution. There may be other solutions. Keystrokes for other calculators may be slightly different.

101. [3] [2] [3] [.] [2] [4] [+] [2] [7] [.] [2] [5] [4] [3] [ENTER] $\{350.4943\} \Rightarrow 350.49$

103. [5] [5] [.] [7] [7] [4] [4] [3] [−] [.] [5] [6] [8] [2] [4] [5] [ENTER]
$\{55.206185\} \Rightarrow 55.21$

105. [2] [5] [.] [2] [5] [×] [1] [3] [2] [.] [1] [7] [9] [ENTER] $\{3337.51975\} \Rightarrow 3{,}337.52$

107. [4] [.] [5] [6] [9] [4] [3] [2] [3] [÷] [.] [4] [5] [6] [ENTER] $\{10.02068487\} \Rightarrow 10.02$

109. $43\frac{1}{2} - 12\frac{1}{3} = 43 + \frac{1}{2} - 12 - \frac{1}{3} = \frac{258}{6} + \frac{3}{6} - \frac{72}{6} - \frac{2}{6} = \frac{187}{6} = 31\frac{1}{6}$ acres

111. $14 \cdot 3\frac{1}{4} = \frac{14}{1} \cdot \frac{13}{4} = \frac{14 \cdot 13}{1 \cdot 4} = \frac{7 \cdot \overset{1}{\cancel{2}} \cdot 13}{1 \cdot 2 \cdot \underset{1}{\cancel{2}}} = \frac{91}{2} = 45\frac{1}{2}$ yd

113. $187.75 - 46.8 - 72.5 = \$68.45$ million

115. 24% of 42,500 $= 0.24(42{,}500) = \$10{,}200$

117. $0.23(17{,}500) = 4{,}025$ defective $\Rightarrow 17{,}500 - 4025 = 13{,}475$ acceptable units

119. $0.12(18{,}700{,}000) = 2{,}244{,}000$ increase $\Rightarrow$ sales $= 18{,}700{,}000 + 2{,}244{,}000 = \$20{,}944{,}000$

121. # gallons $= 15{,}675.2 \div 25.5 = 614.7137255 \Rightarrow$ cost $= 614.7137255(2.87) \approx \$1{,}764.23$

123. Area = length · width $= (253.5 \text{ ft})(178.5 \text{ ft}) = 45{,}249.75 \text{ ft}^2$
Drums of sealer $= 45{,}249.75 \div 4{,}000 \approx 11.3 \Rightarrow$ needs 12 drums; Cost $= 12(97.50) = \$1{,}170$

125. Standard $= 37.50(2{,}530) = \$94{,}875$; High-capacity $= 57.35(1{,}670) = \$95{,}774.50$
The high-capacity order will produce the greater profit.

127. Silage per cow $= 0.57(12{,}000) = 6840$ pounds; $30(6{,}840) = 205{,}200$ lb of silage

129. Regular $= 1{,}730 + 36(107.75) = 1{,}730 + 3{,}879 = 5{,}609$
High $= 4{,}170 + 36(57.50) = 4{,}170 + 2{,}070 = 6{,}240$
The high-efficiency furnace will be more expensive after 3 years.

131-135. Answers may vary.

137. No. Each proper fraction is less than 1. When a number is multiplied by a number less than 1, the result is smaller than the original number.

Exercise 1.3 (page 35)

1.

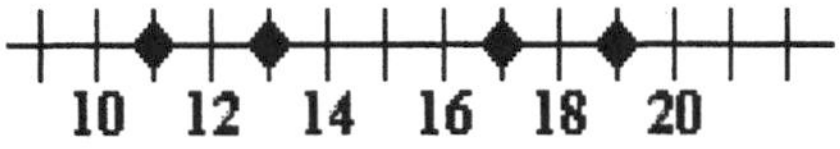

3. 17 is a prime number.

5. exponent

7. multiplication

9. $P = 4s$

11. $P = 2l + 2w$

13. $P = a + b + c$

15. $P = a + b + c + d$

17. $C = \pi D$, or $C = 2\pi r$

19. $V = lwh$

21. $V = \frac{1}{3}Bh$

23. $V = \frac{4}{3}\pi r^3$

25. $4^2 = 4 \cdot 4 = 16$

27. $6^2 = 6 \cdot 6 = 36$

29. $\left(\frac{1}{10}\right)^4 = \frac{1}{10} \cdot \frac{1}{10} \cdot \frac{1}{10} \cdot \frac{1}{10} = \frac{1}{10{,}000}$

Problems 31-33 are to be solved using a calculator. The keystrokes needed to solve each problem using a TI-83 graphing calculator appear in each solution. There may be other solutions. Keystrokes for other calculators may be slightly different.

31. [7] [.] [9] [^] [3] [ENTER]
{493.039}

33. [2] [5] [.] [3] [^] [2] [ENTER]
{640.09}

35. $x^2 = x \cdot x$

37. $3z^4 = 3 \cdot z \cdot z \cdot z \cdot z$

39. $(5t)^2 = 5t \cdot 5t$

41. $5(2x)^3 = 5 \cdot 2x \cdot 2x \cdot 2x$

43. $4(3)^2 = 4 \cdot 9 = 36$

45. $(5 \cdot 2)^3 = 10^3 = 1{,}000$

47. $2(3^2) = 2 \cdot 9 = 18$

49. $(3 \cdot 2)^3 = 6^3 = 216$

51. $3 \cdot 5 - 4 = 15 - 4 = 11$

53. $3(5 - 4) = 3(1) = 3$

55. $3 + 5^2 = 3 + 25 = 28$

57. $(3 + 5)^2 = (8)^2 = 64$

59. $2 + 3 \cdot 5 - 4 = 2 + 15 - 4 = 17 - 4 = 13$

61. $64 \div (3 + 1) = 64 \div 4 = 16$

63. $(7 + 9) \div (2 \cdot 4) = 16 \div 8 = 2$

65. $(5 + 7) \div 3 \cdot 4 = 12 \div 3 \cdot 4 = 4 \cdot 4 = 16$

67. $24 \div 4 \cdot 3 + 3 = 6 \cdot 3 + 3 = 18 + 3 = 21$

69. $3^2 + 2(1 + 4) - 2 = 9 + 2(5) - 2 = 9 + 10 - 2 = 19 - 2 = 17$

71. $5^2 - (7-3)^2 = 5^2 - 4^2 = 25 - 16 = 9$

73. $(2 \cdot 3 - 4)^3 = (6-4)^3 = 2^3 = 8$

75. $\frac{3}{5} \cdot \frac{10}{3} + \frac{1}{2} \cdot 12 = \frac{3}{5} \cdot \frac{10}{3} + \frac{1}{2} \cdot \frac{12}{1} = \frac{\overset{1}{\cancel{3}}}{\underset{1}{\cancel{5}}} \cdot \frac{2 \cdot \overset{1}{\cancel{5}}}{\underset{1}{\cancel{3}}} + \frac{1}{\underset{1}{\cancel{2}}} \cdot \frac{6 \cdot \overset{1}{\cancel{2}}}{1} = \frac{2}{1} + \frac{6}{1} = 2 + 6 = 8$

77. $\left[\frac{1}{3} - \left(\frac{1}{2}\right)^2\right]^2 = \left[\frac{1}{3} - \frac{1}{4}\right]^2 = \left[\frac{1 \cdot \mathbf{4}}{3 \cdot \mathbf{4}} - \frac{1 \cdot \mathbf{3}}{4 \cdot \mathbf{3}}\right]^2 = \left[\frac{4}{12} - \frac{3}{12}\right]^2 = \left[\frac{1}{12}\right]^2 = \frac{1}{144}$

Problems 79-85 are to be solved using a calculator. The keystrokes needed to solve each problem using a TI-83 graphing calculator appear in each solution. There may be other solutions. Keystrokes for other calculators may be slightly different.

79. numerator: [( ] [3] [+] [5] [ )] [x^2] [+] [2] [ENTER] {66} fraction $= \frac{66}{6} = 11$

denominator: [2] [×] [( ] [8] [−] [5] [ )] [ENTER] {6}

81. numerator: [( ] [5] [−] [3] [ )] [x^2] [+] [2] [ENTER] {6} fraction $= \frac{6}{6} = 1$

denominator: [4] [x^2] [−] [( ] [8] [+] [2] [ )] [ENTER] {6}

83. numerator: [2] [×] [( ] [4] [+] [2] [×] [( ] [3] [−] [1] [ )] [ )] [ENTER] {16}

denominator: [3] [×] [( ] [3] [×] [( ] [2] [×] [3] [−] [4] [ )] [ )] [ENTER] {18}

fraction $= \frac{16}{18} = \frac{8}{9}$

85. numerator: [3] [×] [7] [−] [5] [×] [( ] [3] [×] [4] [−] [1] [1] [ )] [ENTER] {16}

denominator: [4] [×] [( ] [3] [+] [2] [ )] [−] [3] [x^2] [+] [5] [ENTER] {16}

fraction $= \frac{16}{16} = 1$

87. $39 = (3 \cdot 8) + (5 \cdot 3)$

89. $87 = (3 \cdot 8 + 5) \cdot 3$

91. $P = 4s = 4(4 \text{ in.}) = 16 \text{ in.}$

93. $P = a + b + c = 3 \text{ m} + 5 \text{ m} + 7 \text{ m} = 15 \text{ m}$

95. $A = s^2 = (5 \text{ m})^2 = 25 \text{ m}^2$

97. $A = bh = (6 \text{ ft})(10 \text{ ft}) = 60 \text{ ft}^2$

99. $C = \pi D \approx \frac{22}{7}(28 \text{ m}) = \frac{22}{7} \cdot \frac{28}{1} \text{ m} = \frac{22}{\underset{1}{\cancel{7}}} \cdot \frac{4 \cdot \overset{1}{\cancel{7}}}{1} \text{ m} = 88 \text{ m}$

101. $A = \pi r^2 \approx \frac{22}{7}(21 \text{ ft})^2 = \frac{22}{7}(441 \text{ ft}^2) = \frac{22}{7} \cdot \frac{441}{1} \text{ ft}^2 = \frac{22}{\underset{1}{\cancel{7}}} \cdot \frac{63 \cdot \overset{1}{\cancel{7}}}{1} \text{ ft}^2 = 1{,}386 \text{ ft}^2$

103. $V = \frac{1}{3}Bh = \frac{1}{3}(3\text{ cm})^2(2\text{ cm}) = \frac{1}{3}(9\text{ cm}^2)(2\text{ cm}) = (3\text{ cm}^2)(2\text{ cm}) = 6\text{ cm}^3$

105. $V = \frac{4}{3}\pi r^3 \approx \frac{4}{3}\cdot\frac{22}{7}(6\text{ m})^3 = \frac{88}{21}(216\text{ m}^3) = \frac{88}{21}\cdot\frac{216}{1}\text{ m}^3 \approx 905\text{ m}^3$

107. Cylinder: $V = Bh = \pi(4\text{ cm})^2(14\text{ cm}) \approx \frac{22}{7}\cdot\frac{16}{1}\text{ cm}^2\cdot\frac{14}{1}\text{ cm} = \frac{22\cdot 16\cdot 2\cdot \overset{1}{\cancel{7}}}{\underset{1}{\cancel{7}}\cdot 1\cdot 1}\text{ cm}^3 = 704\text{ cm}^3$

Cone: $V = \frac{1}{3}Bh = \frac{1}{3}\pi(4\text{ cm})^2(21\text{ cm}) \approx \frac{1}{3}\cdot\frac{22}{7}\cdot\frac{16}{1}\text{ cm}^2\cdot\frac{21}{1}\text{ cm} = \frac{22\cdot 16\cdot \overset{1}{\cancel{3}}\cdot \overset{1}{\cancel{7}}}{\underset{1}{\cancel{3}}\cdot \underset{1}{\cancel{7}}\cdot 1\cdot 1}\text{ cm}^3 = 352\text{ cm}^3$

Total $= 704\text{ cm}^3 + 352\text{ cm}^3 = 1{,}056\text{ cm}^3$

109. $P = 4s = 4\left(30\frac{2}{5}\text{ m}\right) = 4\left(30 + \frac{2}{5}\text{ m}\right) = 4\left(\frac{150}{5} + \frac{2}{5}\text{ m}\right) = \frac{4}{1}\left(\frac{152}{5}\text{ m}\right)$

$= \frac{608}{5}\text{ m} = 121\frac{3}{5}\text{ m}$

111. $V = \frac{4}{3}\pi r^3 = \frac{4}{3}\pi(21.35\text{ ft})^3 \approx 40{,}764.51\text{ ft}^3$

113. $V = lwh = (40\text{ ft})(40\text{ ft})(9\text{ ft}) = 14{,}400\text{ ft}^3$; Per student $= 14{,}400\text{ ft}^3 \div 30 = 480\text{ ft}^3$ per student

115. $f = \frac{rs}{(r+s)(n-1)} = \frac{(8)(12)}{(8+12)(1.6-1)} = \frac{96}{(20)(0.6)} = \frac{96}{12} = 8$

117. Answers may vary.

119. Increasing powers produce larger numbers.

Exercise 1.4 (page 45)

1. $5 + 3(7-2) = 5 + 3(5) = 5 + 15 = 20$

3. $5 + 3(7) - 2 = 5 + 21 - 2 = 26 - 2 = 24$

5. arrows

7. subtract, greater

9. add, opposite

11. $4 + 8 = +(4+8) = +12$

13. $(-3) + (-7) = -(3+7) = -10$

15. $6 + (-4) = +(6-4) = +2$

17. $9 + (-11) = -(11-9) = -2$

19. $(-0.4) + 0.9 = +(0.9 - 0.4) = +0.5$

21. $\frac{1}{5} + \left(+\frac{1}{7}\right) = \frac{7}{35} + \left(+\frac{5}{35}\right) = +\left(\frac{7}{35} + \frac{5}{35}\right) = +\frac{12}{35}$

23. $-\frac{1}{2} + \frac{1}{3} = -\frac{3}{6} + \frac{2}{6} = -\left(\frac{3}{6} - \frac{2}{6}\right) = -\frac{1}{6}$

25. $5 + (-4) = +(5-4) = +1$

27. $-1.3 + 3.5 = +(3.5 - 1.3) = +2.2$

29. $5 + [4 + (-2)] = 5 + [2] = 7$

31. $-2+(-4+5)=-2+1=-1$

33. $(-3+5)+2=2+2=4$

35. $-15+(-4+12)=-15+8=-7$

37. $[-4+(-3)]+[2+(-2)]=[-7]+[0]=-7$

39. $-4+(-3+2)+(-3)=-4+(-1)+(-3)=-5+(-3)=-8$

41. $-|8+(-4)|+7=-|4|+7=-4+7=3$

43. $-5.2+|-2.5+(-4)|=-5.2+|-6.5|$
$=-5.2+6.5=1.3$

45. $8-4=8+(-4)=4$

47. $8-(-4)=8+(+4)=12$

49. $0-(-5)=0+(+5)=5$

51. $\frac{5}{3}-\frac{7}{6}=\frac{10}{6}-\frac{7}{6}=\frac{10}{6}+\left(-\frac{7}{6}\right)$
$=\frac{3}{6}=\frac{1}{2}$

53. $-3\frac{1}{2}-5\frac{1}{4}=-\frac{7}{2}-\frac{21}{4}=-\frac{14}{4}-\frac{21}{4}=-\frac{14}{4}+\left(-\frac{21}{4}\right)=-\frac{35}{4}=-8\frac{3}{4}$

55. $-6.7-(-2.5)=-6.7+(+2.5)=-4.2$

57. $8-4=8+(-4)=4$

59. $-10-(-3)=-10+(+3)=-7$

61. $+3-[(-4)-3]=+3-[(-4)+(-3)]=+3-[-7]=+3+[+7]=10$

63. $(5-3)+(3-5)=[5+(-3)]+[3+(-5)]=[2]+[-2]=0$

65. $5-[4+(-2)-5]=5-[4+(-2)+(-5)]=5-[-3]=5+[+3]=8$

67. $\frac{5-(-4)}{3-(-6)}=\frac{5+(+4)}{3+(+6)}=\frac{9}{9}=1$

69. $\frac{-4-2}{-[2+(-3)]}=\frac{-4+(-2)}{-[-1]}=\frac{-6}{+1}=-6$

71. $\left(\frac{5}{2}-3\right)-\left(\frac{3}{2}-5\right)=\left(\frac{5}{2}-\frac{6}{2}\right)-\left(\frac{3}{2}-\frac{10}{2}\right)=\left(-\frac{1}{2}\right)-\left(-\frac{7}{2}\right)=\left(-\frac{1}{2}\right)+\left(+\frac{7}{2}\right)$
$=\frac{6}{2}=3$

73. $\left(\frac{3}{4}-\frac{4}{5}\right)-\left(\frac{2}{3}+\frac{1}{4}\right)=\left(\frac{15}{20}-\frac{16}{20}\right)-\left(\frac{8}{12}+\frac{3}{12}\right)=\left(-\frac{1}{20}\right)-\left(\frac{11}{12}\right)=\left(-\frac{3}{60}\right)-\left(\frac{55}{60}\right)$
$=-\left(\frac{3}{60}+\frac{55}{60}\right)$
$=-\frac{58}{60}=-\frac{29}{30}$

75. $(3.7-8.25)-(3.75+2.5)=[3.7+(-8.25)]-(6.25)=(-4.55)-(6.25)=-(4.55+6.25)$
$=-10.8$

Problems 77-79 are to be solved using a calculator. The keystrokes needed to solve each problem using a TI-83 graphing calculator appear in each solution. There may be other solutions. Keystrokes for other calculators may be slightly different.

77. [2] [.] [3] [4] [^] [3] [−] [3] [.] [4] [7] [+] [.] [7] [2] [x^2] [ENTER]
$\{9.861304\} \Rightarrow 9.9$

79. [2] [.] [3] [4] [x^2] [−] [3] [.] [4] [7] [x^2] [−] [.] [7] [2] [x^2] [ENTER]
$\{-7.0837\} \Rightarrow -7.1$

81. $(-575) + (+400) = -175$
She still owes \$175.

83. $(+13) + (-4) = +9$

85. $(-14) + 10 = -4°$

87. $1700 - (-300) = 2000$ years

89. $(-2{,}300) + (1{,}750) + (1{,}875) = +1{,}325$ m

91. $32{,}000 - 28{,}000 = 4{,}000$ ft

93. $+32 - (+27) = 5°$

95. $12{,}153 - 23 + 57 = 12{,}187$

97. $500 \cdot 2 - 300 = 1000 - 300 = 700$ shares

99. $437.45 + 25.17 + 37.93 + 45.26 - 17.13 - 83.44 - 22.58 = \$422.66$

101. $115{,}000 - 78 - 446 - 216 - 7{,}612.32 - 23{,}445.11 + 223 = \$83{,}425.57$

103. Answers may vary.

105. The answers agree if the two numbers have the same sign. The answers do not agree if the numbers have opposite signs.

Exercise 1.5 (page 53)

1. $30 \cdot 37\frac{1}{2} = \frac{30}{1} \cdot \frac{75}{2} = \frac{15 \cdot \cancel{2}^{1} \cdot 75}{1 \cdot \cancel{2}_{1}} = 1{,}125$ lb

3. $5^3 - 8(3)^2 = 125 - 8(9) = 125 - 72 = 53$

5. positive

7. positive

9. positive

11. a

13. 0

15. $(+6)(+8) = 48$

17. $(-8)(-7) = 56$

19. $(+12)(-12) = -144$

21. $(-32)(-14) = 448$

23. $\left(\frac{1}{2}\right)(-32) = -\frac{1}{2} \cdot \frac{32}{1} = -\frac{32}{2} = -16$

25. $\left(-\frac{3}{4}\right)\left(-\frac{8}{3}\right) = +\frac{3}{4} \cdot \frac{8}{3} = \frac{24}{12} = 2$

27. $(-3)\left(-\frac{1}{3}\right) = +\frac{3}{1} \cdot \frac{1}{3} = \frac{3}{3} = 1$

29. $(3)(-4)(-6) = (-12)(-6) = 72$

SECTION 1.5

31. $(-2)(3)(4) = (-6)(4) = -24$

33. $(2)(-5)(-6)(-7) = (-10)(-6)(-7) = (+60)(-7) = -420$

35. $(-2)(-2)(-2)(-3)(-4) = (+4)(-2)(-3)(-4) = (-8)(-3)(-4) = (+24)(-4) = -96$

37. $(-2)^2 = (-2)(-2) = 4$

39. $(-4)^3 = (-4)(-4)(-4) = (+16)(-4) = -64$

41. $(-7)^2 = (-7)(-7) = 49$

43. $-(-3)^2 = -(-3)(-3) = -(+9) = -9$

45. $(-1)(2^3) = (-1)(8) = -8$

47. $2 + (-1)(-3) = 2 + 3 = 5$

49. $(-1+2)(-3) = 1(-3) = -3$

51. $[-1-(-3)][-1+(-3)] = [-1+3][-4] = [2][-4] = -8$

53. $-1(2) + (2)(-3) = -2 + (-6) = -8$

55. $(-1)(2)(-3) + 6 = (-2)(-3) + 6 = 6 + 6 = 12$

57. $(-1)^2[2-(-3)] = (-1)(-1)[2+(+3)] = 1[5] = 5$

59. $(-1)^3(-2)^2 + (-3)^2 = (-1)(-1)(-1)(-2)(-2) + (-3)(-3) = -4 + 9 = 5$

61. $\dfrac{80}{-20} = -4$

63. $\dfrac{-110}{-55} = 2$

65. $\dfrac{-160}{40} = -4$

67. $\dfrac{320}{-16} = -20$

69. $\dfrac{8-12}{-2} = \dfrac{-4}{-2} = 2$

71. $\dfrac{20-25}{7-12} = \dfrac{-5}{-5} = 1$

73. $\dfrac{3(4)}{-2} = \dfrac{12}{-2} = -6$

75. $\dfrac{5(-18)}{3} = \dfrac{-90}{3} = -30$

77. $\dfrac{4+(-18)}{-2} = \dfrac{-14}{-2} = 7$

79. $\dfrac{(-2)(5)(4)}{-3+1} = \dfrac{-40}{-2} = 20$

81. $\dfrac{1}{2} - \dfrac{2}{3} = \dfrac{3}{6} + \left(-\dfrac{4}{6}\right) = -\dfrac{1}{6}$

83. $\dfrac{1}{2} - \dfrac{2}{3} - \dfrac{3}{4} = \dfrac{6}{12} + \left(-\dfrac{8}{12}\right) + \left(-\dfrac{9}{12}\right) = -\dfrac{11}{12}$

85. $\left(\dfrac{1}{2} - \dfrac{2}{3}\right)\left(\dfrac{1}{2} + \dfrac{2}{3}\right) = \left(\dfrac{3}{6} - \dfrac{4}{6}\right)\left(\dfrac{3}{6} + \dfrac{4}{6}\right) = \left(-\dfrac{1}{6}\right)\left(\dfrac{7}{6}\right) = -\dfrac{7}{36}$

87. $\left(\dfrac{1}{3} - \dfrac{1}{2}\right)\left(\dfrac{2}{3} - \dfrac{1}{2}\right) = \left(\dfrac{2}{6} - \dfrac{3}{6}\right)\left(\dfrac{4}{6} - \dfrac{3}{6}\right) = \left(-\dfrac{1}{6}\right)\left(\dfrac{1}{6}\right) = -\dfrac{1}{36}$

Problems 89-91 are to be solved using a calculator. The keystrokes needed to solve each problem using a TI-83 graphing calculator appear in each solution. There may be other solutions. Keystrokes for other calculators may be slightly different.

89. [( ] [( ] [(–)] [6] [ )] [^] [3] [+] [4] [×] [( ] [(–)] [3] [ )] [^] [3] [ )] [÷] [( ] [4] [–] [6] [ )] [x^2] [ENTER] {−81}

91. (4 × ((−) 6) x^2 × (−) 3 + 4 x^2 × (−) 6) ÷ (2 × (−) 6 − 2 × (−) 3) ENTER {88}

93. $(+2)(+3) = +6$

95. $(-30)(+15) = -450$

97. $(+23)(-120) = -2{,}760$

99. $\frac{-18}{-3} = +6$

101. **a.** $75(-32) = -\$2400$ **b.** $57(-17) = -\$969$ **c.** $87(-12) = -\$1044$
d. $(-2400) + (-969) + (-1044) = -\$4413$

103. $\frac{(+26)+(+35)+(+17)+(-25)+(-31)+(-12)+(-24)}{7} = \frac{-14}{7} = -2$ per day

105. $613.50(18) = \$11{,}043 \Rightarrow$ enough \$

107. **Answers may vary.**

109. If the quotient is undefined, then the denominator must equal 0, and the product of the two numbers is 0.

111. If x^5 is negative, then x must be negative.

Exercise 1.6 (page 61)

1. $0.14 \cdot 3{,}800 = 532$

3. $\frac{-4+(7-9)}{(-9-7)+4} = \frac{-4+(-2)}{-16+4} = \frac{-6}{-12} = \frac{1}{2}$

5. sum

7. multiplication

9. algebraic

11. variables

13. $x + y$

15. $x(2y)$

17. $y - x$

19. $\frac{y}{x}$

21. $z + \frac{x}{y}$

23. $z - xy$

25. $3xy$

27. $\frac{x+y}{y+z}$

29. $xy + \frac{y}{z}$

31. the sum of x and 3

33. the quotient obtained when x is divided by y

35. the product of 2, x and y (or twice the product of x and y)

37. the quotient obtained when 5 is divided by the sum of x and y

39. the quotient obtained when the sum of 3 and x is divided by y

41. the product of x, y and the sum of x and y

43. $x + z = 8 + 2 = 10$

45. $y - z = 4 - 2 = 2$

47. $yz - 3 = (4)(2) - 3 = 5$

49. $\frac{xy}{z} = \frac{(8)(4)}{2} = 16$

51. $x + y = (-2) + 5 = 3$

53. $xyz = (-2)(5)(-3) = (-10)(-3) = 30$

55. $\frac{yz}{x} = \frac{(5)(-3)}{-2} = \frac{-15}{-2} = \frac{15}{2}$

57. $\frac{3(x+z)}{y} = \frac{3[(-2)+(-3)]}{5} = \frac{3[-5]}{5} = \frac{-15}{5} = -3$

59. $\frac{x(y+z)-25}{x^2+y^2} = \frac{-2[5+(-3)]-25}{(-2)^2+5^2} = \frac{-2[2]-25}{4+25} = \frac{-4-25}{29} = \frac{-29}{29} = -1$

61. $\frac{3(x+z^2)+4}{y(x-z)} = \frac{3\left[(-2)+(-3)^2\right]+4}{5[(-2)-(-3)]} = \frac{3(-2+9)+4}{5(1)} = \frac{3(7)+4}{5} = \frac{21+4}{5} = \frac{25}{5} = 5$

63. $6d$: 1 term; coef. $= 6$

65. $-xy - 4t + 35$: 3 terms; coef. $= -1$

67. $3ab + bc - cd - ef$: 4 terms; coef. $= 3$

69. $-4xyz + 7xy - z$: 3 terms; coef. $= -4$

71. $3x + 4y + 2z + 2$: 4 terms; coef. $= 3$

73. 3rd term: $19x$; factors: 19, x

75. 1st term: $29xyz$; factors: 29, x, y, z

77. 1st term: $3xyz$; factors: 3, x, y, z

79. 3rd term: $17xz$; factors: 17, x, z

81. coefficients: 5, 1 and 8

83. x and y are common to the 1st and 3rd terms.

85. coefficients: 3, 1 and 25; $3 \cdot 1 \cdot 25 = 75$

87. x and y are common to the 1st and 3rd terms.

89. $c + 4$

91. **a.** $h - 20$ **b.** $c + 20$

93. $35{,}000a$

95. $500 - x$

97. $3d + 5$

99. $\frac{N(N-1)}{2} = \frac{10{,}000(10{,}000-1)}{2} = \frac{10{,}000(9{,}999)}{2} = \frac{99{,}990{,}000}{2} = 49{,}995{,}000$ comparisons

101-103. Answers may vary.

105. $37x \Rightarrow 37(2x)$; $37(2x) = 2(37x)$
$37x$ is doubled.

Exercise 1.7 (page 69)

1. $x + y^2 \geq z$

3. $|x| \geq \boxed{0}$

5. positive

7. real

9. $a + b = b + \underline{a}$

11. $(a+b) + c = a + \underline{(b+c)}$

13. $a(b+c) = ab + \underline{ac}$

15. $a \cdot 1 = \underline{a}$

17. element, multiplication

19. $a, \dfrac{1}{a}$

21. $x + y = 12 + (-2) = 10$

23. $xy = 12(-2) = -24$

25. $x^2 = 12^2 = 144$

27. $\dfrac{x}{y^2} = \dfrac{12}{(-2)^2} = \dfrac{12}{4} = 3$

29. $x + y = 5 + 7 = 12$
$y + x = 7 + 5 = 12$

31. $3x + 2y = 3(5) + 2(7) = 15 + 14 = 29$
$2y + 3x = 2(7) + 3(5) = 14 + 15 = 29$

33. $x(x+y) = 5(5+7) = 5(12) = 60$
$(x+y)x = (5+7)5 = (12)5 = 60$

35. $(x+y)+z = [2+(-3)]+1 = -1+1 = \boxed{0}$; $x+(y+z) = 2+(-3+1) = 2+(-2) = \boxed{0}$

37. $(xz)y = [2(1)](-3) = [2](-3) = \boxed{-6}$; $x(yz) = 2[-3(1)] = 2[-3] = \boxed{-6}$

39. $x^2(yz^2) = 2^2[-3(1)^2] = 4[-3(1)] = 4[-3] = \boxed{-12}$
$(x^2y)z^2 = [2^2(-3)](1)^2 = [4(-3)](1) = [-12](1) = \boxed{-12}$

41. $4(x+2) = 4x + 8$

43. $2(z-3) = 2z - 6$

45. $3(x+y) = 3x + 3y$

47. $x(x+3) = x \cdot x + x \cdot 3 = x^2 + 3x$

49. $-x(a+b) = (-x)a + (-x)b = -ax - bx$

51. $-4(x^2+x) = (-4)x^2 + (-4)x$
$= -4x^2 - 4x$

53. $-5(t+2) = -5t + (-5)(2) = -5t - 10$

55. $-2a(x-a) = -2ax + (-2a)(-a)$
$= -2ax + 2a^2$

57. additive inverse: -2
multiplicative inverse: $\dfrac{1}{2}$

59. additive inverse: $-\dfrac{1}{3}$
multiplicative inverse: 3

61. additive inverse: 0
multiplicative inverse: none

63. additive inverse: $\dfrac{5}{2}$
multiplicative inverse: $-\dfrac{2}{5}$

65. additive inverse: 0.2
multiplicative inverse: -5

67. additive inverse: $-\dfrac{4}{3}$
multiplicative inverse: $\dfrac{3}{4}$

69. commutative property of addition

71. commutative property of multiplication

73. distributive property

75. commutative property of addition

77. identity for multiplication

79. additive inverse

81. $3(x+2) = 3x + 3(2) = 3x + 6$

83. $y^2x = xy^2$

85. $(x+y)z = (y+x)z$

87. $(xy)z = x(yz)$

89. $0 + x = x$

91. **Answers may vary.**

93. Closure for addition would not be true (odd number plus odd number equals even number).
Closure for multiplication would be true (odd number times odd number equals odd number).
There would be no additive identity (0 is an even number).
There would be a multiplicative identity, 1 (1 is an odd number).

Chapter 1 Summary (page 71)

1. natural: $1, 2, 3, 4, 5$

2. prime: $2, 3, 5$

3. odd, natural: $1, 3, 5$

4. composite: 4

5. integers: $-6, 0, 5$

6. rational: $-6, -\frac{2}{3}, 0, 2.6, 5$

7. prime: 5

8. real: $-6, -\frac{2}{3}, 0, \sqrt{2}, 2.6, \pi, 5$

9. even integers: $-6, 0$

10. odd integers: 5

11. not rational: $\sqrt{2}, \pi$

12. negative numbers: $-6, -\frac{2}{3}$

13. $-5 \; \boxed{\phantom{<}} \; 12 - 12$
$-5 \; \boxed{<} \; 0$

14. $\frac{24}{6} \; \boxed{\phantom{<}} \; 5$
$4 \; \boxed{<} \; 5$

15. $13 - 13 \; \boxed{} \; 5 - \frac{25}{5}$
$0 \; \boxed{} \; 5 - 5$
$0 \; \boxed{=} \; 0$

16. $\frac{21}{7} \; \boxed{} \; -33$
$3 \; \boxed{>} \; -33$

17. $-(-8) = +8$

18. $-(12 - 4) = -(8) = -8$

19.

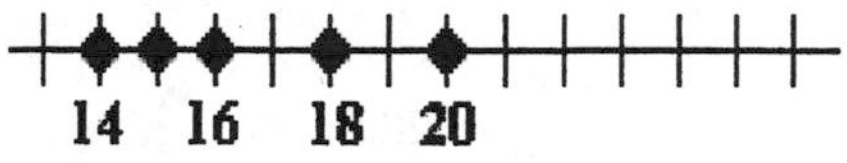

20.

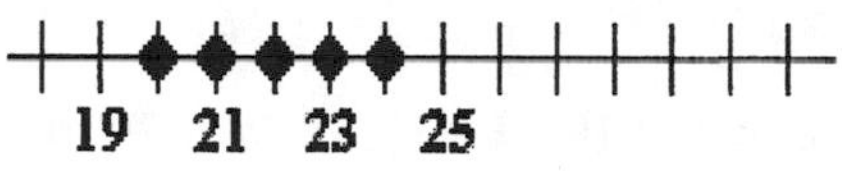

21.

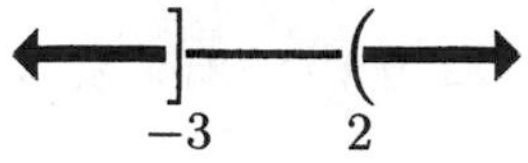

22.

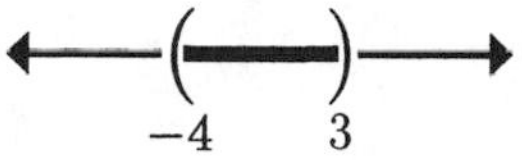

23. $|53 - 42| = |11| = 11$

24. $|-31| = 31$

25. $\frac{45}{27}=\frac{5\cdot\overset{1}{\cancel{9}}}{3\cdot\underset{1}{\cancel{9}}}=\frac{5}{3}$

26. $\frac{121}{11}=\frac{11\cdot\overset{1}{\cancel{11}}}{1\cdot\underset{1}{\cancel{11}}}=\frac{11}{1}=11$

27. $\frac{31}{15}\cdot\frac{10}{62}=\frac{\overset{1}{\cancel{31}}\cdot\overset{1}{\cancel{2}}\cdot\overset{1}{\cancel{5}}}{3\cdot\underset{1}{\cancel{5}}\cdot\underset{1}{\cancel{2}}\cdot\underset{1}{\cancel{31}}}=\frac{1}{3}$

28. $\frac{25}{36}\cdot\frac{12}{15}\cdot\frac{3}{5}=\frac{\overset{1}{\cancel{5}}\cdot\overset{1}{\cancel{5}}\cdot\overset{1}{\cancel{12}}\cdot\overset{1}{\cancel{3}}}{\underset{1}{\cancel{3}}\cdot\underset{1}{\cancel{12}}\cdot 3\cdot\underset{1}{\cancel{5}}\cdot\underset{1}{\cancel{5}}}=\frac{1}{3}$

29. $\frac{18}{21}\div\frac{6}{7}=\frac{18}{21}\cdot\frac{7}{6}=\frac{\overset{1}{\cancel{3}}\cdot\overset{1}{\cancel{6}}\cdot\overset{1}{\cancel{7}}}{\underset{1}{\cancel{3}}\cdot\underset{1}{\cancel{7}}\cdot\underset{1}{\cancel{6}}}=\frac{1}{1}=1$

30. $\frac{14}{24}\div\frac{7}{12}\div\frac{2}{5}=\frac{14}{24}\cdot\frac{12}{7}\cdot\frac{5}{2}$

$=\frac{\overset{1}{\cancel{2}}\cdot\overset{1}{\cancel{7}}\cdot\overset{1}{\cancel{12}}\cdot 5}{\underset{1}{\cancel{2}}\cdot\underset{1}{\cancel{12}}\cdot\underset{1}{\cancel{7}}\cdot 2}=\frac{5}{2}$

31. $\frac{7}{12}+\frac{9}{12}=\frac{7+9}{12}=\frac{16}{12}=\frac{4\cdot\overset{1}{\cancel{4}}}{3\cdot\underset{1}{\cancel{4}}}=\frac{4}{3}$

32. $\frac{13}{24}-\frac{5}{24}=\frac{13-5}{24}=\frac{8}{24}=\frac{\overset{1}{\cancel{8}}}{3\cdot\underset{1}{\cancel{8}}}=\frac{1}{3}$

33. $\frac{1}{3}+\frac{1}{7}=\frac{1\cdot\mathbf{7}}{3\cdot\mathbf{7}}+\frac{1\cdot\mathbf{3}}{7\cdot\mathbf{3}}=\frac{7}{21}+\frac{3}{21}$

$=\frac{7+3}{21}=\frac{10}{21}$

34. $\frac{5}{7}+\frac{4}{9}=\frac{5\cdot\mathbf{9}}{7\cdot\mathbf{9}}+\frac{4\cdot\mathbf{7}}{9\cdot\mathbf{7}}=\frac{45}{63}+\frac{28}{63}$

$=\frac{45+28}{63}=\frac{73}{63}$

35. $\frac{2}{3}-\frac{1}{7}=\frac{2\cdot\mathbf{7}}{3\cdot\mathbf{7}}-\frac{1\cdot\mathbf{3}}{7\cdot\mathbf{3}}=\frac{14}{21}-\frac{3}{21}$

$=\frac{14-3}{21}=\frac{11}{21}$

36. $\frac{4}{5}-\frac{2}{3}=\frac{4\cdot\mathbf{3}}{5\cdot\mathbf{3}}-\frac{2\cdot\mathbf{5}}{3\cdot\mathbf{5}}=\frac{12}{15}-\frac{10}{15}$

$=\frac{12-10}{15}=\frac{2}{15}$

37. $3\frac{2}{3}+5\frac{1}{4}=\frac{11}{3}+\frac{21}{4}=\frac{11\cdot\mathbf{4}}{3\cdot\mathbf{4}}+\frac{21\cdot\mathbf{3}}{4\cdot\mathbf{3}}=\frac{44}{12}+\frac{63}{12}=\frac{44+63}{12}=\frac{107}{12}=8\frac{11}{12}$

38. $7\frac{5}{12}-4\frac{1}{2}=\frac{89}{12}-\frac{9}{2}=\frac{89}{12}-\frac{9\cdot\mathbf{6}}{2\cdot\mathbf{6}}=\frac{89}{12}-\frac{54}{12}=\frac{89-54}{12}=\frac{35}{12}=2\frac{11}{12}$

39. $32.71+15.9=48.61$

40. $27.92-14.93=12.99$

41. $5.3\cdot 3.5=18.55$

42. $21.83\div 5.9=3.7$

43. $2.7(4.92-3.18)=2.7(1.74)\approx 4.70$

44. $\frac{3.3+2.5}{0.22}=\frac{5.8}{0.22}\approx 26.36$

45. $\frac{12.5}{14.7-11.2}=\frac{12.5}{3.5}\approx 3.57$

46. $(3-0.7)(3.63-2)=(2.3)(1.63)\approx 3.75$

47. $17\frac{1}{2}+15\frac{3}{4}=17+\frac{2}{4}+15+\frac{3}{4}=32+\frac{5}{4}=32+1\frac{1}{4}=33\frac{1}{4}$

$100-33\frac{1}{4}=100-33-\frac{1}{4}=67-\frac{1}{4}=66\frac{3}{4}$ acres left

48. avg. $= \dfrac{5.2+4.7+9.5+8}{4} = \dfrac{27.4}{4} = 6.85$ hours

49. $0.15(380) = 57$

50. Front/Back: $2(2.7+2.7+4.2) = 2(9.6) = 19.2$ ft
Top/Bottom: $2(1.2+1.2+4.2) = 2(6.6) = 13.2$ ft
Sides: $2(1.2+2.7) = 2(3.9) = 7.8$ ft
TOTAL $= 19.2 + 13.2 + 7.8 = 40.2$ ft

51. $3^4 = 3\cdot 3\cdot 3\cdot 3 = 81$

52. $\left(\dfrac{2}{3}\right)^2 = \dfrac{2}{3}\cdot\dfrac{2}{3} = \dfrac{4}{9}$

53. $(0.5)^2 = (0.5)(0.5) = 0.25$

54. $5^2 + 2^3 = 5\cdot 5 + 2\cdot 2\cdot 2 = 25 + 8 = 33$

55. $3^2 + 4^2 = 9 + 16 = 25$

56. $(3+4)^2 = 7^2 = 49$

57. $A = \dfrac{1}{2}bh = \dfrac{1}{2}\left(6\dfrac{1}{2}\text{ ft}\right)(7\text{ ft}) = \dfrac{1}{2}\cdot\dfrac{13}{2}\cdot\dfrac{7}{1}\text{ ft}^2 = \dfrac{91}{4}\text{ ft}^2 = 22\dfrac{3}{4}\text{ ft}^2$

58. $V = Bh = \pi r^2 h = \pi\left(\dfrac{32.1}{2}\text{ ft}\right)^2(18.7\text{ ft}) = \pi(257.6025\text{ ft}^2)(18.7\text{ ft}) \approx 15{,}133.6\text{ ft}^3$

59. $5 + 3^3 = 5 + 27 = 32$

60. $7\cdot 2 - 7 = 14 - 7 = 7$

61. $4 + (8\div 4) = 4 + 2 = 6$

62. $(4+8)\div 4 = 12\div 4 = 3$

63. $5^3 - \dfrac{81}{3} = 125 - 27 = 98$

64. $(5-2)^2 + 5^2 + 2^2 = 3^2 + 5^2 + 2^2 = 9 + 25 + 4 = 38$

65. $\dfrac{4\cdot 3 + 3^4}{31} = \dfrac{12+81}{31} = \dfrac{93}{31} = 3$

66. $\dfrac{4}{3}\cdot\dfrac{9}{2} + \dfrac{1}{2}\cdot 18 = \dfrac{2\cdot \cancel{2}^{1}\cdot 3\cdot \cancel{3}^{1}}{\cancel{3}_{1}\cdot \cancel{2}_{1}} + \dfrac{1\cdot 9\cdot \cancel{2}^{1}}{\cancel{2}_{1}\cdot 1} = \dfrac{6}{1} + \dfrac{9}{1} = 15$

67. $8^2 - 6 = 64 - 6 = 58$

68. $(8-6)^2 = 2^2 = 4$

69. $\dfrac{6+8}{6-4} = \dfrac{14}{2} = 7$

70. $\dfrac{6(8)-12}{4+8} = \dfrac{48-12}{12} = \dfrac{36}{12} = 3$

71. $2^2 + 2(3)^2 = 4 + 2(9) = 4 + 18 = 22$

72. $\dfrac{2^2+3}{2^3-1} = \dfrac{4+3}{8-1} = \dfrac{7}{7} = 1$

73. $(+7) + (+8) = +(7+8) = 15$

74. $(-25) + (-32) = -(25+32) = -57$

75. $(-2.7) + (-3.8) = -(2.7+3.8) = -6.5$

76. $\dfrac{1}{3} + \dfrac{1}{6} = \dfrac{2}{6} + \dfrac{1}{6} = \dfrac{3}{6} = \dfrac{1}{2}$

77. $(+12)+(-24)=-(24-12)=-12$

78. $(-44)+(+60)=+(60-44)=16$

79. $3.7+(-2.5)=+(3.7-2.5)=1.2$

80. $-5.6+(+2.06)=-(5.6-2.06)$
$=-3.54$

81. $15-(-4)=15+(+4)=19$

82. $-12-(-13)=-12+(+13)=1$

83. $[-5+(-5)]-(-5)=[-10]+(+5)=-5$

84. $1-[5-(-3)]=1-[5+(+3)]$
$=1-[8]=-7$

85. $\frac{5}{6}-\left(-\frac{2}{3}\right)=\frac{5}{6}+\frac{2}{3}=\frac{5}{6}+\frac{4}{6}=\frac{9}{6}=\frac{3}{2}$

86. $\frac{2}{3}-\left(\frac{1}{3}-\frac{2}{3}\right)=\frac{2}{3}-\left(-\frac{1}{3}\right)$
$=\frac{2}{3}+\frac{1}{3}=\frac{3}{3}=1$

87. $\left|\frac{3}{7}-\left(-\frac{4}{7}\right)\right|=\left|\frac{3}{7}+\frac{4}{7}\right|=\left|\frac{7}{7}\right|=|1|=1$

88. $\frac{3}{7}-\left|-\frac{4}{7}\right|=\frac{3}{7}-\left(+\frac{4}{7}\right)$
$=\frac{3}{7}+\left(-\frac{4}{7}\right)=-\frac{1}{7}$

89. $(+3)(+4)=+12$

90. $(-5)(-12)=+60$

91. $\left(-\frac{3}{14}\right)\left(-\frac{7}{6}\right)=+\frac{3}{14}\cdot\frac{7}{6}$
$=\frac{\overset{1}{\cancel{3}}\cdot\overset{1}{\cancel{7}}}{2\cdot\underset{1}{\cancel{7}}\cdot 2\cdot\underset{1}{\cancel{3}}}=+\frac{1}{4}$

92. $(3.75)(0.37)=+1.3875$

93. $5(-7)=-35$

94. $(-15)(7)=-105$

95. $\left(-\frac{1}{2}\right)\left(\frac{4}{3}\right)=-\frac{1}{2}\cdot\frac{4}{3}=-\frac{1\cdot 2\cdot\overset{1}{\cancel{2}}}{\underset{1}{\cancel{2}}\cdot 3}=-\frac{2}{3}$

96. $(-12.2)(3.7)=-45.14$

97. $\frac{+25}{+5}=+5$

98. $\frac{-14}{-2}=+7$

99. $\frac{(-2)(-7)}{4}=\frac{+14}{4}=+\frac{7\cdot\overset{1}{\cancel{2}}}{2\cdot\underset{1}{\cancel{2}}}=+\frac{7}{2}$

100. $\frac{-22.5}{-3.75}=+6$

101. $\frac{-25}{5}=-5$

102. $\frac{(-3)(-4)}{(-6)}=\frac{+12}{-6}=-2$

103. $\left(\frac{-10}{2}\right)^2 - (-1)^3 = (-5)^2 - (-1)^3$
$= 25 - (-1)$
$= 25 + 1 = 26$

104. $\frac{[-3+(-4)]^2}{10+(-3)} = \frac{[-7]^2}{7} = \frac{49}{7} = 7$

105. $\left(\frac{-3+(-3)}{3}\right)\left(\frac{-15}{5}\right) = \left(\frac{-6}{3}\right)\left(\frac{-15}{5}\right)$
$= (-2)(-3) = 6$

106. $\frac{-2-(-8)}{5+(-1)} = \frac{-2+(+8)}{4} = \frac{6}{4} = \frac{3}{2}$

107. xz

108. $x + 2y$

109. $2(x+y)$

110. $x - yz$

111. the product of 3, x and y

112. 5 decreased by the product of y and z

113. 5 less than the product of y and z

114. the quotient obtained when the sum of x, y and z is divided by twice their product

115. $y + z = -3 + (-1) = -4$

116. $x + y = 2 + (-3) = -1$

117. $x + (y + z) = 2 + [-3 + (-1)]$
$= 2 + [-4] = -2$

118. $x - y = 2 - (-3) = 2 + (+3) = 5$

119. $x - (y - z) = 2 - [-3 - (-1)]$
$= 2 - [-3 + (+1)]$
$= 2 - [-2] = 2 + (+2) = 4$

120. $(x - y) - z = [2 - (-3)] - (-1)$
$= [2 + (+3)] + (+1)$
$= 5 + 1 = 6$

121. $xy = (2)(-3) = -6$

122. $yz = (-3)(-1) = 3$

123. $x(x + z) = 2[2 + (-1)] = 2[1] = 2$

124. $xyz = (2)(-3)(-1) = (-6)(-1) = 6$

125. $y^2z + x = (-3)^2(-1) + 2$
$= 9(-1) + 2 = -9 + 2 = -7$

126. $yz^3 + (xy)^2 = (-3)(-1)^3 + [2(-3)]^2$
$= (-3)(-1) + [-6]^2$
$= 3 + 36 = 39$

127. $\frac{xy}{z} = \frac{(2)(-3)}{(-1)} = \frac{-6}{-1} = 6$

128. $\frac{|xy|}{3z} = \frac{|2(-3)|}{3(-1)} = \frac{|-6|}{-3} = \frac{6}{-3} = -2$

129. three terms

130. 7

131. 1

132. $2 + 4 + 3 = 9$

133. closure property

134. commutative property of multiplication

135. associative property of addition

136. distributive property

137. commutative property of addition

138. associative property of multiplication

139. commutative property of addition

140. identity for multiplication

141. additive inverse

142. identity for addition

Chapter 1 Test (page 76)

1. $31, 37, 41, 43, 47$

2. 2

3.

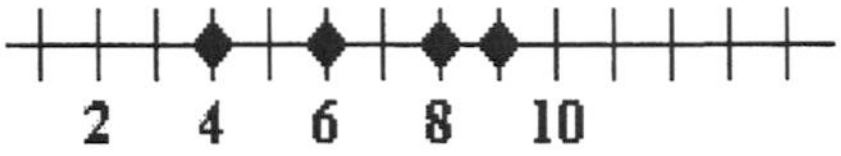

4.

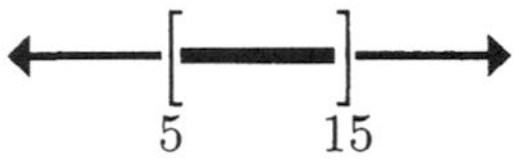

5. $-|23| = -(+23) = -23$

6. $-|7| + |-7| = -(+7) + (+7) = -7 + 7 = 0$

7. $3(4-2)\ \square\ -2(2-5)$
$3(2)\ \square\ -2(-3)$
$6\ \boxed{=}\ 6$

8. $1 + 4 \cdot 3\ \square\ -2(-7)$
$1 + 12\ \square\ +14$
$13\ \boxed{<}\ 14$

9. $25\% \text{ of } 136\ \square\ \frac{1}{2} \text{ of } 66$
$0.25(136)\ \square\ \frac{1}{2}(66)$
$34\ \boxed{>}\ 33$

10. $-13.7\ \square\ -|-13.7|$
$-13.7\ \square\ -(+13.7)$
$-13.7\ \boxed{=}\ -13.7$

11. $\frac{26}{40} = \frac{13 \cdot \cancel{2}^{1}}{20 \cdot \cancel{2}_{1}} = \frac{13}{20}$

12. $\frac{7}{8} \cdot \frac{24}{21} = \frac{\cancel{7}^{1} \cdot \cancel{3}^{1} \cdot \cancel{8}^{1}}{\cancel{8}_{1} \cdot \cancel{3}_{1} \cdot \cancel{7}_{1}} = \frac{1}{1} = 1$

13. $\frac{18}{35} \div \frac{9}{14} = \frac{18}{35} \cdot \frac{14}{9} = \frac{2 \cdot \cancel{9}^{1} \cdot 2 \cdot \cancel{7}^{1}}{5 \cdot \cancel{7}_{1} \cdot \cancel{9}_{1}} = \frac{4}{5}$

14. $\frac{24}{16} + 3 = \frac{3 \cdot \cancel{8}^{1}}{2 \cdot \cancel{8}_{1}} + \frac{3}{1} = \frac{3}{2} + \frac{3 \cdot \mathbf{2}}{1 \cdot \mathbf{2}} = \frac{3}{2} + \frac{6}{2} = \frac{3+6}{2} = \frac{9}{2}\ \left(\text{or } 4\frac{1}{2}\right)$

15. $\frac{17-5}{36} - \frac{2(13-5)}{12} = \frac{12}{36} - \frac{2(8)}{12} = \frac{12}{36} - \frac{16}{12} = \frac{\cancel{12}^{1}}{3 \cdot \cancel{12}_{1}} - \frac{4 \cdot \cancel{4}^{1}}{3 \cdot \cancel{4}_{1}} = \frac{1}{3} - \frac{4}{3} = \frac{1-4}{3} = \frac{-3}{3} = -1$

16. $\frac{|-7-(-6)|}{-7-|-6|} = \frac{|-7+(+6)|}{-7-(+6)} = \frac{|-1|}{-7+(-6)} = \frac{1}{-13} = -\frac{1}{13}$

17. $0.17(457) = 77.69 \approx 77.7$

18. $A = lw = (23.56 \text{ ft})(12.8 \text{ ft}) = 301.568 \text{ ft}^2$
$\approx 301.57 \text{ ft}^2$

19. $A = \frac{1}{2}bh = \frac{1}{2}(16\text{ cm})(8\text{ cm}) = \frac{1}{2}(128\text{ cm}^2)$
$= 64\text{ cm}^2$

20. $V = Bh = \pi r^2 h = \pi(7\text{ in.})^2(10\text{ in.})$
$= \pi(49\text{ in.}^2)(10\text{ in.})$
$= \pi(490\text{ in.}^3) \approx 1{,}539\text{ in.}^3$

21. $xy + z = (-2)(3) + 4 = -6 + 4 = -2$

22. $x(y+z) = -2(3+4) = -2(7) = -14$

23. $\frac{z+4y}{2x} = \frac{4+4(3)}{2(-2)} = \frac{4+12}{-4} = \frac{16}{-4} = -4$

24. $|x^3 - z| = |(-2)^3 - 4| = |-8-4| = |-12|$
$= 12$

25. $x^3 + y^2 + z = (-2)^3 + (3)^2 + 4$
$= -8 + 9 + 4 = 5$

26. $|x| - 3|y| - 4|z| = |-2| - 3|3| - 4|4|$
$= 2 - 3(3) - 4(4)$
$= 2 - 9 - 16 = -23$

27. $\frac{xy}{x+y}$

28. $5y - (x+y)$

29. $x(12+12) + y(7+7) = 24x + 14y$

30. $12a + 8b$

31. 3

32. 4 terms

33. $3(x+2) = 3x + 6$

34. $-p(r-t) = (-p)r + (-p)(-t) = -pr + pt$

35. 0

36. 5

37. commutative property of multiplication

38. distributive property

39. commutative property of addition

40. multiplicative inverse property

Exercise 2.1 (page 89)

1. $\frac{4}{5} + \frac{2}{3} = \frac{4 \cdot 3}{5 \cdot 3} + \frac{2 \cdot 5}{3 \cdot 5} = \frac{12}{15} + \frac{10}{15} = \frac{22}{15}$

3. $\frac{5}{9} \div \frac{3}{5} = \frac{5}{9} \cdot \frac{5}{3} = \frac{25}{27}$

5. $2 + 3 \cdot 4 = 2 + 12 = 14$

7. $3 + 4^3(-5) = 3 + 64(-5) = 3 + (-320)$
$= -317$

9. equation

11. equivalent

13. equal

15. equal

17. regular price

19. 100

21. $x = 2$
equation

23. $7x < 8$
not an equation

25. $x + 7 = 0$
equation

27. $1 + 1 = 3$
equation

29.
$$\begin{aligned} x+2&=3\\ 1+2&\stackrel{?}{=}3\\ 3&=3 \end{aligned}$$
1 is a solution.

31.
$$\begin{aligned} a-7&=0\\ -7-7&\stackrel{?}{=}0\\ -14&\neq 0 \end{aligned}$$
-7 is not a solution.

33.
$$\begin{aligned} \frac{y}{7}&=4\\ \frac{28}{7}&\stackrel{?}{=}4\\ 4&=4 \end{aligned}$$
28 is a solution.

35.
$$\begin{aligned} \frac{x}{5}&=x\\ \frac{0}{5}&\stackrel{?}{=}0\\ 0&=0 \end{aligned}$$
0 is a solution.

37.
$$\begin{aligned} 3k+5&=5k-1\\ 3(3)+5&\stackrel{?}{=}5(3)-1\\ 9+5&\stackrel{?}{=}15-1\\ 14&=14 \end{aligned}$$
3 is a solution.

39.
$$\begin{aligned} \frac{5+x}{10}-x&=\frac{1}{2}\\ \frac{5+0}{10}-0&\stackrel{?}{=}\frac{1}{2}\\ \frac{5}{10}-0&\stackrel{?}{=}\frac{1}{2}\\ \frac{1}{2}&=\frac{1}{2} \end{aligned}$$
0 is a solution.

41.
$$\begin{aligned} x+7&=13\\ x+7-\mathbf{7}&=13-\mathbf{7}\\ x&=6 \end{aligned}\qquad \begin{aligned} x+7&=13\\ 6+7&\stackrel{?}{=}13\\ 13&=13 \end{aligned}$$

43.
$$\begin{aligned} y-7&=12\\ y-7+\mathbf{7}&=12+\mathbf{7}\\ y&=19 \end{aligned}\qquad \begin{aligned} y-7&=12\\ 19-7&\stackrel{?}{=}12\\ 12&=12 \end{aligned}$$

45.
$$\begin{aligned} 1&=y-5\\ 1+\mathbf{5}&=y-5+\mathbf{5}\\ 6&=y \end{aligned}\qquad \begin{aligned} 1&=y-5\\ 1&\stackrel{?}{=}6-5\\ 1&=1 \end{aligned}$$

47.
$$\begin{aligned} p-404&=115\\ p-404+\mathbf{404}&=115+\mathbf{404}\\ p&=519 \end{aligned}$$
Check:
$$\begin{aligned} p-404&=115\\ 519-404&\stackrel{?}{=}115\\ 115&=115 \end{aligned}$$

49.
$$\begin{aligned} -37+z&=37\\ -37+\mathbf{37}+z&=37+\mathbf{37}\\ z&=74 \end{aligned}$$
Check:
$$\begin{aligned} -37+z&=37\\ -37+74&\stackrel{?}{=}37\\ 37&=37 \end{aligned}$$

51.
$$\begin{aligned} -57&=b-29\\ -57+\mathbf{29}&=b-29+\mathbf{29}\\ -28&=b \end{aligned}$$
Check:
$$\begin{aligned} -57&=b-29\\ -57&\stackrel{?}{=}-28-29\\ -57&=-57 \end{aligned}$$

53.
$$\begin{aligned} \frac{4}{3}&=-\frac{2}{3}+x\\ \frac{4}{3}+\mathbf{\frac{2}{3}}&=-\frac{2}{3}+\mathbf{\frac{2}{3}}+x\\ \frac{6}{3}&=x,\text{ or } x=2 \end{aligned}$$
Check:
$$\begin{aligned} \frac{4}{3}&=-\frac{2}{3}+x\\ \frac{4}{3}&\stackrel{?}{=}-\frac{2}{3}+2\\ \frac{4}{3}&\stackrel{?}{=}-\frac{2}{3}+\frac{6}{3}\\ \frac{4}{3}&=\frac{4}{3} \end{aligned}$$

55.

$$d + \frac{2}{3} = \frac{3}{2}$$
$$d + \frac{2}{3} - \mathbf{\frac{2}{3}} = \frac{3}{2} - \mathbf{\frac{2}{3}}$$
$$d = \frac{9}{6} - \frac{4}{6} = \frac{5}{6}$$

Check:
$$d + \frac{2}{3} = \frac{3}{2}$$
$$\frac{5}{6} + \frac{2}{3} \stackrel{?}{=} \frac{3}{2}$$
$$\frac{5}{6} + \frac{4}{6} \stackrel{?}{=} \frac{3}{2}$$
$$\frac{9}{6} \stackrel{?}{=} \frac{3}{2}$$
$$\frac{3}{2} = \frac{3}{2}$$

57.

$$-\frac{3}{5} = x - \frac{2}{5}$$
$$-\frac{3}{5} + \mathbf{\frac{2}{5}} = x - \frac{2}{5} + \mathbf{\frac{2}{5}}$$
$$-\frac{1}{5} = x$$

Check:
$$-\frac{3}{5} = x - \frac{2}{5}$$
$$-\frac{3}{5} \stackrel{?}{=} -\frac{1}{5} - \frac{2}{5}$$
$$-\frac{3}{5} = -\frac{3}{5}$$

59.

$$r - \frac{1}{5} = \frac{3}{10}$$
$$r - \frac{1}{5} + \mathbf{\frac{1}{5}} = \frac{3}{10} + \mathbf{\frac{1}{5}}$$
$$r = \frac{3}{10} + \frac{2}{10} = \frac{5}{10} = \frac{1}{2}$$

Check:
$$r - \frac{1}{5} = \frac{3}{10}$$
$$\frac{1}{2} - \frac{1}{5} \stackrel{?}{=} \frac{3}{10}$$
$$\frac{5}{10} - \frac{2}{10} \stackrel{?}{=} \frac{3}{10}$$
$$\frac{3}{10} = \frac{3}{10}$$

61.

$$\frac{x}{5} = 5 \qquad \frac{x}{5} = 5$$
$$\mathbf{5} \cdot \frac{x}{5} = \mathbf{5} \cdot 5 \qquad \frac{25}{5} \stackrel{?}{=} 5$$
$$x = 25 \qquad 5 = 5$$

63.

$$\frac{x}{32} = -2 \qquad \frac{x}{32} = -2$$
$$\mathbf{32} \cdot \frac{x}{32} = \mathbf{32} \cdot (-2) \qquad \frac{-64}{32} \stackrel{?}{=} -2$$
$$x = -64 \qquad -2 = -2$$

65.

$$\frac{b}{3} = 5 \qquad \frac{b}{3} = 5$$
$$\mathbf{3} \cdot \frac{b}{3} = \mathbf{3} \cdot 5 \qquad \frac{15}{3} \stackrel{?}{=} 5$$
$$b = 15 \qquad 5 = 5$$

67.

$$-3 = \frac{s}{11} \qquad -3 = \frac{s}{11}$$
$$\mathbf{11} \cdot (-3) = \mathbf{11} \cdot \frac{s}{11} \qquad -3 \stackrel{?}{=} \frac{-33}{11}$$
$$-33 = s \qquad -3 = -3$$

69.

$$p + 0.27 = 3.57$$
$$p + 0.27 - \mathbf{0.27} = 3.57 - \mathbf{0.27}$$
$$p = 3.3$$

Check:
$$p + 0.27 = 3.57$$
$$3.3 + 0.27 \stackrel{?}{=} 3.57$$
$$3.57 = 3.57$$

71.

$$y - 2.63 = -8.21$$
$$y - 2.63 + \mathbf{2.63} = -8.21 + \mathbf{2.63}$$
$$y = -5.58$$

Check:
$$y - 2.63 = -8.21$$
$$-5.58 - 2.63 \stackrel{?}{=} -8.21$$
$$-8.21 = -8.21$$

73.
$$6x = 18 \qquad \frac{6x}{6} = \frac{18}{6} \qquad x = 3$$
$$6x = 18 \qquad 6(3) \stackrel{?}{=} 18 \qquad 18 = 18$$

75.
$$-4x = 36 \qquad \frac{-4x}{-4} = \frac{36}{-4} \qquad x = -9$$
$$-4x = 36 \qquad -4(-9) \stackrel{?}{=} 36 \qquad 36 = 36$$

77.
$$4t = 108 \qquad \frac{4t}{4} = \frac{108}{4} \qquad t = 27$$
$$4t = 108 \qquad 4(27) \stackrel{?}{=} 108 \qquad 108 = 108$$

79.
$$11x = -121 \qquad \frac{11x}{11} = \frac{-121}{11} \qquad x = -11$$
$$11x = -121 \qquad 11(-11) \stackrel{?}{=} -121 \qquad -121 = -121$$

81.
$$2x = \frac{1}{7} \qquad \frac{1}{2} \cdot 2x = \frac{1}{2} \cdot \frac{1}{7} \qquad x = \frac{1}{14}$$
$$2x = \frac{1}{7} \qquad 2 \cdot \frac{1}{14} \stackrel{?}{=} \frac{1}{7} \qquad \frac{2}{14} \stackrel{?}{=} \frac{1}{7} \qquad \frac{1}{7} = \frac{1}{7}$$

83.
$$5x = \frac{5}{8} \qquad \frac{1}{5} \cdot 5x = \frac{1}{5} \cdot \frac{5}{8} \qquad x = \frac{1}{8}$$
$$5x = \frac{5}{8} \qquad 5 \cdot \frac{1}{8} \stackrel{?}{=} \frac{5}{8} \qquad \frac{5}{8} = \frac{5}{8}$$

85.
$$\frac{z}{7} = 14 \qquad \mathbf{7} \cdot \frac{z}{7} = \mathbf{7} \cdot 14 \qquad z = 98$$
$$\frac{z}{7} = 14 \qquad \frac{98}{7} \stackrel{?}{=} 14 \qquad 14 = 14$$

87.
$$\frac{w}{7} = \frac{5}{7} \qquad \mathbf{7} \cdot \frac{w}{7} = \mathbf{7} \cdot \frac{5}{7} \qquad w = 5$$
$$\frac{w}{7} = \frac{5}{7} \qquad \frac{5}{7} = \frac{5}{7}$$

89.
$$\frac{s}{-3} = -\frac{5}{6} \qquad \mathbf{-6} \cdot \frac{s}{-3} = \mathbf{-6} \cdot \left(-\frac{5}{6}\right) \qquad 2s = 5 \qquad \frac{2s}{2} = \frac{5}{2} \qquad s = \frac{5}{2}$$
$$\frac{s}{-3} = -\frac{5}{6} \qquad \frac{\frac{5}{2}}{-3} \stackrel{?}{=} -\frac{5}{6} \qquad \frac{\frac{5}{2} \cdot 2}{-3 \cdot 2} \stackrel{?}{=} -\frac{5}{6} \qquad \frac{5}{-6} = -\frac{5}{6}$$

91.
$$-32z = 64 \qquad \frac{-32z}{-32} = \frac{64}{-32} \qquad z = -2$$
$$-32z = 64 \qquad -32(-2) \stackrel{?}{=} 64 \qquad 64 = 64$$

93.
$$18z = -9 \qquad \frac{18z}{18} = \frac{-9}{18} \qquad z = -\frac{1}{2}$$
$$18z = -9 \qquad 18\left(-\frac{1}{2}\right) \stackrel{?}{=} -9 \qquad -9 = -9$$

95.
$$0.25x = 1228 \qquad \frac{0.25x}{0.25} = \frac{1228}{0.25} \qquad x = 4912$$
$$0.25x = 1228 \qquad 0.25(4912) \stackrel{?}{=} 1228 \qquad 4912 = 4912$$

97.
$$\frac{b}{3} = \frac{1}{3} \qquad \mathbf{3} \cdot \frac{b}{3} = \mathbf{3} \cdot \frac{1}{3} \qquad b = 1$$
$$\frac{b}{3} = \frac{1}{3} \qquad \frac{1}{3} = \frac{1}{3}$$

99.
$$-1.2w = -102 \qquad \frac{-1.2w}{-1.2} = \frac{-102}{-1.2} \qquad w = 85$$
$$-1.2w = -102 \qquad -1.2(85) \stackrel{?}{=} -102 \qquad -102 = -102$$

101.

$$\frac{u}{5} = -\frac{3}{10}$$
$$\mathbf{10} \cdot \frac{u}{5} = \mathbf{10} \cdot \left(-\frac{3}{10}\right)$$
$$2u = -3$$
$$\frac{2u}{2} = \frac{-3}{2}$$
$$u = -\frac{3}{2}$$

$$\frac{u}{5} = -\frac{3}{10}$$
$$\frac{-\frac{3}{2}}{5} \stackrel{?}{=} -\frac{3}{10}$$
$$\frac{-\frac{3}{2} \cdot 2}{5 \cdot 2} \stackrel{?}{=} -\frac{3}{10}$$
$$\frac{-3}{10} = -\frac{3}{10}$$

103.

$$\frac{p}{0.2} = 12$$
$$\mathbf{0.2} \cdot \frac{p}{0.2} = \mathbf{0.2} \cdot 12$$
$$p = 2.4$$

$$\frac{p}{0.2} = 12$$
$$\frac{2.4}{0.2} \stackrel{?}{=} 12$$
$$12 = 12$$

105.

$$rb = a$$
$$40\% \cdot 200 = a$$
$$0.40(200) = a$$
$$80 = a$$

107.

$$rb = a$$
$$50\% \cdot 38 = a$$
$$0.50(38) = a$$
$$19 = a$$

109.

$$rb = a$$
$$15\% \cdot b = 48$$
$$0.15b = 48$$
$$\frac{0.15b}{0.15} = \frac{48}{0.15}$$
$$b = 320$$

111.

$$rb = a$$
$$35\% \cdot b = 133$$
$$0.35b = 133$$
$$\frac{0.35b}{0.35} = \frac{133}{0.35}$$
$$b = 380$$

113.

$$rb = a$$
$$28\% \cdot b = 42$$
$$0.28b = 42$$
$$\frac{0.28b}{0.28} = \frac{42}{0.28}$$
$$b = 150$$

115.

$$rb = a$$
$$r(357.5) = 71.5$$
$$\frac{r(357.5)}{357.5} = \frac{71.5}{357.5}$$
$$r = 0.20$$
$$r = 20\%$$

117.

$$rb = a$$
$$r(4) = 0.32$$
$$\frac{r(4)}{4} = \frac{0.32}{4}$$
$$r = 0.08$$
$$r = 8\%$$

119.

$$rb = a$$
$$r(17) = 34$$
$$\frac{r(17)}{17} = \frac{34}{17}$$
$$r = 2.00$$
$$r = 200\%$$

121. Let r = the regular price. Then

$$\boxed{\text{Sale price}} = \boxed{\text{Regular price}} - \boxed{\text{Markdown}}$$
$$7995 = r - 1350$$
$$7995 + \mathbf{1350} = r - 1350 + \mathbf{1350}$$
$$9345 = r$$

The regular price is \$9,345.

123. Let w = the wholesale price. Then

$$\boxed{\text{Retail price}} = \boxed{\text{Wholesale price}} + \boxed{\text{Markup}}$$
$$175 = w + 85$$
$$175 - \mathbf{85} = w + 85 - \mathbf{85}$$
$$90 = w$$

The wholesale price is \$90.

125.

$$A = p + i$$
$$5010 = 4750 + i$$
$$5010 - \mathbf{4750} = 4750 - \mathbf{4750} + i$$
$$260 = i$$

The deposit earned \$260 in interest.

127.

$$v = p + a$$
$$150000 = p + 57000$$
$$150000 - \mathbf{57000} = p + 57000 - \mathbf{57000}$$
$$93000 = p$$

The original price was \$93,000.

129.

$$c = p + t$$
$$512 = 317 + t$$
$$512 - \mathbf{317} = 317 - \mathbf{317} + t$$
$$195 = t$$

The carpet cost \$195 to install.

131. Let $c =$ the condominium price. Then

$$\boxed{\text{Condo price}} = \boxed{\text{House price}} - 57595$$
$$c = 202744 - 57595$$
$$c = 145149$$

The price of the condominium is $145,149.

133. Let $x =$ the original number in the audience.

$$\frac{1}{3} \cdot \boxed{\text{Original audience}} = \boxed{\text{Number who left}}$$
$$\frac{1}{3}x = 78$$
$$\mathbf{3} \cdot \frac{1}{3}x = \mathbf{3} \cdot 78$$
$$x = 234$$

There were originally 234 in the audience.

135. Let $x =$ the number in the senior class.

$$\frac{1}{7} \cdot \boxed{\text{Number in class}} = \boxed{\text{Number off-campus}}$$
$$\tfrac{1}{7}x = 217$$
$$\mathbf{7} \cdot \tfrac{1}{7}x = \mathbf{7} \cdot 217$$
$$x = 1519$$

There are 1,519 in the senior class.

137. Let $r =$ the percentage not pleased.

$$a = 9200 - 4140 = 5060$$
$$rb = a$$
$$r(9200) = 5060$$
$$\frac{r(9200)}{9200} = \frac{5060}{9200}$$
$$r = 0.55$$
$$r = 55\%$$

55% of those surveyed were not pleased.

139. Let $x =$ the selling price.

$$rb = a$$
$$0.05(x) = 13.50$$
$$\frac{0.05x}{0.05} = \frac{13.50}{0.05}$$
$$x = 270$$

The selling price was $270.

141. Let $x =$ the total number of patients.

$$rb = a$$
$$0.18(x) = 1008$$
$$\frac{0.18x}{0.18} = \frac{1008}{0.18}$$
$$x = 5600$$

The hospital treated 5,600 patients.

143. Answers may vary.

145.

$$A_{\text{circle}} = A_{\text{square}}$$
$$\pi r^2 = s^2$$
$$\pi(4.5)^2 = 8^2$$
$$\pi(20.25) = 64$$
$$\frac{\pi(20.25)}{20.25} = \frac{64}{20.25}$$
$$\pi \approx 3.16$$

Exercise 2.2 (page 99)

1. $P = 2l + 2w = 2(8.5\text{ cm}) + 2(16.5\text{ cm}) = 17\text{ cm} + 33\text{ cm} = 50\text{ cm}$

3. $A = \frac{1}{2}h(b + d) = \frac{1}{2}(8.5\text{ in.})(6.7\text{ in.} + 12.2\text{ in.}) = \frac{1}{2}(8.5\text{ in.})(18.9\text{ in.}) = \frac{1}{2}(160.65\text{ in.}^2) = 80.325\text{ in.}^2$

5. cost

7. percent

9.
$$\begin{aligned} 5x - 1 &= 4 \\ 5x - 1 + 1 &= 4 + 1 \\ 5x &= 5 \\ \frac{5x}{5} &= \frac{5}{5} \\ x &= 1 \end{aligned} \qquad \begin{aligned} 5x - 1 &= 4 \\ 5(1) - 1 &\stackrel{?}{=} 4 \\ 5 - 1 &\stackrel{?}{=} 4 \\ 4 &= 4 \end{aligned}$$

11.
$$\begin{aligned} 6x + 2 &= -4 \\ 6x + 2 - 2 &= -4 - 2 \\ 6x &= -6 \\ \frac{6x}{6} &= \frac{-6}{6} \\ x &= -1 \end{aligned} \qquad \begin{aligned} 6x + 2 &= -4 \\ 6(-1) + 2 &\stackrel{?}{=} -4 \\ -6 + 2 &\stackrel{?}{=} -4 \\ -4 &= -4 \end{aligned}$$

13.
$$\begin{aligned} 3x - 8 &= 1 \\ 3x - 8 + 8 &= 1 + 8 \\ 3x &= 9 \\ \frac{3x}{3} &= \frac{9}{3} \\ x &= 3 \end{aligned} \qquad \begin{aligned} 3x - 8 &= 1 \\ 3(3) - 8 &\stackrel{?}{=} 1 \\ 9 - 8 &\stackrel{?}{=} 1 \\ 1 &= 1 \end{aligned}$$

15.
$$\begin{aligned} 11x + 17 &= -5 \\ 11x + 17 - 17 &= -5 - 17 \\ 11x &= -22 \\ \frac{11x}{11} &= \frac{-22}{11} \\ x &= -2 \end{aligned} \qquad \begin{aligned} 11x + 17 &= -5 \\ 11(-2) + 17 &\stackrel{?}{=} -5 \\ -22 + 17 &\stackrel{?}{=} -5 \\ -5 &= -5 \end{aligned}$$

17.
$$\begin{aligned} 43t + 72 &= 158 \\ 43t + 72 - 72 &= 158 - 72 \\ 43t &= 86 \\ \frac{43t}{43} &= \frac{86}{43} \\ t &= 2 \end{aligned} \qquad \begin{aligned} 43t + 72 &= 158 \\ 43(2) + 72 &\stackrel{?}{=} 158 \\ 86 + 72 &\stackrel{?}{=} 158 \\ 158 &= 158 \end{aligned}$$

19.
$$\begin{aligned} -47 - 21s &= 58 \\ -47 + 47 - 21s &= 58 + 47 \\ -21s &= 105 \\ \frac{-21s}{-21} &= \frac{105}{-21} \\ s &= -5 \end{aligned} \qquad \begin{aligned} -47 - 21s &= 58 \\ -47 - 21(-5) &\stackrel{?}{=} 58 \\ -47 + 105 &\stackrel{?}{=} 58 \\ 58 &= 58 \end{aligned}$$

21.
$$\begin{aligned} 2y - \frac{5}{3} &= \frac{4}{3} \\ 2y - \frac{5}{3} + \frac{5}{3} &= \frac{4}{3} + \frac{5}{3} \\ 2y &= \frac{9}{3} \\ 2y &= 3 \\ \frac{2y}{2} &= \frac{3}{2} \\ y &= \frac{3}{2} \end{aligned} \qquad \begin{aligned} 2y - \frac{5}{3} &= \frac{4}{3} \\ 2\left(\frac{3}{2}\right) - \frac{5}{3} &\stackrel{?}{=} \frac{4}{3} \\ 3 - \frac{5}{3} &\stackrel{?}{=} \frac{4}{3} \\ \frac{9}{3} - \frac{5}{3} &\stackrel{?}{=} \frac{4}{3} \\ \frac{4}{3} &= \frac{4}{3} \end{aligned}$$

23.
$$\begin{aligned} -0.4y - 12 &= -20 \\ -0.4y - 12 + 12 &= -20 + 12 \\ -0.4y &= -8 \\ \frac{-0.4y}{-0.4} &= \frac{-8}{-0.4} \\ y &= 20 \end{aligned}$$
$$\begin{aligned} -0.4y - 12 &= -20 \\ -0.4(20) - 12 &\stackrel{?}{=} -20 \\ -8 - 12 &\stackrel{?}{=} -20 \\ -20 &= -20 \end{aligned}$$

25.
$$\begin{aligned} \frac{x}{3} - 3 &= -2 \\ \frac{x}{3} - 3 + 3 &= -2 + 3 \\ \frac{x}{3} &= 1 \\ 3 \cdot \frac{x}{3} &= 3 \cdot 1 \\ x &= 3 \end{aligned}$$
$$\begin{aligned} \frac{x}{3} - 3 &= -2 \\ \frac{3}{3} - 3 &\stackrel{?}{=} -2 \\ 1 - 3 &\stackrel{?}{=} -2 \\ -2 &= -2 \end{aligned}$$

27.
$$\begin{aligned} \frac{z}{9} + 5 &= -1 \\ \frac{z}{9} + 5 - 5 &= -1 - 5 \\ \frac{z}{9} &= -6 \\ 9 \cdot \frac{z}{9} &= 9(-6) \\ z &= -54 \end{aligned}$$
$$\begin{aligned} \frac{z}{9} + 5 &= -1 \\ \frac{-54}{9} + 5 &\stackrel{?}{=} -1 \\ -6 + 5 &\stackrel{?}{=} -1 \\ -1 &= -1 \end{aligned}$$

29.
$$\begin{aligned} \frac{b}{3} + 5 &= 2 \\ \frac{b}{3} + 5 - 5 &= 2 - 5 \\ \frac{b}{3} &= -3 \\ 3 \cdot \frac{b}{3} &= 3(-3) \\ b &= -9 \end{aligned}$$
$$\begin{aligned} \frac{b}{3} + 5 &= 2 \\ \frac{-9}{3} + 5 &\stackrel{?}{=} 2 \\ -3 + 5 &\stackrel{?}{=} 2 \\ 2 &= 2 \end{aligned}$$

31.
$$\begin{aligned} \frac{s}{11} + 9 &= 6 \\ \frac{s}{11} + 9 - 9 &= 6 - 9 \\ \frac{s}{11} &= -3 \\ 11 \cdot \frac{s}{11} &= 11(-3) \\ s &= -33 \end{aligned}$$
$$\begin{aligned} \frac{s}{11} + 9 &= 6 \\ \frac{-33}{11} + 9 &\stackrel{?}{=} 6 \\ -3 + 9 &\stackrel{?}{=} 6 \\ 6 &= 6 \end{aligned}$$

33.
$$\begin{aligned} \frac{k}{5} - \frac{1}{2} &= \frac{3}{2} \\ \frac{k}{5} - \frac{1}{2} + \frac{1}{2} &= \frac{3}{2} + \frac{1}{2} \\ \frac{k}{5} &= \frac{4}{2} \\ \frac{k}{5} &= 2 \\ 5 \cdot \frac{k}{5} &= 5 \cdot 2 \\ k &= 10 \end{aligned}$$
$$\begin{aligned} \frac{k}{5} - \frac{1}{2} &= \frac{3}{2} \\ \frac{10}{5} - \frac{1}{2} &\stackrel{?}{=} \frac{3}{2} \\ 2 - \frac{1}{2} &\stackrel{?}{=} \frac{3}{2} \\ \frac{4}{2} - \frac{1}{2} &\stackrel{?}{=} \frac{3}{2} \\ \frac{3}{2} &= \frac{3}{2} \end{aligned}$$

35.
$$\begin{aligned} \frac{w}{16} + \frac{5}{4} &= 1 \\ \frac{w}{16} + \frac{5}{4} - \frac{5}{4} &= 1 - \frac{5}{4} \\ \frac{w}{16} &= \frac{4}{4} - \frac{5}{4} \\ \frac{w}{16} &= -\frac{1}{4} \\ 16 \cdot \frac{w}{16} &= 16\left(-\frac{1}{4}\right) \\ w &= -4 \end{aligned}$$
$$\begin{aligned} \frac{w}{16} + \frac{5}{4} &= 1 \\ \frac{-4}{16} + \frac{5}{4} &\stackrel{?}{=} 1 \\ -\frac{1}{4} + \frac{5}{4} &\stackrel{?}{=} 1 \\ \frac{4}{4} &\stackrel{?}{=} 1 \\ 1 &= 1 \end{aligned}$$

37.
$$\begin{aligned} \frac{b+5}{3} &= 11 \\ 3 \cdot \frac{b+5}{3} &= 3 \cdot 11 \\ b + 5 &= 33 \\ b + 5 - 5 &= 33 - 5 \\ b &= 28 \end{aligned}$$
$$\begin{aligned} \frac{b+5}{3} &= 11 \\ \frac{28+5}{3} &\stackrel{?}{=} 11 \\ \frac{33}{3} &\stackrel{?}{=} 11 \\ 11 &= 11 \end{aligned}$$

39.
$$\begin{aligned} \frac{r+7}{3} &= 4 \\ 3 \cdot \frac{r+7}{3} &= 3 \cdot 4 \\ r + 7 &= 12 \\ r + 7 - 7 &= 12 - 7 \\ r &= 5 \end{aligned}$$
$$\begin{aligned} \frac{r+7}{3} &= 4 \\ \frac{5+7}{3} &\stackrel{?}{=} 4 \\ \frac{12}{3} &\stackrel{?}{=} 4 \\ 4 &= 4 \end{aligned}$$

41.
$$\begin{aligned}\frac{u-2}{5}&=1\\ 5\cdot\frac{u-2}{5}&=5\cdot 1\\ u-2&=5\\ u-2+2&=5+2\\ u&=7\end{aligned}\qquad\begin{aligned}\frac{u-2}{5}&=1\\ \frac{7-2}{5}&\stackrel{?}{=}1\\ \frac{5}{5}&\stackrel{?}{=}1\\ 1&=1\end{aligned}$$

43.
$$\begin{aligned}\frac{x-4}{4}&=-3\\ 4\cdot\frac{x-4}{4}&=4(-3)\\ x-4&=-12\\ x-4+4&=-12+4\\ x&=-8\end{aligned}\qquad\begin{aligned}\frac{x-4}{4}&=-3\\ \frac{-8-4}{4}&\stackrel{?}{=}-3\\ \frac{-12}{4}&\stackrel{?}{=}-3\\ -3&=-3\end{aligned}$$

45.
$$\begin{aligned}\frac{3x}{2}-6&=9\\ \frac{3x}{2}-6+6&=9+6\\ \frac{3x}{2}&=15\\ 2\cdot\frac{3x}{2}&=2\cdot 15\\ 3x&=30\\ \frac{3x}{3}&=\frac{30}{3}\\ x&=10\end{aligned}\qquad\begin{aligned}\frac{3x}{2}-6&=9\\ \frac{3(10)}{2}-6&\stackrel{?}{=}9\\ \frac{30}{2}-6&\stackrel{?}{=}9\\ 15-6&\stackrel{?}{=}9\\ 9&=9\end{aligned}$$

47.
$$\begin{aligned}\frac{3y}{2}+5&=11\\ \frac{3y}{2}+5-5&=11-5\\ \frac{3y}{2}&=6\\ 2\cdot\frac{3y}{2}&=2\cdot 6\\ 3y&=12\\ \frac{3y}{3}&=\frac{12}{3}\\ y&=4\end{aligned}\qquad\begin{aligned}\frac{3y}{2}+5&=11\\ \frac{3(4)}{2}+5&\stackrel{?}{=}11\\ \frac{12}{2}+5&\stackrel{?}{=}11\\ 6+5&\stackrel{?}{=}11\\ 11&=11\end{aligned}$$

49.
$$\begin{aligned}\frac{2x}{3}+\frac{1}{2}&=3\\ \frac{2x}{3}+\frac{1}{2}-\frac{1}{2}&=3-\frac{1}{2}\\ \frac{2x}{3}&=\frac{5}{2}\\ 3\cdot\frac{2x}{3}&=3\cdot\frac{5}{2}\\ 2x&=\frac{15}{2}\\ \frac{1}{2}\cdot 2x&=\frac{1}{2}\cdot\frac{15}{2}\\ x&=\frac{15}{4}\end{aligned}\qquad\begin{aligned}\frac{2x}{3}+\frac{1}{2}&=3\\ \frac{2\left(\frac{15}{4}\right)}{3}+\frac{1}{2}&\stackrel{?}{=}3\\ \frac{\frac{15}{2}}{3}+\frac{1}{2}&\stackrel{?}{=}3\\ \frac{15}{6}+\frac{1}{2}&\stackrel{?}{=}3\\ \frac{5}{2}+\frac{1}{2}&\stackrel{?}{=}3\\ \frac{6}{2}&\stackrel{?}{=}3\\ 3&=3\end{aligned}$$

51.
$$\begin{aligned}\frac{3x}{4}-\frac{2}{5}&=2\\ \frac{3x}{4}-\frac{2}{5}+\frac{2}{5}&=2+\frac{2}{5}\\ \frac{3x}{4}&=\frac{12}{5}\\ 4\cdot\frac{3x}{4}&=4\cdot\frac{12}{5}\\ 3x&=\frac{48}{5}\\ \frac{1}{3}\cdot 3x&=\frac{1}{3}\cdot\frac{48}{5}\\ x&=\frac{16}{5}\end{aligned}\qquad\begin{aligned}\frac{3x}{4}-\frac{2}{5}&=2\\ \frac{3\left(\frac{16}{5}\right)}{4}-\frac{2}{5}&\stackrel{?}{=}2\\ \frac{\frac{48}{5}}{4}-\frac{2}{5}&\stackrel{?}{=}2\\ \frac{48}{20}-\frac{2}{5}&\stackrel{?}{=}2\\ \frac{12}{5}-\frac{2}{5}&\stackrel{?}{=}2\\ \frac{10}{5}&\stackrel{?}{=}2\\ 2&=2\end{aligned}$$

53.
$$\begin{aligned}\frac{3x-12}{2}&=9\\ 2\cdot\frac{3x-12}{2}&=2\cdot 9\\ 3x-12&=18\\ 3x-12+12&=18+12\\ 3x&=30\\ \frac{3x}{3}&=\frac{30}{3}\\ x&=10\end{aligned}\qquad\begin{aligned}\frac{3x-12}{2}&=9\\ \frac{3(10)-12}{2}&\stackrel{?}{=}9\\ \frac{30-12}{2}&\stackrel{?}{=}9\\ \frac{18}{2}&\stackrel{?}{=}9\\ 9&=9\end{aligned}$$

55.
$$\begin{aligned}\frac{5k-8}{9}&=1\\ 9\cdot\frac{5k-8}{9}&=9\cdot 1\\ 5k-8&=9\\ 5k-8+8&=9+8\\ 5k&=17\\ \frac{5k}{5}&=\frac{17}{5}\\ k&=\frac{17}{5}\end{aligned}\qquad\begin{aligned}\frac{5k-8}{9}&=1\\ \frac{5\cdot\frac{17}{5}-8}{9}&\stackrel{?}{=}1\\ \frac{17-8}{9}&\stackrel{?}{=}1\\ \frac{9}{9}&\stackrel{?}{=}1\\ 1&=1\end{aligned}$$

57.

$$\begin{aligned} \frac{3z+2}{17} &= 0 \\ 17\cdot\frac{3z+2}{17} &= 17\cdot 0 \\ 3z+2 &= 0 \\ 3z+2-2 &= 0-2 \\ 3z &= -2 \\ \frac{3z}{3} &= \frac{-2}{3} \\ z &= -\frac{2}{3} \end{aligned}$$

$$\begin{aligned} \frac{3z+2}{17} &= 0 \\ \frac{3(-\frac{2}{3})+2}{17} &\stackrel{?}{=} 0 \\ \frac{-2+2}{17} &\stackrel{?}{=} 0 \\ \frac{0}{17} &\stackrel{?}{=} 0 \\ 0 &= 0 \end{aligned}$$

59.

$$\begin{aligned} \frac{17k-28}{21}+\frac{4}{3} &= 0 \\ \frac{17k-28}{21}+\frac{4}{3}-\frac{4}{3} &= 0-\frac{4}{3} \\ \frac{17k-28}{21} &= -\frac{4}{3} \\ 21\cdot\frac{17k-28}{21} &= 21\left(-\frac{4}{3}\right) \\ 17k-28 &= -28 \\ 17k-28+28 &= -28+28 \\ 17k &= 0 \\ \frac{17k}{17} &= \frac{0}{17} \\ k &= 0 \end{aligned}$$

$$\begin{aligned} \frac{17k-28}{21}+\frac{4}{3} &= 0 \\ \frac{17\cdot 0-28}{21}+\frac{4}{3} &\stackrel{?}{=} 0 \\ \frac{0-28}{21}+\frac{4}{3} &\stackrel{?}{=} 0 \\ -\frac{4}{3}+\frac{4}{3} &\stackrel{?}{=} 0 \\ 0 &= 0 \end{aligned}$$

61.

$$\begin{aligned} -\frac{x}{3}-\frac{1}{2} &= -\frac{5}{2} \\ -\frac{x}{3}-\frac{1}{2}+\frac{1}{2} &= -\frac{5}{2}+\frac{1}{2} \\ -\frac{x}{3} &= -\frac{4}{2} \\ -\frac{x}{3} &= -2 \\ -3\left(-\frac{x}{3}\right) &= -3(-2) \\ x &= 6 \end{aligned}$$

$$\begin{aligned} -\frac{x}{3}-\frac{1}{2} &= -\frac{5}{2} \\ -\frac{6}{3}-\frac{1}{2} &\stackrel{?}{=} -\frac{5}{2} \\ -2-\frac{1}{2} &\stackrel{?}{=} -\frac{5}{2} \\ -\frac{4}{2}-\frac{1}{2} &\stackrel{?}{=} -\frac{5}{2} \\ -\frac{5}{2} &= -\frac{5}{2} \end{aligned}$$

63.

$$\begin{aligned} \frac{9-5w}{15} &= \frac{2}{5} \\ 15\cdot\frac{9-5w}{15} &= 15\cdot\frac{2}{5} \\ 9-5w &= 6 \\ 9-9-5w &= 6-9 \\ -5w &= -3 \\ \frac{-5w}{-5} &= \frac{-3}{-5} \\ w &= \frac{3}{5} \end{aligned}$$

$$\begin{aligned} \frac{9-5w}{15} &= \frac{2}{5} \\ \frac{9-5(\frac{3}{5})}{15} &\stackrel{?}{=} \frac{2}{5} \\ \frac{9-3}{15} &\stackrel{?}{=} \frac{2}{5} \\ \frac{6}{15} &\stackrel{?}{=} \frac{2}{5} \\ \frac{2}{5} &= \frac{2}{5} \end{aligned}$$

65.

$$\frac{2.4x+4.8}{1.2}=8$$
$$1.2\cdot\frac{2.4x+4.8}{1.2}=1.2(8)$$
$$2.4x+4.8=9.6$$
$$2.4x+4.8-4.8=9.6-4.8$$
$$2.4x=4.8$$
$$\frac{2.4x}{2.4}=\frac{4.8}{2.4}$$
$$x=2$$

$$\frac{2.4x+4.8}{1.2}=8$$
$$\frac{2.4(2)+4.8}{1.2}\stackrel{?}{=}8$$
$$\frac{4.8+4.8}{1.2}\stackrel{?}{=}8$$
$$\frac{9.6}{1.2}\stackrel{?}{=}8$$
$$8=8$$

67.

$$\frac{2.1x-0.13}{0.8}+2.5=0.5$$
$$\frac{2.1x-0.13}{0.8}+2.5-2.5=0.5-2.5$$
$$\frac{2.1x-0.13}{0.8}=-2$$
$$0.8\cdot\frac{2.1x-0.13}{0.8}=0.8(-2)$$
$$2.1x-0.13=-1.6$$
$$2.1x-0.13+0.13=-1.6+0.13$$
$$2.1x=-1.47$$
$$\frac{2.1x}{2.1}=\frac{-1.47}{2.1}$$
$$x=-0.7$$

$$\frac{2.1x-0.13}{0.8}+2.5=0.5$$
$$\frac{2.1(-0.7)-0.13}{0.8}+2.5\stackrel{?}{=}0.5$$
$$\frac{-1.47-0.13}{0.8}+2.5\stackrel{?}{=}0.5$$
$$\frac{-1.6}{0.8}+2.5\stackrel{?}{=}0.5$$
$$-2+2.5\stackrel{?}{=}0.5$$
$$0.5=0.5$$

69. Let $x=$ the original number.
Then $3x-6=$ the other number.

$$\boxed{\text{The other number}}=9$$
$$3x-6=9$$
$$3x-6+6=9+6$$
$$3x=15$$
$$\frac{3x}{3}=\frac{15}{3}$$
$$x=5$$

The original number is 5.

71. Let $x=$ her former rent.
Then $2x-100=$ the new rent.

$$\boxed{\text{The new rent}}=400$$
$$2x-100=400$$
$$2x-100+100=400+100$$
$$2x=500$$
$$\frac{2x}{2}=\frac{500}{2}$$
$$x=250$$

Her former rent was \$250.

73. Let $x =$ the number of days.
Then $16 + 12x =$ the total cost.

$$\boxed{\text{The total cost}} = 100$$
$$16 + 12x = 100$$
$$16 - 16 + 12x = 100 - 16$$
$$12x = 84$$
$$\frac{12x}{12} = \frac{84}{12}$$
$$x = 7$$

The owner was gone for 7 days.

75. Let $x =$ the # of minutes (after the 1st).
Then $0.85 + 0.27x =$ the total cost.

$$\boxed{\text{The total cost}} = 8.50$$
$$0.85 + 0.27x = 8.50$$
$$0.85 - 0.85 + 0.27x = 8.50 - 0.85$$
$$0.27x = 7.65$$
$$\frac{0.27x}{0.27} = \frac{7.65}{0.27}$$
$$x \approx 28.3$$

She can talk for 28 minutes **after the first minute**, for a total of 29 minutes.

77. Let $x =$ the money from ticket sales.
Then $1500 + 0.20x =$ the total income.

$$\boxed{\text{The total income}} = 2980$$
$$1500 + 0.20x = 2980$$
$$1500 - 1500 + 0.20x = 2980 - 1500$$
$$0.20x = 1480$$
$$\frac{0.20x}{0.20} = \frac{1480}{0.20}$$
$$x = 7400$$

The total ticket sales were $7,400.

79. Let $x =$ the score on the fifth exam.

$$\boxed{\text{Average score}} = 90$$
$$\frac{85 + 80 + 95 + 78 + x}{5} = 90$$
$$\frac{338 + x}{5} = 90$$
$$5 \cdot \frac{338 + x}{5} = 5 \cdot 90$$
$$338 + x = 450$$
$$338 - 338 + x = 450 - 338$$
$$x = 112$$

It is impossible to receive an A.

81. Let $x =$ the regular price.
Then $0.80x =$ the sale price.

$$\boxed{\text{Final price}} = 0.90 \cdot \boxed{\text{Sale price}}$$
$$36 = 0.90(0.80x)$$
$$36 = 0.72x$$
$$\frac{36}{0.72} = \frac{0.72x}{0.72}$$
$$50 = x$$

The original price was $50.

83. For a purchase of $100:

$$\boxed{\text{Markdown}} = \boxed{\text{Percent markdown}} \cdot \boxed{\text{Regular price}}$$
$$15 = r \cdot 100$$
$$\frac{15}{100} = \frac{r \cdot 100}{100}$$
$$0.15 = r$$
$$15\% = r$$

For a purchase of $250:

$$\boxed{\text{Markdown}} = \boxed{\text{Percent markdown}} \cdot \boxed{\text{Regular price}}$$
$$15 = r \cdot 250$$
$$\frac{15}{250} = \frac{r \cdot 250}{250}$$
$$0.06 = r$$
$$6\% = r$$

The range of the percent discount is from 6% to 15%.

85. **Answers may vary.**

87.
$$\begin{aligned}\frac{7x+\#}{22}&=\frac{1}{2}\\ \frac{7(1)+\#}{22}&=\frac{1}{2}\\ \frac{7+\#}{22}&=\frac{1}{2}\\ 22\cdot\frac{7+\#}{22}&=22\cdot\frac{1}{2}\\ 7+\#&=11\\ 7-7+\#&=11-7\\ \#&=4\end{aligned}$$

The original equation was $\frac{7x+4}{22}=\frac{1}{2}$.

Exercise 2.3 (page 106)

1. $x^2z(y^3-z)=(-3)^2(0)\left[(-5)^3-0\right]=0$

3. $\frac{x-y^2}{2y-1+x}=\frac{-3-(-5)^2}{2(-5)-1+(-3)}=\frac{-3-(+25)}{-10-1+(-3)}=\frac{-28}{-14}=2$

5. $\frac{6}{7}-\frac{5}{8}=\frac{6\cdot\mathbf{8}}{7\cdot\mathbf{8}}-\frac{5\cdot\mathbf{7}}{8\cdot\mathbf{7}}=\frac{48}{56}-\frac{35}{56}=\frac{48-35}{56}=\frac{13}{56}$

7. $\frac{6}{7}\div\frac{5}{8}=\frac{6}{7}\cdot\frac{8}{5}=\frac{48}{35}$

9. variables, like

11. identity

13. $3x+17x=(3+17)x=20x$

15. $8x^2-5x^2=(8-5)x^2=3x^2$

17. $9x+3y\Rightarrow$ not like terms

19. $$\begin{aligned}3(x+2)+4x&=3\cdot x+3\cdot 2+4x\\&=3x+6+4x=7x+6\end{aligned}$$

21. $$\begin{aligned}5(z-3)+2z&=5\cdot z-5\cdot 3+2z\\&=5z-15+2z=7z-15\end{aligned}$$

23. $$\begin{aligned}12(x+11)-11&=12\cdot x+12\cdot 11-11\\&=12x+132-11\\&=12x+121\end{aligned}$$

25. $8(y+7)-2(y-3)=8\cdot y+8\cdot 7+(-2)\cdot y+(-2)(-3)=8y+56-2y+6=6y+62$

27. $2x+4(y-x)+3y=2x+4\cdot y+4(-x)+3y=2x+4y-4x+3y=-2x+7y$

29. $$\begin{aligned}(x+2)-(x-y)=1(x+2)-1(x-y)=1\cdot x+1\cdot 2+(-1)\cdot x+(-1)(-y)&=x+2-x+y\\&=y+2\end{aligned}$$

31. $2\left(4x+\frac{9}{2}\right)-3\left(x+\frac{2}{3}\right)=2\cdot 4x+2\cdot\frac{9}{2}+(-3)\cdot x+(-3)\cdot\frac{2}{3}=8x+9-3x-2=5x+7$

33. $8(x+3)-3x = 8\cdot x + 8\cdot 3 - 3x = 8x + 24 - 3x = 5x + 24$

35. $3.2(m+1.3) - 2.5(m-7.2) = 3.2m + 4.16 - 2.5m + 18 = 0.7m + 22.16$

37.
$$\begin{aligned} 3x+2 &= 2x \\ 3x-3x+2 &= 2x-3x \\ 2 &= -x \\ -1(2) &= -1(-x) \\ -2 &= x \end{aligned} \qquad \begin{aligned} 3x+2 &= 2x \\ 3(-2)+2 &\stackrel{?}{=} 2(-2) \\ -6+2 &\stackrel{?}{=} -4 \\ -4 &= -4 \end{aligned}$$

39.
$$\begin{aligned} 5x-3 &= 4x \\ 5x-5x-3 &= 4x-5x \\ -3 &= -x \\ -1(-3) &= -1(-x) \\ 3 &= x \end{aligned} \qquad \begin{aligned} 5x-3 &= 4x \\ 5(3)-3 &\stackrel{?}{=} 4(3) \\ 15-3 &\stackrel{?}{=} 12 \\ 12 &= 12 \end{aligned}$$

41.
$$\begin{aligned} 9y-3 &= 6y \\ 9y-9y-3 &= 6y-9y \\ -3 &= -3y \\ \frac{-3}{-3} &= \frac{-3y}{-3} \\ 1 &= y \end{aligned} \qquad \begin{aligned} 9y-3 &= 6y \\ 9(1)-3 &\stackrel{?}{=} 6(1) \\ 9-3 &\stackrel{?}{=} 6 \\ 6 &= 6 \end{aligned}$$

43.
$$\begin{aligned} 8y-7 &= y \\ 8y-8y-7 &= y-8y \\ -7 &= -7y \\ \frac{-7}{-7} &= \frac{-7y}{-7} \\ 1 &= y \end{aligned} \qquad \begin{aligned} 8y-7 &= y \\ 8(1)-7 &\stackrel{?}{=} 1 \\ 8-7 &\stackrel{?}{=} 1 \\ 1 &= 1 \end{aligned}$$

45.
$$\begin{aligned} 9-23w &= 4w \\ 9-23w+23w &= 4w+23w \\ 9 &= 27w \\ \frac{9}{27} &= \frac{27w}{27} \\ \frac{1}{3} &= w \end{aligned} \qquad \begin{aligned} 9-23w &= 4w \\ 9-23\left(\frac{1}{3}\right) &\stackrel{?}{=} 4\left(\frac{1}{3}\right) \\ \frac{27}{3}-\frac{23}{3} &\stackrel{?}{=} \frac{4}{3} \\ \frac{4}{3} &= \frac{4}{3} \end{aligned}$$

47.
$$\begin{aligned} 22-3r &= 8r \\ 22-3r+3r &= 8r+3r \\ 22 &= 11r \\ \frac{22}{11} &= \frac{11r}{11} \\ 2 &= r \end{aligned} \qquad \begin{aligned} 22-3r &= 8r \\ 22-3(2) &\stackrel{?}{=} 8(2) \\ 22-6 &\stackrel{?}{=} 16 \\ 16 &= 16 \end{aligned}$$

49.
$$\begin{aligned} 3(a+2) &= 4a \\ 3a+6 &= 4a \\ 3a-3a+6 &= 4a-3a \\ 6 &= a \end{aligned} \qquad \begin{aligned} 3(a+2) &= 4a \\ 3(6+2) &\stackrel{?}{=} 4(6) \\ 3(8) &\stackrel{?}{=} 24 \\ 24 &= 24 \end{aligned}$$

51.
$$\begin{aligned} 5(b+7) &= 6b \\ 5b+35 &= 6b \\ 5b-5b+35 &= 6b-5b \\ 35 &= b \end{aligned} \qquad \begin{aligned} 5(b+7) &= 6b \\ 5(35+7) &\stackrel{?}{=} 6(35) \\ 5(42) &\stackrel{?}{=} 210 \\ 210 &= 210 \end{aligned}$$

53.

$$\begin{aligned} 2+3(x-5) &= 4(x-1) \\ 2+3x-15 &= 4x-4 \\ 3x-13 &= 4x-4 \\ 3x-3x-13 &= 4x-3x-4 \\ -13 &= x-4 \\ -13+4 &= x-4+4 \\ -9 &= x \end{aligned}$$

$$\begin{aligned} 2+3(x-5) &= 4(x-1) \\ 2+3(-9-5) &\stackrel{?}{=} 4(-9-1) \\ 2+3(-14) &\stackrel{?}{=} 4(-10) \\ 2+(-42) &\stackrel{?}{=} -40 \\ -40 &= -40 \end{aligned}$$

55.

$$\begin{aligned} 10x+3(2-x) &= 5(x+2)-4 \\ 10x+6-3x &= 5x+10-4 \\ 7x+6 &= 5x+6 \\ 7x-5x+6 &= 5x-5x+6 \\ 2x+6 &= 6 \\ 2x+6-6 &= 6-6 \\ 2x &= 0 \\ \frac{2x}{2} &= \frac{0}{2} \\ x &= 0 \end{aligned}$$

$$\begin{aligned} 10x+3(2-x) &= 5(x+2)-4 \\ 10(0)+3(2-0) &\stackrel{?}{=} 5(0+2)-4 \\ 0+3(2) &\stackrel{?}{=} 5(2)-4 \\ 0+6 &\stackrel{?}{=} 10-4 \\ 6 &= 6 \end{aligned}$$

57.

$$\begin{aligned} 3(a+2) &= 2(a-7) \\ 3a+6 &= 2a-14 \\ 3a-2a+6 &= 2a-2a-14 \\ a+6 &= -14 \\ a+6-6 &= -14-6 \\ a &= -20 \end{aligned}$$

$$\begin{aligned} 3(a+2) &= 2(a-7) \\ 3(-20+2) &\stackrel{?}{=} 2(-20-7) \\ 3(-18) &\stackrel{?}{=} 2(-27) \\ -54 &= -54 \end{aligned}$$

59.

$$\begin{aligned} 9(x+11)+5(13-x) &= 0 \\ 9x+99+65-5x &= 0 \\ 4x+164 &= 0 \\ 4x+164-164 &= 0-164 \\ 4x &= -164 \\ \frac{4x}{4} &= \frac{-164}{4} \\ x &= -41 \end{aligned}$$

$$\begin{aligned} 9(x+11)+5(13-x) &= 0 \\ 9(-41+11)+5[13-(-41)] &\stackrel{?}{=} 0 \\ 9(-30)+5[13+41] &\stackrel{?}{=} 0 \\ -270+5(54) &\stackrel{?}{=} 0 \\ -270+270 &\stackrel{?}{=} 0 \\ 0 &= 0 \end{aligned}$$

61.

$$\begin{aligned}
\frac{3(t-7)}{2} &= t-6 \\
2\cdot\frac{3(t-7)}{2} &= 2(t-6) \\
3(t-7) &= 2(t-6) \\
3t-21 &= 2t-12 \\
3t-2t-21 &= 2t-2t-12 \\
t-21 &= -12 \\
t-21+21 &= -12+21 \\
t &= 9
\end{aligned}$$

$$\begin{aligned}
\frac{3(t-7)}{2} &= t-6 \\
\frac{3(9-7)}{2} &\stackrel{?}{=} 9-6 \\
\frac{3(2)}{2} &\stackrel{?}{=} 9-6 \\
\frac{6}{2} &\stackrel{?}{=} 3 \\
3 &= 3
\end{aligned}$$

63.

$$\begin{aligned}
\frac{5(2-s)}{3} &= s+6 \\
3\cdot\frac{5(2-s)}{3} &= 3(s+6) \\
5(2-s) &= 3(s+6) \\
10-5s &= 3s+18 \\
10-5s+5s &= 3s+5s+18 \\
10 &= 8s+18 \\
10-18 &= 8s+18-18 \\
-8 &= 8s \\
\frac{-8}{8} &= \frac{8s}{8} \\
-1 &= s
\end{aligned}$$

$$\begin{aligned}
\frac{5(2-s)}{3} &= s+6 \\
\frac{5[2-(-1)]}{3} &\stackrel{?}{=} -1+6 \\
\frac{5(2+1)}{3} &\stackrel{?}{=} 5 \\
\frac{5(3)}{3} &\stackrel{?}{=} 5 \\
\frac{15}{3} &\stackrel{?}{=} 5 \\
5 &= 5
\end{aligned}$$

65.

$$\begin{aligned}
\frac{4(2x-10)}{3} &= 2(x-4) \\
3\cdot\frac{4(2x-10)}{3} &= 3\cdot 2(x-4) \\
4(2x-10) &= 6(x-4) \\
8x-40 &= 6x-24 \\
8x-6x-40 &= 6x-6x-24 \\
2x-40 &= -24 \\
2x-40+40 &= -24+40 \\
2x &= 16 \\
\frac{2x}{2} &= \frac{16}{2} \\
x &= 8
\end{aligned}$$

$$\begin{aligned}
\frac{4(2x-10)}{3} &= 2(x-4) \\
\frac{4[2(8)-10]}{3} &\stackrel{?}{=} 2(8-4) \\
\frac{4(16-10)}{3} &\stackrel{?}{=} 2(4) \\
\frac{4(6)}{3} &\stackrel{?}{=} 8 \\
\frac{24}{3} &\stackrel{?}{=} 8 \\
8 &= 8
\end{aligned}$$

67.

$$\begin{aligned} 3.1(x-2) &= 1.3x+2.8 \\ 3.1x-6.2 &= 1.3x+2.8 \\ 3.1x-1.3x-6.2 &= 1.3x-1.3x+2.8 \\ 1.8x-6.2 &= 2.8 \\ 1.8x-6.2+6.2 &= 2.8+6.2 \\ 1.8x &= 9.0 \\ \frac{1.8x}{1.8} &= \frac{9.0}{1.8} \\ x &= 5 \end{aligned}$$

$$\begin{aligned} 3.1(x-2) &= 1.3x+2.8 \\ 3.1(5-2) &\stackrel{?}{=} 1.3(5)+2.8 \\ 3.1(3) &\stackrel{?}{=} 6.5+2.8 \\ 9.3 &= 9.3 \end{aligned}$$

69.

$$\begin{aligned} 2.7(y+1) &= 0.3(3y+33) \\ 2.7y+2.7 &= 0.9y+9.9 \\ 2.7y-0.9y+2.7 &= 0.9y-0.9y+9.9 \\ 1.8y+2.7 &= 9.9 \\ 1.8y+2.7-2.7 &= 9.9-2.7 \\ 1.8y &= 7.2 \\ \frac{1.8y}{1.8} &= \frac{7.2}{1.8} \\ y &= 4 \end{aligned}$$

$$\begin{aligned} 2.7(y+1) &= 0.3(3y+33) \\ 2.7(4+1) &= 0.3[3(4)+33] \\ 2.7(5) &\stackrel{?}{=} 0.3(12+33) \\ 13.5 &\stackrel{?}{=} 0.3(45) \\ 13.5 &= 13.5 \end{aligned}$$

71.

$$\begin{aligned} 19.1x-4(x+0.3) &= -46.5 \\ 19.1x-4x-1.2 &= -46.5 \\ 15.1x-1.2 &= -46.5 \\ 15.1x-1.2+1.2 &= -46.5+1.2 \\ 15.1x &= -45.3 \\ \frac{15.1x}{15.1} &= \frac{-45.3}{15.1} \\ x &= -3 \end{aligned}$$

$$\begin{aligned} 19.1x-4(x+0.3) &= -46.5 \\ 19.1(-3)-4(-3+0.3) &\stackrel{?}{=} -46.5 \\ -57.3-4(-2.7) &\stackrel{?}{=} -46.5 \\ -57.3+10.8 &\stackrel{?}{=} -46.5 \\ -46.5 &= -46.5 \end{aligned}$$

73.

$$\begin{aligned} 14.3(x+2)+13.7(x-3) &= 15.5 \\ 14.3x+28.6+13.7x-41.1 &= 15.5 \\ 28.0x-12.5 &= 15.5 \\ 28x-12.5+12.5 &= 15.5+12.5 \\ 28x &= 28 \\ \frac{28x}{28} &= \frac{28}{28} \\ x &= 1 \end{aligned}$$

$$\begin{aligned} 14.3(x+2)+13.7(x-3) &= 15.5 \\ 14.3(1+2)+13.7(1-3) &\stackrel{?}{=} 15.5 \\ 14.3(3)+13.7(-2) &\stackrel{?}{=} 15.5 \\ 42.9-27.4 &\stackrel{?}{=} 15.5 \\ 15.5 &= 15.5 \end{aligned}$$

75.
$$\begin{aligned}
\frac{3.7(2.3x-2.7)}{1.5} &= 5.2(x-1.2)\\
1.5\ \frac{3.7(2.3x-2.7)}{1.5} &= 1.5(5.2)(x-1.2)\\
3.7(2.3x-2.7) &= 7.8(x-1.2)\\
8.51x-9.99 &= 7.8x-9.36\\
8.51x-7.8x-9.99 &= 7.8x-7.8x-9.36\\
0.71x-9.99 &= -9.36\\
0.71x-9.99+9.99 &= -9.36+9.99\\
0.71x &= 0.63\\
\frac{0.71x}{0.71} &= \frac{0.63}{0.71}\\
x &\approx 0.887 \approx 0.9
\end{aligned}$$

77.
$$\begin{aligned}
8x+3(2-x) &= 5(x+2)-4\\
8x+6-3x &= 5x+10-4\\
5x+6 &= 5x+6\\
5x-5x+6 &= 5x-5x+6\\
6 &= 6
\end{aligned}$$
Identity

79.
$$\begin{aligned}
2(s+2) &= 2(s+1)+3\\
2s+4 &= 2s+2+3\\
2s+4 &= 2s+5\\
2s-2s+4 &= 2s-2s+5\\
4 &\neq 5
\end{aligned}$$
Contradiction

81.
$$\begin{aligned}
\frac{2(t-1)}{6}-2 &= \frac{t+2}{6}\\
6\left[\frac{2(t-1)}{6}-2\right] &= 6\cdot\frac{t+2}{6}\\
6\cdot\frac{2(t-1)}{6}-6\cdot 2 &= t+2\\
2(t-1)-12 &= t+2\\
2t-2-12 &= t+2\\
2t-14 &= t+2\\
2t-t-14 &= t-t+2\\
t-14 &= 2\\
t-14+14 &= 2+14\\
t &= 16
\end{aligned}$$

83.
$$\begin{aligned}
2(3z+4) &= 2(3z-2)+13\\
6z+8 &= 6z-4+13\\
6z+8 &= 6z+9\\
6z-6z+8 &= 6z-6z+9\\
8 &\neq 9
\end{aligned}$$
Contradiction

85.
$$2(y-3)-\frac{y}{2}=\frac{3}{2}(y-4)$$
$$2\left[2(y-3)-\frac{y}{2}\right]=2\cdot\frac{3}{2}(y-4)$$
$$2\cdot 2(y-3)-2\cdot\frac{y}{2}=3(y-4)$$
$$4(y-3)-y=3y-12$$
$$4y-12-y=3y-12$$
$$3y-12=3y-12$$
$$3y-3y-12=3y-3y-12$$
$$-12=-12$$
Identity

87.
$$\frac{3x+14}{2}=x-2+\frac{x+18}{2}$$
$$2\cdot\frac{3x+14}{2}=2\left[x-2+\frac{x+18}{2}\right]$$
$$3x+14=2x-2\cdot 2+2\cdot\frac{x+18}{2}$$
$$3x+14=2x-4+x+18$$
$$3x+14=3x+14$$
$$3x-3x+14=3x-3x+14$$
$$14=14$$
Identity

89. **Answers may vary.**

91. **Answers may vary.**

93. Let $x=$ the number.
$$x=2x$$
$$x-x=2x-x$$
$$0=x$$
The number is 0.

Exercise 2.4 (page 114)

1. $V=\frac{1}{3}Bh=\frac{1}{3}s^2h=\frac{1}{3}(10\text{ cm})^2(6\text{ cm})=\frac{1}{3}(100\text{ cm}^2)(6\text{ cm})=\frac{1}{3}(600\text{ cm}^3)=200\text{ cm}^3$

3. $3(x+2)+4(x-3)=3\cdot x+3\cdot 2+4\cdot x-4\cdot 3=3x+6+4x-12=7x-6$

5. $\frac{1}{2}(x+1)-\frac{1}{2}(x+4)=\frac{1}{2}\cdot x+\frac{1}{2}\cdot 1-\frac{1}{2}\cdot x-\frac{1}{2}\cdot 4=\frac{1}{2}x+\frac{1}{2}-\frac{1}{2}x-\frac{4}{2}=-\frac{3}{2}$

7. $A=P+Prt=1200+1200(0.08)(3)=1200+288=\$1488$

9. $2l+2w$

11. vertex

13. complementary

15. $180°$

17. Let $x=$ the length of one part.
Then $2x=$ the length of the other part.

[Length of first part] + [Length of second part] = 12
$$x+2x=12$$
$$3x=12$$
$$\frac{3x}{3}=\frac{12}{3}$$
$$x=4$$
The parts are 4 feet and 8 feet long.

19. [Sum of 3 lengths] = 30
$$x+10+2x+x=30$$
$$4x+10=30$$
$$4x+10-10=30-10$$
$$4x=20$$
$$\frac{4x}{4}=\frac{20}{4}$$
$$x=5$$
The sections are 15, 10, and 5 feet long.

21.

$$\boxed{\text{Sum of 3 lengths}} = 24$$
$$x + x + 4 + x + 2 = 24$$
$$3x + 6 = 24$$
$$3x + 6 - 6 = 24 - 6$$
$$3x = 18$$
$$\frac{3x}{3} = \frac{18}{3}$$
$$x = 6$$

The sections are 6, 10, and 8 feet long.

23. Let x = the number of hardcover books.
Then $11x$ = the number of paperbacks.

$$\boxed{\text{Number of hardcover books}} + \boxed{\text{Number of paperbacks}} = 114{,}000$$
$$x + 11x = 114{,}000$$
$$12x = 114{,}000$$
$$\frac{12x}{12} = \frac{114{,}000}{12}$$
$$x = 9500$$

There were 9500 hardcover books and 104,500 paperbacks sold.

25.

$$x + 40° = 50°$$
$$x + 40° - 40° = 50° - 40°$$
$$x = 10°$$

27.

$$x + 21° = 180°$$
$$x + 21° - 21° = 180° - 21°$$
$$x = 159°$$

29.

$$x + 12° = 59°$$
$$x + 12° - 12° = 59° - 12°$$
$$x = 47°$$

31.

$$x + 63° = 90°$$
$$x + 63° - 63° = 90° - 63°$$
$$x = 27°$$

33. Let x = the measure of the complement.

$$x + 37° = 90°$$
$$x + 37° - 37° = 90° - 37°$$
$$x = 53°$$

35. Let x = the measure of the complement.

$$x + 40° = 90°$$
$$x + 40° - 40° = 90° - 40°$$
$$x = 50°$$

Let x = the measure of the supplement of the complement.

$$x + 50° = 180°$$
$$x + 50° - 50° = 180° - 50°$$
$$x = 130°$$

37.

$$\boxed{\text{Sum of three sides}} = 57$$
$$x + x + x = 57$$
$$3x = 57$$
$$\frac{3x}{3} = \frac{57}{3}$$
$$x = 19$$

Each side is 19 feet long.

39. Let L = the length of the pool.
Then $L - 11$ = the width of the pool.

$$\boxed{\text{Perimeter}} = 94$$
$$2L + 2(L - 11) = 94$$
$$2L + 2L - 22 = 94$$
$$4L - 22 = 94$$
$$4L - 22 + 22 = 94 + 22$$
$$4L = 116$$
$$\frac{4L}{4} = \frac{116}{4}$$
$$L = 29$$

The dimensions are 29 m by 18 m.

41. Let $w =$ the width of the picture.
Then $2w + 5 =$ the length of the picture.

$$\boxed{\text{Perimeter}} = 112$$
$$2w + 2(2w + 5) = 112$$
$$2w + 4w + 10 = 112$$
$$6w + 10 = 112$$
$$6w + 10 - 10 = 112 - 10$$
$$6w = 102$$
$$\frac{6w}{6} = \frac{102}{6}$$
$$w = 17$$

The dimensions are 17 in. by 39 in.

43. Let $a =$ the measure of each angle.

$$\boxed{\text{Sum of angle measures}} = 180$$
$$a + a + a = 180$$
$$3a = 180$$
$$\frac{3a}{3} = \frac{180}{3}$$
$$a = 60$$

Each angle measures 60°.

45. Let $x =$ amount invested at 5%.

$$I = PRT$$
$$300 = x(0.05)(1)$$
$$\frac{300}{0.05} = \frac{0.05x}{0.05}$$

$6000 = x \Rightarrow$ He invested \$6,000.

47. Let $x =$ amount in 9% fund. Then $24000 - x =$ amount in 14% fund.

$$\boxed{\text{Interest at 9\%}} + \boxed{\text{Interest at 14\%}} = \boxed{\text{Total interest}}$$
$$0.09x + 0.14(24000 - x) = 3135$$
$$9x + 14(24000 - x) = 313500$$
$$9x + 336000 - 14x = 313500$$
$$-5x + 336000 = 313500$$
$$-5x + 336000 - 336000 = 313500 - 336000$$
$$-5x = -22500$$
$$\frac{-5x}{-5} = \frac{-22500}{-5}$$

$x = 4500 \Rightarrow$ \$4,500 was invested at 9%, and \$19,500 was invested at 14%.

49. Let $x =$ amount invested in each account.

$$\boxed{\text{Interest at 8\%}} + \boxed{\text{Interest at 11\%}} = \boxed{\text{Total interest}}$$
$$0.08x + 0.11x = 712.50$$
$$8x + 11x = 71250$$
$$19x = 71250$$
$$\frac{19x}{19} = \frac{71250}{19}$$
$$x = 3750$$

\$3,750 is invested in each account.

51. Let $x =$ amount invested at 7%.

$$\boxed{\text{Interest at 6\%}} + \boxed{\text{Interest at 7\%}} = \boxed{\text{Total interest}}$$
$$0.06(15000) + 0.07x = 1250$$
$$6(15000) + 7x = 125000$$
$$90000 + 7x = 125000$$
$$90000 - 90000 + 7x = 125000 - 90000$$
$$7x = 35000$$
$$\frac{7x}{7} = \frac{35000}{7}$$
$$x = 5000$$

\$5,000 should be invested at 7%.

53. Let $r =$ the fund rate and $r + 0.01 =$ the CD rate.
The client invests \$21,000 in CDs, and \$10,500 in the fund.

$$\boxed{\text{CD interest}} = \boxed{\text{Fund interest}} + 840$$
$$(r + 0.01)21000 = r(10500) + 840$$
$$21000r + 210 = 10500r + 840$$
$$21000r - 10500r + 210 = 10500r - 10500r + 840$$
$$10500r + 210 = 840$$
$$10500r + 210 - 210 = 840 - 210$$
$$10500r = 630$$
$$\frac{10500r}{10500} = \frac{630}{10500}$$
$$r = 0.06 \Rightarrow \text{The rates are 6\% and 7\%.}$$

55. **Answers may vary.**

57. Pairs of vertical angles have equal measures.

Exercise 2.5 (page 123)

1. $3 + 4(-5) = 3 + (-20) = -17$

3. $2^3 - 3^2 = 2 \cdot 2 \cdot 2 - 3 \cdot 3 = 8 - 9 = -1$

5.
$$-2x + 3 = 9$$
$$-2x + 3 - 3 = 9 - 3$$
$$-2x = 6$$
$$\frac{-2x}{-2} = \frac{6}{-2}$$
$$x = -3$$

7.
$$\frac{2}{3}p + 1 = 5$$
$$\frac{2}{3}p + 1 - 1 = 5 - 1$$
$$\frac{2}{3}p = 4$$
$$\frac{3}{2} \cdot \frac{2}{3}p = \frac{3}{2} \cdot 4$$
$$p = \frac{12}{2}$$
$$p = 6$$

9. $d = rt$

11. $v = pn$

13. Let $t =$ time for cars to meet.

	r	t	d
Car 1 (A to B)	50	t	$50t$
Car 2 (B to A)	55	t	$55t$

$$\boxed{\text{Distance for car 1}} + \boxed{\text{Distance for car 2}} = 315$$
$$50t + 55t = 315$$
$$105t = 315$$
$$\frac{105t}{105} = \frac{315}{105}$$
$$t = 3$$

The cars meet after 3 hours.

15. Let $t =$ days for crews to meet.

	r	t	d
Crew 1	1.5	t	$1.5t$
Crew 2	1.2	t	$1.2t$

$$\boxed{\text{Distance for crew 1}} + \boxed{\text{Distance for crew 2}} = 9.45$$
$$1.5t + 1.2t = 9.45$$
$$2.7t = 9.45$$
$$\frac{2.7t}{2.7} = \frac{9.45}{2.7}$$
$$t = 3.5$$

The crews meet after 3.5 days.

17. Let $t =$ time for cars to be 715 miles apart.

	r	t	d
Car 1 (going east)	60	t	$60t$
Car 2 (going west)	50	t	$50t$

$$\boxed{\text{Distance 1}} + \boxed{\text{Distance 2}} = 715$$
$$60t + 50t = 715$$
$$110t = 715$$
$$\frac{110t}{110} = \frac{715}{110}$$
$$t = 6.5$$

They will be 715 miles apart after 6.5 hours.

19. Let $t =$ time for boys to be 2 miles apart.

	r	t	d
Boy 1	3	t	$3t$
Boy 2	4	t	$4t$

$$\boxed{\text{Distance 1}} + \boxed{\text{Distance 2}} = 2$$
$$3t + 4t = 2$$
$$7t = 2$$
$$\frac{7t}{7} = \frac{2}{7}$$
$$t = \frac{2}{7}$$

They will lose contact after $\frac{2}{7}$ hour.

21.

	r	t	d
Car	60 mph	t	$60t$
Bus	50 mph	$t+2$	$50(t+2)$

$$\boxed{\text{Car distance}} = \boxed{\text{Bus distance}}$$
$$60t = 50(t+2)$$
$$60t = 50t + 100$$
$$60t - 50t = 50t - 50t + 100$$
$$10t = 100$$
$$\frac{10t}{10} = \frac{100}{10}$$
$$t = 10$$

The car will overtake the bus after 10 hours.

23. Let $t =$ time for cars to be 82.5 miles apart.

	r	t	d
Car 1	42	t	$42t$
Car 2	53	t	$53t$

$$\boxed{\text{Distance 2}} - \boxed{\text{Distance 1}} = 82.5$$
$$53t - 42t = 82.5$$
$$11t = 82.5$$
$$\frac{11t}{11} = \frac{82.5}{11}$$
$$t = 7.5$$

They will be 82.5 miles apart after 7.5 hours.

25. Let $r =$ rate of slow train. Then $r + 20 =$ rate of fast train.

	r	t	d
Slow train	r	3	$3r$
Fast train	$r+20$	3	$3(r+20)$

$$\boxed{\text{Slow dist.}} + \boxed{\text{Fast dist.}} = 330$$
$$3r + 3(r+20) = 330$$
$$3r + 3r + 60 = 330$$
$$6r + 60 = 330$$
$$6r + 60 - 60 = 330 - 60$$
$$6r = 270$$
$$\frac{6r}{6} = \frac{270}{6}$$
$$r = 45$$

The rates are 45 mph and 65 mph.

27. Let $t =$ slower time.
Then $5 - t =$ faster time.

	r	t	d
1st part	40	t	$40t$
2nd part	50	$5-t$	$50(5-t)$

$$\boxed{\text{1st dist.}} + \boxed{\text{2nd dist.}} = 210$$
$$40t + 50(5-t) = 210$$
$$40t + 250 - 50t = 210$$
$$-10t + 250 = 210$$
$$-10t + 250 - 250 = 210 - 250$$
$$-10t = -40$$
$$\frac{-10t}{-10} = \frac{-40}{-10}$$
$$t = 4$$

The car averaged 40 mph for 4 hours.

29. Let $T =$ total number of liters of solution.

12% of the total = liters of acid

$$0.12T = 0.3$$
$$\frac{0.12T}{0.12} = \frac{0.3}{0.12}$$
$$T = 2.5 \Rightarrow \text{There are 2.5 liters of the solution.}$$

31. Let $x =$ gallons of \$1.15 fuel.

$$\boxed{\text{Value of \$1.15 fuel}} + \boxed{\text{Value of \$1.15 fuel}} = \boxed{\text{Value of mixture}}$$
$$1.15x + 0.85(20) = 1(20 + x)$$
$$115x + 85(20) = 100(20 + x)$$
$$115x + 1700 = 2000 + 100x$$
$$115x - 100x + 1700 = 2000 + 100x - 100x$$
$$15x + 1700 = 2000$$
$$15x + 1700 - 1700 = 2000 - 1700$$
$$15x = 300$$
$$\frac{15x}{15} = \frac{300}{15}$$
$$x = 20 \Rightarrow \text{20 gallons of the \$1.15 fuel should be used.}$$

33. Let $x =$ gallons of 3% solution used.

$$\boxed{\text{Salt in 3\% solution}} + \boxed{\text{Salt in 7\% solution}} = \boxed{\text{Salt in mixture}}$$
$$0.03x + 0.07(50) = 0.05(x + 50)$$
$$3x + 7(50) = 5(x + 50)$$
$$3x + 350 = 5x + 250$$
$$3x - 5x + 350 = 5x - 5x + 250$$
$$-2x + 350 = 250$$
$$-2x + 350 - 350 = 250 - 350$$
$$-2x = -100$$
$$\frac{-2x}{-2} = \frac{-100}{-2}$$
$$x = 50$$

50 gallons of the 3% mixture should be used.

35. Let $x =$ ounces of water (0%) added.

$$\boxed{\text{Amt. in 10\% sol.}} + \boxed{\text{Amt. in water}} = \boxed{\text{Amt. in 8\% sol.}}$$
$$0.10(30) + 0(x) = 0.08(30 + x)$$
$$10(30) + 0 = 8(30 + x)$$
$$300 = 240 + 8x$$
$$300 - 240 = 240 - 240 + 8x$$
$$60 = 8x$$
$$\frac{60}{8} = \frac{8x}{8}$$
$$7.5 = x$$

7.5 ounces of water should be added.

37. Let x = pounds of lemondrops. Then $100 - x$ = pounds of jellybeans.

Value of lemondrops	+	Value of jellybeans	=	Value of mixture

$$1.90x + 1.20(100 - x) = 1.48(100)$$
$$190x + 120(100 - x) = 148(100)$$
$$190x + 12000 - 120x = 14800$$
$$70x + 12000 = 14800$$
$$70x + 12000 - 12000 = 14800 - 12000$$
$$70x = 2800$$
$$\frac{70x}{70} = \frac{2800}{70}$$
$x = 40 \Rightarrow$ 40 lb of lemondrops and 60 lb of jellybeans should be used.

39. Let c = cost of cashews. Then $c - 0.30$ = cost of peanuts. 20 pounds of each are used.

Value of cashews	+	Value of peanuts	=	Value of mixture

$$20c + 20(c - 0.30) = 1.05(40)$$
$$20c + 20c - 6 = 42$$
$$40c - 6 = 42$$
$$40c - 6 + 6 = 42 + 6$$
$$40c = 48$$
$$\frac{40c}{40} = \frac{48}{40}$$
$$c = 1.20$$
A bag of cashews is worth \$1.20.

41. Let x = pounds of regular coffee used.

Value of regular	+	Value of gourmet	=	Value of mixture

$$4(x) + 7(40) = 5(x + 40)$$
$$4x + 280 = 5x + 200$$
$$4x - 5x + 280 = 5x - 5x + 200$$
$$-x + 280 = 200$$
$$-x + 280 - 280 = 200 - 280$$
$$-x = -80$$
$$x = 80$$
80 pounds of regular coffee should be used.

43. Let c = cost of hazelnut beans per pound.

Value of chocolate	+	Value of hazelnut	=	Value of mixture

$$7(2) + c(5) = 6(7)$$
$$14 + 5c = 42$$
$$14 - 14 + 5c = 42 - 14$$
$$5c = 28$$
$c = \frac{28}{5} = \$5.60 \Rightarrow$ The hazelnut beans cost \$5.60 per pound.

45. Answers may vary.

47. Answers will vary.

49. Let x and $x + 2$ represent the integers.

$$x + x + 2 = 16$$
$$2x + 2 = 16$$
$$2x = 14$$
$$x = 7$$

This says that the integers are 7 and 9, but these are not **even** integers. The equation has a solution, but not the problem itself.

51. You cannot mix a 10% solution and a 20% solution and end up with a solution that has a greater concentration (30%) than either of the original solutions.

Exercise 2.6 (page 130)

1. $2x - 5y + 3x = 2x + 3x - 5y = (2+3)x - 5y = 5x - 5y$

3. $$\frac{3}{5}(x+5) - \frac{8}{5}(10+x) = \frac{3}{5}x + \frac{3}{5}\cdot 5 - \frac{8}{5}\cdot 10 - \frac{8}{5}x = \frac{3}{5}x + 3 - 16 - \frac{8}{5}x = \left(\frac{3}{5} - \frac{8}{5}\right)x - 13$$
$$= -\frac{5}{5}x - 13$$
$$= -x - 13$$

5. literal

7. isolate

9. subtract

11. $$E = IR$$
$$\frac{E}{R} = \frac{IR}{R}$$
$$\frac{E}{R} = I, \text{ or } I = \frac{E}{R}$$

13. $$V = lwh$$
$$\frac{V}{lh} = \frac{lwh}{lh}$$
$$\frac{V}{lh} = w, \text{ or } w = \frac{V}{lh}$$

15. $$P = a + b + c$$
$$P - a = a - a + b + c$$
$$P - a = b + c$$
$$P - a - c = b + c - c$$
$$P - a - c = b, \text{ or } b = P - a - c$$

17. $$P = 2l + 2w$$
$$P - 2l = 2l - 2l + 2w$$
$$P - 2l = 2w$$
$$\frac{P-2l}{2} = \frac{2w}{2}$$
$$\frac{P-2l}{2} = w, \text{ or } w = \frac{P-2l}{2}$$

19. $$A = P + Prt$$
$$A - P = P - P + Prt$$
$$A - P = Prt$$
$$\frac{A-P}{Pr} = \frac{Prt}{Pr}$$
$$\frac{A-P}{Pr} = t, \text{ or } t = \frac{A-P}{Pr}$$

21. $$C = 2\pi r$$
$$\frac{C}{2\pi} = \frac{2\pi r}{2\pi}$$
$$\frac{C}{2\pi} = r, \text{ or } r = \frac{C}{2\pi}$$

23. $$K = \frac{wv^2}{2g}$$
$$2g \cdot K = 2g \cdot \frac{wv^2}{2g}$$
$$2gK = wv^2$$
$$\frac{2gK}{v^2} = \frac{wv^2}{v^2}$$
$$\frac{2gK}{v^2} = w, \text{ or } w = \frac{2gK}{v^2}$$

25. $$P = I^2R$$
$$\frac{P}{I^2} = \frac{I^2R}{I^2}$$
$$\frac{P}{I^2} = R, \text{ or } R = \frac{P}{I^2}$$

27.

$$\begin{aligned} K &= \frac{wv^2}{2g} \\ 2g \cdot K &= 2g \cdot \frac{wv^2}{2g} \\ 2gK &= wv^2 \\ \frac{2gK}{2K} &= \frac{wv^2}{2K} \\ g &= \frac{wv^2}{2K} \end{aligned}$$

29.

$$\begin{aligned} F &= \frac{GMm}{d^2} \\ d^2 \cdot F &= d^2 \cdot \frac{GMm}{d^2} \\ d^2F &= GMm \\ \frac{d^2F}{Gm} &= \frac{GMm}{Gm} \\ \frac{d^2F}{Gm} &= M, \text{ or } M = \frac{d^2F}{Gm} \end{aligned}$$

31.

$$\begin{aligned} F &= \frac{GMm}{d^2} \\ d^2 \cdot F &= d^2 \cdot \frac{GMm}{d^2} \\ d^2F &= GMm \\ \frac{d^2F}{F} &= \frac{GMm}{F} \\ d^2 &= \frac{GMm}{F} \end{aligned}$$

33.

$$\begin{aligned} G &= 2(r-1)b \\ \frac{G}{2b} &= \frac{2(r-1)b}{2b} \\ \frac{G}{2b} &= r-1 \\ \frac{G}{2b}+1 &= r-1+1 \\ \frac{G}{2b}+1 &= r, \text{ or } r = \frac{G}{2b}+1 \end{aligned}$$

35.

$$\begin{aligned} d &= rt & t &= \frac{d}{r} \\ \frac{d}{r} &= \frac{rt}{r} & t &= \frac{135}{45} \\ \frac{d}{r} &= t & t &= 3 \end{aligned}$$

37.

$$\begin{aligned} i &= prt & t &= \frac{i}{pr} \\ \frac{i}{pr} &= \frac{prt}{pr} & t &= \frac{12}{100(0.06)} \\ \frac{i}{pr} &= t & t &= \frac{12}{6} = 2 \end{aligned}$$

39.

$$\begin{aligned} P &= a+b+c & c &= P-a-b \\ P-a &= a-a+b+c & c &= 37-15-19 \\ P-a &= b+c & c &= 22-19 \\ P-a-b &= b-b+c & c &= 3 \\ P-a-b &= c \end{aligned}$$

41.

$$\begin{aligned} K &= \frac{1}{2}h(a+b) & h &= \frac{2K}{a+b} \\ 2 \cdot K &= 2 \cdot \frac{1}{2}h(a+b) & h &= \frac{2(48)}{7+5} \\ 2K &= h(a+b) & h &= \frac{96}{12} = 8 \\ \frac{2K}{a+b} &= \frac{h(a+b)}{a+b} \\ \frac{2K}{a+b} &= h \end{aligned}$$

43.

$$\begin{aligned} E &= IR & I &= \frac{E}{R} \\ \frac{E}{R} &= \frac{IR}{R} & I &= \frac{48}{12} \\ \frac{E}{R} &= I & I &= 4 \text{ amperes} \end{aligned}$$

45.
$$C = 2\pi r \qquad \frac{C}{2\pi} = \frac{2\pi r}{2\pi} \qquad \frac{C}{2\pi} = r$$
$$r = \frac{C}{2\pi} \qquad r = \frac{14.32}{2\pi} \qquad r \approx 2.28 \text{ feet}$$

47.
$$P = I^2R \qquad \frac{P}{I^2} = \frac{I^2R}{I^2} \qquad \frac{P}{I^2} = R$$
$$R = \frac{P}{I^2} \qquad R = \frac{2700}{14^2} \qquad R = \frac{2700}{196} = 13.78 \text{ ohms}$$

49.
$$F = \frac{GMm}{d^2}$$
$$d^2 \cdot F = d^2 \cdot \frac{GMm}{d^2}$$
$$d^2F = GMm$$
$$\frac{Fd^2}{GM} = \frac{GMm}{GM}$$
$$\frac{Fd^2}{GM} = m, \text{ or } m = \frac{Fd^2}{GM}$$

51.
$$L = 2D + 3.25(r + R)$$
$$L - 3.25(r + R) = 2D + 3.25(r + R) - 3.25(r + R)$$
$$L - 3.25(r + R) = 2D$$
$$\frac{L - 3.25(r + R)}{2} = \frac{2D}{2}$$
$$\frac{L - 3.25r - 3.25R}{2} = D$$
$$D = \frac{L - 3.25r - 3.25R}{2}$$
$$D = \frac{25 - 3.25(1) - 3.25(3)}{2}$$
$$D = \frac{25 - 3.25 - 9.75}{2} = \frac{12}{2} = 6 \text{ ft}$$

53.
$$C = 0.15(T - C)$$
$$C = 0.15T - 0.15C$$
$$C + 0.15C = 0.15T - 0.15C + 0.15C$$
$$1.15C = 0.15T$$
$$\frac{1.15C}{1.15} = \frac{0.15T}{1.15}$$
$$C \approx 0.1304T$$

The maximum contribution is about 13% of taxable income.

55. **Answers may vary.**

57.
$$E = mc^2$$
$$E = 1(300{,}000)^2 = 90{,}000{,}000{,}000 \text{ joules}$$

Exercise 2.7 (page 138)

1. $3x^2 - 2(y^2 - x^2) = 3x^2 + (-2)y^2 - (-2)x^2 = 3x^2 - 2y^2 + 2x^2 = 5x^2 - 2y^2$

3. $\frac{1}{3}(x + 6) - \frac{4}{3}(x - 9) = \frac{1}{3}x + \frac{1}{3} \cdot 6 - \frac{4}{3}x - \frac{4}{3}(-9) = -\frac{3}{3}x + 2 + 12 = -x + 14$

5. is less than

7. $\geq$

9. inequality

SECTION 2.7

11.
$$\begin{aligned} x+2 &> 5 \\ x+2-2 &> 5-2 \\ x &> 3 \end{aligned}$$

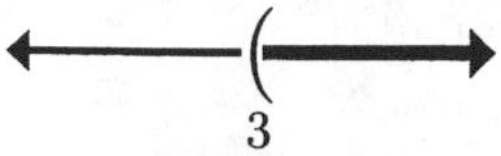

13.
$$\begin{aligned} -x-3 &\le 7 \\ -x-3+3 &\le 7+3 \\ -x &\le 10 \\ -1(-x) &\ge -1(10) \\ x &\ge -10 \end{aligned}$$

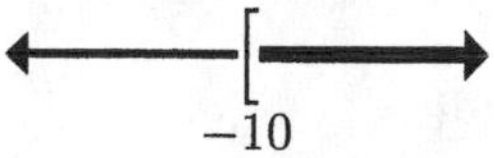

15.
$$\begin{aligned} 3+x &< 2 \\ 3-3+x &< 2-3 \\ x &< -1 \end{aligned}$$

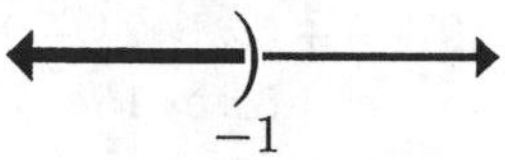

17.
$$\begin{aligned} 2x-3 &\le 5 \\ 2x-3+3 &\le 5+3 \\ 2x &\le 8 \\ \frac{2x}{2} &\le \frac{8}{2} \\ x &\le 4 \end{aligned}$$

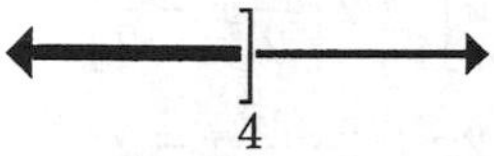

19.
$$\begin{aligned} -3x-7 &> -1 \\ -3x-7+7 &> -1+7 \\ -3x &> 6 \\ \frac{-3x}{-3} &< \frac{6}{-3} \\ x &< -2 \end{aligned}$$

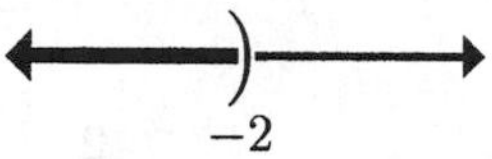

21.
$$\begin{aligned} -4x+1 &> 17 \\ -4x+1-1 &> 17-1 \\ -4x &> 16 \\ \frac{-4x}{-4} &< \frac{16}{-4} \\ x &< -4 \end{aligned}$$

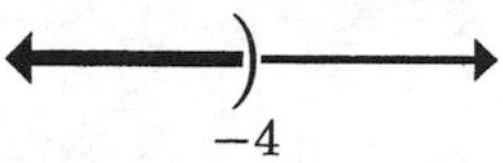

23.
$$\begin{aligned} 2x+9 &\le x+8 \\ 2x-x+9 &\le x-x+8 \\ x+9 &\le 8 \\ x+9-9 &\le 8-9 \\ x &\le -1 \end{aligned}$$

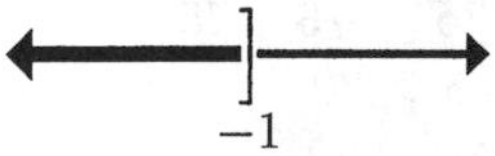

25.
$$\begin{aligned} 9x+13 &\ge 8x \\ 9x-8x+13 &\ge 8x-8x \\ x+13 &\ge 0 \\ x+13-13 &\ge 0-13 \\ x &\ge -13 \end{aligned}$$

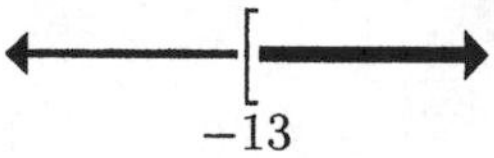

27.
$$\begin{aligned} 8x+4 &> 6x-2 \\ 8x-6x+4 &> 6x-6x-2 \\ 2x+4 &> -2 \\ 2x+4-4 &> -2-4 \\ 2x &> -6 \\ \frac{2x}{2} &> \frac{-6}{2} \\ x &> -3 \end{aligned}$$

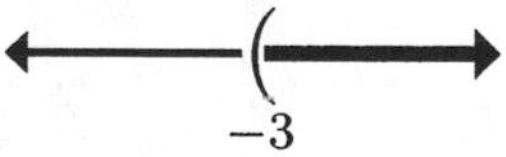

29.
$$\begin{aligned} 5x+7 &< 2x+1 \\ 5x-2x+7 &< 2x-2x+1 \\ 3x+7 &< 1 \\ 3x+7-7 &< 1-7 \\ 3x &< -6 \\ \frac{3x}{3} &< \frac{-6}{3} \\ x &< -2 \end{aligned}$$

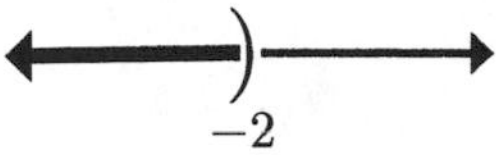

31.
$$\begin{aligned} 7-x &\le 3x-1 \\ 7-x-3x &\le 3x-3x-1 \\ 7-4x &\le -1 \\ 7-7-4x &\le -1-7 \\ -4x &\le -8 \\ \tfrac{-4x}{-4} &\ge \tfrac{-8}{-4} \\ x &\ge 2 \end{aligned}$$

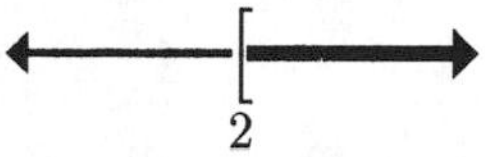

33.

$$\begin{aligned} 9-2x &> 24-7x \\ 9-2x+7x &> 24-7x+7x \\ 9+5x &> 24 \\ 9-9+5x &> 24-9 \\ 5x &> 15 \\ \tfrac{5x}{5} &> \tfrac{15}{5} \\ x &> 3 \end{aligned}$$

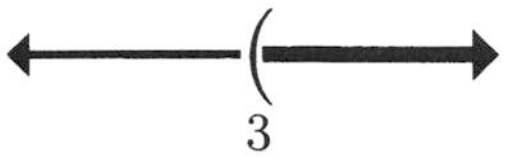

35.

$$\begin{aligned} 3(x-8) &< 5x+6 \\ 3x-24 &< 5x+6 \\ 3x-5x-24 &< 5x-5x+6 \\ -2x-24 &< 6 \\ -2x-24+24 &< 6+24 \\ -2x &< 30 \\ \tfrac{-2x}{-2} &> \tfrac{30}{-2} \\ x &> -15 \end{aligned}$$

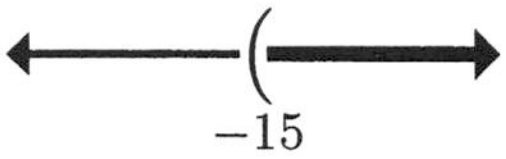

37.

$$\begin{aligned} 8(5-x) &\le 10(8-x) \\ 40-8x &\le 80-10x \\ 40-8x+10x &\le 80-10x+10x \\ 40+2x &\le 80 \\ 40-40+2x &\le 80-40 \\ 2x &\le 40 \\ \tfrac{2x}{2} &\le \tfrac{40}{2} \\ x &\le 20 \end{aligned}$$

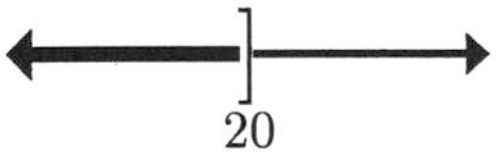

39.

$$\begin{aligned} \frac{5}{2}(7x-15)+x &\ge \frac{13}{2}x-\frac{3}{2} \\ 2\left[\frac{5}{2}(7x-15)+x\right] &\ge 2\left[\frac{13}{2}x-\frac{3}{2}\right] \\ 2\cdot\frac{5}{2}(7x-15)+2x &\ge 2\cdot\frac{13}{2}x-2\cdot\frac{3}{2} \\ 5(7x-15)+2x &\ge 13x-3 \\ 35x-75+2x &\ge 13x-3 \\ 37x-75 &\ge 13x-3 \\ 37x-13x-75 &\ge 13x-13x-3 \\ 24x-75 &\ge -3 \\ 24x-75+75 &\ge -3+75 \\ 24x &\ge 72 \\ \frac{24x}{24} &\ge \frac{72}{24} \\ x &\ge 3 \end{aligned}$$

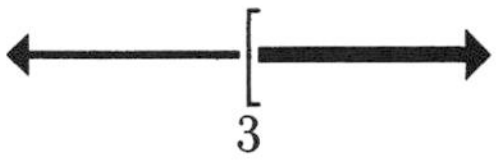

41.

$$\begin{aligned} \frac{3x-3}{2} &< 2x+2 \\ 2\cdot\frac{3x-3}{2} &< 2(2x+2) \\ 3x-3 &< 4x+4 \\ 3x-4x-3 &< 4x-4x+4 \\ -x-3 &< 4 \end{aligned}$$

$$\begin{aligned} -x-3 &< 4 \\ -x-3+3 &< 4+3 \\ -x &< 7 \\ -1(-x) &> -1(7) \\ x &> -7 \end{aligned}$$

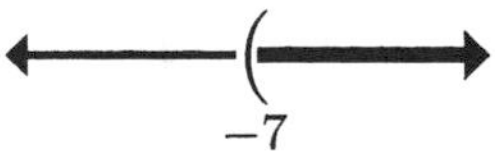

43.

$$\begin{aligned} \frac{2(x+5)}{3} &\le 3x-6 \\ 3\cdot\frac{2(x+5)}{3} &\le 3(3x-6) \\ 2(x+5) &\le 9x-18 \\ 2x+10 &\le 9x-18 \\ 2x-9x+10 &\le 9x-9x-18 \\ -7x+10 &\le -18 \end{aligned}$$

$$\begin{aligned} -7x+10 &\le -18 \\ -7x+10-10 &\le -18-10 \\ -7x &\le -28 \\ \frac{-7x}{-7} &\ge \frac{-28}{-7} \\ x &\ge 4 \end{aligned}$$

45.

$$\begin{aligned} 2 < \ & x-5 < 5 \\ 2+5 < \ & x-5+5 < 5+5 \\ 7 < \ & x < 10 \end{aligned}$$

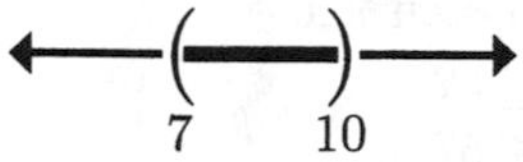

47.

$$\begin{aligned} -5 < \ & x+4 \le 7 \\ -5-4 < \ & x+4-4 \le 7-4 \\ -9 < \ & x \le 3 \end{aligned}$$

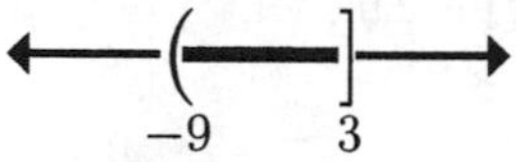

49.

$$\begin{aligned} 0 \le \ & x+10 \le 10 \\ 0-10 \le \ & x+10-10 \le 10-10 \\ -10 \le \ & x \le 0 \end{aligned}$$

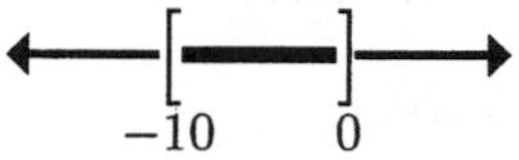

51.

$$\begin{aligned} 4 < \ & -2x < 10 \\ \frac{4}{-2} > \ & \frac{-2x}{2} > \frac{10}{-2} \\ -2 > \ & x > -5 \\ -5 < \ & x < -2 \end{aligned}$$

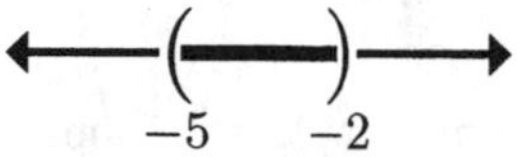

53.

$$\begin{aligned} -3 \le \ & \frac{x}{2} \le 5 \\ 2(-3) \le \ & 2\cdot\frac{x}{2} \le 2(5) \\ -6 \le \ & x \le 10 \end{aligned}$$

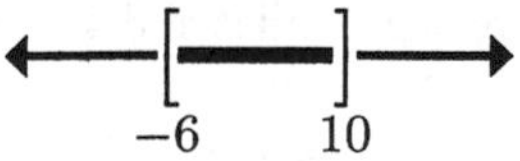

55.

$$\begin{aligned} 3 \le \ & 2x-1 < 5 \\ 3+1 \le \ & 2x-1+1 < 5+1 \\ 4 \le \ & 2x < 6 \\ \frac{4}{2} \le \ & \frac{2x}{2} < \frac{6}{2} \\ 2 \le \ & x < 3 \end{aligned}$$

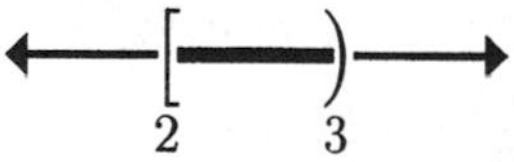

57.

$$\begin{aligned} 0 < \ & 10-5x \le 15 \\ 0-10 < \ & 10-10-5x \le 15-10 \\ -10 < \ & -5x \le 5 \\ \frac{-10}{-5} > \ & \frac{-5x}{-5} \ge \frac{5}{-5} \\ 2 > \ & x \ge -1 \\ -1 \le \ & x < 2 \end{aligned}$$

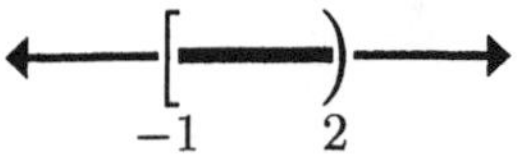

59.

$$\begin{aligned} -6 < \ & 3(x+2) < 9 \\ -6 < \ & 3x+6 < 9 \\ -6-6 < \ & 3x+6-6 < 9-6 \\ -12 < \ & 3x < 3 \\ \frac{-12}{3} < \ & \frac{3x}{3} < \frac{3}{3} \\ -4 < \ & x < 1 \end{aligned}$$

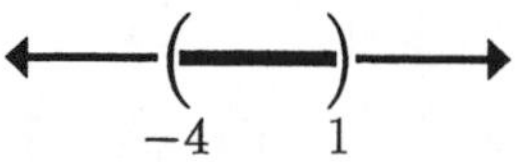

61.

$$\begin{aligned} -4 &< \frac{x-2}{2} < 6 \\ 2(-4) &< 2\cdot\frac{x-2}{2} < 2(6) \\ -8 &< x-2 < 12 \\ -8+2 &< x-2+2 < 12+2 \\ -6 &< x < 14 \end{aligned}$$

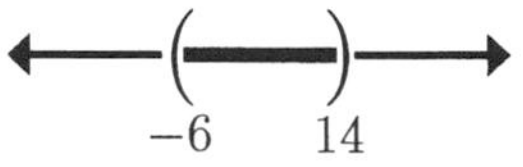

63. Let s = score on last exam.

$$\begin{aligned} \text{Average score} &\geq 80 \\ \frac{68+75+79+s}{4} &\geq 80 \\ 4\cdot\frac{68+75+79+s}{4} &\geq 4(80) \\ 68+75+79+s &\geq 320 \\ 222+s &\geq 320 \\ 222-222+s &\geq 320-222 \\ s &\geq 98 \end{aligned}$$

Her last test score must be at least 98%.

65. Let r = rating of third model.

$$\begin{aligned} \text{Average rating} &\geq 21 \\ \frac{17+19+r}{3} &\geq 21 \\ 3\cdot\frac{17+19+r}{3} &\geq 3(21) \\ 17+19+r &\geq 63 \\ 36+r &\geq 63 \\ 36-36+r &\geq 63-36 \\ r &\geq 27 \end{aligned}$$

It must have a rating of at least 27 mpg.

67. Let s = length of each side.

$$\begin{aligned} 0 &< \text{Perimeter} \leq 57 \\ 0 &< 3s \leq 57 \\ \frac{0}{3} &< \frac{3s}{3} \leq \frac{57}{3} \\ 0 &< s \leq 19 \end{aligned}$$

Each side can be from 0 feet to 19 feet long.

69.

$$\begin{aligned} 470\text{ ft} \leq\ &\text{range in feet} \leq 13{,}143\text{ ft} \\ \frac{470\text{ ft}}{5280} \leq\ &\text{range in miles} \leq \frac{13143\text{ ft}}{5280} \\ 0.1\text{ miles} \leq\ &\text{range in miles} \leq 2.5\text{ miles} \end{aligned}$$

The range is from 0.1 miles to 2.5 miles.

71.

$$\begin{aligned} 17500\text{ ft} <\ &\text{range in feet} < 21700\text{ ft} \\ \frac{17500\text{ ft}}{5280} <\ &\text{range in miles} < \frac{21700\text{ ft}}{5280} \\ 3.3\text{ miles} <\ &\text{range in miles} < 4.1\text{ miles} \end{aligned}$$

The range is between 3.3 miles and 4.1 miles.

73.

$$\begin{aligned} 19^\circ &< C < 22^\circ \\ 19^\circ &< \frac{5}{9}(F-32) < 22^\circ \\ \frac{9}{5}(19^\circ) &< \frac{9}{5}\cdot\frac{5}{9}(F-32) < \frac{9}{5}(22^\circ) \\ 34.2^\circ &< F-32 < 39.6^\circ \\ 34.2^\circ+32 &< F-32+32 < 39.6^\circ+32 \\ 66.2^\circ &< F < 71.6^\circ \end{aligned}$$

The Fahrenheit temperature is between 66.2° and 71.6°.

75.

$$\begin{aligned} 5.9\text{ in.} &< r < 6.1\text{ in.} \\ 2\pi(5.9\text{ in.}) &< 2\pi\cdot r < 2\pi(6.1\text{ in}) \\ 2(3.14)(5.9\text{ in.}) &< 2\pi r < 2(3.14)(6.1\text{ in.}) \\ 37.052\text{ in.} &< C < 38.308\text{ in.} \end{aligned}$$

The circumference can vary between about 37.052 inches and 38.308 inches.

77.
$$150 \text{ lb} < \text{range in lbs} < 190 \text{ lb}$$
$$\frac{150 \text{ lb}}{2.2} < \text{range in kg} < \frac{190 \text{ lb}}{2.2}$$
$$68.18 \text{ kg} < \text{range in kg} < 86.36 \text{ kg}$$
The weight is between 68.18 and 86.36 kg.

79. Let w = width. Then $2w - 3$ = length.
$$\begin{aligned} 24 &< \text{perimeter} < 48 \\ 24 &< 2w + 2(2w - 3) < 48 \\ 24 &< 2w + 4w - 6 < 48 \\ 24 &< 6w - 6 < 48 \\ 24 + 6 &< 6w - 6 + 6 < 48 + 6 \\ 30 &< 6w < 54 \\ \frac{30}{6} &< \frac{6w}{6} < \frac{54}{6} \\ 5 &< w < 9 \end{aligned}$$
The width could be between 5 and 9 feet.

81. **Answers may vary.**

83. This is not correct because x could represent a negative number, and then multiplying both sides of the equation would require the inequality to change from $<$ to $>$.

Chapter 2 Summary (page 142)

1.
$$\begin{aligned} 3x + 7 &= 1 \\ 3(-2) + 7 &\stackrel{?}{=} 1 \\ -6 + 7 &\stackrel{?}{=} 1 \\ 1 &= 1 \end{aligned}$$
-2 is a solution.

2.
$$\begin{aligned} 5 - 2x &= 3 \\ 5 - 2(-1) &\stackrel{?}{=} 3 \\ 5 + 2 &\stackrel{?}{=} 3 \\ 7 &\neq 3 \end{aligned}$$
-1 is not a solution.

3.
$$\begin{aligned} 2(x + 3) &= x \\ 2(-3 + 3) &\stackrel{?}{=} -3 \\ 2(0) &\stackrel{?}{=} -3 \\ 0 &\neq -3 \end{aligned}$$
-3 is not a solution.

4.
$$\begin{aligned} 5(3 - x) &= 2 - 4x \\ 5(3 - 13) &\stackrel{?}{=} 2 - 4(13) \\ 5(-10) &\stackrel{?}{=} 2 - 52 \\ -50 &= -50 \end{aligned}$$
13 is a solution.

5.
$$\begin{aligned} 3(x + 5) &= 2(x - 3) \\ 3(-21 + 5) &\stackrel{?}{=} 2(-21 - 3) \\ 3(-16) &\stackrel{?}{=} 2(-24) \\ -48 &= -48 \end{aligned}$$
-21 is a solution.

6.
$$\begin{aligned} 2(x - 7) &= x + 14 \\ 2(0 - 7) &\stackrel{?}{=} 0 + 14 \\ 2(-7) &\stackrel{?}{=} 14 \\ -14 &\neq 14 \end{aligned}$$
0 is not a solution.

7.
$$\begin{aligned} x - 7 &= -6 \\ x - 7 + 7 &= -6 + 7 \\ x &= 1 \end{aligned} \qquad \begin{aligned} x - 7 &= -6 \\ 1 - 7 &\stackrel{?}{=} -6 \\ -6 &= -6 \end{aligned}$$

8.
$$\begin{aligned} y - 4 &= 5 \\ y - 4 + 4 &= 5 + 4 \\ y &= 9 \end{aligned} \qquad \begin{aligned} y - 4 &= 5 \\ 9 - 4 &\stackrel{?}{=} 5 \\ 5 &= 5 \end{aligned}$$

9.
$$\begin{aligned} p + 4 &= 20 \\ p + 4 - 4 &= 20 - 4 \\ p &= 16 \end{aligned} \qquad \begin{aligned} p + 4 &= 20 \\ 16 + 4 &\stackrel{?}{=} 20 \\ 20 &= 20 \end{aligned}$$

10.
$$\begin{aligned} x + \frac{3}{5} &= \frac{3}{5} \\ x + \frac{3}{5} - \frac{3}{5} &= \frac{3}{5} - \frac{3}{5} \\ x &= 0 \end{aligned} \qquad \begin{aligned} x + \frac{3}{5} &= \frac{3}{5} \\ 0 + \frac{3}{5} &\stackrel{?}{=} \frac{3}{5} \\ \frac{3}{5} &= \frac{3}{5} \end{aligned}$$

11.

$$\begin{aligned} y-\frac{7}{2}&=\frac{1}{2} \\ y-\frac{7}{2}+\frac{7}{2}&=\frac{1}{2}+\frac{7}{2} \\ y&=\frac{8}{2}=4 \end{aligned}$$

$$\begin{aligned} y-\frac{7}{2}&=\frac{1}{2} \\ 4-\frac{7}{2}&\stackrel{?}{=}\frac{1}{2} \\ \frac{8}{2}-\frac{7}{2}&\stackrel{?}{=}\frac{1}{2} \\ \frac{1}{2}&=\frac{1}{2} \end{aligned}$$

12.

$$\begin{aligned} z+\frac{5}{3}&=-\frac{1}{3} \\ z+\frac{5}{3}-\frac{5}{3}&=-\frac{1}{3}-\frac{5}{3} \\ z&=-\frac{6}{3} \\ z&=-2 \end{aligned}$$

$$\begin{aligned} z+\frac{5}{3}&=-\frac{1}{3} \\ -2+\frac{5}{3}&\stackrel{?}{=}-\frac{1}{3} \\ -\frac{6}{3}+\frac{5}{3}&\stackrel{?}{=}-\frac{1}{3} \\ -\frac{1}{3}&=-\frac{1}{3} \end{aligned}$$

13. Let r = the regular price. Then

$$\boxed{\text{Sale price}}=\boxed{\text{Regular price}}-\boxed{\text{Markdown}}$$

$$\begin{aligned} 69.95&=r-35.45 \\ 69.95+35.45&=r-35.45+35.45 \\ 105.40&=r \end{aligned}$$

The regular price is \$105.40.

14. Let w = the wholesale price. Then

$$\boxed{\text{Retail price}}=\boxed{\text{Wholesale price}}+\boxed{\text{Markup}}$$

$$\begin{aligned} 212.95&=w+115.25 \\ 212.95-115.25&=w+115.25-115.25 \\ 97.70&=w \end{aligned}$$

The wholesale price is \$97.70.

15.

$$\begin{aligned} 3x&=15 \\ \frac{3x}{3}&=\frac{15}{3} \\ x&=5 \end{aligned} \qquad \begin{aligned} 3x&=15 \\ 3(5)&\stackrel{?}{=}15 \\ 15&=15 \end{aligned}$$

16.

$$\begin{aligned} 8r&=-16 \\ \frac{8r}{8}&=\frac{-16}{8} \\ r&=-2 \end{aligned} \qquad \begin{aligned} 8r&=-16 \\ 8(-2)&\stackrel{?}{=}-16 \\ -16&=-16 \end{aligned}$$

17.

$$\begin{aligned} 10z&=5 \\ \frac{10z}{10}&=\frac{5}{10} \\ z&=\frac{1}{2} \end{aligned} \qquad \begin{aligned} 10z&=5 \\ 10\left(\frac{1}{2}\right)&\stackrel{?}{=}5 \\ 5&=5 \end{aligned}$$

18.

$$\begin{aligned} 14s&=21 \\ \frac{14s}{14}&=\frac{21}{14} \\ s&=\frac{3}{2} \end{aligned} \qquad \begin{aligned} 14s&=21 \\ 14\left(\frac{3}{2}\right)&\stackrel{?}{=}21 \\ 21&=21 \end{aligned}$$

19.

$$\begin{aligned} \frac{y}{3}&=6 \\ 3\cdot\frac{y}{3}&=3\ 6 \\ y&=18 \end{aligned} \qquad \begin{aligned} \frac{y}{3}&=6 \\ \frac{18}{3}&\stackrel{?}{=}6 \\ 6&=6 \end{aligned}$$

20.

$$\begin{aligned} \frac{w}{7}&=-5 \\ 7\ \frac{w}{7}&=7(-5) \\ w&=-35 \end{aligned} \qquad \begin{aligned} \frac{w}{7}&=-5 \\ \frac{-35}{7}&\stackrel{?}{=}-5 \\ -5&=-5 \end{aligned}$$

21.

$$\begin{aligned} \frac{a}{-7}&=\frac{1}{14} \\ -7\cdot\frac{a}{-7}&=-7\left(\frac{1}{14}\right) \\ a&=-\frac{7}{14} \\ a&=-\frac{1}{2} \end{aligned} \qquad \begin{aligned} \frac{a}{-7}&=\frac{1}{14} \\ \frac{-\frac{1}{2}}{-7}&\stackrel{?}{=}\frac{1}{14} \\ \frac{2\left(-\frac{1}{2}\right)}{2(-7)}&\stackrel{?}{=}\frac{1}{14} \\ \frac{-1}{-14}&=\frac{1}{14} \end{aligned}$$

22.

$$\begin{aligned} \frac{t}{12}&=\frac{1}{2} \\ 12\cdot\frac{t}{12}&=12\left(\frac{1}{2}\right) \\ t&=\frac{12}{2} \\ t&=6 \end{aligned} \qquad \begin{aligned} \frac{t}{12}&=\frac{1}{2} \\ \frac{6}{12}&\stackrel{?}{=}\frac{1}{2} \\ \frac{1}{2}&=\frac{1}{2} \end{aligned}$$

23.
$$\begin{aligned} rb &= a \\ 35\% \cdot 700 &= a \\ 0.35(700) &= a \\ 245 &= a \end{aligned}$$

24.
$$\begin{aligned} rb &= a \\ 72\% \cdot b &= 936 \\ 0.72b &= 936 \\ \frac{0.72b}{0.72} &= \frac{936}{0.72} \\ b &= 1{,}300 \end{aligned}$$

25.
$$\begin{aligned} rb &= a \\ r \cdot 2300 &= 851 \\ \frac{r \cdot 2300}{2300} &= \frac{851}{2300} \\ r &= 0.37 \\ r &= 37\% \end{aligned}$$

26.
$$\begin{aligned} rb &= a \\ r \cdot 576 &= 72 \\ \frac{r \cdot 576}{576} &= \frac{72}{576} \\ r &= 0.125 \\ r &= 12.5\% \end{aligned}$$

27.
$$\begin{aligned} 5y + 6 &= 21 \\ 5y + 6 - 6 &= 21 - 6 \\ 5y &= 15 \\ \frac{5y}{5} &= \frac{15}{5} \\ y &= 3 \end{aligned} \qquad \begin{aligned} 5y + 6 &= 21 \\ 5(3) + 6 &\stackrel{?}{=} 21 \\ 15 + 6 &\stackrel{?}{=} 21 \\ 21 &= 21 \end{aligned}$$

28.
$$\begin{aligned} 5y - 9 &= 1 \\ 5y - 9 + 9 &= 1 + 9 \\ 5y &= 10 \\ \frac{5y}{5} &= \frac{10}{5} \\ y &= 2 \end{aligned} \qquad \begin{aligned} 5y - 9 &= 1 \\ 5(2) - 9 &\stackrel{?}{=} 1 \\ 10 - 9 &\stackrel{?}{=} 1 \\ 1 &= 1 \end{aligned}$$

29.
$$\begin{aligned} -12z + 4 &= -8 \\ -12z + 4 - 4 &= -8 - 4 \\ -12z &= -12 \\ \frac{-12z}{-12} &= \frac{-12}{-12} \\ z &= 1 \end{aligned} \qquad \begin{aligned} -12z + 4 &= -8 \\ -12(1) + 4 &\stackrel{?}{=} -8 \\ -12 + 4 &\stackrel{?}{=} -8 \\ -8 &= -8 \end{aligned}$$

30.
$$\begin{aligned} 17z + 3 &= 20 \\ 17z + 3 - 3 &= 20 - 3 \\ 17z &= 17 \\ \frac{17z}{17} &= \frac{17}{17} \\ z &= 1 \end{aligned} \qquad \begin{aligned} 17z + 3 &= 20 \\ 17(1) + 3 &\stackrel{?}{=} 20 \\ 17 + 3 &\stackrel{?}{=} 20 \\ 20 &= 20 \end{aligned}$$

31.
$$\begin{aligned} 13 - 13t &= 0 \\ 13 - 13 - 13t &= 0 - 13 \\ -13t &= -13 \\ \frac{-13t}{-13} &= \frac{-13}{-13} \\ t &= 1 \end{aligned} \qquad \begin{aligned} 13 - 13t &= 0 \\ 13 - 13(1) &\stackrel{?}{=} 0 \\ 13 - 13 &\stackrel{?}{=} 0 \\ 0 &= 0 \end{aligned}$$

32.
$$\begin{aligned} 10 + 7t &= -4 \\ 10 - 10 + 7t &= -4 - 10 \\ 7t &= -14 \\ \frac{7t}{7} &= \frac{-14}{7} \\ t &= -2 \end{aligned} \qquad \begin{aligned} 10 + 7t &= -4 \\ 10 + 7(-2) &\stackrel{?}{=} -4 \\ 10 + (-14) &\stackrel{?}{=} -4 \\ -4 &= -4 \end{aligned}$$

33.
$$\begin{aligned} 23a - 43 &= 3 \\ 23a - 43 + 43 &= 3 + 43 \\ 23a &= 46 \\ \frac{23a}{23} &= \frac{46}{23} \\ a &= 2 \end{aligned} \qquad \begin{aligned} 23a - 43 &= 3 \\ 23(2) - 43 &\stackrel{?}{=} 3 \\ 46 - 43 &\stackrel{?}{=} 3 \\ 3 &= 3 \end{aligned}$$

34.

$$\begin{aligned} 84 - 21a &= -63 \\ 84 - 84 - 21a &= -63 - 84 \\ -21a &= -147 \\ \frac{-21a}{-21} &= \frac{-147}{-21} \\ a &= 7 \end{aligned} \qquad \begin{aligned} 84 - 21a &= -63 \\ 84 - 21(7) &\stackrel{?}{=} -63 \\ 84 - 147 &\stackrel{?}{=} -63 \\ -63 &= -63 \end{aligned}$$

35.

$$\begin{aligned} 3x + 7 &= 1 \\ 3x + 7 - 7 &= 1 - 7 \\ 3x &= -6 \\ \frac{3x}{3} &= \frac{-6}{3} \\ x &= -2 \end{aligned} \qquad \begin{aligned} 3x + 7 &= 1 \\ 3(-2) + 7 &\stackrel{?}{=} 1 \\ -6 + 7 &\stackrel{?}{=} 1 \\ 1 &= 1 \end{aligned}$$

36.

$$\begin{aligned} 7 - 9x &= 16 \\ 7 - 7 - 9x &= 16 - 7 \\ -9x &= 9 \\ \frac{-9x}{-9} &= \frac{9}{-9} \\ x &= -1 \end{aligned} \qquad \begin{aligned} 7 - 9x &= 16 \\ 7 - 9(-1) &\stackrel{?}{=} 16 \\ 7 + 9 &\stackrel{?}{=} 16 \\ 16 &= 16 \end{aligned}$$

37.

$$\begin{aligned} \frac{b+3}{4} &= 2 \\ 4 \cdot \frac{b+3}{4} &= 4 \cdot 2 \\ b + 3 &= 8 \\ b + 3 - 3 &= 8 - 3 \\ b &= 5 \end{aligned} \qquad \begin{aligned} \frac{b+3}{4} &= 2 \\ \frac{5+3}{4} &\stackrel{?}{=} 2 \\ \frac{8}{4} &\stackrel{?}{=} 2 \\ 2 &= 2 \end{aligned}$$

38.

$$\begin{aligned} \frac{b-7}{2} &= -2 \\ 2 \cdot \frac{b-7}{2} &= 2(-2) \\ b - 7 &= -4 \\ b - 7 + 7 &= -4 + 7 \\ b &= 3 \end{aligned} \qquad \begin{aligned} \frac{b-7}{2} &= -2 \\ \frac{3-7}{2} &\stackrel{?}{=} -2 \\ \frac{-4}{2} &\stackrel{?}{=} -2 \\ -2 &= -2 \end{aligned}$$

39.

$$\begin{aligned} \frac{x-8}{5} &= 1 \\ 5 \cdot \frac{x-8}{5} &= 5(1) \\ x - 8 &= 5 \\ x - 8 + 8 &= 5 + 8 \\ x &= 13 \end{aligned} \qquad \begin{aligned} \frac{x-8}{5} &= 1 \\ \frac{13-8}{5} &\stackrel{?}{=} 1 \\ \frac{5}{5} &\stackrel{?}{=} 1 \\ 1 &= 1 \end{aligned}$$

40.

$$\begin{aligned} \frac{x+10}{2} &= -1 \\ 2 \cdot \frac{x+10}{2} &= 2(-1) \\ x + 10 &= -2 \\ x + 10 - 10 &= -2 - 10 \\ x &= -12 \end{aligned} \qquad \begin{aligned} \frac{x+10}{2} &= -1 \\ \frac{-12+10}{2} &\stackrel{?}{=} -1 \\ \frac{-2}{2} &\stackrel{?}{=} -1 \\ -1 &= -1 \end{aligned}$$

41.

$$\begin{aligned} \frac{2y-2}{4} &= 2 \\ 4 \cdot \frac{2y-2}{4} &= 4(2) \\ 2y - 2 &= 8 \\ 2y - 2 + 2 &= 8 + 2 \\ 2y &= 10 \\ \frac{2y}{2} &= \frac{10}{2} \\ y &= 5 \end{aligned} \qquad \begin{aligned} \frac{2y-2}{4} &= 2 \\ \frac{2(5)-2}{4} &\stackrel{?}{=} 2 \\ \frac{10-2}{4} &\stackrel{?}{=} 2 \\ \frac{8}{4} &\stackrel{?}{=} 2 \\ 2 &= 2 \end{aligned}$$

42.

$$\begin{aligned} \frac{3y+12}{11} &= 3 \\ 11 \cdot \frac{3y+12}{11} &= 11(3) \\ 3y + 12 &= 33 \\ 3y + 12 - 12 &= 33 - 12 \\ 3y &= 21 \\ \frac{3y}{3} &= \frac{21}{3} \\ y &= 7 \end{aligned} \qquad \begin{aligned} \frac{3y+12}{11} &= 3 \\ \frac{3(7)+12}{11} &\stackrel{?}{=} 3 \\ \frac{21+12}{11} &\stackrel{?}{=} 3 \\ \frac{33}{11} &\stackrel{?}{=} 3 \\ 3 &= 3 \end{aligned}$$

43.

$$\begin{aligned} \frac{x}{2}+7&=11 \\ \frac{x}{2}+7-7&=11-7 \\ \frac{x}{2}&=4 \\ 2\cdot\frac{x}{2}&=2\cdot 4 \\ x&=8 \end{aligned} \qquad \begin{aligned} \frac{x}{2}+7&=11 \\ \frac{8}{2}+7&\stackrel{?}{=}11 \\ 4+7&\stackrel{?}{=}11 \\ 11&=11 \end{aligned}$$

44.

$$\begin{aligned} \frac{r}{3}-3&=7 \\ \frac{r}{3}-3+3&=7+3 \\ \frac{r}{3}&=10 \\ 3\cdot\frac{r}{3}&=3\cdot 10 \\ r&=30 \end{aligned} \qquad \begin{aligned} \frac{r}{3}-3&=7 \\ \frac{30}{3}-3&\stackrel{?}{=}7 \\ 10-3&\stackrel{?}{=}7 \\ 7&=7 \end{aligned}$$

45.

$$\begin{aligned} \frac{a}{2}+\frac{9}{4}&=6 \\ \frac{a}{2}+\frac{9}{4}-\frac{9}{4}&=\frac{24}{4}-\frac{9}{4} \\ \frac{a}{2}&=\frac{15}{4} \\ 2\cdot\frac{a}{2}&=2\cdot\frac{15}{4} \\ a&=\frac{15}{2} \end{aligned} \qquad \begin{aligned} \frac{a}{2}+\frac{9}{4}&=6 \\ \frac{\frac{15}{2}}{2}+\frac{9}{4}&\stackrel{?}{=}6 \\ \frac{2\cdot\frac{15}{2}}{2\cdot 2}+\frac{9}{4}&\stackrel{?}{=}6 \\ \frac{15}{4}+\frac{9}{4}&\stackrel{?}{=}6 \\ \frac{24}{4}&\stackrel{?}{=}6 \\ 6&=6 \end{aligned}$$

46.

$$\begin{aligned} \frac{x}{8}-2.3&=3.2 \\ \frac{x}{8}-2.3+2.3&=3.2+2.3 \\ \frac{x}{8}&=5.5 \\ 8\cdot\frac{x}{8}&=8(5.5) \\ x&=44 \end{aligned} \qquad \begin{aligned} \frac{x}{8}-2.3&=3.2 \\ \frac{44}{8}-2.3&\stackrel{?}{=}3.2 \\ 5.5-2.3&\stackrel{?}{=}3.2 \\ 3.2&=3.2 \end{aligned}$$

47. Let $x=$ the regular price.

$$\boxed{\text{Sale price}}=\boxed{\text{Regular price}}-\boxed{\text{Markdown}}$$

$$\begin{aligned} 240&=x-0.25x \\ 240&=0.75x \\ \frac{240}{0.75}&=\frac{0.75x}{0.75} \\ 320&=x \end{aligned}$$

The regular price is \$320.

48. Let $r=$ the sales tax rate.

$$\boxed{\text{Total price}}=\boxed{\text{Price before tax}}+\boxed{\text{Sales tax}}$$

$$\begin{aligned} 40.47&=38+38r \\ 40.47-38&=38-38+38r \\ 2.47&=38r \\ \frac{2.47}{38}&=\frac{38r}{38} \\ 0.065&=r \end{aligned}$$

The sales tax rate is 6.5%.

49. Let $r=$ the % increase.

$$\boxed{\text{New price}}=\boxed{\text{Original price}}+\boxed{\text{Increase}}$$

$$\begin{aligned} 1100&=560+560r \\ 1100-560&=560-560+560r \\ 540&=560r \\ \frac{540}{560}&=\frac{560r}{560} \\ 0.964&=r \end{aligned}$$

The percent increase is 96.4%.

50. Let $r=$ the % discount.

$$\boxed{\text{Markdown}}=\boxed{\text{Percent markdown}}\cdot\boxed{\text{Regular price}}$$

$$\begin{aligned} 465-215&=r\cdot 465 \\ 250&=465r \\ \frac{250}{465}&=\frac{465r}{465} \\ 0.538&=r \end{aligned}$$

The percent markdown is 53.8%.

51. $5x+9x=(5+9)x=14x$

52. $7a+12a=(7+12)a=19a$

53. $18b-13b=(18-13)b=5b$

54. $21x-23x=(21-23)x=-2x$

55. $5y - 7y = (5-7)y = -2y$

56. $19x - 19 \Rightarrow$ not like terms

57. $$\begin{aligned} 7(x+2)+2(x-7) &= 7x+14+2x-14 \\ &= 7x+2x+14-14 \\ &= (7+2)x+0 = 9x \end{aligned}$$

58. $$\begin{aligned} 2(3-x)+x-6x &= 6-2x+x-6x \\ &= 6+(-2+1-6)x \\ &= 6-7x \end{aligned}$$

59. $$\begin{aligned} y^2+3\left(y^2-2\right) &= y^2+3y^2-6 \\ &= (1+3)y^2-6 \\ &= 4y^2-6 \end{aligned}$$

60. $$\begin{aligned} 2x^2-2\left(x^2-2\right) &= 2x^2-2x^2+4 \\ &= (2-2)x^2+4 \\ &= 0x^2+4 = 4 \end{aligned}$$

61. $$\begin{aligned} 2x-19 &= 2-x \\ 2x+x-19 &= 2-x+x \\ 3x-19 &= 2 \\ 3x-19+19 &= 2+19 \\ 3x &= 21 \\ \frac{3x}{3} &= \frac{21}{3} \\ x &= 7 \end{aligned}$$

$$\begin{aligned} 2x-19 &= 2-x \\ 2(7)-19 &\stackrel{?}{=} 2-7 \\ 14-19 &\stackrel{?}{=} -5 \\ -5 &= -5 \end{aligned}$$

62. $$\begin{aligned} 5b-19 &= 2b+20 \\ 5b-2b-19 &= 2b-2b+20 \\ 3b-19 &= 20 \\ 3b-19+19 &= 20+19 \\ 3b &= 39 \\ \frac{3b}{3} &= \frac{39}{3} \\ b &= 13 \end{aligned}$$

$$\begin{aligned} 5b-19 &= 2b+20 \\ 5(13)-19 &\stackrel{?}{=} 2(13)+20 \\ 65-19 &\stackrel{?}{=} 26+20 \\ 46 &= 46 \end{aligned}$$

63. $$\begin{aligned} 3x+20 &= 5-2x \\ 3x+2x+20 &= 5-2x+2x \\ 5x+20 &= 5 \\ 5x+20-20 &= 5-20 \\ 5x &= -15 \\ \frac{5x}{5} &= \frac{-15}{5} \\ x &= -3 \end{aligned}$$

$$\begin{aligned} 3x+20 &= 5-2x \\ 3(-3)+20 &\stackrel{?}{=} 5-2(-3) \\ -9+20 &\stackrel{?}{=} 5+6 \\ 11 &= 11 \end{aligned}$$

64.

$$\begin{aligned} 0.9x + 10 &= 0.7x + 1.8 \\ 0.9x - 0.7x + 10 &= 0.7x - 0.7x + 1.8 \\ 0.2x + 10 &= 1.8 \\ 0.2x + 10 - 10 &= 1.8 - 10 \\ 0.2x &= -8.2 \\ \frac{0.2x}{0.2} &= \frac{-8.2}{0.2} \\ x &= -41 \end{aligned}$$

$$\begin{aligned} 0.9x + 10 &= 0.7x + 1.8 \\ 0.9(-41) + 10 &\stackrel{?}{=} 0.7(-41) + 1.8 \\ -36.9 + 10 &\stackrel{?}{=} -28.7 + 1.8 \\ -26.9 &= -26.9 \end{aligned}$$

65.

$$\begin{aligned} 10(t-3) &= 3(t+11) \\ 10t - 30 &= 3t + 33 \\ 10t - 3t - 30 &= 3t - 3t + 33 \\ 7t - 30 &= 33 \\ 7t - 30 + 30 &= 33 + 30 \\ 7t &= 63 \\ \frac{7t}{7} &= \frac{63}{7} \\ t &= 9 \end{aligned}$$

$$\begin{aligned} 10(t-3) &= 3(t+11) \\ 10(9-3) &\stackrel{?}{=} 3(9+11) \\ 10(6) &\stackrel{?}{=} 3(20) \\ 60 &= 60 \end{aligned}$$

66.

$$\begin{aligned} 2(5x-7) &= 2(x-35) \\ 10x - 14 &= 2x - 70 \\ 10x - 2x - 14 &= 2x - 2x - 70 \\ 8x - 14 &= -70 \\ 8x - 14 + 14 &= -70 + 14 \\ 8x &= -56 \\ \frac{8x}{8} &= \frac{-56}{8} \\ x &= -7 \end{aligned}$$

$$\begin{aligned} 2(5x-7) &= 2(x-35) \\ 2[5(-7)-7] &\stackrel{?}{=} 2(-7-35) \\ 2[-35-7] &\stackrel{?}{=} 2(-42) \\ 2(-42) &\stackrel{?}{=} 2(-42) \\ -84 &= -84 \end{aligned}$$

67.

$$\begin{aligned} \frac{3u-6}{5} &= 3 \\ 5 \cdot \frac{3u-6}{5} &= 5(3) \\ 3u - 6 &= 15 \\ 3u - 6 + 6 &= 15 + 6 \\ 3u &= 21 \\ \frac{3u}{3} &= \frac{21}{3} \\ u &= 7 \end{aligned}$$

$$\begin{aligned} \frac{3u-6}{5} &= 3 \\ \frac{3(7)-6}{5} &\stackrel{?}{=} 3 \\ \frac{21-6}{5} &\stackrel{?}{=} 3 \\ \frac{15}{5} &\stackrel{?}{=} 3 \\ 3 &= 3 \end{aligned}$$

68.
$$\begin{aligned} \frac{5v-35}{3} &= -5 \\ 3\cdot\frac{5v-35}{3} &= 3(-5) \\ 5v-35 &= -15 \\ 5v-35+35 &= -15+35 \\ 5v &= 20 \\ \frac{5v}{5} &= \frac{20}{5} \\ v &= 4 \end{aligned}$$

$$\begin{aligned} \frac{5v-35}{3} &= -5 \\ \frac{5(4)-35}{3} &\stackrel{?}{=} -5 \\ \frac{20-35}{3} &\stackrel{?}{=} -5 \\ \frac{-15}{3} &\stackrel{?}{=} -5 \\ -5 &= -5 \end{aligned}$$

69.
$$\begin{aligned} \frac{7x-28}{4} &= -21 \\ 4\cdot\frac{7x-28}{4} &= 4(-21) \\ 7x-28 &= -84 \\ 7x-28+28 &= -84+28 \\ 7x &= -56 \\ \frac{7x}{7} &= \frac{-56}{7} \\ x &= -8 \end{aligned}$$

$$\begin{aligned} \frac{7x-28}{4} &= -21 \\ \frac{7(-8)-28}{4} &\stackrel{?}{=} -21 \\ \frac{-56-28}{4} &\stackrel{?}{=} -21 \\ \frac{-84}{4} &\stackrel{?}{=} -21 \\ -21 &= -21 \end{aligned}$$

70.
$$\begin{aligned} \frac{27+9y}{5} &= -27 \\ 5\cdot\frac{27+9y}{5} &= 5(-27) \\ 27+9y &= -135 \\ 27-27+9y &= -135-27 \\ 9y &= -162 \\ \frac{9y}{9} &= \frac{-162}{9} \\ y &= -18 \end{aligned}$$

$$\begin{aligned} \frac{27+9y}{5} &= -27 \\ \frac{27+9(-18)}{5} &\stackrel{?}{=} -27 \\ \frac{27+(-162)}{5} &\stackrel{?}{=} -27 \\ \frac{-135}{5} &\stackrel{?}{=} -27 \\ -27 &= -27 \end{aligned}$$

71.
$$\begin{aligned} 2x-5 &= x-5+x \\ 2x-5 &= 2x-5 \\ 2x-2x-5 &= 2x-2x-5 \\ -5 &= -5 \end{aligned}$$

Identity

72.
$$\begin{aligned} -3(a+1)-a &= -4a+3 \\ -3a-3-a &= -4a+3 \\ -4a-3 &= -4a+3 \\ -4a+4a-3 &= -4a+4a+3 \\ -3 &\neq 3 \end{aligned}$$

Contradiction

73.
$$\begin{aligned} 2(x-1)+4 &= 4(1+x)-(2x+2) \\ 2x-2+4 &= 4+4x-2x-2 \\ 2x+2 &= 2x+2 \\ 2x-2x+2 &= 2x-2x+2 \\ 2 &= 2 \end{aligned}$$
Identity

74. Let $x =$ the length of one part. Then $2x-7 =$ the length of the other part.

$\boxed{\text{Length of first part}} + \boxed{\text{Length of second part}} = 8$
$$\begin{aligned} x+2x-7 &= 8 \\ 3x-7 &= 8 \\ 3x-7+7 &= 8+7 \\ 3x &= 15 \\ \frac{3x}{3} &= \frac{15}{3} \\ x &= 5 \end{aligned}$$
One part should be cut 5 feet long.

75.
$$\begin{aligned} x+47° &= 62° \\ x+47°-47° &= 62°-47° \\ x &= 15° \end{aligned}$$

76.
$$\begin{aligned} x+135° &= 180° \\ x+135°-135° &= 180°-135° \\ x &= 45° \end{aligned}$$

77. Let $x =$ the measure of the complement.
$$\begin{aligned} x+69° &= 90° \\ x+69°-69° &= 90°-69° \\ x &= 21° \end{aligned}$$

78. Let $x =$ the measure of the supplement.
$$\begin{aligned} x+69° &= 180° \\ x+69°-69° &= 180°-69° \\ x &= 111° \end{aligned}$$

79. Let $w =$ the width of the picture. Then $2w+3 =$ the length of the picture.

$\boxed{\text{Perimeter}} = 84$
$$\begin{aligned} 2w+2(2w+3) &= 84 \\ 2w+4w+6 &= 84 \\ 6w+6 &= 84 \\ 6w+6-6 &= 84-6 \\ 6w &= 78 \\ \frac{6w}{6} &= \frac{78}{6} \end{aligned}$$
$w = 13 \Rightarrow$ The width is 13 inches.

80. Let $x =$ amount in 7% CD. Then $27000-x =$ amount in 9% fund.

$\boxed{\text{Interest at 7\%}} + \boxed{\text{Interest at 9\%}} = \boxed{\text{Total interest}}$
$$\begin{aligned} 0.07x+0.09(27000-x) &= 2110 \\ 7x+9(27000-x) &= 211000 \\ 7x+243000-9x &= 211000 \\ -2x+243000 &= 211000 \\ -2x+243000-243000 &= 211000-243000 \\ -2x &= -32000 \\ \frac{-2x}{-2} &= \frac{-32000}{-2} \end{aligned}$$
$x = 16000 \Rightarrow$ \$16,000 is invested at 7%, and \$11,000 is invested at 9%.

81. Let $t =$ time for friends to meet.

	r	t	d
Walk	3	t	$3t$
Bike	12	t	$12t$

$$\boxed{\text{Distance walked}} + \boxed{\text{Distance biked}} = 5$$
$$3t + 12t = 5$$
$$15t = 5$$
$$\frac{15t}{15} = \frac{5}{15}$$
$$t = \frac{1}{3}$$

They meet after $\frac{1}{3}$ hour (or 20 minutes).

82. Let $t =$ time to be 90 miles apart.

	r	t	d
1st Team	20	t	$20t$
2nd Team	25	t	$25t$

$$\boxed{\text{Distance of 1st Team}} + \boxed{\text{Distance of 2nd Team}} = 90$$
$$20t + 25t = 90$$
$$45t = 90$$
$$\frac{45t}{45} = \frac{90}{45}$$
$$t = 2$$

They will lose contact in 2 hours.

83. Let $t =$ time needed to be 75 miles apart.

	r	t	d
Bus	65	t	$65t$
Truck	55	t	$55t$

$$\boxed{\text{Bus Distance}} - \boxed{\text{Truck Distance}} = 75$$
$$65t - 55t = 75$$
$$10t = 75$$
$$\frac{10t}{10} = \frac{75}{10}$$
$$t = 7.5$$

They will be 75 miles apart after 7.5 hours.

84. Let $x =$ liters of 1% butterfat milk used.

$$\boxed{\text{Butterfat in 4\% milk}} + \boxed{\text{Butterfat in 1\% milk}} = \boxed{\text{Butterfat in 2\% mixture}}$$
$$0.04(12) + 0.01x = 0.02(12 + x)$$
$$4(12) + 1x = 2(12 + x)$$
$$48 + x = 24 + 2x$$
$$x - 2x + 48 = 24 + 2x - 2x$$
$$-x + 48 = 24$$
$$-x + 48 - 48 = 24 - 48$$
$$-x = -24$$
$$x = 24$$

24 liters of the 1% milk should be used.

85. Let $x =$ liters of 10% solution used.

$$\boxed{\text{Acid in 5\% solution}} + \boxed{\text{Acid in 10\% solution}} = \boxed{\text{Acid in 7\% solution}}$$
$$0.05(2) + 0.10x = 0.07(2 + x)$$
$$0.1 + 0.1x = 0.14 + 0.07x$$
$$10 + 10x = 14 + 7x$$
$$10x - 7x + 10 = 14 + 7x - 7x$$
$$3x + 10 = 14$$
$$3x + 10 - 10 = 14 - 10$$
$$3x = 4$$
$$\frac{3x}{3} = \frac{4}{3}$$
$$x = \tfrac{4}{3} = 1\tfrac{1}{3}$$

$1\frac{1}{3}$ liters of the 10% solution should be used.

86. Let $x =$ pounds of 90¢ candy.
Then $20 - x =$ pounds of \$1.50 candy.

$$\boxed{\text{Value of 90¢ candy}} + \boxed{\text{Value of \$1.50 candy}} = \boxed{\text{Value of mixture}}$$
$$0.90x + 1.50(20 - x) = 1.20(20)$$
$$90x + 150(20 - x) = 120(20)$$
$$90x + 3000 - 150x = 2400$$
$$-60x + 3000 = 2400$$
$$-60x + 3000 - 3000 = 2400 - 3000$$
$$-60x = -600$$
$$\frac{-60x}{-60} = \frac{-600}{-60}$$
$$x = 10$$

10 lb of each should be used.

87.
$$\begin{aligned} E &= IR \\ \frac{E}{I} &= \frac{IR}{I} \\ \frac{E}{I} &= R, \text{ or } R = \frac{E}{I} \end{aligned}$$

88.
$$\begin{aligned} i &= prt \\ \frac{i}{pr} &= \frac{prt}{pr} \\ \frac{i}{pr} &= t, \text{ or } t = \frac{i}{pr} \end{aligned}$$

89.
$$\begin{aligned} P &= I^2R \\ \frac{P}{I^2} &= \frac{I^2R}{I^2} \\ \frac{P}{I^2} &= R, \text{ or } R = \frac{P}{I^2} \end{aligned}$$

90.
$$\begin{aligned} d &= rt \\ \frac{d}{t} &= \frac{rt}{t} \\ \frac{d}{t} &= r, \text{ or } r = \frac{d}{t} \end{aligned}$$

91.
$$\begin{aligned} V &= lwh \\ \frac{V}{lw} &= \frac{lwh}{lw} \\ \frac{V}{lw} &= h, \text{ or } h = \frac{V}{lw} \end{aligned}$$

92.
$$\begin{aligned} y &= mx + b \\ y - b &= mx + b - b \\ y - b &= mx \\ \frac{y-b}{x} &= \frac{mx}{x} \\ \frac{y-b}{x} &= m, \text{ or } m = \frac{y-b}{x} \end{aligned}$$

93.
$$\begin{aligned} V &= \pi r^2 h \\ \frac{V}{\pi r^2} &= \frac{\pi r^2 h}{\pi r^2} \\ \frac{V}{\pi r^2} &= h, \text{ or } h = \frac{V}{\pi r^2} \end{aligned}$$

94.
$$\begin{aligned} a &= 2\pi rh \\ \frac{a}{2\pi h} &= \frac{2\pi rh}{2\pi h} \\ \frac{a}{2\pi h} &= r, \text{ or } r = \frac{a}{2\pi h} \end{aligned}$$

95.
$$\begin{aligned} F &= \frac{GMm}{d^2} \\ d^2 \cdot F &= d^2 \cdot \frac{GMm}{d^2} \\ d^2F &= GMm \\ \frac{d^2F}{Mm} &= \frac{GMm}{Mm} \\ \frac{d^2F}{Mm} &= G, \text{ or } G = \frac{d^2F}{Mm} \end{aligned}$$

96.
$$\begin{aligned} P &= \frac{RT}{mV} \\ mV \cdot P &= mV \cdot \frac{RT}{mV} \\ mVP &= RT \\ \frac{mVP}{VP} &= \frac{RT}{VP} \\ m &= \frac{RT}{VP} \end{aligned}$$

97.
$$\begin{aligned} 3x + 2 &< 5 \\ 3x + 2 - 2 &< 5 - 2 \\ 3x &< 3 \\ \frac{3x}{3} &< \frac{3}{3} \\ x &< 1 \end{aligned}$$

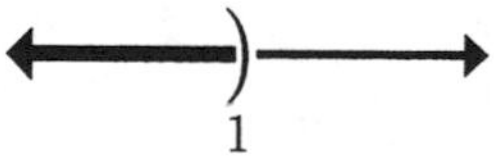

98.
$$\begin{aligned} -5x - 8 &> 7 \\ -5x - 8 + 8 &> 7 + 8 \\ -5x &> 15 \\ \frac{-5x}{-5} &< \frac{15}{-5} \\ x &< -3 \end{aligned}$$

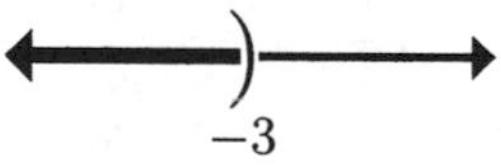

99.
$$\begin{aligned} 5x - 3 &\geq 2x + 9 \\ 5x - 2x - 3 &\geq 2x - 2x + 9 \\ 3x - 3 &\geq 9 \\ 3x - 3 + 3 &\geq 9 + 3 \\ 3x &\geq 12 \\ \frac{3x}{3} &\geq \frac{12}{3} \\ x &\geq 4 \end{aligned}$$

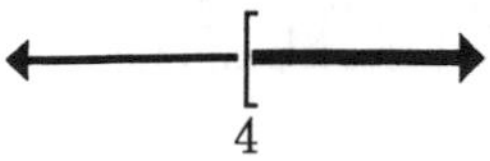

100.
$$\begin{aligned} 7x + 1 &\leq 8x - 5 \\ 7x - 8x + 1 &\leq 8x - 8x - 5 \\ -x + 1 &\leq -5 \\ -x + 1 - 1 &\leq -5 - 1 \\ -x &\leq -6 \\ -(-x) &\geq -(-6) \\ x &\geq 6 \end{aligned}$$

101.
$$\begin{aligned} 5(3-x) &\le 3(x-3) \\ 15-5x &\le 3x-9 \\ 15-5x-3x &\le 3x-3x-9 \\ 15-8x &\le -9 \\ 15-15-8x &\le -9-15 \\ -8x &\le -24 \\ \frac{-8x}{-8} &\ge \frac{-24}{-8} \\ x &\ge 3 \end{aligned}$$

102.
$$\begin{aligned} 3(5-x) &\ge 2x \\ 15-3x &\ge 2x \\ 15-3x+3x &\ge 2x+3x \\ 15 &\ge 5x \\ \frac{15}{5} &\ge \frac{5x}{5} \\ 3 &\ge x, \text{ or } x \le 3 \end{aligned}$$

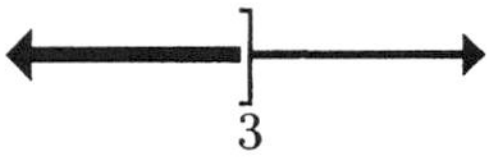

103.
$$\begin{aligned} 8 < {} & x+2 < 13 \\ 8-2 < {} & x+2-2 < 13-2 \\ 6 < {} & x < 11 \end{aligned}$$

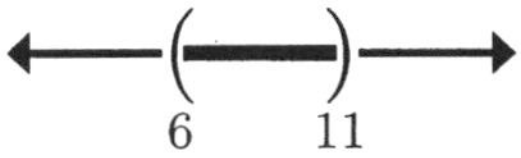

104.
$$\begin{aligned} 0 \le {} & 2-2x < 4 \\ 0-2 \le {} & 2-2-2x < 4-2 \\ -2 \le {} & -2x < 2 \\ \frac{-2}{-2} \ge {} & \frac{-2x}{-2} > \frac{2}{-2} \\ 1 \ge {} & x > -1 \\ -1 < {} & x \le 1 \end{aligned}$$

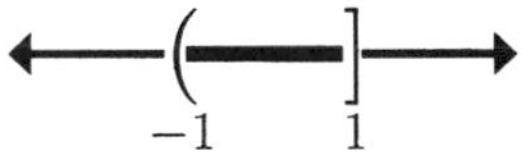

105. Let l = length. Then $l-6$ = width.
$$\begin{aligned} \text{perimeter} &\le 68 \\ 2l+2(l-6) &\le 68 \\ 2l+2l-12 &\le 68 \\ 4l-12 &\le 68 \\ 4l-12+12 &\le 68+12 \\ 4l &\le 80 \\ \frac{4l}{4} &\le \frac{80}{4} \\ l &\le 20 \end{aligned}$$
The length must be no more than 20 feet.

Chapter 2 Test (page 146)

1.
$$\begin{aligned} 5x+3 &= -2 \\ 5(-1)+3 &\stackrel{?}{=} -2 \\ -5+3 &\stackrel{?}{=} -2 \\ -2 &= -2 \end{aligned}$$
−1 is a solution.

2.
$$\begin{aligned} 3(x+2) &= 2x \\ 3(-6+2) &\stackrel{?}{=} 2(-6) \\ 3(-4) &\stackrel{?}{=} -12 \\ -12 &= -12 \end{aligned}$$
−6 is a solution.

3.
$$\begin{aligned} -3(2-x) &= 0 \\ -3[2-(-2)] &\stackrel{?}{=} 0 \\ -3[4] &\stackrel{?}{=} 0 \\ -12 &\ne 0 \end{aligned}$$
−2 is not a solution.

4.
$$\begin{aligned} 3(x+2) &= 2x+7 \\ 3(1+2) &\stackrel{?}{=} 2(1)+7 \\ 3(3) &\stackrel{?}{=} 2+7 \\ 9 &= 9 \end{aligned}$$
1 is a solution.

5.
$$\begin{aligned} x+17 &= -19 \\ x+17-17 &= -19-17 \\ x &= -36 \end{aligned}$$

6.
$$\begin{aligned} a-15 &= 32 \\ a-15+15 &= 32+15 \\ a &= 47 \end{aligned}$$

7.
$$\begin{aligned} 12x &= -144 \\ \frac{12x}{12} &= \frac{-144}{12} \\ x &= -12 \end{aligned}$$

8.
$$\begin{aligned} \frac{x}{7} &= -1 \\ 7 \cdot \frac{x}{7} &= 7(-1) \\ x &= -7 \end{aligned}$$

9.
$$\begin{aligned} 8x+2 &= -14 \\ 8x+2-2 &= -14-2 \\ 8x &= -16 \\ \frac{8x}{8} &= \frac{-16}{8} \\ x &= -2 \end{aligned}$$

10.
$$\begin{aligned} 3 &= 5-2x \\ 3-5 &= 5-5-2x \\ -2 &= -2x \\ \frac{-2}{-2} &= \frac{-2x}{-2} \\ 1 &= x \end{aligned}$$

11.
$$\begin{aligned} \frac{2x-5}{3} &= 3 \\ 3 \cdot \frac{2x-5}{3} &= 3(3) \\ 2x-5 &= 9 \\ 2x-5+5 &= 9+5 \\ 2x &= 14 \\ \tfrac{2x}{2} &= \tfrac{14}{2} \\ x &= 7 \end{aligned}$$

12.
$$\begin{aligned} \frac{3x-18}{2} &= 6x \\ 2 \cdot \frac{3x-18}{2} &= 2(6x) \\ 3x-18 &= 12x \\ 3x-3x-18 &= 12x-3x \\ -18 &= 9x \\ \tfrac{-18}{9} &= \tfrac{9x}{9} \\ -2 &= x \end{aligned}$$

13.
$$\begin{aligned} 23-5(x+10) &= -12 \\ 23-5x-50 &= -12 \\ -5x-27 &= -12 \\ -5x-27+27 &= -12+27 \\ -5x &= 15 \\ \frac{-5x}{-5} &= \frac{15}{-5} \\ x &= -3 \end{aligned}$$

14.
$$\begin{aligned} \frac{7}{8}(x-4) &= 5x-\frac{7}{2} \\ 8 \cdot \frac{7}{8}(x-4) &= 8\left(5x-\frac{7}{2}\right) \\ 7(x-4) &= 8(5x)-8 \cdot \tfrac{7}{2} \\ 7x-28 &= 40x-28 \\ 7x-40x-28 &= 40x-40x-28 \\ -33x-28 &= -28 \\ -33x-28+28 &= -28+28 \\ -33x &= 0 \\ x &= 0 \end{aligned}$$

15. $x+5(x-3) = x+5x-15 = 6x-15$

16. $3x-5(2-x) = 3x-10+5x = 8x-10$

17. $-3(x+3)+3(x-3) = -3x-9+3x-9 = -3x+3x-9-9 = -18$

18. $-4(2x-5)-7(4x+1) = -8x+20-28x-7 = -36x+13$

19.
$$x + 45° = 120°$$
$$x + 45° - 45° = 120° - 45°$$
$$x = 75°$$

20. Let x = the measure of the supplement.
$$x + 105° = 180°$$
$$x + 105° - 105° = 180° - 105°$$
$$x = 75°$$

21. Let x = amount at 6%. Then $10000 - x$ = amount at 5%.

[Interest at 6%] + [Interest at 5%] = [Total interest]

$$0.06x + 0.05(10000 - x) = 560$$
$$6x + 5(10000 - x) = 56000$$
$$6x + 50000 - 5x = 56000$$
$$x + 50000 = 56000$$
$$x + 50000 - 50000 = 56000 - 50000$$
$x = 6000 \Rightarrow$ \$6,000 is invested at 6%, and \$4,000 is invested at 5%.

22. Let t = time to meet.

	r	t	d
Car	65	t	$65t$
Truck	55	t	$55t$

[Car Dist] + [Truck Dist] = 72

$$65t + 55t = 72$$
$$120t = 72$$
$$\frac{120t}{120} = \frac{72}{120}$$
$$t = \frac{3}{5}$$

They meet after $\frac{3}{5}$ of an hour.

23. Let x = liters of water used.

[Amt. in 10% sol.] + [Amt. in water] = [Amt. in 8% sol.]

$$0.10(30) + 0(x) = 0.08(30 + x)$$
$$10(30) + 0 = 8(30 + x)$$
$$300 = 240 + 8x$$
$$300 - 240 = 8x$$
$$60 = 8x$$
$$\frac{60}{8} = \frac{8x}{8}$$
$$\frac{15}{2} = x$$

$7\frac{1}{2}$ liters of water should be used.

24. Let x = pounds of peanuts used.

[Value of cashews] + [Value of peanuts] = [Value of mixture]

$$6(20) + 3x = 4(20 + x)$$
$$120 + 3x = 80 + 4x$$
$$120 + 3x - 4x = 80 + 4x - 4x$$
$$-x + 120 = 80$$
$$-x + 120 - 120 = 80 - 120$$
$$-x = -40$$
$x = 40$; 40 pounds of peanuts should be used.

25.
$$d = rt$$
$$\frac{d}{r} = \frac{rt}{r}$$
$\frac{d}{r} = t$, or $t = \frac{d}{r}$

26.
$$P = 2l + 2w$$
$$P - 2w = 2l + 2w - 2w$$
$$P - 2w = 2l$$
$$\frac{P - 2w}{2} = \frac{2l}{2}$$
$\frac{P - 2w}{2} = l$, or $l = \frac{P - 2w}{2}$

27.
$$A = 2\pi rh$$
$$\frac{A}{2\pi r} = \frac{2\pi rh}{2\pi r}$$
$\frac{A}{2\pi r} = h$, or $h = \frac{A}{2\pi r}$

28.
$$A = P + Prt$$
$$A - P = P - P + Prt$$
$$A - P = Prt$$
$$\frac{A-P}{Pt} = \frac{Prt}{Pt}$$
$$\frac{A-P}{Pt} = r \text{, or } r = \frac{A-P}{Pt}$$

29.
$$8x - 20 \geq 4$$
$$8x - 20 + 20 \geq 4 + 20$$
$$8x \geq 24$$
$$\frac{8x}{8} \geq \frac{24}{8}$$
$$x \geq 3$$

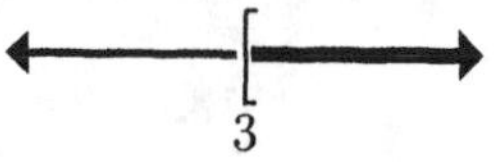

30.
$$x - 2(x + 7) > 14$$
$$x - 2x - 14 > 14$$
$$-x - 14 > 14$$
$$-x - 14 + 14 > 14 + 14$$
$$-x > 28$$
$$-(-x) < -(28)$$
$$x < -28$$

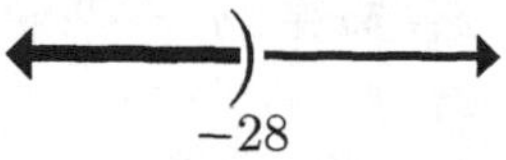

31.
$$-4 \leq 2(x+1) < 10$$
$$-4 \leq 2x + 2 < 10$$
$$-4 - 2 \leq 2x + 2 - 2 < 10 - 2$$
$$-6 \leq 2x < 8$$
$$\frac{-6}{2} \leq \frac{2x}{2} < \frac{8}{2}$$
$$-3 \leq x < 4$$

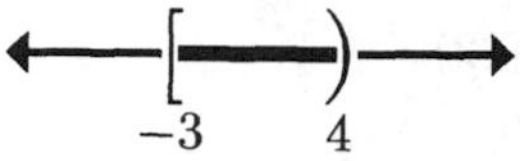

32.
$$-2 < 5(x-1) \leq 10$$
$$-2 < 5x - 5 \leq 10$$
$$-2 + 5 < 5x - 5 + 5 \leq 10 + 5$$
$$3 < 5x \leq 15$$
$$\frac{3}{5} < \frac{5x}{5} \leq \frac{15}{5}$$
$$\frac{3}{5} < x \leq 3$$

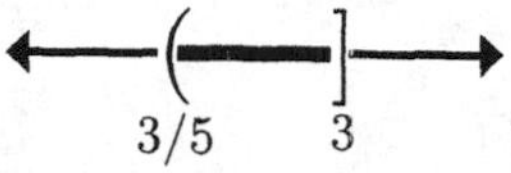

Cumulative Review Exercises (page 147)

1. $\frac{27}{9} = 3$: integer, rational number, real number, positive number

2. $-0.25 = -\frac{1}{4}$: rational number, real number, negative number

3.

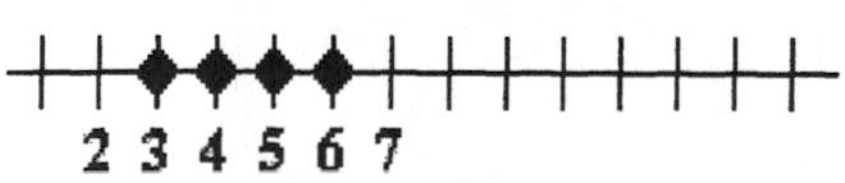

4.

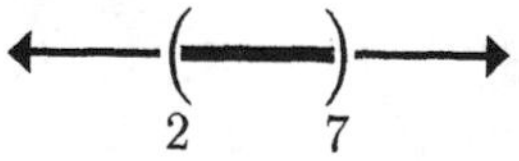

5. $\frac{|-3|-|3|}{|-3-3|} = \frac{(+3)-(+3)}{|-6|} = \frac{0}{+6} = 0$

6. $\frac{5}{7} \cdot \frac{14}{3} = \frac{5}{\cancel{7}_1} \cdot \frac{2 \cdot \cancel{7}^1}{3} = \frac{10}{3}$

7. $2\frac{3}{5} + 5\frac{1}{2} = \frac{13}{5} + \frac{11}{2} = \frac{13 \cdot 2}{5 \cdot 2} + \frac{11 \cdot 5}{2 \cdot 5} = \frac{26}{10} + \frac{55}{10} = \frac{81}{10} = 8\frac{1}{10}$

8. $35.7 - 0.05 = 35.65$

9. $(3x - 2y)z = [3(-5) - 2(3)]0 = 0$

10. $\frac{x - 3y + |z|}{2 - x} = \frac{-5 - 3(3) + |0|}{2 - (-5)} = \frac{-5 - 9 + 0}{2 + 5} = \frac{-14}{7} = -2$

11.
$$x^2 - y^2 + z^2 = (-5)^2 - (3)^2 + 0^2$$
$$= 25 - 9 + 0 = 16$$

12. $\frac{x}{y} + \frac{y+2}{3-z} = \frac{-5}{3} + \frac{3+2}{3-0} = -\frac{5}{3} + \frac{5}{3} = 0$

13.
$$\begin{aligned} rb &= a \\ 0.075(330) &= a \\ 24.75 &= a \end{aligned}$$

14.
$$\begin{aligned} rb &= a \\ 0.32b &= 1688 \\ \tfrac{0.32b}{0.32} &= \tfrac{1688}{0.32} \\ b &= 5{,}275 \end{aligned}$$

15. 2nd term: $5x^2y$; coefficient: 5

16. 3rd term: $37y$; factors: 37, y

17. $3x - 5x + 2y = -2x + 2y$

18.
$$\begin{aligned} 3(x-7) + 2(8-x) &= 3x - 21 + 16 - 2x \\ &= x - 5 \end{aligned}$$

19. $2x^2y^3 - 4x^2y^3 = -2x^2y^3$

20. $2(3 - x) + 5(x + 2) = 6 - 2x + 5x + 10 = 3x + 16$

21.
$$\begin{aligned} 3(x-5) + 2 &= 2x \\ 3x - 15 + 2 &= 2x \\ 3x - 13 &= 2x \\ 3x - 3x - 13 &= 2x - 3x \\ -13 &= -x \\ 13 &= x \end{aligned}$$

22.
$$\begin{aligned} \tfrac{x-5}{3} - 5 &= 7 \\ \tfrac{x-5}{3} - 5 + 5 &= 7 + 5 \\ \tfrac{x-5}{3} &= 12 \\ 3 \cdot \tfrac{x-5}{3} &= 3(12) \\ x - 5 &= 36 \\ x - 5 + 5 &= 36 + 5 \\ x &= 41 \end{aligned}$$

23.
$$\begin{aligned} \frac{2x-1}{5} &= \frac{1}{2} \\ 10 \cdot \frac{2x-1}{5} &= 10 \cdot \frac{1}{2} \\ 2(2x-1) &= 5 \\ 4x - 2 &= 5 \\ 4x - 2 + 2 &= 5 + 2 \\ 4x &= 7 \\ \tfrac{4x}{4} &= \tfrac{7}{4} \\ x &= \tfrac{7}{4} \end{aligned}$$

24.
$$\begin{aligned} 2(a-5) - (3a+1) &= 0 \\ 2a - 10 - 3a - 1 &= 0 \\ -a - 11 &= 0 \\ -a + a - 11 &= 0 + a \\ -11 &= a \end{aligned}$$

25. Let x = Dealer's invoice.

[Price] = [Dealer's invoice] + [Markup]

$$\begin{aligned} 23499 &= x + 0.03x \\ 23499 &= 1.03x \\ \tfrac{23499}{1.03} &= \tfrac{1.03x}{1.03} \\ 22814.56 &= r \end{aligned}$$

The dealer's invoice was \$22,814.56.

26. Let x = original sofa price.

[Price] = [New sofa price] + [New chair price]

$$\begin{aligned} 780 &= x - 0.35x + 300 - 0.35(300) \\ 780 &= 0.65x + 300 - 105 \\ 780 &= 0.65x + 195 \\ 780 - 195 &= 0.65x + 195 - 195 \\ 585 &= 0.65x \\ \tfrac{585}{0.65} &= \tfrac{0.65x}{0.65} \\ 900 &= x \end{aligned}$$

The original price of the sofa was \$900.

27. Let $x =$ original car price.

$$13725.25 = x + 0.085x$$
$$13725.25 = 1.085x$$
$$\frac{13725.25}{1.085} = \frac{1.085x}{1.085}$$
$$12650 = x$$

The original price was $12,650.

28. Let x and $3x =$ the unknown amounts.

$\boxed{\text{Cement}} + \boxed{\text{Gravel}} = \boxed{\text{Concrete}}$

$$x + 3x = 500$$
$$4x = 500$$
$$\tfrac{4x}{4} = \tfrac{500}{4}$$
$$x = 125 \text{ lbs cement}$$

29. Let x and $2x =$ the lengths.

$$\text{Total length} = 35$$
$$14 + x + 2x = 35$$
$$14 + 3x = 35$$
$$14 - 14 + 3x = 35 - 14$$
$$3x = 21$$
$$\tfrac{3x}{3} = \tfrac{21}{3}$$
$$x = 7 \text{ ft}$$

The section will not span the doorway.

30. Let $x =$ the length of one panel.

$$\boxed{\text{Total length}} = 18$$
$$x + x + 3.4 = 18$$
$$2x + 3.4 = 18$$
$$2x + 3.4 - 3.4 = 18 - 3.4$$
$$2x = 14.6$$
$$\frac{2x}{2} = \frac{14.6}{2}$$
$$x = 7.3$$

The lengths are 7.3 and 10.7 feet.

31. Let $n =$ number of kwh used.

$$17.50 + 0.18n = 43.96$$
$$17.50 - 17.50 + 0.18n = 43.96 - 17.50$$
$$0.18n = 26.46$$
$$\frac{0.18n}{0.18} = \frac{26.46}{0.18}$$
$$n = 147$$

147 kwh were used that month.

32. Let $n =$ number of feet required.

$$35 + 1.50n = 162.50$$
$$35 - 35 + 1.50n = 162.50 - 35$$
$$1.50n = 127.50$$
$$\frac{1.50n}{1.50} = \frac{127.50}{1.50}$$
$$n = 85$$

85 ft of gutters were required.

33.

$$A = \frac{1}{2}h(b + B)$$
$$2A = 2 \cdot \frac{1}{2}h(b + B)$$
$$2A = h(b + B)$$
$$\tfrac{2A}{b+B} = \tfrac{h(b+B)}{b+B}$$
$$\tfrac{2A}{b+B} = h, \text{ or } h = \tfrac{2A}{b+B}$$

34.

$$y = mx + b$$
$$y - b = mx + b - b$$
$$y - b = mx$$
$$\tfrac{y-b}{m} = \tfrac{mx}{m}$$
$$\tfrac{y-b}{m} = x, \text{ or } x = \tfrac{y-b}{m}$$

35. $4^2 - 5^2 = 16 - 25 = -9$

36. $(4 - 5)^2 = (-1)^2 = 1$

37. $5(4^3 - 2^3) = 5(64 - 8) = 5(56) = 280$

38.

$$-2(5^4 - 7^3) = -2(625 - 343)$$
$$= -2(282) = -564$$

39.
$$\begin{aligned} 8(4+x) &> 10(6+x) \\ 32+8x &> 60+10x \\ 32+8x-10x &> 60+10x-10x \\ 32-2x &> 60 \\ 32-32-2x &> 60-32 \\ -2x &> 28 \\ \frac{-2x}{-2} &< \frac{28}{-2} \\ x &< -14 \end{aligned}$$

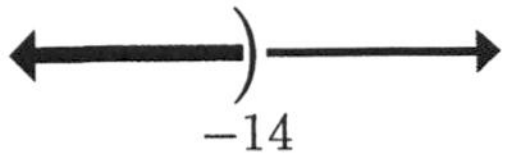

40.
$$\begin{aligned} -9 <\ & 3(x+2) \le 3 \\ -9 <\ & 3x+6 \le 3 \\ -9-6 <\ & 3x+6-6 \le 3-6 \\ -15 <\ & 3x \le -3 \\ \frac{-15}{3} <\ & \frac{3x}{3} \le \frac{-3}{3} \\ -5 <\ & x \le -1 \end{aligned}$$

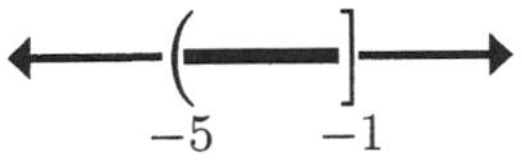

Exercise 3.1 (page 158)

1. $-3-3(-5) = -3-(-15) = -3+15$
$= 12$

3. opposite of -8: 8

5.
$$\begin{aligned} -4x+7 &= -21 \\ -4x+7-7 &= -21-7 \\ -4x &= -28 \\ \frac{-4x}{-4} &= \frac{-28}{-4} \\ x &= 7 \end{aligned}$$

7.
$$\begin{aligned} (x+1)(x+y)^2 &= (-2+1)[-2+(-5)]^2 \\ &= (-1)(-7)^2 \\ &= -1(+49) = -49 \end{aligned}$$

9. ordered pair

11. origin

13. rectangular coordinate

15. no

17. origin, left, up

19. II

21.

x	y
4	3 or -3
0	5 or -5
-3	4 or -4
5 or -5	0
-4	-3 or 3
0	-5 or 5
3	-4 or 4

23. The point $(-10, 60)$ indicates that 10 minutes before the workout started, her heart rate was 60 beats per minute.

25. The point with an x-coordinate of 30 is $(30, 150)$, so her heart rate was 150 beats per minute one half-hour after starting.

27. The points on the graph with a y-coordinate of 100 have x-coordinates of approximately 5 and 50, so her heart rate was 100 beats per minute after about 5 and 50 minutes.

29. Before the workout, her heart rate was 60 beats per minute. After the workout, her heart rate was about 70 beats per minute, or about 10 beats per minute higher.

31.

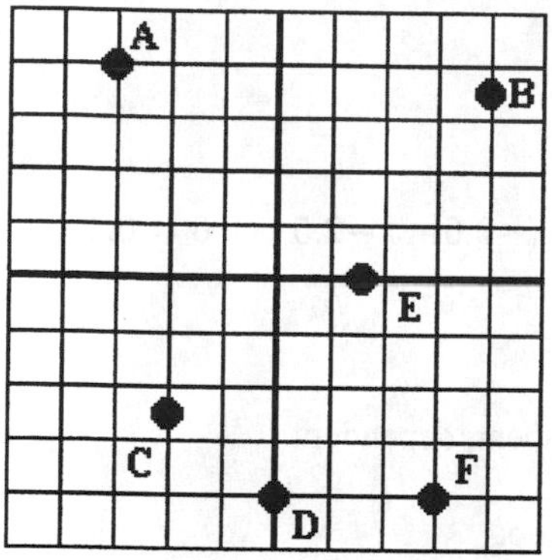

33.

City	Ordered Pair
Carbondale	$(5\frac{1}{2}, J)$
Champaign	(7, D)
Chicago	$(8\frac{1}{2}, B)$
Peoria	$(5\frac{3}{4}, C)$
Rockford	$(5\frac{3}{4}, A)$
Springfield	$(4\frac{1}{2}, E)$
St. Louis	$(4\frac{1}{4}, H)$

35. **a.** The highest point on the graph for the 60° angle is $(-2, 7)$, for a height of 7 ft. The highest point on the graph for the 30° angle is (0, 3), for a height of 3 ft. The difference is 4 ft.

b. The farthest point on the graph for the 60° angle is (3, 0), for a final location of 3 ft. The farthest point on the graph for the 30° angle is (7, 0), for a final location of 7 ft. The difference is 4 ft.

37. **a.** To find the charge for a 1-day rental, find the y-coordinate of the point with an x-coordinate of 1. This is the point (1, 2). The charge will be $2.

b. To find the charge for a 2-day rental, find the y-coordinate of the point with an x-coordinate of 2. This is the point (2, 4). The charge will be $4.

c. To find the charge for a 5-day rental, find the y-coordinate of the point with an x-coordinate of 5. This is the point (5, 7). The charge will be $7.

d. To find the charge for a 7-day rental, find the y-coordinate of the point with an x-coordinate of 7. This is the point (7, 9). The charge will be $9.

39.

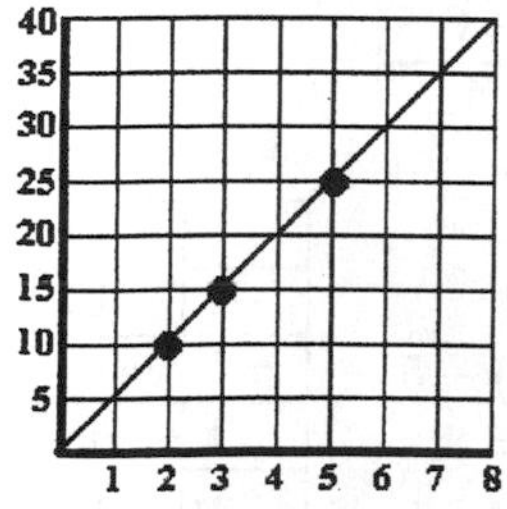

a. $(7, 35) \Rightarrow 35$ miles
b. $(4, 20) \Rightarrow 4$ gallons
c. $(6.5, 32.5) \Rightarrow 32.5$ miles

41.

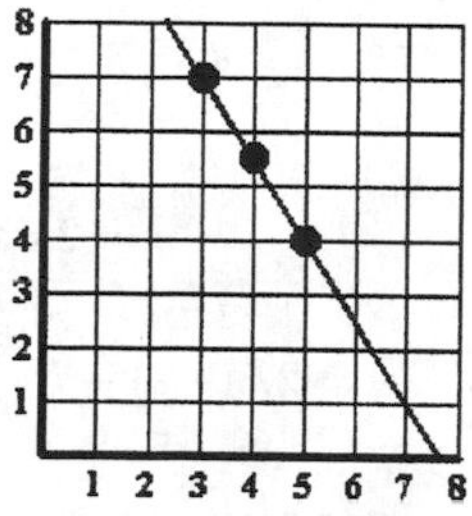

a. A 3-year old car is worth $7000.
b. $(7, 1) \Rightarrow \$1000$
c. $(6, 2500) \Rightarrow 6$ years

43-47. Answers may vary.

Exercise 3.2 (page 173)

1.
$$\begin{aligned} \frac{x}{8} &= -12 \\ 8 \cdot \frac{x}{8} &= 8(-12) \\ x &= -96 \end{aligned}$$

3. expression

5.
$$\begin{aligned} rb &= a \\ 0.005(250) &= a \\ 1.25 &= a \end{aligned}$$

7. $-2.5 - (-2.6) = -2.5 + 2.6 = 0.1$

9. two

11. independent, dependent

13. linear

15. y-intercept

17.
$$\begin{aligned} x - 2y &= -4 \\ 4 - 2(4) &\stackrel{?}{=} -4 \\ 4 - 8 &\stackrel{?}{=} -4 \\ -4 &= -4 \Rightarrow (4,4) \text{ is a solution.} \end{aligned}$$

19.
$$\begin{aligned} y &= \frac{2}{3}x + 5 \\ 12 &\stackrel{?}{=} \frac{2}{3}(6) + 5 \\ 12 &\stackrel{?}{=} 4 + 5 \\ 12 &\neq 9 \Rightarrow (6,12) \text{ is not a solution.} \end{aligned}$$

21. $y = x - 3$

$x = 0$: $y = 0 - 3$, $y = -3$

$x = 1$: $y = 1 - 3$, $y = -2$

$x = -2$: $y = -2 - 3$, $y = -5$

$x = -4$: $y = -4 - 3$, $y = -7$

x	y
0	−3
1	−2
−2	−5
−4	−7

23. $y = -2x$

$x = 0$: $y = -2(0)$, $y = 0$

$x = 1$: $y = -2(1)$, $y = -2$

$x = 3$: $y = -2(3)$, $y = -6$

$x = -2$: $y = -2(-2)$, $y = 4$

input	*output*
0	0
1	−2
3	−6
−2	4

25. $y = 2.5x - 5$

$x = -4$: $y = 2.5(-4) - 5$, $y = -10 - 5$, $y = -15$

$x = -2$: $y = 2.5(-2) - 5$, $y = -5 - 5$, $y = -10$

$x = 0$: $y = 2.5(0) - 5$, $y = 0 - 5$, $y = -5$

$x = 4$: $y = 2.5(4) - 5$, $y = 10 - 5$, $y = 5$

x	y
−4	−15
−2	−10
0	−5
4	5

27. $y = 2x$

$x = 0$	$x = 1$	$x = -1$
$y = 2(0)$	$y = 2(1)$	$y = 2(-1)$
$y = 0$	$y = 2$	$y = -2$

x	y
0	0
1	2
−1	−2

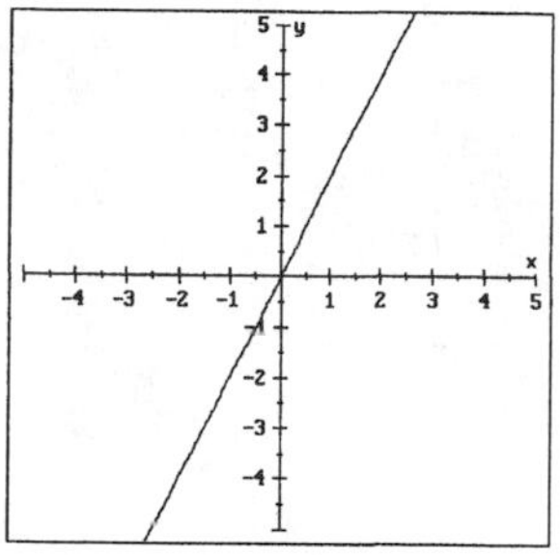

29. $y = 2x - 1$

$x = 0$	$x = 1$	$x = -1$
$y = 2(0) - 1$	$y = 2(1) - 1$	$y = 2(-1) - 1$
$y = 0 - 1$	$y = 2 - 1$	$y = -2 - 1$
$y = -1$	$y = 1$	$y = -3$

x	y
0	−1
1	1
−1	−3

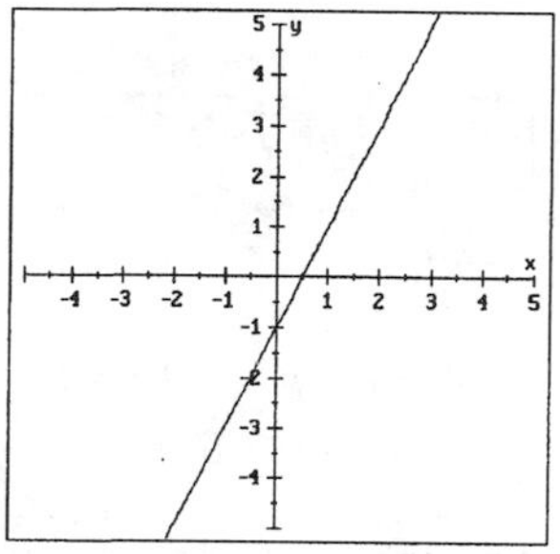

31. $y = 1.2x - 2$

$x = 0$	$x = 5$	$x = -5$
$y = 1.2(0) - 2$	$y = 1.2(5) - 2$	$y = 1.2(-5) - 2$
$y = 0 - 2$	$y = 6 - 2$	$y = -6 - 2$
$y = -2$	$y = 4$	$y = -8$

x	y
0	−2
5	4
−5	−8

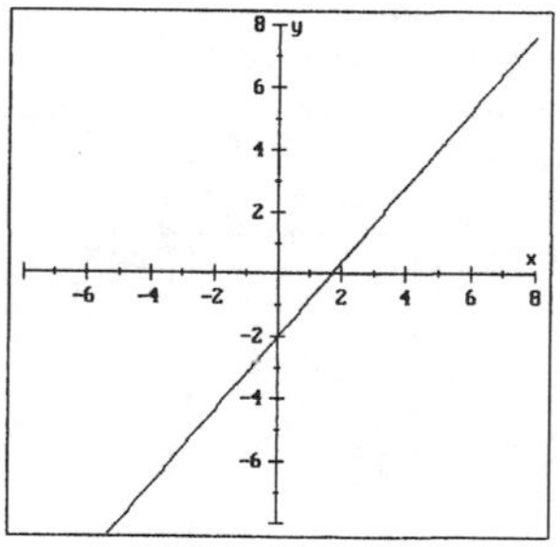

33. $y = \frac{x}{2} - 2$

$x = 0$	$x = 2$	$x = -2$
$y = \frac{0}{2} - 2$	$y = \frac{2}{2} - 2$	$y = \frac{-2}{2} - 2$
$y = 0 - 2$	$y = 1 - 2$	$y = -1 - 2$
$y = -2$	$y = -1$	$y = -3$

x	y
0	−2
2	−1
−2	−3

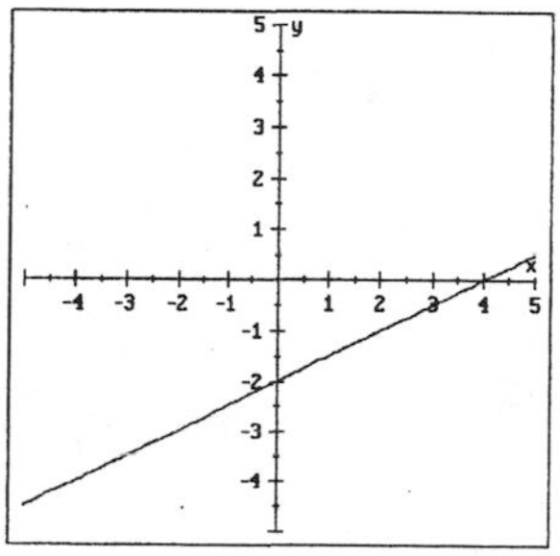

35.

$$x + y = 7$$

$$\begin{array}{cc} x = 0 & y = 0 \\ \hline 0 + y = 7 & x + 0 = 7 \\ y = 7 & x = 7 \\ (0, 7) & (7, 0) \end{array}$$

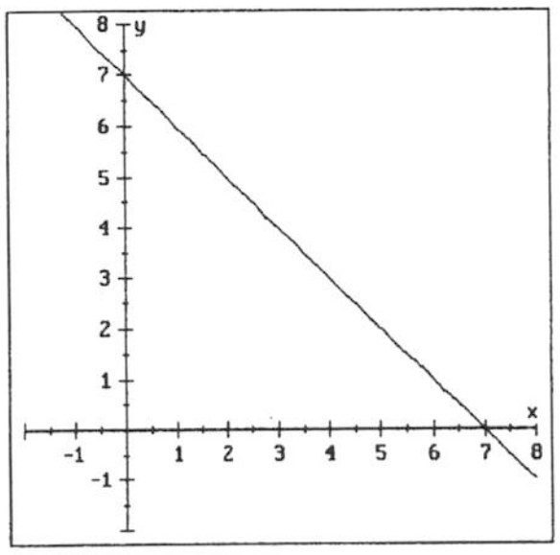

37.

$$x - y = 7$$

$$\begin{array}{cc} x = 0 & y = 0 \\ \hline 0 - y = 7 & x - 0 = 7 \\ -y = 7 & x = 7 \\ y = -7 & (7, 0) \\ (0, -7) & \end{array}$$

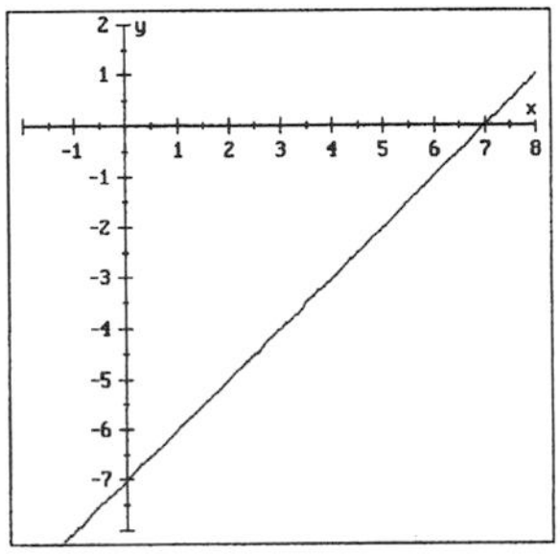

39.

$$y = -2x + 5$$

$$2x + y = 5$$

$$\begin{array}{cc} x = 0 & y = 0 \\ \hline 2(0) + y = 5 & 2x + 0 = 5 \\ 0 + y = 5 & 2x = 5 \\ y = 5 & x = \frac{5}{2} \\ (0, 5) & \left(\frac{5}{2}, 0\right) \end{array}$$

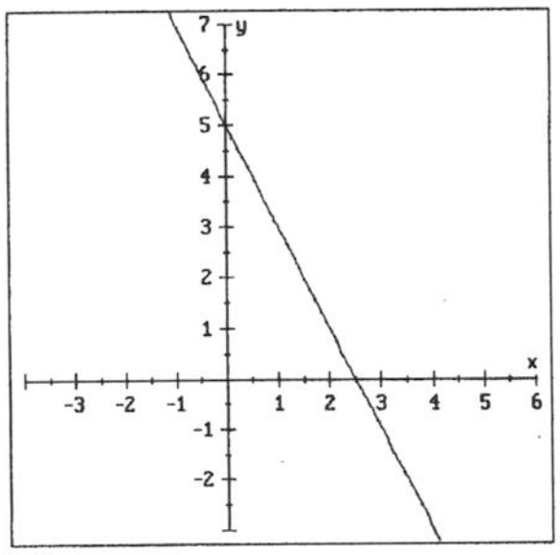

41.

$$2x + 3y = 12$$

$$\begin{array}{cc} x = 0 & y = 0 \\ \hline 2(0) + 3y = 12 & 2x + 3(0) = 12 \\ 0 + 3y = 12 & 2x + 0 = 12 \\ 3y = 12 & 2x = 12 \\ y = 4 & x = 6 \\ (0, 4) & (6, 0) \end{array}$$

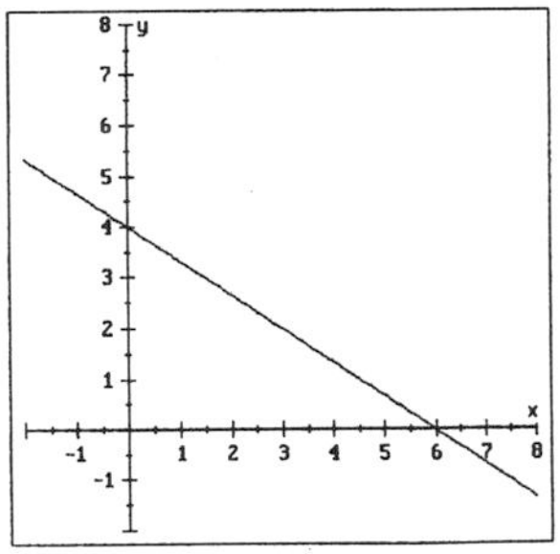

43. $y = -5$

horizontal, y-coordinate $= -5$

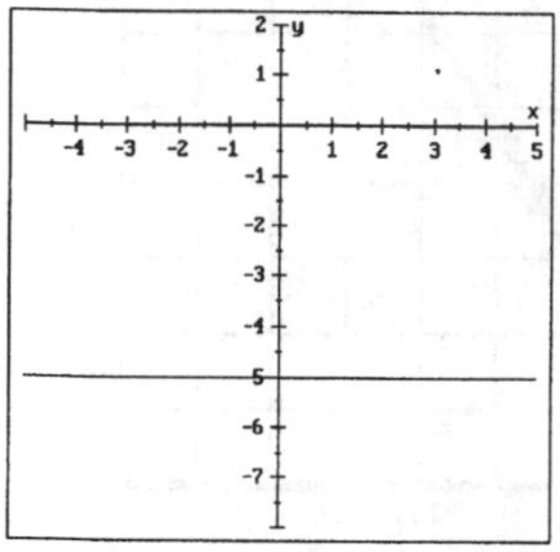

45. $x = 5$

vertical, x-coordinate $= 5$

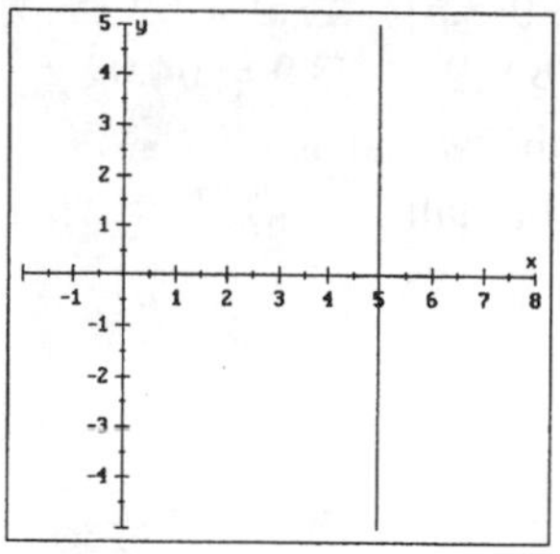

47. $y = 0$

horizontal, y-coordinate $= 0$

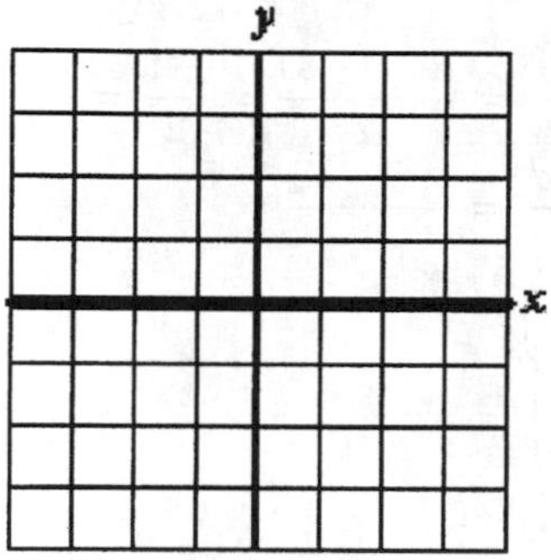

49. $2x = 5$

$x = \frac{5}{2} = 2\frac{1}{2}$

vertical, x-coordinate $= \frac{5}{2}$

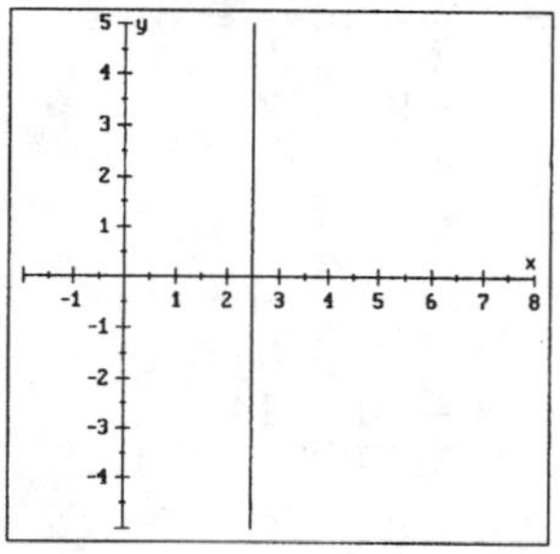

51. **a.** $c = 50 + 25u$

b.

u	c
4	$50 + 25(4) = 150$
8	$50 + 25(8) = 250$
14	$50 + 25(14) = 400$

c. The service fee is \$50.

d. cost for 18 units = \$500

cost for 12 units = \$350

Total cost = \$850

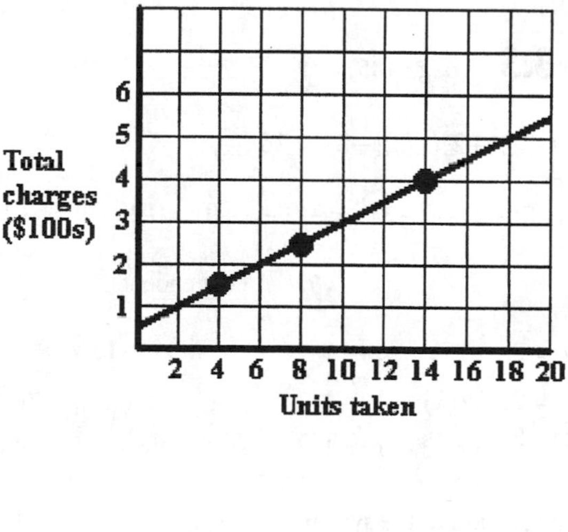

53. **a.**

r	h
7	$3.9(7) + 28.9 = 56.2$
8.5	$3.9(8.5) + 28.9 = 62.1$
9	$3.9(9) + 28.9 = 64.0$

b. ...taller the woman is.

c. 58.1 inches tall

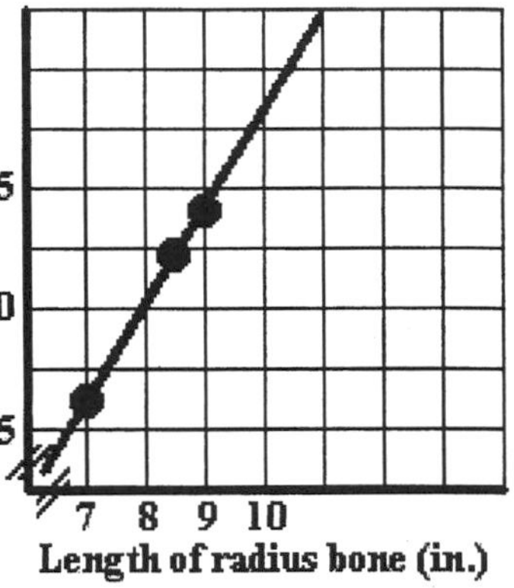

55-59. Answers may vary.

61. $x_M = \frac{a+c}{2} = \frac{5+7}{2} = \frac{12}{2} = 6$

$y_M = \frac{b+d}{2} = \frac{3+9}{2} = \frac{12}{2} = 6$

$M(6,6)$

63. $x_M = \frac{a+c}{2} = \frac{2+(-3)}{2} = \frac{-1}{2} = -\frac{1}{2}$

$y_M = \frac{b+d}{2} = \frac{-7+12}{2} = \frac{5}{2}$

$M\left(-\frac{1}{2}, \frac{5}{2}\right)$

65. $x_M = \frac{a+c}{2} = \frac{4+10}{2} = \frac{14}{2} = 7$

$y_M = \frac{b+d}{2} = \frac{6+6}{2} = \frac{12}{2} = 6$

$M(7,6)$

Exercise 3.3 (page 185)

1. $(-2)^4 = (-2)(-2)(-2)(-2) = 16$

3. $3x - x^2 = 3(-3) - (-3)^2 = -9 - (+9) = -9 - 9 = -18$

5. system

7. independent

9. inconsistent

11.

$x + y = 2$	$2x - y = 1$
$1 + 1 \stackrel{?}{=} 2$	$2(1) - 1 \stackrel{?}{=} 1$
$2 = 2$	$2 - 1 \stackrel{?}{=} 1$
	$1 = 1$

$(1,1)$ is a solution to the system.

13.

$2x + y = 4$	$x + y = 1$
$2(3) + (-2) \stackrel{?}{=} 4$	$3 + (-2) \stackrel{?}{=} 1$
$6 + (-2) \stackrel{?}{=} 4$	$1 = 1$
$4 = 4$	

$(3,-2)$ is a solution to the system.

15.

$$\begin{aligned} 2x - 3y &= -7 & 4x - 5y &= 25 \\ 2(4) - 3(5) &\stackrel{?}{=} -7 & 4(4) - 5(5) &\stackrel{?}{=} 25 \\ 8 - 15 &\stackrel{?}{=} -7 & 16 - 25 &\stackrel{?}{=} 25 \\ -7 &= -7 & -9 &\neq 25 \end{aligned}$$

$(4, 5)$ is not a solution to the system.

17.

$$\begin{aligned} 4x + 5y &= -23 & -3x + 2y &= 0 \\ 4(-2) + 5(-3) &\stackrel{?}{=} -23 & -3(-2) + 2(-3) &\stackrel{?}{=} 0 \\ -8 + (-15) &\stackrel{?}{=} -23 & 6 + (-6) &\stackrel{?}{=} 0 \\ -23 &= -23 & 0 &= 0 \end{aligned}$$

$(-2, -3)$ is a solution to the system.

19.

$$\begin{aligned} 2x + y &= 4 & 4x - 3y &= 11 \\ 2(\tfrac{1}{2}) + 3 &\stackrel{?}{=} 4 & 4(\tfrac{1}{2}) - 3(3) &\stackrel{?}{=} 11 \\ 1 + 3 &\stackrel{?}{=} 4 & 2 - 9 &\stackrel{?}{=} 11 \\ 4 &= 4 & -7 &\neq 11 \end{aligned}$$

$(\frac{1}{2}, 3)$ is not a solution to the system.

21.

$$\begin{aligned} 5x - 4y &= -6 & 8y &= 10x + 12 \\ 5(-\tfrac{2}{5}) - 4(\tfrac{1}{4}) &\stackrel{?}{=} -6 & 8(\tfrac{1}{4}) &\stackrel{?}{=} 10(-\tfrac{2}{5}) + 12 \\ -2 - 1 &\stackrel{?}{=} -6 & 2 &\stackrel{?}{=} -4 + 12 \\ -3 &\neq -6 & 2 &\neq 8 \end{aligned}$$

$(-\frac{2}{5}, \frac{1}{4})$ is not a solution to the system.

23. $x + y = 2$ $\quad$ $x - y = 0$

x	y
0	2
1	1

x	y
0	0
1	1

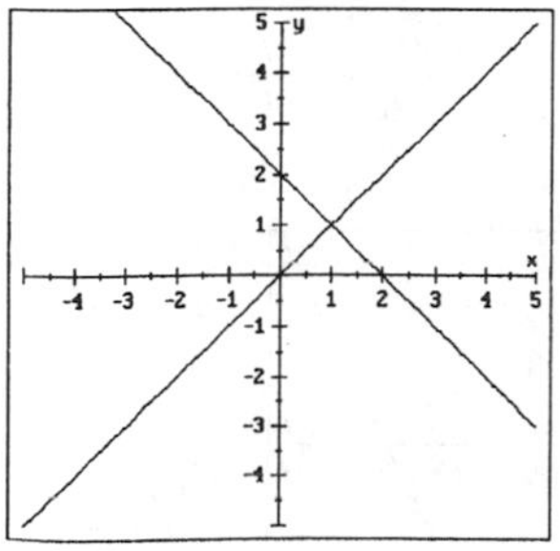

$(1, 1)$

25. $x + y = 2$ $\quad$ $x - y = 4$

x	y
0	2
1	1

x	y
0	-4
1	-3

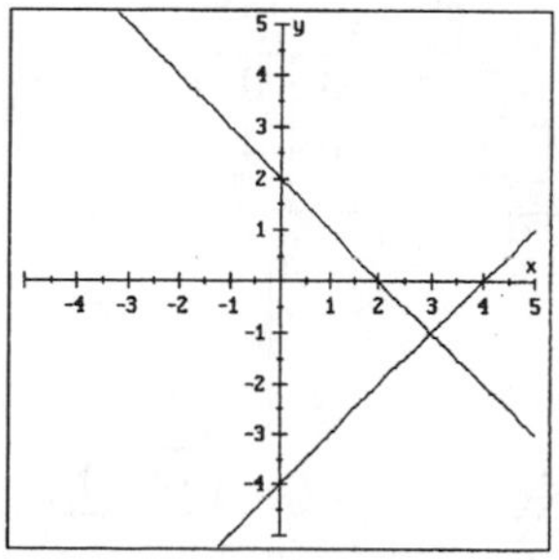

$(3, -1)$

27. $y = 2x \quad x + y = 0$

x	y
0	0
1	2

x	y
2	-2
1	-1

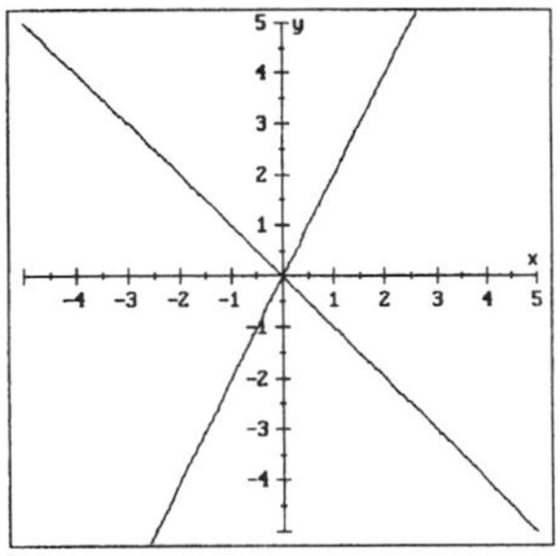

$\boxed{(0,0)}$

29. $3x + 2y = -8 \quad 2x - 3y = -1$

x	y
0	-4
2	-7

x	y
1	1
-2	-1

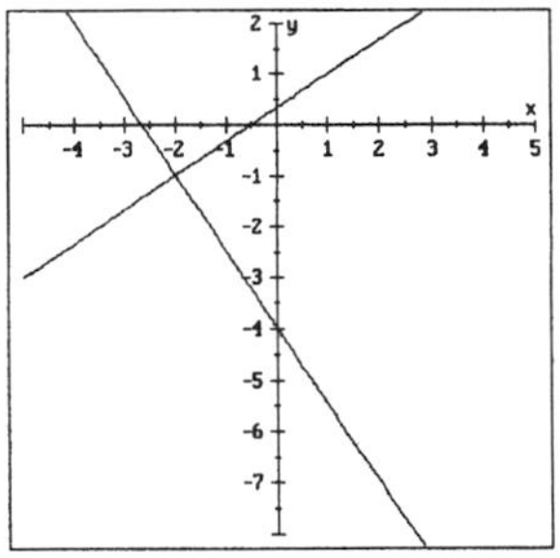

$\boxed{(-2,-1)}$

31. $4x - 2y = 8 \quad y = 2x - 4$

x	y
0	-4
1	-2

x	y
0	-4
1	-2

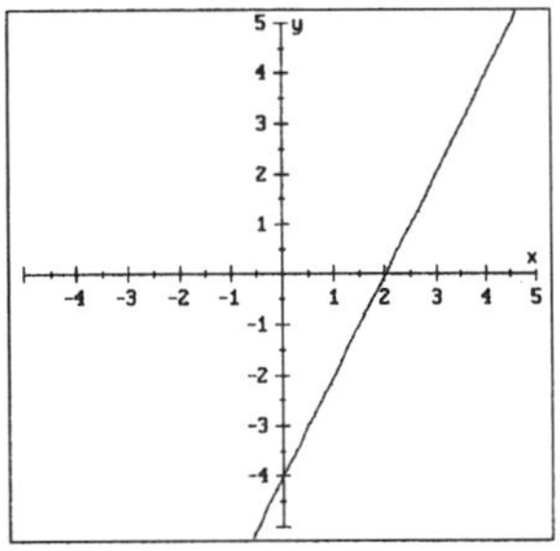

dependent equations

33. $2x - 3y = -18 \quad 3x + 2y = -1$

x	y
-3	4
-6	2

x	y
1	-2
-1	1

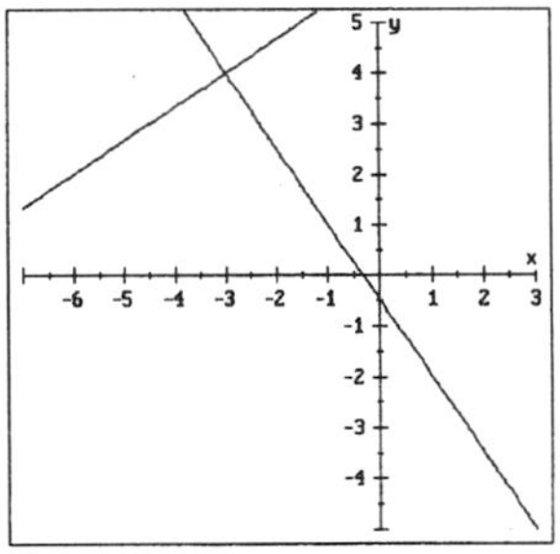

$\boxed{(-3,4)}$

35. $4x = 3(4 - y) \quad 2y = 4(3 - x)$

$4x = 12 - 3y \quad 2y = 12 - 4x$

x	y
0	4
3	0

x	y
3	0
0	6

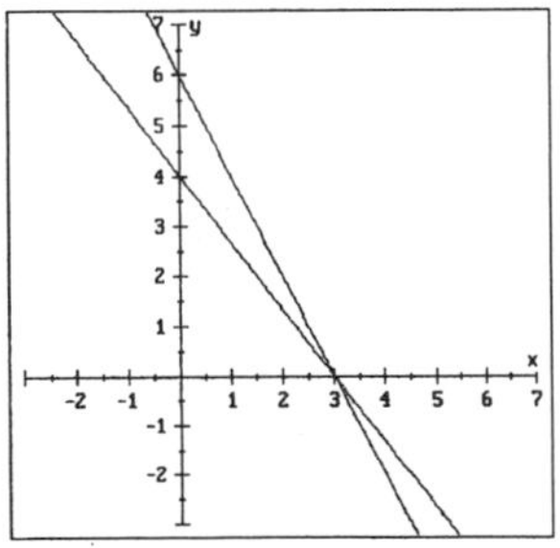

$\boxed{(3,0)}$

37. $x + 2y = -4 \quad x - \frac{1}{2}y = 6$

$2x - y = 12$

x	y
0	-2
2	-3

x	y
4	-4
6	0

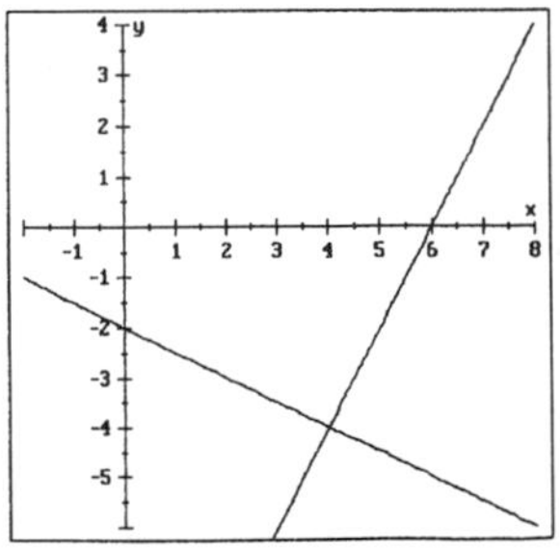

$\boxed{(4,-4)}$

39. $-\frac{3}{4}x + y = 3$ $\quad$ $\frac{1}{4}x + y = -1$

$-3x + 4y = 12$ $\quad$ $x + 4y = -4$

x	y
0	3
-4	0

x	y
0	-1
-4	0

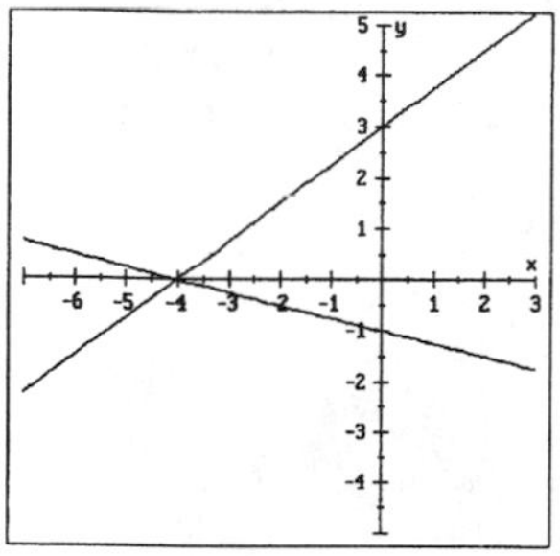

$\boxed{(-4, 0)}$

41. $\frac{1}{2}x + \frac{1}{4}y = 0$ $\quad$ $\frac{1}{4}x - \frac{3}{8}y = -2$

$2x + y = 0$ $\quad$ $2x - 3y = -16$

x	y
0	0
1	-2

x	y
-2	4
-5	2

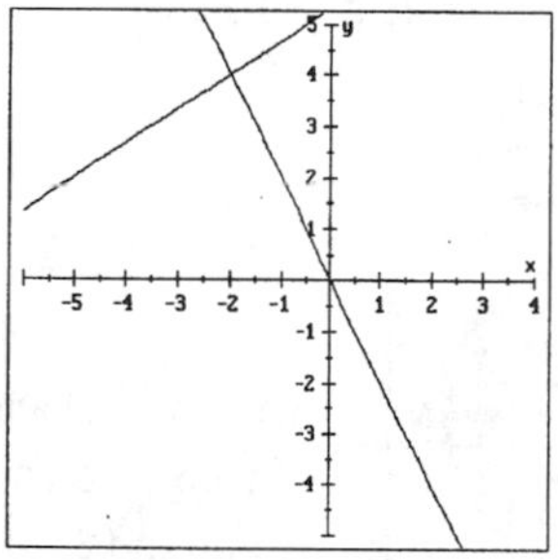

$\boxed{(-2, 4)}$

43. $\frac{1}{3}x - \frac{1}{2}y = \frac{1}{6}$ $\quad$ $\frac{2}{5}x + \frac{1}{2}y = \frac{13}{10}$

$2x - 3y = 1$ $\quad$ $4x + 5y = 13$

x	y
5	3
2	1

x	y
-3	5
2	1

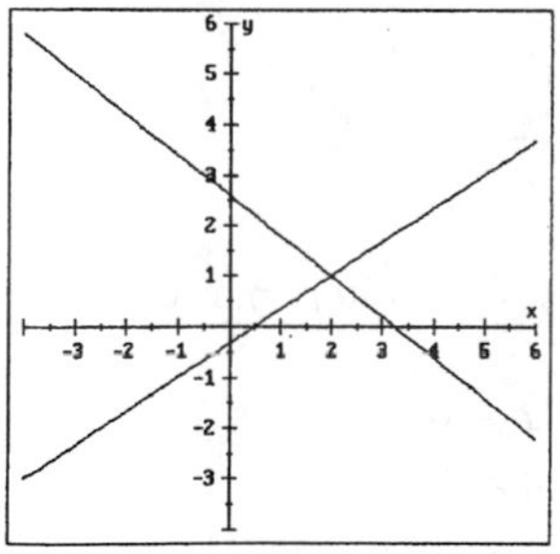

$\boxed{(2, 1)}$

45. $\begin{cases} y = 4 - x \\ y = 2 + x \end{cases}$

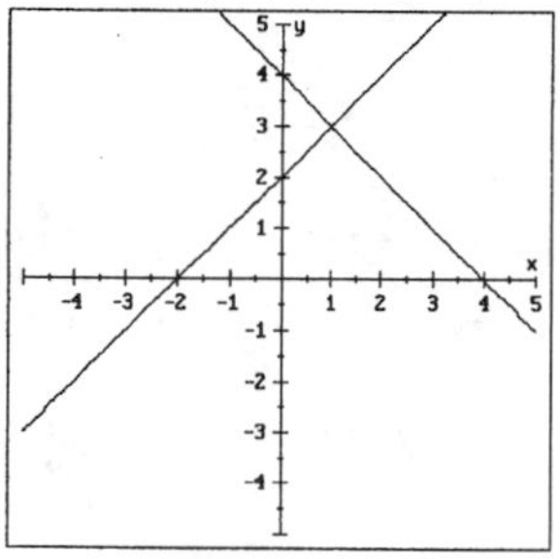

$\boxed{(1, 3)}$

47. $\begin{cases} 3x - 6y = 4 & \Rightarrow y = \frac{1}{2}x - \frac{2}{3} \\ 2x + y = 1 & \Rightarrow y = -2x + 1 \end{cases}$

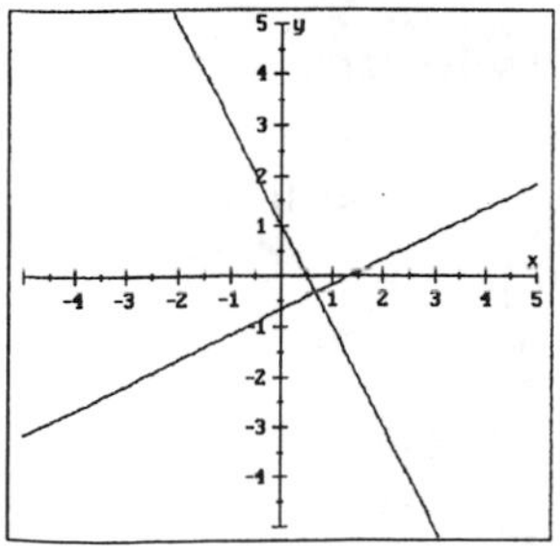

$\boxed{(0.67, -0.33)}$

49. **a.** Donors outnumbered patients.
b. 2003; about 4100
c. Patients outnumber donors.

51. **a.** Houston, New Orleans, St. Augustine
b. St. Louis, Memphis, New Orleans
c. New Orleans

53. $y = -\frac{1}{2}x + 3$

x	y
4	1
2	2

$3y = 2x + 2$

x	y
2	2
-1	0

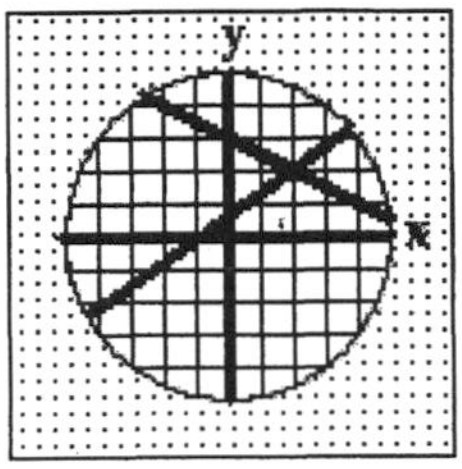

There is a chance of a collision at the point (2, 2).

55-57. Answers may vary.

Exercise 3.4 (page 193)

1. $y^2 - x^2 = (3)^2 - (-2)^2 = 9 - 4 = 5$

3. $\frac{3x - 2y}{2x + y} = \frac{3(-2) - 2(3)}{2(-2) + 3} = \frac{-6 - 6}{-4 + 3} = \frac{-12}{-1} = 12$

5. $-x(3y - 4) = -(-2)[3(3) - 4] = -(-2)(5) = 10$

7. y, terms

9. remove

11. infinitely many

13. $\begin{cases} (1) & y = 2x \\ (2) & x + y = 6 \end{cases}$

Substitute $y = 2x$ from (1) into (2):

$$\begin{aligned} x + y &= 6 \\ x + 2x &= 6 \\ 3x &= 6 \\ x &= 2 \end{aligned}$$

Substitute and solve for y:

$y = 2x$

$y = 2(2) = 4$

$\boxed{x = 2, y = 4}$

15. $\begin{cases} (1) & y = 2x - 6 \\ (2) & 2x + y = 6 \end{cases}$

Substitute $y = 2x - 6$ from (1) into (2):

$$\begin{aligned} 2x + y &= 6 \\ 2x + 2x - 6 &= 6 \\ 4x &= 12 \\ x &= 3 \end{aligned}$$

Substitute and solve for y:

$y = 2x - 6$

$y = 2(3) - 6 = 6 - 6 = 0$

$\boxed{x = 3, y = 0}$

17. $\begin{cases} (1) & y = 2x + 5 \\ (2) & x + 2y = -5 \end{cases}$

Substitute $y = 2x + 5$ from (1) into (2):

$$\begin{aligned} x + 2y &= -5 \\ x + 2(2x + 5) &= -5 \\ x + 4x + 10 &= -5 \\ 5x &= -15 \\ x &= -3 \end{aligned}$$

Substitute and solve for y:

$y = 2x + 5$

$y = 2(-3) + 5 = -6 + 5 = -1$

$\boxed{x = -3, y = -1}$

19. $\begin{cases} (1) & 2a + 4b = -24 \\ (2) & a = 20 - 2b \end{cases}$

Substitute $a = 20 - 2b$ from (2) into (1):

$$\begin{aligned} 2a + 4b &= -24 \\ 2(20 - 2b) + 4b &= -24 \\ 40 - 4b + 4b &= -24 \\ 40 &\neq -24 \end{aligned}$$

$\boxed{\text{inconsistent system}}$

21. $\begin{cases} (1) & 2a = 3b - 13 \\ (2) & b = 2a + 7 \end{cases}$

Substitute $b = 2a + 7$ from (2) into (1):

$$\begin{aligned} 2a &= 3b - 13 \\ 2a &= 3(2a + 7) - 13 \\ 2a &= 6a + 21 - 13 \\ -4a &= 8 \\ a &= -2 \end{aligned}$$

Substitute and solve for b:

$b = 2a + 7$

$b = 2(-2) + 7 = -4 + 7 = 3$

$\boxed{a = -2, b = 3}$

23. $\begin{cases} 2x + y = 4 & \Rightarrow (1) \quad y = -2x + 4 \\ 4x + y = 5 & \Rightarrow (2) \quad 4x + y = 5 \end{cases}$

Substitute $y = -2x + 4$ from (1) into (2):

$$\begin{aligned} 4x + y &= 5 \\ 4x + (-2x + 4) &= 5 \\ 4x - 2x + 4 &= 5 \\ 2x &= 1 \\ x &= \frac{1}{2} \end{aligned}$$

Substitute and solve for y:

$y = -2x + 4$

$y = -2\left(\frac{1}{2}\right) + 4 = -1 + 4 = 3$

$\boxed{x = \frac{1}{2}, y = 3}$

25. $\begin{cases} r + 3s = 9 & \Rightarrow (1) \quad r = -3s + 9 \\ 3r + 2s = 13 & \Rightarrow (2) \quad 3r + 2s = 13 \end{cases}$

Substitute $r = -3s + 9$ from (1) into (2):

$$\begin{aligned} 3r + 2s &= 13 \\ 3(-3s + 9) + 2s &= 13 \\ -9s + 27 + 2s &= 13 \\ -7s &= -14 \\ s &= 2 \end{aligned}$$

Substitute and solve for r:

$r = -3s + 9$

$r = -3(2) + 9 = -6 + 9 = 3$

$\boxed{r = 3, s = 2}$

27. $\begin{cases} 4x + 5y = 2 & \Rightarrow (1) \quad 4x + 5y = 2 \\ 3x - y = 11 & \Rightarrow (2) \quad y = 3x - 11 \end{cases}$

Substitute $y = 3x - 11$ from (2) into (1):

$$\begin{aligned} 4x + 5y &= 2 \\ 4x + 5(3x - 11) &= 2 \\ 4x + 15x - 55 &= 2 \\ 19x &= 57 \\ x &= 3 \end{aligned}$$

Substitute and solve for y:

$y = 3x - 11$

$y = 3(3) - 11 = 9 - 11 = -2$

$\boxed{x = 3, y = -2}$

29. $\begin{cases} 2x + y = 0 & \Rightarrow (1) \quad y = -2x \\ 3x + 2y = 1 & \Rightarrow (2) \quad 3x + 2y = 1 \end{cases}$

Substitute $y = -2x$ from (1) into (2):

$$\begin{aligned} 3x + 2y &= 1 \\ 3x + 2(-2x) &= 1 \\ 3x - 4x &= 1 \\ -x &= 1 \\ x &= -1 \end{aligned}$$

Substitute and solve for y:

$y = -2x$

$y = -2(-1) = 2$

$\boxed{x = -1, y = 2}$

31. $\begin{cases} 3x + 4y = -7 & \Rightarrow (1) \quad 3x + 4y = -7 \\ 2y - x = -1 & \Rightarrow (2) \quad x = 2y + 1 \end{cases}$

Substitute $x = 2y + 1$ from (2) into (1):

$$\begin{aligned} 3x + 4y &= -7 \\ 3(2y + 1) + 4y &= -7 \\ 6y + 3 + 4y &= -7 \\ 10y &= -10 \\ y &= -1 \end{aligned}$$

Substitute and solve for x:

$x = 2y + 1$

$x = 2(-1) + 1 = -2 + 1 = -1$

$\boxed{x = -1, y = -1}$

33. $\begin{cases} 9x = 3y + 12 & \Rightarrow (1) \quad 9x = 3y + 12 \\ 4 = 3x - y & \Rightarrow (2) \quad y = 3x - 4 \end{cases}$

Substitute $y = 3x - 4$ from (2) into (1):

$9x = 3y + 12$

$9x = 3(3x - 4) + 12$

$9x = 9x - 12 + 12$

$0 = 0$ $\boxed{\text{dependent equations}}$

35. $\begin{cases} (1) \quad 2x + 3y = 5 & \Rightarrow 2x = 5 - 3y \quad \Rightarrow x = \dfrac{5 - 3y}{2} \\ (2) \quad 3x + 2y = 5 \end{cases}$

Substitute $x = \frac{5-3y}{2}$ from (1) into (2):

$$\begin{aligned} 3x + 2y &= 5 \\ 3 \cdot \frac{5 - 3y}{2} + 2y &= 5 \\ 3(5 - 3y) + 4y &= 10 \\ 15 - 9y + 4y &= 10 \\ -5y &= -5 \\ y &= 1 \end{aligned}$$

Substitute and solve for x:

$$x = \frac{5 - 3y}{2}$$

$$x = \frac{5 - 3(1)}{2}$$

$$x = \frac{5 - 3}{2} = \frac{2}{2} = 1$$

Solution:

$\boxed{x = 1, y = 1}$

37. $\begin{cases} (1) \quad a = \dfrac{3}{2}b + 5 \\ (2) \quad 2a - 3b = 8 \end{cases}$

Substitute $a = \frac{3}{2}b + 5$ from (1) into (2):

$$\begin{aligned} 2a - 3b &= 8 \\ 2\left(\frac{3}{2}b + 5\right) - 3b &= 8 \\ 3b + 10 - 3b &= 8 \\ 10 &\neq 8 \end{aligned}$$

$\boxed{\text{inconsistent system}}$

39. $\begin{cases} (1) \quad 2x + 5y = -2 \quad \Rightarrow 2x = -2 - 5y \quad \Rightarrow x = \dfrac{-2-5y}{2} \\ (2) \quad 4x + 3y = 10 \end{cases}$

Substitute $x = \dfrac{-2-5y}{2}$ from (1) into (2):

$$\begin{aligned} 4x + 3y &= 10 \\ 4 \cdot \frac{-2-5y}{2} + 3y &= 10 \\ 4(-2-5y) + 6y &= 20 \\ -8 - 20y + 6y &= 20 \\ -14y &= 28 \\ y &= -2 \end{aligned}$$

Substitute and solve for x:

$$\begin{aligned} x &= \frac{-2-5y}{2} \\ x &= \frac{-2-5(-2)}{2} \\ x &= \frac{-2+10}{2} = \frac{8}{2} = 4 \end{aligned}$$

Solution: $\boxed{x = 4, y = -2}$

41. $\begin{cases} (1) \quad 2x - 3y = -3 \quad \Rightarrow 2x = 3y - 3 \quad \Rightarrow x = \dfrac{3y-3}{2} \\ (2) \quad 3x + 5y = -14 \end{cases}$

Substitute $x = \dfrac{3y-3}{2}$ from (1) into (2):

$$\begin{aligned} 3x + 5y &= -14 \\ 3 \cdot \frac{3y-3}{2} + 5y &= -14 \\ 3(3y-3) + 10y &= -28 \\ 9y - 9 + 10y &= -28 \\ 19y &= -19 \\ y &= -1 \end{aligned}$$

Substitute and solve for x:

$$\begin{aligned} x &= \frac{3y-3}{2} \\ x &= \frac{3(-1)-3}{2} \\ x &= \frac{-3-3}{2} = \frac{-6}{2} = -3 \end{aligned}$$

Solution: $\boxed{x = -3, y = -1}$

43. $\begin{cases} (1) \quad 7x - 2y = -1 \quad \Rightarrow 7x = 2y - 1 \quad \Rightarrow x = \dfrac{2y-1}{7} \\ (2) \quad -5x + 2y = -1 \end{cases}$

Substitute $x = \dfrac{2y-1}{7}$ from (1) into (2):

$$\begin{aligned} -5x + 2y &= -1 \\ -5 \cdot \frac{2y-1}{7} + 2y &= -1 \\ -5(2y-1) + 14y &= -7 \\ -10y + 5 + 14y &= -7 \\ 4y &= -12 \\ y &= -3 \end{aligned}$$

Substitute and solve for x:

$$\begin{aligned} x &= \frac{2y-1}{7} \\ x &= \frac{2(-3)-1}{7} \\ x &= \frac{-6-1}{7} = \frac{-7}{7} = -1 \end{aligned}$$

Solution: $\boxed{x = -1, y = -3}$

45. $\begin{cases} (1) \quad 2a + 3b = 2 \quad \Rightarrow 2a = 2 - 3b \quad \Rightarrow a = \dfrac{2-3b}{2} \\ (2) \quad 8a - 3b = 3 \end{cases}$

Substitute $a = \dfrac{2-3b}{2}$ from (1) into (2):

$$\begin{aligned} 8a - 3b &= 3 \\ 8 \cdot \frac{2-3b}{2} - 3b &= 3 \\ 4(2-3b) - 3b &= 3 \\ 8 - 12b - 3b &= 3 \\ -15b &= -5 \\ b &= \frac{1}{3} \end{aligned}$$

Substitute and solve for a:

$$a = \frac{2-3b}{2}$$
$$a = \frac{2-3\left(\frac{1}{3}\right)}{2}$$
$$a = \frac{2-1}{2} = \frac{1}{2}$$

Solution: $\boxed{a = \frac{1}{2}, b = \frac{1}{3}}$

47. $\begin{cases} (1) \quad y - x = 3x \quad \Rightarrow y = 4x \\ (2) \quad 2(x+y) = 14 - y \quad \Rightarrow 2x + 2y = 14 - y \quad \Rightarrow 2x + 3y = 14 \end{cases}$

Substitute $y = 4x$ from (1) into (2):

$$\begin{aligned} 2x + 3y &= 14 \\ 2x + 3(4x) &= 14 \\ 2x + 12x &= 14 \\ 14x &= 14 \\ x &= 1 \end{aligned}$$

Substitute and solve for y:

$$y = 4x$$
$$y = 4(1) = 4$$

Solution: $\boxed{x = 1, y = 4}$

49. $\begin{cases} (1) \quad 3(x-1) + 3 = 8 + 2y \quad \Rightarrow 3x - 3 + 3 = 8 + 2y \quad \Rightarrow 3x = 2y + 8 \quad \Rightarrow x = \dfrac{2y+8}{3} \\ (2) \quad 2(x+1) = 4 + 3y \quad \Rightarrow 2x + 2 = 4 + 3y \quad \Rightarrow 2x - 3y = 2 \end{cases}$

Substitute $x = \dfrac{2y+8}{3}$ from (1) into (2):

$$\begin{aligned} 2x - 3y &= 2 \\ 2 \cdot \frac{2y+8}{3} - 3y &= 2 \\ 2(2y+8) - 9y &= 6 \\ 4y + 16 - 9y &= 6 \\ -5y &= -10 \\ y &= 2 \end{aligned}$$

Substitute and solve for x:

$$x = \frac{2y+8}{3}$$
$$x = \frac{2(2)+8}{3}$$
$$x = \frac{4+8}{3} = \frac{12}{3} = 4$$

Solution: $\boxed{x = 4, y = 2}$

51. $\begin{cases} (1) & 6a = 5(3+b+a) - a \Rightarrow 6a = 15 + 5b + 5a - a \Rightarrow 2a = 5b + 15 \Rightarrow a = \dfrac{5b+15}{2} \\ (2) & 3(a-b) + 4b = 5(1+b) \Rightarrow 3a - 3b + 4b = 5 + 5b \Rightarrow 3a - 4b = 5 \end{cases}$

Substitute $a = \dfrac{5b+15}{2}$ from (1) into (2):

$$3a - 4b = 5$$
$$3 \cdot \frac{5b+15}{2} - 4b = 5$$
$$3(5b+15) - 8b = 10$$
$$15b + 45 - 8b = 10$$
$$7b = -35$$
$$b = -5$$

Substitute and solve for a:

$$a = \frac{5b+15}{2}$$
$$a = \frac{5(-5)+15}{2}$$
$$a = \frac{-25+15}{2} = \frac{-10}{2} = -5$$

Solution: $\boxed{a = -5, b = -5}$

53. $\begin{cases} (1) & \frac{1}{2}x + \frac{1}{2}y = -1 \Rightarrow x + y = -2 \Rightarrow x = -y - 2 \\ (2) & \frac{1}{3}x - \frac{1}{2}y = -4 \Rightarrow 2x - 3y = -24 \end{cases}$

Substitute $x = -y - 2$ from (1) into (2):

$$2x - 3y = -24$$
$$2(-y-2) - 3y = -24$$
$$-2y - 4 - 3y = -24$$
$$-5y = -20$$
$$y = 4$$

Substitute and solve for x:

$$x = -y - 2$$
$$x = -4 - 2 = -6$$

Solution: $\boxed{x = -6, y = 4}$

55. $\begin{cases} (1) & 5x = \frac{1}{2}y - 1 \Rightarrow 10x = y - 2 \\ (2) & \frac{1}{4}y = 10x - 1 \Rightarrow y = 40x - 4 \end{cases}$

Substitute $y = 40x - 4$ from (2) into (1):

$$10x = y - 2$$
$$10x = 40x - 4 - 2$$
$$-30x = -6$$
$$x = \frac{1}{5}$$

Substitute and solve for y:

$$y = 40x - 4$$
$$y = 40\left(\frac{1}{5}\right) - 4$$
$$y = 8 - 4 = 4$$

Solution: $\boxed{x = \frac{1}{5}, y = 4}$

57. $\begin{cases} (1) & \dfrac{6x-1}{3} - \dfrac{5}{3} = \dfrac{3y+1}{2} \Rightarrow [\times 6] \Rightarrow 2(6x-1) - 2(5) = 3(3y+1) \\ & \qquad 12x - 2 - 10 = 9y + 3 \Rightarrow 12x = 9y + 15 \\ (2) & \dfrac{1+5y}{4} + \dfrac{x+3}{4} = \dfrac{17}{2} \Rightarrow [\times 4] \Rightarrow 1 + 5y + x + 3 = 2(17) \Rightarrow x = 30 - 5y \end{cases}$

Substitute $x = 30 - 5y$ from (2) into (1):

$$12x = 9y + 15$$
$$12(30 - 5y) = 9y + 15$$
$$360 - 60y = 9y + 15$$
$$-69y = -345$$
$$y = 5$$

Substitute and solve for x:

$$x = 30 - 5y$$
$$x = 30 - 5(5)$$
$$x = 30 - 25 = 5$$

Solution: $\boxed{x = 5, y = 5}$

59. $\begin{cases} (1) & x + y = 90 \\ (2) & y = 2x \end{cases}$

Substitute $y = 2x$ from (2) into (1):

$$\begin{aligned} x + y &= 90 \\ x + 2x &= 90 \\ 3x &= 90 \\ x &= 30 \end{aligned}$$

Substitute and solve for y:

$y = 2x$

$y = 2(30) = 60$

$\boxed{x = 30°, y = 60°}$

61-63. Answers may vary.

Exercise 3.5 (page 200)

1.
$$\begin{aligned} 8(3x - 5) - 12 &= 4(2x + 3) \\ 24x - 40 - 12 &= 8x + 12 \\ 24x - 52 &= 8x + 12 \\ 24x - 8x - 52 &= 8x - 8x + 12 \\ 16x - 52 + 52 &= 12 + 52 \\ 16x &= 64 \\ \frac{16x}{16} &= \frac{64}{16} \\ x &= 4 \end{aligned}$$

3.
$$\begin{aligned} x - 2 &= \frac{x + 2}{3} \\ 3(x - 2) &= 3 \cdot \frac{x + 2}{3} \\ 3x - 6 &= x + 2 \\ 3x - x - 6 &= x - x + 2 \\ 2x - 6 + 6 &= 2 + 6 \\ 2x &= 8 \\ \frac{2x}{2} &= \frac{8}{2} \\ x &= 4 \end{aligned}$$

5.
$$\begin{aligned} 7x - 9 &\le 5 \\ 7x - 9 + 9 &\le 5 + 9 \\ 7x &\le 14 \\ \frac{7x}{7} &\le \frac{14}{7} \\ x &\le 2 \end{aligned}$$

7. coefficient

9. general

11. 15

13.
$$\begin{aligned} x + y &= 5 \\ x - y &= -3 \\ \hline 2x \phantom{{}+y} &= 2 \\ x \phantom{{}+y} &= 1 \end{aligned}$$

Substitute and solve for y:

$x + y = 5$

$1 + y = 5$

$y = 4$ $\boxed{(1, 4)}$

15.
$$\begin{aligned} x - y &= -5 \\ x + y &= 1 \\ \hline 2x \phantom{{}+y} &= -4 \\ x \phantom{{}+y} &= -2 \end{aligned}$$

Substitute and solve for y:

$x + y = 1$

$-2 + y = 1$

$y = 3$ $\boxed{(-2, 3)}$

17.
$$\begin{aligned} 2x + y &= -1 \\ -2x + y &= 3 \\ \hline 2y &= 2 \\ y &= 1 \end{aligned}$$
Substitute and solve for x:
$$\begin{aligned} 2x + y &= -1 \\ 2x + 1 &= -1 \\ 2x &= -2 \\ x &= -1 \quad \boxed{(-1, 1)} \end{aligned}$$

19.
$$\begin{aligned} 2x - 3y &= -11 \\ 3x + 3y &= 21 \\ \hline 5x &= 10 \\ x &= 2 \end{aligned}$$
Substitute and solve for y:
$$\begin{aligned} 3x + 3y &= 21 \\ 3(2) + 3y &= 21 \\ 6 + 3y &= 21 \\ 3y &= 15 \\ y &= 5 \quad \boxed{(2, 5)} \end{aligned}$$

21.
$$\begin{aligned} 2x + y &= -2 \\ -2x - 3y &= -6 \\ \hline -2y &= -8 \\ y &= 4 \end{aligned}$$
Substitute and solve for x:
$$\begin{aligned} 2x + y &= -2 \\ 2x + 4 &= -2 \\ 2x &= -6 \\ x &= -3 \quad \boxed{(-3, 4)} \end{aligned}$$

23.
$$\begin{aligned} 4x + 3y &= 24 \\ 4x - 3y &= -24 \\ \hline 8x &= 0 \\ x &= 0 \end{aligned}$$
Substitute and solve for y:
$$\begin{aligned} 4x + 3y &= 24 \\ 4(0) + 3y &= 24 \\ 0 + 3y &= 24 \\ 3y &= 24 \\ y &= 8 \quad \boxed{(0, 8)} \end{aligned}$$

25.
$$\begin{aligned} x + y = 5 &\Rightarrow \times(-1) \quad -x - y = -5 \\ x + 2y = 8 & \qquad\qquad\quad\ \ x + 2y = 8 \\ & \qquad\qquad\qquad\quad\ \ \ y = 3 \end{aligned}$$
$$\begin{aligned} x + 2y &= 8 \\ x + 2(3) &= 8 \\ x + 6 &= 8 \\ x &= 2 \end{aligned}$$
Solution: $\boxed{(2, 3)}$

27.
$$\begin{aligned} 2x + y = 4 &\Rightarrow \times(-1) \quad -2x - y = -4 \\ 2x + 3y = 0 & \qquad\qquad\quad\ \ 2x + 3y = 0 \\ & \qquad\qquad\qquad\quad\ \ 2y = -4 \\ & \qquad\qquad\qquad\quad\ \ \ y = -2 \end{aligned}$$
$$\begin{aligned} 2x + y &= 4 \\ 2x + (-2) &= 4 \\ 2x &= 6 \\ x &= 3 \end{aligned}$$
Solution: $\boxed{(3, -2)}$

29.
$$\begin{aligned} 3x + 29 = 5y &\Rightarrow 3x - 5y = -29 \Rightarrow \times(-1) \quad -3x + 5y = 29 \\ 4y - 34 = -3x &\Rightarrow 3x + 4y = 34 \qquad\qquad\qquad\quad\ \ 3x + 4y = 34 \\ & \qquad\qquad\qquad\qquad\qquad\qquad\qquad\quad 9y = 63 \\ & \qquad\qquad\qquad\qquad\qquad\qquad\qquad\quad\ \ y = 7 \end{aligned}$$
$$\begin{aligned} 3x + 4y &= 34 \\ 3x + 4(7) &= 34 \\ 3x + 28 &= 34 \\ 3x &= 6 \\ x &= 2 \end{aligned}$$
Solution: $\boxed{(2, 7)}$

31.
$$\begin{aligned} 2x = 3(y-2) &\Rightarrow 2x = 3y - 6 \Rightarrow 2x - 3y = -6 \Rightarrow \times(-1) \quad -2x + 3y = 6 \\ 2(x+4) = 3y &\Rightarrow 2x + 8 = 3y \Rightarrow 2x - 3y = -8 \qquad\qquad\qquad\quad 2x - 3y = -8 \\ & \qquad\qquad\qquad\qquad\qquad\qquad\qquad\qquad\qquad\qquad\quad 0 \neq -2 \end{aligned}$$
$\boxed{\text{inconsistent system}}$

33.
$$\begin{array}{rcl} -2(x+1)=3(y-2) \Rightarrow & -2x-2=3y-6 & \Rightarrow -2x-3y=-4 \\ 3(y+2)=6-2(x-2) \Rightarrow & 3y+6=6-2x+4 & \Rightarrow \quad 2x+3y=4 \\ & & \qquad 0=0 \end{array}$$
$\boxed{\text{dependent equations}}$

35.
$$\begin{array}{l} 4(x+1)=17-3(y-1) \Rightarrow 4x+4=17-3y+3 \Rightarrow 4x+3y=16 \\ 2(x+2)+3(y-1)=9 \Rightarrow 2x+4+3y-3=9 \Rightarrow 2x+3y=8 \end{array}$$

$$\begin{array}{lll} 4x+3y=16 \Rightarrow \times(-1) & -4x-3y=-16 & 4x+3y=16 \\ 2x+3y=8 & 2x+3y=8 & 4(4)+3y=16 \\ & -2x=-8 & 16+3y=16 \\ & x=4 & 3y=0 \\ & & y=0 \end{array}$$
Solution: $\boxed{(4,0)}$

37.
$$\begin{array}{lll} 2x+y=10 \Rightarrow \times(-2) & -4x-2y=-20 & 2x+y=10 \\ x+2y=10 & x+2y=10 & 2\left(\frac{10}{3}\right)+y=10 \\ & -3x=-10 & \frac{20}{3}+y=\frac{30}{3} \\ & x=\frac{10}{3} & y=\frac{10}{3} \end{array}$$
Solution: $\boxed{\left(\frac{10}{3},\frac{10}{3}\right)}$

39.
$$\begin{array}{lll} 2x-y=16 \Rightarrow \times(2) & 4x-2y=32 & 3x+2y=3 \\ 3x+2y=3 & 3x+2y=3 & 3(5)+2y=3 \\ & 7x=35 & 15+2y=3 \\ & x=5 & 2y=-12 \\ & & y=-6 \end{array}$$
Solution: $\boxed{(5,-6)}$

41.
$$\begin{array}{lll} 2x+3y=2 \Rightarrow \times(-2) & -4x-6y=-4 & 2x+3y=2 \\ 4x-9y=-1 & 4x-9y=-1 & 2x+3\left(\frac{1}{3}\right)=2 \\ & -15y=-5 & 2x+1=2 \\ & y=\frac{-5}{-15}=\frac{1}{3} & 2x=1 \\ & & x=\frac{1}{2} \end{array}$$
Solution: $\boxed{\left(\frac{1}{2},\frac{1}{3}\right)}$

43.
$$\begin{array}{lll} 4x+5y=-20 \Rightarrow \times(4) & 16x+20y=-80 & 4x+5y=-20 \\ 5x-4y=-25 \Rightarrow \times(5) & 25x-20y=-125 & 4(-5)+5y=-20 \\ & 41x=-205 & -20+5y=-20 \\ & x=-5 & 5y=0 \\ & & y=0 \end{array}$$
Solution: $\boxed{(-5,0)}$

45.
$$\begin{array}{llll} 6x=-3y \Rightarrow & 6x+3y=0 & 6x+3y=0 & 6x=-3y \\ 5y=2x+12 \Rightarrow & -2x+5y=12 \Rightarrow \times(3) & -6x+15y=36 & 6x=-3(2) \\ & & 18y=36 & 6x=-6 \\ & & y=2 & x=-1 \end{array}$$
Solution: $\boxed{(-1,2)}$

47.

$$\begin{aligned} 4(2x-y) &= 18 &&\Rightarrow 8x-4y=18 &&\Rightarrow 8x-4y=18\\ 3(x-3) &= 2y-1 &&\Rightarrow 3x-9=2y-1 &&\Rightarrow 3x-2y=8 \end{aligned}$$

$$\begin{aligned} 8x-4y &= 18 \\ 3x-2y &= 8 \Rightarrow \times(-2) \end{aligned} \qquad \begin{aligned} 8x-4y &= 18\\ -6x+4y &= -16\\ \hline 2x &= 2\\ x &= 1 \end{aligned} \qquad \begin{aligned} 8x-4y &= 18\\ 8(1)-4y &= 18\\ 8-4y &= 18\\ -4y &= 10\\ y &= -\tfrac{10}{4} = -\tfrac{5}{2} \end{aligned}$$

Solution: $\boxed{(1, -\frac{5}{2})}$

49.

$$\begin{aligned} 3(x-2y) &= 12 &&\Rightarrow 3x-6y=12 && 3x-6y=12 && 3x-6y=12\\ x &= 2(y+2) &&\Rightarrow x=2y+4 \Rightarrow && x-2y=4 \Rightarrow \times(-3) && -3x+6y=-12\\ &&&&&& 0=0 \end{aligned}$$

dependent equations

51.

$$\begin{aligned} \frac{3}{5}x+\frac{4}{5}y &= 1 \Rightarrow \times(5) & 3x+4y &= 5 \Rightarrow \times(2) & 6x+8y &= 10\\ -\frac{1}{4}x+\frac{3}{8}y &= 1 \Rightarrow \times(8) & -2x+3y &= 8 \Rightarrow \times(3) & -6x+9y &= 24\\ &&&& 17y &= 34\\ &&&& y &= 2 \end{aligned} \qquad \begin{aligned} 3x+4y &= 5\\ 3x+4(2) &= 5\\ 3x+8 &= 5\\ 3x &= -3\\ x &= -1 \end{aligned}$$

Solution: $\boxed{(-1,2)}$

53.

$$\begin{aligned} \frac{3}{5}x+y &= 1 \Rightarrow \times(5) & 3x+5y &= 5\\ \frac{4}{5}x-y &= -1 \Rightarrow \times(5) & 4x-5y &= -5\\ && 7x &= 0\\ && x &= 0 \end{aligned} \qquad \begin{aligned} 3x+5y &= 5\\ 3(0)+5y &= 5\\ 5y &= 5\\ y &= 1 \end{aligned}$$

Solution: $\boxed{(0,1)}$

55.

$$\begin{aligned} \frac{x}{2}-\frac{y}{3} &= -2 \Rightarrow \times(6) & 3x-2y &= -12 \Rightarrow & 3x-2y &= -12\\ \frac{2x-3}{2}+\frac{6y+1}{3} &= \frac{17}{6} \Rightarrow \times(6) & 3(2x-3)+2(6y+1) &= 17 \Rightarrow & 6x-9+12y+2 &= 17 \end{aligned}$$

$$\begin{aligned} 3x-2y &= -12 \Rightarrow \times(-2)\\ 6x+12y &= 24 \end{aligned} \qquad \begin{aligned} -6x+4y &= 24\\ 6x+12y &= 24\\ \hline 16y &= 48\\ y &= 3 \end{aligned} \qquad \begin{aligned} 3x-2y &= -12\\ 3x-2(3) &= -12\\ 3x-6 &= -12\\ 3x &= -6\\ x &= -2 \end{aligned}$$

Solution: $\boxed{(-2,3)}$

57. $\frac{x-3}{2}+\frac{y+5}{3}=\frac{11}{6} \Rightarrow \times(6) \quad 3(x-3)+2(y+5)=11 \Rightarrow 3x-9+2y+10=11$

$\frac{x+3}{3}-\frac{5}{12}=\frac{y+3}{4} \Rightarrow \times(12) \quad 4(x+3)-5=3(y+3) \Rightarrow 4x+12-5=3y+9$

$$\begin{aligned} 3x+2y&=10 \Rightarrow \times(3) \\ 4x-3y&=\ \ 2 \Rightarrow \times(2) \end{aligned} \qquad \begin{aligned} 9x+6y&=30 \\ 8x-6y&=\ \ 4 \\ 17x\qquad &=34 \\ x\qquad &=\ \ 2 \end{aligned} \qquad \begin{aligned} 3x+2y&=10 \\ 3(2)+2y&=10 \\ 6+2y&=10 \\ 2y&=4 \\ y&=2 \end{aligned}$$

Solution: $\boxed{(2,2)}$

59. $$\begin{aligned} 2(x+y)&=10 \Rightarrow & 2x+2y&=10 \Rightarrow \times(5) \\ 5(x-y)&=5 \Rightarrow & 5x-5y&=5 \Rightarrow \times(2) \end{aligned} \qquad \begin{aligned} 10x+10y&=50 \\ 10x-10y&=10 \\ 20x\qquad &=60 \\ x\qquad &=\ \ 3 \end{aligned} \qquad \begin{aligned} 2x+2y&=10 \\ 2(3)+2y&=10 \\ 6+2y&=10 \\ 2y&=4 \\ y&=2 \end{aligned}$$

Solution: $\boxed{x=3,\ y=2}$

61. **Answers may vary.**

63. $(1,4)$

Exercise 3.6 (page 210)

1. $x<4$

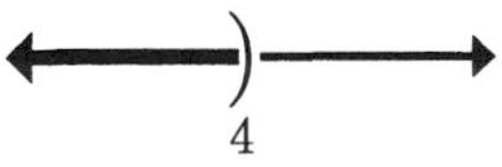

3. $-1<x\le 2$

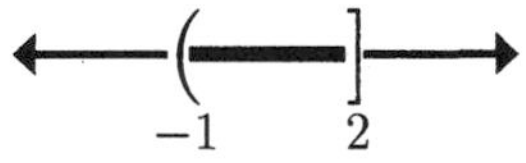

5. $8\cdot 8\cdot 8\cdot c=8^3c$

7. $a\cdot a\cdot b\cdot b=a^2b^2$

9. variable

11. system

13. Let x and y represent the integers.
$$\begin{cases} (1) & x=2y \\ (2) & x+y=96 \end{cases}$$

Substitute $x=2y$ into (2):
$$\begin{aligned} x+y&=96 \\ 2y+y&=96 \\ 3y&=96 \\ y&=32 \end{aligned}$$

Substitute; solve for x:
$x=2y$
$x=2(32)=64$
The integers are 32 and 64.

15. Let x and y represent the integers.
$$\begin{aligned} 3x+\ \ y&=29 \Rightarrow \times(-2) \\ x+2y&=18 \end{aligned} \qquad \begin{aligned} -6x-2y&=-58 \\ x+2y&=\ \ 18 \\ -5x\qquad &=-40 \\ x\qquad &=\ \ \ 8 \end{aligned} \qquad \begin{aligned} x+2y&=18 \\ 8+2y&=18 \\ 2y&=10 \\ y&=5 \end{aligned}$$

The integers are 5 and 8.

17. Let $x =$ number of horses and $y =$ number of cows.

$$\begin{cases} (1) & y = 5x \\ (2) & x + y = 168 \end{cases}$$

Substitute $y = 5x$ from (1) into (2):

$$\begin{aligned} x + y &= 168 \\ x + 5x &= 168 \\ 6x &= 168 \\ x &= 28 \end{aligned}$$

Substitute and solve for y:

$$\begin{aligned} y &= 5x \\ y &= 5(28) = 140 \end{aligned}$$

He has 140 cows.

19. Let $x =$ cost of a gallon of paint and $y =$ cost of a brush.

$$\begin{aligned} 8x + 3y = 135 &\Rightarrow \times\ (2) & 16x + 6y &= 270 \\ 6x + 2y = 100 &\Rightarrow \times\ (-3) & -18x - 6y &= -300 \\ & & -2x &= -30 \\ & & x &= 15 \end{aligned}$$

$$\begin{aligned} 6x + 2y &= 100 \\ 6(15) + 2y &= 100 \\ 90 + 2y &= 100 \\ 2y &= 10 \\ y &= 5 \end{aligned}$$

The paint costs \$15 per gallon, while each brush costs \$5.

21. Let $x =$ cost of contact cleaner and $y =$ cost of soaking solution.

$$\begin{aligned} 2x + 3y = 29.40 &\Rightarrow \times\ (-2) & -4x - 6y &= -58.80 \\ 3x + 2y = 28.60 &\Rightarrow \times\ (3) & 9x + 6y &= 85.80 \\ & & 5x &= 27 \\ & & x &= 5.40 \end{aligned}$$

$$\begin{aligned} 2x + 3y &= 29.40 \\ 2(5.40) + 3y &= 29.40 \\ 10.80 + 3y &= 29.40 \\ 3y &= 18.60 \\ y &= 6.20 \end{aligned}$$

The contact cleaner costs \$5.40, while the soaking solution costs \$6.20.

23. Let $x =$ the length of one piece.
Let $y =$ the other length.

$$\begin{cases} (1) & x = y + 5 \\ (2) & x + y = 25 \end{cases}$$

Substitute $x = y + 5$ into (2):

$$\begin{aligned} x + y &= 25 \\ y + 5 + y &= 25 \\ 2y &= 20 \\ y &= 10 \end{aligned}$$

Substitute; solve for x:

$$\begin{aligned} x &= y + 5 \\ x &= 10 + 5 = 15 \end{aligned}$$

The lengths are 10 ft and 15 ft.

25. Let $x =$ Lena's amount.
Let $y =$ Chayla's amount.

$$\begin{cases} (1) & x = y + 50000 \\ (2) & x + y = 250000 \end{cases}$$

Substitute $x = y + 50000$ into (2):

$$\begin{aligned} x + y &= 250000 \\ y + 50000 + y &= 250000 \\ 2y &= 200000 \\ y &= 100000 \end{aligned}$$

Chayla will get \$100,000.

27. Let $x =$ time for causes.
Let $y =$ time for outcome.

$$\begin{cases} (1) & x = 4y \\ (2) & x + y = 30 \end{cases}$$

Substitute $x = 4y$ into (2):

$$\begin{aligned} x + y &= 30 \\ 4y + y &= 30 \\ 5y &= 30 \\ y &= 6 \end{aligned}$$

Substitute; solve for x:

$$\begin{aligned} x &= 4y \\ x &= 4(6) = 24 \end{aligned}$$

24 minutes will be devoted to the causes, and 6 minutes to the outcome.

29. Let $x =$ deaths from cancer.
Let $y =$ deaths from accidents.

$$\begin{cases} (1) & x = 6y \\ (2) & x + y = 630000 \end{cases}$$

Substitute $x = 6y$ into (2):

$$\begin{aligned} x + y &= 630000 \\ 6y + y &= 630000 \\ 7y &= 630000 \\ y &= 90000 \end{aligned}$$

Substitute; solve for x:

$$\begin{aligned} x &= 6y \\ x &= 6(90000) = 540000 \end{aligned}$$

90,000 die from accidents and 540,000 die from cancer.

31. Let $w =$ the width.
Let $l =$ the length.

$$\begin{cases} (1) & l = w + 5 \\ (2) & 2l + 2w = 110 \end{cases}$$

Substitute $l = w + 5$ into (2):

$$\begin{aligned} 2l + 2w &= 110 \\ 2(w + 5) + 2w &= 110 \\ 2w + 10 + 2w &= 110 \\ 4w &= 100 \\ w &= 25 \end{aligned}$$

Substitute; solve for l:

$$\begin{aligned} l &= w + 5 \\ l &= 25 + 5 = 30 \end{aligned}$$

The dimensions are 25 ft by 30 ft.

33. Let $x =$ the length.
Let $y =$ the width.

$$\begin{cases} (1) & x = 2y + 2 \\ (2) & 2x + 2y = 34 \end{cases}$$

Substitute $x = 2y + 2$ into (2):

$$\begin{aligned} 2x + 2y &= 34 \\ 2(2y + 2) + 2y &= 34 \\ 4y + 4 + 2y &= 34 \\ 6y &= 30 \\ y &= 5 \end{aligned}$$

Substitute; solve for x:

$$\begin{aligned} x &= 2y + 2 \\ x &= 2(5) + 2 = 12 \end{aligned}$$

The area is $(5)(12) = 60$ ft^2.

35. Let $c =$ the cost.
Let $n =$ number of years.

$$\begin{cases} (1) & c = 2250 + 412n \\ (2) & c = 1715 + 466n \end{cases}$$

Substitute:

$$\begin{aligned} 2250 + 412n &= 1715 + 466n \\ -54n &= -535 \\ n &\approx 9.9 \end{aligned}$$

The break point is about 9.9 years.

37. Since 7 years is less than the break point, choose the furnace with the lower up-front cost (80+).

39. Let $x =$ Bill's principal and $y =$ Janette's principal.

$$\begin{aligned} x + y &= 5000 \Rightarrow \times(-5) & -5x - 5y &= -25000 \\ 0.05x + 0.07y &= 310 \Rightarrow \times(100) & 5x + 7y &= 31000 \\ & & 2y &= 6000 \\ & & y &= 3000 \end{aligned}$$

$$\begin{aligned} x + y &= 5000 \\ x + 3000 &= 5000 \\ x &= 2000 \end{aligned}$$

Bill invested \$2000.

41. Let $x =$ number of student tickets and $y =$ number of nonstudent tickets.

$$\begin{aligned} x + y &= 350 \Rightarrow \times(-1) & -x - y &= -350 \\ x + 2y &= 450 & x + 2y &= 450 \\ & & y &= 100 \end{aligned}$$

$$\begin{aligned} x + y &= 350 \\ x + 100 &= 350 \\ x &= 250 \end{aligned}$$

There were 250 student tickets sold.

43. Let b = speed of boat in still water and c = speed of current.

	d	r	t	Equation $(d = r \cdot t)$
Downstream	24	$b+c$	2	$24 = (b+c)\cdot 2 \Rightarrow 2b+2c = 24$
Upstream	24	$b-c$	3	$24 = (b-c)\cdot 3 \Rightarrow 3b-3c = 24$

$$\begin{aligned} 2b+2c=24 &\Rightarrow \times\,(3) & 6b+6c &= 72 \\ 3b-3c=24 &\Rightarrow \times\,(2) & 6b-6c &= 48 \\ & & 12b &= 120 \\ & & b &= 10 \end{aligned}$$

The speed of the boat is 10 mph in still water.

45. Let p = airspeed of plane and w = speed of wind.

	d	r	t	Equation $(d = r \cdot t)$
With wind	600	$p+w$	2	$600 = (p+w)\cdot 2 \Rightarrow 2p+2w = 600$
Against wind	600	$p-w$	3	$600 = (p-w)\cdot 3 \Rightarrow 3p-3w = 600$

$$\begin{aligned} 2p+2w=600 &\Rightarrow \times\,(3) & 6p+6w &= 1800 \\ 3p-3w=600 &\Rightarrow \times\,(2) & 6p-6w &= 1200 \\ & & 12p &= 3000 \\ & & p &= 250 \end{aligned}$$

$$\begin{aligned} 2p+2w &= 600 \\ 2(250)+2w &= 600 \\ 500+2w &= 600 \\ 2w &= 100 \\ w &= 50 \end{aligned}$$

The speed of the wind is 50 mph.

47. Let x = liters of first solution and y = liters of second solution.

	Fractional part that is alcohol	Number of liters of solution	Number of liters of alcohol
First solution	0.40	x	$0.40x$
Second solution	0.55	y	$0.55y$
Final solution	0.50	15	$0.50(15) = 7.5$

$$\begin{aligned} x + y &= 15 \\ 0.40x+0.55y &= 7.5 \end{aligned}$$

$$\begin{aligned} x+y = 15 &\Rightarrow \times\,(-40) & -40x-40y &= -600 \\ 0.40x+0.55y = 7.5 \Rightarrow \times\,(100) \quad 40x+55y = 750 & & 40x+55y &= 750 \\ & & 15y &= 150 \\ & & y &= 10 \end{aligned}$$

$$\begin{aligned} x+y &= 15 \\ x+10 &= 15 \\ x &= 5 \end{aligned}$$

The chemist should use 5 L of the 40% solution and 10 L of the 55% solution.

49. Let x = pounds of peanuts and y = pounds of cashews.

	Cost per pound	Number of pounds	Total value
Peanuts	3	x	$3x$
Cashews	6	y	$6y$
Mixture	4	48	$4(48) = 192$

$$\begin{aligned} x+y=48 &\Rightarrow \times\,(-3) & -3x-3y &= -144 \\ 3x+6y=192 & & 3x+6y &= 192 \\ & & 3y &= 48 \\ & & y &= 16 \end{aligned}$$

$$\begin{aligned} x+y &= 48 \\ x+16 &= 48 \\ x &= 32 \end{aligned}$$

The merchant should use 32 pounds of peanuts and 16 pounds of cashews.

51. Let $x =$ number of inexpensive radios and $y =$ number of expensive radios.

$$\begin{aligned} x + y &= 25 \Rightarrow \times(-87) \\ 87x + 119y &= 2495 \end{aligned} \qquad \begin{aligned} -87x - 87y &= -2175 \\ 87x + 119y &= 2495 \\ \hline 32y &= 320 \\ y &= 10 \end{aligned} \qquad \begin{aligned} x + y &= 25 \\ x + 10 &= 25 \\ x &= 15 \end{aligned}$$

There were 15 inexpensive radios sold.

53. Let $r =$ the lower rate. Then $r + 0.01 =$ the higher rate.

$\boxed{\text{Interest on \$950}} + \boxed{\text{Interest on \$1,200}} = \boxed{\text{Total interest}}$

$$\begin{aligned} 950r + 1200(r + 0.01) &= 205.50 \\ 950r + 1200r + 12 &= 205.50 \\ 2150r &= 193.50 \\ r &= 0.09 \Rightarrow \text{The lower rate is 9\%.} \end{aligned}$$

55. $\begin{cases} (1) & p = -\frac{1}{2}q + 1300 \\ (2) & p = \frac{1}{3}q + \frac{1400}{3} \end{cases}$

Substitute:

$$\begin{aligned} -\tfrac{1}{2}q + 1300 &= \tfrac{1}{3}q + \tfrac{1400}{3} \\ 6\left(-\tfrac{1}{2}q + 1300\right) &= 6\left(\tfrac{1}{3}q + \tfrac{1400}{3}\right) \\ -3q + 7800 &= 2q + 2800 \\ -5q &= -5000 \\ q &= 1000 \end{aligned}$$

Substitute $q = 1000$ and solve for p:

$$\begin{aligned} p &= -\tfrac{1}{2}q + 1300 \\ &= -\tfrac{1}{2}(1000) + 1300 \\ &= -500 + 1300 = 800 \end{aligned}$$

The equilibrium price is $800.

57. **Answers may vary.**

Exercise 3.7 (page 222)

1. $$\begin{aligned} 3x + 5 &= 14 \\ 3x &= 9 \\ x &= 3 \end{aligned}$$

3. $$\begin{aligned} A &= P + Prt \\ A - P &= Prt \\ \frac{A - P}{Pr} &= \frac{Prt}{Pr} \\ \frac{A - P}{Pr} &= t, \text{ or } t = \frac{A - P}{Pr} \end{aligned}$$

5. $2a + 5(a - 3) = 2a + 5a - 15 = 7a - 15$

7. $$\begin{aligned} 4(b - a) + 3b + 2a &= 4b - 4a + 3b + 2a \\ &= -2a + 7b \end{aligned}$$

9. inequality

11. boundary

13. inequalities

15. doubly shaded

17. **a.** $$\begin{aligned} 5x - 3y &\geq 0 \\ 5(1) - 3(1) &\overset{?}{\geq} 0 \\ 5 - 3 &\overset{?}{\geq} 0 \\ 2 &\geq 0 \end{aligned}$$

$(1, 1)$ is a solution.

b. $$\begin{aligned} 5x - 3y &\geq 0 \\ 5(-2) - 3(-3) &\overset{?}{\geq} 0 \\ -10 + 9 &\overset{?}{\geq} 0 \\ -1 &\ngeq 0 \end{aligned}$$

$(-2, -3)$ is not a solution.

17. **c.**

$$5x - 3y \geq 0$$
$$5(0) - 3(0) \overset{?}{\geq} 0$$
$$0 - 0 \overset{?}{\geq} 0$$
$$0 \geq 0$$

$(0, 0)$ is a solution.

d.

$$5x - 3y \geq 0$$
$$5\left(\frac{1}{5}\right) - 3\left(\frac{4}{3}\right) \overset{?}{\geq} 0$$
$$1 - 4 \overset{?}{\geq} 0$$
$$-3 \not\geq 0$$

$\left(\frac{1}{5}, \frac{4}{3}\right)$ is not a solution.

19. **a.** does not include boundary

b. does include boundary

21. Test point: $(0, 0)$

$$y \leq x + 2$$
$$0 \overset{?}{\leq} 0 + 2$$
$$0 \leq 2$$

Shade half-plane with test point.

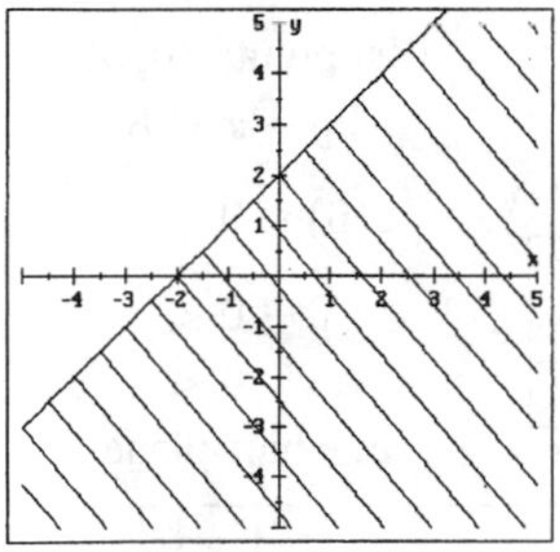

23. Test point: $(0, 0)$

$$y > 2x - 4$$
$$0 \overset{?}{>} 2(0) - 4$$
$$0 \overset{?}{>} 0 - 4$$
$$0 > -4$$

Shade half-plane with test point.

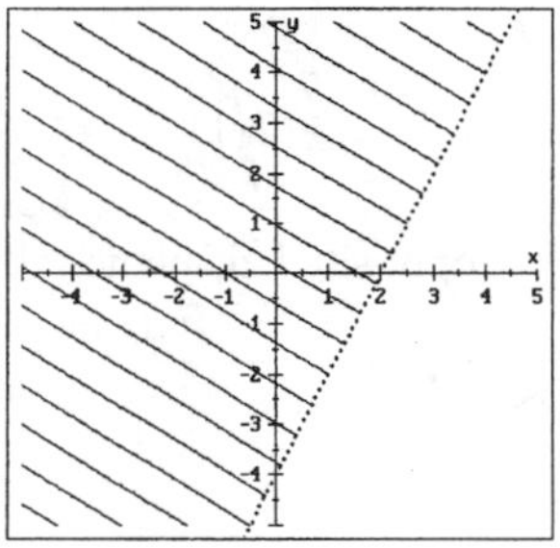

25. Test point: $(0, 0)$

$$x - 2y \leq 4$$
$$0 - 2(0) \overset{?}{\leq} 4$$
$$0 - 0 \overset{?}{\leq} 4$$
$$0 \leq 4$$

Shade half-plane with test point.

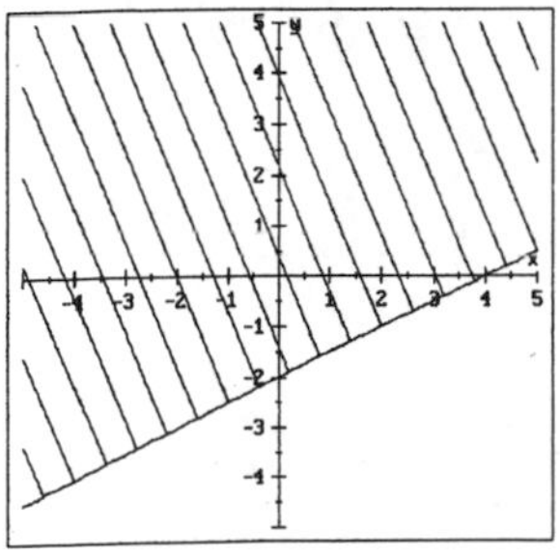

27. Test point: $(1, 1)$

$$y \leq 4x$$
$$1 \overset{?}{\leq} 4(1)$$
$$1 \leq 4$$

Shade half-plane with test point.

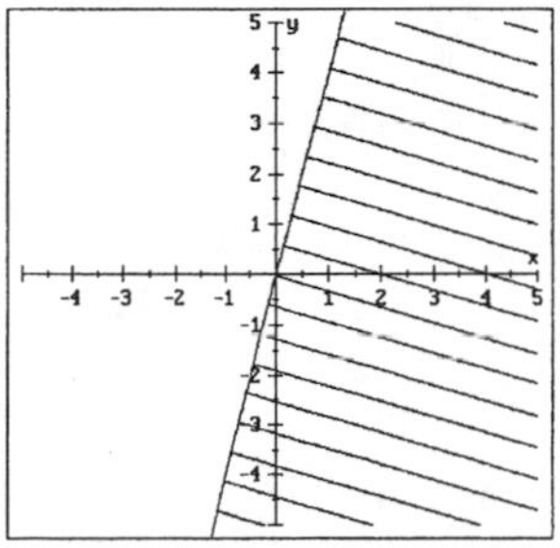

29. Boundary (solid) $y = 3 - x$

x	y
0	3
3	0

Test point: $(0,0)$

$y \geq 3 - x$

$0 \overset{?}{\geq} 3 - 0$

$0 \not\geq 3$

opposite half-plane

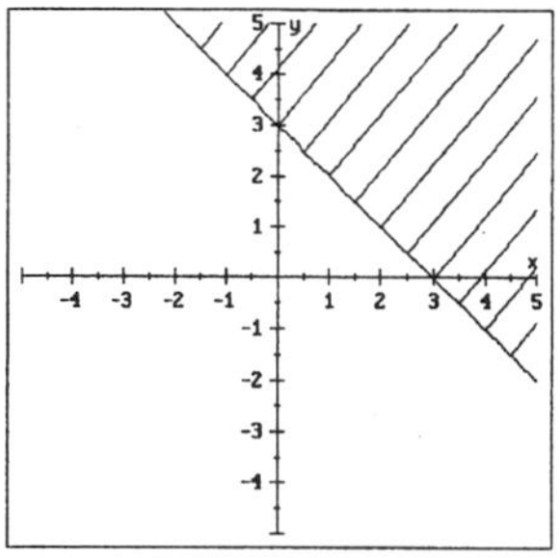

31. Boundary (dotted) $y = 2 - 3x$

x	y
0	2
1	-1

Test point: $(0,0)$

$y < 2 - 3x$

$0 \overset{?}{<} 2 - 3(0)$

$0 \overset{?}{<} 2 - 0$

$0 < 2$

same half-plane

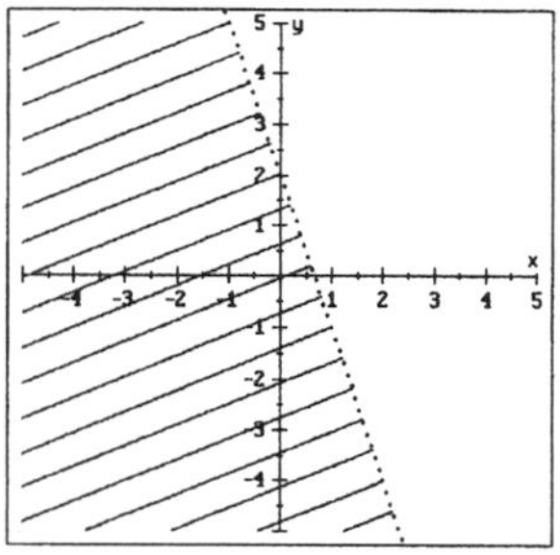

33. Boundary (solid) $y = 2x$

x	y
0	0
2	4

Test point: $(1,1)$

$y \geq 2x$

$1 \overset{?}{\geq} 2(1)$

$1 \not\geq 2$

opposite half-plane

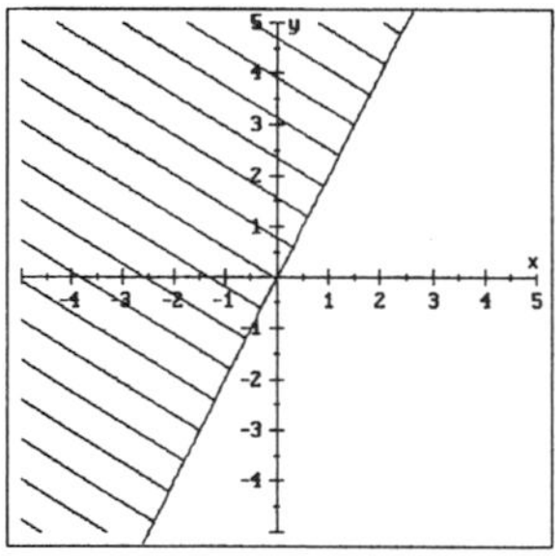

35. Boundary (dotted) $2y - x = 8$

x	y
0	4
-8	0

Test point: $(0,0)$

$2y - x < 8$

$2(0) - 0 \overset{?}{<} 8$

$0 - 0 \overset{?}{<} 8$

$0 < 8$

same half-plane

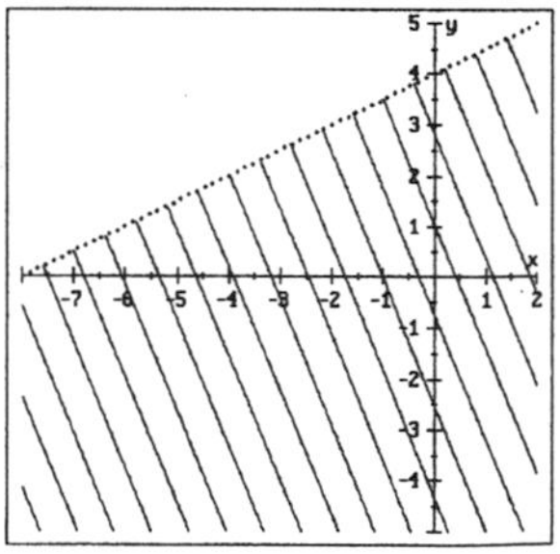

37. Boundary (dotted)

$3x - 4y = 12$

x	y
0	-3
4	0

Test point: $(0, 0)$

$3x - 4y > 12$

$3(0) - 4(0) \overset{?}{>} 12$

$0 - 0 \overset{?}{>} 12$

$0 \not> 12$

opposite half-plane

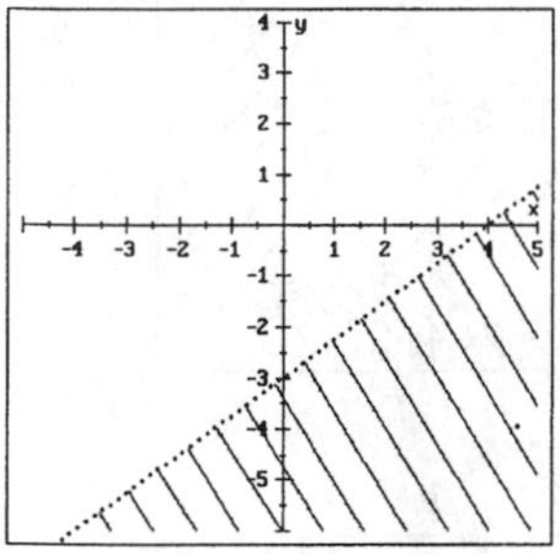

39. Boundary (solid)

$5x + 4y = 20$

x	y
0	5
4	0

Test point: $(0, 0)$

$5x + 4y \geq 20$

$5(0) + 4(0) \overset{?}{\geq} 20$

$0 + 0 \overset{?}{\geq} 20$

$0 \not\geq 20$

opposite half-plane

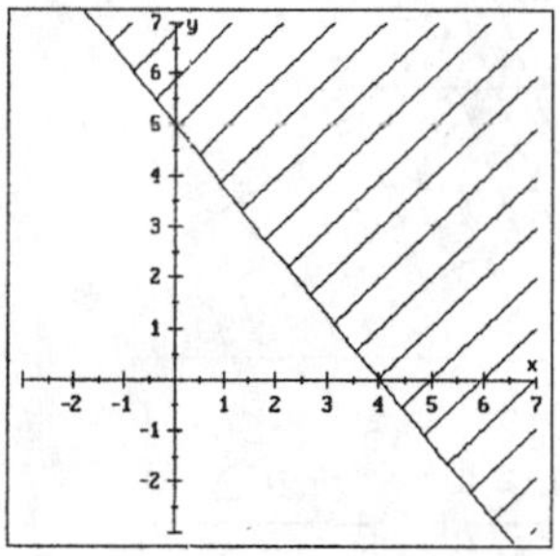

41. Boundary (dotted)

$x = 2$

x	y
2	0
2	4

Test point: $(0, 0)$

$x < 2$

$0 < 2$

same half-plane

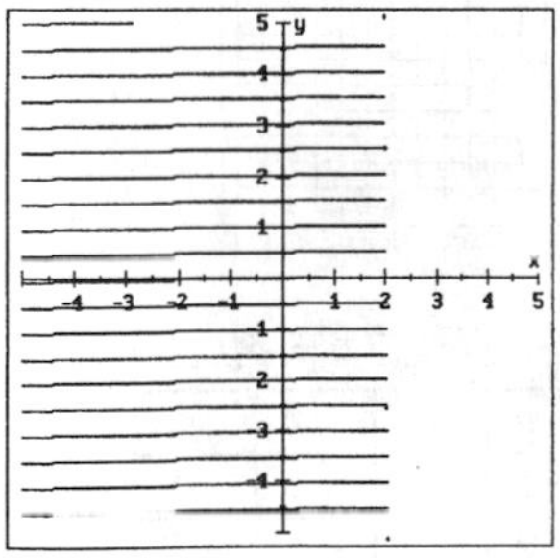

43. Boundary (solid)

$y = 1$

x	y
0	1
4	1

Test point: $(0, 0)$

$y \leq 1$

$0 \leq 1$

same half-plane

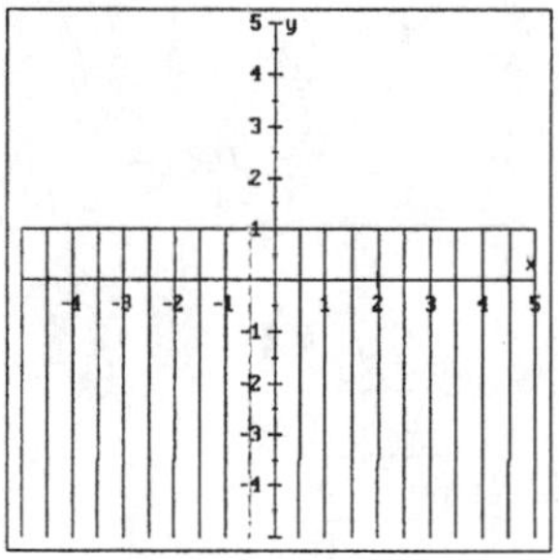

45. $x + 2y \le 3$ $\quad 2x - y \ge 1$

x	y
1	1
3	0

x	y
0	-1
2	3

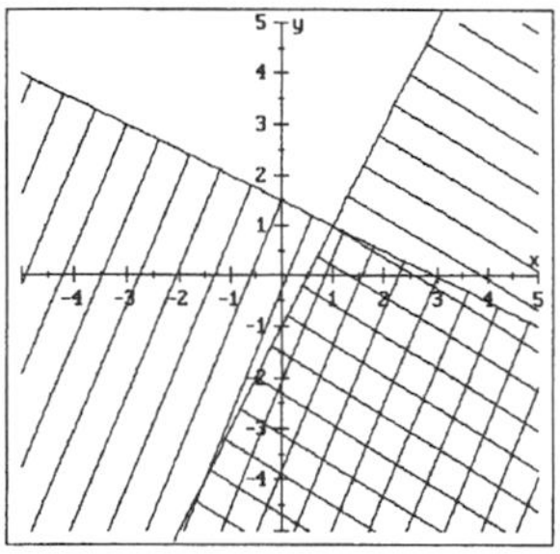

47. $x + y < -1$ $\quad x - y > -1$

x	y
0	-1
-1	0

x	y
0	1
-1	0

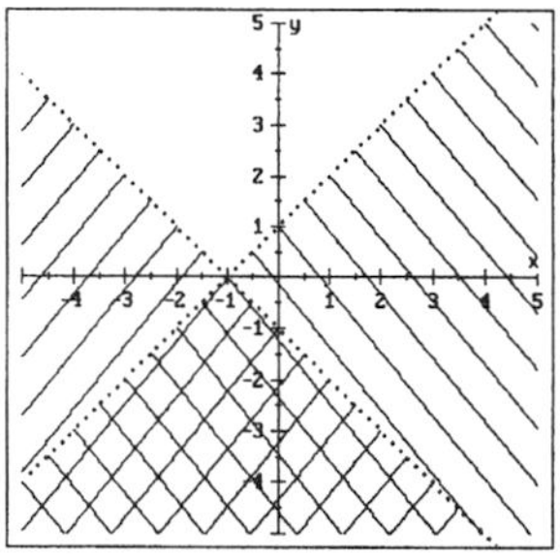

49. $2x - y < 4$ $\quad x + y \ge -1$

x	y
0	-4
2	0

x	y
0	-1
-1	0

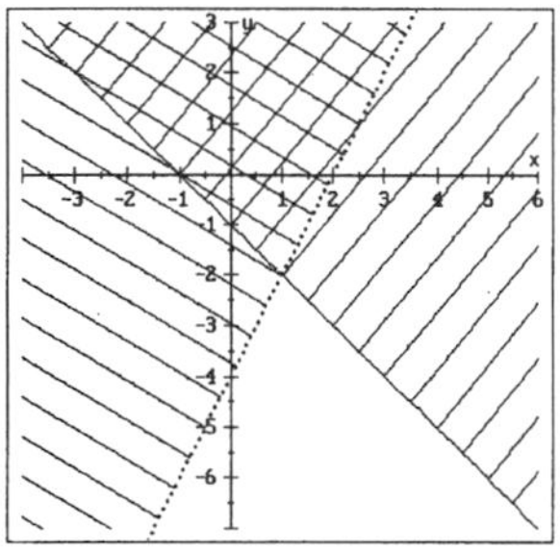

51. $x > 2$ $\quad y \le 3$

x	y
2	0
2	2

x	y
0	3
1	3

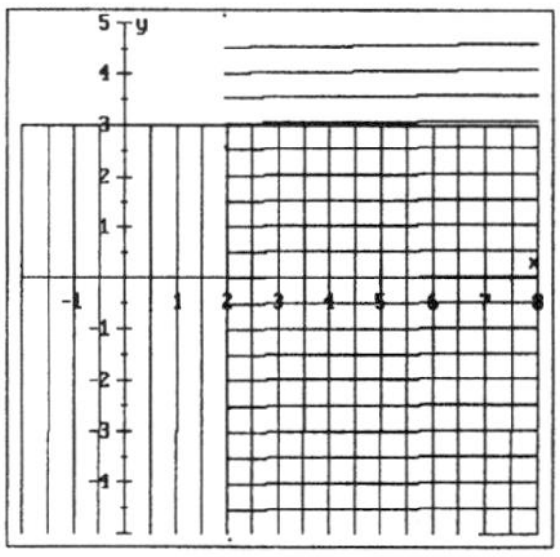

53. $x + y < 1$ $\quad x + y > 3$

x	y
0	1
1	0

x	y
0	3
3	0

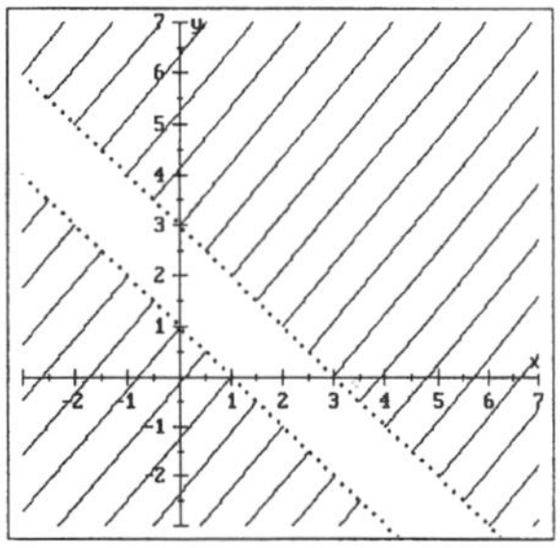

(no solutions)

55. $3x + 4y > -7$ $\quad 2x - 3y \ge 1$

x	y
-1	-1
-5	2

x	y
2	1
-1	-1

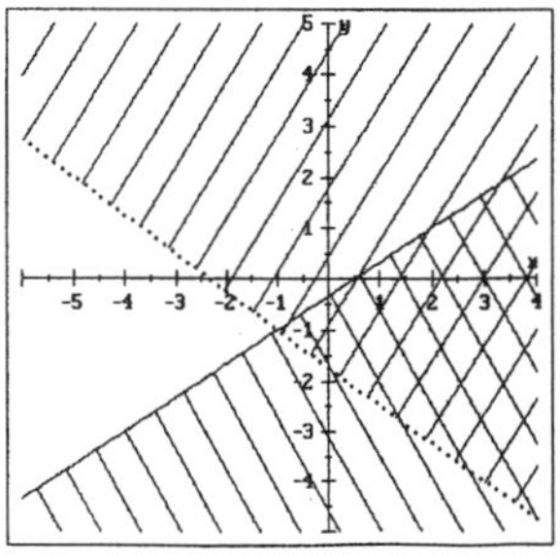

57. $2x - 4y > -6$ $\quad 3x + y \geq 5$

x	y
1	2
-3	0

x	y
0	5
1	2

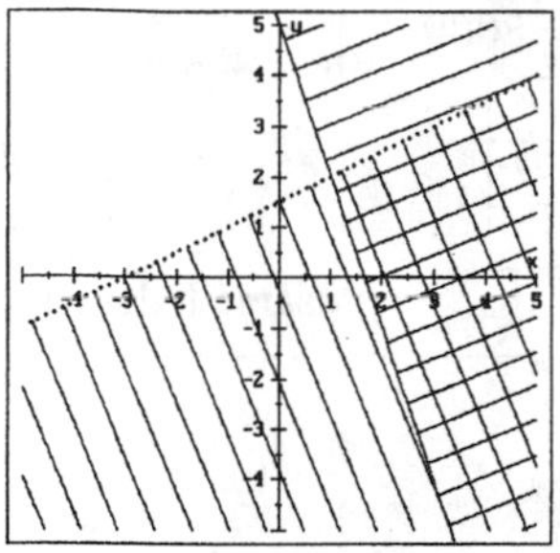

59. $3x - y \leq -4$ $\quad 3y > -2(x+5)$

x	y
0	4
-1	1

x	y
1	-4
-5	0

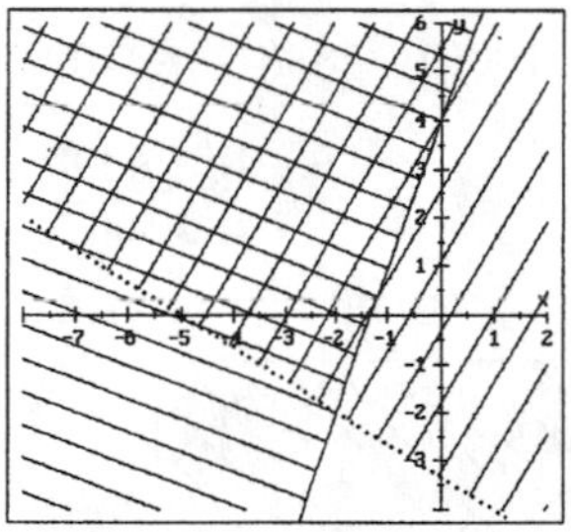

61. $\dfrac{x}{2} + \dfrac{y}{3} \geq 2$ $\quad \dfrac{x}{2} - \dfrac{y}{2} < -1$

$3x + 2y \geq 12$ $\quad x - y < -2$

x	y
0	6
4	0

x	y
0	2
-2	0

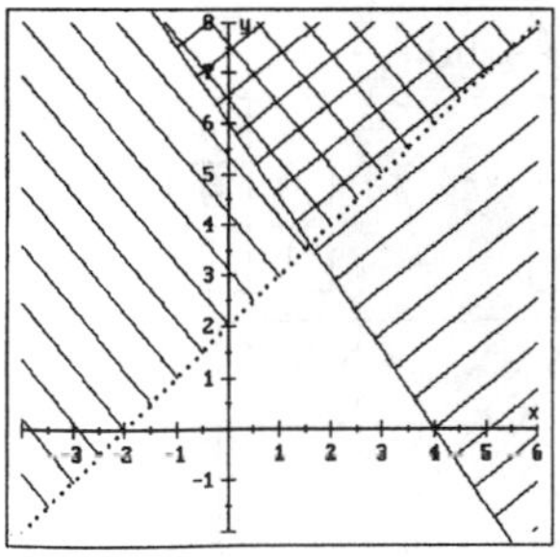

63. $\boxed{\text{Cake cost}} + \boxed{\text{Pie cost}} \leq 120$

$3x + 4y \leq 120$

Boundary (solid)

$3x + 4y = 120$

x	y
0	30
40	0

Test point: $(0,0)$

$$3x + 4y \leq 120$$

$$3(0) + 4(0) \stackrel{?}{\leq} 120$$

$$0 + 0 \stackrel{?}{\leq} 120$$

$$0 \leq 120$$

same half-plane

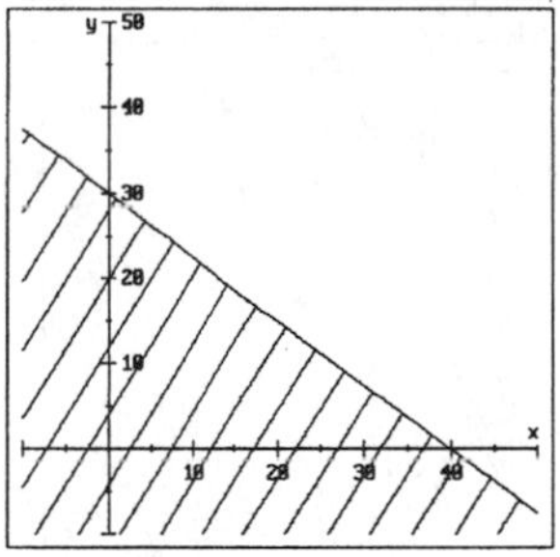

Solutions: $(10,10), (20,10), (10,20)$

65. $\boxed{\text{Leather}} + \boxed{\text{Nylon}} \geq 4400$

$100x + 88y \geq 4400$

Boundary (solid)

x	y
0	50
44	0

Test point: $(0, 0)$

$100x + 88y \geq 4400$

$100(0) + 88(0) \overset{?}{\geq} 4400$

$0 + 0 \overset{?}{\geq} 4400$

$0 \not\geq 4400$

opposite half-plane

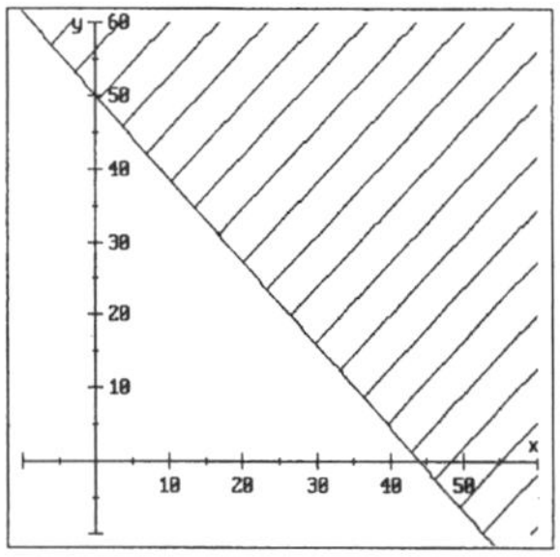

Solutions: $(50, 50), (30, 40), (40, 40)$

67. $\boxed{\text{R. stock}} + \boxed{\text{M. stock}} \leq 8000$

$40x + 50y \leq 8000$

Boundary (solid)

$40x + 50y = 8000$

x	y
0	160
200	0

Test point: $(0, 0)$

$40x + 50y \leq 8000$

$40(0) + 50(0) \overset{?}{\leq} 8000$

$0 + 0 \overset{?}{\leq} 8000$

$0 \leq 8000$

same half-plane

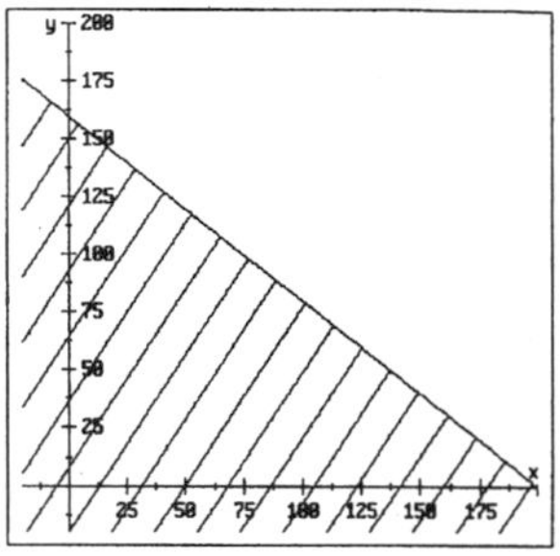

Solutions: $(80, 40), (80, 80), (120, 40)$

69. $10x + 15y \geq 30$ $\quad$ $10x + 15y \leq 60$

x	y
0	2
3	0

x	y
0	4
6	0

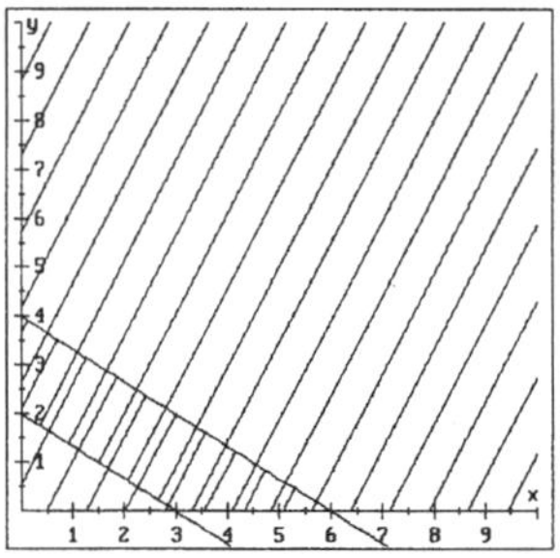

Solutions: $(1, 2), (4, 1)$

71. $150x + 100y \leq 900$ $\quad$ $y > x$

x	y
0	9
6	0

x	y
0	0
1	1

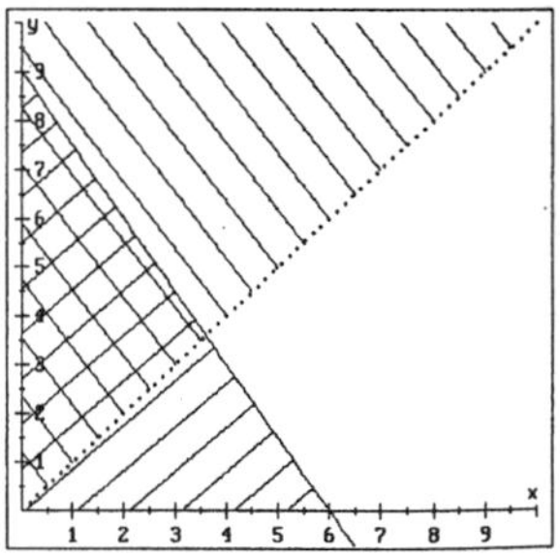

Solutions: $(2, 4), (1, 5)$

73-79. Answers may vary.

Chapter 3 Summary (page 228)

1-6.

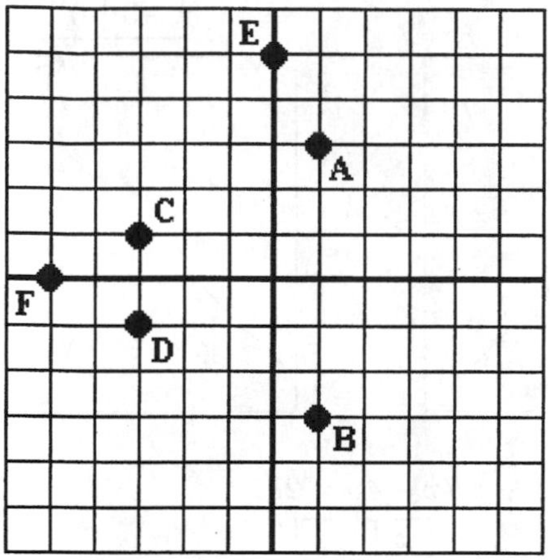

7. $(3, 1)$
8. $(-4, 5)$
9. $(-3, -4)$
10. $(2, -3)$
11. $(0, 0)$
12. $(0, 4)$
13. $(-5, 0)$
14. $(0, -3)$

15.
$$3x - 4y = 12$$
$$3(2) - 4(1) \stackrel{?}{=} 12$$
$$6 - 4 \stackrel{?}{=} 12$$
$$2 \neq 12$$
$(2, 1)$ is not a solution.

16.
$$3x - 4y = 12$$
$$3(3) - 4\left(-\frac{3}{4}\right) \stackrel{?}{=} 12$$
$$9 + 3 \stackrel{?}{=} 12$$
$$12 = 12$$
$\left(3, -\frac{3}{4}\right)$ is a solution.

17. $y = x - 5$

x	y
0	-5
2	-3

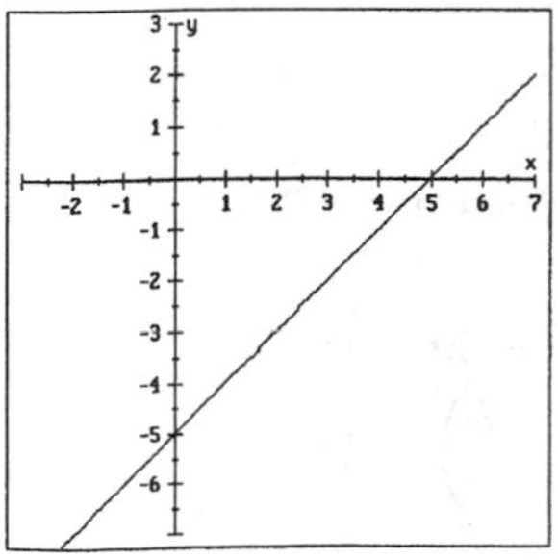

18. $y = 2x + 1$

x	y
0	1
2	5

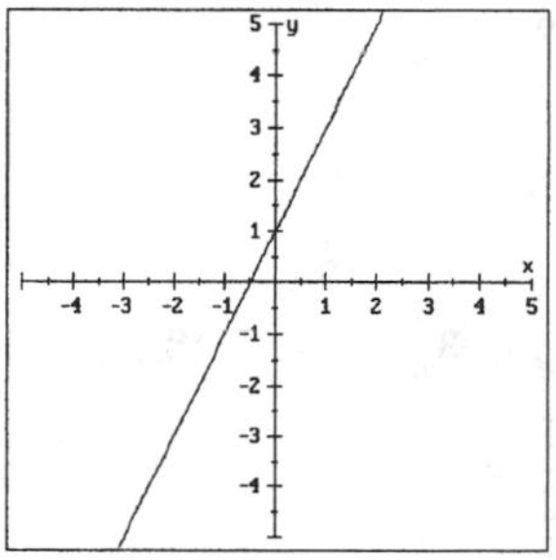

19. $y = \frac{x}{2} + 2$

x	y
0	2
2	3

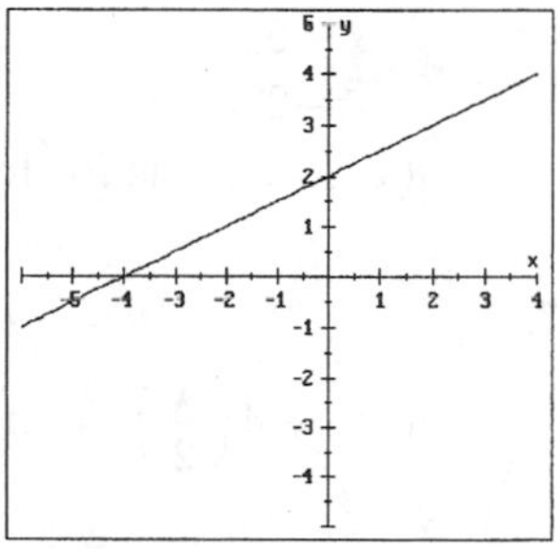

20. $y = 3$

x	y
0	3
2	3

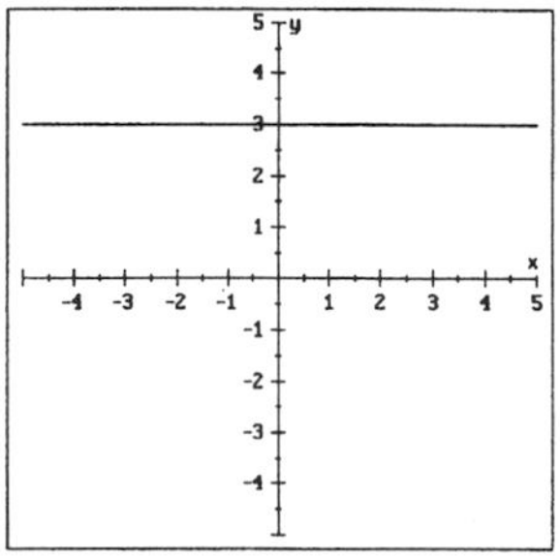

21. $x + y = 4$

x	y
0	4
4	0

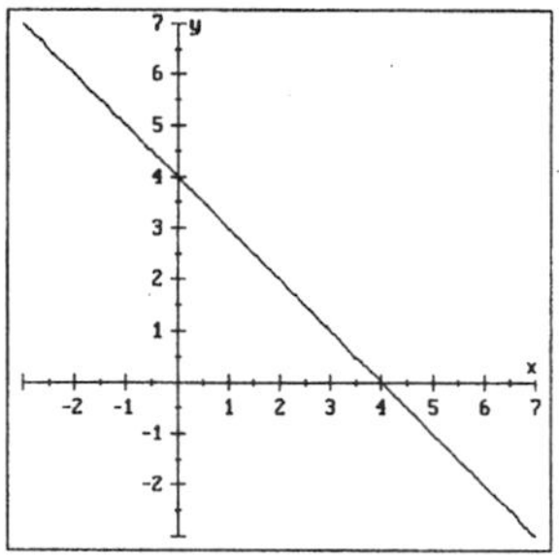

22. $x - y = -3$

x	y
0	3
-3	0

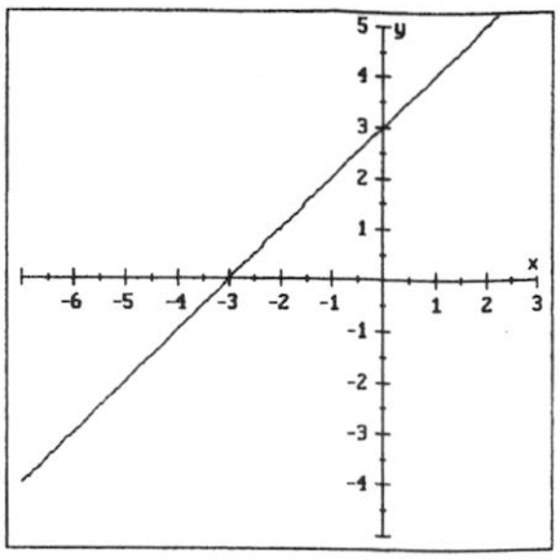

23. $3x + 5y = 15$

x	y
0	3
5	0

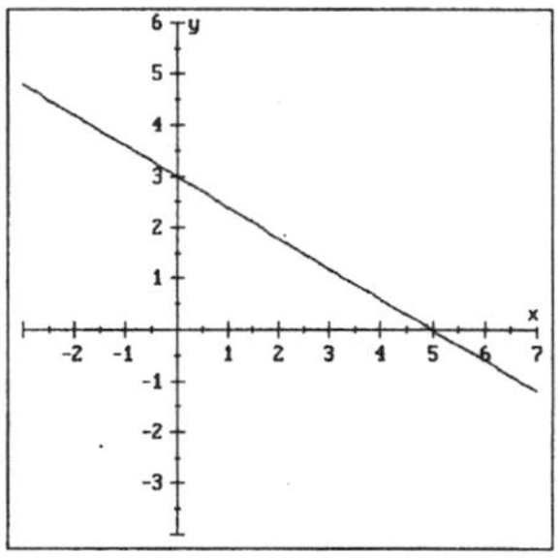

24. $7x - 4y = 28$

x	y
0	-7
4	0

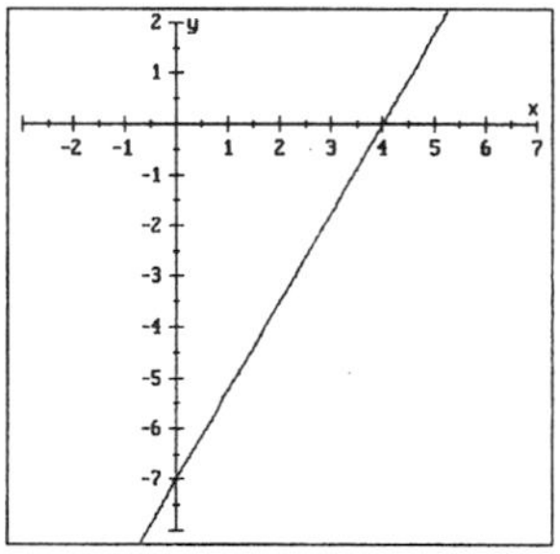

25.

$$\begin{array}{rr} 3x - y = -2 & 2x + 3y = 17 \\ 3(1) - 5 \stackrel{?}{=} -2 & 2(1) + 3(5) \stackrel{?}{=} 17 \\ 3 - 5 \stackrel{?}{=} -2 & 2 + 15 \stackrel{?}{=} 17 \\ -2 = -2 & 17 = 17 \end{array}$$

$(1, 5)$ is a solution to the system.

26.

$$\begin{array}{rr} 5x + 3y = 2 & -3x + 2y = 16 \\ 5(-2) + 3(4) \stackrel{?}{=} 2 & -3(-2) + 2(4) \stackrel{?}{=} 16 \\ -10 + 12 \stackrel{?}{=} 2 & 6 + 8 \stackrel{?}{=} 16 \\ 2 = 2 & 14 \neq 16 \end{array}$$

$(-2, 4)$ is not a solution to the system.

27.

$$\begin{array}{rr} 2x + 4y = 30 & \frac{x}{4} - y = 3 \\ 2(14) + 4\left(\frac{1}{2}\right) \stackrel{?}{=} 30 & \frac{14}{4} - \frac{1}{2} \stackrel{?}{=} 3 \\ 28 + 2 \stackrel{?}{=} 30 & \frac{7}{2} - \frac{1}{2} \stackrel{?}{=} 3 \\ 30 = 30 & \frac{6}{2} = 3 \end{array}$$

$\left(14, \frac{1}{2}\right)$ is a solution to the system.

28.

$$\begin{array}{rr} 4x - 6y = 18 & \frac{x}{3} + \frac{y}{2} = \frac{5}{6} \\ 4\left(\frac{7}{2}\right) - 6\left(-\frac{2}{3}\right) \stackrel{?}{=} 18 & \frac{\frac{7}{2}}{3} + \frac{-\frac{2}{3}}{2} \stackrel{?}{=} \frac{5}{6} \\ 14 + 4 \stackrel{?}{=} 18 & \frac{7}{6} - \frac{2}{6} \stackrel{?}{=} \frac{5}{6} \\ 18 = 18 & \frac{5}{6} = \frac{5}{6} \end{array}$$

$\left(\frac{7}{2}, -\frac{2}{3}\right)$ is a solution to the system.

29. $x + y = 7$ $\quad 2x - y = 5$

x	y
0	7
7	0

x	y
2	−1
3	1

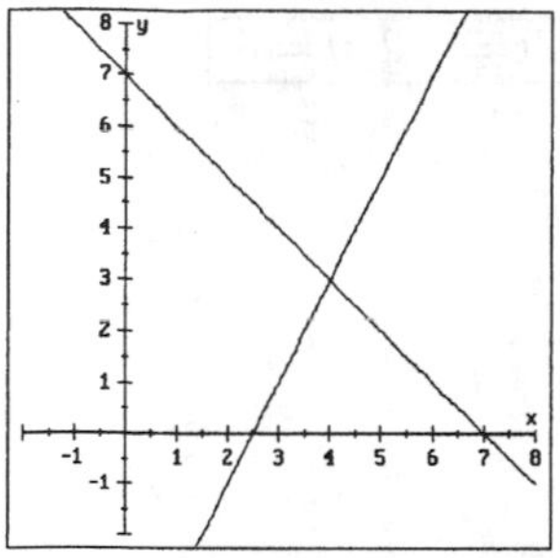

$\boxed{(4, 3)}$

30. $\dfrac{x}{3} + \dfrac{y}{5} = -1$ $\quad x - 3y = -3$

x	y
0	−5
−3	0

x	y
0	1
−3	0

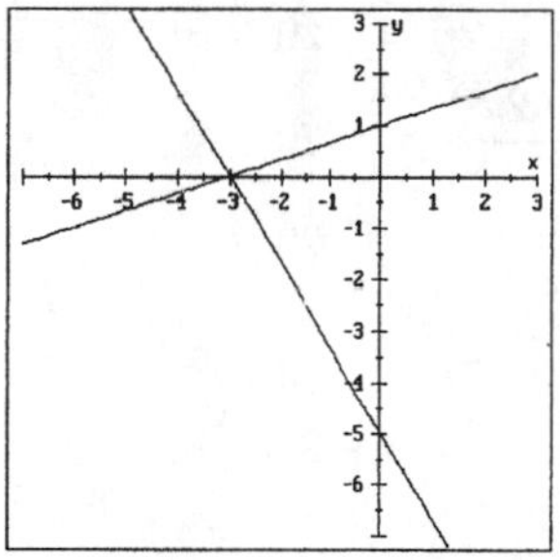

$\boxed{(-3, 0)}$

31. $3x + 6y = 6$ $\quad x + 2y = 2$

x	y
0	1
2	0

x	y
0	1
2	0

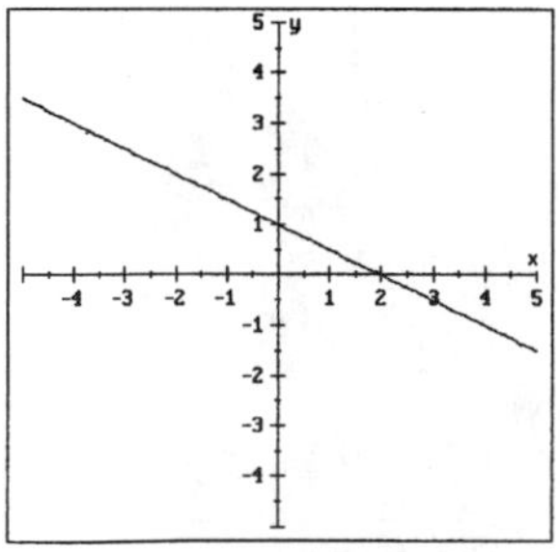

(dependent equations)

32. $6x + 3y = 12$ $\quad 2x + y = 2$

x	y
0	4
2	0

x	y
0	2
1	0

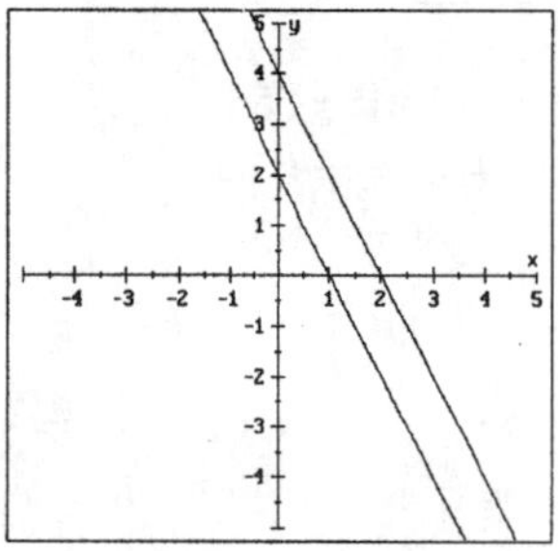

(inconsistent system)

33. $\begin{cases} (1) \quad x = 3y + 5 \\ (2) \quad 5x - 4y = 3 \end{cases}$

Substitute $x = 3y + 5$ from (1):

$$5x - 4y = 3$$
$$5(3y + 5) - 4y = 3$$
$$15y + 25 - 4y = 3$$
$$11y = -22$$
$$y = -2$$

Substitute and solve for x:

$$x = 3y + 5$$
$$x = 3(-2) + 5$$
$$x = -6 + 5 = -1$$

Solution:

$\boxed{x = -1, y = -2}$

34. $\begin{cases} 3x - \dfrac{2y}{5} = 2(x-2) & \Rightarrow 15x - 2y = 10(x-2) & \Rightarrow 5x - 2y = -20 \ (1) \\ 2x - 3 = 3 - 2y & \Rightarrow 2x + 2y = 6 & \Rightarrow 2x + 2y = 6 \ (2) \end{cases}$

Substitute $y = \dfrac{6-2x}{2}$ from (2):

$$\begin{aligned} 5x - 2y &= -20 \\ 5x - 2\left(\frac{6-2x}{2}\right) &= -20 \\ 5x - (6-2x) &= -20 \\ 5x - 6 + 2x &= -20 \\ 7x &= -14 \\ x &= -2 \end{aligned}$$

Substitute and solve for y:

$$\begin{aligned} y &= \frac{6-2x}{2} \\ y &= \frac{6-2(-2)}{2} \\ y &= \frac{6+4}{2} = \frac{10}{2} = 5 \end{aligned}$$

Solution: $\boxed{x = -2, y = 5}$

35. $\begin{cases} 8x + 5y = 3 \ (1) \\ 5x - 8y = 13 \ (2) \end{cases}$

Substitute $y = \dfrac{3-8x}{5}$ from (1):

$$\begin{aligned} 5x - 8y &= 13 \\ 5x - 8\left(\frac{3-8x}{5}\right) &= 13 \\ 25x - 8(3-8x) &= 65 \\ 25x - 24 + 64x &= 65 \\ 89x &= 89 \\ x &= 1 \end{aligned}$$

Substitute and solve for y:

$$\begin{aligned} y &= \frac{3-8x}{5} \\ y &= \frac{3-8(1)}{5} \\ y &= \frac{3-8}{5} = \frac{-5}{5} = -1 \end{aligned}$$

Solution: $\boxed{x = 1, y = -1}$

36. $\begin{cases} 6(x+2) = y - 1 & \Rightarrow 6x + 12 = y - 1 & \Rightarrow 6x - y = -13 \ (1) \\ 5(y-1) = x + 2 & \Rightarrow 5y - 5 = x + 2 & \Rightarrow -x + 5y = 7 \ (2) \end{cases}$

Substitute $y = 6x + 13$ from (1):

$$\begin{aligned} -x + 5y &= 7 \\ -x + 5(6x+13) &= 7 \\ -x + 30x + 65 &= 7 \\ 29x &= -58 \\ x &= -2 \end{aligned}$$

Substitute and solve for y:

$$\begin{aligned} y &= 6x + 13 \\ y &= 6(-2) + 13 \\ y &= -12 + 13 = 1 \end{aligned}$$

Solution: $\boxed{x = -2, y = 1}$

37.
$$\begin{aligned} 2x + y &= 1 \\ \underline{5x - y} &\underline{= 20} \\ 7x &= 21 \\ x &= 3 \end{aligned}$$

$$\begin{aligned} 2x + y &= 1 \\ 2(3) + y &= 1 \\ 6 + y &= 1 \\ y &= -5 \end{aligned}$$

Solution: $\boxed{(3, -5)}$

38.
$x + 8y = 7 \Rightarrow \times\ (-1)$

$\underline{x - 4y = 1}$

$$\begin{aligned} -x - 8y &= -7 \\ \underline{x - 4y} &\underline{= 1} \\ -12y &= -6 \\ y &= \tfrac{1}{2} \end{aligned}$$

$$\begin{aligned} x + 8y &= 7 \\ x + 8\left(\tfrac{1}{2}\right) &= 7 \\ x + 4 &= 7 \\ x &= 3 \end{aligned}$$

Solution: $\boxed{(3, \tfrac{1}{2})}$

39. $5x + y = 2 \Rightarrow \times(-2)$ $\quad -10x - 2y = -4$
$3x + 2y = 11$ $\quad 3x + 2y = 11$

$-7x = 7$
$x = -1$

$5x + y = 2$
$5(-1) + y = 2$
$-5 + y = 2$
$y = 7$

Solution: $\boxed{(-1, 7)}$

40. $x + y = 3 \Rightarrow x + y = 3 \Rightarrow \times(-1)$ $\quad -x - y = -3$
$3x = 2 - y \Rightarrow 3x + y = 2$ $\quad 3x + y = 2$

$2x = -1$
$x = -\frac{1}{2}$

$x + y = 3$
$-\frac{1}{2} + y = \frac{6}{2}$
$y = \frac{7}{2}$

Solution: $\boxed{\left(-\frac{1}{2}, \frac{7}{2}\right)}$

41. $11x + 3y = 27 \Rightarrow \times(4)$ $\quad 44x + 12y = 108$
$8x + 4y = 36 \Rightarrow \times(-3)$ $\quad -24x - 12y = -108$

$20x = 0$
$x = 0$

$11x + 3y = 27$
$11(0) + 3y = 27$
$0 + 3y = 27$
$3y = 27$
$y = 9$

Solution: $\boxed{(0, 9)}$

42. $9x + 3y = 5 \Rightarrow 9x + 3y = 5$ $\quad 9x + 3y = 5$
$3x = 4 - y \Rightarrow 3x + y = 4 \Rightarrow \times(-3)$ $\quad -9x - 3y = -12$

$0 \neq -7$

$\boxed{\text{inconsistent system}}$

43. $9x + 3y = 5$ $\quad 9x + 3y = 5$
$3x + y = \frac{5}{3} \Rightarrow \times(-3)$ $\quad -9x - 3y = -5$

$0 = 0$

$\boxed{\text{dependent equations}}$

44. $\dfrac{x}{3} + \dfrac{y+2}{2} = 1 \Rightarrow \times(6)$ $\quad 2x + 3(y+2) = 6 \Rightarrow$ $\quad 2x + 3y + 6 = 6$
$\dfrac{x+8}{8} + \dfrac{y-3}{3} = 0 \Rightarrow \times(24)$ $\quad 3(x+8) + 8(y-3) = 0 \Rightarrow$ $\quad 3x + 24 + 8y - 24 = 0$

$2x + 3y = 0 \Rightarrow \times(3)$ $\quad 6x + 9y = 0$
$3x + 8y = 0 \Rightarrow \times(-2)$ $\quad -6x - 16y = 0$

$-7y = 0$
$y = 0$

$2x + 3y = 0$
$2x + 3(0) = 0$
$2x = 0$
$x = 0$

Solution: $\boxed{(0, 0)}$

45. Let x and y represent the integers.

$$\begin{cases} (1) & x = 5y \\ (2) & x + y = 18 \end{cases}$$

Substitute $x = 5y$ into (2):
$x + y = 18$
$5y + y = 18$
$6y = 18$
$y = 3$

Substitute; solve for x:
$x = 5y$
$x = 5(3) = 15$
The integers are 3 and 15.

46. Let $w =$ the width.
Let $l =$ the length.

$$\begin{cases} (1) & l = 3w \\ (2) & 2l + 2w = 24 \end{cases}$$

Substitute $l = 3w$ into (2):

$$\begin{aligned} 2l + 2w &= 24 \\ 2(3w) + 2w &= 24 \\ 6w + 2w &= 24 \\ 8w &= 24 \\ w &= 3 \end{aligned}$$

Substitute; solve for l:

$l = 3w$

$l = 3(3) = 9$

The dimensions are 3 ft by 9 ft.

47. Let $x =$ the cost of a grapefruit.
Let $y =$ the cost of an orange.

$$\begin{cases} (1) & x = y + 15 \\ (2) & x + y = 85 \end{cases}$$

Substitute $x = y + 15$ into (2):

$$\begin{aligned} x + y &= 85 \\ y + 15 + y &= 85 \\ 2y &= 70 \\ y &= 35 \end{aligned}$$

Substitute; solve for x:

$x = y + 15$

$x = 35 + 15 = 50$

A grapefruit costs 50¢.

48. Let $x =$ the electric bill.
Let $y =$ the gas bill.

$$\begin{cases} (1) & x = y - 23 \\ (2) & x + y = 109 \end{cases}$$

Substitute $x = y - 23$ into (2):

$$\begin{aligned} x + y &= 109 \\ y - 23 + y &= 109 \\ 2y &= 132 \\ y &= 66 \end{aligned}$$

The gas bill was \$66.

49. Let $x =$ cost of a gallon of milk and $y =$ cost of one dozen eggs.

$$\begin{aligned} 2x + 3y = 6.80 &\Rightarrow \times(-2) & -4x - 6y &= -13.60 \\ 3x + 2y = 7.35 &\Rightarrow \times(3) & 9x + 6y &= 22.05 \\ & & 5x &= 8.45 \\ & & x &= 1.69 \end{aligned}$$

A gallon of milk costs \$1.69.

50. Let $x =$ principal at 10% and $y =$ principal at 6%.

$$\begin{aligned} x + y &= 3000 & & & x + y &= 3000 \Rightarrow \times(-6) & -6x - 6y &= -18000 \\ 0.10x + 0.06y &= 270 \Rightarrow \times(100) & & & 10x + 6y &= 27000 & 10x + 6y &= 27000 \\ & & & & & & 4x &= 9000 \\ & & & & & & x &= 2250 \end{aligned}$$

$$\begin{aligned} x + y &= 3000 \\ 2250 + y &= 3000 \\ y &= 750 \end{aligned}$$

He invested \$750 at the 6% rate.

51. Let $b =$ speed of boat in still water and $c =$ speed of current.

	d	r	t	Equation $(d = r \cdot t)$
Downstream	56	$b + c$	4	$56 = (b + c) \cdot 4 \Rightarrow 4b + 4c = 56$
Upstream	56	$b - c$	7	$56 = (b - c) \cdot 7 \Rightarrow 7b - 7c = 56$

$$\begin{aligned} 4b + 4c = 56 &\Rightarrow \times\ (7) & 28b + 28c &= 392 \\ 7b - 7c = 56 &\Rightarrow \times\ (4) & 28b - 28c &= 224 \\ & & 56b &= 616 \\ & & b &= 11 \end{aligned}$$

$$\begin{aligned} 4b + 4c &= 56 \\ 4(11) + 4c &= 56 \\ 44 + 4c &= 56 \\ 4c &= 12 \\ c &= 3 \end{aligned}$$

The speed of the current is 3 mph.

52. Boundary (solid)

$y = x + 2$

x	y
0	2
-2	0

Test point: $(0, 0)$

$y \geq x + 2$

$0 \overset{?}{\geq} 0 + 2$

$0 \not\geq 2$

opposite half-plane

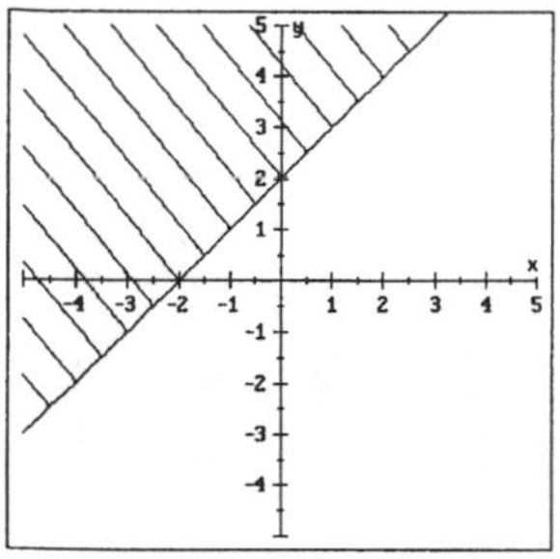

53. Boundary (dotted)

$x = 3$

x	y
3	0
3	4

Test point: $(0, 0)$

$x < 3$

$0 < 3$

same half-plane

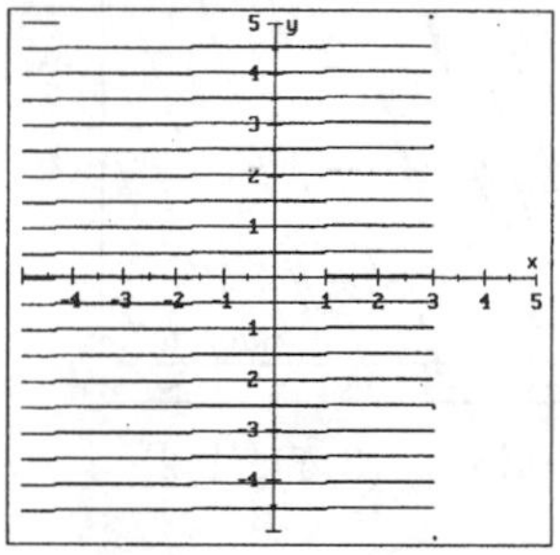

54. $5x + 3y < 15$ $\quad$ $3x - y > 3$

x	y
0	5
3	0

x	y
0	-3
1	0

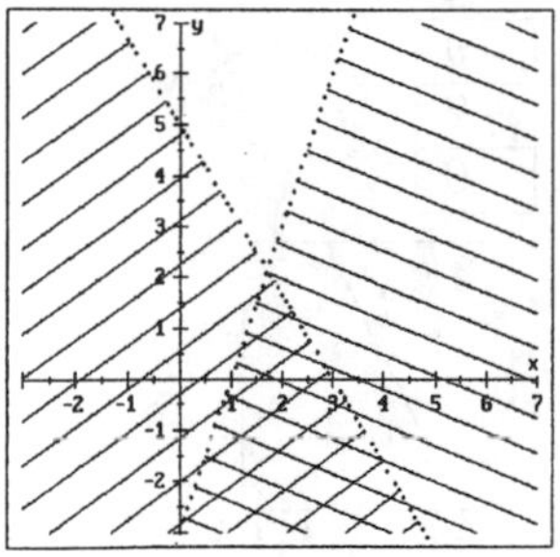

55. $5x - 3y \geq 5$ $\quad$ $3x + 2y \geq 3$

x	y
1	0
4	5

x	y
1	0
3	-3

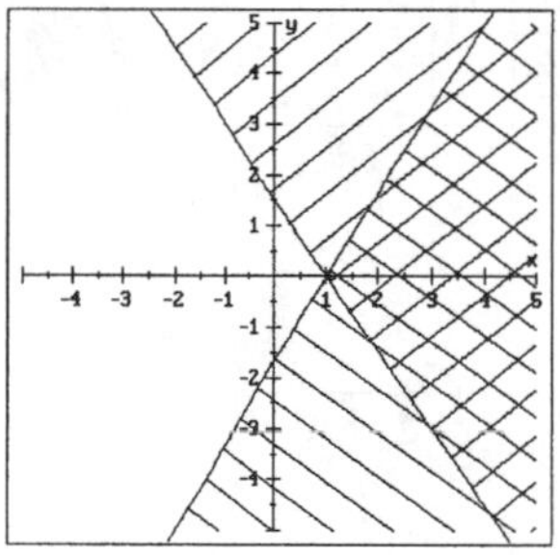

56. $x \geq 3y$ $\quad$ $y < 3x$

x	y
0	0
3	1

x	y
0	0
1	3

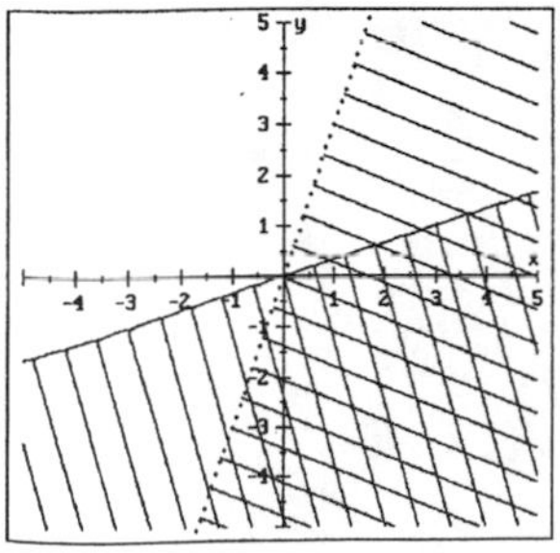

57. $x \geq 0$ $\quad$ $x \leq 3$

x	y
0	0
0	4

x	y
3	1
3	0

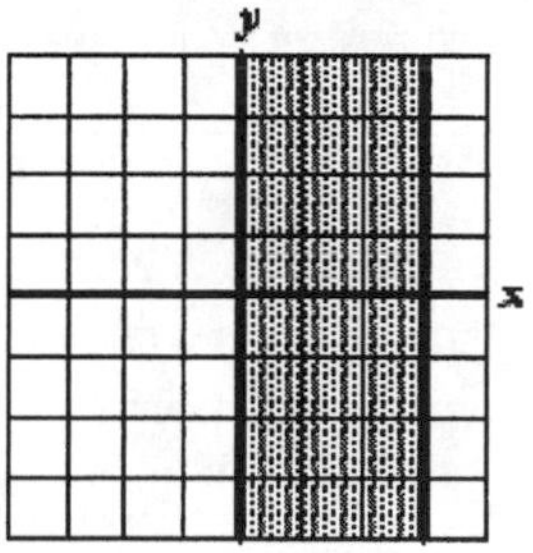

58. $10x + 20y \geq 40 \quad 10x + 20y \leq 60$

x	y
0	2
4	0

x	y
0	3
6	0

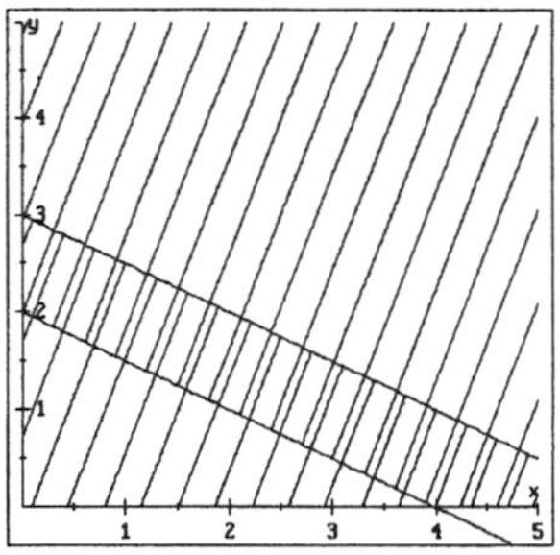

Solutions: $(3, 1), (1, 2)$

Chapter 3 Test (page 232)

1.

$$y = \frac{x}{2} + 1$$

x	y
0	1
2	3

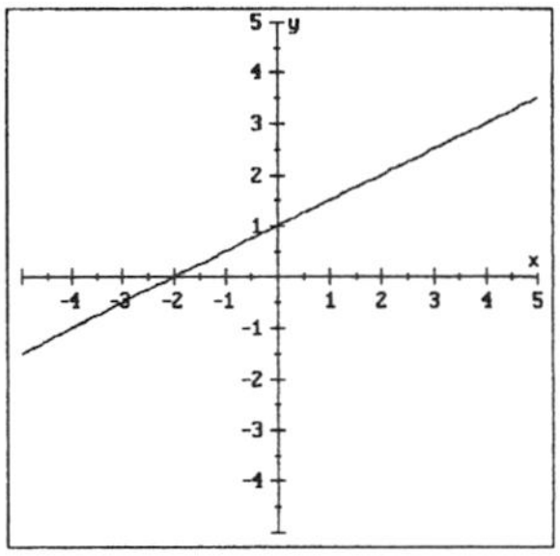

2.

$$2(x+1) - y = 4$$
$$2x + 2 - y = 4$$
$$2x - y = 2$$

x	y
0	-2
1	0

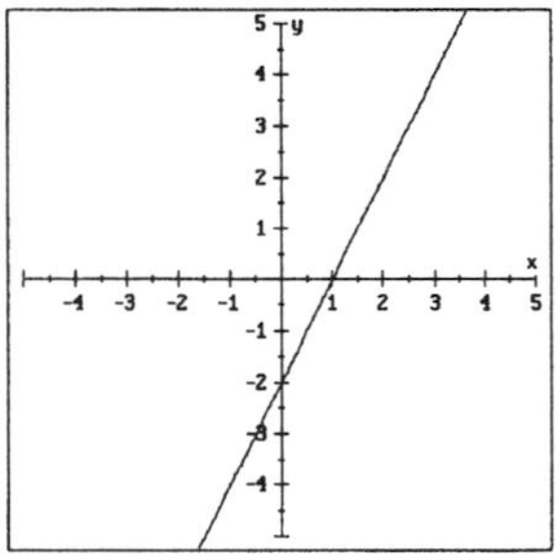

3. $x = 1$

x	y
1	2
1	0

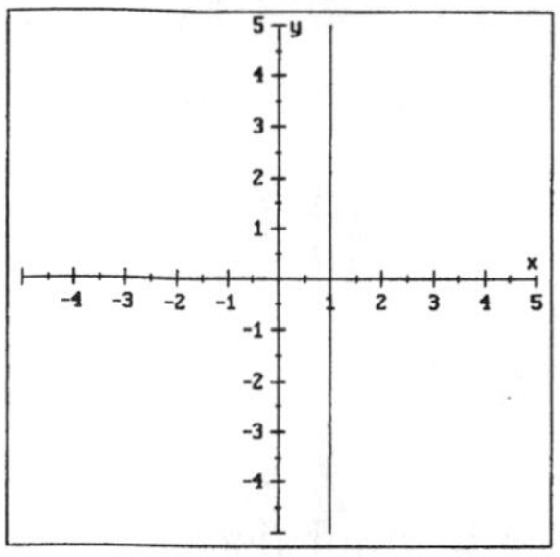

4.
$$\begin{aligned} 2y &= 8 \\ y &= 4 \end{aligned}$$

x	y
0	4
2	4

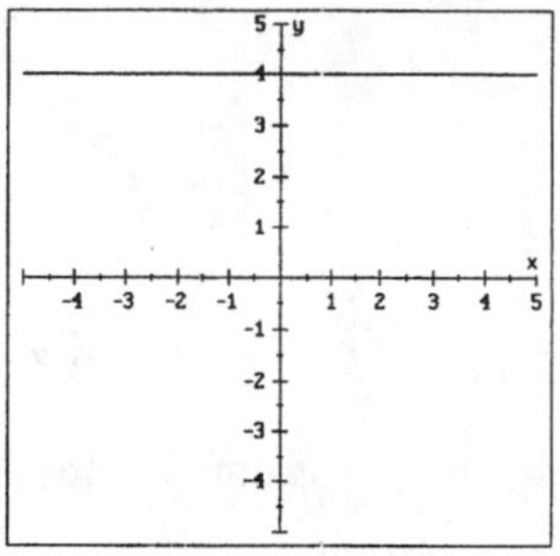

5.
$$\begin{aligned} 3x - 2y &= 12 & 2x + 3y &= -5 \\ 3(2) - 2(-3) &\stackrel{?}{=} 12 & 2(2) + 3(-3) &\stackrel{?}{=} -5 \\ 6 + 6 &\stackrel{?}{=} 12 & 4 - 9 &\stackrel{?}{=} -5 \\ 12 &= 12 & -5 &= -5 \end{aligned}$$

$(2, -3)$ is a solution to the system.

6.
$$\begin{aligned} 4x + y &= -9 & 2x - 3y &= -7 \\ 4(-2) + (-1) &\stackrel{?}{=} -9 & 2(-2) - 3(-1) &\stackrel{?}{=} -7 \\ -8 - 1 &\stackrel{?}{=} -9 & -4 + 3 &\stackrel{?}{=} -7 \\ -9 &= -9 & -1 &\neq -7 \end{aligned}$$

$(-2, -1)$ is not a solution to the system.

7. $3x + y = 7 \qquad x - 2y = 0$

x	y
0	7
2	1

x	y
0	0
4	2

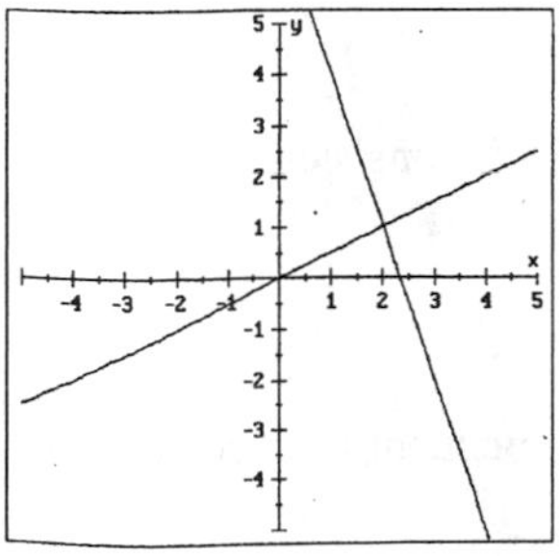

$(2, 1)$

8. $x + \dfrac{y}{2} = 1 \qquad y = 1 - 3x$

x	y
0	2
1	0

x	y
0	1
1	-2

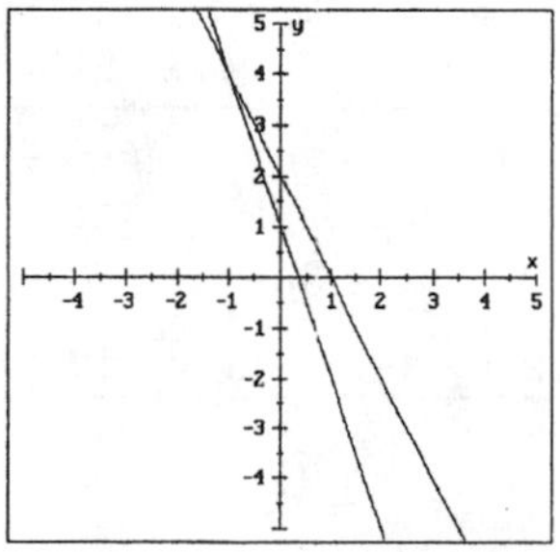

$(-1, 4)$

9. $\begin{cases} (1) & y = x - 1 \\ (2) & x + y = -7 \end{cases}$

Substitute $y = x - 1$ from (1):

$$x + y = -7$$
$$x + x - 1 = -7$$
$$2x = -6$$
$$x = -3$$

Substitute and solve for y:

$$y = x - 1$$
$$y = -3 - 1$$
$$y = -4$$

Solution: $\boxed{x = -3, y = -4}$

10. $\begin{cases} (1) & \frac{x}{6} + \frac{y}{10} = 3 \quad \Rightarrow 5x + 3y = 90 \quad \Rightarrow x = \frac{90 - 3y}{5} \\ (2) & \frac{5x}{16} - \frac{3y}{16} = \frac{15}{8} \quad \Rightarrow 5x - 3y = 30 \end{cases}$

Substitute $x = \frac{90 - 3y}{5}$ from (1) into (2):

$$5x - 3y = 30$$
$$5\left(\frac{90 - 3y}{5}\right) - 3y = 30$$
$$90 - 3y - 3y = 30$$
$$-6y = -60$$
$$y = 10$$

Substitute and solve for x:

$$x = \frac{90 - 3y}{5}$$
$$x = \frac{90 - 3(10)}{5}$$
$$x = \frac{90 - 30}{5} = \frac{60}{5} = 12$$

Solution: $\boxed{x = 12, y = 10}$

11.

$$\begin{array}{rl} 3x - y = & 2 \\ 2x + y = & 8 \\ \hline 5x \quad = & 10 \\ x \quad = & 2 \end{array}$$

$$2x + y = 8$$
$$2(2) + y = 8$$
$$4 + y = 8$$
$$y = 4$$

Solution: $\boxed{(2, 4)}$

12.

$$\begin{array}{rl} 4x + \;\; 3 = -3y \\ \frac{-x}{7} + \frac{4y}{21} = \quad 1 \Rightarrow \times (21) \\ \hline \end{array} \quad \begin{array}{rl} 4x + 3y = -3 \Rightarrow \times (3) \\ -3x + 4y = \;\; 21 \Rightarrow \times (4) \\ \hline \end{array} \quad \begin{array}{rl} 12x + \;\; 9y = & -9 \\ -12x + 16y = & 84 \\ \hline 25y = & 75 \\ y = & 3 \end{array}$$

$$4x + 3y = -3$$
$$4x + 3(3) = -3$$
$$4x + 9 = -3$$
$$4x = -12$$
$$x = -3$$

$\boxed{(-3, 3)}$

13.

$$\begin{array}{rl} 2x + 3(y - 2) = 0 \Rightarrow 2x + 3y - 6 = 0 \Rightarrow & 2x + 3y = 6 \\ -3y = 2(x - 4) \Rightarrow \quad -3y = 2x - 8 \Rightarrow & -2x - 3y = -8 \\ \hline & 0 \neq -2 \end{array} \Rightarrow \text{inconsistent}$$

14.

$$\begin{array}{rl} \frac{x}{3} + y - 4 = 0 \Rightarrow x + 3y - 12 = 0 \Rightarrow & x + 3y = 12 \\ -3y = x - 12 \Rightarrow -x - 3y = -12 \Rightarrow & -x - 3y = -12 \\ \hline & 0 = 0 \end{array} \Rightarrow \text{dependent (and consistent)}$$

15. Let x and y represent the numbers.

$$\begin{cases} (1) & x = 3y + 2 \\ (2) & x + y = -18 \end{cases}$$

Substitute $x = 3y + 2$.

$$\begin{aligned} x + y &= -18 \\ 3y + 2 + y &= -18 \\ 4y &= -20 \\ y &= -5 \end{aligned}$$

Substitute; solve for x:

$$\begin{aligned} x &= 3y + 2 \\ x &= 3(-5) + 2 \\ x &= -15 + 2 = -13 \end{aligned}$$

The numbers are -5 and -13.
The product is $(-5)(-13) = 65$.

16. Let x = number of adult tickets and y = number of child tickets.

$$\begin{aligned} x + y &= 7 \Rightarrow \times(-14) & -14x - 14y &= -98 \\ 21x + 14y &= 119 & 21x + 14y &= 119 \\ & & 7x &= 21 \\ & & x &= 3 \end{aligned}$$

He bought 3 adult tickets.

17. Let x = principal at 8% and y = principal at 9%.

$$\begin{aligned} x + y &= 10000 & x + y &= 10000 \Rightarrow \times(-9) & -9x - 9y &= -90000 \\ 0.08x + 0.09y &= 840 \Rightarrow \times(100) & 8x + 9y &= 84000 & 8x + 9y &= 84000 \\ & & & & -x &= -6000 \\ & & & & x &= 6000 \end{aligned}$$

$$\begin{aligned} x + y &= 10000 \\ 6000 + y &= 10000 \\ y &= 4000 \end{aligned}$$

She invested \$4000 at the 9% rate.

18. Let b = speed of boat in still water and c = speed of current.

	d	r	t	Equation $(d = r \cdot t)$
Downstream	8	$b + c$	2	$8 = (b + c) \cdot 2 \Rightarrow 2b + 2c = 8$
Upstream	8	$b - c$	4	$8 = (b - c) \cdot 4 \Rightarrow 4b - 4c = 8$

$$\begin{aligned} 2b + 2c &= 8 \Rightarrow \times(2) & 4b + 4c &= 16 \\ 4b - 4c &= 8 \Rightarrow & 4b - 4c &= 8 \\ & & 8b &= 24 \\ & & b &= 3 \end{aligned}$$

$$\begin{aligned} 2b + 2c &= 8 \\ 2(3) + 2c &= 8 \\ 6 + 2c &= 8 \\ 2c &= 2 \\ c &= 1 \end{aligned}$$

The speed of the current is 1 mph.

19. $x + y < 3$ $\quad x - y < 1$

x	y
0	3
3	0

x	y
0	−1
1	0

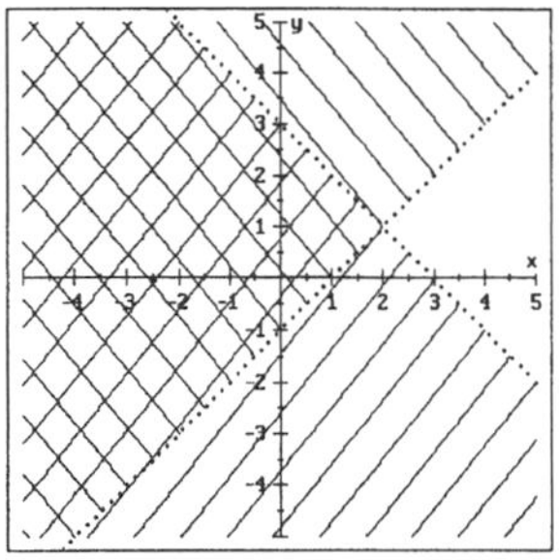

20. $2x + 3y \le 6$ $\quad x \ge 2$

x	y
0	2
3	0

x	y
2	0
2	3

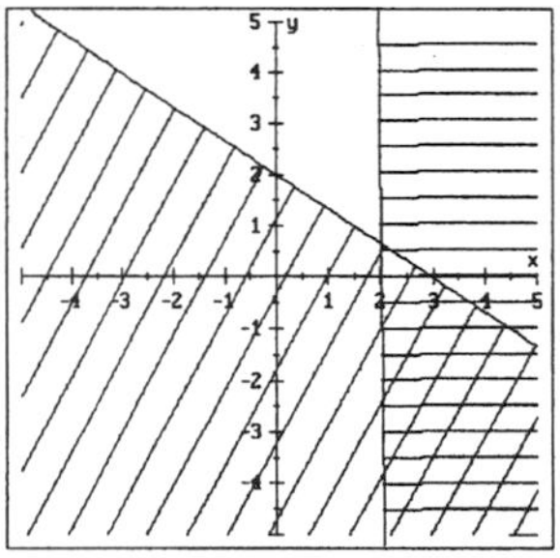

Exercise 4.1 (page 241)

1. 

3. the product of 3 and the sum of x and y

5. $|2x| + 3$

7. $-5; 3$

9. $(3x)(3x)(3x)(3x)$

11. $y \cdot y \cdot y \cdot y \cdot y$

13. $x^n y^n$

15. a^{bc}

17. Answers may vary.

19. Base: 4
Exponent: 3

21. Base: x
Exponent: 5

23. Base: $2y$
Exponent: 3

25. Base: x
Exponent: 4

27. Base: x
Exponent: 1

29. Base: x
Exponent: 3

31. $5^3 = 5 \cdot 5 \cdot 5$

33. $x^7 = x \cdot x \cdot x \cdot x \cdot x \cdot x \cdot x$

35. $-4x^5 = -4 \cdot x \cdot x \cdot x \cdot x \cdot x$

37. $(3t)^5 = (3t)(3t)(3t)(3t)(3t)$

39. $2 \cdot 2 \cdot 2 = 2^3$

41. $x \cdot x \cdot x \cdot x = x^4$

43. $(2x)(2x)(2x) = (2x)^3$

45. $-4t \cdot t \cdot t \cdot t = -4t^4$

47. $5^4 = 5 \cdot 5 \cdot 5 \cdot 5 = 625$

49. $2^2 + 3^2 = 4 + 9 = 13$

51. $5^4 - 4^3 = 625 - 64 = 561$

53. $-5(3^4 + 4^3) = -5(81 + 64)$
$= -5(145) = -725$

55. $x^4x^3 = x^{4+3} = x^7$

57. $x^5x^5 = x^{5+5} = x^{10}$

59. $tt^2 = t^{1+2} = t^3$

61. $a^3a^4a^5 = a^{3+4+5} = a^{12}$

63. $y^3(y^2y^4) = y^3y^{2+4} = y^3y^6 = y^{3+6} = y^9$

65. $4x^2(3x^5) = 4 \cdot 3x^2x^5 = 12x^{2+5} = 12x^7$

67. $(-y^2)(4y^3) = -1 \cdot 4y^2y^3 = -4y^{2+3} = -4y^5$

69. $6x^3(-x^2)(-x^4) = 6(-1)(-1)x^3x^2x^4 = 6x^{3+2+4} = 6x^9$

71. $(3^2)^4 = 3^{2\cdot4} = 3^8$

73. $(y^5)^3 = y^{5\cdot3} = y^{15}$

75. $(a^3)^7 = a^{3\cdot7} = a^{21}$

77. $(x^2x^3)^5 = (x^{2+3})^5 = (x^5)^5 = x^{5\cdot5} = x^{25}$

79. $(3zz^2z^3)^5 = (3z^{1+2+3})^5 = (3z^6)^5 = 3^5(z^6)^5 = 243z^{6\cdot5} = 243z^{30}$

81. $(x^5)^2(x^7)^3 = x^{5\cdot2}x^{7\cdot3} = x^{10}x^{21} = x^{10+21} = x^{31}$

83. $(r^3r^2)^4(r^3r^5)^2 = (r^{3+2})^4(r^{3+5})^2 = (r^5)^4(r^8)^2 = r^{5\cdot4}r^{8\cdot2} = r^{20}r^{16} = r^{20+16} = r^{36}$

85. $(s^3)^3(s^2)^2(s^5)^4 = s^{3\cdot3}s^{2\cdot2}s^{5\cdot4} = s^9s^4s^{20} = s^{9+4+20} = s^{33}$

87. $(xy)^3 = x^3y^3$

89. $(r^3s^2)^2 = (r^3)^2(s^2)^2 = r^{3\cdot2}s^{2\cdot2} = r^6s^4$

91. $(4ab^2)^2 = 4^2a^2(b^2)^2 = 16a^2b^{2\cdot2} = 16a^2b^4$

93. $(-2r^2s^3t)^3 = (-2)^3(r^2)^3(s^3)^3t^3 = -8r^{2\cdot3}s^{3\cdot3}t^3 = -8r^6s^9t^3$

95. $\left(\frac{a}{b}\right)^3 = \frac{a^3}{b^3}$

97. $\left(\frac{x^2}{y^3}\right)^5 = \frac{(x^2)^5}{(y^3)^5} = \frac{x^{2\cdot5}}{y^{3\cdot5}} = \frac{x^{10}}{y^{15}}$

99. $\left(\frac{-2a}{b}\right)^5 = \frac{(-2a)^5}{b^5} = \frac{(-2)^5a^5}{b^5} = \frac{-32a^5}{b^5}$

101. $\left(\frac{b^2}{3a}\right)^3 = \frac{(b^2)^3}{(3a)^3} = \frac{b^{2\cdot3}}{3^3a^3} = \frac{b^6}{27a^3}$

103. $\frac{x^5}{x^3} = x^{5-3} = x^2$

105. $\frac{y^3y^4}{yy^2} = \frac{y^{3+4}}{y^{1+2}} = \frac{y^7}{y^3} = y^{7-3} = y^4$

107. $\frac{12a^2a^3a^4}{4(a^4)^2} = \frac{12a^{2+3+4}}{4a^{4\cdot2}} = \frac{12a^9}{4a^8} = \frac{12}{4}a^{9-8} = 3a^1 = 3a$

109. $\frac{(ab^2)^3}{(ab)^2} = \frac{a^3(b^2)^3}{a^2b^2} = \frac{a^3b^{2\cdot3}}{a^2b^2} = \frac{a^3b^6}{a^2b^2} = a^{3-2}b^{6-2} = a^1b^4 = ab^4$

111. $\frac{20(r^4s^3)^4}{6(rs^3)^3} = \frac{20(r^4)^4(s^3)^4}{6r^3(s^3)^3} = \frac{20r^{4\cdot4}s^{3\cdot4}}{6r^3s^{3\cdot3}} = \frac{20r^{16}s^{12}}{6r^3s^9} = \frac{20}{6}r^{16-3}s^{12-9} = \frac{10}{3}r^{13}s^3 = \frac{10r^{13}s^3}{3}$

113. $\frac{17(x^4y^3)^8}{34(x^5y^2)^4} = \frac{17(x^4)^8(y^3)^8}{34(x^5)^4(y^2)^4} = \frac{17x^{4\cdot8}y^{3\cdot8}}{34x^{5\cdot4}y^{2\cdot4}} = \frac{17x^{32}y^{24}}{34x^{20}y^8} = \frac{17}{34}x^{32-20}y^{24-8} = \frac{1}{2}x^{12}y^{16} = \frac{x^{12}y^{16}}{2}$

115. $\left(\frac{y^3y}{2yy^2}\right)^3 = \left(\frac{y^{3+1}}{2y^{1+2}}\right)^3 = \left(\frac{y^4}{2y^3}\right)^3 = \left(\frac{1}{2}y^{4-3}\right)^3 = \left(\frac{1}{2}y^1\right)^3 = \left(\frac{1}{2}\right)^3 (y^1)^3 = \frac{1}{8}y^{1\cdot 3} = \frac{y^3}{8}$

117. $\left(\frac{-2r^3r^3}{3r^4r}\right)^3 = \left(\frac{-2r^{3+3}}{3r^{4+1}}\right)^3 = \left(\frac{-2r^6}{3r^5}\right)^3 = \left(\frac{-2}{3}r^{6-5}\right)^3 = \frac{(-2)^3}{3^3}(r^1)^3 = \frac{-8}{27}r^{1\cdot 3} = -\frac{8r^3}{27}$

119.

Bounce	1	2	3	4
Height	$\frac{1}{2}(32)$	$\frac{1}{2}\left[\frac{1}{2}(32)\right] = 32\left(\frac{1}{2}\right)^2$	$\frac{1}{2}\left[32\left(\frac{1}{2}\right)^2\right] = 32\left(\frac{1}{2}\right)^3$	$\frac{1}{2}\left[32\left(\frac{1}{2}\right)^3\right] = 32\left(\frac{1}{2}\right)^4$

$32\left(\frac{1}{2}\right)^4 = 32\left(\frac{1^4}{2^4}\right) = 32\left(\frac{1}{16}\right) = \frac{32}{16} = 2 \text{ ft}$

121.

Years	7	14	21	28
Value	2,000	4,000	8,000	16,000

123. $A = P(1+r)^t = 8000(1+0.06)^{30} = 8000(1.06)^{30} \approx 8000(5.743491) \approx \$45{,}947.93$

125. Answers may vary.

127. No. $2^3 = 8, 3^2 = 9 \Rightarrow 2^3 \neq 3^2$

Exercise 4.2 (page 246)

1. $\frac{3a^2+4b+8}{a+2b^2} = \frac{3(-2)^2+4(3)+8}{-2+2(3)^2} = \frac{3(4)+12+8}{-2+2(9)} = \frac{12+12+8}{-2+18} = \frac{32}{16} = 2$

3.
$$\begin{aligned} 5\left(x-\frac{1}{2}\right) &= \frac{7}{2} \\ 5x - \frac{5}{2} &= \frac{7}{2} \\ 5x - \frac{5}{2} + \frac{5}{2} &= \frac{7}{2} + \frac{5}{2} \\ 5x &= \frac{12}{2} \\ 5x &= 6 \\ \frac{5x}{5} &= \frac{6}{5} \\ x &= \frac{6}{5} \end{aligned}$$

5.
$$\begin{aligned} P &= L + \frac{s}{f}i \\ P - L &= \frac{s}{f}i \\ f(P-L) &= f\cdot\frac{s}{f}i \\ f(P-L) &= si \\ \frac{f(P-L)}{i} &= \frac{si}{i} \\ \frac{f(P-L)}{i} &= s, \text{ or } s = \frac{f(P-L)}{i} \end{aligned}$$

7. 1

9. 1

11. $2^5 \cdot 2^{-2} = 2^{5+(-2)} = 2^3 = 8$

13. $4^{-3}\cdot 4^{-2}\cdot 4^5 = 4^{-3+(-2)+5} = 4^0 = 1$

15. $\frac{3^5\cdot 3^{-2}}{3^3} = \frac{3^{5+(-2)}}{3^3} = \frac{3^3}{3^3} = 3^{3-3} = 3^0 = 1$

17. $\frac{2^5\cdot 2^7}{2^6\cdot 2^{-3}} = \frac{2^{5+7}}{2^{6+(-3)}} = \frac{2^{12}}{2^3} = 2^{12-3} = 2^9 = 512$

19. $2x^0 = 2 \cdot x^0 = 2 \cdot 1 = 2$

21. $(-x)^0 = 1$

23. $\left(\dfrac{a^2b^3}{ab^4}\right)^0 = 1$

25. $\dfrac{x^0 - 5x^0}{2x^0} = \dfrac{1-5(1)}{2(1)} = \dfrac{1-5}{2} = \dfrac{-4}{2} = -2$

27. $x^{-2} = \dfrac{1}{x^2}$

29. $b^{-5} = \dfrac{1}{b^5}$

31. $(2y)^{-4} = \dfrac{1}{(2y)^4} = \dfrac{1}{2^4y^4} = \dfrac{1}{16y^4}$

33. $(-5p)^{-3} = \dfrac{1}{(-5p)^3} = \dfrac{1}{(-5)^3p^3} = -\dfrac{1}{125p^3}$

35. $(4t)^{-3} = \dfrac{1}{(4t)^3} = \dfrac{1}{4^3t^3} = \dfrac{1}{64t^3}$

37. $(ab^2)^{-3} = \dfrac{1}{(ab^2)^3} = \dfrac{1}{a^3(b^2)^3} = \dfrac{1}{a^3b^6}$

39. $(x^2y)^{-2} = \dfrac{1}{(x^2y)^2} = \dfrac{1}{(x^2)^2y^2} = \dfrac{1}{x^4y^2}$

41. $\dfrac{y^4}{y^5} = y^{4-5} = y^{-1} = \dfrac{1}{y}$

43. $\dfrac{p^3}{p^6} = p^{3-6} = p^{-3} = \dfrac{1}{p^3}$

45. $\dfrac{(r^2)^3}{(r^3)^4} = \dfrac{r^6}{r^{12}} = r^{6-12} = r^{-6} = \dfrac{1}{r^6}$

47. $\dfrac{y^4y^3}{y^4y^{-2}} = \dfrac{y^7}{y^2} = y^{7-2} = y^5$

49. $\dfrac{a^4a^{-2}}{a^2a^0} = \dfrac{a^2}{a^2} = a^{2-2} = a^0 = 1$

51. $(ab^2)^{-2} = \dfrac{1}{(ab^2)^2} = \dfrac{1}{a^2(b^2)^2} = \dfrac{1}{a^2b^4}$

53. $(x^2y)^{-3} = \dfrac{1}{(x^2y)^3} = \dfrac{1}{(x^2)^3y^3} = \dfrac{1}{x^6y^3}$

55. $(x^{-4}x^3)^3 = (x^{-1})^3 = x^{-3} = \dfrac{1}{x^3}$

57. $(y^3y^{-2})^{-2} = (y^1)^{-2} = y^{-2} = \dfrac{1}{y^2}$

59. $(a^{-2}b^{-3})^{-4} = (a^{-2})^{-4}(b^{-3})^{-4} = a^8b^{12}$

61. $(-2x^3y^{-2})^{-5} = (-2)^{-5}(x^3)^{-5}(y^{-2})^{-5} = \dfrac{1}{(-2)^5}x^{-15}y^{10} - -\dfrac{y^{10}}{32x^{15}}$

63. $\left(\dfrac{a^3}{a^{-4}}\right)^2 = \left(a^{3-(-4)}\right)^2 = (a^7)^2 = a^{14}$

65. $\left(\dfrac{b^5}{b^{-2}}\right)^{-2} = \left(b^{5-(-2)}\right)^{-2} = (b^7)^{-2}$
$= b^{-14} = \dfrac{1}{b^{14}}$

67. $\left(\dfrac{4x^2}{3x^{-5}}\right)^4 = \left(\dfrac{4}{3}x^{2-(-5)}\right)^4 = \left(\dfrac{4}{3}x^7\right)^4 = \left(\dfrac{4}{3}\right)^4(x^7)^4 = \dfrac{256}{81}x^{28} = \dfrac{256x^{28}}{81}$

69. $\left(\dfrac{12y^3z^{-2}}{3y^{-4}z^3}\right)^2 = \left(\dfrac{12}{3}y^{3-(-4)}z^{-2-3}\right)^2 = (4y^7z^{-5})^2 = 4^2(y^7)^2(z^{-5})^2 = 16y^{14}z^{-10} = \dfrac{16y^{14}}{z^{10}}$

71. $\left(\frac{2x^3y^{-2}}{4xy^2}\right)^7 = \left(\frac{2}{4}x^{3-1}y^{-2-2}\right)^7 = \left(\frac{1}{2}x^2y^{-4}\right)^7 = \left(\frac{1}{2}\right)^7(x^2)^7(y^{-4})^7 = \frac{1}{128}x^{14}y^{-28} = \frac{x^{14}}{128y^{28}}$

73. $\left(\frac{14u^{-2}v^3}{21u^{-3}v}\right)^4 = \left(\frac{14}{21}u^{-2-(-3)}v^{3-1}\right)^4 = \left(\frac{2}{3}u^1v^2\right)^4 = \left(\frac{2}{3}\right)^4(u^1)^4(v^2)^4 = \frac{16}{81}u^4v^8 = \frac{16u^4v^8}{81}$

75. $\left(\frac{6a^2b^3}{2ab^2}\right)^{-2} = \left(\frac{6}{2}a^{2-1}b^{3-2}\right)^{-2} = (3a^1b^1)^{-2} = \frac{1}{(3ab)^2} = \frac{1}{3^2a^2b^2} = \frac{1}{9a^2b^2}$

77. $\left(\frac{18a^2b^3c^{-4}}{3a^{-1}b^2c}\right)^{-3} = \left(\frac{18}{3}a^{2-(-1)}b^{3-2}c^{-4-1}\right)^{-3} = (6a^3b^1c^{-5})^{-3} = 6^{-3}(a^3)^{-3}(b^1)^{-3}(c^{-5})^{-3}$

$= \frac{1}{6^3}a^{-9}b^{-3}c^{15} = \frac{c^{15}}{216a^9b^3}$

79. $\frac{(2x^{-2}y)^{-3}}{(4x^2y^{-1})^3} = \frac{2^{-3}(x^{-2})^{-3}y^{-3}}{4^3(x^2)^3(y^{-1})^3} = \frac{x^6y^{-3}}{2^3\cdot 4^3x^6y^{-3}} = \frac{1}{8\cdot 64}x^{6-6}y^{-3-(-3)} = \frac{1}{512}x^0y^0 = \frac{1}{512}$

81. $\frac{(17x^5y^{-5}z)^{-3}}{(17x^{-5}y^3z^2)^{-4}} = \frac{17^{-3}(x^5)^{-3}(y^{-5})^{-3}z^{-3}}{17^{-4}(x^{-5})^{-4}(y^3)^{-4}(z^2)^{-4}} = \frac{17^{-3}x^{-15}y^{15}z^{-3}}{17^{-4}x^{20}y^{-12}z^{-8}}$

$= 17^{-3-(-4)}x^{-15-20}y^{15-(-12)}z^{-3-(-8)}$

$= 17^1x^{-35}y^{27}z^5 = \frac{17y^{27}z^5}{x^{35}}$

83. $x^{2m}x^m = x^{2m+m} = x^{3m}$

85. $u^{2m}v^{3n}u^{3m}v^{-3n} = u^{2m+3m}v^{3n+(-3n)}$

$= u^{5m}v^0 = u^{5m}$

87. $y^{3m+2}y^{-m} = y^{3m+2+(-m)} = y^{2m+2}$

89. $\frac{y^{3m}}{y^{2m}} = y^{3m-2m} = y^m$

91. $\frac{x^{3n}}{x^{6n}} = x^{3n-6n} = x^{-3n} = \frac{1}{x^{3n}}$

93. $\left(x^{m+1}\right)^2 = x^{(m+1)\cdot 2} = x^{2m+2}$

95. $\left(x^{3-2n}\right)^{-4} = x^{(3-2n)(-4)} = x^{-12+8n}$

$= x^{8n-12}$

97. $\left(y^{2-n}\right)^{-4} = y^{(2-n)(-4)} = y^{-8+4n}$

$= y^{4n-8}$

Problems 99-103 are to be solved using a calculator. The keystrokes needed to solve each problem using a TI-83 graphing calculator appear in each solution. There may be other solutions. Keystrokes for other calculators may be slightly different.

99. $P = A(1+i)^{-n} = 100{,}000(1+0.07)^{-40}$

[1] [0] [0] [0] [0] [0] [(] [1] [+] [.] [0] [7] [)] [^] [(–)] [4] [0] [=]

The original deposit should be $6,678.04.

101. $P = A(1+i)^{-n} = 100{,}000(1+0.09)^{-40}$

[1][0][0][0][0][0][(][1][+][.][0][9][)][^][(-)][4][0][=]

The original deposit should be \$3,183.76.

103. $P = A(1+i)^{-n} = 1{,}000{,}000(1+0.08)^{-60}$

[1][0][0][0][0][0][0][(][1][+][.][0][8][)][^][(-)][6][0][=]

The original deposit should be \$9,875.85.

105. **Answers may vary.**

107. If $x > 1$, then x raised to a negative power is less than x. If $x = 1$, then x raised to a negative power is equal to x. If $0 < x < 1$, then x raised to a negative power is greater than x.

Exercise 4.3 (page 253)

1. $-5y^{55} = -5(-1)^{55} = -5(-1) = 5$

3. commutative property of addition

5.
$$\begin{aligned} 3(x-4)-6 &= 0 \\ 3x-12-6 &= 0 \\ 3x-18 &= 0 \\ 3x &= 18 \\ x &= 6 \end{aligned}$$

7. scientific notation

9. $23{,}000 = 2.3 \times 10^4$

11. $1{,}700{,}000 = 1.7 \times 10^6$

13. $0.062 = 6.2 \times 10^{-2}$

15. $0.0000051 = 5.1 \times 10^{-6}$

17.
$$\begin{aligned} 42.5 \times 10^2 &= 4.25 \times 10^1 \times 10^2 \\ &= 4.25 \times 10^3 \end{aligned}$$

19.
$$\begin{aligned} 0.25 \times 10^{-2} &= 2.5 \times 10^{-1} \times 10^{-2} \\ &= 2.5 \times 10^{-3} \end{aligned}$$

21. $2.3 \times 10^2 = 230$

23. $8.12 \times 10^5 = 812{,}000$

25. $1.15 \times 10^{-3} = 0.00115$

27. $9.76 \times 10^{-4} = 0.000976$

29. $25 \times 10^6 = 25{,}000{,}000$

31. $0.51 \times 10^{-3} = 0.00051$

33. $(3.4 \times 10^2)(2.1 \times 10^3) = (3.4)(2.1) \times 10^2 \times 10^3 = 7.14 \times 10^5 = 714{,}000$

35. $$\frac{9.3 \times 10^2}{3.1 \times 10^{-2}} = \frac{9.3}{3.1} \times \frac{10^2}{10^{-2}} = 3 \times 10^4 = 30{,}000$$

37. $$\frac{96{,}000}{(12{,}000)(0.00004)} = \frac{9.6 \times 10^4}{(1.2 \times 10^4)(4 \times 10^{-5})} = \frac{9.6 \times 10^4}{4.8 \times 10^{-1}} = \frac{9.6}{4.8} \times \frac{10^4}{10^{-1}} = 2 \times 10^5 = 200{,}000$$

39. $25{,}700{,}000{,}000{,}000 = 2.57 \times 10^{13}$ mi

41. $1.14 \times 10^8 = 114{,}000{,}000$ mi

43. $0.00622 = 6.22 \times 10^{-3}$ mi

45. $3.6 \times 10^7 \text{ mi} \times 5280 \text{ ft/mi} = 19008 \times 10^7 \text{ ft} = 1.9008 \times 10^{11} \text{ ft}$

47. $2.617 \times 10^{11} \text{ bar} \times 42 \text{ gal/bar} = 109.914 \times 10^{11} \text{ gal} = 1.09914 \times 10^{13} \text{ gallons}$

49. $\dfrac{3.3 \times 10^4 \text{ cm}}{\text{sec}} = \dfrac{3.3 \times 10^4 \text{ cm}}{\text{sec}} \cdot \dfrac{1 \text{ m}}{100 \text{ cm}} \cdot \dfrac{1 \text{ km}}{1{,}000 \text{ m}} = \dfrac{3.3 \times 10^4}{10^5} \text{ km/sec} = 3.3 \times 10^{-1} \text{ km/sec}$

51. x-rays, visible light, infrared

53. 1.5×10^{-4} in.; $25{,}000{,}000{,}000{,}000 = 2.5 \times 10^{13}$

55. **Answers may vary.**

57. **Answers may vary.**

Exercise 4.4 (page 264)

1.
$$
\begin{aligned}
5(u-5)+9 &= 2(u+4)\\
5u-25+9 &= 2u+8\\
5u-16 &= 2u+8\\
3u-16 &= 8\\
3u &= 24\\
u &= 8
\end{aligned}
$$

3.
$$
\begin{aligned}
-4(3y+2) &\le 28\\
-12y-8 &\le 28\\
-12y &\le 36\\
\frac{-12y}{-12} &\ge \frac{36}{-12}\\
y &\ge -3
\end{aligned}
$$

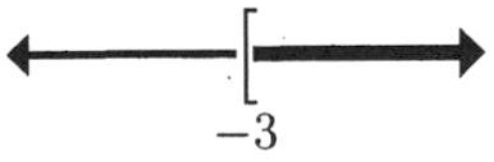

5. $(x^2x^4)^3 = (x^6)^3 = x^{18}$

7. $\left(\dfrac{y^2y^5}{y^4}\right)^3 = \left(\dfrac{y^7}{y^4}\right)^3 = (y^3)^3 = y^9$

9. algebraic

11. polynomial

13. trinomial

15. degree

17. function

19. domain

21. yes

23. yes

25. no; fractional exponent on x

27. binomial

29. trinomial

31. monomial

33. binomial

35. trinomial

37. none of these (4 terms)

39. 4th

41. 3rd

43. 8th $(3+5=8)$

45. 6th $(3+1+2=6)$

47. 12th

49. 0th

51. $5x-3 = 5(2)-3 = 10-3 = 7$

53. $5x-3 = 5(-1)-3 = -5-3 = -8$

55. $-x^2-4 = -(0)^2-4 = -0-4 = -4$

57. $-x^2-4 = -(-1)^2-4 = -1-4 = -5$

59. $f(0) = 5(0) + 1 = 0 + 1 = 1$

61. $f\left(-\frac{1}{2}\right) = 5\left(-\frac{1}{2}\right) + 1 = -\frac{5}{2} + \frac{2}{2} = -\frac{3}{2}$

63. $f(0) = (0)^2 - 2(0) + 3 = 0 - 0 + 3 = 3$

65. $f(-2) = (-2)^2 - 2(-2) + 3$
$= 4 + 4 + 3 = 11$

67. $f(0.5) = (0.5)^2 - 2(0.5) + 3 = 0.25 - 1 + 3 = 2.25$

69.

x	$x^2 - 3$
-2	$(-2)^2 - 3 = 4 - 3 = 1$
-1	$(-1)^2 - 3 = 1 - 3 = -2$
0	$(0)^2 - 3 = 0 - 3 = -3$
1	$(1)^2 - 3 = 1 - 3 = -2$
2	$(2)^2 - 3 = 4 - 3 = 1$

71.

x	$x^3 + 2$
-2	$(-2)^3 + 2 = -8 + 2 = -6$
-1	$(-1)^3 + 2 = -1 + 2 = 1$
0	$(0)^3 + 2 = 0 + 2 = 2$
1	$(1)^3 + 2 = 1 + 2 = 3$
2	$(2)^3 + 2 = 8 + 2 = 10$

73. $f(x) = x^2 - 1$

x	$f(x)$
-2	3
-1	0
0	-1
1	0
2	3

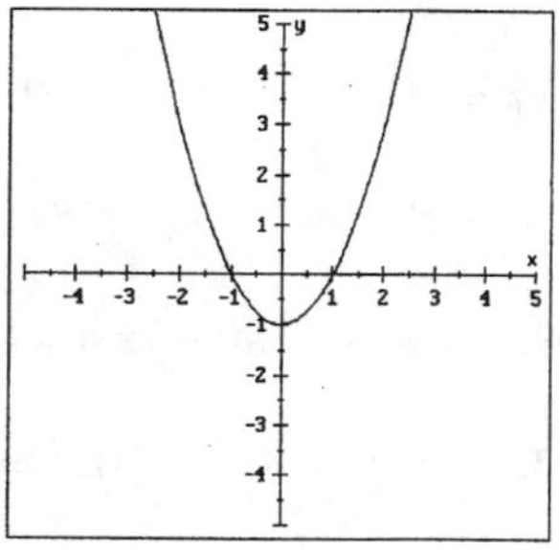

75. $f(x) = x^3 + 2$

x	$f(x)$
-2	-6
-1	1
0	2
1	3
2	10

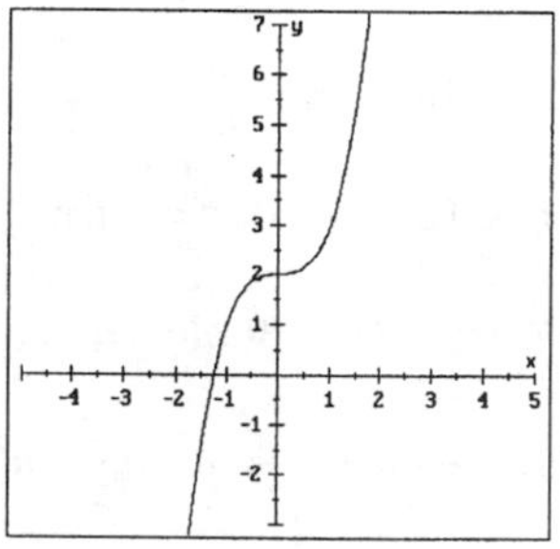

77. $h = -16t^2 + 64t$
$h = -16(2)^2 + 64(2)$
$h = -16(4) + 128$
$h = -64 + 128 = 64$ ft

79. $f(d) = -0.08d^2 + 100d$
$f(815) = -0.08(815)^2 + 100(815)$
$= -0.08(664225) + 81500$
$= -53138 + 81500$
$= \$28,362$

81. $d = 0.04v^2 + 0.9v$
$d = 0.04(30)^2 + 0.9(30)$
$d = 0.04(900) + 27$
$d = 36 + 27 = 63$ ft

83. **Answers may vary.**

85. There are many possible polynomials. One is $2x - 2$.

Exercise 4.5 (page 270)

1. $ab + cd = (3)(-2) + (-1)(2)$
$= -6 + (-2) = -8$

3. $a(b + c) = 3[-2 + (-1)] = 3(-3) = -9$

SECTION 4.5

5. $-4(2x-9) \geq 12$

$-8x+36 \geq 12$

$-8x \geq -24$

$\frac{-8x}{-8} \leq \frac{-24}{-8}$

$x \leq 3$

7. monomial

9. coefficients, variables

11. like terms

13. like terms; $3y + 4y = 7y$

15. unlike terms

17. like terms
$3x^3 + 4x^3 + 6x^3 = 13x^3$

19. like terms; $-5x^3y^2 + 13x^3y^2 = 8x^3y^2$

21. like terms
$-23t^6 + 32t^6 + 56t^6 = 65t^6$

23. unlike terms

25. $4y + 5y = 9y$

27. $-2x + 3x = 1x = x$

29. $-8t^2 - 4t^2 = -12t^2$

31. $15x^2 + 10x^2 = 25x^2$

33. $32u^3 - 16u^3 = 16u^3$

35. $18x^5y^2 - 11x^5y^2 = 7x^5y^2$

37. $22ab^2 - 30ab^2 = -8ab^2$

39. $3rst + 4rst + 7rst = 14rst$

41. $-4a^2bc + 5a^2bc - 7a^2bc = -6a^2bc$

43. $(3x)^2 - 4x^2 + 10x^2 = 9x^2 - 4x^2 + 10x^2$
$= 15x^2$

45. $5x^2y^2 + 2(xy)^2 - (3x^2)y^2 = 5x^2y^2 + 2x^2y^2 - 3x^2y^2 = 4x^2y^2$

47. $(-3x^2y)^4 + (4x^4y^2)^2 - 2x^8y^4 = 81x^8y^4 + 16x^8y^4 - 2x^8y^4 = 95x^8y^4$

49. $(3x+7) + (4x-3) = 3x + 7 + 4x - 3$
$= 7x + 4$

51. $(4a+3) - (2a-4) = 4a + 3 - 2a + 4$
$= 2a + 7$

53. $(2x+3y) + (5x-10y) = 2x + 3y + 5x - 10y = 7x - 7y$

55. $(-8x-3y) - (11x+y) = -8x - 3y - 11x - y = -19x - 4y$

57. $(3x^2-3x-2) + (3x^2+4x-3) = 3x^2 - 3x - 2 + 3x^2 + 4x - 3 = 6x^2 + x - 5$

59. $(2b^2+3b-5) - (2b^2-4b-9) = 2b^2 + 3b - 5 - 2b^2 + 4b + 9 = 7b + 4$

61. $(2x^2-3x+1) - (4x^2-3x+2) + (2x^2+3x+2)$
$= 2x^2 - 3x + 1 - 4x^2 + 3x - 2 + 2x^2 + 3x + 2 = 3x + 1$

63. $2(x+3) + 3(x+3) = 2x + 6 + 3x + 9 = 5x + 15$

65. $-8(x-y)+11(x-y)=-8x+8y+11x-11y=3x-3y$

67. $2(x^2-5x-4)-3(x^2-5x-4)+6(x^2-5x-4)$
$=2x^2-10x-8-3x^2+15x+12+6x^2-30x-24=5x^2-25x-20$

69.
$$\begin{array}{rr} & 3x^2+4x+\ \ 5 \\ + & \underline{2x^2-3x+\ \ 6} \\ & 5x^2+\ \ x+11 \end{array}$$

71.
$$\begin{array}{rr} & 2x^3-3x^2+4x-7 \\ + & \underline{-9x^3-4x^2-5x+6} \\ & -7x^3-7x^2-\ \ x-1 \end{array}$$

73.
$$\begin{array}{rr} & -3x^2y+4xy+25y^2 \\ + & \underline{5x^2y-3xy-12y^2} \\ & 2x^2y+\ \ xy+13y^2 \end{array}$$

75.
$$\begin{array}{rr} & 3x^2+4x-5 \\ - & \underline{-2x^2-2x+3} \\ & 3x^2+4x-5 \\ + & \underline{2x^2+2x-3} \\ & 5x^2+6x-8 \end{array}$$

77.
$$\begin{array}{rr} & 4x^3+4x^2-3x+10 \\ - & \underline{5x^3-2x^2-4x-\ \ 4} \\ & 4x^3+4x^2-3x+10 \\ + & \underline{-5x^3+2x^2+4x+\ \ 4} \\ & -x^3+6x^2+\ \ x+14 \end{array}$$

79.
$$\begin{array}{rr} & -2x^2y^2-\ \ 4xy+12y^2 \\ - & \underline{10x^2y^2+\ \ 9xy-24y^2} \\ & -2x^2y^2-\ \ 4xy+12y^2 \\ + & \underline{-10x^2y^2-\ \ 9xy+24y^2} \\ & -12x^2y^2-13xy+36y^2 \end{array}$$

81. $\left[(2x^2-3x+4)+(3x^2-2)\right]+(x^2+x-3)$
$=\left[2x^2-3x+4+3x^2-2\right]+x^2+x-3$
$=5x^2-3x+2+x^2+x-3=6x^2-2x-1$

83. $\left[(3t^3+t^2)+(-t^3+6t-3)\right]-(t^3-2t^2+2)$
$=\left[3t^3+t^2-t^3+6t-3\right]-t^3+2t^2-2$
$=2t^3+t^2+6t-3-t^3+2t^2-2=t^3+3t^2+6t-5$

85. $\left[(-2x^2-7x+1)+(-4x^2+8x-1)\right]+(3x^2+4x-7)$
$=\left[-2x^2-7x+1-4x^2+8x-1\right]+3x^2+4x-7$
$=-6x^2+x+3x^2+4x-7=-3x^2+5x-7$

87. $2(x+3)+4(x-2)=2x+6+4x-8=6x-2$

89. $-2(x^2+7x-1)-3(x^2-2x+7)=-2x^2-14x+2-3x^2+6x-21=-5x^2-8x-19$

91. $2(2y^2-2y+2)-4(3y^2-4y-1)+4(y^3-y^2-y)$
$=4y^2-4y+4-12y^2+16y+4+4y^3-4y^2-4y=4y^3-12y^2+8y+8$

93. $2(a^2b^2-ab)-3(ab+2ab^2)+(b^2-ab+a^2b^2)=2a^2b^2-2ab-3ab-6ab^2+b^2-ab+a^2b^2$
$=3a^2b^2-6ab+b^2-6ab^2$

95. $-4(x^2y^2 + xy^3 + xy^2z) - 2(x^2y^2 - 4xy^2z) - 2(8xy^3 - y)$
$$= -4x^2y^2 - 4xy^3 - 4xy^2z - 2x^2y^2 + 8xy^2z - 16xy^3 + 2y$$
$$= -6x^2y^2 + 4xy^2z - 20xy^3 + 2y$$

97. $y = 900x + 105{,}000$
$y = 900(10) + 105{,}000$
$y = 9{,}000 + 105{,}000 = \$114{,}000$

99. $y = 1{,}000x + 120{,}000$
$y = 1{,}000(12) + 120{,}000$
$y = 12{,}000 + 120{,}000 = \$132{,}000$

101. a. $y = 900x + 105{,}000$
$y = 900(20) + 105{,}000$
$y = 18{,}000 + 105{,}000 = \$123{,}000$

$y = 1{,}000x + 120{,}000$
$y = 1{,}000(20) + 120{,}000$
$y = 20{,}000 + 120{,}000 = \$140{,}000$

$\$123{,}000 + \$140{,}000 = \$263{,}000$

101. b. $y = 1{,}900x + 225{,}000 = 1{,}900(20) + 225{,}000 = 38{,}000 + 225{,}000 = \$263{,}000$

103. $y = -1{,}100x + 6{,}600$

105. $y = (-1{,}100x + 6{,}600) + (-1{,}700x + 9{,}200) = -2{,}800x + 15{,}800$

107. Answers may vary.

109. $P(x+h) + P(x) = [3(x+h) - 5] + [3x - 5] = 3x + 3h - 5 + 3x - 5 = 6x + 3h - 10$

111. $P(x) - Q(x) = (x^{23} + 5x^2 + 73) - (x^{23} + 4x^2 + 73) = x^{23} + 5x^2 + 73 - x^{23} - 4x^2 - 73 = x^2$
$P(7) - Q(7) = 7^2 = 49$

Exercise 4.6 (page 280)

1. distributive property

3. commutative property of multiplication

5.
$$\frac{5}{3}(5y+6) - 10 = 0$$
$$3\left[\frac{5}{3}(5y+6) - 10\right] = 3(0)$$
$$3 \cdot \frac{5}{3}(5y+6) - 3 \cdot 10 = 0$$
$$5(5y+6) - 30 = 0$$
$$25y + 30 - 30 = 0$$
$$25y = 0$$
$$y = 0$$

7. monomial

9. trinomial

11. $(2x)(3x) = 6x^2$

13. $(5)(3x) = 15x$

15. $(3x^2)(4x^3) = (3)(4)x^2x^3 = 12x^5$

17. $(-5t^3)(2t^4) = (-5)(2)t^3t^4 = -10t^7$

19. $(3b^2)(-2b)(4b^3) = (3)(-2)(4)b^2bb^3 = -24b^6$

21. $(2x^2y^3)(3x^3y^2) = (2)(3)x^2x^3y^3y^2 = 6x^5y^5$

23. $(x^2y^5)(x^2z^5)(-3y^2z^3) = (1)(1)(-3)x^2x^2y^5y^2z^5x^3 = -3x^4y^7z^8$

25. $(x^2y^3)^5 = (x^2)^5(y^3)^5 = x^{10}y^{15}$

27. $(x^5y^2)^3 = (x^5)^3(y^2)^3 = x^{15}y^6$

29. $(a^2b^3c)^5 = (a^2)^5(b^3)^5c^5 = a^{10}b^{15}c^5$

31. $(a^3b^2c)(abc^3)^2 = (a^3b^2c)\left[a^2b^2(c^3)^2\right] = (a^3b^2c)(a^2b^2c^6) = a^3a^2b^2b^2cc^6 = a^5b^4c^7$

33. $3(x+4) = 3(x) + 3(4) = 3x + 12$

35. $-4(t+7) = -4(t) + (-4)(7) = -4t - 28$

37. $3x(x-2) = 3x(x) - 3x(2) = 3x^2 - 6x$

39. $-2x^2(3x^2 - x) = -2x^2(3x^2) - (-2x^2)(x) = -6x^4 + 2x^3$

41. $3xy(x+y) = 3xy(x) + 3xy(y) = 3x^2y + 3xy^2$

43. $2x^2(3x^2 + 4x - 7) = 2x^2(3x^2) + 2x^2(4x) - 2x^2(7) = 6x^4 + 8x^3 - 14x^2$

45. $\frac{1}{4}x^2(8x^5 - 4) = \frac{1}{4}x^2(8x^5) - \frac{1}{4}x^2(4) = 2x^7 - x^2$

47. $-\frac{2}{3}r^2t^2(9r - 3t) = -\frac{2}{3}r^2t^2(9r) - \left(-\frac{2}{3}r^2t^2\right)(3t) = -6r^3t^2 + 2r^2t^3$

49. $(3xy)(-2x^2y^3)(x+y) = (-6x^3y^4)(x+y) = -6x^3y^4(x) + (-6x^3y^4)(y) = -6x^4y^4 - 6x^3y^5$

51. $(a+4)(a+5) = a(a) + a(5) + 4(a) + 4(5) = a^2 + 5a + 4a + 20 = a^2 + 9a + 20$

53. $(3x-2)(x+4) = 3x(x) + 3x(4) + (-2)(x) + (-2)(4) = 3x^2 + 12x - 2x - 8 = 3x^2 + 10x - 8$

55. $(2a+4)(3a-5) = 2a(3a) - 2a(5) + 4(3a) - 4(5) = 6a^2 - 10a + 12a - 20 = 6a^2 + 2a - 20$

57. $(3x-5)(2x+1) = 3x(2x) + 3x(1) + (-5)(2x) + (-5)(1) = 6x^2 + 3x - 10x - 5 = 6x^2 - 7x - 5$

59. $(x+3)(2x-3) = x(2x) - x(3) + 3(2x) - 3(3) = 2x^2 - 3x + 6x - 9 = 2x^2 + 3x - 9$

61. $(2s+3t)(3s-t) = 2s(3s) - 2s(t) + 3t(3s) - 3t(t) = 6s^2 - 2st + 9st - 3t^2 = 6s^2 + 7st - 3t^2$

63. $(x+y)(x+z) = x(x) + x(z) + y(x) + y(z) = x^2 + xz + xy + yz$

65. $(u+v)(u+2t) = u(u) + u(2t) + v(u) + v(2t) = u^2 + 2tu + uv + 2tv$

67. $(-2r-3s)(2r+7s) = (-2r)(2r) + (-2r)(7s) + (-3s)(2r) + (-3s)(7s) = -4r^2 - 14rs - 6rs - 21s^2 = -4r^2 - 20rs - 21s^2$

69.
$$\begin{array}{r} 4x + 3 \\ x + 2 \\ \hline 4x^2 + 3x \\ 8x + 6 \\ \hline 4x^2 + 11x + 6 \end{array}$$

71.
$$\begin{array}{r} 4x - 2y \\ 3x + 5y \\ \hline 12x^2 - 6xy \\ 20xy - 10y^2 \\ \hline 12x^2 + 14xy - 10y^2 \end{array}$$

73.
$$\begin{array}{r} x^2 + x + 1 \\ x - 1 \\ \hline x^3 + x^2 + x \\ - x^2 - x - 1 \\ \hline x^3 - 1 \end{array}$$

75. $(2x+1)\left(x^2+3x-1\right) = (2x+1)\left(x^2\right) + (2x+1)(3x) - (2x+1)(1)$
$= 2x\left(x^2\right) + 1\left(x^2\right) + 2x(3x) + 1(3x) - 2x(1) - 1(1)$
$= 2x^3 + x^2 + 6x^2 + 3x - 2x - 1 = 2x^3 + 7x^2 + x - 1$

77. $(x+4)(x+4) = x(x) + x(4) + 4(x) + 4(4) = x^2 + 4x + 4x + 16 = x^2 + 8x + 16$

79. $(t-3)(t-3) = t(t) - t(3) + (-3)(t) - (-3)(3) = t^2 - 3t - 3t + 9 = t^2 - 6t + 9$

81. $(r+4)(r-4) = r(r) - r(4) + 4(r) - 4(4) = r^2 - 4r + 4r - 16 = r^2 - 16$

83. $(x+5)^2 = (x+5)(x+5) = x(x) + x(5) + 5(x) + 5(5) = x^2 + 5x + 5x + 25 = x^2 + 10x + 25$

85. $(2s+1)(2s+1) = 2s(2s) + 2s(1) + 1(2s) + 1(1) = 4s^2 + 2s + 2s + 1 = 4s^2 + 4s + 1$

87. $(4x+5)(4x-5) = 4x(4x) - 4x(5) + 5(4x) - 5(5) = 16x^2 - 20x + 20x - 25 = 16x^2 - 25$

89. $(x-2y)^2 = (x-2y)(x-2y) = x(x) - x(2y) + (-2y)(x) - (-2y)(2y)$
$= x^2 - 2xy - 2xy + 4y^2 = x^2 - 4xy + 4y^2$

91. $(2a-3b)^2 = (2a-3b)(2a-3b) = 2a(2a) - 2a(3b) + (-3b)(2a) - (-3b)(3b)$
$= 4a^2 - 6ab - 6ab + 9b^2 = 4a^2 - 12ab + 9b^2$

93. $(4x+5y)(4x-5y) = 4x(4x) - 4x(5y) + 5y(4x) - 5y(5y)$
$= 16x^2 - 20xy + 20xy - 25y^2 = 16x^2 - 25y^2$

95. $2(x-4)(x+1) = 2[x(x) + x(1) + (-4)(x) + (-4)(1)] = 2\left[x^2 + x - 4x - 4\right]$
$= 2\left[x^2 - 3x - 4\right] = 2x^2 - 6x - 8$

97. $3a(a+b)(a-b) = 3a[a(a) - a(b) + b(a) - b(b)] = 3a\left[a^2 - ab + ab - b^2\right]$
$= 3a\left[a^2 - b^2\right] = 3a^3 - 3ab^2$

99. $(4t+3)\left(t^2+2t+3\right) = 4t\left(t^2\right) + 4t(2t) + 4t(3) + 3\left(t^2\right) + 3(2t) + 3(3)$
$= 4t^3 + 8t^2 + 12t + 3t^2 + 6t + 9 = 4t^3 + 11t^2 + 18t + 9$

101. $(-3x+y)\left(x^2 - 8xy + 16y^2\right)$
$= -3x\left(x^2\right) - (-3x)(8xy) + (-3x)\left(16y^2\right) + y\left(x^2\right) - y(8xy) + y\left(16y^2\right)$
$= -3x^3 + 24x^2y - 48xy^2 + x^2y - 8xy^2 + 16y^3 = -3x^3 + 25x^2y - 56xy^2 + 16y^3$

103. $(x-2y)\left(x^2+2xy+4y^2\right)$
$$= x\left(x^2\right)+x(2xy)+x\left(4y^2\right)+(-2y)\left(x^2\right)+(-2y)(2xy)+(-2y)\left(4y^2\right)$$
$$= x^3+2x^2y+4xy^2-2x^2y-4xy^2-8y^3 = x^3-8y^3$$

105. $2t(t+2)+3t(t-5) = [2t(t)+2t(2)]+[3t(t)-3t(5)] = \left[2t^2+4t\right]+\left[3t^2-15t\right]$
$$= 2t^2+4t+3t^2-15t = 5t^2-11t$$

107. $(b+2)(b-2)+2b(b+1) = [b(b)-b(2)+2(b)-2(2)]+[2b(b)+2b(1)]$
$$= \left[b^2-2b+2b-4\right]+\left[2b^2+2b\right]$$
$$= \left[b^2-4\right]+\left[2b^2+2b\right] = b^2-4+2b^2+2b = 3b^2+2b-4$$

109. $3xy(x+y)-2x(xy-x) = [3xy(x)+3xy(y)]-[(2x)(xy)-(2x)(x)]$
$$= \left[3x^2y+3xy^2\right]-\left[2x^2y-2x^2\right]$$
$$= 3x^2y+3xy^2-2x^2y+2x^2 = x^2y+3xy^2+2x^2$$

111. $(x+y)(x-y)+x(x+y) = [x(x)-x(y)+y(x)-y(y)]+[x(x)+x(y)]$
$$= \left[x^2-xy+xy-y^2\right]+\left[x^2+xy\right]$$
$$= \left[x^2-y^2\right]+\left[x^2+xy\right] = x^2-y^2+x^2+xy = 2x^2+xy-y^2$$

113. $(x+2)^2-(x-2)^2 = [(x+2)(x+2)]-[(x-2)(x-2)]$
$$= [x(x)+x(2)+2(x)+2(2)]-[x(x)-x(2)+(-2)(x)-(-2)(2)]$$
$$= \left[x^2+2x+2x+4\right]-\left[x^2-2x-2x+4\right]$$
$$= \left[x^2+4x+4\right]-\left[x^2-4x+4\right] = x^2+4x+4-x^2+4x-4 = 8x$$

115. $(2s-3)(s+2)+(3s+1)(s-3)$
$$= [2s(s)+2s(2)+(-3)(s)+(-3)(2)]+[3s(s)-3s(3)+1(s)-1(3)]$$
$$= \left[2s^2+4s-3s-6\right]+\left[3s^2-9s+s-3\right]$$
$$= \left[2s^2+s-6\right]+\left[3s^2-8s-3\right] = 2s^2+s-6+3s^2-8s-3 = 5s^2-7s-9$$

117.
$$(s-4)(s+1) = s^2+5$$
$$s^2+s-4s-4 = s^2+5$$
$$s^2-3s-4 = s^2+5$$
$$-3s-4 = 5$$
$$-3s = 9$$
$$s = -3$$

119.
$$z(z+2) = (z+4)(z-4)$$
$$z^2+2z = z^2-4z+4z-16$$
$$z^2+2z = z^2-16$$
$$2z = -16$$
$$z = -8$$

121.
$$(x+4)(x-4) = (x-2)(x+6)$$
$$x^2-4x+4x-16 = x^2+6x-2x-12$$
$$x^2-16 = x^2+4x-12$$
$$-16 = 4x-12$$
$$-4 = 4x$$
$$-1 = x$$

123.
$$\begin{aligned}(a-3)^2 &= (a+3)^2\\(a-3)(a-3) &= (a+3)(a+3)\\a^2-3a-3a+9 &= a^2+3a+3a+9\\a^2-6a+9 &= a^2+6a+9\\-6a+9 &= 6a+9\\-6a &= 6a\\12a &= 0\\a &= 0\end{aligned}$$

125.
$$\begin{aligned}4+(2y-3)^2 &= (2y-1)(2y+3)\\4+(2y-3)(2y-3) &= (2y-1)(2y+3)\\4+4y^2-6y-6y+9 &= 4y^2+6y-2y-3\\4y^2-12y+13 &= 4y^2+4y-3\\-12y+13 &= 4y-3\\-12y &= 4y-16\\-16y &= -16\\y &= 1\end{aligned}$$

127. Let $r =$ the smaller radius.
Then $r+3 =$ the larger radius.
$$\begin{aligned}\boxed{\text{Larger area}} - \boxed{\text{Smaller area}} &= 15\pi\\\pi(r+3)^2-\pi r^2 &= 15\pi\\\pi(r+3)(r+3)-\pi r^2 &= 15\pi\\\pi\left(r^2+3r+3r+9\right)-\pi r^2 &= 15\pi\\\pi\left(r^2+6r+9\right)-\pi r^2 &= 15\pi\\\pi r^2+6\pi r+9\pi-\pi r^2 &= 15\pi\\6\pi r+9\pi &= 15\pi\\6\pi r &= 6\pi\\\frac{6\pi r}{6\pi} &= \frac{6\pi}{6\pi}\\r &= 1\end{aligned}$$
The larger radius $= 1+3 = 4$ m.

129. Let $s =$ the side of the softball field. Then $s+30 =$ the side of the baseball field.
$$\begin{aligned}\boxed{\text{Larger area}} - \boxed{\text{Smaller area}} &= 4500\\(s+30)^2-s^2 &= 4500\\(s+30)(s+30)-s^2 &= 4500\\s^2+30s+30s+900-s^2 &= 4500\\60s+900 &= 4500\\60s &= 3600\\s &= 60\end{aligned}$$
The baseball field has a side of $60+30=90$ feet.

131. **Answers may vary.**

133. **Answers may vary.**

Exercise 4.7 (page 286)

1. binomial

3. none of these

5. 2

7. polynomial

9. two

11. $\frac{a}{b}$

13. $\frac{5}{15} = \frac{5(1)}{5(3)} = \frac{1}{3}$

15. $\frac{-125}{75} = -\frac{25(5)}{25(3)} = -\frac{5}{3}$

17. $\frac{120}{160} = \frac{40(3)}{40(4)} = \frac{3}{4}$

19. $\frac{-3612}{-3612} = 1$

21. $\frac{-90}{360} = -\frac{90(1)}{90(4)} = -\frac{1}{4}$

23. $\frac{5880}{2660} = \frac{140(42)}{140(19)} = \frac{42}{19}$

25. $\frac{xy}{yz} = \frac{x}{z}$

27. $\frac{r^3s^2}{rs^3} = r^{3-1}s^{2-3} = r^2s^{-1} = \frac{r^2}{s}$

29. $\dfrac{8x^3y^2}{4xy^3} = \dfrac{8}{4}x^{3-1}y^{2-3} = 2x^2y^{-1} = \dfrac{2x^2}{y}$

31. $\dfrac{12u^5v}{-4u^2v^3} = \dfrac{12}{-4}u^{5-2}v^{1-3} = -3u^3v^{-2}$
$= -\dfrac{3u^3}{v^2}$

33. $\dfrac{-16r^3y^2}{-4r^2y^4} = \dfrac{-16}{-4}r^{3-2}y^{2-4} = 4ry^{-2}$
$= \dfrac{4r}{y^2}$

35. $\dfrac{-65rs^2t}{15r^2s^3t} = \dfrac{-65}{15}r^{1-2}s^{2-3}t^{1-1}$
$= -\dfrac{13}{3}r^{-1}s^{-1}t^0 = -\dfrac{13}{3rs}$

37. $\dfrac{x^2x^3}{xy^6} = \dfrac{x^5}{xy^6} = \dfrac{x^{5-1}}{y^6} = \dfrac{x^4}{y^6}$

39. $\dfrac{(a^3b^4)^3}{ab^4} = \dfrac{a^9b^{12}}{ab^4} = a^{9-1}b^{12-4} = a^8b^8$

41. $\dfrac{15(r^2s^3)^2}{-5(rs^5)^3} = \dfrac{15r^4s^6}{-5r^3s^{15}} = \dfrac{15}{-5}r^{4-3}s^{6-15} = -3r^1s^{-9} = -\dfrac{3r}{s^9}$

43. $\dfrac{-32(x^3y)^3}{128(x^2y^2)^3} = \dfrac{-32x^9y^3}{128x^6y^6} = \dfrac{-32}{128}x^{9-6}y^{3-6} = -\dfrac{1}{4}x^3y^{-3} = -\dfrac{x^3}{4y^3}$

45. $\dfrac{(5a^2b)^3}{(2a^2b^2)^3} = \dfrac{125a^6b^3}{8a^6b^6} = \dfrac{125}{8}a^{6-6}b^{3-6} = \dfrac{125}{8}a^0b^{-3} = \dfrac{125}{8b^3}$

47. $\dfrac{-(3x^3y^4)^3}{-(9x^4y^5)^2} = \dfrac{-27x^9y^{12}}{-81x^8y^{10}} = \dfrac{-27}{-81}x^{9-8}y^{12-10} = \dfrac{1}{3}xy^2 = \dfrac{xy^2}{3}$

49. $\dfrac{(a^2a^3)^4}{(a^4)^3} = \dfrac{(a^5)^4}{a^{12}} = \dfrac{a^{20}}{a^{12}} = a^{20-12} = a^8$

51. $\dfrac{(z^3z^{-4})^3}{(z^{-3})^2} = \dfrac{(z^{-1})^3}{z^{-6}} = \dfrac{z^{-3}}{z^{-6}} = z^{-3-(-6)} = z^3$

53. $\dfrac{(a^2b)^3(ab^2)^2}{(3a^3b^2)^4} = \dfrac{(a^6b^3)(a^2b^4)}{81a^{12}b^8} = \dfrac{a^8b^7}{81a^{12}b^8} = \dfrac{1}{81}a^{8-12}b^{7-8} = \dfrac{1}{81}a^{-4}b^{-1} = \dfrac{1}{81a^4b}$

55. $\dfrac{6x+9y}{3xy} = \dfrac{6x}{3xy} + \dfrac{9y}{3xy} = \dfrac{2}{y} + \dfrac{3}{x}$

57. $\dfrac{xy+6}{3y} = \dfrac{xy}{3y} + \dfrac{6}{3y} = \dfrac{x}{3} + \dfrac{2}{y}$

59. $\dfrac{5x-10y}{25xy} = \dfrac{5x}{25xy} - \dfrac{10y}{25xy} = \dfrac{1}{5y} - \dfrac{2}{5x}$

61. $\dfrac{3x^2+6y^3}{3x^2y^2} = \dfrac{3x^2}{3x^2y^2} + \dfrac{6y^3}{3x^2y^2} = \dfrac{1}{y^2} + \dfrac{2y}{x^2}$

63. $\dfrac{15a^3b^2-10a^2b^3}{5a^2b^2} = \dfrac{15a^3b^2}{5a^2b^2} - \dfrac{10a^2b^3}{5a^2b^2}$
$= 3a - 2b$

65. $\dfrac{4x-2y+8z}{4xy} = \dfrac{4x}{4xy} - \dfrac{2y}{4xy} + \dfrac{8z}{4xy}$
$= \dfrac{1}{y} - \dfrac{1}{2x} + \dfrac{2z}{xy}$

67. $\dfrac{12x^3y^2-8x^2y-4x}{4xy} = \dfrac{12x^3y^2}{4xy} - \dfrac{8x^2y}{4xy} - \dfrac{4x}{4xy} = 3x^2y - 2x - \dfrac{1}{y}$

69. $\dfrac{-25x^2y+30xy^2-5xy}{-5xy}=\dfrac{-25x^2y}{-5xy}+\dfrac{30xy^2}{-5xy}-\dfrac{5xy}{-5xy}=5x-6y+1$

71. $\dfrac{5x(4x-2y)}{2y}=\dfrac{20x^2-10xy}{2y}=\dfrac{20x^2}{2y}-\dfrac{10xy}{2y}=\dfrac{10x^2}{y}-5x$

73. $\dfrac{(-2x)^3+(3x^2)^2}{6x^2}=\dfrac{-8x^3+9x^4}{6x^2}=\dfrac{-8x^3}{6x^2}+\dfrac{9x^4}{6x^2}=-\dfrac{4x}{3}+\dfrac{3x^2}{2}$

75. $\dfrac{4x^2y^2-2(x^2y^2+xy)}{2xy}=\dfrac{4x^2y^2-2x^2y^2-2xy}{2xy}=\dfrac{2x^2y^2-2xy}{2xy}=\dfrac{2x^2y^2}{2xy}-\dfrac{2xy}{2xy}=xy-1$

77. $\dfrac{(3x-y)(2x-3y)}{6xy}=\dfrac{6x^2-9xy-2xy+3y^2}{6xy}=\dfrac{6x^2-11xy+3y^2}{6xy}=\dfrac{6x^2}{6xy}-\dfrac{11xy}{6xy}+\dfrac{3y^2}{6xy}$

$=\dfrac{x}{y}-\dfrac{11}{6}+\dfrac{y}{2x}$

79. $\dfrac{(a+b)^2-(a-b)^2}{2ab}=\dfrac{(a+b)(a+b)-(a-b)(a-b)}{2ab}$

$=\dfrac{[a^2+ab+ab+b^2]-[a^2-ab-ab+b^2]}{2ab}$

$=\dfrac{[a^2+2ab+b^2]-[a^2-2ab+b^2]}{2ab}$

$=\dfrac{a^2+2ab+b^2-a^2+2ab-b^2}{2ab}=\dfrac{4ab}{2ab}=2$

81. $l=\dfrac{P-2w}{2}$

$l=\dfrac{P}{2}-\dfrac{2w}{2}$

$l=\dfrac{P}{2}-w$

They are the same.

83. $C=\dfrac{0.15x+12}{x}$

$C=\dfrac{0.15x}{x}+\dfrac{12}{x}$

$C=0.15+\dfrac{12}{x}$

They are the same.

85. **Answers may vary.**

87. $\dfrac{x^{500}-x^{499}}{x^{499}}=\dfrac{x^{500}}{x^{499}}-\dfrac{x^{499}}{x^{499}}=x-1$

Let $x=501$: $x-1=501-1=500$

Exercise 4.8 (page 292)

1. $21, 22, 24, 25, 26, 27, 28$

3. $|a-b|=|-2-3|=|-5|=5$

5. $-|a^2-b^2|=-\left|(-2)^2-3^2\right|=-|4-9|=-|-5|=-(+5)=-5$

7. $3(2x^2-4x+5)+2(x^2+3x-7)=6x^2-12x+15+2x^2+6x-14=8x^2-6x+1$

SECTION 4.8

9. divisor; dividend

11. remainder

13. $4x^3 - 2x^2 + 7x + 6$

15. $6x^4 - x^3 + 2x^2 + 9x$

17. $0x^3$ and $0x$

19.
$$\begin{array}{r} x+2 \\ x+5\overline{)x^2+7x+10} \\ \underline{x^2+5x} \\ 2x+10 \\ \underline{2x+10} \\ 0 \end{array}$$

21.
$$\begin{array}{r} a+5 \\ a-4\overline{)a^2+a-20} \\ \underline{a^2-4a} \\ 5a-20 \\ \underline{5a-20} \\ 0 \end{array}$$

23.
$$\begin{array}{r} x+2 \\ x+2\overline{)x^2+4x+4} \\ \underline{x^2+2x} \\ 2x+4 \\ \underline{2x+4} \\ 0 \end{array}$$

25.
$$\begin{array}{r} y+12 \\ y+1\overline{)y^2+13y+12} \\ \underline{y^2+y} \\ 12y+12 \\ \underline{12y+12} \\ 0 \end{array}$$

27.
$$\begin{array}{r} a+b \\ a+b\overline{)a^2+2ab+b^2} \\ \underline{a^2+ab} \\ ab+b^2 \\ \underline{ab+b^2} \\ 0 \end{array}$$

29.
$$\begin{array}{r} 3a-2 \\ 2a+3\overline{)6a^2+5a-6} \\ \underline{6a^2+9a} \\ -4a-6 \\ \underline{-4a-6} \\ 0 \end{array}$$

31.
$$\begin{array}{r} b+3 \\ 3b+2\overline{)3b^2+11b+6} \\ \underline{3b^2+2b} \\ 9b+6 \\ \underline{9b+6} \\ 0 \end{array}$$

33.
$$\begin{array}{r} x-3y \\ 2x-y\overline{)2x^2-7xy+3y^2} \\ \underline{2x^2-xy} \\ -6xy+3y^2 \\ \underline{-6xy+3y^2} \\ 0 \end{array}$$

35.
$$\begin{array}{r} a+2b \\ a+b\overline{)a^2+3ab+2b^2} \\ \underline{a^2+ab} \\ 2ab+2b^2 \\ \underline{2ab+2b^2} \\ 0 \end{array}$$

37.
$$\begin{array}{r} 2x+1 \\ 5x+3\overline{)10x^2+11x+3} \\ \underline{10x^2+6x} \\ 5x+3 \\ \underline{5x+3} \\ 0 \end{array}$$

39.
$$\begin{array}{r} x-7 \\ 2x+4\overline{)2x^2-10x-28} \\ \underline{2x^2+4x} \\ -14x-28 \\ \underline{-14x-28} \\ 0 \end{array}$$

41.
$$\begin{array}{r} 3x+2y \\ 2x-y\overline{)6x^2+xy-2y^2} \\ \underline{6x^2-3xy} \\ 4xy-2y^2 \\ \underline{4xy-2y^2} \\ 0 \end{array}$$

43.
$$\begin{array}{r} 2x-y \\ x+3y\overline{)2x^2+5xy-3y^2} \\ \underline{2x^2+6xy} \\ -xy-3y^2 \\ \underline{-xy-3y^2} \\ 0 \end{array}$$

45.
$$\begin{array}{r} x+5y \\ 3x-2y\overline{)3x^2+13xy-10y^2} \\ \underline{3x^2-2xy} \\ 15xy-10y^2 \\ \underline{15xy-10y^2} \\ 0 \end{array}$$

47.
$$\begin{array}{r} x-5y \\ 4x+y\overline{)4x^2-19xy-5y^2} \\ \underline{4x^2+xy} \\ -20xy-5y^2 \\ \underline{-20xy-5y^2} \\ 0 \end{array}$$

49.
$$\begin{array}{r} x^2+2x-1 \\ 2x+3\overline{)2x^3+7x^2+4x-3} \\ \underline{2x^3+3x^2} \\ 4x^2+4x \\ \underline{4x^2+6x} \\ -2x-3 \\ \underline{-2x-3} \\ 0 \end{array}$$

51.
$$\begin{array}{r}
2x^2 + 2x + 1 \\
3x+2\;\overline{)\,6x^3 + 10x^2 + 7x + 2} \\
\underline{6x^3 + 4x^2} \\
6x^2 + 7x \\
\underline{6x^2 + 4x} \\
3x + 2 \\
\underline{3x + 2} \\
0
\end{array}$$

53.
$$\begin{array}{r}
x^2 + xy + y^2 \\
2x+y\;\overline{)\,2x^3 + 3x^2y + 3xy^2 + y^3} \\
\underline{2x^3 + x^2y} \\
2x^2y + 3xy^2 \\
\underline{2x^2y + xy^2} \\
2xy^2 + y^3 \\
\underline{2xy^2 + y^3} \\
0
\end{array}$$

55.
$$\begin{array}{r}
x + 1 + \frac{-1}{2x+3} \\
2x+3\;\overline{)\,2x^2 + 5x + 2} \\
\underline{2x^2 + 3x} \\
2x + 2 \\
\underline{2x + 3} \\
-1
\end{array}$$

57.
$$\begin{array}{r}
2x + 2 + \frac{-3}{2x+1} \\
2x+1\;\overline{)\,4x^2 + 6x - 1} \\
\underline{4x^2 + 2x} \\
4x - 1 \\
\underline{4x + 2} \\
-3
\end{array}$$

59.
$$\begin{array}{r}
x^2 + 2x + 1 \\
x+1\;\overline{)\,x^3 + 3x^2 + 3x + 1} \\
\underline{x^3 + x^2} \\
2x^2 + 3x \\
\underline{2x^2 + 2x} \\
x + 1 \\
\underline{x + 1} \\
0
\end{array}$$

61.
$$\begin{array}{r}
x^2 + 2x - 1 + \frac{6}{2x+3} \\
2x+3\;\overline{)\,2x^3 + 7x^2 + 4x + 3} \\
\underline{2x^3 + 3x^2} \\
4x^2 + 4x \\
\underline{4x^2 + 6x} \\
-2x + 3 \\
\underline{-2x - 3} \\
6
\end{array}$$

63.
$$\begin{array}{r}
2x^2 + 8x + 14 + \frac{31}{x-2} \\
x-2\;\overline{)\,2x^3 + 4x^2 - 2x + 3} \\
\underline{2x^3 - 4x^2} \\
8x^2 - 2x \\
\underline{8x^2 - 16x} \\
14x + 3 \\
\underline{14x - 28} \\
31
\end{array}$$

65.
$$\begin{array}{r}
x + 1 \\
x-1\;\overline{)\,x^2 + 0x - 1} \\
\underline{x^2 - x} \\
x - 1 \\
\underline{x - 1} \\
0
\end{array}$$

67.
$$\begin{array}{r}
2x - 3 \\
2x+3\;\overline{)\,4x^2 + 0x - 9} \\
\underline{4x^2 + 6x} \\
-6x - 9 \\
\underline{-6x - 9} \\
0
\end{array}$$

69.
$$\begin{array}{r}
x^2 - x + 1 \\
x+1\;\overline{)\,x^3 + 0x^2 + 0x + 1} \\
\underline{x^3 + x^2} \\
-x^2 + 0x \\
\underline{-x^2 - x} \\
x + 1 \\
\underline{x + 1} \\
0
\end{array}$$

71.
$$\begin{array}{r}
a^2 - 3a + 10 + \frac{-30}{a+3} \\
a+3\;\overline{)\,a^3 + 0a^2 + a + 0} \\
\underline{a^3 + 3a^2} \\
-3a^2 + a \\
\underline{-3a^2 - 9a} \\
10a + 0 \\
\underline{10a + 30} \\
-30
\end{array}$$

73.
$$\begin{array}{r}
5x^2 - x + 4 + \frac{16}{3x-4} \\
3x-4\;\overline{)\,15x^3 - 23x^2 + 16x + 0} \\
\underline{15x^3 - 20x^2} \\
-3x^2 + 16x \\
\underline{-3x^2 + 4x} \\
12x + 0 \\
\underline{12x - 16} \\
16
\end{array}$$

75.

$$\begin{array}{r} 2t^2 - 3tu + u^2 + \frac{u^3}{5t-2u} \\ 5t - 2u \overline{\big)\, 10t^3 - 19t^2u + 11tu^2 - u^3} \\ \underline{10t^3 - 4t^2u \qquad\qquad\qquad} \\ -15t^2u + 11tu^2 \qquad \\ \underline{-15t^2u + 6tu^2 \qquad} \\ 5tu^2 - u^3 \\ \underline{5tu^2 - 2u^3} \\ u^3 \end{array}$$

77. **Answers may vary.**

79. $x^2 - 2x$ is added to the dividend when it should be subtracted.

Chapter 4 Summary (page 295)

1. $(-3x)^4 = (-3x)(-3x)(-3x)(-3x)$

2. $\left(\frac{1}{2}pq\right)^3 = \left(\frac{1}{2}pq\right)\left(\frac{1}{2}pq\right)\left(\frac{1}{2}pq\right)$

3. $5^3 = 5 \cdot 5 \cdot 5 = 125$

4. $3^5 = 3 \cdot 3 \cdot 3 \cdot 3 \cdot 3 = 243$

5. $(-8)^2 = (-8)(-8) = 64$

6. $-8^2 = -1 \cdot 8 \cdot 8 = -64$

7. $3^2 + 2^2 = 9 + 4 = 13$

8. $(3+2)^2 = (5)^2 = 25$

9. $x^3x^2 = x^{3+2} = x^5$

10. $x^2x^7 = x^{2+7} = x^9$

11. $(y^7)^3 = y^{7 \cdot 3} = y^{21}$

12. $(x^{21})^2 = x^{21 \cdot 2} = x^{42}$

13. $(ab)^3 = a^3b^3$

14. $(3x)^4 = 3^4x^4 = 81x^4$

15. $b^3b^4b^5 = b^{3+4+5} = b^{12}$

16. $-z^2(z^3y^2) = -z^{2+3}y^2 = -y^2z^5$

17. $(16s)^2s = 16^2s^2s = 256s^3$

18. $-3y(y^5) = -3y^6$

19. $(x^2x^3)^3 = (x^5)^3 = x^{15}$

20. $(2x^2y)^2 = 2^2(x^2)^2y^2 = 4x^4y^2$

21. $\frac{x^7}{x^3} = x^{7-3} = x^4$

22. $\left(\frac{x^2y}{xy^2}\right)^2 = \frac{x^4y^2}{x^2y^4} = x^2y^{-2} = \frac{x^2}{y^2}$

23. $\frac{8(y^2x)^2}{4(yx^2)^2} = \frac{8y^4x^2}{4y^2x^4} = \frac{8}{4}y^2x^{-2} = \frac{2y^2}{x^2}$

24. $\frac{(5y^2z^3)^3}{25(yz)^5} = \frac{125y^6z^9}{25y^5z^5} = 5yz^4$

25. $x^0 = 1$

26. $(3x^2y^2)^0 = 1$

27. $(3x^0)^2 = (3 \cdot 1)^2 = 3^2 = 9$

28. $(3x^2y^0)^2 = (3x^2 \cdot 1)^2 = (3x^2)^2 = 9x^4$

29. $x^{-3} = \frac{1}{x^3}$

30. $x^{-2}x^3 = x^1 = x$

31. $y^4y^{-3} = y^1 = y$

32. $\frac{x^3}{x^{-7}} = x^{3-(-7)} = x^{10}$

33. $\left(x^{-3}x^4\right)^{-2} = \left(x^1\right)^{-2} = x^{-2} = \frac{1}{x^2}$

34. $\left(a^{-2}b\right)^{-3} = a^6b^{-3} = \frac{a^6}{b^3}$

35. $\left(\frac{x^2}{x}\right)^{-5} = (x)^{-5} = \frac{1}{x^5}$

36. $\left(\frac{15z^4}{5z^3}\right)^{-2} = (3z)^{-2} = \frac{1}{(3z)^2} = \frac{1}{9z^2}$

37. $728 = 7.28 \times 10^2$

38. $9{,}370 = 9.37 \times 10^3$

39. $0.0136 = 1.36 \times 10^{-2}$

40. $0.00942 = 9.42 \times 10^{-3}$

41. $7.73 = 7.73 \times 10^0$

42. $753 \times 10^3 = 7.53 \times 10^2 \times 10^3$
$= 7.53 \times 10^5$

43. $0.018 \times 10^{-2} = 1.8 \times 10^{-2} \times 10^{-2}$
$= 1.8 \times 10^{-4}$

44. $600 \times 10^2 = 6.00 \times 10^2 \times 10^2$
$= 6.00 \times 10^4$

45. $7.26 \times 10^5 = 726{,}000$

46. $3.91 \times 10^{-4} = 0.000391$

47. $2.68 \times 10^0 = 2.68$

48. $5.76 \times 10^1 = 57.6$

49. $739 \times 10^{-2} = 7.39$

50. $0.437 \times 10^{-3} = 0.000437$

51. $\frac{(0.00012)(0.00004)}{0.00000016} = \frac{(1.2 \times 10^{-4})(4 \times 10^{-5})}{1.6 \times 10^{-7}} = \frac{4.8 \times 10^{-9}}{1.6 \times 10^{-7}} = 3 \times 10^{-2} = 0.03$

52. $\frac{(4{,}800)(20{,}000)}{600{,}000} = \frac{(4.8 \times 10^3)(2 \times 10^4)}{6 \times 10^5} = \frac{9.6 \times 10^7}{6 \times 10^5} = 1.6 \times 10^2 = 160$

53. $13x^7$: monomial, degree $= 7$

54. $5^3x + x^2$: binomial, degree $= 2$

55. $-3x^5 + x - 1$: trinomial, degree $= 5$

56. $9x + 21x^3y^2$: binomial, degree $= 3 + 2 = 5$

57. $3x + 2 = 3(3) + 2 = 9 + 2 = 11$

58. $3x + 2 = 3(0) + 2 = 0 + 2 = 2$

59. $3x + 2 = 3(-2) + 2 = -6 + 2 = -4$

60. $3x + 2 = 3\left(\frac{2}{3}\right) + 2 = 2 + 2 = 4$

61. $5x^4 - x = 5(3)^4 - 3 = 5(81) - 3$
$= 405 - 3 = 402$

62. $5x^4 - x = 5(0)^4 - 0 = 5(0) - 0$
$= 0 - 0 = 0$

63. $5x^4 - x = 5(-2)^4 - (-2) = 5(16) + 2$
$= 80 + 2 = 82$

64. $5x^4 - x = 5(-0.3)^4 - (-0.3)$
$= 5(0.0081) + 0.3$
$= 0.0405 + 3 = 0.3405$

65. $f(0) = 0^2 - 4 = 0 - 4 = -4$

66. $f(5) = 5^2 - 4 = 25 - 4 = 21$

67. $f(-2) = (-2)^2 - 4 = 4 - 4 = 0$

68. $f\left(\frac{1}{2}\right) = \left(\frac{1}{2}\right)^2 - 4 = \frac{1}{4} - 4 = -\frac{15}{4}$

69. $f(x) = x^2 - 5$

x	-2	-1	0	1	2
$f(x)$	-1	-4	-5	-4	-1

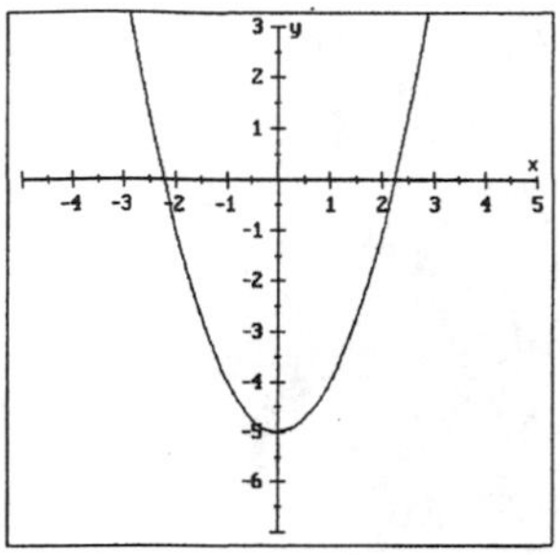

70. $f(x) = x^3 - 2$

x	-2	-1	0	1	2
$f(x)$	-10	-3	-2	-1	6

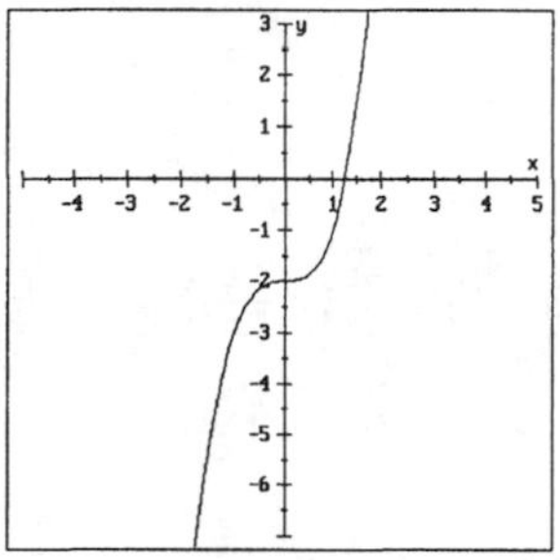

71. $3x + 5x - x = 7x$

72. $3x + 2y$: not like terms

73. $(xy)^2 + 3x^2y^2 = x^2y^2 + 3x^2y^2 = 4x^2y^2$

74. $-2x^2yz + 3x^2yz = x^2yz$

75. $(3x^2 + 2x) + (5x^2 - 8x) = 3x^2 + 2x + 5x^2 - 8x = 8x^2 - 6x$

76. $(7a^2 + 2a - 5) - (3a^2 - 2a + 1) = 7a^2 + 2a - 5 - 3a^2 + 2a - 1 = 4a^2 + 4a - 6$

77. $3(9x^2 + 3x + 7) - 2(11x^2 - 5x + 9) = 27x^2 + 9x + 21 - 22x^2 + 10x - 18 = 5x^2 + 19x + 3$

78. $4(4x^3 + 2x^2 - 3x - 8) - 5(2x^3 - 3x + 8) = 16x^3 + 8x^2 - 12x - 32 - 10x^3 + 15x - 40$
$= 6x^3 + 8x^2 + 3x - 72$

79. $(2x^2y^3)(5xy^2) = 10x^3y^5$

80. $(xyz^3)(x^3z)^2 = xyz^3x^6z^2 = x^7yz^5$

81. $5(x + 3) = 5x + 15$

82. $3(2x + 4) = 6x + 12$

83. $x^2(3x^2 - 5) = 3x^4 - 5x^2$

84. $2y^2(y^2 + 5y) = 2y^4 + 10y^3$

85. $-x^2y(y^2 - xy) = -x^2y^3 + x^3y^2$

86. $-3xy(xy - x) = -3x^2y^2 + 3x^2y$

87. $(x + 3)(x + 2) = x^2 + 2x + 3x + 6$
$= x^2 + 5x + 6$

88. $(2x + 1)(x - 1) = 2x^2 - 2x + x - 1$
$= 2x^2 - x - 1$

89. $(3a - 3)(2a + 2) = 6a^2 + 6a - 6a - 6$
$= 6a^2 - 6$

90. $6(a - 1)(a + 1) = 6(a^2 + a - a - 1)$
$= 6(a^2 - 1) = 6a^2 - 6$

91. $(a-b)(2a+b) = 2a^2 + ab - 2ab - b^2$
$= 2a^2 - ab - b^2$

92. $(3x-y)(2x+y) = 6x^2 + 3xy - 2xy - y^2$
$= 6x^2 + xy - y^2$

93. $(x+3)(x+3) = x^2 + 3x + 3x + 9$
$= x^2 + 6x + 9$

94. $(x+5)(x-5) = x^2 - 5x + 5x - 25$
$= x^2 - 25$

95. $(y-2)(y+2) = y^2 + 2y - 2y - 4$
$= y^2 - 4$

96. $(x+4)^2 = (x+4)(x+4)$
$= x^2 + 4x + 4x + 16$
$= x^2 + 8x + 16$

97. $(x-3)^2 = (x-3)(x-3) = x^2 - 3x - 3x + 9 = x^2 - 6x + 9$

98. $(y-1)^2 = (y-1)(y-1) = y^2 - y - y + 1 = y^2 - 2y + 1$

99. $(2y+1)^2 = (2y+1)(2y+1) = 4y^2 + 2y + 2y + 1 = 4y^2 + 4y + 1$

100. $(y^2+1)(y^2-1) = y^4 - y^2 + y^2 - 1 = y^4 - 1$

101. $(3x+1)\left(x^2+2x+1\right) = 3x^3 + 6x^2 + 3x + x^2 + 2x + 1 = 3x^3 + 7x^2 + 5x + 1$

102. $(2a-3)\left(4a^2+6a+9\right) = 8a^3 + 12a^2 + 18a - 12a^2 - 18a - 27 = 8a^3 - 27$

103.
$$\begin{aligned} x^2 + 3 &= x(x+3) \\ x^2 + 3 &= x^2 + 3x \\ 3 &= 3x \\ 1 &= x \end{aligned}$$

104.
$$\begin{aligned} x^2 + x &= (x+1)(x+2) \\ x^2 + x &= x^2 + 2x + x + 2 \\ x^2 + x &= x^2 + 3x + 2 \\ x &= 3x + 2 \\ -2x &= 2 \\ x &= -1 \end{aligned}$$

105.
$$\begin{aligned} (x+2)(x-5) &= (x-4)(x-1) \\ x^2 - 5x + 2x - 10 &= x^2 - x - 4x + 4 \\ x^2 - 3x - 10 &= x^2 - 5x + 4 \\ -3x - 10 &= -5x + 4 \\ -3x &= -5x + 14 \\ 2x &= 14 \\ x &= 7 \end{aligned}$$

106.
$$\begin{aligned} (x-1)(x-2) &= (x-3)(x+1) \\ x^2 - 2x - x + 2 &= x^2 + x - 3x - 3 \\ x^2 - 3x + 2 &= x^2 - 2x - 3 \\ -3x + 2 &= -2x - 3 \\ -3x &= -2x - 5 \\ -x &= -5 \\ x &= 5 \end{aligned}$$

107.
$$\begin{aligned} x^2 + x(x+2) &= x(2x+1) + 1 \\ x^2 + x^2 + 2x &= 2x^2 + x + 1 \\ 2x^2 + 2x &= 2x^2 + x + 1 \\ 2x &= x + 1 \\ x &= 1 \end{aligned}$$

108.
$$\begin{aligned} (x+5)(3x+1) &= x^2 + (2x-1)(x-5) \\ 3x^2 + 16x + 5 &= 3x^2 - 11x + 5 \\ 16x + 5 &= -11x + 5 \\ 16x &= -11x \\ 27x &= 0 \\ x &= 0 \end{aligned}$$

109. $\dfrac{3x+6y}{2xy}=\dfrac{3x}{2xy}+\dfrac{6y}{2xy}=\dfrac{3}{2y}+\dfrac{3}{x}$

110. $\dfrac{14xy-21x}{7xy}=\dfrac{14xy}{7xy}-\dfrac{21x}{7xy}=2-\dfrac{3}{y}$

111. $\dfrac{15a^2bc+20ab^2c-25abc^2}{-5abc}=\dfrac{15a^2bc}{-5abc}+\dfrac{20ab^2c}{-5abc}-\dfrac{25abc^2}{-5abc}=-3a-4b+5c$

112.
$$\begin{aligned}\frac{(x+y)^2+(x-y)^2}{-2xy}&=\frac{(x+y)(x+y)+(x-y)(x-y)}{-2xy}\\&=\frac{x^2+xy+xy+y^2+x^2-xy-xy+y^2}{-2xy}\\&=\frac{2x^2+2y^2}{-2xy}=\frac{2x^2}{-2xy}+\frac{2y^2}{-2xy}=-\frac{x}{y}-\frac{y}{x}\end{aligned}$$

113.
$$\begin{array}{r|l} & x+1+\frac{3}{x+2}\\ \hline x+2 & x^2+3x+5\\ & \underline{x^2+2x}\\ & \quad x+5\\ & \quad \underline{x+2}\\ & \qquad 3\end{array}$$

114.
$$\begin{array}{r|l} & x-5\\ \hline x-1 & x^2-6x+5\\ & \underline{x^2-x}\\ & -5x+5\\ & \underline{-5x+5}\\ & \qquad 0\end{array}$$

115.
$$\begin{array}{r|l} & 2x+1\\ \hline x+3 & 2x^2+7x+3\\ & \underline{2x^2+6x}\\ & \quad x+3\\ & \quad \underline{x+3}\\ & \qquad 0\end{array}$$

116.
$$\begin{array}{r|l} & x+5+\frac{3}{3x-1}\\ \hline 3x-1 & 3x^2+14x-2\\ & \underline{3x^2-x}\\ & \quad 15x-2\\ & \quad \underline{15x-5}\\ & \qquad 3\end{array}$$

117.
$$\begin{array}{r|l} & 3x^2+2x+1+\frac{2}{2x-1}\\ \hline 2x-1 & 6x^3+x^2+0x+1\\ & \underline{6x^3-3x^2}\\ & \quad 4x^2+0x\\ & \quad \underline{4x^2-2x}\\ & \qquad 2x+1\\ & \qquad \underline{2x-1}\\ & \qquad\quad 2\end{array}$$

118.
$$\begin{array}{r|l} & 3x^2-x-4\\ \hline 3x+1 & 9x^3+0x^2-13x-4\\ & \underline{9x^3+3x^2}\\ & \quad -3x^2-13x\\ & \quad \underline{-3x^2-x}\\ & \qquad -12x-4\\ & \qquad \underline{-12x-4}\\ & \qquad\quad 0\end{array}$$

Chapter 4 Test (page 299)

1. $2xxxyyyy=2x^3y^4$

2. $3^2+5^3=9+125=134$

3. $y^2(yy^3)=y^2y^4=y^6$

4. $\left(-3b^2\right)\left(2b^3\right)\left(-b^2\right)=(-3)(2)(-1)b^2b^3b^2=6b^7$

5. $(2x^3)^5(x^2)^3=32x^{15}x^6=32x^{21}$

6. $(2rr^2r^3)^3=(2r^6)^3=8r^{18}$

7. $3x^0=3(1)=3$

8. $2y^{-5}y^2=2y^{-3}=\dfrac{2}{y^3}$

9. $\dfrac{y^2}{yy^{-2}}=\dfrac{y^2}{y^{-1}}=y^3$

10. $\left(\dfrac{a^2b^{-1}}{4a^3b^{-2}}\right)^{-3} = \left(\dfrac{1}{4}a^{-1}b^{1}\right)^{-3} = \left(\dfrac{1}{4}\right)^{-3}a^3b^{-3} = \dfrac{64a^3}{b^3}$

11. $28{,}000 = 2.8 \times 10^4$

12. $0.0025 = 2.5 \times 10^{-3}$

13. $7.4 \times 10^3 = 7{,}400$

14. $9.3 \times 10^{-5} = 0.000093$

15. $3x^2 + 2$: binomial

16. $2 + 3 + 5 = 10$th degree

17. $x^2 + x + 2 = (-2)^2 + (-2) - 2$
$= 4 - 2 - 2 = 0$

18. $f(x) = x^2 + 2$

x	$f(x)$
-2	6
-1	3
0	2
1	3
2	6

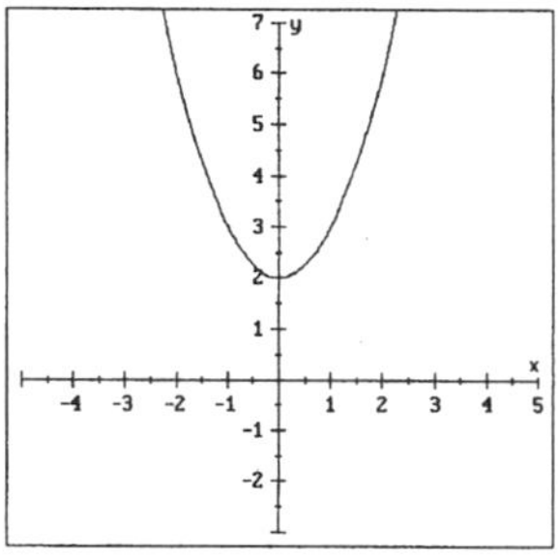

19. $-6(x - y) + 2(x + y) - 3(x + 2y) = -6x + 6y + 2x + 2y - 3x - 6y = -7x + 2y$

20. $-2(x^2 + 3x - 1) - 3(x^2 - x + 2) + 5(x^2 + 2) = -2x^2 - 6x + 2 - 3x^2 + 3x - 6 + 5x^2 + 10$
$= -3x + 6$

21.
$$\begin{array}{rr} & 3x^3 + 4x^2 - x - 7 \\ + & 2x^3 - 2x^2 + 3x + 2 \\ \hline & 5x^3 + 2x^2 + 2x - 5 \end{array}$$

22.
$$\begin{array}{rr} & 2x^2 - 7x + 3 \\ - & 3x^2 - 2x - 1 \\ \hline \end{array} \quad \Rightarrow \quad \begin{array}{rr} & 2x^2 - 7x + 3 \\ + & -3x^2 + 2x + 1 \\ \hline & -x^2 - 5x + 4 \end{array}$$

23. $(-2x^3)(2x^2y) = -4x^5y$

24. $3y^2(y^2 - 2y + 3) = 3y^4 - 6y^3 + 9y^2$

25. $(2x - 5)(3x + 4) = 6x^2 + 8x - 15x - 20 = 6x^2 - 7x - 20$

26. $(2x - 3)(x^2 - 2x + 4) = 2x^3 - 4x^2 + 8x - 3x^2 + 6x - 12 = 2x^3 - 7x^2 + 14x - 12$

27.
$$\begin{aligned} (a+2)^2 &= (a-3)^2 \\ (a+2)(a+2) &= (a-3)(a-3) \\ a^2 + 4a + 4 &= a^2 - 6a + 9 \\ 4a + 4 &= -6a + 9 \\ 10a &= 5 \\ a &= \frac{5}{10} = \frac{1}{2} \end{aligned}$$

28. $\dfrac{8x^2y^3z^4}{16x^3y^2z^4} = \dfrac{1}{2}x^{-1}y^1z^0 = \dfrac{y}{2x}$

29. $\dfrac{6a^2 - 12b^2}{24ab} = \dfrac{6a^2}{24ab} - \dfrac{12b^2}{24ab}$
$\qquad = \dfrac{a}{4b} - \dfrac{b}{2a}$

30.
$$\begin{array}{r}
x - 2 \\
2x+3 \overline{\smash{)}\, 2x^2 - x - 6} \\
\underline{2x^2 + 3x} \\
-4x - 6 \\
\underline{-4x - 6} \\
0
\end{array}$$

Cumulative Review Exercises (page 300)

1. $5 + 3 \cdot 2 = 5 + 6 = 11$

2. $3 \cdot 5^2 - 4 = 3 \cdot 25 - 4 = 75 - 4 = 71$

3. $\dfrac{3x - y}{xy} = \dfrac{3(2) - (-5)}{2(-5)} = \dfrac{6 + 5}{-10} = -\dfrac{11}{10}$

4. $\dfrac{x^2 - y^2}{x + y} = \dfrac{2^2 - (-5)^2}{2 + (-5)} = \dfrac{4 - 25}{-3} = \dfrac{-21}{-3} = 7$

5.
$$\begin{aligned}
\tfrac{4}{5}x + 6 &= 18 \\
\tfrac{4}{5}x &= 12 \\
5 \cdot \tfrac{4}{5}x &= 5(12) \\
4x &= 60 \\
x &= 15
\end{aligned}$$

6.
$$\begin{aligned}
x - 2 &= \frac{x+2}{3} \\
3(x - 2) &= 3 \cdot \frac{x+2}{3} \\
3x - 6 &= x + 2 \\
2x - 6 &= 2 \\
2x &= 8 \\
x &= 4
\end{aligned}$$

7.
$$\begin{aligned}
2(5x + 2) &= 3(3x - 2) \\
10x + 4 &= 9x - 6 \\
x + 4 &= -6 \\
x &= -10
\end{aligned}$$

8.
$$\begin{aligned}
4(y + 1) &= -2(4 - y) \\
4y + 4 &= -8 + 2y \\
2y + 4 &= -8 \\
2y &= -12 \\
y &= -6
\end{aligned}$$

9.
$$\begin{aligned}
5x - 3 &> 7 \\
5x &> 10 \\
x &> 2
\end{aligned}$$

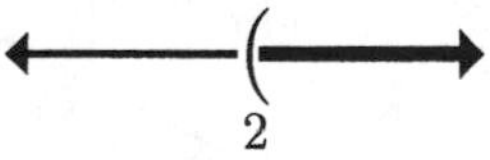

10.
$$\begin{aligned}
7x - 9 &< 5 \\
7x &< 14 \\
x &< 2
\end{aligned}$$

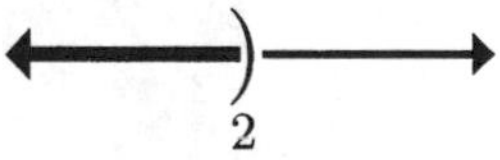

11.
$$\begin{array}{rcl}
-2 < & -x + 3 & < 5 \\
-5 < & -x & < 2 \\
\dfrac{-5}{-1} > & \dfrac{-x}{-1} & > \dfrac{2}{-1} \\
5 > & x & > -2 \\
-2 < & x & < 5
\end{array}$$

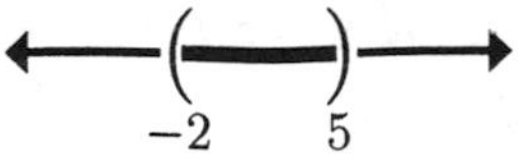

12.
$$\begin{array}{rcl}
0 \le & \dfrac{4 - x}{3} & \le 2 \\
3(0) \le & 3 \cdot \dfrac{4 - x}{3} & \le 3(2) \\
0 \le & 4 - x & \le 6 \\
-4 \le & -x & \le 2 \\
\dfrac{-4}{-1} \ge & \dfrac{-x}{-1} & \ge \dfrac{2}{-1} \\
4 \ge & x & \ge -2 \\
-2 \le & x & \le 4
\end{array}$$

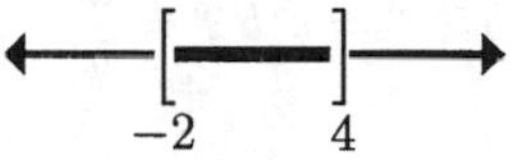

13.
$$\begin{aligned} A &= p + prt \\ A - p &= p - p + prt \\ A - p &= prt \\ \frac{A-p}{pt} &= \frac{prt}{pt} \\ \frac{A-p}{pt} &= r, \text{ or } r = \frac{A-p}{pt} \end{aligned}$$

14.
$$\begin{aligned} A &= \frac{1}{2}bh \\ 2A &= 2 \cdot \frac{1}{2}bh \\ 2A &= bh \\ \frac{2A}{b} &= \frac{bh}{b} \\ \frac{2A}{b} &= h, \text{ or } h = \frac{2A}{b} \end{aligned}$$

15.
$$3x - 4y = 12$$

$x = 0$	$y = 0$
$3(0) - 4y = 12$	$3x - 4(0) = 12$
$0 - 4y = 12$	$3x - 0 = 12$
$-4y = 12$	$3x = 12$
$y = -3$	$x = 4$

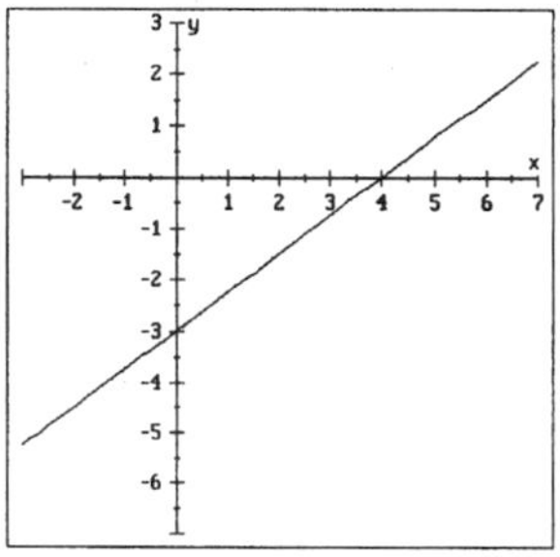

16.
$$y - 2 = \frac{1}{2}(x - 4)$$

$x = 0$	$x = 4$
$y - 2 = \frac{1}{2}(0 - 4)$	$y - 2 = \frac{1}{2}(4 - 4)$
$y - 2 = \frac{1}{2}(-4)$	$y - 2 = \frac{1}{2}(0)$
$y - 2 = -2$	$y - 2 = 0$
$y = 0$	$y = 2$

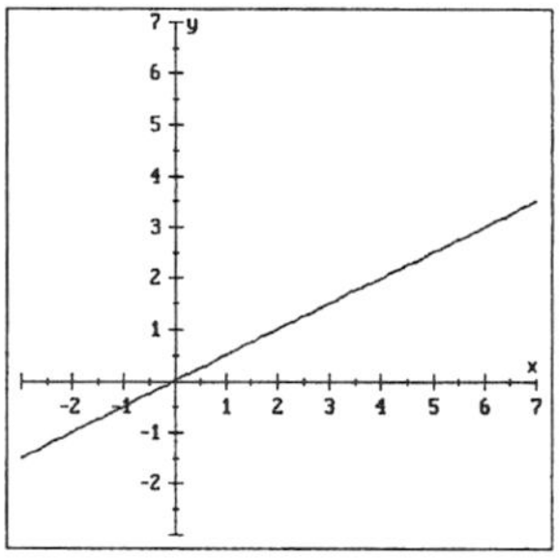

17. $f(0) = 5(0) - 2 = 0 - 2 = -2$

18. $f(3) = 5(3) - 2 = 15 - 2 = 13$

19. $f(-2) = 5(-2) - 2 = -10 - 2 = -12$

20. $f\left(\frac{1}{5}\right) = 5\left(\frac{1}{5}\right) - 2 = 1 - 2 = -1$

21. $(y^3y^5)y^6 = y^8y^6 = y^{14}$

22. $\dfrac{x^3y^4}{x^2y^3} = x^{3-2}y^{4-3} = x^1y^1 = xy$

23. $\dfrac{a^4b^{-3}}{a^{-3}b^3} = a^{4-(-3)}b^{-3-3} = a^7b^{-6} = \dfrac{a^7}{b^6}$

24.
$$\begin{aligned} \left(\frac{-x^{-2}y^3}{x^{-3}y^2}\right)^2 &= \left(-x^{-2-(-3)}y^{3-2}\right)^2 \\ &= \left(-x^1y^1\right)^2 = x^2y^2 \end{aligned}$$

25. $(3x^2 + 2x - 7) - (2x^2 - 2x + 7) = 3x^2 + 2x - 7 - 2x^2 + 2x - 7 = x^2 + 4x - 14$

26. $(3x - 7)(2x + 8) = 6x^2 + 24x - 14x - 56 = 6x^2 + 10x - 56$

27. $(x-2)(x^2+2x+4) = x^3+2x^2+4x-2x^2-4x-8 = x^3-8$

28.
$$\begin{array}{r|l} & 2x + 1 \\ \hline x-3 & 2x^2-5x-3 \\ & \underline{2x^2-6x} \\ & \quad x-3 \\ & \quad \underline{x-3} \\ & \quad\quad 0 \end{array}$$

29. $(1.6\times 10^2)(3\times 10^{16}) = 4.8\times 10^{18}$ m

30.
$$\begin{aligned} A &= 2lw+2wd+2ld \\ 202 &= 2(9)(5)+2(5)d+2(9)d \\ 202 &= 90+28d \\ 112 &= 28d \Rightarrow d = 4 \text{ inches} \end{aligned}$$

31.
$$\begin{aligned} A &= \pi(R+r)(R-r) \\ &= \pi(17+3)(17-3) \\ &= \pi(20)(14) \\ &= 280\pi \approx 879.6 \text{ in.}^2 \end{aligned}$$

32. Let r = the regular price. Then an employee can purchase the TV for $0.75r$.
Purchase price + Sales tax = Total price
$$\begin{aligned} 0.75r + 0.08(0.75r) &= 414.72 \\ 0.81r &= 414.72 \\ \tfrac{0.81r}{0.81} &= \tfrac{414.72}{0.81} \\ r &= 512 \Rightarrow \text{The regular price is \$512.} \end{aligned}$$

Exercise 5.1 (page 309)

1.
$$\begin{aligned} 3x-2(x+1) &= 5 \\ 3x-2x-2 &= 5 \\ x &= 7 \end{aligned}$$

3.
$$\begin{aligned} \frac{2x-7}{5} &= 3 \\ 5\cdot\frac{2x-7}{5} &= 5(3) \\ 2x-7 &= 15 \\ 2x &= 22 \\ x &= 11 \end{aligned}$$

5. prime

7. largest (or greatest)

9. $a+b$

11. $12 = 2\cdot 6 = 2\cdot(2\cdot 3) = 2^2\cdot 3$

13. $15 = 3\cdot 5$

15. $40 = 4\cdot 10 = (2\cdot 2)\cdot(2\cdot 5) = 2^3\cdot 5$

17. $98 = 2\cdot 49 = 2\cdot(7\cdot 7) = 2\cdot 7^2$

19. $225 = 15\cdot 15 = (3\cdot 5)\cdot(3\cdot 5) = 3^2\cdot 5^2$

21. $288 = 24\cdot 12 = (4\cdot 6)\cdot(4\cdot 3) = (2^2\cdot 2\cdot 3)\cdot(2^2\cdot 3) = 2^5\cdot 3^2$

23. $4a+12 = \boxed{4}(a+3)$

25. $r^4+r^2 = r^2(\boxed{r^2}+1)$

27. $4y^2+8y-2xy = 2y(2y+\boxed{4}-\boxed{x})$

29. $a(x+y)+b(x+y) = (x+y)\boxed{(a+b)}$

31. $(r-s)p-(r-s)q = (r-s)\boxed{(p-q)}$

33. $3x+6 = 3(x+2)$

35. $4x - 8 = 4(x - 2)$

37. $xy - xz = x(y - z)$

39. $t^3 + 2t^2 = t^2(t + 2)$

41. $r^4 + r^2 = r^2(r^2 + 1)$

43. $a^3b^3z^3 - a^2b^3z^2 = a^2b^3z^2(az - 1)$

45. $24x^2y^3z^4 + 8xy^2z^3 = 8xy^2z^3(3xyz + 1)$

47. $12uvw^3 - 18uv^2w^2 = 6uvw^2(2w - 3v)$

49. $3x + 3y - 6z = 3(x + y - 2z)$

51. $ab + ac - ad = a(b + c - d)$

53. $4y^2 + 8y - 2xy = 2y(2y + 4 - x)$

55. $12r^2 - 3rs + 9r^2s^2 = 3r(4r - s + 3rs^2)$

57. $abx - ab^2x + abx^2 = abx(1 - b + x)$

59. $4x^2y^2z^2 - 6xy^2z^2 + 12xyz^2 = 2xyz^2(2xy - 3y + 6)$

61. $70a^3b^2c^2 + 49a^2b^3c^3 - 21a^2b^2c^2 = 7a^2b^2c^2(10a + 7bc - 3)$

63. $-x - 2 = -(x + 2)$

65. $-a - b = -(a + b)$

67. $-2x + 5y = -(2x - 5y)$

69. $-2a + 3b = -(2a - 3b)$

71. $-3m - 4n + 1 = -(3m + 4n - 1)$

73. $-3xy + 2z + 5w = -(3xy - 2z - 5w)$

75. $-3ab - 5ac + 9bc = -(3ab + 5ac - 9bc)$

77. $-3x^2y - 6xy^2 = -3xy(x + 2y)$

79. $-4a^2b^3 + 12a^3b^2 = -4a^2b^2(b - 3a)$

81. $-8a^5b^2 - 8a^3b^4 = -8a^3b^2(a^2 + b^2)$

83. $-4a^2b^2c^2 + 14a^2b^2c - 10ab^2c^2 = -2ab^2c(2ac - 7a + 5c)$

85. $-14a^6b^6 + 49a^2b^3 - 21ab = -7ab(2a^5b^5 - 7ab^2 + 3)$

87. $-5a^2b^3c + 15a^3b^4c^2 - 25a^4b^3c = -5a^2b^3c(1 - 3abc + 5a^2)$

89. $(x + y)2 + (x + y)b = (x + y)(2 + b)$

91. $3(x + y) - a(x + y) = (x + y)(3 - a)$

93. $3(r - 2s) - x(r - 2s) = (r - 2s)(3 - x)$

95. $(x - 3)^2 + (x - 3) = (x - 3)(x - 3) + 1(x - 3) = (x - 3)(x - 3 + 1) = (x - 3)(x - 2)$

97. $2x(a^2 + b) + 2y(a^2 + b) = (a^2 + b)(2x + 2y) = 2(a^2 + b)(x + y)$

99. $3x^2(r + 3s) - 6y^2(r + 3s) = (r + 3s)(3x^2 - 6y^2) = 3(r + 3s)(x^2 - 2y^2)$

101. $3x(a + b + c) - 2y(a + b + c) = (a + b + c)(3x - 2y)$

103. $14x^2y(r + 2s - t) - 21xy(r + 2s - t) = (r + 2s - t)(14x^2y - 21xy) = 7xy(r + 2s - t)(2x - 3)$

105. $(x + 3)(x + 1) - y(x + 1) = (x + 1)(x + 3 - y)$

SECTION 5.1

107. $(3x-y)(x^2-2)+(x^2-2)=(3x-y)(x^2-2)+1(x^2-2)=(x^2-2)(3x-y+1)$

109. $2x+2y+ax+ay=2(x+y)+a(x+y)=(x+y)(2+a)$

111. $7r+7s-kr-ks=7(r+s)-k(r+s)=(r+s)(7-k)$

113. $xr+xs+yr+ys=x(r+s)+y(r+s)=(r+s)(x+y)$

115. $2ax+2bx+3a+3b=2x(a+b)+3(a+b)=(a+b)(2x+3)$

117. $2ab+2ac+3b+3c=2a(b+c)+3(b+c)=(b+c)(2a+3)$

119. $2x^2+2xy-3x-3y=2x(x+y)-3(x+y)=(x+y)(2x-3)$

121. $3tv-9tw+uv-3uw=3t(v-3w)+u(v-3w)=(v-3w)(3t+u)$

123. $9mp+3mq-3np-nq=3m(3p+q)-n(3p+q)=(3p+q)(3m-n)$

125. $mp-np-m+n=p(m-n)-1(m-n)=(m-n)(p-1)$

127. $x(a-b)+y(b-a)=x(a-b)-y(a-b)=(a-b)(x-y)$

129. $ax^3+bx^3+2ax^2y+2bx^2y=x^2[ax+bx+2ay+2by]=x^2[x(a+b)+2y(a+b)]$
$=x^2(a+b)(x+2y)$

131. $4a^2b+12a^2-8ab-24a=4a[ab+3a-2b-6]=4a[a(b+3)-2(b+3)]$
$=4a(b+3)(a-2)$

133. $x^3+2x^2+x+2=x^2(x+2)+1(x+2)=(x+2)(x^2+1)$

135. $x^3y-x^2y-xy^2+y^2=y[x^3-x^2-xy+y]=y[x^2(x-1)-y(x-1)]$
$=y(x-1)(x^2-y)$

137. $2r-bs-2s+br=2r-2s+br-bs=2(r-s)+b(r-s)=(r-s)(2+b)$

139. $ax+by+bx+ay=ax+ay+bx+by=a(x+y)+b(x+y)=(x+y)(a+b)$

141. $ac+bd-ad-bc=ac-bc-ad+bd=c(a-b)-d(a-b)=(a-b)(c-d)$

143. $ar^2-brs+ars-br^2=r[ar-bs+as-br]=r[ar-br+as-bs]=r[r(a-b)+s(a-b)]$
$=r(a-b)(r+s)$

145. $ba+3+a+3b=ba+a+3b+3=a(b+1)+3(b+1)=(b+1)(a+3)$

147. $pr+qs-ps-qr=pr-qr-ps+qs=r(p-q)-s(p-q)=(p-q)(r-s)$

149. Answers may vary.

151. Answers may vary.

153. $ax + ay + bx + by = a(x+y) + b(x+y) = (x+y)(a+b)$
$ax + bx + ay + by = x(a+b) + y(a+b) = (a+b)(x+y)$

Exercise 5.2 (page 315)

1. $$\frac{p}{w} + \frac{v^2}{2g} + h = k$$
$$\frac{p}{w} = k - h - \frac{v^2}{2g}$$
$$w \cdot \frac{p}{w} = w\left(k - h - \frac{v^2}{2g}\right)$$
$$p = w\left(k - h - \frac{v^2}{2g}\right)$$

3. difference of two squares

5. $(p-q)$

7. $x^2 - 9 = (x+3)\boxed{(x-3)}$

9. $4m^2 - 9n^2 = (2m+3n)\boxed{(2m-3n)}$

11. $x^2 - 16 = x^2 - 4^2 = (x+4)(x-4)$

13. $y^2 - 49 = y^2 - 7^2 = (y+7)(y-7)$

15. $4y^2 - 49 = (2y)^2 - 7^2 = (2y+7)(2y-7)$

17. $9x^2 - y^2 = (3x)^2 - y^2 = (3x+y)(3x-y)$

19. $25t^2 - 36u^2 = (5t)^2 - (6u)^2$
$= (5t+6u)(5t-6u)$

21. $16a^2 - 25b^2 = (4a)^2 - (5b)^2$
$= (4a+5b)(4a-5b)$

23. $a^2 + b^2$: sum of squares $\Rightarrow$ prime

25. $a^4 - 4b^2 = (a^2)^2 - (2b)^2$
$= (a^2+2b)(a^2-2b)$

27. $49y^2 - 225z^4 = (7y)^2 - (15z^2)^2 = (7y+15z^2)(7y-15z^2)$

29. $196x^4 - 169y^2 = (14x^2)^2 - (13y)^2 = (14x^2+13y)(14x^2-13y)$

31. $8x^2 - 32y^2 = 8[x^2 - 4y^2] = 8[x^2 - (2y)^2] = 8(x+2y)(x-2y)$

33. $2a^2 - 8y^2 = 2[a^2 - 4y^2] = 2[a^2 - (2y)^2] = 2(a+2y)(a-2y)$

35. $3r^2 - 12s^2 = 3[r^2 - 4s^2] = 3[r^2 - (2s)^2] = 3(r+2s)(r-2s)$

37. $x^3 - xy^2 = x(x^2 - y^2) = x(x+y)(x-y)$

39. $4a^2x - 9b^2x = x[4a^2 - 9b^2] = x[(2a)^2 - (3b)^2] = x(2a+3b)(2a-3b)$

41. $3m^3 - 3mn^2 = 3m(m^2 - n^2) = 3m(m+n)(m-n)$

SECTION 5.2

43. $4x^4 - x^2y^2 = x^2(4x^2 - y^2) = x^2\left[(2x)^2 - y^2\right] = x^2(2x+y)(2x-y)$

45. $2a^3b - 242ab^3 = 2ab[a^2 - 121b^2] = 2ab\left[a^2 - (11b)^2\right] = 2ab(a+11b)(a-11b)$

47. $x^4 - 81 = \left(x^2+9\right)\left(x^2-9\right) = \left(x^2+9\right)(x+3)(x-3)$

49. $a^4 - 16 = \left(a^2+4\right)\left(a^2-4\right) = \left(a^2+4\right)(a+2)(a-2)$

51. $a^4 - b^4 = (a^2+b^2)(a^2-b^2) = (a^2+b^2)(a+b)(a-b)$

53. $81r^4 - 256s^4 = (9r^2+16s^2)(9r^2-16s^2) = (9r^2+16s^2)(3r+4s)(3r-4s)$

55. $a^4 - b^8 = (a^2+b^4)(a^2-b^4) = (a^2+b^4)(a+b^2)(a-b^2)$

57. $x^8 - y^8 = (x^4+y^4)(x^4-y^4) = (x^4+y^4)(x^2+y^2)(x^2-y^2) = (x^4+y^4)(x^2+y^2)(x+y)(x-y)$

59. $2x^4 - 2y^4 = 2(x^4-y^4) = 2(x^2+y^2)(x^2-y^2) = 2(x^2+y^2)(x+y)(x-y)$

61. $a^4b - b^5 = b(a^4-b^4) = b(a^2+b^2)(a^2-b^2) = b(a^2+b^2)(a+b)(a-b)$

63. $48m^4n - 243n^5 = 3n\left(16m^4 - 81n^4\right) = 3n\left(4m^2+9n^2\right)\left(4m^2-9n^2\right) = 3n\left(4m^2+9n^2\right)(2m+3n)(2m-3n)$

65. $3a^5y + 6ay^5 = 3ay(a^4+2y^4)$

67. $3a^{10} - 3a^2b^4 = 3a^2(a^8-b^4) = 3a^2(a^4+b^2)(a^4-b^2) = 3a^2(a^4+b^2)(a^2+b)(a^2-b)$

69. $2x^8y^2 - 32y^6 = 2y^2(x^8-16y^4) = 2y^2(x^4+4y^2)(x^4-4y^2) = 2y^2(x^4+4y^2)(x^2+2y)(x^2-2y)$

71. $a^6b^2 - a^2b^6c^4 = a^2b^2\left(a^4 - b^4c^4\right) = a^2b^2\left(a^2+b^2c^2\right)\left(a^2-b^2c^2\right) = a^2b^2\left(a^2+b^2c^2\right)(a+bc)(a-bc)$

73. $a^2b^7 - 625a^2b^3 = a^2b^3(b^4-625) = a^2b^3(b^2+25)(b^2-25) = a^2b^3(b^2+25)(b+5)(b-5)$

75. $243r^5s - 48rs^5 = 3rs\left(81r^4 - 16s^4\right) = 3rs\left(9r^2+4s^2\right)\left(9r^2-4s^2\right) = 3rs\left(9r^2+4s^2\right)(3r+2s)(3r-2s)$

77. $16(x-y)^2 - 9 = [4(x-y)]^2 - 3^2 = [4(x-y)+3][4(x-y)-3] = (4x-4y+3)(4x-4y-3)$

79. $a^3 - 9a + 3a^2 - 27 = a(a^2-9) + 3(a^2-9) = (a^2-9)(a+3) = (a+3)(a-3)(a+3)$

81. $y^3 - 16y - 3y^2 + 48 = y(y^2-16) - 3(y^2-16) = (y^2-16)(y-3) = (y+4)(y-4)(y-3)$

83. $3x^3 - 12x + 3x^2 - 12 = 3\left[x^3 - 4x + x^2 - 4\right] = 3\left[x(x^2-4) + 1(x^2-4)\right] = 3\left(x^2-4\right)(x+1) = 3(x+2)(x-2)(x+1)$

85. $3m^3 - 3mn^2 + 3am^2 - 3an^2 = 3\left[m^3 - mn^2 + am^2 - an^2\right] = 3\left[m\left(m^2 - n^2\right) + a\left(m^2 - n^2\right)\right]$
$= 3\left(m^2 - n^2\right)(m + a)$
$= 3(m + n)(m - n)(m + a)$

87. $2m^3n^2 - 32mn^2 + 8m^2 - 128 = 2\left[m^3n^2 - 16mn^2 + 4m^2 - 64\right]$
$= 2\left[mn^2\left(m^2 - 16\right) + 4\left(m^2 - 16\right)\right]$
$= 2\left(m^2 - 16\right)\left(mn^2 + 4\right) = 2(m + 4)(m - 4)\left(mn^2 + 4\right)$

89. **Answers may vary.**

91. $399 \cdot 401 = (400 - 1)(400 + 1) = 400^2 - 1^2 = 160{,}000 - 1 = 159{,}999$

Exercise 5.3 (page 324)

1. $x - 3 > 5$
$x > 8$

3. $-3x - 5 \geq 4$
$-3x \geq 9$
$\frac{-3x}{-3} \leq \frac{9}{-3}$
$x \leq -3$

5. $\frac{3(x - 1)}{4} < 12$
$4 \cdot \frac{3x - 3}{4} < 4(12)$
$3x - 3 < 48$
$3x < 51$
$x < 17$

7. $-2 < x \leq 4$

9. $x^2 + 2xy + y^2 = (x + y)^2$

11. $y^2 + 6y + 8 = \left(y + \boxed{4}\right)\left(y + \boxed{2}\right)$

13. $x^2 - xy - 2y^2 = \left(x + \boxed{y}\right)\left(x - \boxed{2y}\right)$

15. $x^2 + 3x + 2 = (x + 2)(x + 1)$

17. $z^2 + 12z + 11 = (z + 1)(z + 11)$

19. $a^2 - 4a - 5 = (a - 5)(a + 1)$

21. $t^2 - 9t + 14 = (t - 2)(t - 7)$

23. $u^2 + 10u + 15$: prime

25. $y^2 - y - 30 = (y - 6)(y + 5)$

27. $a^2 + 6a - 16 = (a + 8)(a - 2)$

29. $t^2 - 5t - 50 = (t - 10)(t + 5)$

31. $r^2 - 9r - 12$: prime

33. $y^2 + 2yz + z^2 = (y + z)(y + z)$

35. $x^2 + 4xy + 4y^2 = (x + 2y)(x + 2y)$

37. $m^2 + 3mn - 10n^2 = (m + 5n)(m - 2n)$

39. $a^2 - 4ab - 12b^2 = (a - 6b)(a + 2b)$

41. $u^2 + 2uv - 15v^2 = (u + 5v)(u - 3v)$

43. $$\begin{aligned} -x^2 - 7x - 10 &= -(x^2 + 7x + 10) \\ &= -(x + 5)(x + 2) \end{aligned}$$

45. $$\begin{aligned} -y^2 - 2y + 15 &= -(y^2 + 2y - 15) \\ &= -(y + 5)(y - 3) \end{aligned}$$

47. $$\begin{aligned} -t^2 - 15t + 34 &= -(t^2 + 15t - 34) \\ &= -(t + 17)(t - 2) \end{aligned}$$

49. $$\begin{aligned} -r^2 + 14r - 40 &= -(r^2 - 14r + 40) \\ &= -(r - 4)(r - 10) \end{aligned}$$

51. $$\begin{aligned} -a^2 - 4ab - 3b^2 &= -(a^2 + 4ab + 3b^2) \\ &= -(a + 3b)(a + b) \end{aligned}$$

53. $$\begin{aligned} -x^2 + 6xy + 7y^2 &= -(x^2 - 6xy - 7y^2) \\ &= -(x - 7y)(x + y) \end{aligned}$$

55. $$\begin{aligned} 4 - 5x + x^2 &= x^2 - 5x + 4 \\ &= (x - 4)(x - 1) \end{aligned}$$

57. $$\begin{aligned} 10y + 9 + y^2 &= y^2 + 10y + 9 \\ &= (y + 9)(y + 1) \end{aligned}$$

59. $$\begin{aligned} c^2 - 5 + 4c &= c^2 + 4c - 5 \\ &= (c + 5)(c - 1) \end{aligned}$$

61. $$\begin{aligned} -r^2 + 2s^2 + rs &= -r^2 + rs + 2s^2 \\ &= -(r^2 - rs - 2s^2) \\ &= -(r - 2s)(r + s) \end{aligned}$$

63. $$\begin{aligned} 4rx + r^2 + 3x^2 &= r^2 + 4rx + 3x^2 \\ &= (r + 3x)(r + x) \end{aligned}$$

65. $$\begin{aligned} -3ab + a^2 + 2b^2 &= a^2 - 3ab + 2b^2 \\ &= (a - 2b)(a - b) \end{aligned}$$

67. $$\begin{aligned} 2x^2 + 10x + 12 &= 2(x^2 + 5x + 6) \\ &= 2(x + 2)(x + 3) \end{aligned}$$

69. $$\begin{aligned} 3y^3 + 6y^2 + 3y &= 3y(y^2 + 2y + 1) \\ &= 3y(y + 1)(y + 1) \end{aligned}$$

71. $$\begin{aligned} -5a^2 + 25a - 30 &= -5(a^2 - 5a + 6) \\ &= -5(a - 3)(a - 2) \end{aligned}$$

73. $3z^2 - 15tz + 12t^2 = 3(z^2 - 5tz + 4t^2) = 3(z - 4t)(z - t)$

75. $12xy + 4x^2y - 72y = 4y(3x + x^2 - 18) = 4y(x^2 + 3x - 18) = 4y(x + 6)(x - 3)$

77. $-4x^2y - 4x^3 + 24xy^2 = -4x(xy + x^2 - 6y^2) = -4x(x^2 + xy - 6y^2) = -4x(x + 3y)(x - 2y)$

79. $x^2 + 4x + 4 - y^2 = (x + 2)(x + 2) - y^2 = (x + 2)^2 - y^2 = (x + 2 + y)(x + 2 - y)$

81. $b^2 - 6b + 9 - c^2 = (b - 3)(b - 3) - c^2 = (b - 3)^2 - c^2 = (b - 3 + c)(b - 3 - c)$

83. $a^2 + 2ab + b^2 - 4 = (a + b)(a + b) - 4 = (a + b)^2 - 2^2 = (a + b + 2)(a + b - 2)$

85. $$\begin{aligned} b^2 - y^2 - 4y - 4 = b^2 - (y^2 + 4y + 4) = b^2 - (y + 2)(y + 2) &= b^2 - (y + 2)^2 \\ &= [b + (y + 2)][b - (y + 2)] \\ &= (b + y + 2)(b - y - 2) \end{aligned}$$

87. $x^2 + 6x + 9 = (x+3)(x+3)$

89. $y^2 - 8y + 16 = (y-4)(y-4)$

91. $t^2 + 20t + 100 = (t+10)(t+10)$

93. $u^2 - 18u + 81 = (u-9)(u-9)$

95. $x^2 + 4xy + 4y^2 = (x+2y)(x+2y)$

97. $r^2 - 10rs + 25s^2 = (r-5s)(r-5s)$

99. **Answers may vary.**

101. Both answers check. Both may be factored more completely:
$(2x+6)(x+7) = 2(x+3)(x+7)$
$(x+3)(2x+14) = (x+3)\ \ 2(x+7)$

Exercise 5.4 (page 332)

1.
$$
\begin{aligned}
l &= f + (n-1)d \\
l &= f + nd - d \\
l - f + d &= nd \\
\frac{l - f + d}{d} &= \frac{nd}{d} \\
\frac{l - f + d}{d} &= n
\end{aligned}
$$

3. descending

5. opposites

7. $6x^2 + 7x + 2 = (2x+1)(3x+\boxed{2})$

9. $6x^2 + x - 2 = (3x+\boxed{2})(2x-\boxed{1})$

11. $12x^2 - 7xy + y^2 = (3x-\boxed{y})(4x-\boxed{y})$

13. $2a^2 + 5a + 3 = (2a+3)(a+1)$

15. $2x^2 - 3x + 1 = (2x-1)(x-1)$

17. $3a^2 + 13a + 4 = (3a+1)(a+4)$

19. $4z^2 + 13z + 3 = (z+3)(4z+1)$

21. $6y^2 + 7y + 2 = (3y+2)(2y+1)$

23. $6x^2 - 7x + 2 = (2x-1)(3x-2)$

25. $3a^2 - 4a - 4 = (3a+2)(a-2)$

27. $2x^2 - 3x - 2 = (2x+1)(x-2)$

29. $2m^2 + 5m - 12 = (2m-3)(m+4)$

31. $10y^2 - 3y - 1 = (5y+1)(2y-1)$

33. $12y^2 - 5y - 2 = (4y+1)(3y-2)$

35. $5t^2 + 13t + 6 = (5t+3)(t+2)$

37. $16m^2 - 14m + 3 = (8m-3)(2m-1)$

39. $3x^2 - 4xy + y^2 = (3x-y)(x-y)$

41. $2u^2 + uv - 3v^2 = (2u+3v)(u-v)$

43. $4a^2 - 4ab + b^2 = (2a-b)(2a-b)$

45. $6r^2 + rs - 2s^2 = (2r-s)(3r+2s)$

47. $4x^2 + 8xy + 3y^2 = (2x+y)(2x+3y)$

49. $4a^2 - 15ab + 9b^2 = (4a-3b)(a-3b)$

51.
$$
\begin{aligned}
-13x + 3x^2 - 10 &= 3x^2 - 13x - 10 \\
&= (3x+2)(x-5)
\end{aligned}
$$

53.
$$
\begin{aligned}
15 + 8a^2 - 26a &= 8a^2 - 26a + 15 \\
&= (4a-3)(2a-5)
\end{aligned}
$$

55. $12y^2 + 12 - 25y = 12y^2 - 25y + 12$
$= (3y-4)(4y-3)$

57. $3x^2 + 6 + x = 3x^2 + x + 6 \Rightarrow$ prime

59. $2a^2 + 3b^2 + 5ab = 2a^2 + 5ab + 3b^2$
$= (2a+3b)(a+b)$

61. $pq + 6p^2 - q^2 = 6p^2 + pq - q^2$
$= (2p+q)(3p-q)$

63. $b^2 + 4a^2 + 16ab = 4a^2 + 16ab + b^2$
PRIME

65. $12x^2 + 10y^2 - 23xy = 12x^2 - 23xy + 10y^2$
$= (3x-2y)(4x-5y)$

67. $-21mn - 10n^2 + 10m^2 = 10m^2 - 21mn - 10n^2 = (5m+2n)(2m-5n)$

69. $4x^2 + 10x - 6 = 2(2x^2 + 5x - 3)$
$= 2(2x-1)(x+3)$

71. $y^3 + 13y^2 + 12y = y(y^2 + 13y + 12)$
$= y(y+12)(y+1)$

73. $6x^3 - 15x^2 - 9x = 3x(2x^2 - 5x - 3)$
$= 3x(2x+1)(x-3)$

75. $30r^5 + 63r^4 - 30r^3 = 3r^3(10r^2 + 21r - 10)$
$= 3r^3(5r-2)(2r+5)$

77. $4a^2 - 4ab - 8b^2 = 4(a^2 - ab - 2b^2)$
$= 4(a-2b)(a+b)$

79. $8x^2 - 12xy - 8y^2 = 4(2x^2 - 3xy - 2y^2)$
$= 4(2x+y)(x-2y)$

81. $-16m^3n - 20m^2n^2 - 6mn^3 = -2mn(8m^2 + 10mn + 3n^2) = -2mn(4m+3n)(2m+n)$

83. $-28u^3v^3 + 26u^2v^4 - 6uv^5 = -2uv^3(14u^2 - 13uv + 3v^2) = -2uv^3(7u-3v)(2u-v)$

85. $4x^2 + 12x + 9 = (2x+3)(2x+3)$
$= (2x+3)^2$

87. $9x^2 + 12x + 4 = (3x+2)(3x+2)$
$= (3x+2)^2$

89. $16x^2 - 8xy + y^2 = (4x-y)(4x-y) = (4x-y)^2$

91. $4x^2 + 4xy + y^2 - 16 = (4x^2 + 4xy + y^2) - 16 = (2x+y)(2x+y) - 16$
$= (2x+y)^2 - 4^2 = (2x+y+4)(2x+y-4)$

93. $9 - a^2 - 4ab - 4b^2 = 9 - (a^2 + 4ab + 4b^2) = 9 - (a+2b)(a+2b)$
$= 3^2 - (a+2b)^2$
$= (3 + (a+2b))(3 - (a+2b))$
$= (3+a+2b)(3-a-2b)$

95. $9p^2 + 1 + 6p - q^2 = 9p^2 + 6p + 1 - q^2 = (3p+1)(3p+1) - q^2 = (3p+1)^2 - q^2$
$= (3p+1+q)(3p+1-q)$

97. **Answers may vary.**

99. $(6x+1)(x+6) = 6x^2+37x+6$; $(6x-1)(x-6) = 6x^2-37x+6 \Rightarrow b = \pm 37$
$(6x+2)(x+3) = 6x^2+20x+6$; $(6x-2)(x-3) = 6x^2-20x+6 \Rightarrow b = \pm 20$
$(6x+3)(x+2) = 6x^2+15x+6$; $(6x-3)(x-2) = 6x^2-15x+6 \Rightarrow b = \pm 15$
$(6x+6)(x+1) = 6x^2+12x+6$; $(6x-6)(x-1) = 6x^2-12x+6 \Rightarrow b = \pm 12$
$(2x+1)(3x+6) = 6x^2+15x+6$; $(2x-1)(3x-6) = 6x^2-15x+6 \Rightarrow b = \pm 15$
$(2x+2)(3x+3) = 6x^2+12x+6$; $(2x-2)(3x-3) = 6x^2-12x+6 \Rightarrow b = \pm 12$
$(2x+3)(3x+2) = 6x^2+13x+6$; $(2x-3)(3x-2) = 6x^2-13x+6 \Rightarrow b = \pm 13$
$(2x+6)(3x+1) = 6x^2+20x+6$; $(2x-6)(3x-1) = 6x^2-20x+6 \Rightarrow b = \pm 20$
$b = \pm 12, \pm 13, \pm 15, \pm 20, \pm 37$

Exercise 5.5 (page 337)

1. $1 \times 10^{-13} = 0.0000000000001$ cm

3. $x^3+y^3 = (x+y)\boxed{(x^2-xy+y^2)}$

5. $$\begin{aligned} y^3+1 &= y^3+1^3 \\ &= (y+1)\left(y^2-1y+1^2\right) \\ &= (y+1)\left(y^2-y+1\right) \end{aligned}$$

7. $$\begin{aligned} a^3-27 &= a^3-3^3 \\ &= (a-3)\left(a^2+3a+3^2\right) \\ &= (a-3)\left(a^2+3a+9\right) \end{aligned}$$

9. $$\begin{aligned} 8+x^3 &= 2^3+x^3 \\ &= (2+x)\left(2^2-2x+x^2\right) \\ &= (2+x)\left(4-2x+x^2\right) \end{aligned}$$

11. $$\begin{aligned} 64-z^3 &= 4^3-z^3 \\ &= (4-z)\left(4^2+4z+z^2\right) \\ &= (4-z)\left(16+4z+z^2\right) \end{aligned}$$

13. $s^3-t^3 = (s-t)(s^2+st+t^2)$

15. $125p^3-q^3 = (5p)^3-q^3 = (5p-q)\left((5p)^2+5pq+q^2\right) = (5p-q)(25p^2+5pq+q^2)$

17. $$\begin{aligned} 27x^3+y^3 &= (3x)^3+y^3 \\ &= (3x+y)\left((3x)^2-3xy+y^2\right) \\ &= (3x+y)\left(9x^2-3xy+y^2\right) \end{aligned}$$

19. $$\begin{aligned} a^3+8b^3 &= a^3+(2b)^3 \\ &= (a+2b)\left(a^2-2ab+(2b)^2\right) \\ &= (a+2b)\left(a^2-2ab+4b^2\right) \end{aligned}$$

21. $64x^3-27 = (4x)^3-3^3 = (4x-3)\left((4x)^2+3(4x)+3^2\right) = (4x-3)(16x^2+12x+9)$

23. $$\begin{aligned} 27x^3-125y^3 = (3x)^3-(5y)^3 &= (3x-5y)\left((3x)^2+(3x)(5y)+(5y)^2\right) \\ &= (3x-5y)\left(9x^2+15xy+25y^2\right) \end{aligned}$$

25. $a^6-b^3 = (a^2)^3-b^3 = (a^2-b)\left((a^2)^2+a^2b+b^2\right) = (a^2-b)(a^4+a^2b+b^2)$

27. $x^9+y^6 = (x^3)^3+(y^2)^3 = (x^3+y^2)\left((x^3)^2-x^3y^2+(y^2)^2\right) = (x^3+y^2)(x^6-x^3y^2+y^4)$

29. $2x^3+54 = 2(x^3+27) = 2(x^3+3^3) = 2(x+3)(x^2-3x+3^2) = 2(x+3)(x^2-3x+9)$

SECTION 5.5

31. $-x^3 + 216 = -(x^3 - 216) = -(x^3 - 6^3) = -(x-6)(x^2 + 6x + 6^2) = -(x-6)(x^2 + 6x + 36)$

33. $$\begin{aligned} 64m^3x - 8n^3x = 8x(8m^3 - n^3) = 8x((2m)^3 - n^3) &= 8x(2m-n)((2m)^2 + 2mn + n^2) \\ &= 8x(2m-n)(4m^2 + 2mn + n^2) \end{aligned}$$

35. $$\begin{aligned} x^4y + 216xy^4 = xy(x^3 + 216y^3) = xy(x^3 + (6y)^3) &= xy(x+6y)(x^2 - 6xy + (6y)^2) \\ &= xy(x+6y)(x^2 - 6xy + 36y^2) \end{aligned}$$

37. $$\begin{aligned} 81r^4s^2 - 24rs^5 = 3rs^2(27r^3 - 8s^3) &= 3rs^2((3r)^3 - (2s)^3) \\ &= 3rs^2(3r-2s)((3r)^2 + (3r)(2s) + (2s)^2) \\ &= 3rs^2(3r-2s)(9r^2 + 6rs + 4s^2) \end{aligned}$$

39. $$\begin{aligned} 125a^6b^2 + 64a^3b^5 = a^3b^2(125a^3 + 64b^3) &= a^3b^2((5a)^3 + (4b)^3) \\ &= a^3b^2(5a+4b)((5a)^2 - (5a)(4b) + (4b)^2) \\ &= a^3b^2(5a+4b)(25a^2 - 20ab + 16b^2) \end{aligned}$$

41. $$\begin{aligned} y^7z - yz^4 = yz(y^6 - z^3) = yz\left((y^2)^3 - z^3\right) &= yz(y^2 - z)((y^2) + y^2z + z^2) \\ &= yz(y^2 - z)(y^4 + y^2z + z^2) \end{aligned}$$

43. $$\begin{aligned} 2mp^4 + 16mpq^3 = 2mp(p^3 + 8q^3) = 2mp(p^3 + (2q)^3) &= 2mp(p+2q)(p^2 - 2pq + (2q)^2) \\ &= 2mp(p+2q)(p^2 - 2pq + 4q^2) \end{aligned}$$

45. $$\begin{aligned} x^6 - 1 = (x^3)^2 - 1^2 = (x^3+1)(x^3-1) &= (x^3 + 1^3)(x^3 - 1^3) \\ &= (x+1)(x^2 - 1x + 1^2)(x-1)(x^2 + 1x + 1^2) \\ &= (x+1)(x^2 - x + 1)(x-1)(x^2 + x + 1) \end{aligned}$$

47. $$\begin{aligned} x^{12} - y^6 = (x^6)^2 - (y^3)^2 &= (x^6 + y^3)(x^6 - y^3) \\ &= \left((x^2)^3 + y^3\right)\left((x^2)^3 - y^3\right) \\ &= (x^2 + y)\left((x^2)^2 - x^2y + y^2\right)(x^2 - y)\left((x^2)^2 + x^2y + y^2\right) \\ &= (x^2 + y)(x^4 - x^2y + y^2)(x^2 - y)(x^4 + x^2y + y^2) \end{aligned}$$

49. $3(x^3 + y^3) - z(x^3 + y^3) = (x^3 + y^3)(3 - z) = (x+y)(x^2 - xy + y^2)(3 - z)$

51. $$\begin{aligned} (m^3 + 8n^3) + (m^3x + 8n^3x) &= 1(m^3 + 8n^3) + x(m^3 + 8n^3) \\ &= (m^3 + 8n^3)(1 + x) \\ &= (m+2n)(m^2 - 2mn + (2n)^2)(1+x) \\ &= (m+2n)(m^2 - 2mn + 4n^2)(1+x) \end{aligned}$$

53. $(a^4+27a)-(a^3b+27b)=a(a^3+27)-b(a^3+27)=(a^3+27)(a-b)$
$=(a^3+3^3)(a-b)$
$=(a+3)(a^2-3a+3^2)(a-b)$
$=(a+3)(a^2-3a+9)(a-b)$

55. $y^3(y^2-1)-27(y^2-1)=(y^2-1)(y^3-27)=(y^2-1^2)(y^3-3^3)$
$=(y+1)(y-1)(y-3)(y^2+3y+9)$

57. **Answers may vary.**

59. $a^3-b^3=11^3-7^3=1331-343=\boxed{988}$
$(a-b)(a^2+ab+b^2)=(11-7)(11^2+11\cdot 7+7^2)=4(121+77+49)=4(247)=\boxed{988}$

Exercise 5.6 (page 341)

1. $2(t-5)+t=3(2-t)$
$2t-10+t=6-3t$
$3t-10=6-3t$
$6t=16$
$t=\frac{16}{6}=\frac{8}{3}$

3. $5-3(t+1)=t+2$
$5-3t-3=t+2$
$-3t+2=t+2$
$-4t=0$
$t=\frac{0}{-4}=0$

5. factors

7. binomials

9. $6x+3=3(2x+1)$

11. $x^2-6x-7=(x-7)(x+1)$

13. $6t^2+7t-3=(2t+3)(3t-1)$

15. $4x^2-25=(2x)^2-5^2=(2x+5)(2x-5)$

17. $t^2-2t+1=(t-1)(t-1)$

19. $a^3-8=a^3-2^3=(a-2)(a^2+2a+2^2)$
$=(a-2)(a^2+2a+4)$

21. $x^2y^2-2x^2-y^2+2=x^2(y^2-2)-1(y^2-2)=(y^2-2)(x^2-1)=(y^2-2)(x^2-1^2)$
$=(y^2-2)(x+1)(x-1)$

23. $70p^4q^3-35p^4q^2+49p^5q^2=7p^4q^2(10q-5+7p)$

25. $2ab^2+8ab-24a=2a(b^2+4b-12)$
$=2a(b+6)(b-2)$

27. $-8p^3q^7-4p^2q^3=-4p^2q^3(2pq^4+1)$

29. $4a^2-4ab+b^2-9=(4a^2-4ab+b^2)-9=(2a-b)^2-3^2=(2a-b+3)(2a-b-3)$

31. $x^2+7x+1\Rightarrow$ prime

33. $-2x^5 + 128x^2 = -2x^2(x^3 - 64) = -2x^2(x^3 - 4^3) = -2x^2(x-4)(x^2+4x+16)$

35. $14t^3 - 40t^2 + 6t^4 = 2t^2(7t - 20 + 3t^2) = 2t^2(3t^2 + 7t - 20) = 2t^2(3t-5)(t+4)$

37. $a^2(x-a) - b^2(x-a) = (x-a)(a^2-b^2) = (x-a)(a+b)(a-b)$

39. $$\begin{aligned} 8p^6 - 27q^6 &= \left(2p^2\right)^3 - \left(3q^2\right)^3 = \left(2p^2 - 3q^2\right)\left(\left(2p^2\right)^2 + \left(2p^2\right)\left(3q^2\right) + \left(3q^2\right)^2\right) \\ &= \left(2p^2 - 3q^2\right)\left(4p^4 + 6p^2q^2 + 9q^4\right) \end{aligned}$$

41. $$\begin{aligned} 125p^3 - 64y^3 &= (5p)^3 - (4y)^3 = (5p - 4y)\left((5p)^2 + (5p)(4y) + (4y)^2\right) \\ &= (5p - 4y)\left(25p^2 + 20py + 16y^2\right) \end{aligned}$$

43. $-16x^4y^2z + 24x^5y^3z^4 - 15x^2y^3z^7 = -x^2y^2z(16x^2 - 24x^3yz^3 + 15yz^6)$

45. $81p^4 - 16q^4 = \left(9p^2\right)^2 - \left(4q^2\right)^2 = \left(9p^2 + 4q^2\right)\left(9p^2 - 4q^2\right) = \left(9p^2 + 4q^2\right)(3p + 2q)(3p - 2q)$

47. $4x^2 + 9y^2 \Rightarrow$ prime

49. $$\begin{aligned} 54x^3 + 250y^6 = 2\left(27x^3 + 125y^6\right) &= 2\left((3x)^3 + \left(5y^2\right)^3\right) \\ &= 2\left(3x + 5y^2\right)\left((3x)^2 - (3x)\left(5y^2\right) + \left(5y^2\right)^2\right) \\ &= 2\left(3x + 5y^2\right)\left(9x^2 - 15xy^2 + 25y^4\right) \end{aligned}$$

51. $10r^2 - 13r - 4 \Rightarrow$ prime

53. $$\begin{aligned} 21t^3 - 10t^2 + t &= t\left(21t^2 - 10t + 1\right) \\ &= t(3t-1)(7t-1) \end{aligned}$$

55. $$\begin{aligned} x^5 - x^3y^2 + x^2y^3 - y^5 = x^3\left(x^2 - y^2\right) + y^3\left(x^2 - y^2\right) &= \left(x^2 - y^2\right)\left(x^3 + y^3\right) \\ &= (x+y)(x-y)(x+y)\left(x^2 - xy + y^2\right) \end{aligned}$$

57. $$\begin{aligned} 2a^2c - 2b^2c + 4a^2d - 4b^2d &= 2\left(a^2c - b^2c + 2a^2d - 2b^2d\right) \\ &= 2\left(c\left(a^2 - b^2\right) + 2d\left(a^2 - b^2\right)\right) \\ &= 2\left(a^2 - b^2\right)(c + 2d) = 2(a+b)(a-b)(c+2d) \end{aligned}$$

59. Answers may vary.

61. Answers may vary.

Exercise 5.7 (page 347)

1. $u^3u^2u^4 = u^{3+2+4} = u^9$

3. $\dfrac{a^3b^4}{a^2b^5} = a^1b^{-1} = \dfrac{a}{b}$

5. quadratic

7. second

9.
$$\begin{aligned} (x-2)(x+3)&=0 \\ x-2=0 \quad \textbf{or} \quad x+3&=0 \\ x=2 \qquad\qquad x&=-3 \end{aligned}$$

11.
$$\begin{aligned} (x-4)(x+1)&=0 \\ x-4=0 \quad \textbf{or} \quad x+1&=0 \\ x=4 \qquad\qquad x&=-1 \end{aligned}$$

13.
$$\begin{aligned} (2x-5)(3x+6)&=0 \\ 2x-5=0 \quad \textbf{or} \quad 3x+6&=0 \\ 2x=5 \qquad\qquad 3x&=-6 \\ x=\tfrac{5}{2} \qquad\qquad x&=-2 \end{aligned}$$

15.
$$\begin{aligned} (x-1)(x+2)(x-3)&=0 \\ x-1=0 \quad \textbf{or} \quad x+2=0 \quad \textbf{or} \quad x-3&=0 \\ x=1 \qquad\qquad x=-2 \qquad\qquad x&=3 \end{aligned}$$

17.
$$\begin{aligned} x^2-3x&=0 \\ x(x-3)&=0 \\ x=0 \quad \textbf{or} \quad x-3&=0 \\ x=0 \qquad\qquad x&=3 \end{aligned}$$

19.
$$\begin{aligned} 2x^2-5x&=0 \\ x(2x-5)&=0 \\ x=0 \quad \textbf{or} \quad 2x-5&=0 \\ x=0 \qquad\qquad 2x&=5 \\ x=0 \qquad\qquad x&=\tfrac{5}{2} \end{aligned}$$

21.
$$\begin{aligned} x^2-7x&=0 \\ x(x-7)&=0 \\ x=0 \quad \textbf{or} \quad x-7&=0 \\ x=0 \qquad\qquad x&=7 \end{aligned}$$

23.
$$\begin{aligned} 3x^2+8x&=0 \\ x(3x+8)&=0 \\ x=0 \quad \textbf{or} \quad 3x+8&=0 \\ x=0 \qquad\qquad 3x&=-8 \\ x=0 \qquad\qquad x&=-\tfrac{8}{3} \end{aligned}$$

25.
$$\begin{aligned} 8x^2-16x&=0 \\ 8x(x-2)&=0 \\ 8x=0 \quad \textbf{or} \quad x-2&=0 \\ x=0 \qquad\qquad x&=2 \end{aligned}$$

27.
$$\begin{aligned} 10x^2+2x&=0 \\ 2x(5x+1)&=0 \\ 2x=0 \quad \textbf{or} \quad 5x+1&=0 \\ x=0 \qquad\qquad 5x&=-1 \\ x=0 \qquad\qquad x&=-\tfrac{1}{5} \end{aligned}$$

29.
$$\begin{aligned} x^2-25&=0 \\ (x+5)(x-5)&=0 \\ x+5=0 \quad \textbf{or} \quad x-5&=0 \\ x=-5 \qquad\qquad x&=5 \end{aligned}$$

31.
$$\begin{aligned} y^2-49&=0 \\ (y+7)(y-7)&=0 \\ y+7=0 \quad \textbf{or} \quad y-7&=0 \\ y=-7 \qquad\qquad y&=7 \end{aligned}$$

33.
$$\begin{aligned} 4x^2-1&=0 \\ (2x+1)(2x-1)&=0 \\ 2x+1=0 \quad \textbf{or} \quad 2x-1&=0 \\ 2x=-1 \qquad\qquad 2x&=1 \\ x=-\tfrac{1}{2} \qquad\qquad x&=\tfrac{1}{2} \end{aligned}$$

35.
$$\begin{aligned} 9y^2-4&=0 \\ (3y+2)(3y-2)&=0 \\ 3y+2=0 \quad \textbf{or} \quad 3y-2&=0 \\ 3y=-2 \qquad\qquad 3y&=2 \\ y=-\tfrac{2}{3} \qquad\qquad y&=\tfrac{2}{3} \end{aligned}$$

37.
$$\begin{aligned} x^2 &= 49 \\ x^2 - 49 &= 0 \\ (x+7)(x-7) &= 0 \end{aligned}$$
$$x+7=0 \quad \textbf{or} \quad x-7=0$$
$$x=-7 \qquad\qquad x=7$$

39.
$$\begin{aligned} 4x^2 &= 81 \\ 4x^2 - 81 &= 0 \\ (2x+9)(2x-9) &= 0 \end{aligned}$$
$$2x+9=0 \quad \textbf{or} \quad 2x-9=0$$
$$2x=-9 \qquad\qquad 2x=9$$
$$x=-\tfrac{9}{2} \qquad\qquad x=\tfrac{9}{2}$$

41.
$$\begin{aligned} x^2 - 13x + 12 &= 0 \\ (x-1)(x-12) &= 0 \end{aligned}$$
$$x-1=0 \quad \textbf{or} \quad x-12=0$$
$$x=1 \qquad\qquad x=12$$

43.
$$\begin{aligned} x^2 - 2x - 15 &= 0 \\ (x+3)(x-5) &= 0 \end{aligned}$$
$$x+3=0 \quad \textbf{or} \quad x-5=0$$
$$x=-3 \qquad\qquad x=5$$

45.
$$\begin{aligned} x^2 - 4x - 21 &= 0 \\ (x+3)(x-7) &= 0 \end{aligned}$$
$$x+3=0 \quad \textbf{or} \quad x-7=0$$
$$x=-3 \qquad\qquad x=7$$

47.
$$\begin{aligned} x^2 + 8 - 9x &= 0 \\ x^2 - 9x + 8 &= 0 \\ (x-8)(x-1) &= 0 \end{aligned}$$
$$x-8=0 \quad \textbf{or} \quad x-1=0$$
$$x=8 \qquad\qquad x=1$$

49.
$$\begin{aligned} a^2 + 8a &= -15 \\ a^2 + 8a + 15 &= 0 \\ (a+3)(a+5) &= 0 \end{aligned}$$
$$a+3=0 \quad \textbf{or} \quad a+5=0$$
$$a=-3 \qquad\qquad a=-5$$

51.
$$\begin{aligned} 2y - 8 &= -y^2 \\ y^2 + 2y - 8 &= 0 \\ (y+4)(y-2) &= 0 \end{aligned}$$
$$y+4=0 \quad \textbf{or} \quad y-2=0$$
$$y=-4 \qquad\qquad y=2$$

53.
$$\begin{aligned} 2x^2 - 5x + 2 &= 0 \\ (2x-1)(x-2) &= 0 \end{aligned}$$
$$2x-1=0 \quad \textbf{or} \quad x-2=0$$
$$2x=1 \qquad\qquad x=2$$
$$x=\tfrac{1}{2} \qquad\qquad x=2$$

55.
$$\begin{aligned} 6x^2 + x - 2 &= 0 \\ (2x-1)(3x+2) &= 0 \end{aligned}$$
$$2x-1=0 \quad \textbf{or} \quad 3x+2=0$$
$$2x=1 \qquad\qquad 3x=-2$$
$$x=\tfrac{1}{2} \qquad\qquad x=-\tfrac{2}{3}$$

57.
$$\begin{aligned} 5p^2 - 6p + 1 &= 0 \\ (5p-1)(p-1) &= 0 \end{aligned}$$
$$5p-1=0 \quad \textbf{or} \quad p-1=0$$
$$5p=1 \qquad\qquad p=1$$
$$p=\tfrac{1}{5} \qquad\qquad p=1$$

59.
$$\begin{aligned} 14m^2 + 23m + 3 &= 0 \\ (7m+1)(2m+3) &= 0 \end{aligned}$$
$$7m+1=0 \quad \textbf{or} \quad 2m+3=0$$
$$7m=-1 \qquad\qquad 2m=-3$$
$$m=-\tfrac{1}{7} \qquad\qquad m=-\tfrac{3}{2}$$

61.
$$\begin{aligned} (x-1)\left(x^2+5x+6\right) &= 0 \\ (x-1)(x+2)(x+3) &= 0 \end{aligned}$$
$$x-1=0 \quad \textbf{or} \quad x+2=0 \quad \textbf{or} \quad x+3=0$$
$$x=1 \qquad\qquad x=-2 \qquad\qquad x=-3$$

63.
$$(x+3)(x^2+2x-15)=0$$
$$(x+3)(x+5)(x-3)=0$$
$$x+3=0 \quad \textbf{or} \quad x+5=0 \quad \textbf{or} \quad x-3=0$$
$$x=-3 \qquad\qquad x=-5 \qquad\qquad x=3$$

65.
$$(x-5)(2x^2+x-3)=0$$
$$(x-5)(2x+3)(x-1)=0$$
$$x-5=0 \quad \textbf{or} \quad 2x+3=0 \quad \textbf{or} \quad x-1=0$$
$$x=5 \qquad\qquad 2x=-3 \qquad\qquad x=1$$
$$x=-\tfrac{3}{2}$$

67.
$$(p^2-81)(p+2)=0$$
$$(p+9)(p-9)(p+2)=0$$
$$p+9=0 \quad \textbf{or} \quad p-9=0 \quad \textbf{or} \quad p+2=0$$
$$p=-9 \qquad\qquad p=9 \qquad\qquad p=-2$$

69.
$$x^3+3x^2+2x=0$$
$$x(x^2+3x+2)=0$$
$$x(x+2)(x+1)=0$$
$$x=0 \quad \textbf{or} \quad x+2=0 \quad \textbf{or} \quad x+1=0$$
$$x=0 \qquad\qquad x=-2 \qquad\qquad x=-1$$

71.
$$x^3-27x-6x^2=0$$
$$x(x^2-27-6x)=0$$
$$x(x^2-6x-27)=0$$
$$x(x-9)(x+3)=0$$
$$x=0 \quad \textbf{or} \quad x-9=0 \quad \textbf{or} \quad x+3=0$$
$$x=0 \qquad\qquad x=9 \qquad\qquad x=-3$$

73.
$$3x^2-8x=3$$
$$3x^2-8x-3=0$$
$$(3x+1)(x-3)=0$$
$$3x+1=0 \quad \textbf{or} \quad x-3=0$$
$$3x=-1 \qquad\qquad x=3$$
$$x=-\tfrac{1}{3} \qquad\qquad x=3$$

75.
$$15x^2-2=7x$$
$$15x^2-7x-2=0$$
$$(3x-2)(5x+1)=0$$
$$3x-2=0 \quad \textbf{or} \quad 5x+1=0$$
$$3x=2 \qquad\qquad 5x=-1$$
$$x=\tfrac{2}{3} \qquad\qquad x=-\tfrac{1}{5}$$

77.

$$\begin{aligned} x(6x+5) &= 6 \\ 6x^2+5x &= 6 \\ 6x^2+5x-6 &= 0 \\ (3x-2)(2x+3) &= 0 \end{aligned}$$

$3x-2=0$ **or** $2x+3=0$
$3x=2$ $\quad 2x=-3$
$x=\frac{2}{3}$ $\quad x=-\frac{3}{2}$

79.

$$\begin{aligned} (x+1)(8x+1) &= 18x \\ 8x^2+9x+1 &= 18x \\ 8x^2-9x+1 &= 0 \\ (x-1)(8x-1) &= 0 \end{aligned}$$

$x-1=0$ **or** $8x-1=0$
$x=1$ $\quad 8x=1$
$x=1$ $\quad x=\frac{1}{8}$

81.

$$\begin{aligned} 2x(3x^2+10x) &= -6x \\ 6x^3+20x^2 &= -6x \\ 6x^3+20x^2+6x &= 0 \\ 2x(3x^2+10x+3) &= 0 \\ 2x(3x+1)(x+3) &= 0 \end{aligned}$$

$2x=0$ **or** $3x+1=0$ **or** $x+3=0$
$x=0$ $\quad 3x=-1$ $\quad x=-3$
$x=0$ $\quad x=-\frac{1}{3}$ $\quad x=-3$

83.

$$\begin{aligned} x^3+7x^2 &= x^2-9x \\ x^3+6x^2+9x &= 0 \\ x(x^2+6x+9) &= 0 \\ x(x+3)(x+3) &= 0 \end{aligned}$$

$x=0$ **or** $x+3=0$ **or** $x+3=0$
$x=0$ $\quad x=-3$ $\quad x=-3$

85.

$$\begin{aligned} x^3+1.3x^2-0.3x &= 0 \\ 10x^3+13x^2-3x &= 0 \\ x(10x^2+13x-3) &= 0 \\ x(5x-1)(2x+3) &= 0 \end{aligned}$$

$x=0$ **or** $5x-1=0$ **or** $2x+3=0$
$x=0$ $\quad 5x=1$ $\quad 2x=-3$
$x=0$ $\quad x=\frac{1}{5}$ $\quad x=-\frac{3}{2}$

87. **Answers may vary.**

89. **Answers may vary.**

91.

$$\begin{aligned} 3a^2+9a-2a-6 &= 0 \\ 3a(a+3)-2(a+3) &= 0 \\ (a+3)(3a-2) &= 0 \end{aligned}$$

$a+3=0$ **or** $3a-2=0$
$a=-3$ $\quad 3a=2$
$\quad a=\frac{2}{3}$

$$\begin{aligned} 3a^2+9a-2a-6 &= 0 \\ 3a^2+7a-6 &= 0 \\ (a+3)(3a-2) &= 0 \end{aligned}$$

$a+3=0$ **or** $3a-2=0$
$a=-3$ $\quad 3a=2$
$\quad a=\frac{2}{3}$

Exercise 5.8 (page 352)

1.

$$\begin{aligned} -2(5z+2) &= 3(2-3z) \\ -10z-4 &= 6-9z \\ -4 &= 6+z \\ -10 &= z \end{aligned}$$

3. Let $w =$ the width and $3w =$ the length.

$$\begin{aligned} \text{Perimeter} &= 120 \\ 2(w)+2(3w) &= 120 \\ 2w+6w &= 120 \\ 8w &= 120 \\ w &= 15 \end{aligned}$$

The area $= 15(45) = 675$ cm^2.

5. analyze

7. Let $x =$ the first positive integer. Then $x + 2 =$ the other positive integer.

$$x(x+2) = 35$$
$$x^2 + 2x = 35$$
$$x^2 + 2x - 35 = 0$$
$$(x+7)(x-5) = 0$$
$$x + 7 = 0 \quad \textbf{or} \quad x - 5 = 0$$
$$x = -7 \qquad\qquad x = 5$$

The answer of -7 is not a positive integer, so the integers are 5 and 7.

9. Let $x =$ the composite integer.

$$x^2 + 4 = 10x - 5$$
$$x^2 - 10x + 9 = 0$$
$$(x-9)(x-1) = 0$$
$$x - 9 = 0 \quad \textbf{or} \quad x - 1 = 0$$
$$x = 9 \qquad\qquad x = 1$$

The answer of 1 is not a composite integer, so the integer is 9.

11. Let $v = 144$ and $h = 0$:

$$h = vt - 16t^2$$
$$0 = 144t - 16t^2$$
$$0 = 16t(9 - t)$$
$$16t = 0 \quad \textbf{or} \quad 9 - t = 0$$
$$t = 0 \qquad\qquad 9 = t$$

Since $t = 0$ is when the object was first thrown, it will hit the ground in 9 seconds.

13. Let $v = 220$ and $h = 600$:

$$h = vt - 16t^2$$
$$600 = 220t - 16t^2$$
$$16t^2 - 220t + 600 = 0$$
$$4\left(4t^2 - 55t + 150\right) = 0$$
$$4(4t - 15)(t - 10) = 0$$
$$4t - 15 = 0 \quad \textbf{or} \quad t - 10 = 0$$
$$4t = 15 \qquad\qquad t = 10$$
$$t = \tfrac{15}{4} \qquad\qquad t = 10$$

The cannonball will be at a height of 600 feet after $\frac{15}{4}$ seconds and after 10 seconds.

15.

$$h = -16t^2 + 64$$
$$0 = -16t^2 + 64$$
$$16t^2 - 64 = 0$$
$$16\left(t^2 - 4\right) = 0$$
$$16(t+2)(t-2) = 0$$
$$t + 2 = 0 \quad \textbf{or} \quad t - 2 = 0$$
$$t = -2 \qquad\qquad t = 2$$

The value of $t = -2$ does not make sense, so the dive lasts 2 seconds.

17.

$$75^2 = h^2 + 72^2$$
$$5625 = h^2 + 5184$$
$$0 = h^2 - 441$$
$$0 = (h+21)(h-21)$$
$$h + 21 = 0 \quad \textbf{or} \quad h - 21 = 0$$
$$h = -21 \qquad\qquad h = 21$$

The value of $h = -21$ does not make sense, so the camper started at a height of 21 feet.

19. Area = Length · Width

$$\begin{aligned} 36 &= (2w+1)w \\ 36 &= 2w^2 + w \\ 0 &= 2w^2 + w - 36 \\ 0 &= (2w+9)(w-4) \end{aligned}$$

$$\begin{array}{lcl} 2w+9=0 & \textbf{or} & w-4=0 \\ 2w=-9 & & w=4 \\ w=-\frac{9}{2} & & w=4 \end{array}$$

Since the answer $w = -\frac{9}{2}$ does not make sense, the dimensions are 4 m by 9 m.

21. Let w = the width and $w + 2$ = the length.

Area = Length · Width

$$\begin{aligned} 143 &= (w+2)w \\ 143 &= w^2 + 2w \\ 0 &= w^2 + 2w - 143 \\ 0 &= (w+13)(w-11) \end{aligned}$$

$$\begin{array}{lcl} w+13=0 & \textbf{or} & w-11=0 \\ w=-13 & & w=11 \end{array}$$

Since the answer $w = -13$ does not make sense, the dimensions are 11 ft by 13 ft and the perimeter is 48 ft.

23. Let b = the base and $5b - 2$ = the height.

$$\begin{aligned} A &= \frac{1}{2}bh \\ 36 &= \frac{1}{2}b(5b-2) \\ 72 &= b(5b-2) \\ 72 &= 5b^2 - 2b \\ 0 &= 5b^2 - 2b - 72 \\ 0 &= (5b+18)(b-4) \end{aligned}$$

$$\begin{array}{lcl} 5b+18=0 & \textbf{or} & b-4=0 \\ 5b=-18 & & b=4 \\ h=-\frac{18}{5} & & b=4 \end{array}$$

Since the answer $b = -\frac{18}{5}$ does not make sense, the base is 4 in. and the height is 18 in.

25. Let x = the base and the height.

$$\boxed{\text{Base}} + \boxed{\text{Height}} = \boxed{\text{Area}} - 6$$

$$\begin{aligned} x + x &= \tfrac{1}{2}(x)(x) - 6 \\ 2x &= \tfrac{1}{2}x^2 - 6 \\ 0 &= \tfrac{1}{2}x^2 - 2x - 6 \\ 0 &= x^2 - 4x - 12 \\ 0 &= (x-6)(x+2) \end{aligned}$$

$$\begin{array}{lcl} x-6=0 & \textbf{or} & x+2=0 \\ x=6 & & x=-2 \end{array}$$

The answer $x = -2$ does not make sense, so the area is $\frac{1}{2}(6)(6) = 18$ square units.

27.

$$\boxed{\text{Large area}} - \boxed{\text{Small area}} = \boxed{\text{Border area}}$$

$$\begin{aligned} (10+2w)(25+2w) - (10)(25) &= 74 \\ 250 + 70w + 4w^2 - 250 &= 74 \\ 4w^2 + 70w - 74 &= 0 \\ 2\left(2w^2 + 35w - 37\right) &= 0 \\ 2(2w+37)(w-1) &= 0 \end{aligned}$$

$$\begin{array}{lcl} 2w+37=0 & \textbf{or} & w-1=0 \\ 2w=-37 & & w=1 \\ w=-\frac{37}{2} & & w=1 \end{array}$$

Since the answer $w = -\frac{37}{2}$ does not make sense, the width should be 1 meter.

29. Let w = the width and $2w + 1$ = the height.

$$\begin{aligned} V &= lwh \\ 210 &= 10w(2w+1) \\ 210 &= 20w^2 + 10w \\ 0 &= 20w^2 + 10w - 210 \\ 0 &= 10\left(2w^2 + w - 21\right) \\ 0 &= 10(2w+7)(w-3) \end{aligned}$$

$$\begin{array}{lcl} 2w+7=0 & \textbf{or} & w-3=0 \\ 2w=-7 & & w=3 \\ w=-\frac{7}{2} & & w=3 \end{array}$$

Since the answer $w = -\frac{7}{2}$ does not make sense, the width is 3 cm.

31. Let $x =$ one edge and $x - 3 =$ the other.

$$V = \frac{Bh}{3}$$
$$84 = \frac{x(x-3)(9)}{3}$$
$$252 = 9x^2 - 27x$$
$$0 = 9x^2 - 27x - 252$$
$$0 = 9(x+4)(x-7)$$
$$x + 4 = 0 \quad \textbf{or} \quad x - 7 = 0$$
$$x = -4 \qquad\qquad x = 7$$

Since the answer $x = -4$ does not make sense, the base is 7 cm by 4 cm.

33.

$$C = \tfrac{1}{2}(n^2 - n)$$
$$6 = \tfrac{1}{2}(n^2 - n)$$
$$2(66) = 2 \cdot \tfrac{1}{2}(n^2 - n)$$
$$132 = n^2 - n$$
$$0 = n^2 - n - 132$$
$$0 = (n+11)(n-12)$$
$$n + 11 = 0 \quad \textbf{or} \quad n - 12 = 0$$
$$n = -11 \qquad\qquad n = 12$$

The value of $n = -11$ does not make sense, so 12 telephones are needed.

35. Area #1 $= \pi r^2 = \pi(20 \text{ m})^2 = 400\pi \text{ m}^2$

Area #2 $= \pi r^2 = \pi(21 \text{ m})^2 = 441\pi \text{ m}^2$

Total Area $= 400\pi + 441\pi \text{ m}^2 = 841\pi \text{ m}^2$

37.

$$a^2 + b^2 = c^2$$
$$x^2 + (x+2)^2 = (x+4)^2$$
$$x^2 + x^2 + 4x + 4 = x^2 + 8x + 16$$
$$2x^2 + 4x + 4 = x^2 + 8x + 16$$
$$x^2 - 4x - 12 = 0$$
$$(x-6)(x+2) = 0$$
$$x - 6 = 0 \quad \textbf{or} \quad x + 2 = 0$$
$$x = 6 \qquad\qquad x = -2$$

The value of $x = -2$ does not make sense, so $x = 6$, and the height was 16 ft.

39. **Answers may vary.**

41. Let $w =$ the width and $w + 2 =$ the length.

$$\text{Area} = 18$$
$$w(w+2) = 18$$
$$w^2 + 2w = 18$$
$$w^2 + 2w - 18 = 0$$

This is a prime trinomial, so it cannot be factored in order to solve.

Chapter 5 Summary (page 356)

1. $35 = 5 \cdot 7$

2. $45 = 9 \cdot 5 = 3^2 \cdot 5$

3. $96 = 12 \cdot 8 = 4 \cdot 3 \cdot 2^3 = 2^2 \cdot 3 \cdot 2^3 = 2^5 \cdot 3$

4. $102 = 2 \cdot 51 = 2 \cdot 3 \cdot 17$

5. $87 = 3 \cdot 29$

6. $99 = 9 \cdot 11 = 3^2 \cdot 11$

7. $2{,}050 = 50 \cdot 41 = 25 \cdot 2 \cdot 41 = 2 \cdot 5^2 \cdot 41$

8. $4{,}096 = 64 \cdot 64 = 2^6 \cdot 2^6 = 2^{12}$

9. $3x + 9y = 3(x + 3y)$

10. $5ax^2 + 15a = 5a(x^2 + 3)$

11. $7x^2 + 14x = 7x(x+2)$

12. $3x^2 - 3x = 3x(x-1)$

13. $2x^3 + 4x^2 - 8x = 2x(x^2 + 2x - 4)$

14. $ax + ay - az = a(x + y - z)$

15. $ax + ay - a = a(x + y - 1)$

16. $x^2yz + xy^2z = xyz(x+y)$

17. $(x+y)a + (x+y)b = (x+y)(a+b)$

18. $(x+y)^2 + (x+y) = (x+y)(x+y) + 1(x+y) = (x+y)(x+y+1)$

19. $2x^2(x+2) + 6x(x+2) = (x+2)(2x^2+6x) = (x+2)(2x)(x+3) = 2x(x+2)(x+3)$

20. $$\begin{aligned}3x(y+z) - 9x(y+z)^2 &= 3x[(y+z) - 3(y+z)(y+z)] = 3x(y+z)(1-3(y+z))\\ &= 3x(y+z)(1-3y-3z)\end{aligned}$$

21. $3p + 9q + ap + 3aq = 3(p+3q) + a(p+3q) = (p+3q)(3+a)$

22. $ar - 2as + 7r - 14s = a(r-2s) + 7(r-2s) = (r-2s)(a+7)$

23. $$\begin{aligned}x^2 + ax + bx + ab &= x(x+a) + b(x+a)\\ &= (x+a)(x+b)\end{aligned}$$

24. $$\begin{aligned}xy + 2x - 2y - 4 &= x(y+2) - 2(y+2)\\ &= (y+2)(x-2)\end{aligned}$$

25. $xa + yb + ya + xb = xa + ya + xb + yb = a(x+y) + b(x+y) = (x+y)(a+b)$

26. $x^2 - 9 = x^2 - 3^2 = (x+3)(x-3)$

27. $x^2y^2 - 16 = (xy+4)(xy-4)$

28. $(x+2)^2 - y^2 = (x+2+y)(x+2-y)$

29. $z^2 - (x+y)^2 = [z + (x+y)][z - (x+y)] = (z+x+y)(z-x-y)$

30. $6x^2y - 24y^3 = 6y(x^2 - 4y^2) = 6y\left[x^2 - (2y)^2\right] = 6y(x+2y)(x-2y)$

31. $(x+y)^2 - z^2 = [(x+y)+z][(x+y)-z] = (x+y+z)(x+y-z)$

32. $x^2 + 10x + 21 = (x+3)(x+7)$

33. $x^2 + 4x - 21 = (x+7)(x-3)$

34. $x^2 + 2x - 24 = (x+6)(x-4)$

35. $x^2 - 4x - 12 = (x-6)(x+2)$

36. $2x^2 - 5x - 3 = (2x+1)(x-3)$

37. $3x^2 - 14x - 5 = (3x+1)(x-5)$

38. $6x^2 + 7x - 3 = (2x+3)(3x-1)$

39. $$\begin{aligned}6x^2 + 3x - 3 &= 3(2x^2 + x - 1)\\ &= 3(2x-1)(x+1)\end{aligned}$$

40. $$\begin{aligned}6x^3 + 17x^2 - 3x &= x(6x^2 + 17x - 3)\\ &= x(x+3)(6x-1)\end{aligned}$$

41. $$\begin{aligned}4x^3 - 5x^2 - 6x &= x(4x^2 - 5x - 6)\\ &= x(x-2)(4x+3)\end{aligned}$$

42. $12x - 4x^3 - 2x^2 = -4x^3 - 2x^2 + 12x = -2x(2x^2 + x - 6) = -2x(2x - 3)(x + 2)$

43. $-4a^3 + 4a^2b + 24ab^2 = -4a(a^2 - ab - 6b^2) = -4a(a - 3b)(a + 2b)$

44. $c^3 - 27 = c^3 - 3^3 = (c - 3)(c^2 + 3c + 9)$

45. $d^3 + 8 = d^3 + 2^3 = (d + 2)(d^2 - 2d + 4)$

46.
$$\begin{aligned} 2x^3 + 54 &= 2(x^3 + 27) \\ &= 2(x^3 + 3^3) \\ &= 2(x + 3)(x^2 - 3x + 9) \end{aligned}$$

47.
$$\begin{aligned} 2ab^4 - 2ab &= 2ab(b^3 - 1) \\ &= 2ab(b^3 - 1^3) \\ &= 2ab(b - 1)(b^2 + b + 1) \end{aligned}$$

48.
$$\begin{aligned} 3x^2y - xy^2 - 6xy + 2y^2 = y(3x^2 - xy - 6x + 2y) &= y[x(3x - y) - 2(3x - y)] \\ &= y(3x - y)(x - 2) \end{aligned}$$

49.
$$\begin{aligned} 5x^2 + 10x - 15xy - 30y = 5(x^2 + 2x - 3xy - 6y) &= 5[x(x + 2) - 3y(x + 2)] \\ &= 5(x + 2)(x - 3y) \end{aligned}$$

50.
$$\begin{aligned} 2a^2x + 2abx + a^3 + a^2b = a(2ax + 2bx + a^2 + ab) &= a[2x(a + b) + a(a + b)] \\ &= a(a + b)(2x + a) \end{aligned}$$

51.
$$\begin{aligned} x^2 + 2ax + a^2 - y^2 = (x^2 + 2ax + a^2) - y^2 &= (x + a)(x + a) - y^2 \\ &= (x + a)^2 - y^2 = (x + a + y)(x + a - y) \end{aligned}$$

52. $x^2 - 4 + bx + 2b = x^2 - 2^2 + bx + 2b = (x + 2)(x - 2) + b(x + 2) = (x + 2)(x - 2 + b)$

53.
$$\begin{aligned} ax^6 - ay^6 = a(x^6 - y^6) = a\left[(x^3)^2 - (y^3)^2\right] &= a(x^3 + y^3)(x^3 - y^3) \\ &= a(x + y)(x^2 - xy + y^2)(x - y)(x^2 + xy + y^2) \end{aligned}$$

54.
$$\begin{aligned} &x^2 + 2x = 0 \\ &x(x + 2) = 0 \\ &x = 0 \quad \textbf{or} \quad x + 2 = 0 \\ &x = 0 \qquad\qquad x = -2 \end{aligned}$$

55.
$$\begin{aligned} &2x^2 - 6x = 0 \\ &2x(x - 3) = 0 \\ &2x = 0 \quad \textbf{or} \quad x - 3 = 0 \\ &x = 0 \qquad\qquad x = 3 \end{aligned}$$

56.
$$\begin{aligned} &3x^2 = 2x \\ &3x^2 - 2x = 0 \\ &x(3x - 2) = 0 \\ &x = 0 \quad \textbf{or} \quad 3x - 2 = 0 \\ &\qquad\qquad 3x = 2 \\ &\qquad\qquad x = \tfrac{2}{3} \end{aligned}$$

57.
$$\begin{aligned} &5x^2 + 25x = 0 \\ &5x(x + 5) = 0 \\ &5x = 0 \quad \textbf{or} \quad x + 5 = 0 \\ &x = \tfrac{0}{5} \qquad\qquad x = -5 \\ &x = 0 \end{aligned}$$

58.
$$\begin{aligned} &x^2 - 9 = 0 \\ &(x + 3)(x - 3) = 0 \\ &x + 3 = 0 \quad \textbf{or} \quad x - 3 = 0 \\ &x = -3 \qquad\qquad x = 3 \end{aligned}$$

59.
$$\begin{aligned} &x^2 - 25 = 0 \\ &(x + 5)(x - 5) = 0 \\ &x + 5 = 0 \quad \textbf{or} \quad x - 5 = 0 \\ &x = -5 \qquad\qquad x = 5 \end{aligned}$$

60.
$$\begin{aligned} a^2 - 7a + 12 &= 0 \\ (a-3)(a-4) &= 0 \end{aligned}$$
$$a - 3 = 0 \quad \textbf{or} \quad a - 4 = 0$$
$$a = 3 \qquad\qquad a = 4$$

61.
$$\begin{aligned} x^2 - 2x - 15 &= 0 \\ (x+3)(x-5) &= 0 \end{aligned}$$
$$x + 3 = 0 \quad \textbf{or} \quad x - 5 = 0$$
$$x = -3 \qquad\qquad x = 5$$

62.
$$\begin{aligned} 2x - x^2 + 24 &= 0 \\ x^2 - 2x - 24 &= 0 \\ (x+4)(x-6) &= 0 \end{aligned}$$
$$x + 4 = 0 \quad \textbf{or} \quad x - 6 = 0$$
$$x = -4 \qquad\qquad x = 6$$

63.
$$\begin{aligned} 16 + x^2 - 10x &= 0 \\ x^2 - 10x + 16 &= 0 \\ (x-2)(x-8) &= 0 \end{aligned}$$
$$x - 2 = 0 \quad \textbf{or} \quad x - 8 = 0$$
$$x = 2 \qquad\qquad x = 8$$

64.
$$\begin{aligned} 2x^2 - 5x - 3 &= 0 \\ (2x+1)(x-3) &= 0 \end{aligned}$$
$$2x + 1 = 0 \quad \textbf{or} \quad x - 3 = 0$$
$$2x = -1 \qquad\qquad x = 3$$
$$x = -\tfrac{1}{2} \qquad\qquad x = 3$$

65.
$$\begin{aligned} 2x^2 + x - 3 &= 0 \\ (2x+3)(x-1) &= 0 \end{aligned}$$
$$2x + 3 = 0 \quad \textbf{or} \quad x - 1 = 0$$
$$2x = -3 \qquad\qquad x = 1$$
$$x = -\tfrac{3}{2} \qquad\qquad x = 1$$

66.
$$\begin{aligned} 4x^2 &= 1 \\ 4x^2 - 1 &= 0 \\ (2x+1)(2x-1) &= 0 \end{aligned}$$
$$2x + 1 = 0 \quad \textbf{or} \quad 2x - 1 = 0$$
$$2x = -1 \qquad\qquad 2x = 1$$
$$x = -\tfrac{1}{2} \qquad\qquad x = \tfrac{1}{2}$$

67.
$$\begin{aligned} 9x^2 &= 4 \\ 9x^2 - 4 &= 0 \\ (3x+2)(3x-2) &= 0 \end{aligned}$$
$$3x + 2 = 0 \quad \textbf{or} \quad 3x - 2 = 0$$
$$3x = -2 \qquad\qquad 3x = 2$$
$$x = -\tfrac{2}{3} \qquad\qquad x = \tfrac{2}{3}$$

68.
$$\begin{aligned} x^3 - 7x^2 + 12x &= 0 \\ x\left(x^2 - 7x + 12\right) &= 0 \\ x(x-3)(x-4) &= 0 \end{aligned}$$
$$x = 0 \quad \textbf{or} \quad x - 3 = 0 \quad \textbf{or} \quad x - 4 = 0$$
$$x = 0 \qquad\qquad x = 3 \qquad\qquad x = 4$$

69.
$$\begin{aligned} x^3 + 5x^2 + 6x &= 0 \\ x\left(x^2 + 5x + 6\right) &= 0 \\ x(x+2)(x+3) &= 0 \end{aligned}$$
$$x = 0 \quad \textbf{or} \quad x + 2 = 0 \quad \textbf{or} \quad x + 3 = 0$$
$$x = 0 \qquad\qquad x = -2 \qquad\qquad x = -3$$

70.
$$2x^3 + 5x^2 = 3x$$
$$2x^3 + 5x^2 - 3x = 0$$
$$x(2x^2 + 5x - 3) = 0$$
$$x(2x - 1)(x + 3) = 0$$
$x = 0$ **or** $2x - 1 = 0$ **or** $x + 3 = 0$
$x = 0$ $\quad 2x = 1$ $\quad x = -3$
$x = 0$ $\quad x = \frac{1}{2}$ $\quad x = -3$

71.
$$3x^3 - 2x = x^2$$
$$3x^3 - x^2 - 2x = 0$$
$$x(3x^2 - x - 2) = 0$$
$$x(3x + 2)(x - 1) = 0$$
$x = 0$ **or** $3x + 2 = 0$ **or** $x - 1 = 0$
$x = 0$ $\quad 3x = -2$ $\quad x = 1$
$x = 0$ $\quad x = -\frac{2}{3}$ $\quad x = 1$

72. Let $x =$ one number. Then
$12 - x =$ the other number.
$$x(12 - x) = 35$$
$$12x - x^2 = 35$$
$$0 = x^2 - 12x + 35$$
$$0 = (x - 5)(x - 7)$$
$x - 5 = 0$ **or** $x - 7 = 0$
$x = 5$ $\quad x = 7$
The numbers are 5 and 7.

73. Let $x =$ the positive number.
$$3x^2 + 5x = 2$$
$$3x^2 + 5x - 2 = 0$$
$$(3x - 1)(x + 2) = 0$$
$3x - 1 = 0$ **or** $x + 2 = 0$
$3x = 1$ $\quad x = -2$
$x = \frac{1}{3}$ $\quad x = -2$
The answer of -2 is not positive,
so the number is $\frac{1}{3}$.

74. Let $w =$ the width and $w + 2 =$ the length.
Area = Length · Width
$$48 = (w + 2)w$$
$$0 = w^2 + 2w - 48$$
$$0 = (w + 8)(w - 6)$$
$w + 8 = 0$ **or** $w - 6 = 0$
$w = -8$ $\quad w = 6$
Since the answer $w = -8$ does not make sense, the dimensions are 6 ft by 8 ft.

75. Let $w =$ the width and $2w + 3 =$ the length.
Area = Length · Width
$$27 = (2w + 3)w$$
$$0 = 2w^2 + 3w - 27$$
$$0 = (2w + 9)(w - 3)$$
$2w + 9 = 0$ **or** $w - 3 = 0$
$2w = -9$ $\quad w = 3$
$w = -\frac{9}{2}$ $\quad w = 3$
Since the answer $w = -\frac{9}{2}$ does not make sense, the dimensions are 3 ft by 9 ft.

76. Let $w =$ the width and $w + 3 =$ the length.

$$\begin{aligned} \text{Area} &= \text{Perimeter} \\ (w+3)w &= 2w + 2(w+3) \\ w^2 + 3w &= 2w + 2w + 6 \\ w^2 + 3w &= 4w + 6 \\ w^2 - w - 6 &= 0 \\ (w+2)(w-3) &= 0 \end{aligned}$$

$$w + 2 = 0 \quad \textbf{or} \quad w - 3 = 0$$
$$w = -2 \qquad\qquad w = 3$$

Since the answer $w = -2$ does not make sense, the dimensions are 3 ft by 6 ft.

Chapter 5 Test (page 359)

1. $196 = 14 \cdot 14 = 2 \cdot 7 \cdot 2 \cdot 7 = 2^2 \cdot 7^2$

2. $111 = 3 \cdot 37$

3. $60ab^2c^3 + 30a^3b^2c - 25a = 5a(12b^2c^3 + 6a^2b^2c - 5)$

4. $3x^2(a+b) - 6xy(a+b) = 3x[x(a+b) - 2y(a+b)] = 3x(a+b)(x-2y)$

5. $ax + ay + bx + by = a(x+y) + b(x+y)$
$= (x+y)(a+b)$

6. $x^2 - 25 = x^2 - 5^2 = (x+5)(x-5)$

7. $3a^2 - 27b^2 = 3(a^2 - 9b^2) = 3\left(a^2 - (3b)^2\right) = 3(a+3b)(a-3b)$

8. $16x^4 - 81y^4 = \left(4x^2\right)^2 - \left(9y^2\right)^2 = \left(4x^2 + 9y^2\right)\left(4x^2 - 9y^2\right)$
$= \left(4x^2 + 9y^2\right)\left((2x)^2 - (3y)^2\right)$
$= \left(4x^2 + 9y^2\right)(2x + 3y)(2x - 3y)$

9. $x^2 + 4x + 3 = (x+3)(x+1)$

10. $x^2 - 9x - 22 = (x-11)(x+2)$

11. $x^2 + 10xy + 9y^2 = (x+y)(x+9y)$

12. $6x^2 - 30xy + 24y^2 = 6\left(x^2 - 5xy + 4y^2\right)$
$= 6(x-4y)(x-y)$

13. $3x^2 + 13x + 4 = (3x+1)(x+4)$

14. $2a^2 + 5a - 12 = (2a-3)(a+4)$

15. $2x^2 + 3xy - 2y^2 = (2x-y)(x+2y)$

16. $12 - 25x + 12x^2 = 12x^2 - 25x + 12$
$= (4x-3)(3x-4)$

17. $12a^2 + 6ab - 36b^2 = 6\left(2a^2 + ab - 6b^2\right)$
$= 6(2a-3b)(a+2b)$

18. $x^3 - 64 = x^3 - 4^3 = (x-4)\left(x^2 + 4x + 16\right)$

19. $216 + 8a^3 = 8(27 + a^3) = 8(3^3 + a^3) = 8(3+a)(9 - 3a + a^2)$

20. $x^9z^3 - y^3z^6 = z^3(x^9 - y^3z^3) = z^3\left((x^3)^3 - (yz)^3\right) = z^3(x^3 - yz)\left((x^3)^2 + x^3yz + (yz)^2\right)$

$= z^3(x^3 - yz)(x^6 + x^3yz + y^2z^2)$

21.
$$x^2 + 3x = 0$$
$$x(x+3) = 0$$
$x = 0$ **or** $x + 3 = 0$
$x = 0$ $\quad x = -3$

22.
$$2x^2 + 5x + 3 = 0$$
$$(2x+3)(x+1) = 0$$
$2x + 3 = 0$ **or** $x + 1 = 0$
$2x = -3$ $\quad x = -1$
$x = -\frac{3}{2}$ $\quad x = -1$

23.
$$9y^2 - 81 = 0$$
$$9(y^2 - 9) = 0$$
$$9(y+3)(y-3) = 0$$
$y + 3 = 0$ **or** $y - 3 = 0$
$y = -3$ $\quad y = 3$

24.
$$-3(y-6) + 2 = y^2 + 2$$
$$-3y + 18 + 2 = y^2 + 2$$
$$-3y + 20 = y^2 + 2$$
$$0 = y^2 + 3y - 18$$
$$0 = (y+6)(y-3)$$
$y + 6 = 0$ **or** $y - 3 = 0$
$y = -6$ $\quad y = 3$

25.
$$10x^2 - 13x = 9$$
$$10x^2 - 13x - 9 = 0$$
$$(2x+1)(5x-9) = 0$$
$2x + 1 = 0$ **or** $5x - 9 = 0$
$2x = -1$ $\quad 5x = 9$
$x = -\frac{1}{2}$ $\quad x = \frac{9}{5}$

26.
$$10x^2 - x = 9$$
$$10x^2 - x - 9 = 0$$
$$(10x+9)(x-1) = 0$$
$10x + 9 = 0$ **or** $x - 1 = 0$
$10x = -9$ $\quad x = 1$
$x = -\frac{9}{10}$ $\quad x = 1$

27.
$$10x^2 + 43x = 9$$
$$10x^2 + 43x - 9 = 0$$
$$(2x+9)(5x-1) = 0$$
$2x + 9 = 0$ **or** $5x - 1 = 0$
$2x = -9$ $\quad 5x = 1$
$x = -\frac{9}{2}$ $\quad x = \frac{1}{5}$

28.
$$10x^2 - 89x = 9$$
$$10x^2 - 89x - 9 = 0$$
$$(10x+1)(x-9) = 0$$
$10x + 1 = 0$ **or** $x - 9 = 0$
$10x = -1$ $\quad x = 9$
$x = -\frac{1}{10}$ $\quad x = 9$

29. Let $v = 192$ and $h = 0$:
$$h = vt - 16t^2$$
$$0 = 192t - 16t^2$$
$$0 = 16t(12 - t)$$
$16t = 0$ **or** $12 - t = 0$
$t = 0$ $\quad 12 = t$

Since $t = 0$ is when the cannonball was fired, it will hit the ground in 12 seconds.

30. Let $h =$ the height and $h + 2 =$ the base.
$$A = \frac{1}{2}bh$$
$$40 = \frac{1}{2}(h+2)h$$
$$80 = (h+2)h$$
$$0 = h^2 + 2h - 80$$
$$0 = (h+10)(h-8)$$
$h + 10 = 0$ **or** $h - 8 = 0$
$h = -10$ $\quad h = 8$

Since the answer $h = -10$ does not make sense, the height is 8 m and the base is 10 m.

Exercise 6.1 (page 368)

1. If a, b and c are real numbers, then $(a+b)+c=a+(b+c)$.

3. 0 is the additive identity.

5. $\frac{5}{3}$ is the additive inverse of $-\frac{5}{3}$.

7. numerator

9. 0

11. negatives (or opposites)

13. $\frac{a}{b}$

15. factor; common

17. $y-2=0$
$y=2$

19. $x^2-x-2=0$
$(x+1)(x-2)=0$
$x+1=0$ **or** $x-2=0$
$x=-1$ $\quad$ $x=2$

21. $2m^2-m-3=0$
$(2m-3)(m+1)=0$
$2m-3=0$ **or** $m+1=0$
$2m=3$ $\quad$ $m=-1$
$m=\frac{3}{2}$ $\quad$ $m=-1$

23. $\frac{8}{10}=\frac{2\cdot 4}{2\cdot 5}=\frac{4}{5}$

25. $\frac{28}{35}=\frac{7\cdot 4}{7\cdot 5}=\frac{4}{5}$

27. $\frac{8}{52}=\frac{4\cdot 2}{4\cdot 13}=\frac{2}{13}$

29. $\frac{10}{45}=\frac{5\cdot 2}{5\cdot 9}=\frac{2}{9}$

31. $\frac{-18}{54}=-\frac{18\cdot 1}{18\cdot 3}=-\frac{1}{3}$

33. $\frac{4x}{2}=\frac{2\cdot 2x}{2\cdot 1}=2x$

35. $\frac{-6x}{18}=-\frac{6\cdot x}{6\cdot 3}=-\frac{x}{3}$

37. $\frac{45}{9a}=\frac{9\cdot 5}{9\cdot a}=\frac{5}{a}$

39. $\frac{7+3}{5z}=\frac{10}{5z}=\frac{5\cdot 2}{5\cdot z}=\frac{2}{z}$

41. $\frac{(3+4)a}{24-3}=\frac{7a}{21}=\frac{7\cdot a}{7\cdot 3}=\frac{a}{3}$

43. $\frac{2x}{3x}=\frac{x\cdot 2}{x\cdot 3}=\frac{2}{3}$

45. $\frac{6x^2}{4x^2}=\frac{2x^2\cdot 3}{2x^2\cdot 2}=\frac{3}{2}$

47. $\frac{2x^2}{3y}\Rightarrow$ simplest form

49. $\frac{15x^2y}{5xy^2}=\frac{5xy\cdot 3x}{5xy\cdot y}=\frac{3x}{y}$

51. $\frac{28x}{32y}=\frac{4\cdot 7x}{4\cdot 8y}=\frac{7x}{8y}$

53. $\frac{x+3}{3(x+3)}=\frac{1(x+3)}{3(x+3)}=\frac{1}{3}$

55. $\frac{5x+35}{x+7}=\frac{5(x+7)}{1(x+7)}=5$

57. $\frac{x^2+3x}{2x+6}=\frac{x(x+3)}{2(x+3)}=\frac{x}{2}$

59. $\frac{15x-3x^2}{25y-5xy}=\frac{3x(5-x)}{5y(5-x)}=\frac{3x}{5y}$

61. $\frac{6a-6b+6c}{9a-9b+9c}=\frac{6(a-b+c)}{9(a-b+c)}=\frac{6}{9}=\frac{2}{3}$

63. $\frac{x-7}{7-x}=\frac{-(7-x)}{7-x}=-1$

65. $\frac{6x-3y}{3y-6x}=\frac{-(3y-6x)}{3y-6x}=-1$

67. $\dfrac{a+b-c}{c-a-b} = \dfrac{-(c-a-b)}{c-a-b} = -1$

69. $\dfrac{x^2+3x+2}{x^2+x-2} = \dfrac{(x+2)(x+1)}{(x+2)(x-1)} = \dfrac{x+1}{x-1}$

71. $\dfrac{x^2-8x+15}{x^2-x-6} = \dfrac{(x-5)(x-3)}{(x+2)(x-3)} = \dfrac{x-5}{x+2}$

73. $\dfrac{2x^2-8x}{x^2-6x+8} = \dfrac{2x(x-4)}{(x-2)(x-4)} = \dfrac{2x}{x-2}$

75. $\dfrac{xy+2x^2}{2xy+y^2} = \dfrac{x(y+2x)}{y(2x+y)} = \dfrac{x}{y}$

77. $\dfrac{x^2+3x+2}{x^3+x^2} = \dfrac{(x+2)(x+1)}{x^2(x+1)} = \dfrac{x+2}{x^2}$

79. $\dfrac{x^2-8x+16}{x^2-16} = \dfrac{(x-4)(x-4)}{(x+4)(x-4)} = \dfrac{x-4}{x+4}$

81. $\dfrac{2x^2-8}{x^2-3x+2} = \dfrac{2(x^2-4)}{(x-1)(x-2)} = \dfrac{2(x+2)(x-2)}{(x-1)(x-2)} = \dfrac{2(x+2)}{x-1}$

83. $\dfrac{x^2-2x-15}{x^2+2x-15} = \dfrac{(x-5)(x+3)}{(x+5)(x-3)}$

85. $\dfrac{x^2-3(2x-3)}{9-x^2} = \dfrac{x^2-6x+9}{3^2-x^2} = \dfrac{(x-3)(x-3)}{(3+x)(3-x)} = -\dfrac{x-3}{x+3}$

87. $\dfrac{4(x+3)+4}{3(x+2)+6} = \dfrac{4x+12+4}{3x+6+6} = \dfrac{4x+16}{3x+12} = \dfrac{4(x+4)}{3(x+4)} = \dfrac{4}{3}$

89. $\dfrac{x^2-9}{(2x+3)-(x+6)} = \dfrac{x^2-9}{2x+3-x-6} = \dfrac{x^2-9}{x-3} = \dfrac{(x+3)(x-3)}{x-3} = x+3$

91. $\dfrac{x^3+1}{x^2-x+1} = \dfrac{x^3+1^3}{x^2-x+1} = \dfrac{(x+1)(x^2-1x+1^2)}{x^2-x+1} = \dfrac{(x+1)(x^2-x+1)}{x^2-x+1} = x+1$

93. $\dfrac{2a^3-16}{2a^2+4a+8} = \dfrac{2(a^3-8)}{2(a^2+2a+4)} = \dfrac{a^3-2^3}{a^2+2a+4} = \dfrac{(a-2)(a^2+2a+2^2)}{a^2+2a+4}$

$= \dfrac{(a-2)(a^2+2a+4)}{a^2+2a+4} = a-2$

95. $\dfrac{ab+b+2a+2}{ab+a+b+1} = \dfrac{b(a+1)+2(a+1)}{a(b+1)+1(b+1)} = \dfrac{(a+1)(b+2)}{(b+1)(a+1)} = \dfrac{b+2}{b+1}$

97. $\dfrac{xy+3y+3x+9}{x^2-9} = \dfrac{y(x+3)+3(x+3)}{(x+3)(x-3)} = \dfrac{(x+3)(y+3)}{(x+3)(x-3)} = \dfrac{y+3}{x-3}$

99. **Answers may vary.**

101. $\dfrac{3-x}{3+x} = \dfrac{-(x-3)}{x+3} = -\dfrac{x-3}{x+3}$

Exercise 6.2 (page 375)

1. $2x^3y^2(-3x^2y^4z) = -6x^5y^6z$

3. $(3y)^{-4} = \dfrac{1}{(3y)^4} = \dfrac{1}{81y^4}$

5. $\dfrac{x^{3m}}{x^{4m}} = x^{-m} = \dfrac{1}{x^m}$

7. $-4(y^3 - 4y^2 + 3y - 2) + 6(-2y^2 + 4) - 4(-2y^3 - y)$
$$= -4y^3 + 16y^2 - 12y + 8 - 12y^2 + 24 + 8y^3 + 4y$$
$$= 4y^3 + 4y^2 - 8y + 32$$

9. numerator

11. numerators; denominators

13. 1

15. divisor; multiply

17. $\dfrac{5}{7} \cdot \dfrac{9}{13} = \dfrac{5 \cdot 9}{7 \cdot 13} = \dfrac{45}{91}$

19. $\dfrac{25}{35} \cdot \dfrac{-21}{55} = \dfrac{-5 \cdot 5 \cdot 3 \cdot 7}{7 \cdot 5 \cdot 5 \cdot 11} = -\dfrac{3}{11}$

21. $\dfrac{2}{3} \cdot \dfrac{15}{2} \cdot \dfrac{1}{7} = \dfrac{2 \cdot 3 \cdot 5}{3 \cdot 2 \cdot 7} = \dfrac{5}{7}$

23. $\dfrac{3x}{y} \cdot \dfrac{y}{2} = \dfrac{3x \cdot y}{y \cdot 2} = \dfrac{3x}{2}$

25. $\dfrac{5y}{7} \cdot \dfrac{7x}{5z} = \dfrac{5 \cdot y \cdot 7 \cdot x}{7 \cdot 5 \cdot z} = \dfrac{xy}{z}$

27. $\dfrac{7z}{9z} \cdot \dfrac{4z}{2z} = \dfrac{7 \cdot z \cdot 2 \cdot 2 \cdot z}{9 \cdot z \cdot 2 \cdot z} = \dfrac{14}{9}$

29. $\dfrac{2x^2y}{3xy} \cdot \dfrac{3xy^2}{2} = \dfrac{6x^3y^3}{6xy} = x^2y^2$

31. $\dfrac{8x^2y^2}{4x^2} \cdot \dfrac{2xy}{2y} = \dfrac{16x^3y^3}{8x^2y} = 2xy^2$

33. $\dfrac{-2xy}{x^2} \cdot \dfrac{3xy}{2} = -\dfrac{6x^2y^2}{2x^2} = -3y^2$

35. $\dfrac{ab^2}{a^2b} \cdot \dfrac{b^2c^2}{abc} \cdot \dfrac{abc^2}{a^3c^2} = \dfrac{a^2b^5c^4}{a^6b^2c^3} = \dfrac{b^3c}{a^4}$

37. $\dfrac{10r^2st^3}{6rs^2} \cdot \dfrac{3r^3t}{2rst} \cdot \dfrac{2s^3t^4}{5s^2t^3} = \dfrac{60r^5s^4t^8}{60r^2s^5t^4} = \dfrac{r^3t^4}{s}$

39. $\dfrac{z+7}{7} \cdot \dfrac{z+2}{z} = \dfrac{(z+7)(z+2)}{7z}$

41. $\dfrac{x-2}{2} \cdot \dfrac{2x}{x-2} = \dfrac{2x(x-2)}{2(x-2)} = x$

43. $\dfrac{x+5}{5} \cdot \dfrac{x}{x+5} = \dfrac{x(x+5)}{5(x+5)} = \dfrac{x}{5}$

45. $\dfrac{(x+1)^2}{x+1} \cdot \dfrac{x+2}{x+1} = \dfrac{(x+2)(x+1)^2}{(x+1)^2}$
$$= x+2$$

47. $\dfrac{2x+6}{x+3} \cdot \dfrac{3}{4x} = \dfrac{2(x+3)}{x+3} \cdot \dfrac{3}{4x}$
$$= \frac{6(x+3)}{4x(x+3)} = \frac{3}{2x}$$

49. $\dfrac{x^2-x}{x} \cdot \dfrac{3x-6}{3x-3} = \dfrac{x(x-1)}{x} \cdot \dfrac{3(x-2)}{3(x-1)} = \dfrac{3x(x-1)(x-2)}{3x(x-1)} = x-2$

51. $\dfrac{7y-14}{y-2} \cdot \dfrac{x^2}{7x} = \dfrac{7(y-2)}{y-2} \cdot \dfrac{x^2}{7x} = \dfrac{7x^2(y-2)}{7x(y-2)} = x$

53. $\dfrac{x^2+x-6}{5x}\cdot\dfrac{5x-10}{x+3}=\dfrac{(x+3)(x-2)}{5x}\cdot\dfrac{5(x-2)}{x+3}=\dfrac{5(x+3)(x-2)^2}{5x(x+3)}=\dfrac{(x-2)^2}{x}$

55. $\dfrac{m^2-2m-3}{2m+4}\cdot\dfrac{m^2-4}{m^2+3m+2}=\dfrac{(m-3)(m+1)}{2(m+2)}\cdot\dfrac{(m+2)(m-2)}{(m+2)(m+1)}$

$=\dfrac{(m-3)(m+1)(m+2)(m-2)}{2(m+2)(m+2)(m+1)}=\dfrac{(m-3)(m-2)}{2(m+2)}$

57. $\dfrac{x^2+7xy+12y^2}{x^2+2xy-8y^2}\cdot\dfrac{x^2-xy-2y^2}{x^2+4xy+3y^2}=\dfrac{(x+4y)(x+3y)}{(x+4y)(x-2y)}\cdot\dfrac{(x-2y)(x+y)}{(x+3y)(x+y)}$

$=\dfrac{(x+4y)(x+3y)(x-2y)(x+y)}{(x+4y)(x-2y)(x+3y)(x+y)}=1$

59. $\dfrac{ax+bx+ay+by}{x^3-y^3}\cdot\dfrac{x^2+xy+y^2}{ax+bx}=\dfrac{x(a+b)+y(a+b)}{(x-y)(x^2+xy+y^2)}\cdot\dfrac{x^2+xy+y^2}{x(a+b)}$

$=\dfrac{(a+b)(x+y)}{(x-y)(x^2+xy+y^2)}\cdot\dfrac{x^2+xy+y^2}{x(a+b)}$

$=\dfrac{(a+b)(x+y)(x^2+xy+y^2)}{(x-y)(x^2+xy+y^2)\cdot x(a+b)}=\dfrac{x+y}{x(x-y)}$

61. $\dfrac{x^2-y^2}{y^2-xy}\cdot\dfrac{yx^3-y^4}{ax+ay+bx+by}=\dfrac{(x+y)(x-y)}{y(y-x)}\cdot\dfrac{y(x^3-y^3)}{a(x+y)+b(x+y)}$

$=-\dfrac{(x+y)}{y}\cdot\dfrac{y(x-y)(x^2+xy+y^2)}{(x+y)(a+b)}$

$=-\dfrac{(x+y)\cdot y(x-y)(x^2+xy+y^2)}{y(x+y)(a+b)}$

$=-\dfrac{(x-y)(x^2+xy+y^2)}{a+b}$

63. $\dfrac{abc^2}{a+1}\cdot\dfrac{c}{a^2b^2}\cdot\dfrac{a^2+a}{ac}=\dfrac{abc^2}{a+1}\cdot\dfrac{c}{a^2b^2}\cdot\dfrac{a(a+1)}{ac}=\dfrac{a^2bc^3(a+1)}{a^3b^2c(a+1)}=\dfrac{c^2}{ab}$

65. $\dfrac{3x^2+5x+2}{x^2-9}\cdot\dfrac{x-3}{x^2-4}\cdot\dfrac{x^2+5x+6}{6x+4}=\dfrac{(3x+2)(x+1)}{(x+3)(x-3)}\cdot\dfrac{x-3}{(x+2)(x-2)}\cdot\dfrac{(x+2)(x+3)}{2(3x+2)}$

$=\dfrac{(3x+2)(x+1)(x-3)(x+2)(x+3)}{2(x+3)(x-3)(x+2)(x-2)(3x+2)}=\dfrac{x+1}{2(x-2)}$

67. $\dfrac{1}{3}\div\dfrac{1}{2}=\dfrac{1}{3}\cdot\dfrac{2}{1}=\dfrac{2}{3}$

69. $\dfrac{21}{14}\div\dfrac{5}{2}=\dfrac{21}{14}\cdot\dfrac{2}{5}=\dfrac{3\cdot7\cdot2}{2\cdot7\cdot5}=\dfrac{3}{5}$

71. $\dfrac{2}{y}\div\dfrac{4}{3}=\dfrac{2}{y}\cdot\dfrac{3}{4}=\dfrac{2\cdot3}{y\cdot2\cdot2}=\dfrac{3}{2y}$

73. $\dfrac{3x}{2}\div\dfrac{x}{2}=\dfrac{3x}{2}\cdot\dfrac{2}{x}=\dfrac{3\cdot x\cdot2}{2\cdot x}=3$

75. $\dfrac{3x}{y} \div \dfrac{2x}{4} = \dfrac{3x}{y} \cdot \dfrac{4}{2x} = \dfrac{3 \cdot x \cdot 2 \cdot 2}{y \cdot 2 \cdot x} = \dfrac{6}{y}$

77. $\dfrac{4x}{3x} \div \dfrac{2y}{9y} = \dfrac{4x}{3x} \cdot \dfrac{9y}{2y} = \dfrac{2 \cdot 2 \cdot x \cdot 3 \cdot 3 \cdot y}{3 \cdot x \cdot 2 \cdot y}$
$= 6$

79. $\dfrac{x^2}{3} \div \dfrac{2x}{4} = \dfrac{x^2}{3} \cdot \dfrac{4}{2x} = \dfrac{x \cdot x \cdot 2 \cdot 2}{3 \cdot 2 \cdot x} = \dfrac{2x}{3}$

81. $\dfrac{x^2y}{3xy} \div \dfrac{xy^2}{6y} = \dfrac{x^2y}{3xy} \cdot \dfrac{6y}{xy^2} = \dfrac{6x^2y^2}{3x^2y^3} = \dfrac{2}{y}$

83. $\dfrac{x+2}{3x} \div \dfrac{x+2}{2} = \dfrac{x+2}{3x} \cdot \dfrac{2}{x+2} = \dfrac{2(x+2)}{3x(x+2)} = \dfrac{2}{3x}$

85. $\dfrac{(z-2)^2}{3z^2} \div \dfrac{z-2}{6z} = \dfrac{(z-2)^2}{3z^2} \cdot \dfrac{6z}{z-2} = \dfrac{6z(z-2)^2}{3z^2(z-2)} = \dfrac{2(z-2)}{z}$

87. $\dfrac{(z-7)^2}{z+2} \div \dfrac{z(z-7)}{5z^2} = \dfrac{(z-7)^2}{z+2} \cdot \dfrac{5z^2}{z(z-7)} = \dfrac{5z^2(z-7)^2}{z(z+2)(z-7)} = \dfrac{5z(z-7)}{z+2}$

89. $\dfrac{x^2-4}{3x+6} \div \dfrac{x-2}{x+2} = \dfrac{x^2-4}{3x+6} \cdot \dfrac{x+2}{x-2} = \dfrac{(x+2)(x-2)(x+2)}{3(x+2)(x-2)} = \dfrac{x+2}{3}$

91. $\dfrac{x^2-1}{3x-3} \div \dfrac{x+1}{3} = \dfrac{x^2-1}{3x-3} \cdot \dfrac{3}{x+1} = \dfrac{(x+1)(x-1)3}{3(x-1)(x+1)} = 1$

93. $\dfrac{5x^2+13x-6}{x+3} \div \dfrac{5x^2-17x+6}{x-2} = \dfrac{5x^2+13x-6}{x+3} \cdot \dfrac{x-2}{5x^2-17x+6}$
$= \dfrac{(5x-2)(x+3)(x-2)}{(x+3)(5x-2)(x-3)} = \dfrac{x-2}{x-3}$

95. $\dfrac{2x^2+8x-42}{x-3} \div \dfrac{2x^2+14x}{x^2+5x} = \dfrac{2(x^2+4x-21)}{x-3} \cdot \dfrac{x^2+5x}{2x^2+14x}$
$= \dfrac{2(x+7)(x-3)(x)(x+5)}{(x-3)(2x)(x+7)} = x+5$

97. $\dfrac{ab+4a+2b+8}{b^2+4b+16} \div \dfrac{b^2-16}{b^3-64} = \dfrac{ab+4a+2b+8}{b^2+4b+16} \cdot \dfrac{b^3-64}{b^2-16}$
$= \dfrac{a(b+4)+2(b+4)}{b^2+4b+16} \cdot \dfrac{b^3-4^3}{(b+4)(b-4)}$
$= \dfrac{(b+4)(a+2)}{b^2+4b+16} \cdot \dfrac{(b-4)(b^2+4b+16)}{(b+4)(b-4)}$
$= \dfrac{(b+4)(a+2)(b-4)(b^2+4b+16)}{(b^2+4b+16)(b+4)(b-4)} = a+2$

99. $$\frac{p^3-p^2q+pq^2}{mp-mq+np-nq} \div \frac{q^3+p^3}{q^2-p^2} = \frac{p^3-p^2q+pq^2}{mp-mq+np-nq}\cdot\frac{q^2-p^2}{q^3+p^3}$$
$$= \frac{p(p^2-pq+q^2)}{m(p-q)+n(p-q)}\cdot\frac{(q+p)(q-p)}{(q+p)(q^2-pq+p^2)}$$
$$= \frac{p(p^2-pq+q^2)}{(p-q)(m+n)}\cdot\frac{(q+p)(q-p)}{(q+p)(q^2-pq+p^2)}$$
$$= \frac{p(p^2-pq+q^2)(q+p)(q-p)}{(p-q)(m+n)(q+p)(q^2-pq+p^2)} = -\frac{p}{m+n}$$

101. $$\frac{x}{3}\cdot\frac{9}{4}\div\frac{x^2}{6} = \frac{x}{3}\cdot\frac{9}{4}\cdot\frac{6}{x^2} = \frac{x\cdot3\cdot3\cdot2\cdot3}{3\cdot2\cdot2\cdot x\cdot x} = \frac{9}{2x}$$

103. $$\frac{x^2}{18}\div\frac{x^3}{6}\div\frac{12}{x^2} = \frac{x^2}{18}\cdot\frac{6}{x^3}\cdot\frac{x^2}{12} = \frac{6x^4}{216x^3} = \frac{x}{36}$$

105. $$\frac{x^2-1}{x^2-9}\cdot\frac{x+3}{x+2}\div\frac{5}{x+2} = \frac{x^2-1}{x^2-9}\cdot\frac{x+3}{x+2}\cdot\frac{x+2}{5} = \frac{(x+1)(x-1)(x+3)(x+2)}{5(x+3)(x-3)(x+2)} = \frac{(x+1)(x-1)}{5(x-3)}$$

107. $$\frac{x^2-4}{2x+6}\div\frac{x+2}{4}\cdot\frac{x+3}{x-2} = \frac{x^2-4}{2x+6}\cdot\frac{4}{x+2}\cdot\frac{x+3}{x-2} = \frac{4(x+2)(x-2)(x+3)}{2(x+3)(x+2)(x-2)} = 2$$

109. $$\frac{x-x^2}{x^2-4}\left(\frac{2x+4}{x+2}\div\frac{5}{x+2}\right) = \frac{x-x^2}{x^2-4}\left(\frac{2x+4}{x+2}\cdot\frac{x+2}{5}\right) = \frac{x-x^2}{x^2-4}\left(\frac{2(x+2)(x+2)}{5(x+2)}\right)$$
$$= \frac{x(1-x)}{(x+2)(x-2)}\cdot\frac{2(x+2)}{5} = \frac{2x(1-x)}{5(x-2)}$$

111. $$\frac{y^2}{x+1}\cdot\frac{x^2+2x+1}{x^2-1}\div\frac{3y}{xy-y} = \frac{y^2}{x+1}\cdot\frac{(x+1)(x+1)}{(x+1)(x-1)}\cdot\frac{xy-y}{3y} = \frac{y^2(x+1)(x+1)(y)(x-1)}{(x+1)(x+1)(x-1)(3y)} = \frac{y^2}{3}$$

113. $$\frac{x^2+x-6}{x^2-4}\cdot\frac{x^2+2x}{x-2}\div\frac{x^2+3x}{x+2} = \frac{x^2+x-6}{x^2-4}\cdot\frac{x^2+2x}{x-2}\cdot\frac{x+2}{x^2+3x}$$
$$= \frac{(x+3)(x-2)(x)(x+2)(x+2)}{(x+2)(x-2)(x-2)(x)(x+3)} = \frac{x+2}{x-2}$$

115. Answers may vary.

117. Answers may vary.

119. You always get the original value of x after simplifying.

Exercise 6.3 (page 387)

1. $49 = 7 \cdot 7 = 7^2$

3. $136 = 4 \cdot 34 = 2 \cdot 2 \cdot 2 \cdot 17 = 2^3 \cdot 17$

5. $102 = 2 \cdot 51 = 2 \cdot 3 \cdot 17$

7. $144 = 16 \cdot 9 = 2 \cdot 2 \cdot 2 \cdot 2 \cdot 3 \cdot 3 = 2^4 \cdot 3^2$

9. LCD

11. numerators; common denominator

13. $\frac{1}{3} + \frac{1}{3} = \frac{1+1}{3} = \frac{2}{3}$

15. $\frac{2}{9} + \frac{1}{9} = \frac{2+1}{9} = \frac{3}{9} = \frac{1}{3}$

17. $\frac{2x}{y} + \frac{2x}{y} = \frac{2x+2x}{y} = \frac{4x}{y}$

19. $\frac{4}{7y} + \frac{10}{7y} = \frac{4+10}{7y} = \frac{14}{7y} = \frac{2}{y}$

21. $\frac{y+2}{5z} + \frac{y+4}{5z} = \frac{y+2+y+4}{5z} = \frac{2y+6}{5z}$

23. $\frac{3x-5}{x-2} + \frac{6x-13}{x-2} = \frac{3x-5+6x-13}{x-2} = \frac{9x-18}{x-2} = \frac{9(x-2)}{x-2} = 9$

25. $\frac{5}{7} - \frac{4}{7} = \frac{5-4}{7} = \frac{1}{7}$

27. $\frac{35}{72} - \frac{44}{72} = \frac{35-44}{72} = \frac{-9}{72} = -\frac{1}{8}$

29. $\frac{2x}{y} - \frac{x}{y} = \frac{2x-x}{y} = \frac{x}{y}$

31. $\frac{9y}{3x} - \frac{6y}{3x} = \frac{9y-6y}{3x} = \frac{3y}{3x} = \frac{y}{x}$

33. $\frac{6x-5}{3xy} - \frac{3x-5}{3xy} = \frac{6x-5-(3x-5)}{3xy} = \frac{6x-5-3x+5}{3xy} = \frac{3x}{3xy} = \frac{1}{y}$

35. $\frac{3y-2}{y+3} - \frac{2y-5}{y+3} = \frac{3y-2-(2y-5)}{y+3} = \frac{3y-2-2y+5}{y+3} = \frac{y+3}{y+3} = 1$

37. $\frac{13x}{15} + \frac{12x}{15} - \frac{5x}{15} = \frac{13x+12x-5x}{15}$
$= \frac{20x}{15} = \frac{4x}{3}$

39. $\frac{x}{3y} + \frac{2x}{3y} - \frac{x}{3y} = \frac{x+2x-x}{3y} = \frac{2x}{3y}$

41. $\frac{3x}{y+2} - \frac{3y}{y+2} + \frac{x+y}{y+2} = \frac{3x-3y+x+y}{y+2} = \frac{4x-2y}{y+2} = \frac{2(2x-y)}{y+2}$

43. $\frac{x+1}{x-2} - \frac{2(x-3)}{x-2} + \frac{3(x+1)}{x-2} = \frac{x+1-2(x-3)+3(x+1)}{x-2} = \frac{x+1-2x+6+3x+3}{x-2}$
$= \frac{2x+10}{x-2} = \frac{2(x+5)}{x-2}$

45. $\frac{25}{4} = \frac{25 \cdot 5}{4 \cdot 5} = \frac{125}{20}$

47. $\frac{8}{x} = \frac{8 \cdot xy}{x \cdot xy} = \frac{8xy}{x^2y}$

49. $\dfrac{3x}{x+1}=\dfrac{3x(x+1)}{(x+1)(x+1)}=\dfrac{3x(x+1)}{(x+1)^2}$

51. $\dfrac{2y}{x}=\dfrac{2y(x+1)}{x(x+1)}=\dfrac{2y(x+1)}{x^2+x}$

53. $\dfrac{z}{z-1}=\dfrac{z(z+1)}{(z-1)(z+1)}=\dfrac{z(z+1)}{z^2-1}$

55. $\dfrac{2}{x+1}=\dfrac{2(x+2)}{(x+1)(x+2)}=\dfrac{2(x+2)}{x^2+3x+2}$

57. $2x=2\cdot x$
$6x=2\cdot 3\cdot x$
$\text{LCD}=2\cdot 3\cdot x=6x$

59. $3x=3\cdot x$
$6y=2\cdot 3\cdot y$
$9xy=3^2\cdot x\cdot y$
$\text{LCD}=2\cdot 3^2\cdot x\cdot y=18xy$

61. $x^2-1=(x+1)(x-1)$
$x+1=x+1$
$\text{LCD}=(x+1)(x-1)=x^2-1$

63. $x^2+6x=x(x+6)$
$x+6=x+6$
$x=x$
$\text{LCD}=x(x+6)=x^2+6x$

65. $x^2-4x-5=(x+1)(x-5)$; $x^2-25=(x+5)(x-5)$; $\text{LCD}=(x+1)(x-5)(x+5)$

67. $\dfrac{1}{2}+\dfrac{2}{3}=\dfrac{1\cdot 3}{2\cdot 3}+\dfrac{2\cdot 2}{3\cdot 2}=\dfrac{3}{6}+\dfrac{4}{6}=\dfrac{7}{6}$

69. $\dfrac{2y}{9}+\dfrac{y}{3}=\dfrac{2y}{9}+\dfrac{y\cdot 3}{3\cdot 3}=\dfrac{2y}{9}+\dfrac{3y}{9}=\dfrac{5y}{9}$

71. $\dfrac{21x}{14}-\dfrac{5x}{21}=\dfrac{21x\cdot 3}{14\cdot 3}-\dfrac{5x\cdot 2}{21\cdot 2}$
$=\dfrac{63x}{42}-\dfrac{10x}{42}=\dfrac{53x}{42}$

73. $\dfrac{4x}{3}+\dfrac{2x}{y}=\dfrac{4x\cdot y}{3\cdot y}+\dfrac{2x\cdot 3}{y\cdot 3}$
$=\dfrac{4xy}{3y}+\dfrac{6x}{3y}=\dfrac{4xy+6x}{3y}$

75. $\dfrac{2}{x}-3x=\dfrac{2}{x}-\dfrac{3x}{1}=\dfrac{2}{x}-\dfrac{3x\cdot x}{1\cdot x}=\dfrac{2}{x}-\dfrac{3x^2}{x}=\dfrac{2-3x^2}{x}$

77. $\dfrac{y+2}{5y}+\dfrac{y+4}{15y}=\dfrac{(y+2)3}{5y\cdot 3}+\dfrac{y+4}{15y}=\dfrac{3y+6}{15y}+\dfrac{y+4}{15y}=\dfrac{4y+10}{15y}$

79. $\dfrac{x+5}{xy}-\dfrac{x-1}{x^2y}=\dfrac{(x+5)x}{xy\cdot x}-\dfrac{x-1}{x^2y}=\dfrac{x^2+5x}{x^2y}-\dfrac{x-1}{x^2y}=\dfrac{x^2+4x+1}{x^2y}$

81. $\dfrac{x}{x+1}+\dfrac{x-1}{x}=\dfrac{x\cdot x}{(x+1)x}+\dfrac{(x-1)(x+1)}{x(x+1)}=\dfrac{x^2}{x(x+1)}+\dfrac{x^2-1}{x(x+1)}=\dfrac{2x^2-1}{x(x+1)}$

83. $\dfrac{x-1}{x}+\dfrac{y+1}{y}=\dfrac{(x-1)y}{x\cdot y}+\dfrac{(y+1)x}{y\cdot x}=\dfrac{xy-y}{xy}+\dfrac{xy+x}{xy}=\dfrac{2xy+x-y}{xy}$

85. $\dfrac{x}{x-2}+\dfrac{4+2x}{x^2-4}=\dfrac{x}{x-2}+\dfrac{2(x+2)}{(x+2)(x-2)}=\dfrac{x}{x-2}+\dfrac{2}{x-2}=\dfrac{x+2}{x-2}$

87. $$\frac{x+1}{x-1}+\frac{x-1}{x+1}=\frac{(x+1)(x+1)}{(x-1)(x+1)}+\frac{(x-1)(x-1)}{(x+1)(x-1)}=\frac{x^2+2x+1}{(x+1)(x-1)}+\frac{x^2-2x+1}{(x+1)(x-1)}$$
$$=\frac{2x^2+2}{(x+1)(x-1)}$$

89. $$\frac{2x+2}{x-2}-\frac{2x}{2-x}=\frac{2x+2}{x-2}+\frac{2x}{x-2}=\frac{4x+2}{x-2}=\frac{2(2x+1)}{x-2}$$

91. $$\frac{2x}{x^2-3x+2}+\frac{2x}{x-1}-\frac{x}{x-2}=\frac{2x}{(x-2)(x-1)}+\frac{2x}{x-1}-\frac{x}{x-2}$$
$$=\frac{2x}{(x-2)(x-1)}+\frac{2x(x-2)}{(x-1)(x-2)}-\frac{x(x-1)}{(x-2)(x-1)}$$
$$=\frac{2x+2x(x-2)-x(x-1)}{(x-2)(x-1)}$$
$$=\frac{2x+2x^2-4x-x^2+x}{(x-2)(x-1)}$$
$$=\frac{x^2-x}{(x-2)(x-1)}=\frac{x(x-1)}{(x-2)(x-1)}=\frac{x}{x-2}$$

93. $$\frac{2x}{x-1}+\frac{3x}{x+1}-\frac{x+3}{x^2-1}=\frac{2x}{x-1}+\frac{3x}{x+1}-\frac{x+3}{(x+1)(x-1)}$$
$$=\frac{2x(x+1)}{(x-1)(x+1)}+\frac{3x(x-1)}{(x+1)(x-1)}-\frac{x+3}{(x+1)(x-1)}$$
$$=\frac{2x(x+1)+3x(x-1)-(x+3)}{(x+1)(x-1)}$$
$$=\frac{2x^2+2x+3x^2-3x-x-3}{(x+1)(x-1)}$$
$$=\frac{5x^2-2x-3}{(x+1)(x-1)}=\frac{(5x+3)(x-1)}{(x+1)(x-1)}=\frac{5x+3}{x+1}$$

95. $$\frac{x+1}{2x+4}-\frac{x^2}{2x^2-8}=\frac{x+1}{2(x+2)}-\frac{x^2}{2(x+2)(x-2)}=\frac{(x+1)(x-2)}{2(x+2)(x-2)}-\frac{x^2}{2(x+2)(x-2)}$$
$$=\frac{x^2-2x+x-2-x^2}{2(x+2)(x-2)}$$
$$=\frac{-x-2}{2(x+2)(x-2)}$$
$$=\frac{-(x+2)}{2(x+2)(x-2)}=-\frac{1}{2(x-2)}$$

97. $\frac{-a}{3a^2-27}+\frac{1}{3a+9}=\frac{-a}{3(a^2-9)}+\frac{1}{3(a+3)}=\frac{-a}{3(a+3)(a-3)}+\frac{1}{3(a+3)}$
$=\frac{-a}{3(a+3)(a-3)}+\frac{1(a-3)}{3(a+3)(a-3)}$
$=\frac{-a+1(a-3)}{3(a+3)(a-3)}$
$=\frac{-a+a-3}{3(a+3)(a-3)}$
$=\frac{-3}{3(a+3)(a-3)}=\frac{-1}{(a+3)(a-3)}$

99. **Answers may vary.**

101. **Answers may vary.**

103. The subtraction needs to be distributed in the numerator of the 2nd fraction.

105. $\frac{a}{b}+\frac{c}{d}=\frac{a(d)}{b(d)}+\frac{c(b)}{d(b)}$
$=\frac{ad}{bd}+\frac{bc}{bd}=\frac{ad+bc}{bd}$

Exercise 6.4 (page 394)

1. $t^3t^4t^2=t^{3+4+2}=t^9$

3. $-2r(r^3)^2=-2rr^6=-2r^7$

5. $\left(\frac{3r}{4r^3}\right)^4=\left(\frac{3}{4r^2}\right)^4=\frac{3^4}{(4r^2)^4}=\frac{81}{256r^8}$

7. $\left(\frac{6r^{-2}}{2r^3}\right)^{-2}=(3r^{-5})^{-2}=\frac{r^{10}}{9}$

9. complex fraction

11. single; divide

13. $\frac{\frac{2}{3}}{\frac{3}{4}}=\frac{2}{3}\div\frac{3}{4}=\frac{2}{3}\cdot\frac{4}{3}=\frac{8}{9}$

15. $\frac{\frac{4}{5}}{\frac{32}{15}}=\frac{4}{5}\div\frac{32}{15}=\frac{4}{5}\cdot\frac{15}{32}=\frac{3}{8}$

17. $\frac{\frac{2}{3}+1}{\frac{1}{3}+1}=\frac{(\frac{2}{3}+1)3}{(\frac{1}{3}+1)3}=\frac{\frac{2}{3}\cdot 3+1\cdot 3}{\frac{1}{3}\cdot 3+1\cdot 3}$
$=\frac{2+3}{1+3}=\frac{5}{4}$

19. $\frac{\frac{1}{2}+\frac{3}{4}}{\frac{3}{2}+\frac{1}{4}}=\frac{(\frac{1}{2}+\frac{3}{4})4}{(\frac{3}{2}+\frac{1}{4})4}=\frac{\frac{1}{2}\cdot 4+\frac{3}{4}\cdot 4}{\frac{3}{2}\cdot 4+\frac{1}{4}\cdot 4}$
$=\frac{2+3}{6+1}=\frac{5}{7}$

21. $\frac{\frac{x}{y}}{\frac{1}{x}}=\frac{x}{y}\div\frac{1}{x}=\frac{x}{y}\cdot\frac{x}{1}=\frac{x^2}{y}$

23. $\frac{\frac{5t^2}{9x^2}}{\frac{3t}{x^2t}}=\frac{5t^2}{9x^2}\div\frac{3t}{x^2t}=\frac{5t^2}{9x^2}\cdot\frac{x^2t}{3t}$
$=\frac{5x^2t^3}{27x^2t}=\frac{5t^2}{27}$

25. $\frac{\frac{1}{x}-3}{\frac{5}{x}+2}=\frac{(\frac{1}{x}-3)x}{(\frac{5}{x}+2)x}=\frac{\frac{1}{x}\cdot x-3\cdot x}{\frac{5}{x}\cdot x+2\cdot x}$
$=\frac{1-3x}{5+2x}$

SECTION 6.4

27. $$\frac{\frac{2}{x}+2}{\frac{4}{x}+2}=\frac{\left(\frac{2}{x}+2\right)x}{\left(\frac{4}{x}+2\right)x}=\frac{\frac{2}{x}\cdot x+2\cdot x}{\frac{4}{x}\cdot x+2\cdot x}=\frac{2+2x}{4+2x}=\frac{2(1+x)}{2(2+x)}=\frac{1+x}{2+x}$$

29. $$\frac{\frac{3y}{x}-y}{y-\frac{y}{x}}=\frac{\left(\frac{3y}{x}-y\right)x}{\left(y-\frac{y}{x}\right)x}=\frac{\frac{3y}{x}\cdot x-y\cdot x}{y\cdot x-\frac{y}{x}\cdot x}=\frac{3y-xy}{xy-y}=\frac{y(3-x)}{y(x-1)}=\frac{3-x}{x-1}$$

31. $$\frac{\frac{2}{a+2}+1}{\frac{3}{a+2}}=\frac{\left(\frac{2}{a+2}+1\right)(a+2)}{\left(\frac{3}{a+2}\right)(a+2)}=\frac{\frac{2}{a+2}(a+2)+1(a+2)}{3}=\frac{2+a+2}{3}=\frac{a+4}{3}$$

33. $$\frac{\frac{1}{x+1}}{1+\frac{1}{x+1}}=\frac{\left(\frac{1}{x+1}\right)(x+1)}{\left(1+\frac{1}{x+1}\right)(x+1)}=\frac{\frac{1}{x+1}(x+1)}{1(x+1)+\frac{1}{x+1}(x+1)}=\frac{1}{x+1+1}=\frac{1}{x+2}$$

35. $$\frac{\frac{x}{x+2}}{\frac{x}{x+2}+x}=\frac{\left(\frac{x}{x+2}\right)(x+2)}{\left(\frac{x}{x+2}+x\right)(x+2)}=\frac{\frac{x}{x+2}(x+2)}{\frac{x}{x+2}(x+2)+x(x+2)}=\frac{x}{x+x^2+2x}=\frac{x}{x^2+3x}$$
$$=\frac{x}{x(x+3)}=\frac{1}{x+3}$$

37. $$\frac{1}{\frac{1}{x}+\frac{1}{y}}=\frac{1(xy)}{\left(\frac{1}{x}+\frac{1}{y}\right)xy}=\frac{xy}{\frac{1}{x}\cdot xy+\frac{1}{y}\cdot xy}=\frac{xy}{y+x}$$

39. $$\frac{\frac{2}{x}}{\frac{2}{y}-\frac{4}{x}}=\frac{\frac{2}{x}(xy)}{\left(\frac{2}{y}-\frac{4}{x}\right)(xy)}=\frac{\frac{2}{x}(xy)}{\frac{2}{y}(xy)-\frac{4}{x}(xy)}=\frac{2y}{2x-4y}=\frac{2y}{2(x-2y)}=\frac{y}{x-2y}$$

41. $$\frac{\frac{3}{x}+\frac{2x}{y}}{\frac{4}{x}}=\frac{\left(\frac{3}{x}+\frac{2x}{y}\right)(xy)}{\frac{4}{x}(xy)}=\frac{\frac{3}{x}(xy)+\frac{2x}{y}(xy)}{4y}=\frac{3y+2x^2}{4y}$$

43. $$\frac{3+\frac{3}{x-1}}{3-\frac{3}{x}}=\frac{\left(3+\frac{3}{x-1}\right)(x)(x-1)}{\left(3-\frac{3}{x}\right)(x)(x-1)}=\frac{3x(x-1)+\frac{3}{x-1}(x)(x-1)}{3x(x-1)-\frac{3}{x}(x)(x-1)}=\frac{3x^2-3x+3x}{3x^2-3x-3(x-1)}$$
$$=\frac{3x^2}{3x^2-6x+3}$$
$$=\frac{3x^2}{3(x^2-2x+1)}$$
$$=\frac{x^2}{x^2-2x+1}=\frac{x^2}{(x-1)^2}$$

45. $$\frac{\frac{3}{x}+\frac{4}{x+1}}{\frac{2}{x+1}-\frac{3}{x}}=\frac{\left(\frac{3}{x}+\frac{4}{x+1}\right)(x)(x+1)}{\left(\frac{2}{x+1}-\frac{3}{x}\right)(x)(x+1)}=\frac{\frac{3}{x}(x)(x+1)+\frac{4}{x+1}(x)(x+1)}{\frac{2}{x+1}(x)(x+1)-\frac{3}{x}(x)(x+1)}$$
$$=\frac{3(x+1)+4x}{2x-3(x+1)}=\frac{3x+3+4x}{2x-3x-3}=\frac{7x+3}{-x-3}$$

47. $$\frac{\frac{2}{x}-\frac{3}{x+1}}{\frac{2}{x+1}-\frac{3}{x}}=\frac{\left(\frac{2}{x}-\frac{3}{x+1}\right)(x)(x+1)}{\left(\frac{2}{x+1}-\frac{3}{x}\right)(x)(x+1)}=\frac{2(x+1)-3x}{2x-3(x+1)}$$
$$=\frac{2x+2-3x}{2x-3x-3}=\frac{-x+2}{-x-3}=\frac{-(x-2)}{-(x+3)}=\frac{x-2}{x+3}$$

49. $$\frac{\frac{m}{m+2}-\frac{2}{m-1}}{\frac{3}{m+2}+\frac{m}{m-1}}=\frac{\left(\frac{m}{m+2}-\frac{2}{m-1}\right)(m+2)(m-1)}{\left(\frac{3}{m+2}+\frac{m}{m-1}\right)(m+2)(m-1)}=\frac{\frac{m}{m+2}(m+2)(m-1)-\frac{2}{m-1}(m+2)(m-1)}{\frac{3}{m+2}(m+2)(m-1)+\frac{m}{m-1}(m+2)(m-1)}$$
$$=\frac{m(m-1)-2(m+2)}{3(m-1)+m(m+2)}$$
$$=\frac{m^2-m-2m-4}{3m-3+m^2+2m}$$
$$=\frac{m^2-3m-4}{m^2+5m-3}=\frac{(m-4)(m+1)}{m^2+5m-3}.$$

51. $$\frac{\frac{2}{x+2}}{\frac{3}{x-3}+\frac{1}{x}}=\frac{\left(\frac{2}{x+2}\right)(x)(x+2)(x-3)}{\left(\frac{3}{x-3}+\frac{1}{x}\right)(x)(x+2)(x-3)}=\frac{2x(x-3)}{\frac{3}{x-3}(x)(x+2)(x-3)+\frac{1}{x}(x)(x+2)(x-3)}$$
$$=\frac{2x^2-6x}{3x(x+2)+(x+2)(x-3)}$$
$$=\frac{2x^2-6x}{3x^2+6x+x^2-3x+2x-6}$$
$$=\frac{2x^2-6x}{4x^2+5x-6}=\frac{2x(x-3)}{(4x-3)(x+2)}$$

53. $$\frac{\frac{1}{x}+\frac{2}{x+1}}{\frac{2}{x-1}-\frac{1}{x}}=\frac{\left(\frac{1}{x}+\frac{2}{x+1}\right)(x)(x+1)(x-1)}{\left(\frac{2}{x-1}-\frac{1}{x}\right)(x)(x+1)(x-1)}=\frac{\frac{1}{x}(x)(x+1)(x-1)+\frac{2}{x+1}(x)(x+1)(x-1)}{\frac{2}{x-1}(x)(x+1)(x-1)-\frac{1}{x}(x)(x+1)(x-1)}$$
$$=\frac{(x+1)(x-1)+2x(x-1)}{2x(x+1)-(x+1)(x-1)}$$
$$=\frac{x^2-1+2x^2-2x}{2x^2+2x-(x^2-1)}$$
$$=\frac{3x^2-2x-1}{2x^2+2x-x^2+1}$$
$$=\frac{3x^2-2x-1}{x^2+2x+1}=\frac{(3x+1)(x-1)}{(x+1)(x+1)}$$

55. $$\frac{\frac{1}{y^2+y}-\frac{1}{xy+x}}{\frac{1}{xy+x}-\frac{1}{y^2+y}}=\frac{\frac{1}{y(y+1)}-\frac{1}{x(y+1)}}{\frac{1}{x(y+1)}-\frac{1}{y(y+1)}}=\frac{\left(\frac{1}{y(y+1)}-\frac{1}{x(y+1)}\right)(x)(y)(y+1)}{\left(\frac{1}{x(y+1)}-\frac{1}{y(y+1)}\right)(x)(y)(y+1)}$$
$$=\frac{1x-1y}{1y-1x}=\frac{x-y}{-(x-y)}=-1$$

57. $$\frac{x^{-2}}{y^{-1}}=\frac{\frac{1}{x^2}}{\frac{1}{y}}=\frac{1}{x^2}\div\frac{1}{y}=\frac{1}{x^2}\cdot\frac{y}{1}=\frac{y}{x^2}$$

59. $\dfrac{1+x^{-1}}{x^{-1}-1}=\dfrac{1+\frac{1}{x}}{\frac{1}{x}-1}=\dfrac{(1+\frac{1}{x})x}{(\frac{1}{x}-1)x}=\dfrac{1x+\frac{1}{x}\cdot x}{\frac{1}{x}\cdot x-1x}=\dfrac{x+1}{1-x}$

61. $\dfrac{a^{-2}+a}{a+1}=\dfrac{\frac{1}{a^2}+a}{a+1}=\dfrac{(\frac{1}{a^2}+a)a^2}{(a+1)a^2}=\dfrac{1+a^3}{a^3+a^2}=\dfrac{(1+a)(1-a+a^2)}{a^2(a+1)}=\dfrac{a^2-a+1}{a^2}$

63. $\dfrac{2x^{-1}+4x^{-2}}{2x^{-2}+x^{-1}}=\dfrac{\frac{2}{x}+\frac{4}{x^2}}{\frac{2}{x^2}+\frac{1}{x}}=\dfrac{(\frac{2}{x}+\frac{4}{x^2})x^2}{(\frac{2}{x^2}+\frac{1}{x})x^2}=\dfrac{2x+4}{2+x}=\dfrac{2(x+2)}{x+2}=2$

65. $\dfrac{1-25y^{-2}}{1+10y^{-1}+25y^{-2}}=\dfrac{1-\frac{25}{y^2}}{1+\frac{10}{y}+\frac{25}{y^2}}=\dfrac{\left(1-\frac{25}{y^2}\right)y^2}{\left(1+\frac{10}{y}+\frac{25}{y^2}\right)y^2}=\dfrac{y^2-25}{y^2+10y+25}$

$=\dfrac{(y+5)(y-5)}{(y+5)(y+5)}=\dfrac{y-5}{y+5}$

67. **Answers may vary.**

69. **Answers may vary.**

Exercise 6.5 (page 401)

1. $x^2+4x=x(x+4)$

3. $2x^2+x-3=(2x+3)(x-1)$

5. $x^4-16=(x^2)^2-4^2=(x^2+4)(x^2-4)=(x^2+4)(x+2)(x-2)$

7. extraneous

9. LCD

11. xy

13.
$$\begin{aligned}\frac{x}{2}+4&=\frac{3x}{2}\\ 2\left(\frac{x}{2}+4\right)&=2\left(\frac{3x}{2}\right)\\ x+8&=3x\\ 8&=2x\\ 4&=x\end{aligned}$$

15.
$$\begin{aligned}\frac{2y}{5}-8&=\frac{4y}{5}\\ 5\left(\frac{2y}{5}-8\right)&=5\left(\frac{4y}{5}\right)\\ 2y-40&=4y\\ -40&=2y\\ -20&=y\end{aligned}$$

17.
$$\begin{aligned}\frac{x}{3}+1&=\frac{x}{2}\\ 6\left(\frac{x}{3}+1\right)&=6\left(\frac{x}{2}\right)\\ 2x+6&=3x\\ 6&=x\end{aligned}$$

19.
$$\begin{aligned}\frac{x}{5}-\frac{x}{3}&=-8\\ 15\left(\frac{x}{5}-\frac{x}{3}\right)&=15(-8)\\ 3x-5x&=-120\\ -2x&=-120\\ x&=60\end{aligned}$$

21.
$$\begin{aligned}\frac{3a}{2}+\frac{a}{3}&=-22\\ 6\left(\frac{3a}{2}+\frac{a}{3}\right)&=6(-22)\\ 9a+2a&=-132\\ 11a&=-132\\ a&=-12\end{aligned}$$

23.
$$\begin{aligned}\frac{x-3}{3}+2x&=-1\\ 3\left(\frac{x-3}{3}+2x\right)&=3(-1)\\ x-3+6x&=-3\\ 7x&=0\\ x&=0\end{aligned}$$

25.
$$\begin{aligned} \frac{z-3}{2} &= z+2 \\ 2\left(\frac{z-3}{2}\right) &= 2(z+2) \\ z-3 &= 2z+4 \\ -7 &= z \end{aligned}$$

27.
$$\begin{aligned} \frac{5(x+1)}{8} &= x+1 \\ 8\left(\frac{5(x+1)}{8}\right) &= 8(x+1) \\ 5(x+1) &= 8(x+1) \\ 5x+5 &= 8x+8 \\ -3x &= 3 \\ x &= -1 \end{aligned}$$

29.
$$\begin{aligned} \frac{c-4}{4} &= \frac{c+4}{8} \\ 8\left(\frac{c-4}{4}\right) &= 8\left(\frac{c+4}{8}\right) \\ 2(c-4) &= c+4 \\ 2c-8 &= c+4 \\ c &= 12 \end{aligned}$$

31.
$$\begin{aligned} \frac{x+1}{3}+\frac{x-1}{5} &= \frac{2}{15} \\ 15\left(\frac{x+1}{3}+\frac{x-1}{5}\right) &= 15\left(\frac{2}{15}\right) \\ 5(x+1)+3(x-1) &= 2 \\ 5x+5+3x-3 &= 2 \\ 8x &= 0 \\ x &= 0 \end{aligned}$$

33.
$$\begin{aligned} \frac{3x-1}{6}-\frac{x+3}{2} &= \frac{3x+4}{3} \\ 12\left(\frac{3x-1}{6}-\frac{x+3}{2}\right) &= 12\left(\frac{3x+4}{3}\right) \\ 2(3x-1)-6(x+3) &= 4(3x+4) \\ 6x-2-6x-18 &= 12x+16 \\ -36 &= 12x \\ -3 &= x \end{aligned}$$

35.
$$\begin{aligned} \frac{3}{x}+2 &= 3 \\ x\left(\frac{3}{x}+2\right) &= x(3) \\ 3+2x &= 3x \\ 3 &= x \end{aligned}$$

37.
$$\begin{aligned} \frac{5}{a}-\frac{4}{a} &= 8+\frac{1}{a} \\ a\left(\frac{5}{a}-\frac{4}{a}\right) &= a\left(8+\frac{1}{a}\right) \\ 5-4 &= 8a+1 \\ 0 &= 8a \\ 0 &= a \end{aligned}$$

0 does not check. ⇒ no solutions

39.
$$\begin{aligned} \frac{2}{y+1}+5 &= \frac{12}{y+1} \\ (y+1)\left(\frac{2}{y+1}+5\right) &= (y+1)\left(\frac{12}{y+1}\right) \\ 2+5(y+1) &= 12 \\ 2+5y+5 &= 12 \\ 5y &= 5 \\ y &= 1 \end{aligned}$$

41.
$$\begin{aligned} \frac{1}{x-1}+\frac{3}{x-1} &= 1 \\ (x-1)\left(\frac{1}{x-1}+\frac{3}{x-1}\right) &= (x-1)(1) \\ 1+3 &= x-1 \\ 5 &= x \end{aligned}$$

43.
$$\begin{aligned} \frac{a^2}{a+2}-\frac{4}{a+2} &= a \\ (a+2)\left(\frac{a^2}{a+2}-\frac{4}{a+2}\right) &= (a+2)a \\ a^2-4 &= a^2+2a \\ -4 &= 2a \\ -2 &= a \end{aligned}$$

-2 does not check ⇒ no solutions

45.
$$\begin{aligned} \frac{x}{x-5}-\frac{5}{x-5} &= 3 \\ (x-5)\left(\frac{x}{x-5}-\frac{5}{x-5}\right) &= (x-5)3 \\ x-5 &= 3x-15 \\ -2x &= -10 \\ x &= 5 \end{aligned}$$

5 does not check ⇒ no solutions

47.
$$\begin{aligned}\frac{3r}{2}-\frac{3}{r}&=\frac{3r}{2}+3\\2r\left(\frac{3r}{2}-\frac{3}{r}\right)&=2r\left(\frac{3r}{2}+3\right)\\3r^2-6&=3r^2+6r\\-6&=6r\\-1&=r\end{aligned}$$

49.
$$\begin{aligned}\frac{1}{3}+\frac{2}{x-3}&=1\\3(x-3)\left(\frac{1}{3}+\frac{2}{x-3}\right)&=3(x-3)(1)\\x-3+3(2)&=3x-9\\x-3+6&=3x-9\\12&=2x\\6&=x\end{aligned}$$

51.
$$\begin{aligned}\frac{u}{u-1}+\frac{1}{u}&=\frac{u^2+1}{u^2-u}\\\frac{u}{u-1}+\frac{1}{u}&=\frac{u^2+1}{u(u-1)}\\u(u-1)\left(\frac{u}{u-1}+\frac{1}{u}\right)&=u(u-1)\left(\frac{u^2+1}{u(u-1)}\right)\\u^2+u-1&=u^2+1\\u-1&=1\\u&=2\end{aligned}$$

53.
$$\begin{aligned}\frac{3}{x-2}+\frac{1}{x}&=\frac{2(3x+2)}{x^2-2x}\\\frac{3}{x-2}+\frac{1}{x}&=\frac{6x+4}{x(x-2)}\\x(x-2)\left(\frac{3}{x-2}+\frac{1}{x}\right)&=x(x-2)\left(\frac{6x+4}{x(x-2)}\right)\\3x+x-2&=6x+4\\4x-2&=6x+4\\-2x&=6\\x&=-3\end{aligned}$$

55.
$$\begin{aligned}\frac{7}{q^2-q-2}+\frac{1}{q+1}&=\frac{3}{q-2}\\\frac{7}{(q-2)(q+1)}+\frac{1}{q+1}&=\frac{3}{q-2}\\(q-2)(q+1)\left(\frac{7}{(q-2)(q+1)}+\frac{1}{q+1}\right)&=(q-2)(q+1)\left(\frac{3}{q-2}\right)\\7+q-2&=3(q+1)\\q+5&=3q+3\\-2q&=-2\\q&=1\end{aligned}$$

57.

$$\frac{3y}{3y-6}+\frac{8}{y^2-4}=\frac{2y}{2y+4}$$
$$\frac{3y}{3(y-2)}+\frac{8}{(y+2)(y-2)}=\frac{2y}{2(y+2)}$$
$$\frac{y}{y-2}+\frac{8}{(y+2)(y-2)}=\frac{y}{y+2}$$
$$(y+2)(y-2)\left(\frac{y}{y-2}+\frac{8}{(y+2)(y-2)}\right)=(y+2)(y-2)\left(\frac{y}{y+2}\right)$$
$$y(y+2)+8=y(y-2)$$
$$y^2+2y+8=y^2-2y$$
$$8=-4y$$
$$-2=y: \quad -2 \text{ does not check} \Rightarrow \text{no solutions}$$

59.

$$y+\frac{2}{3}=\frac{2y-12}{3y-9}$$
$$y+\frac{2}{3}=\frac{2(y-6)}{3(y-3)}$$
$$3(y-3)\left(y+\frac{2}{3}\right)=3(y-3)\left(\frac{2y-12}{3(y-3)}\right)$$
$$3y(y-3)+2(y-3)=2y-12$$
$$3y^2-9y+2y-6=2y-12$$
$$3y^2-9y+6=0$$
$$3\left(y^2-3y+2\right)=0$$
$$3(y-2)(y-1)=0$$
$$y-2=0 \quad \textbf{or} \quad y-1=0$$
$$y=2 \qquad\qquad y=1$$

61.

$$\frac{5}{4y+12}-\frac{3}{4}=\frac{5}{4y+12}-\frac{y}{4}$$
$$\frac{5}{4(y+3)}-\frac{3}{4}=\frac{5}{4(y+3)}-\frac{y}{4}$$
$$4(y+3)\left(\frac{5}{4(y+3)}-\frac{3}{4}\right)=4(y+3)\left(\frac{5}{4(y+3)}-\frac{y}{4}\right)$$
$$5-3(y+3)=5-y(y+3)$$
$$5-3y-9=5-y^2-3y$$
$$y^2-9=0$$
$$(y+3)(y-3)=0$$
$$y+3=0 \quad \textbf{or} \quad y-3=0 \quad \Rightarrow y=-3 \text{ does not check, so the only solution is } y=3.$$
$$y=-3 \qquad\qquad y=3$$

63.

$$\frac{x}{x-1}-\frac{12}{x^2-x}=\frac{-1}{x-1}$$
$$\frac{x}{x-1}-\frac{12}{x(x-1)}=\frac{-1}{x-1}$$
$$x(x-1)\left(\frac{x}{x-1}-\frac{12}{x(x-1)}\right)=x(x-1)\left(\frac{-1}{x-1}\right)$$
$$x^2-12=-x$$
$$x^2+x-12=0$$
$$(x+4)(x-3)=0$$
$$x+4=0 \quad \textbf{or} \quad x-3=0$$
$$x=-4 \qquad\qquad x=3$$

65.

$$\frac{z-4}{z-3}=\frac{z+2}{z+1}$$
$$(z-3)(z+1)\left(\frac{z-4}{z-3}\right)=(z-3)(z+1)\left(\frac{z+2}{z+1}\right)$$
$$(z+1)(z-4)=(z-3)(z+2)$$
$$z^2-3z-4=z^2-z-6$$
$$-2z=-2$$
$$z=1$$

67.

$$\frac{n}{n^2-9}+\frac{n+8}{n+3}=\frac{n-8}{n-3}$$
$$\frac{n}{(n+3)(n-3)}+\frac{n+8}{n+3}=\frac{n-8}{n-3}$$
$$(n+3)(n-3)\left(\frac{n}{(n+3)(n-3)}+\frac{n+8}{n+3}\right)=(n+3)(n-3)\left(\frac{n-8}{n-3}\right)$$
$$n+(n-3)(n+8)=(n+3)(n-8)$$
$$n+n^2+5n-24=n^2-5n-24$$
$$11n=0$$
$$n=0$$

69.

$$\frac{b+2}{b+3}+1=\frac{-7}{b-5}$$
$$(b+3)(b-5)\left(\frac{b+2}{b+3}+1\right)=(b+3)(b-5)\left(\frac{-7}{b-5}\right)$$
$$(b-5)(b+2)+(b+3)(b-5)=-7(b+3)$$
$$b^2-3b-10+b^2-2b-15=-7b-21$$
$$2b^2+2b-4=0$$
$$2\left(b^2+b-2\right)=0$$
$$2(b+2)(b-1)=0$$
$$b+2=0 \quad \textbf{or} \quad b-1=0$$
$$b=-2 \qquad\qquad b=1$$

71.
$$\begin{aligned}\frac{1}{a}+\frac{1}{b}&=1\\ ab\left(\frac{1}{a}+\frac{1}{b}\right)&=ab(1)\\ b+a&=ab\\ b&=ab-a\\ b&=a(b-1)\\ \frac{b}{b-1}&=a\end{aligned}$$

73.
$$\begin{aligned}\frac{1}{f}&=\frac{1}{d_1}+\frac{1}{d_2}\\ fd_1d_2\cdot\frac{1}{f}&=fd_1d_2\left(\frac{1}{d_1}+\frac{1}{d_2}\right)\\ d_1d_2&=fd_2+fd_1\\ d_1d_2&=f(d_2+d_1)\\ \frac{d_1d_2}{d_2+d_1}&=f\end{aligned}$$

75. **Answers may vary.**

77.
$$\begin{aligned}x&=\frac{1}{x}\\ x(x)&=x\left(\frac{1}{x}\right)\\ x^2&=1\\ x^2-1&=0\\ (x+1)(x-1)&=0\end{aligned}$$
$x+1=0$ **or** $x-1=0$
$x=-1$ $\quad$ $x=1$

The numbers 1 and -1 are equal to their own reciprocals.

Exercise 6.6 (page 407)

1.
$$\begin{aligned}x^2-5x-6&=0\\ (x-6)(x+1)&=0\end{aligned}$$
$x-6=0$ **or** $x+1=0$
$x=6$ $\quad$ $x=-1$

3.
$$\begin{aligned}(t+2)(t^2+7t+12)&=0\\ (t+2)(t+4)(t+3)&=0\end{aligned}$$
$t+2=0$ **or** $t+4=0$ **or** $t+3=0$
$t=-2$ $\quad$ $t=-4$ $\quad$ $t=-3$

5.
$$\begin{aligned}y^3-y^2&=0\\ y^2(y-1)&=0\end{aligned}$$
$y=0$ **or** $y=0$ **or** $y-1=0$
$y=0$ $\quad$ $y=0$ $\quad$ $y=1$

7.
$$\begin{aligned}(x^2-1)(x^2-4)&=0\\ (x+1)(x-1)(x+2)(x-2)&=0\end{aligned}$$
$x+1=0$ **or** $x-1=0$ **or** $x+2=0$ **or** $x-2=0$
$x=-1$ $\quad$ $x=1$ $\quad$ $x=-2$ $\quad$ $x=2$

9. **Answers may vary.**

11. Let $x =$ the number.

$$\frac{2(3)}{4+x} = 1$$
$$(4+x)\left(\frac{6}{4+x}\right) = (4+x)(1)$$
$$6 = 4+x$$
$$2 = x \Rightarrow \text{The number is 2.}$$

13. Let $x =$ the number.

$$\frac{3+x}{4+2x} = \frac{4}{7}$$
$$7(4+2x)\left(\frac{3+x}{4+2x}\right) = 7(4+2x)\left(\frac{4}{7}\right)$$
$$7(x+3) = 4(4+2x)$$
$$7x+21 = 16+8x$$
$$5 = x$$

The number is 5.

15. Let $x =$ the number. $\Rightarrow x + \frac{1}{x} = \frac{13}{6}$

$$6x\left(x+\frac{1}{x}\right) = 6x\left(\frac{13}{6}\right)$$
$$6x^2 + 6 = 13x$$
$$6x^2 - 13x + 6 = 0$$
$$(2x-3)(3x-2) = 0$$

$2x - 3 = 0$ **or** $3x - 2 = 0$

$2x = 3$ $\quad$ $3x = 2$

$x = \frac{3}{2}$ $\quad$ $x = \frac{2}{3}$ $\quad$ The numbers are $\frac{3}{2}$ and $\frac{2}{3}$.

17. Let $x =$ minutes for both to grade the set.

Teacher in 1 minute	+	Aide in 1 minute	=	Total in 1 minute

$$\frac{1}{45} + \frac{1}{90} = \frac{1}{x}$$
$$90x\left(\frac{1}{45}+\frac{1}{90}\right) = 90x\left(\frac{1}{x}\right)$$
$$2x + x = 90$$
$$3x = 90$$
$$x = 30$$

It will take them 30 minutes to grade the set of quizzes.

19. Let $x =$ hours for both pipes to fill the pool.

1st pipe in 1 hour	+	2nd pipe in 1 hour	=	Total in 1 hour

$$\frac{1}{5} + \frac{1}{4} = \frac{1}{x}$$
$$20x\left(\frac{1}{5}+\frac{1}{4}\right) = 20x\left(\frac{1}{x}\right)$$
$$4x + 5x = 20$$
$$9x = 20$$
$$x = \frac{20}{9}$$

The pool can be filled in $2\frac{2}{9}$ hours.

21. Let $x =$ days for both working together.

Roofer in 1 day	+	Owner in 1 day	=	Total in 1 day

$$\frac{1}{4} + \frac{1}{7} = \frac{1}{x}$$
$$28x\left(\frac{1}{4}+\frac{1}{7}\right) = 28x\left(\frac{1}{x}\right)$$
$$7x + 4x = 28$$

$$7x + 4x = 28$$
$$11x = 28$$
$$x = \frac{28}{11}$$

The pool can be filled in $2\frac{6}{11}$ days.

23. Let $r =$ the rate for the heron.

	d	r	t
Goose	120	$r+10$	$\frac{120}{r+10}$
Heron	80	r	$\frac{80}{r}$

Time for goose = Time for heron

$$\frac{120}{r+10} = \frac{80}{r}$$
$$r(r+10)\left(\frac{120}{r+10}\right) = r(r+10)\left(\frac{80}{r}\right)$$
$$120r = 80(r+10)$$
$$120r = 80r + 800$$
$$40r = 800$$
$$20 = r$$

The heron's speed is 20 mph, while the goose's speed is 30 mph.

25. Let $r =$ the rate at which he walks.

	d	r	t
Walks	8	r	$\frac{8}{r}$
Rides	28	$r+10$	$\frac{28}{r+10}$

Time he walks = Time he rides

$$\frac{8}{r} = \frac{28}{r+10}$$
$$r(r+10)\left(\frac{8}{r}\right) = r(r+10)\left(\frac{28}{r+10}\right)$$
$$8(r+10) = 28r$$
$$8r + 80 = 28r$$
$$80 = 20r$$
$$4 = r$$

He walks 4 miles per hour, so it will take him $\frac{30}{4} = 7\frac{1}{2}$ hours to walk 30 miles.

27. Let $r =$ the speed of the current.

	d	r	t
Downstream	22	$18+r$	$\frac{22}{18+r}$
Upstream	14	$18-r$	$\frac{14}{18-r}$

Time downstream = Time upstream

$$\frac{22}{18+r} = \frac{14}{18-r}$$
$$(18+r)(18-r)\left(\frac{22}{18+r}\right) = (18+r)(18-r)\left(\frac{14}{18-r}\right)$$
$$22(18-r) = 14(18+r)$$
$$396 - 22r = 252 + 14r$$
$$144 = 36r$$
$$4 = r$$

The current has a speed of 4 miles per hour.

29. Let $r =$ the lower interest rate.

	I	P	r
Lower rate CD	175	$\frac{175}{r}$	r
Higher rate CD	200	$\frac{200}{r+.01}$	$r+.01$

Lower rate principal = Higher rate principal

$$\frac{175}{r} = \frac{200}{r+.01}$$
$$r(r+.01)\left(\frac{175}{r}\right) = r(r+.01)\left(\frac{200}{r+.01}\right)$$
$$175(r+.01) = 200r$$
$$175r + 1.75 = 200r$$
$$1.75 = 25r$$
$$0.07 = r$$

The rates are 7% and 8%.

31. Let $x =$ the number who contributed.

$$\boxed{\text{Original share}} = \boxed{\text{Share with more workers}} + 2$$

$$\frac{35}{x} = \frac{35}{x+2} + 2$$

$$x(x+2)\left(\frac{35}{x}\right) = x(x+2)\left(\frac{35}{x+2} + 2\right)$$

$$35(x+2) = 35x + 2x(x+2)$$

$$35x + 70 = 35x + 2x^2 + 4x$$

$$0 = 2x^2 + 4x - 70$$

$$0 = 2(x^2 + 2x - 35)$$

$$0 = 2(x+7)(x-5)$$

$x + 7 = 0$ **or** $x - 5 = 0$

$x = -7$ $\qquad$ $x = 5$

Since the answer cannot be negative, 5 workers must have contributed.

33. Let $x =$ the number bought.

Then each cost $\frac{120}{x}$.

$$\boxed{\text{Number bought}} \cdot \boxed{\text{Amount charged for each}} = 120$$

$$(x+10) \cdot \left(\frac{120}{x} - 1\right) = 120$$

$$120 - x + \frac{1200}{x} - 10 = 120$$

$$\frac{1200}{x} - x - 10 = 0$$

$$x\left(\frac{1200}{x} - x - 10\right) = 0$$

$$-x^2 - 10x + 1200 = 0$$

$$-(x^2 + 10x - 1200) = 0$$

$$-(x+40)(x-30) = 0$$

$x + 40 = 0$ **or** $x - 30 = 0$

$x = -40$ $\qquad$ $x = 30$

The store can buy 30 at the regular price.

35. Let $r =$ the still-water speed.

$$\boxed{\text{Time upstream}} + \boxed{\text{Time downstream}} = 5$$

$$\frac{60}{r-5} + \frac{60}{r+5} = 5$$

$$(r+5)(r-5)\left(\frac{60}{r-5} + \frac{60}{r+5}\right) = 5(r+5)(r-5)$$

$$60(r+5) + 60(r-5) = 5(r^2 - 25)$$

$$60r + 300 + 60r - 300 = 5r^2 - 125$$

$$0 = 5r^2 - 120r - 125$$

$$0 = 5(r^2 - 24r - 25)$$

$$0 = 5(r+1)(r-25)$$

$r + 1 = 0$ **or** $r - 25 = 0$

$r = -1$ $\qquad$ $r = 25$

The still-water speed should be 25 miles per hour.

	d	r	t
Upstream	60	$r-5$	$\frac{60}{r-5}$
Downstream	60	$r+5$	$\frac{60}{r+5}$

37. **Answers may vary.**

39. **Answers may vary.**

Exercise 6.7 (page 413)

1.

$$2x + 4 = 38$$

$$2x = 34$$

$$x = 17$$

3.

$$3(x+2) = 24$$

$$3x + 6 = 24$$

$$3x = 18$$

$$x = 6$$

5. $2x+6=2(x+3)$

7. $2x^2-x-6=(2x+3)(x-2)$

9. quotient

11. equal

13. **Answers may vary.**

15. $\frac{5}{7}$

17. $\frac{17}{34}=\frac{1}{2}$

19. $\frac{22}{33}=\frac{2}{3}$

21. $\frac{7}{24.5}=\frac{14}{49}=\frac{2}{7}$

23. $\frac{4\text{ oz}}{12\text{ oz}}=\frac{1}{3}$

25. $\frac{12\text{ min}}{1\text{ hr}}=\frac{12\text{ min}}{60\text{ min}}=\frac{1}{5}$

27. $\frac{3\text{ days}}{1\text{ week}}=\frac{3\text{ days}}{7\text{ days}}=\frac{3}{7}$

29. $\frac{18\text{ months}}{2\text{ years}}=\frac{18\text{ months}}{24\text{ months}}=\frac{3}{4}$

31. $750+652+188+125+110=\$1{,}825$

33. $\frac{110}{1825}=\frac{22}{365}$

35. $995+1245+1680+4580+225=\$8{,}725$

37. $\frac{1680}{8725}=\frac{336}{1745}$

39. $\frac{125}{2000}=\frac{1}{16}$

41. $\frac{\$53.55}{17\text{ gal}}=\$3.15/\text{gal}$

43. $\frac{84¢}{12\text{ oz}}=7¢/\text{oz}$

45. $\frac{89¢}{6\text{ oz}}\approx 14.873¢/\text{oz};\ \frac{119¢}{8\text{ oz}}=14.875¢/\text{oz}$
The 6-oz can is the better buy.

47. $\frac{54\text{ pg}}{40\text{ min}}=1.35\,\frac{\text{pg}}{\text{min}};\ \frac{80\text{ pg}}{62\text{ min}}\approx 1.29\,\frac{\text{pg}}{\text{min}}$
The 1st student read faster.

49. $\frac{11880\text{ gal}}{27\text{ min}}=440\text{ gal/min}$

51. $\frac{36.75-35}{35}=\frac{1.75}{35}=0.05=5\%$

53. $\frac{325\text{ mi}}{5\text{ hr}}=65\text{ mi/hr}$

55. $\frac{1235\text{ mi}}{51.3\text{ gal}}\approx 24\,\frac{\text{mi}}{\text{gal}};\ \frac{1456\text{ mi}}{55.78\text{ gal}}\approx 26\,\frac{\text{mi}}{\text{gal}}$
The 2nd car had the better mpg rating.

57. **Answers may vary.**

59. $\frac{17}{19}=\frac{17\cdot 21}{19\cdot 21}=\frac{357}{399};\ \frac{19}{21}=\frac{19\cdot 19}{21\cdot 19}=\frac{361}{399}$

Exercise 6.8 (page 423)

1. $\frac{9}{10}=0.9=90\%$

3. $33\frac{1}{3}\%=\frac{100}{3}\%=\frac{100}{3}\cdot\frac{1}{100}=\frac{1}{3}$

5. $rb = a$
$0.30(1600) = a$
$480 = a$

7. $rb = a$
$0.25(98) = a$
$24.50 = a$
$98 - 24.50 = \$73.50$

9. proportion; ratios

11. means

13. shape

15. ad; bc

17. triangle

19. $\frac{9}{7} \stackrel{?}{=} \frac{81}{70}$
$9 \cdot 70 \stackrel{?}{=} 7 \cdot 81$
$630 \neq 567$
not a proportion

21. $\frac{-7}{3} \stackrel{?}{=} \frac{14}{-6}$
$-7(-6) \stackrel{?}{=} 3 \cdot 14$
$42 = 42$
proportion

23. $\frac{9}{19} \stackrel{?}{=} \frac{38}{80}$
$9 \cdot 80 \stackrel{?}{=} 19 \cdot 38$
$720 \neq 722$
not a proportion

25. $\frac{10.4}{3.6} \stackrel{?}{=} \frac{41.6}{14.4}$
$10.4(14.4) \stackrel{?}{=} 3.6(41.6)$
$149.76 = 149.76$
proportion

27. $\frac{2}{3} = \frac{x}{6}$
$2(6) = 3x$
$12 = 3x$
$4 = x$

29. $\frac{5}{10} = \frac{3}{c}$
$5c = 10(3)$
$5c = 30$
$c = 6$

31. $\frac{-6}{x} = \frac{8}{4}$
$-6(4) = 8x$
$-24 = 8x$
$-3 = x$

33. $\frac{x}{3} = \frac{9}{3}$
$x(3) = 3(9)$
$3x = 27$
$x = 9$

35. $\frac{x+1}{5} = \frac{3}{15}$
$15(x+1) = 5(3)$
$15x + 15 = 15$
$15x = 0$
$x = 0$

37. $\frac{x+3}{12} = \frac{-7}{6}$
$6(x+3) = 12(-7)$
$6x + 18 = -84$
$6x = -102$
$x = -17$

39. $\frac{4-x}{13} = \frac{11}{26}$
$26(4-x) = 13(11)$
$104 - 26x = 143$
$-26x = 39$
$x = -\frac{39}{26} = -\frac{3}{2}$

41. $\frac{2x+1}{18} = \frac{14}{3}$
$3(2x+1) = 18(14)$
$6x + 3 = 252$
$6x = 249$
$x = \frac{249}{6} = \frac{83}{2}$

43. $\frac{3p-2}{12} = \frac{p+1}{3}$
$3(3p-2) = 12(p+1)$
$9p - 6 = 12p + 12$
$-3p = 18$
$p = -6$

45. Let $c =$ the cost of 51 pints of yogurt.

$$\frac{3 \text{ pints}}{51 \text{ pints}} = \frac{\text{cost of 3 pints}}{\text{cost of 51 pints}}$$
$$\frac{3}{51} = \frac{1}{c}$$
$$c \cdot 3 = 51 \cdot 1$$
$$3c = 51$$
$$c = 17$$

The 51 pints will cost \$17.

47. Let $c =$ the cost of 39 packets.

$$\frac{3 \text{ packets}}{39 \text{ packets}} = \frac{\text{cost of 3 packets}}{\text{cost of 39 packets}}$$
$$\frac{3}{39} = \frac{50}{c}$$
$$c \cdot 3 = 39 \cdot 50$$
$$3c = 1{,}950$$
$$c = 650 \text{ cents} = \$6.50$$

The 39 packets will cost \$6.50.

49. Let $d =$ the # of drops of pure essence needed with 56 drops of alcohol.

$$\frac{3 \text{ pure}}{7 \text{ alcohol}} = \frac{\text{drops of pure}}{56 \text{ drops of alcohol}}$$
$$\frac{3}{7} = \frac{d}{56}$$
$$3 \cdot 56 = 7d$$
$$168 = 7d$$
$$24 = d$$

24 drops of pure essence will be needed.

51. Let $f =$ the cups of flour needed for 12 dozen cookies.

$$\frac{1\frac{1}{4} \text{ C flour}}{3\frac{1}{2} \text{ dozen}} = \frac{\text{C flour}}{12 \text{ dozen}}$$
$$\frac{1\frac{1}{4}}{3\frac{1}{2}} = \frac{f}{12}$$
$$1\frac{1}{4} \cdot 12 = 3\frac{1}{2}f$$
$$\frac{5}{4} \cdot 12 = \frac{7}{2}f$$
$$15 = \frac{7}{2}f$$
$$30 = 7f$$
$$\frac{30}{7} = f$$
$$4\frac{2}{7} = f$$

$4\frac{2}{7}\left(\approx 4\frac{1}{4}\right)$ cups of flour will be needed.

53. Let $n =$ the number defective.

$$\frac{5 \text{ defective}}{100 \text{ parts}} = \frac{\text{number defective}}{940 \text{ parts}}$$
$$\frac{5}{100} = \frac{n}{940}$$
$$5 \cdot 940 = 100n$$
$$4700 = 100n$$
$$47 = n$$

There will be 47 defective parts.

55. Let $g =$ gallons of gas needed for 315 miles.

$$\frac{42 \text{ miles}}{315 \text{ miles}} = \frac{\text{gas for 42 miles}}{\text{gas for 315 miles}}$$
$$\frac{42}{315} = \frac{1}{g}$$
$$g \cdot 42 = 315 \cdot 1$$
$$42g = 315$$
$$g = \frac{315}{42} = \frac{15}{2} = 7\frac{1}{2}$$

To go 315 miles, $7\frac{1}{2}$ gallons of gas are needed.

57. Let $p =$ the amount he was paid last week.

$$\frac{\text{30 hours}}{\text{40 hours}} = \frac{\text{pay for 30 hours}}{\text{pay for 40 hours}}$$

$$\frac{30}{40} = \frac{p}{412}$$

$$412 \cdot 30 = 40 \cdot p$$

$$12{,}360 = 40p$$

$$309 = p$$

He was paid \$309 last week.

59. Let $l =$ the length of a real caboose.

$$\frac{\text{169 feet}}{\text{1 foot}} = \frac{\text{length of a real caboose}}{\text{length of a model caboose}}$$

$$\frac{169}{1} = \frac{l}{3.5}$$

$$169(3.5) = 1 \cdot l$$

$$591.5 = l$$

A real engine is 591.5 inches $(49' \; 3.5'')$long.

61. Let $t =$ the number of teachers required.

$$\frac{\text{3 teachers}}{\text{50 students}} = \frac{\text{number of teachers}}{\text{number of students}}$$

$$\frac{3}{50} = \frac{t}{2{,}700}$$

$$2{,}700 \cdot 3 = 50 \cdot t$$

$$8{,}100 = 50t$$

$$162 = t$$

A total of 162 teachers are needed.

63. Let $x =$ the amount of oil to be added.

$$\frac{50}{1} = \frac{\text{ounces of gasoline}}{\text{ounces of oil}}$$

$$\frac{50}{1} = \frac{6 \cdot 128}{x}$$

$$\frac{50}{1} = \frac{768}{x}$$

$$x \cdot 50 = 1 \cdot 768$$

$$50x = 768$$

$$x = \frac{768}{50} \approx 15.36 \text{ ounces}$$

The directions are close.

65. $$\frac{\text{height of man}}{\text{man's shadow}} = \frac{\text{height of tree}}{\text{tree's shadow}}$$

$$\frac{6}{4} = \frac{h}{26}$$

$$6 \cdot 26 = 4h$$

$$156 = 4h$$

$$39 = h$$

The tree is 39 feet tall.

67. $$\frac{75}{32} = \frac{w}{20}$$

$$75 \cdot 20 = 32w$$

$$1500 = 32w$$

$$\frac{1500}{32} = w, \text{ or } w = 46\frac{7}{8}$$

The river is $46\frac{7}{8}$ feet wide.

69. $$\frac{x}{1350} = \frac{5}{1}$$

$$1 \cdot x = 1350 \cdot 5$$

$$x = 6750$$

The plane will lose 6,750 feet in altitude.

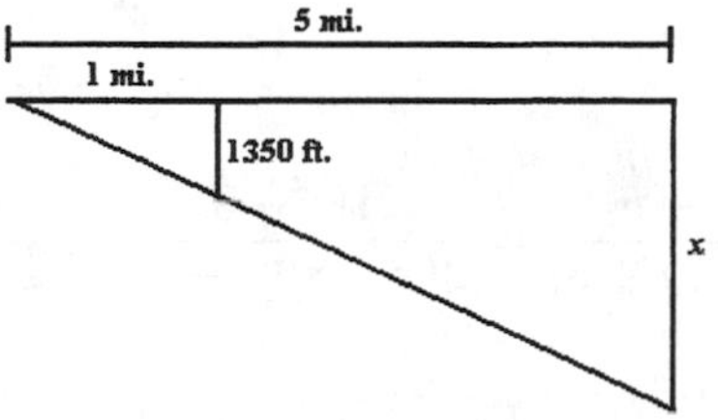

71.
$$\frac{x}{750} = \frac{52800}{2500}$$
$$2500 \cdot x = 750 \cdot 52800$$
$$2500x = 39{,}600{,}000$$
$$x = 15{,}840$$
The road will rise 15,840 ft.

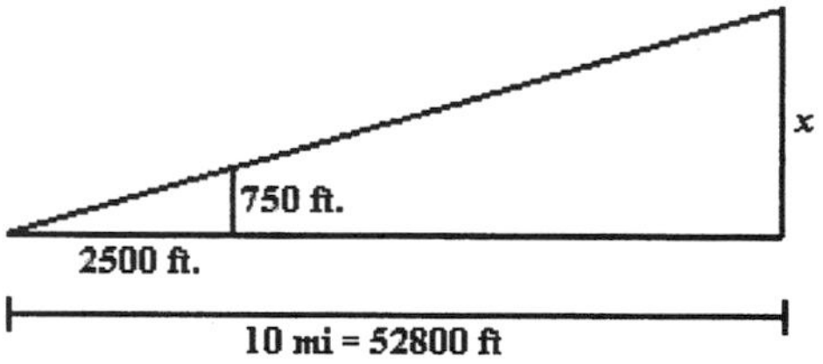

73. **Answers may vary.**

75. If $\frac{a}{b} = \frac{c}{d}$, then $ad = bc$. Thus, $\frac{a}{b} = \frac{a(b+d)}{b(b+d)} = \frac{ab+ad}{b(b+d)} = \frac{ab+bc}{b(b+d)} = \frac{b(a+c)}{b(b+d)} = \frac{a+c}{b+d}$.

Chapter 6 Summary (page 427)

1.
$$(x+3)(x-3) = 0$$
$x+3=0$ **or** $x-3=0$
$x=-3$ $\quad$ $x=3$

2.
$$x^2+x-6=0$$
$$(x+3)(x-2)=0$$
$x+3=0$ **or** $x-2=0$
$x=-3$ $\quad$ $x=2$

3. $\frac{10}{25} = \frac{5\cdot 2}{5\cdot 5} = \frac{2}{5}$

4. $\frac{-12}{18} = -\frac{6\cdot 2}{6\cdot 3} = -\frac{2}{3}$

5. $\frac{-51}{153} = -\frac{51\cdot 1}{51\cdot 3} = -\frac{1}{3}$

6. $\frac{105}{45} = \frac{15\cdot 7}{15\cdot 3} = \frac{7}{3}$

7. $\frac{3x^2}{6x^3} = \frac{1}{2x}$

8. $\frac{5xy^2}{2x^2y^2} = \frac{5}{2x}$

9. $\frac{x^2}{x^2+x} = \frac{x^2}{x(x+1)} = \frac{x}{x+1}$

10. $\frac{x+2}{x^2+2x} = \frac{x+2}{x(x+2)} = \frac{1}{x}$

11. $\frac{6xy}{3xy} = 2$

12. $\frac{8x^2y}{2x(4xy)} = \frac{8x^2y}{8x^2y} = 1$

13. $\frac{3p-2}{2-3p} = -1$

14. $\frac{x^2-x-56}{x^2-5x-24} = \frac{(x-8)(x+7)}{(x-8)(x+3)} = \frac{x+7}{x+3}$

15. $\frac{2x^2-16x}{2x^2-18x+16} = \frac{2x(x-8)}{2(x^2-9x+8)} = \frac{2x(x-8)}{2(x-8)(x-1)} = \frac{x}{x-1}$

16. $\frac{a^2+2a+ab+2b}{a^2+2ab+b^2} = \frac{a(a+2)+b(a+2)}{(a+b)(a+b)} = \frac{(a+2)(a+b)}{(a+b)(a+b)} = \frac{a+2}{a+b}$

17. $\frac{3xy}{2x} \cdot \frac{4x}{2y^2} = \frac{12x^2y}{4xy^2} = \frac{3x}{y}$

18.
$$\frac{3x}{x^2-x} \cdot \frac{2x-2}{x^2} = \frac{3x(2x-2)}{(x^2-x)x^2} = \frac{6x(x-1)}{x^3(x-1)}$$
$$= \frac{6}{x^2}$$

19. $$\frac{x^2+3x+2}{x^2+2x}\cdot\frac{x}{x+1}=\frac{(x^2+3x+2)x}{(x^2+2x)(x+1)}=\frac{x(x+2)(x+1)}{x(x+2)(x+1)}=1$$

20. $$\frac{x^2+x}{3x-15}\cdot\frac{6x-30}{x^2+2x+1}=\frac{(x^2+x)(6x-30)}{(3x-15)(x^2+2x+1)}=\frac{x(x+1)\cdot 6(x-5)}{3(x-5)\cdot(x+1)(x+1)}=\frac{2x}{x+1}$$

21. $$\frac{3x^2}{5x^2y}\div\frac{6x}{15xy^2}=\frac{3x^2}{5x^2y}\cdot\frac{15xy^2}{6x}=\frac{45x^3y^2}{30x^3y}=\frac{3y}{2}$$

22. $$\frac{x^2+5x}{x^2+4x-5}\div\frac{x^2}{x-1}=\frac{x^2+5x}{x^2+4x-5}\cdot\frac{x-1}{x^2}=\frac{x(x+5)(x-1)}{x^2(x+5)(x-1)}=\frac{1}{x}$$

23. $$\frac{x^2-x-6}{2x-1}\div\frac{x^2-2x-3}{2x^2+x-1}=\frac{x^2-x-6}{2x-1}\cdot\frac{2x^2+x-1}{x^2-2x-3}=\frac{(x-3)(x+2)(2x-1)(x+1)}{(2x-1)(x-3)(x+1)}$$
$$=x+2$$

24. $$\frac{x^2-3x}{x^2-x-6}\div\frac{x^2-x}{x^2+x-2}=\frac{x^2-3x}{x^2-x-6}\cdot\frac{x^2+x-2}{x^2-x}=\frac{x(x-3)\cdot(x+2)(x-1)}{(x+2)(x-3)\cdot x(x-1)}=1$$

25. $$\frac{x^2+4x+4}{x^2+x-6}\left(\frac{x-2}{x-1}\div\frac{x+2}{x^2+2x-3}\right)=\frac{x^2+4x+4}{x^2+x-6}\left(\frac{x-2}{x-1}\cdot\frac{x^2+2x-3}{x+2}\right)$$
$$=\frac{(x+2)(x+2)}{(x+3)(x-2)}\left(\frac{(x-2)(x+3)(x-1)}{(x-1)(x+2)}\right)$$
$$=\frac{(x+2)(x+2)(x-2)(x+3)(x-1)}{(x+3)(x-2)(x-1)(x+2)}=x+2$$

26. $$\frac{x}{x+y}+\frac{y}{x+y}=\frac{x+y}{x+y}=1$$

27. $$\frac{3x}{x-7}-\frac{x-2}{x-7}=\frac{3x-x+2}{x-7}=\frac{2x+2}{x-7}$$

28. $$\frac{x}{x-1}+\frac{1}{x}=\frac{(x)(x)}{(x-1)(x)}+\frac{1(x-1)}{x(x-1)}=\frac{x^2}{x(x-1)}+\frac{x-1}{x(x-1)}=\frac{x^2+x-1}{x(x-1)}$$

29. $$\frac{1}{7}-\frac{1}{x}=\frac{1(x)}{7(x)}-\frac{1(7)}{x(7)}=\frac{x}{7x}-\frac{7}{7x}=\frac{x-7}{7x}$$

30. $$\frac{3}{x+1}-\frac{2}{x}=\frac{3(x)}{(x+1)(x)}-\frac{2(x+1)}{x(x+1)}=\frac{3x}{x(x+1)}-\frac{2x+2}{x(x+1)}=\frac{3x-2x-2}{x(x+1)}=\frac{x-2}{x(x+1)}$$

31. $$\frac{x+2}{2x}-\frac{2-x}{x^2}=\frac{(x+2)x}{2x(x)}-\frac{(2-x)2}{x^2(2)}=\frac{x^2+2x}{2x^2}-\frac{4-2x}{2x^2}=\frac{x^2+2x-4+2x}{2x^2}$$
$$=\frac{x^2+4x-4}{2x^2}$$

32. $\frac{x}{x+2}+\frac{3}{x}-\frac{4}{x^2+2x}=\frac{x}{x+2}+\frac{3}{x}-\frac{4}{x(x+2)}=\frac{x(x)}{x(x+2)}+\frac{3(x+2)}{x(x+2)}-\frac{4}{x(x+2)}$

$=\frac{x^2}{x(x+2)}+\frac{3x+6}{x(x+2)}-\frac{4}{x(x+2)}$

$=\frac{x^2+3x+2}{x(x+2)}=\frac{(x+2)(x+1)}{x(x+2)}=\frac{x+1}{x}$

33. $\frac{2}{x-1}-\frac{3}{x+1}+\frac{x-5}{x^2-1}=\frac{2}{x-1}-\frac{3}{x+1}+\frac{x-5}{(x+1)(x-1)}$

$=\frac{2(x+1)}{(x-1)(x+1)}-\frac{3(x-1)}{(x+1)(x-1)}+\frac{x-5}{(x+1)(x-1)}$

$=\frac{2(x+1)-3(x-1)+x-5}{(x+1)(x-1)}$

$=\frac{2x+2-3x+3+x-5}{(x+1)(x-1)}=\frac{0}{(x+1)(x-1)}=0$

34. $\frac{\frac{3}{2}}{\frac{2}{3}}=\frac{3}{2}\div\frac{2}{3}=\frac{3}{2}\cdot\frac{3}{2}=\frac{9}{4}$

35. $\frac{\frac{3}{2}+1}{\frac{2}{3}+1}=\frac{(\frac{3}{2}+1)6}{(\frac{2}{3}+1)6}=\frac{\frac{3}{2}\cdot 6+1\cdot 6}{\frac{2}{3}\cdot 6+1\cdot 6}=\frac{9+6}{4+6}=\frac{15}{10}=\frac{3}{2}$

36. $\frac{\frac{1}{x}+1}{\frac{1}{x}-1}=\frac{(\frac{1}{x}+1)x}{(\frac{1}{x}-1)x}=\frac{\frac{1}{x}\cdot x+1\cdot x}{\frac{1}{x}\cdot x-1\cdot x}$

$=\frac{1+x}{1-x}$

37. $\frac{1+\frac{3}{x}}{2-\frac{1}{x^2}}=\frac{(1+\frac{3}{x})x^2}{(2-\frac{1}{x^2})x^2}=\frac{x^2+3x}{2x^2-1}$

38. $\frac{\frac{2}{x-1}+\frac{x-1}{x+1}}{\frac{1}{x^2-1}}=\frac{\frac{2}{x-1}+\frac{x-1}{x+1}}{\frac{1}{(x+1)(x-1)}}=\frac{(\frac{2}{x-1}+\frac{x-1}{x+1})(x+1)(x-1)}{\left(\frac{1}{(x+1)(x-1)}\right)(x+1)(x-1)}=\frac{2(x+1)+(x-1)(x-1)}{1}$

$=2x+2+x^2-2x+1=x^2+3$

39. $\frac{\frac{a}{b}+c}{\frac{b}{a}+c}=\frac{(\frac{a}{b}+c)ab}{(\frac{b}{a}+c)ab}=\frac{a^2+abc}{b^2+abc}=\frac{a(a+bc)}{b(b+ac)}$

40.
$$\frac{3}{x}=\frac{2}{x-1}$$
$$x(x-1)\left(\frac{3}{x}\right)=x(x-1)\left(\frac{2}{x-1}\right)$$
$$3(x-1)=2x$$
$$3x-3=2x$$
$x=3\Rightarrow$ The answer checks.

41.
$$\begin{aligned}
\frac{5}{x+4} &= \frac{3}{x+2} \\
(x+4)(x+2)\left(\frac{5}{x+4}\right) &= (x+4)(x+2)\left(\frac{3}{x+2}\right) \\
5(x+2) &= 3(x+4) \\
5x+10 &= 3x+12 \\
2x &= 2 \\
x &= 1 \Rightarrow \text{The answer checks.}
\end{aligned}$$

42.
$$\begin{aligned}
\frac{2}{3x} + \frac{1}{x} &= \frac{5}{9} \\
9x\left(\frac{2}{3x} + \frac{1}{x}\right) &= 9x\left(\frac{5}{9}\right) \\
6+9 &= 5x \\
15 &= 5x \\
3 &= x \Rightarrow \text{The answer checks.}
\end{aligned}$$

43.
$$\begin{aligned}
\frac{2x}{x+4} &= \frac{3}{x-1} \\
(x+4)(x-1)\left(\frac{2x}{x+4}\right) &= (x+4)(x-1)\left(\frac{3}{x-1}\right) \\
2x(x-1) &= 3(x+4) \\
2x^2 - 2x &= 3x+12 \\
2x^2 - 5x - 12 &= 0 \\
(2x+3)(x-4) &= 0
\end{aligned}$$

$$\begin{aligned}
2x+3 &= 0 \quad \textbf{or} \quad & x-4 &= 0 \quad \Rightarrow \text{Both answers check.} \\
2x &= -3 & x &= 4 \\
x &= -\tfrac{3}{2} & x &= 4
\end{aligned}$$

44.
$$\begin{aligned}
\frac{2}{x-1} + \frac{3}{x+4} &= \frac{-5}{x^2+3x-4} \\
\frac{2}{x-1} + \frac{3}{x+4} &= \frac{-5}{(x+4)(x-1)} \\
(x-1)(x+4)\left(\frac{2}{x-1} + \frac{3}{x+4}\right) &= (x-1)(x+4)\left(\frac{-5}{(x+4)(x-1)}\right) \\
2(x+4) + 3(x-1) &= -5 \\
2x+8+3x-3 &= -5 \\
5x &= -10 \\
x &= -2 \Rightarrow \text{The answer checks.}
\end{aligned}$$

45.

$$\frac{4}{x+2}-\frac{3}{x+3}=\frac{6}{x^2+5x+6}$$
$$\frac{4}{x+2}-\frac{3}{x+3}=\frac{6}{(x+2)(x+3)}$$
$$(x+2)(x+3)\left(\frac{4}{x+2}-\frac{3}{x+3}\right)=(x+2)(x+3)\left(\frac{6}{(x+2)(x+3)}\right)$$
$$4(x+3)-3(x+2)=6$$
$$4x+12-3x-6=6$$
$$x=0 \Rightarrow \text{The answer checks.}$$

46.

$$\frac{1}{r}=\frac{1}{r_1}+\frac{1}{r_2}$$
$$rr_1r_2\cdot\frac{1}{r}=rr_1r_2\left(\frac{1}{r_1}+\frac{1}{r_2}\right)$$
$$r_1r_2=rr_2+rr_1$$
$$r_1r_2-rr_1=rr_2$$
$$r_1(r_2-r)=rr_2$$
$$r_1=\frac{rr_2}{r_2-r}$$

47.

$$E=1-\frac{T_2}{T_1}$$
$$T_1E=T_1\left(1-\frac{T_2}{T_1}\right)$$
$$T_1E=T_1-T_2$$
$$T_1E-T_1=-T_2$$
$$T_1(E-1)=-T_2$$
$$T_1=-\frac{T_2}{E-1}=\frac{T_2}{1-E}$$

48.

$$H=\frac{RB}{R+B}$$
$$(R+B)H=(R+B)\left(\frac{RB}{R+B}\right)$$
$$RH+BH=RB$$
$$BH=RB-RH$$
$$BH=R(B-H)$$
$$\frac{BH}{B-H}=R$$

49. Let $x =$ hours for both pipes to empty.

[1st pipe in 1 hour] + [2nd pipe in 1 hour] = [Total in 1 hour]

$$\frac{1}{18}+\frac{1}{20}=\frac{1}{x}$$
$$180x\left(\frac{1}{18}+\frac{1}{20}\right)=180x\left(\frac{1}{x}\right)$$
$$10x+9x=180$$
$$19x=180$$
$$x=\frac{180}{19}$$

It can be emptied in $9\frac{9}{19}$ hours.

50. Let $x =$ days for both working together.

[Painter in 1 day] + [Owner in 1 day] = [Total in 1 day]

$$\frac{1}{10}+\frac{1}{14}=\frac{1}{x}$$
$$70x\left(\frac{1}{10}+\frac{1}{14}\right)=70x\left(\frac{1}{x}\right)$$
$$7x+5x=70$$
$$12x=70$$
$$x=\frac{70}{12}=\frac{35}{6}$$

It can be painted in $5\frac{5}{6}$ days.

51. Let $r =$ the rate at which he jogs.

	d	r	t
Jogs	10	r	$\frac{10}{r}$
Rides	30	$r+10$	$\frac{30}{r+10}$

$$\boxed{\text{Time he jogs}} = \boxed{\text{Time he rides}}$$
$$\frac{10}{r} = \frac{30}{r+10}$$
$$r(r+10)\left(\frac{10}{r}\right) = r(r+10)\left(\frac{30}{r+10}\right)$$
$$10(r+10) = 30r$$
$$10r + 100 = 30r$$
$$100 = 20r$$
$$5 = r$$

He jogs 5 miles per hour.

52. Let $r =$ the speed of the wind.

	d	r	t
Downwind	400	$360+r$	$\frac{400}{360+r}$
Upwind	320	$360-r$	$\frac{320}{360-r}$

$$\boxed{\text{Time downwind}} = \boxed{\text{Time upwind}}$$
$$\frac{400}{360+r} = \frac{320}{360-r}$$
$$(360+r)(360-r)\left(\frac{400}{360+r}\right) = (360+r)(360-r)\left(\frac{320}{360-r}\right)$$
$$400(360-r) = 320(360+r)$$
$$144000 - 400r = 115200 + 320r$$
$$28800 = 720r$$

$40 = r \Rightarrow$ The wind has a speed of 40 miles per hour.

53. $\frac{3}{6} = \frac{1}{2}$

54. $\frac{12x}{15x} = \frac{4}{5}$

55. $\frac{2\text{ ft}}{1\text{ yd}} = \frac{2\text{ ft}}{3\text{ ft}} = \frac{2}{3}$

56. $\frac{5\text{ pt}}{3\text{ qt}} = \frac{5\text{ pt}}{6\text{ pt}} = \frac{5}{6}$

57. $\frac{\$8.79}{3\text{ lb}} = \$2.93/\text{lb}$

58. $\frac{2275\text{ kwh}}{4\text{ weeks}} = 568.75\text{ kwh/week}$

59.
$$\frac{4}{7} \stackrel{?}{=} \frac{20}{34}$$
$$4 \cdot 34 \stackrel{?}{=} 7 \cdot 20$$
$$136 \neq 140$$
not a proportion

60.
$$\frac{5}{7} \stackrel{?}{=} \frac{30}{42}$$
$$5 \cdot 42 \stackrel{?}{=} 7 \cdot 30$$
$$210 = 210$$
proportion

61.
$$\frac{3}{x} = \frac{6}{9}$$
$$3(9) = 6x$$
$$27 = 6x$$
$$\tfrac{27}{6} = x$$
$$\tfrac{9}{2} = x$$

62.
$$\frac{x}{3} = \frac{x}{5}$$
$$5x = 3x$$
$$2x = 0$$
$$x = 0$$

63. $\frac{x-2}{5} = \frac{x}{7}$
$7(x-2) = 5x$
$7x - 14 = 5x$
$2x = 14$
$x = 7$

64. $\frac{4x-1}{18} = \frac{x}{6}$
$6(4x-1) = 18x$
$24x - 6 = 18x$
$6x = 6$
$x = 1$

65. Let $h =$ the height of the pole.
$\frac{\text{height of pole}}{\text{shadow of pole}} = \frac{\text{height of man}}{\text{shadow of man}}$
$\frac{h}{12} = \frac{6}{3.6}$
$3.6h = 12(6)$
$3.6h = 72$
$h = \frac{72}{3.6} = 20 \text{ ft}$

Chapter 6 Test (page 430)

1. $\frac{48x^2y}{54xy^2} = \frac{6 \cdot 8x^2y}{6 \cdot 9xy^2} = \frac{8x}{9y}$

2. $\frac{2x^2 - x - 3}{4x^2 - 9} = \frac{(2x-3)(x+1)}{(2x+3)(2x-3)} = \frac{x+1}{2x+3}$

3. $\frac{3(x+2)-3}{2x-4-(x-5)} = \frac{3x+6-3}{2x-4-x+5}$
$= \frac{3x+3}{x+1} = \frac{3(x+1)}{x+1} = 3$

4. $\frac{12x^2y}{15xyz} \cdot \frac{25y^2z}{16xt} = \frac{3 \cdot 4 \cdot 5 \cdot 5x^2y^3z}{3 \cdot 5 \cdot 4 \cdot 4x^2ytz} = \frac{5y^2}{4t}$

5. $\frac{x^2+3x+2}{3x+9} \cdot \frac{x+3}{x^2-4} = \frac{(x+2)(x+1)}{3(x+3)} \cdot \frac{x+3}{(x+2)(x-2)} = \frac{(x+2)(x+1)(x+3)}{3(x+3)(x+2)(x-2)} = \frac{x+1}{3(x-2)}$

6. $\frac{8x^2y}{25xt} \div \frac{16x^2y^3}{30xyt^3} = \frac{8x^2y}{25xt} \cdot \frac{30xyt^3}{16x^2y^3} = \frac{4 \cdot 2 \cdot 5 \cdot 2 \cdot 3x^3y^2t^3}{5 \cdot 5 \cdot 4 \cdot 4x^3y^3t} = \frac{3t^2}{5y}$

7. $\frac{x^2-x}{3x^2+6x} \div \frac{3x-3}{3x^3+6x^2} = \frac{x^2-x}{3x^2+6x} \cdot \frac{3x^3+6x^2}{3x-3} = \frac{x(x-1)}{3x(x+2)} \cdot \frac{3x^2(x+2)}{3(x-1)}$
$= \frac{3x^3(x-1)(x+2)}{9x(x-1)(x+2)} = \frac{x^2}{3}$

8. $\frac{x^2+xy}{x-y} \cdot \frac{x^2-y^2}{x^2-2x} \div \frac{x^2+2xy+y^2}{x^2-4} = \frac{x^2+xy}{x-y} \cdot \frac{x^2-y^2}{x^2-2x} \cdot \frac{x^2-4}{x^2+2xy+y^2}$
$= \frac{x(x+y)}{x-y} \cdot \frac{(x+y)(x-y)}{x(x-2)} \cdot \frac{(x+2)(x-2)}{(x+y)(x+y)}$
$= \frac{x(x+y)(x+y)(x-y)(x+2)(x-2)}{x(x-y)(x-2)(x+y)(x+y)} = x+2$

9. $$\frac{5x-4}{x-1}+\frac{5x+3}{x-1}=\frac{5x-4+5x+3}{x-1}=\frac{10x-1}{x-1}$$

10. $$\frac{3y+7}{2y+3}-\frac{3(y-2)}{2y+3}=\frac{3y+7-3(y-2)}{2y+3}=\frac{3y+7-3y+6}{2y+3}=\frac{13}{2y+3}$$

11. $$\frac{x+1}{x}+\frac{x-1}{x+1}=\frac{(x+1)(x+1)}{x(x+1)}+\frac{x(x-1)}{x(x+1)}=\frac{(x+1)(x+1)+x(x-1)}{x(x+1)}$$
$$=\frac{x^2+2x+1+x^2-x}{x(x+1)}=\frac{2x^2+x+1}{x(x+1)}$$

12. $$\frac{5x}{x-2}-3=\frac{5x}{x-2}-\frac{3}{1}=\frac{5x}{x-2}-\frac{3(x-2)}{x-2}=\frac{5x-3(x-2)}{x-2}=\frac{5x-3x+6}{x-2}=\frac{2x+6}{x-2}$$

13. $$\frac{\frac{8x^2}{xy^3}}{\frac{4y^3}{x^2y^3}}=\frac{8x^2}{xy^3}\div\frac{4y^3}{x^2y^3}=\frac{8x^2}{xy^3}\cdot\frac{x^2y^3}{4y^3}=\frac{8x^4y^3}{4xy^6}=\frac{2x^3}{y^3}$$

14. $$\frac{1+\frac{y}{x}}{\frac{y}{x}-1}=\frac{\left(1+\frac{y}{x}\right)x}{\left(\frac{y}{x}-1\right)x}=\frac{1\cdot x+\frac{y}{x}\cdot x}{\frac{y}{x}\cdot x-1\cdot x}$$
$$=\frac{x+y}{y-x}$$

15. $$\frac{x}{10}-\frac{1}{2}=\frac{x}{5}$$
$$10\left(\frac{x}{10}-\frac{1}{2}\right)=10\left(\frac{x}{5}\right)$$
$$x-5=2x$$
$$-5=x$$

16. $$3x-\frac{2(x+3)}{3}=16-\frac{x+2}{2}$$
$$6\left(3x-\frac{2x+6}{3}\right)=6\left(16-\frac{x+2}{2}\right)$$
$$18x-2(2x+6)=96-3(x+2)$$
$$18x-4x-12=96-3x-6$$
$$17x=102$$
$$x=6$$

17. $$\frac{7}{x+4}-\frac{1}{2}=\frac{3}{x+4}$$
$$2(x+4)\left(\frac{7}{x+4}-\frac{1}{2}\right)=2(x+4)\left(\frac{3}{x+4}\right)$$
$$2(7)-(x+4)=2(3)$$
$$14-x-4=6$$
$$-x=-4$$
$$x=4$$

The answer checks.

18.

$$H = \frac{RB}{R+B}$$
$$(R+B)H = (R+B)\left(\frac{RB}{R+B}\right)$$
$$RH + BH = RB$$
$$RH = RB - BH$$
$$RH = B(R-H)$$
$$\frac{RH}{R-H} = B$$

19. Let $x =$ hours working together.

$$\boxed{\text{1st worker in 1 hour}} + \boxed{\text{2nd worker in 1 hour}} = \boxed{\text{Total in 1 hour}}$$
$$\frac{1}{7} + \frac{1}{9} = \frac{1}{x}$$
$$63x\left(\frac{1}{7} + \frac{1}{9}\right) = 63x\left(\frac{1}{x}\right)$$
$$9x + 7x = 63$$
$$16x = 63$$
$$x = \frac{63}{16}$$

They can finish in $3\frac{15}{16}$ hours.

20. Let $r =$ the speed of the current.

	d	r	t
Downstream	28	$23+r$	$\frac{28}{23+r}$
Upstream	18	$23-r$	$\frac{18}{23-r}$

$$\boxed{\text{Time downstream}} = \boxed{\text{Time upstream}}$$
$$\frac{28}{23+r} = \frac{18}{23-r}$$
$$(23+r)(23-r)\left(\frac{28}{23+r}\right) = (23+r)(23-r)\left(\frac{18}{23-r}\right)$$
$$28(23-r) = 18(23+r)$$
$$644 - 28r = 414 + 18r$$
$$230 = 46r$$
$$5 = r$$

The current has a speed of 5 miles per hour.

21.

$$\frac{575}{\frac{1}{2}} = \frac{x}{7}$$
$$7(575) = \frac{1}{2}x$$
$$4025 = \frac{1}{2}x$$
$$2(4025) = 2\left(\frac{1}{2}x\right)$$

$8050 = x \Rightarrow$ The plane will lose 8,050 feet of altitude.

22. $\frac{6 \text{ ft}}{3 \text{ yd}} = \frac{6 \text{ ft}}{9 \text{ ft}} = \frac{2}{3}$

23.

$$\frac{3xy}{5xy} \stackrel{?}{=} \frac{3xt}{5xt}$$
$$3xy \cdot 5xt \stackrel{?}{=} 5xy \cdot 3xt$$

$15x^2yt = 15x^2yt \Rightarrow$ proportion

24. $\dfrac{y}{y-1} = \dfrac{y-2}{y}$

$$\begin{aligned} y(y) &= (y-1)(y-2) \\ y^2 &= y^2 - 3y + 2 \\ 3y &= 2 \\ y &= \frac{2}{3} \end{aligned}$$

25. Let $h =$ the height of the tree.

$$\begin{aligned} \frac{\text{height of tree}}{\text{shadow of tree}} &= \frac{\text{height of man}}{\text{shadow of man}} \\ \frac{h}{30} &= \frac{6}{4} \\ 4h &= 30(6) \\ 4h &= 180 \\ h &= 45 \text{ ft} \end{aligned}$$

Cumulative Review Exercises (page 431)

1. $x^2x^5 = x^{2+5} = x^7$

2. $(x^2)^5 = x^{2\cdot 5} = x^{10}$

3. $\dfrac{x^5}{x^2} = x^{5-2} = x^3$

4. $(3x^5)^0 = 1$

5. $(3x^2 - 2x) + (6x^3 - 3x^2 - 1) = 3x^2 - 2x + 6x^3 - 3x^2 - 1 = 6x^3 - 2x - 1$

6. $(4x^3 - 2x) - (2x^3 - 2x^2 - 3x + 1) = 4x^3 - 2x - 2x^3 + 2x^2 + 3x - 1 = 2x^3 + 2x^2 + x - 1$

7. $3(5x^2 - 4x + 3) + 2(-x^2 + 2x - 4) = 15x^2 - 12x + 9 - 2x^2 + 4x - 8 = 13x^2 - 8x + 1$

8. $4(3x^2 - 4x - 1) - 2(-2x^2 + 4x - 3) = 12x^2 - 16x - 4 + 4x^2 - 8x + 6 = 16x^2 - 24x + 2$

9. $(3x^3y^2)(-4x^2y^3) = 3(-4)x^3x^2y^2y^3 = -12x^5y^5$

10. $-5x^2(7x^3 - 2x^2 - 2) = -35x^5 + 10x^4 + 10x^2$

11. $(3x + 1)(2x + 4) = 6x^2 + 12x + 2x + 4 = 6x^2 + 14x + 4$

12. $(5x - 4y)(3x + 2y) = 15x^2 + 10xy - 12xy - 8y^2 = 15x^2 - 2xy - 8y^2$

13.

$$\begin{array}{r} x + 4 \\ x+3 \overline{\big)\, x^2 + 7x + 12} \\ \underline{x^2 + 3x} \\ 4x + 12 \\ \underline{4x + 12} \\ 0 \end{array}$$

14.

$$\begin{array}{r} x^2 + x + 1 \\ 2x-3 \overline{\big)\, 2x^3 - x^2 - x - 3} \\ \underline{2x^3 - 3x^2} \\ 2x^2 - x \\ \underline{2x^2 - 3x} \\ 2x - 3 \\ \underline{2x - 3} \\ 0 \end{array}$$

15. $3x^2y - 6xy^2 = 3xy(x - 2y)$

16. $3(a + b) + x(a + b) = (a + b)(3 + x)$

17.
$$\begin{aligned} 2a + 2b + ab + b^2 &= 2(a + b) + b(a + b) \\ &= (a + b)(2 + b) \end{aligned}$$

18.
$$\begin{aligned} 25p^4 - 16q^2 &= (5p^2)^2 - (4q)^2 \\ &= (5p^2 + 4q)(5p^2 - 4q) \end{aligned}$$

CUMULATIVE REVIEW EXERCISES

19. $x^2 - 11x - 12 = (x-12)(x+1)$

20. $x^2 - xy - 6y^2 = (x-3y)(x+2y)$

21. $6a^2 - 7a - 20 = (2a-5)(3a+4)$

22. $8m^2 - 10mn - 3n^2 = (4m+n)(2m-3n)$

23. $p^3 - 27q^3 = p^3 - (3q)^3 = (p-3q)\left(p^2 + p(3q) + (3q)^2\right) = (p-3q)(p^2 + 3pq + 9q^2)$

24.
$$\begin{aligned} 8r^3 + 64s^3 = 8\left(r^3 + 8s^3\right) = 8\left[r^3 + (2s)^3\right] &= 8(r+2s)\left(r^2 - r\cdot 2s + (2s)^2\right) \\ &= 8(r+2s)\left(r^2 - 2rs + 4s^2\right) \end{aligned}$$

25.
$$\begin{aligned} \frac{4}{5}x + 6 &= 18 \\ \frac{4}{5}x &= 12 \\ 5\cdot\frac{4}{5}x &= 5(12) \\ 4x &= 60 \\ x &= 15 \end{aligned}$$

26.
$$\begin{aligned} 5 - \frac{x+2}{3} &= 7 - x \\ 3\left(5 - \frac{x+2}{3}\right) &= 3(7-x) \\ 15 - (x+2) &= 21 - 3x \\ 15 - x - 2 &= 21 - 3x \\ 2x &= 8 \\ x &= 4 \end{aligned}$$

27.
$$6x^2 - x - 2 = 0$$
$$(2x+1)(3x-2) = 0$$
$$2x + 1 = 0 \quad \textbf{or} \quad 3x - 2 = 0$$
$$2x = -1 \qquad\qquad 3x = 2$$
$$x = -\tfrac{1}{2} \qquad\qquad x = \tfrac{2}{3}$$

28.
$$5x^2 = 10x$$
$$5x^2 - 10x = 0$$
$$5x(x-2) = 0$$
$$5x = 0 \quad \textbf{or} \quad x - 2 = 0$$
$$x = 0 \qquad\qquad x = 2$$

29.
$$x^2 + 3x + 2 = 0$$
$$(x+1)(x+2) = 0$$
$$x + 1 = 0 \quad \textbf{or} \quad x + 2 = 0$$
$$x = -1 \qquad\qquad x = -2$$

30.
$$2y^2 + 5y - 12 = 0$$
$$(2y-3)(y+4) = 0$$
$$2y - 3 = 0 \quad \textbf{or} \quad y + 4 = 0$$
$$2y = 3 \qquad\qquad y = -4$$
$$y = \tfrac{3}{2} \qquad\qquad y = -4$$

31.
$$\begin{aligned} 5x - 3 &> 7 \\ 5x &> 10 \\ x &> 2 \end{aligned}$$

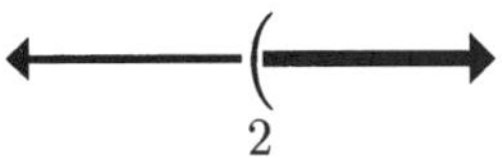

32.
$$\begin{aligned} 7x - 9 &< 5 \\ 7x &< 14 \\ x &< 2 \end{aligned}$$

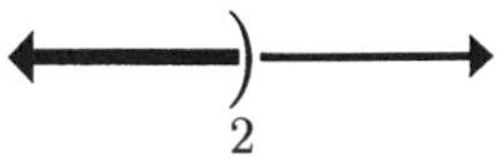

33.
$$\begin{aligned} -2 &< -x+3 < 5 \\ -5 &< -x < 2 \\ \frac{-5}{-1} &> \frac{-x}{-1} > \frac{2}{-1} \\ 5 &> x > -2 \\ -2 &< x < 5 \end{aligned}$$

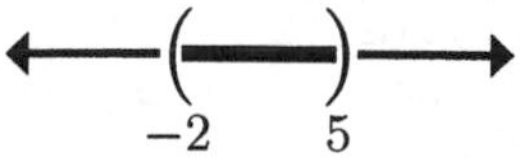

34.
$$\begin{aligned} 0 &\le \frac{4-x}{3} \le 2 \\ 3(0) &\le 3\cdot\frac{4-x}{3} \le 3(2) \\ 0 &\le 4-x \le 6 \\ -4 &\le -x \le 2 \\ \frac{-4}{-1} &\ge \frac{-x}{-1} \ge \frac{2}{-1} \\ 4 &\ge x \ge -2 \\ -2 &\le x \le 4 \end{aligned}$$

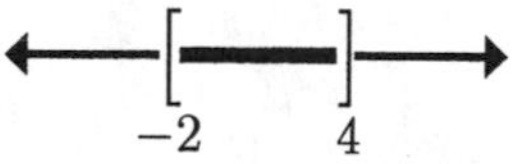

35. $4x - 3y = 12$

x	y
0	−4
3	0

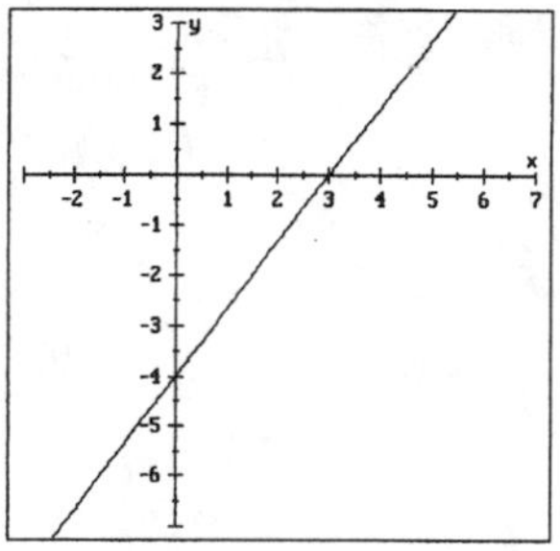

36.
$$\begin{aligned} 3x + 4y &= 4y + 12 \\ 3x &= 12 \\ x &= 4 \end{aligned}$$

x	y
4	0
4	2

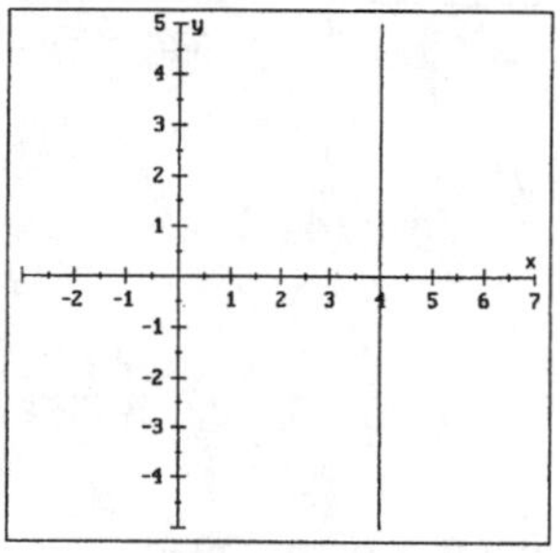

37. $f(0) = 2(0)^2 - 3 = 0 - 3 = -3$

38. $f(3) = 2(3)^2 - 3 = 2(9) - 3 = 18 - 3 = 15$

39. $f(-2) = 2(-2)^2 - 3 = 2(4) - 3$
$= 8 - 3 = 5$

40. $f(2x) = 2(2x)^2 - 3 = 2(4x^2) - 3$
$= 8x^2 - 3$

41. $\dfrac{x^2+2x+1}{x^2-1} = \dfrac{(x+1)(x+1)}{(x+1)(x-1)} = \dfrac{x+1}{x-1}$

42. $\dfrac{x^2+2x-15}{x^2+3x-10} = \dfrac{(x+5)(x-3)}{(x+5)(x-2)} = \dfrac{x-3}{x-2}$

43. $\dfrac{x^2+x-6}{5x-5}\cdot\dfrac{5x-10}{x+3} = \dfrac{(x+3)(x-2)\cdot 5(x-2)}{5(x-1)\cdot(x+3)} = \dfrac{(x-2)^2}{x-1}$

44.
$$\begin{aligned} \frac{p^2-p-6}{3p-9} \div \frac{p^2+6p+9}{p^2-9} = \frac{p^2-p-6}{3p-9}\cdot\frac{p^2-9}{p^2+6p+9} &= \frac{(p-3)(p+2)\cdot(p+3)(p-3)}{3(p-3)\cdot(p+3)(p+3)} \\ &= \frac{(p+2)(p-3)}{3(p+3)} \end{aligned}$$

45. $\dfrac{3x}{x+2} + \dfrac{5x}{x+2} - \dfrac{7x-2}{x+2} = \dfrac{3x+5x-(7x-2)}{x+2} = \dfrac{3x+5x-7x+2}{x+2} = \dfrac{x+2}{x+2} = 1$

46. $$\frac{x-1}{x+1}+\frac{x+1}{x-1}=\frac{(x-1)(x-1)}{(x+1)(x-1)}+\frac{(x+1)(x+1)}{(x-1)(x+1)}=\frac{x^2-2x+1}{(x+1)(x-1)}+\frac{x^2+2x+1}{(x+1)(x-1)}$$
$$=\frac{2x^2+2}{(x+1)(x-1)}$$

47. $$\frac{a+1}{2a+4}-\frac{a^2}{2a^2-8}=\frac{a+1}{2(a+2)}-\frac{a^2}{2(a+2)(a-2)}=\frac{(a+1)(a-2)}{2(a+2)(a-2)}-\frac{a^2}{2(a+2)(a-2)}$$
$$=\frac{a^2-2a+a-2-a^2}{2(a+2)(a-2)}$$
$$=\frac{-a-2}{2(a+2)(a-2)}$$
$$=\frac{-(a+2)}{2(a+2)(a-2)}=-\frac{1}{2(a-2)}$$

48. $$\frac{\frac{1}{x}+\frac{1}{y}}{\frac{1}{x}-\frac{1}{y}}=\frac{\left(\frac{1}{x}+\frac{1}{y}\right)xy}{\left(\frac{1}{x}-\frac{1}{y}\right)xy}=\frac{\frac{1}{x}\cdot xy+\frac{1}{y}\cdot xy}{\frac{1}{x}\cdot xy-\frac{1}{y}\cdot xy}=\frac{y+x}{y-x}$$

Exercise 7.1 (page 444)

1. $$\left(\frac{t^3t^5t^{-6}}{t^2t^{-4}}\right)^{-3}=\left(\frac{t^2t^{-4}}{t^3t^5t^{-6}}\right)^3=\left(\frac{t^{-2}}{t^2}\right)^3=\left(\frac{1}{t^4}\right)^3=\frac{1}{t^{12}}$$

3. Let x = the number of pies made.

$\boxed{\text{Expenses}}=\boxed{\text{Income}}$

$$1200+3.40x=5.95x$$
$$1200=2.55x$$
$470.59=x\Rightarrow$ He must sell at least 471 pies.

5. equation

7. multiplied; divided

9. contradiction

11. is greater than

13. half-open

15. positive

17.
$$2x+1=13$$
$$2x+1-1=13-1$$
$$2x=12$$
$$\frac{2x}{2}=\frac{12}{2}$$
$$x=6$$

19.
$$3(x+1)=15$$
$$3x+3=15$$
$$3x+3-3=15-3$$
$$3x=12$$
$$\frac{3x}{3}=\frac{12}{3}$$
$$x=4$$

21.
$$\begin{aligned} 2r-5&=1-r\\ 2r+r-5&=1-r+r\\ 3r-5+5&=1+5\\ 3r&=6\\ \tfrac{3r}{3}&=\tfrac{6}{3}\\ r&=2 \end{aligned}$$

23.
$$\begin{aligned} 3(2y-4)-6&=3y\\ 6y-12-6&=3y\\ 6y-18&=3y\\ 6y-3y-18&=3y-3y\\ 3y-18+18&=0+18\\ 3y&=18\\ \tfrac{3y}{3}&=\tfrac{18}{3}\\ y&=6 \end{aligned}$$

25.
$$\begin{aligned} 5(5-a)&=37-2a\\ 25-5a&=37-2a\\ 25-5a+2a&=37-2a+2a\\ 25-25-3a&=37-25\\ -3a&=12\\ \tfrac{-3a}{-3}&=\tfrac{12}{-3}\\ a&=-4 \end{aligned}$$

27.
$$\begin{aligned} 4(y+1)&=-2(4-y)\\ 4y+4&=-8+2y\\ 4y-2y+4&=-8+2y-2y\\ 2y+4-4&=-8-4\\ 2y&=-12\\ \tfrac{2y}{2}&=\tfrac{-12}{2}\\ y&=-6 \end{aligned}$$

29.
$$\begin{aligned} 2(a-5)-(3a+1)&=0\\ 2a-10-3a-1&=0\\ -a-11+11&=0+11\\ -a&=11\\ a&=-11 \end{aligned}$$

31.
$$\begin{aligned} \frac{x}{2}-\frac{x}{3}&=4\\ 6\left(\frac{x}{2}-\frac{x}{3}\right)&=6(4)\\ 6\left(\frac{x}{2}\right)-6\left(\frac{x}{3}\right)&=24\\ 3x-2x&=24\\ x&=24 \end{aligned}$$

33.
$$\begin{aligned} \frac{x}{6}+1&=\frac{x}{3}\\ 6\left(\frac{x}{6}+1\right)&=6\left(\frac{x}{3}\right)\\ 6\left(\frac{x}{6}\right)+6(1)&=2x\\ x+6&=2x\\ x-2x+6&=2x-2x\\ -x+6-6&=0-6\\ -x&=-6\\ x&=6 \end{aligned}$$

35.
$$\begin{aligned} \frac{a+1}{3}+\frac{a-1}{5}&=\frac{2}{15}\\ 15\left(\frac{a+1}{3}+\frac{a-1}{5}\right)&=15\left(\frac{2}{15}\right)\\ 15\left(\frac{a+1}{3}\right)+15\left(\frac{a-1}{5}\right)&=2\\ 5(a+1)+3(a-1)&=2\\ 5a+5+3a-3&=2\\ 8a+2-2&=2-2\\ 8a&=0\\ \tfrac{8a}{8}&=\tfrac{0}{8}\\ a&=0 \end{aligned}$$

37.
$$\begin{aligned} 4(2-3t)+6t&=-6t+8\\ 8-12t+6t&=-6t+8\\ 8-6t&=-6t+8\\ -6t+8&=-6t+8 \end{aligned}$$
Identity

39.
$$\frac{a+1}{4}+\frac{2a-3}{4}=\frac{a}{2}-2$$
$$4\left(\frac{a+1}{4}+\frac{2a-3}{4}\right)=4\left(\frac{a}{2}-2\right)$$
$$4\left(\frac{a+1}{4}\right)+4\left(\frac{2a-3}{4}\right)=4\left(\frac{a}{2}\right)+4(-2)$$
$$a+1+2a-3=2a-8$$
$$3a-2=2a-8$$
$$3a-2a-2=2a-2a-8$$
$$a-2+2=-8+2$$
$$a=-6$$

41.
$$3(x-4)+6=-2(x+4)+5x$$
$$3x-12+6=-2x-8+5x$$
$$3x-6=3x-8$$
$$3x-3x-6=3x-3x-8$$
$$-6\neq-8$$
Contradiction

43.
$$y(y+2)=(y+1)^2-1$$
$$y^2+2y=(y+1)(y+1)-1$$
$$y^2+2y=y^2+2y+1-1$$
$$y^2+2y=y^2+2y$$
Identity

45.
$$V=\frac{1}{3}Bh$$
$$3V=3\cdot\frac{1}{3}Bh$$
$$3V=Bh$$
$$\frac{3V}{h}=\frac{Bh}{h}$$
$$\frac{3V}{h}=B\text{, or }B=\frac{3V}{h}$$

47.
$$p=2l+2w$$
$$p-2l=2l-2l+2w$$
$$p-2l=2w$$
$$\frac{p-2l}{2}=\frac{2w}{2}$$
$$\frac{p-2l}{2}=w\text{, or }w=\frac{p-2l}{2}$$

49.
$$z=\frac{x-\mu}{\sigma}$$
$$\sigma z=\sigma\cdot\frac{x-\mu}{\sigma}$$
$$\sigma z=x-\mu$$
$$\sigma z+\mu=x\text{, or }x=z\sigma+\mu$$

51.
$$y=mx+b$$
$$y-b=mx+b-b$$
$$y-b=mx$$
$$\frac{y-b}{m}=\frac{mx}{m}$$
$$\frac{y-b}{m}=x\text{, or }x=\frac{y-b}{m}$$

53.
$$P=L+\frac{s}{f}i$$
$$P-L=\frac{s}{f}i$$
$$f(P-L)=f\cdot\frac{s}{f}i$$
$$f(P-L)=si$$
$$\frac{f(P-L)}{s}=\frac{si}{s}$$
$$\frac{f(P-L)}{s}=i\text{, or }i=\frac{f(P-L)}{s}$$

55.

$$\begin{aligned} 5x - 3 &> 7 \\ 5x - 3 + 3 &> 7 + 3 \\ 5x &> 10 \\ \tfrac{5x}{5} &> \tfrac{10}{5} \\ x &> 2 \end{aligned}$$

solution set: $(2, \infty)$

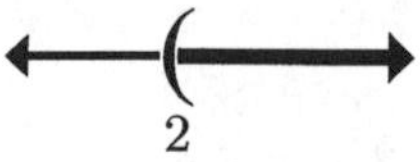

57.

$$\begin{aligned} -3x - 1 &\le 5 \\ -3x - 1 + 1 &\le 5 + 1 \\ -3x &\le 6 \\ \tfrac{-3x}{-3} &\ge \tfrac{6}{-3} \\ x &\ge -2 \end{aligned}$$

solution set: $[-2, \infty)$

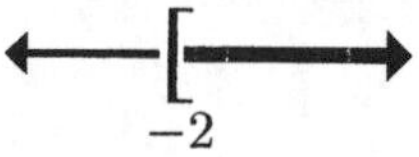

59.

$$\begin{aligned} 8 - 9y &\ge -y \\ 8 - 9y + y &\ge -y + y \\ 8 - 8 - 8y &\ge 0 - 8 \\ -8y &\ge -8 \\ \tfrac{-8y}{-8} &\le \tfrac{-8}{-8} \\ y &\le 1 \end{aligned}$$

solution set: $(-\infty, 1]$

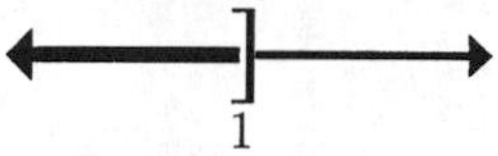

61.

$$\begin{aligned} -3(a + 2) &> 2(a + 1) \\ -3a - 6 &> 2a + 2 \\ -3a - 2a - 6 &> 2a - 2a + 2 \\ -5a - 6 + 6 &> 2 + 6 \\ -5a &> 8 \\ \tfrac{-5a}{-5} &< \tfrac{8}{-5} \\ a &< -\tfrac{8}{5} \end{aligned}$$

solution set: $(-\infty, -\frac{8}{5})$

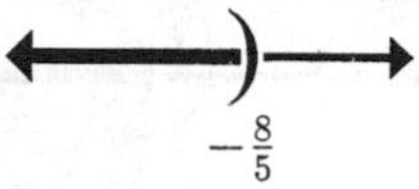

63.

$$\begin{aligned} \frac{1}{2}y + 2 &\ge \frac{1}{3}y - 4 \\ 6\left(\frac{1}{2}y + 2\right) &\ge 6\left(\frac{1}{3}y - 4\right) \\ 3y + 12 &\ge 2y - 24 \\ 3y - 2y + 12 - 12 &\ge 2y - 2y - 24 - 12 \\ y &\ge -36 \end{aligned}$$

solution set: $[-36, \infty)$

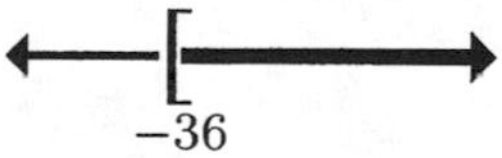

65.

$$\begin{aligned} -2 < -b + 3 &< 5 \\ -2 - 3 < -b + 3 - 3 &< 5 - 3 \\ -5 < -b &< 2 \\ \tfrac{-5}{-1} > \tfrac{-b}{-1} &> \tfrac{2}{-1} \end{aligned}$$

$5 > b > -2$, or $-2 < b < 5$

solution set: $(-2, 5)$

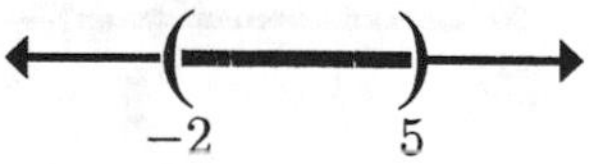

67.

$$\begin{aligned} 15 > 2x - 7 &> 9 \\ 15 + 7 > 2x - 7 + 7 &> 9 + 7 \\ 22 > 2x &> 16 \\ \tfrac{22}{2} > \tfrac{2x}{2} &> \tfrac{16}{2} \end{aligned}$$

$11 > x > 8$, or $8 < x < 11$

solution set: $(8, 11)$

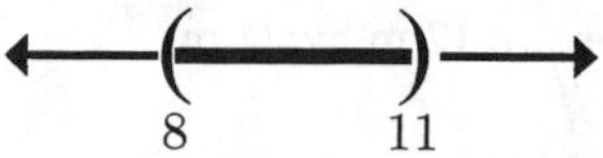

69.

$$\begin{aligned} -6 < -3(x - 4) &\le 24 \\ \frac{-6}{-3} > \frac{-3(x - 4)}{-3} &\ge \frac{24}{-3} \\ 2 > x - 4 &\ge -8 \\ 2 + 4 > x - 4 + 4 &\ge -8 + 4 \end{aligned}$$

$6 > x \ge -4$, or $-4 \le x < 6$

solution set: $[-4, 6)$

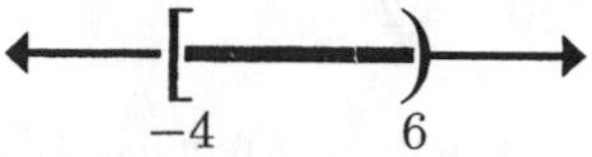

71. $0 \geq \frac{1}{2}x - 4 > 6$

This inequality indicates that $0 \geq 6$ (by the transitive property). Since this is not possible, there is no solution to the inequality.

73.

$$0 \leq \frac{4-x}{3} \leq 2$$
$$3(0) \leq 3\left(\frac{4-x}{3}\right) \leq 3(2)$$
$$0 \leq 4 - x \leq 6$$
$$0 - 4 \leq 4 - 4 - x \leq 6 - 4$$
$$-4 \leq -x \leq 2$$
$$\tfrac{-4}{-1} \geq \tfrac{-x}{-1} \geq \tfrac{2}{-1}$$
$$4 \geq x \geq -2, \text{ or } -2 \leq x \leq 4$$

solution set: $[-2, 4]$

75.

$$\begin{aligned} 3x + 2 &< 8 \\ 3x + 2 - 2 &< 8 - 2 \\ 3x &< 6 \\ x &< 2 \end{aligned} \qquad \text{or} \qquad \begin{aligned} 2x - 3 &> 11 \\ 2x - 3 + 3 &> 11 + 3 \\ 2x &> 14 \\ x &> 7 \end{aligned}$$

If $x < 2$ **or** $x > 7$, then the solution set is $(-\infty, 2) \cup (7, \infty)$.

77.

$$\begin{aligned} -4(x+2) &\geq 12 \\ -4x - 8 + 8 &\geq 12 + 8 \\ -4x &\geq 20 \\ \tfrac{-4x}{-4} &\leq \tfrac{20}{-4} \\ x &\leq -5 \end{aligned} \qquad \text{or} \qquad \begin{aligned} 3x + 8 &< 11 \\ 3x + 8 - 8 &< 11 - 8 \\ 3x &< 3 \\ x &< 1 \end{aligned}$$

If $x \leq -5$ **or** $x < 1$, then $x < 1$. The solution set is $(-\infty, 1)$.

79.

$$x < -3 \quad \text{and} \quad x > 3$$

It is impossible for x to be both less than -3 and greater than 3. There is no solution.

81. Let $x =$ the length of the shorter piece.

Then $2x + 1 =$ the length of the other piece.

[shorter length] + [other length] = [total length]

$$x + 2x + 1 = 22$$
$$3x + 1 = 22$$
$$3x = 21$$
$$x = 7$$

The pieces have lengths of 7 feet and 15 feet.

83. Let $w =$ the width. Then $2w =$ the length.

[Perimeter of garden] = 72

$$w + 2w + w + 2w = 72$$
$$6w = 72$$
$$w = 12$$

The dimension are 12 m by 24 m.

85.
$$\boxed{\text{Total amount of fence}} = 150$$
$$x+x+x+x+x+5+x+5+x = 150$$
$$7x+10 = 150$$
$$7x = 140$$
$$x = 20$$
The dimensions are 20 feet by 45 feet.

87.
$$\text{Price} < 42$$
$$\text{Price} - \text{Cost} < 42 - \text{Cost}$$
$$\text{Profit} < 42 - 27$$
$$\text{Profit} < 15$$

89. Let $x =$ the # of compact discs bought.
$$175 + 8.50x \le 330$$
$$8.50x \le 155$$
$$x \le 18.24$$
He can buy at most 18 discs.

91. **Answers may vary.**

93. The 4 was not distributed correctly in the 2nd line. It should be $4x + 12 = 16$.

Exercise 7.2 (page 451)

1.
$$3(2a-1) = 2a$$
$$6a - 3 = 2a$$
$$4a = 3$$
$$a = \frac{3}{4}$$

3.
$$\frac{5x}{2} - 1 = \frac{x}{3} + 12$$
$$6\left(\frac{5x}{2} - 1\right) = 6\left(\frac{x}{3} + 12\right)$$
$$15x - 6 = 2x + 72$$
$$13x = 78$$
$$x = 6$$

5. x

7. 0

9. $a = -b$

11. $|8| = 8$

13. $|-12| = 12$

15. $-|2| = -2$

17. $-|-30| = -30$

19. $-(-|50|) = -(-50) = 50$

21. $|\pi - 4| = -(\pi - 4) = 4 - \pi$ ($\pi - 4$ is less than zero.)

23. $|2| = 2$; $|5| = 5$; $|2|$ is smaller.

25. $|5| = 5$; $|-8| = 8$; $|5|$ is smaller.

27. $|-2| = 2$; $|10| = 10$; $|-2|$ is smaller.

29. $|-3| = 3$; $-|-4| = -4$; $-|-4|$ is smaller.

31. $-|-5| = -5$; $-|-7| = -7$; $-|-7|$ is smaller.

33. $-x < 0$ if $x > 0$; $|x+1| > 0$; $-x$ is smaller.

35.
$$|x| = 8$$
$$x = 8 \quad \textbf{or} \quad x = -8$$

37.
$$|x-3| = 6$$
$$x - 3 = 6 \quad \textbf{or} \quad x - 3 = -6$$
$$x = 9 \qquad\qquad x = -3$$

39.
$$|2x-3|=5$$
$$2x-3=5 \quad \text{or} \quad 2x-3=-5$$
$$2x=8 \qquad 2x=-2$$
$$x=4 \qquad x=-1$$

41.
$$|3x+2|=16$$
$$3x+2=16 \quad \text{or} \quad 3x+2=-16$$
$$3x=14 \qquad 3x=-18$$
$$x=\tfrac{14}{3} \qquad x=-6$$

43.
$$\left|\tfrac{7}{2}x+3\right|=-5$$
Since an absolute value cannot be negative, there is no solution.

45.
$$\left|\frac{x}{2}-1\right|=3$$
$$\frac{x}{2}-1=3 \quad \text{or} \quad \frac{x}{2}-1=-3$$
$$\frac{x}{2}=4 \qquad \frac{x}{2}=-2$$
$$x=8 \qquad x=-4$$

47.
$$|3-4x|=5$$
$$3-4x=5 \quad \text{or} \quad 3-4x=-5$$
$$-4x=2 \qquad -4x=-8$$
$$x=-\tfrac{1}{2} \qquad x=2$$

49.
$$|3x+24|=0$$
$$3x+24=0 \quad \text{or} \quad 3x+24=-0$$
$$3x=-24 \qquad 3x=-24$$
$$x=-8 \qquad x=-8$$

51.
$$\left|\frac{3x+48}{3}\right|=12$$
$$\frac{3x+48}{3}=12 \quad \text{or} \quad \frac{3x+48}{3}=-12$$
$$3x+48=36 \qquad 3x+48=-36$$
$$3x=-12 \qquad 3x=-84$$
$$x=-4 \qquad x=-28$$

53.
$$|x+3|+7=10$$
$$|x+3|=3$$
$$x+3=3 \quad \text{or} \quad x+3=-3$$
$$x=0 \qquad x=-6$$

55.
$$|0.3x-3|-2=7$$
$$|0.3x-3|=9$$
$$0.3x-3=9 \quad \text{or} \quad 0.3x-3=-9$$
$$0.3x=12 \qquad 0.3x=-6$$
$$x=40 \qquad x=-20$$

57.
$$\left|\frac{3}{5}x-4\right|-2=-2$$
$$\left|\frac{3}{5}x-4\right|=0$$
$$\frac{3}{5}x-4=0 \quad \text{or} \quad \frac{3}{5}x-4=-0$$
$$\frac{3}{5}x=4 \qquad \frac{3}{5}x=4$$
$$x=\tfrac{20}{3} \qquad x=\tfrac{20}{3}$$

59. Graph $y=|0.75x+0.12|$ and $y=12.3$ and find the x-coordinate(s) of any point(s) of intersection:

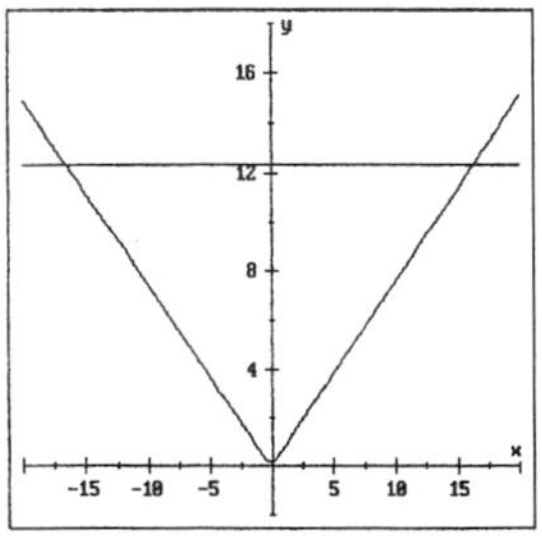

solutions: -16.6, 16.2

61.
$$|2x+1| = |3x+3|$$
$$2x+1 = 3x+3 \quad \textbf{or} \quad 2x+1 = -(3x+3)$$
$$-x = 2 \qquad 2x+1 = -3x-3$$
$$x = -2 \qquad 5x = -4$$
$$x = -\tfrac{4}{5}$$

63.
$$|3x-1| = |x+5|$$
$$3x-1 = x+5 \quad \textbf{or} \quad 3x-1 = -(x+5)$$
$$2x = 6 \qquad 3x-1 = -x-5$$
$$x = 3 \qquad 4x = -4$$
$$x = -1$$

65.
$$|2-x| = |3x+2|$$
$$2-x = 3x+2 \quad \textbf{or} \quad 2-x = -(3x+2)$$
$$-4x = 0 \qquad 2-x = -3x-2$$
$$x = 0 \qquad 2x = -4$$
$$x = -2$$

67.
$$\left|\frac{x}{2}+2\right| = \left|\frac{x}{2}-2\right|$$
$$\frac{x}{2}+2 = \frac{x}{2}-2 \quad \textbf{or} \quad \frac{x}{2}+2 = -\left(\frac{x}{2}-2\right)$$
$$0 = -4 \qquad \frac{x}{2}+2 = -\frac{x}{2}+2$$
(no solution from this part) $\qquad x = 0$

69.
$$\left|x+\tfrac{1}{3}\right| = |x-3|$$
$$x+\tfrac{1}{3} = x-3 \quad \textbf{or} \quad x+\tfrac{1}{3} = -(x-3)$$
$$3x+1 = 3x-9 \qquad x+\tfrac{1}{3} = -x+3$$
$$0 = -10 \qquad 3x+1 = -3x+9$$
(no solution from this part) $\qquad 6x = 8$
$$x = \tfrac{4}{3}$$

71. $|3x+7| = -|8x-2|$
Since an absolute value cannot be negative, there is no solution.

73. **Answers may vary.**

75. $|x| + k = 0$
$|x| = -k$
If this equation has exactly two solutions, then $-k > 0$, or $k < 0$.

77. **Answers may vary.**

79. **Answers may vary.**

Exercise 7.3 (page 458)

1.
$$A = p + prt$$
$$A - p = prt$$
$$\frac{A-p}{pr} = t, \text{ or } t = \frac{A-p}{pr}$$

3.
$$P = 2w + 2l$$
$$P - 2w = 2l$$
$$\frac{P-2w}{2} = l, \text{ or } l = \frac{P-2w}{2}$$

5. $-k < x < k$

7. $x < -k$ or $x > k$

9. $|2x| < 8$

$-8 < 2x < 8$

$\frac{-8}{2} < \frac{2x}{2} < \frac{8}{2}$

$-4 < x < 4$

solution set: $(-4, 4)$

11. $|x + 9| \leq 12$

$-12 \leq x + 9 \leq 12$

$-12 - 9 \leq x + 9 - 9 \leq 12 - 9$

$-21 \leq x \leq 3$

solution set: $[-21, 3]$

13. $|3x + 2| \leq -3$; Since an absolute value can never be negative, this inequality has no solution.

15. $|4x - 1| \leq 7$

$-7 \leq 4x - 1 \leq 7$

$-7 + 1 \leq 4x - 1 + 1 \leq 7 + 1$

$-6 \leq 4x \leq 8$

$\frac{-6}{4} \leq \frac{4x}{4} \leq \frac{8}{4}$

$-\frac{3}{2} \leq x \leq 2$

solution set: $\left[-\frac{3}{2}, 2\right]$

17. $|3 - 2x| < 7$

$-7 < 3 - 2x < 7$

$-7 - 3 < 3 - 3 - 2x < 7 - 3$

$-10 < -2x < 4$

$\frac{-10}{-2} > \frac{-2x}{-2} > \frac{4}{-2}$

$5 > x > -2$, or $-2 < x < 5$

solution set: $(-2, 5)$

19. $|5x| > 5$

$5x < -5$ **or** $5x > 5$

$x < -1$ $\quad$ $x > 1$

solution set: $(-\infty, -1) \cup (1, \infty)$

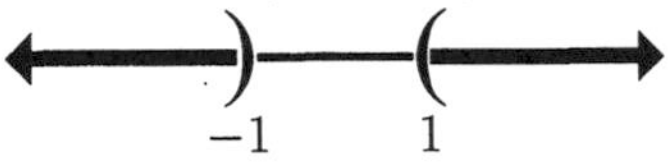

21. $|x - 12| > 24$

$x - 12 < -24$ **or** $x - 12 > 24$

$x < -12$ $\quad$ $x > 36$

solution set: $(-\infty, -12) \cup (36, \infty)$

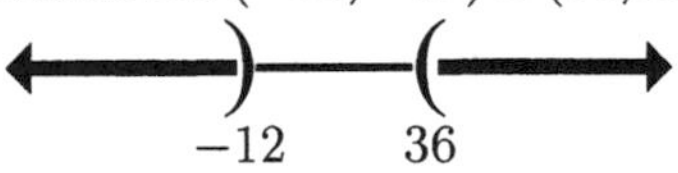

23. $|3x + 2| > 14$

$3x + 2 < -14$ **or** $3x + 2 > 14$

$3x < -16$ $\quad$ $3x > 12$

$x < -\frac{16}{3}$ $\quad$ $x > 4$

solution set: $\left(-\infty, -\frac{16}{3}\right) \cup (4, \infty)$

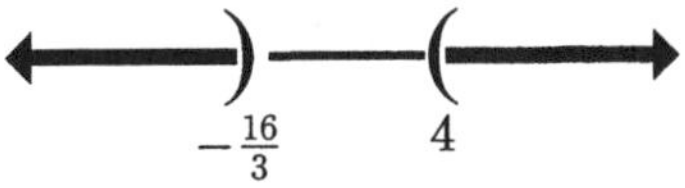

25. $|4x + 3| > -5$

Since an absolute value is always at least 0, this inequality is true for all real numbers.

solution set: $(-\infty, \infty)$

27.

$$|2-3x| \geq 8$$

$$2-3x \leq -8 \quad \textbf{or} \quad 2-3x \geq 8$$

$$-3x \leq -10 \qquad -3x \geq 6$$

$$x \geq \tfrac{10}{3} \qquad x \leq -2$$

solution set: $(-\infty, -2] \cup \left[\frac{10}{3}, \infty\right)$

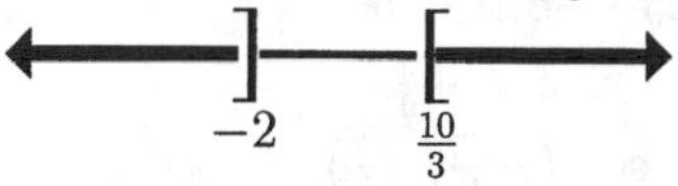

29.

$$-|2x-3| < -7$$

$$|2x-3| > 7$$

$$2x-3 < -7 \quad \textbf{or} \quad 2x-3 > 7$$

$$2x < -4 \qquad 2x > 10$$

$$x < -2 \qquad x > 5$$

solution set: $(-\infty, -2) \cup (5, \infty)$

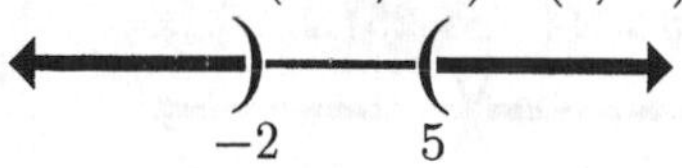

31.

$$|8x-3| > 0$$

$$8x-3 < -0 \quad \textbf{or} \quad 8x-3 > 0$$

$$8x < 3 \qquad 8x > 3$$

$$x < \tfrac{3}{8} \qquad x > \tfrac{3}{8}$$

solution set: $\left(-\infty, \frac{3}{8}\right) \cup \left(\frac{3}{8}, \infty\right)$

33.

$$\left|\frac{x-2}{3}\right| \leq 4$$

$$-4 \leq \frac{x-2}{3} \leq 4$$

$$-12 \leq x-2 \leq 12$$

$$-10 \leq x \leq 14$$

solution set: $[-10, 14]$

35. $|3x+1| + 2 < 6$

$$|3x+1| < 4$$

$$-4 < 3x+1 < 4$$

$$-5 < 3x < 3$$

$$-\tfrac{5}{3} < x < 1$$

solution set: $\left(-\frac{5}{3}, 1\right)$

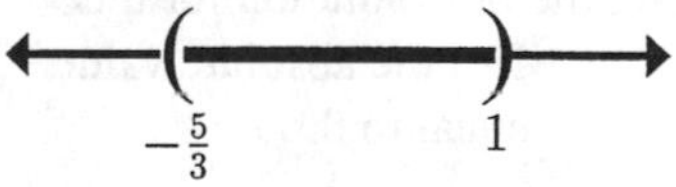

37.

$$3|2x+5| \geq 9$$

$$|2x+5| \geq 3$$

$$2x+5 \leq -3 \quad \textbf{or} \quad 2x+5 \geq 3$$

$$2x \leq -8 \qquad 2x \geq -2$$

$$x \leq -4 \qquad x \geq -1$$

solution set: $(-\infty, -4] \cup [-1, \infty)$

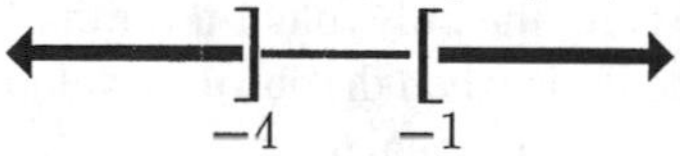

39. $|5x-1| + 4 \leq 0$

$$|5x-1| \leq -4$$

Since an absolute value can never be negative, this inequality has no solution.

41.

$$\left|\tfrac{1}{3}x+7\right| + 5 > 6$$

$$\left|\tfrac{1}{3}x+7\right| > 1$$

$$\tfrac{1}{3}x+7 < -1 \quad \textbf{or} \quad \tfrac{1}{3}x+7 > 1$$

$$x+21 < -3 \qquad x+21 > 3$$

$$x < -24 \qquad x > -18$$

solution set: $(-\infty, -24) \cup (-18, \infty)$

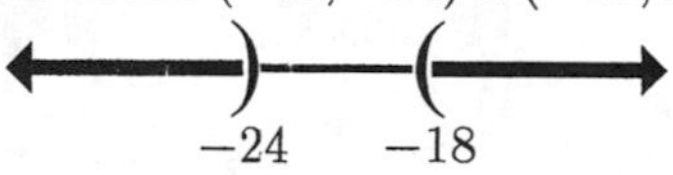

43.

$$\left|\tfrac{1}{5}x-5\right|+4>4$$
$$\left|\tfrac{1}{5}x-5\right|>0$$
$$\tfrac{1}{5}x-5<-0 \quad \textbf{or} \quad \tfrac{1}{5}x-5>0$$
$$\tfrac{1}{5}x<5 \qquad \tfrac{1}{5}x>5$$
$$x<25 \qquad x>25$$

solution set: $(-\infty,25)\cup(25,\infty)$

45.

$$\left|\tfrac{3}{5}x+\tfrac{7}{3}\right|<2$$
$$-2<\tfrac{3}{5}x+\tfrac{7}{3}<2$$
$$15(-2)<15\left(\tfrac{3}{5}x+\tfrac{7}{3}\right)<15(2)$$
$$-30<9x+35<30$$
$$-65<9x<-5$$
$$-\tfrac{65}{9}<x<-\tfrac{5}{9}$$

solution set: $\left(-\tfrac{65}{9},-\tfrac{5}{9}\right)$

47.

$$\left|3\left(\frac{x+4}{4}\right)\right|>0$$
$$3\left(\tfrac{x+4}{4}\right)<-0 \quad \textbf{or} \quad 3\left(\tfrac{x+4}{4}\right)>0$$
$$\tfrac{4}{3}\cdot 3\left(\tfrac{x+4}{4}\right)<\tfrac{4}{3}(-0) \qquad \tfrac{4}{3}\cdot 3\left(\tfrac{x+4}{4}\right)>\tfrac{4}{3}\cdot 0$$
$$x+4<0 \qquad x+4>0$$
$$x<-4 \qquad x>-4$$

solution set: $(-\infty,-4)\cup(-4,\infty)$

49.

$$\left|\tfrac{1}{7}x+1\right|\le 0$$

Since an absolute value can never be less than zero, the only solution for this inequality is when the absolute value is equal to 0.

$$\left|\tfrac{1}{7}x+1\right|=0$$
$$\tfrac{1}{7}x+1=0 \quad \textbf{or} \quad \tfrac{1}{7}x+1=-0$$
$$\tfrac{1}{7}x=-1 \qquad \tfrac{1}{7}x=-1$$
$$x=-7 \qquad x=-7$$

solution set: $[-7,-7]$

51.

$$\left|\frac{x-5}{10}\right|\le 0$$

Since an absolute value can never be less than zero, the only solution for this inequality is when the absolute value is equal to 0.

$$\left|\frac{x-5}{10}\right|=0$$
$$\frac{x-5}{10}=0 \quad \textbf{or} \quad \frac{x-5}{10}=-0$$
$$x-5=0 \qquad x-5=0$$
$$x=5 \qquad x=5$$

solution set: $[5,5]$

53. Find the x-coordinates of all points on the graph of $y = |0.5x + 0.7|$ that are below points on the graph of $y = 2.6$:

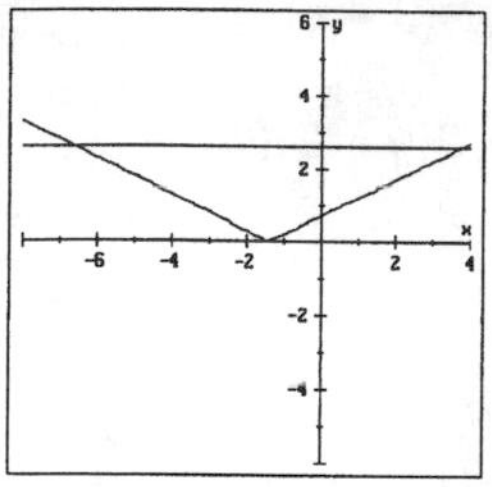

solution set: $(-6.6, 3.8)$

55. Find the x-coordinates of all points on the graph of $y = |2.15x - 3.05|$ that are above points on the graph of $y = 3.8$:

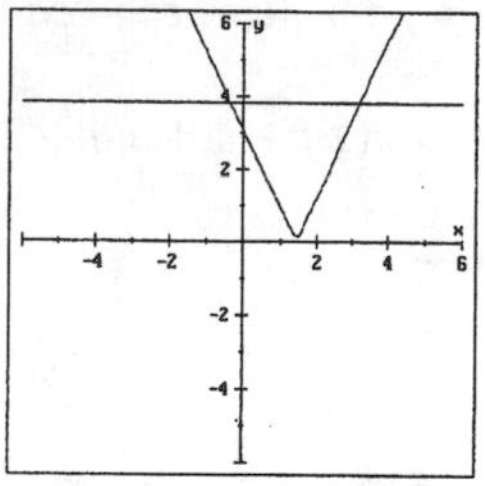

solution set: $(-\infty, -0.4) \cup (3.2, \infty)$

57. **Answers may vary.**

59. $|x| + |y| > |x + y|$ when x and y have different signs.

Exercise 7.4 (page 469)

1. 1.1×10^3 ft/sec = 1,100 ft/sec

3.
$$\begin{aligned} \frac{2}{3}(5t - 3) &= 38 \\ 3 \cdot \frac{2}{3}(5t - 3) &= 3(38) \\ 2(5t - 3) &= 114 \\ 10t - 6 &= 114 \\ 10t &= 120 \\ t &= 12 \end{aligned}$$

5. $ab + ac$

7. perfect

9. $x^2 - xy + y^2$

11. $2x + 8 = \mathbf{2} \cdot x + \mathbf{2} \cdot 4 = \mathbf{2}(x + 4)$

13. $2x^2 - 6x = \mathbf{2x} \cdot x - \mathbf{2x} \cdot 3 = \mathbf{2x}(x - 3)$

15. $15x^2y - 10x^2y^2 = \mathbf{5x^2y} \cdot 3 - \mathbf{5x^2y} \cdot 2y = \mathbf{5x^2y}(3 - 2y)$

17. $13ab^2c^3 - 26a^3b^2c = \mathbf{13ab^2c} \cdot c^2 - \mathbf{13ab^2c} \cdot 2a^2 = \mathbf{13ab^2c}(c^2 - 2a^2)$

19. $27z^3 + 12z^2 + 3z = \mathbf{3z} \cdot 9z^2 + \mathbf{3z} \cdot 4z + \mathbf{3z} \cdot 1 = \mathbf{3z}(9z^2 + 4z + 1)$

21. $24s^3 - 12s^2t + 6st^2 = \mathbf{6s}(4s^2) - \mathbf{6s}(2st) + \mathbf{6s}(t^2) = \mathbf{6s}(4s^2 - 2st + t^2)$

23.
$$\begin{aligned} -3a - 6 &= \mathbf{(-3)}(a) + \mathbf{(-3)}(2) \\ &= \mathbf{-3}(a + 2) \end{aligned}$$

25.
$$\begin{aligned} -6x^2 - 3xy &= \mathbf{(-3x)}(2x) + \mathbf{(-3x)}(y) \\ &= \mathbf{-3x}(2x + y) \end{aligned}$$

27. $-63u^3v^6z^9 + 28u^2v^7z^2 - 21u^3v^3z^4 = -7u^2v^3z^2(9uv^3z^7 - 4v^4 + 3uz^2)$

29.
$$\begin{aligned} x^{n+2} + x^{n+3} &= x^2(x^{n+2-2} + x^{n+3-2}) \\ &= x^2(x^n + x^{n+1}) \end{aligned}$$

31.
$$\begin{aligned} 2y^{n+2} - 3y^{n+3} &= y^n(2y^{n+2-n} - 3y^{n+3-n}) \\ &= y^n(2y^2 - 3y^3) \end{aligned}$$

33. $ax + bx + ay + by = x(a+b) + y(a+b)$
$= (a+b)(x+y)$

35. $x^2 + yx + 2x + 2y = x(x+y) + 2(x+y)$
$= (x+y)(x+2)$

37. $3c - cd + 3d - c^2 = 3c + 3d - c^2 - cd = 3(c+d) - c(c+d) = (c+d)(3-c)$

39. $4y + 8 + 2xy + 4x = 2(2y + 4 + xy + 2x) = 2[2(y+2) + x(y+2)] = 2(y+2)(2+x)$

41.
$$r_1 r_2 = rr_2 + rr_1$$
$$r_1 r_2 - rr_1 = rr_2$$
$$r_1(r_2 - r) = rr_2$$
$$r_1 = \frac{rr_2}{r_2 - r}$$

43.
$$S(1-r) = a - lr$$
$$S - Sr = a - lr$$
$$S - a = Sr - lr$$
$$S - a = r(S - l)$$
$$\frac{S-a}{S-l} = r, \text{ or } r = \frac{S-a}{S-l}$$

45. $x^2 - 4 = x^2 - 2^2 = (x+2)(x-2)$

47. $9y^2 - 64 = (3y)^2 - 8^2 = (3y+8)(3y-8)$

49. $81a^4 - 49b^2 = (9a^2)^2 - (7b)^2$
$= (9a^2 + 7b)(9a^2 - 7b)$

51. $(x+y)^2 - z^2 = [(x+y) + z][(x+y) - z]$
$= (x+y+z)(x+y-z)$

53. $x^4 - y^4 = (x^2 + y^2)(x^2 - y^2) = (x^2 + y^2)(x+y)(x-y)$

55. $2x^2 - 288 = 2(x^2 - 144)$
$= 2(x+12)(x-12)$

57. $2x^3 - 32x = 2x(x^2 - 16)$
$= 2x(x+4)(x-4)$

59. $x^{2m} - y^{4n} = (x^m)^2 - (y^{2n})^2 = (x^m + y^{2n})(x^m - y^{2n})$

61. $a^2 - b^2 + a + b = (a^2 - b^2) + (a+b) = (a+b)(a-b) + 1(a+b)$
$= (a+b)(a-b+1)$

63. $2x + y + 4x^2 - y^2 = (2x+y) + (4x^2 - y^2) = (2x+y)(1) + (2x+y)(2x-y)$
$= (2x+y)(1 + 2x - y)$

65. $x^2 + 5x + 6 = (x+2)(x+3)$

67. $a^2 + 5a - 52 \Rightarrow$ prime

69. $2x^2 - 11x + 5 = (2x-1)(x-5)$

71. $6y^2 + 7y + 2 = (2y+1)(3y+2)$

73. $8a^2 + 6a - 9 = (4a-3)(2a+3)$

75. $a^2 - 3ab - 4b^2 = (a-4b)(a+b)$

77. $2y^2 + yt - 6t^2 = (2y-3t)(y+2t)$

79. $5x^2 + 4x + 1 \Rightarrow$ prime

81. $8x^2 - 10x + 3 = (4x-3)(2x-1)$

83. $3x^2 + 12x - 63 = 3(x^2 + 4x - 21)$
$= 3(x+7)(x-3)$

85. $3x^3 - 10x^2 + 3x = x(3x^2 - 10x + 3)$
$= x(3x-1)(x-3)$

87. $a^2b^2 - 13ab^2 + 22b^2 = b^2(a^2 - 13a + 22) = b^2(a-11)(a-2)$

89. $-a^2+4a+32=-(a^2-4a-32)$
$=-(a-8)(a+4)$

91. $-3x^2+15x-18=-3(x^2-5x+6)$
$=-3(x-2)(x-3)$

93. $-3a^2+ab+2b^2=-(3a^2-ab-2b^2)=-(3a+2b)(a-b)$

95. $-4x^3-9x+12x^2=-x(4x^2-12x+9)=-x(2x-3)(2x-3)=-x(2x-3)^2$

97. $x^4+8x^2+15=(x^2+5)(x^2+3)$

99. $y^4-13y^2+30=(y^2-10)(y^2-3)$

101. $a^4-13a^2+36=(a^2-4)(a^2-9)=(a+2)(a-2)(a+3)(a-3)$

103. $x^{2n}+2x^n+1=(x^n+1)(x^n+1)=(x^n+1)^2$

105. $2a^{6n}-3a^{3n}-2=(2a^{3n}+1)(a^{3n}-2)$

107. $x^{4n}+2x^{2n}y^{2n}+y^{4n}=(x^{2n}+y^{2n})(x^{2n}+y^{2n})=(x^{2n}+y^{2n})^2$

109. $6x^{2n}+7x^n-3=(3x^n-1)(2x^n+3)$

111. $x^2+4x+4-y^2=(x^2+4x+4)-y^2=(x+2)(x+2)-y^2=(x+2)^2-y^2$
$=[(x+2)+y][(x+2)-y]$
$=(x+2+y)(x+2-y)$

113. $x^2+2x+1-9z^2=(x^2+2x+1)-9z^2=(x+1)(x+1)-9z^2$
$=(x+1)^2-(3z)^2$
$=[(x+1)+3z][(x+1)-3z]$
$=(x+1+3z)(x+1-3z)$

115. $c^2-4a^2+4ab-b^2=c^2-(4a^2-4ab+b^2)=c^2-(2a-b)(2a-b)$
$=c^2-(2a-b)^2$
$=[c+(2a-b)][c-(2a-b)]$
$=(c+2a-b)(c-2a+b)$

117. $y^3+1=y^3+1^3$
$=(y+1)(y^2-1y+1^2)$
$=(y+1)(y^2-y+1)$

119. $a^3-27=a^3-3^3$
$=(a-3)(a^2+3a+3^2)$
$=(a-3)(a^2+3a+9)$

121. $8+x^3=2^3+x^3$
$=(2+x)(2^2-2x+x^2)$
$=(2+x)(4-2x+x^2)$

123. $s^3-t^3=(s-t)(s^2+st+t^2)$

125. $27x^3+y^3=(3x)^3+y^3$
$=(3x+y)((3x)^2-3xy+y^2)$
$=(3x+y)(9x^2-3xy+y^2)$

127. $a^3+8b^3=a^3+(2b)^3$
$=(a+2b)(a^2-2ab+(2b)^2)$
$=(a+2b)(a^2-2ab+4b^2)$

129. $27x^3 - 125y^3 = (3x)^3 - (5y)^3 = (3x - 5y)\left((3x)^2 + (3x)(5y) + (5y)^2\right)$
$= (3x - 5y)\left(9x^2 + 15xy + 25y^2\right)$

131. $a^6 - b^3 = (a^2)^3 - b^3 = (a^2 - b)\left((a^2)^2 + a^2b + b^2\right) = (a^2 - b)(a^4 + a^2b + b^2)$

133. $x^9 + y^6 = (x^3)^3 + (y^2)^3 = (x^3 + y^2)\left((x^3)^2 - x^3y^2 + (y^2)^2\right)$
$= (x^3 + y^2)(x^6 - x^3y^2 + y^4)$

135. $2x^3 + 54 = 2(x^3 + 27)$
$= 2(x^3 + 3^3)$
$= 2(x + 3)(x^2 - 3x + 3^2)$
$= 2(x + 3)(x^2 - 3x + 9)$

137. $-x^3 + 216 = -(x^3 - 216)$
$= -(x^3 - 6^3)$
$= -(x - 6)(x^2 + 6x + 6^2)$
$= -(x - 6)(x^2 + 6x + 36)$

139. $64m^3x - 8n^3x = 8x(8m^3 - n^3) = 8x\left((2m)^3 - n^3\right) = 8x(2m - n)\left((2m)^2 + 2mn + n^2\right)$
$= 8x(2m - n)(4m^2 + 2mn + n^2)$

141. $x^4y + 216xy^4 = xy(x^3 + 216y^3) = xy\left(x^3 + (6y)^3\right) = xy(x + 6y)\left(x^2 - 6xy + (6y)^2\right)$
$= xy(x + 6y)(x^2 - 6xy + 36y^2)$

143. $81r^4s^2 - 24rs^5 = 3rs^2(27r^3 - 8s^3) = 3rs^2\left((3r)^3 - (2s)^3\right)$
$= 3rs^2(3r - 2s)\left((3r)^2 + (3r)(2s) + (2s)^2\right)$
$= 3rs^2(3r - 2s)(9r^2 + 6rs + 4s^2)$

145. $125a^6b^2 + 64a^3b^5 = a^3b^2(125a^3 + 64b^3) = a^3b^2\left((5a)^3 + (4b)^3\right)$
$= a^3b^2(5a + 4b)\left((5a)^2 - (5a)(4b) + (4b)^2\right)$
$= a^3b^2(5a + 4b)(25a^2 - 20ab + 16b^2)$

147. $2mp^4 + 16mpq^3 = 2mp(p^3 + 8q^3) = 2mp\left(p^3 + (2q)^3\right) = 2mp(p + 2q)\left(p^2 - 2pq + (2q)^2\right)$
$= 2mp(p + 2q)(p^2 - 2pq + 4q^2)$

149. $3(x^3 + y^3) - z(x^3 + y^3) = (x^3 + y^3)(3 - z) = (x + y)(x^2 - xy + y^2)(3 - z)$

151. $(m^3 + 8n^3) + (m^3x + 8n^3x) = 1(m^3 + 8n^3) + x(m^3 + 8n^3)$
$= (m^3 + 8n^3)(1 + x)$
$= (m + 2n)\left(m^2 - 2mn + (2n)^2\right)(1 + x)$
$= (m + 2n)(m^2 - 2mn + 4n^2)(1 + x)$

153. $$\begin{aligned}(a^4+27a)-(a^3b+27b)=a(a^3+27)-b(a^3+27)&=(a^3+27)(a-b)\\&=(a^3+3^3)(a-b)\\&=(a+3)(a^2-3a+3^2)(a-b)\\&=(a+3)(a^2-3a+9)(a-b)\end{aligned}$$

155. $$\begin{aligned}y^3(y^2-1)-27(y^2-1)=(y^2-1)(y^3-27)&=(y^2-1^2)(y^3-3^3)\\&=(y+1)(y-1)(y-3)(y^2+3y+9)\end{aligned}$$

157. $$\begin{aligned}x^6-1=(x^3)^2-1^2=(x^3+1)(x^3-1)&=(x^3+1^3)(x^3-1^3)\\&=(x+1)(x^2-1x+1^2)(x-1)(x^2+1x+1^2)\\&=(x+1)(x^2-x+1)(x-1)(x^2+x+1)\end{aligned}$$

159. $$\begin{aligned}x^{12}-y^6=(x^6)^2-(y^3)^2&=(x^6+y^3)(x^6-y^3)\\&=\left((x^2)^3+y^3\right)\left((x^2)^3-y^3\right)\\&=(x^2+y)\left((x^2)^2-x^2y+y^2\right)(x^2-y)\left((x^2)^2+x^2y+y^2\right)\\&=(x^2+y)(x^4-x^2y+y^2)(x^2-y)(x^4+x^2y+y^2)\end{aligned}$$

161. Answers may vary.

163. Answers may vary.

165. $x^2-q^2=x^2+0x-q^2$: $a=1, b=0, c=-q^2 \Rightarrow b^2-4ac=0^2-4(1)(-q^2)=4q^2=(2q)^2$.
Since $4q^2$ is a perfect square, the binomial is factorable, and the test works.

167. $$\begin{aligned}x^{32}-y^{32}=(x^{16}+y^{16})(x^{16}-y^{16})&=(x^{16}+y^{16})(x^8+y^8)(x^8-y^8)\\&=(x^{16}+y^{16})(x^8+y^8)(x^4+y^4)(x^4-y^4)\\&=(x^{16}+y^{16})(x^8+y^8)(x^4+y^4)(x^2+y^2)(x^2-y^2)\\&=(x^{16}+y^{16})(x^8+y^8)(x^4+y^4)(x^2+y^2)(x+y)(x-y)\end{aligned}$$

Exercise 7.5 (page 483)

1. 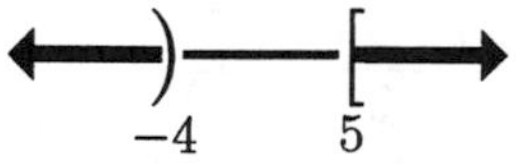

3. $$\begin{aligned}P&=2l+2w\\P-2l&=2w\\\frac{P-2l}{2}&=\frac{2w}{2}\\\frac{P-2l}{2}&=w, \text{ or } w=\frac{P-2l}{2}\end{aligned}$$

5.

$$\begin{aligned} a^4 - 13a^2 + 36 &= 0 \\ (a^2-4)(a^2-9) &= 36 \\ (a+2)(a-2)(a+3)(a-3) &= 36 \end{aligned}$$

$$\begin{array}{ccccccc} a+2=0 & \textbf{or} & a-2=0 & \textbf{or} & a+3=0 & \textbf{or} & a-3=0 \\ a=-2 & & a=2 & & a=-3 & & a=3 \end{array}$$

7. $\dfrac{a}{b}$

9. $\dfrac{ad}{bc}$

11. $\dfrac{12x^3}{3x} = \dfrac{12}{3} \cdot \dfrac{x^3}{x} = 4x^2$

13. $\dfrac{-24x^3y^4}{18x^4y^3} = -\dfrac{24}{18} \cdot \dfrac{x^3y^4}{x^4y^3} = -\dfrac{4y}{3x}$

15. $\dfrac{9y^2(y-z)}{21y(y-z)^2} = \dfrac{3y}{7(y-z)}$

17. $\dfrac{(a-b)(b-c)(c-d)}{(c-d)(b-c)(a-b)} = 1$

19. $\dfrac{x+y}{x^2-y^2} = \dfrac{x+y}{(x+y)(x-y)} = \dfrac{1}{x-y}$

21. $\dfrac{12-3x^2}{x^2-x-2} = \dfrac{3(4-x^2)}{(x-2)(x+1)} = \dfrac{3(2+x)(2-x)}{(x-2)(x+1)} = \dfrac{-3(x+2)(x-2)}{(x-2)(x+1)} = \dfrac{-3(x+2)}{x+1}$

23. $$\begin{aligned} \frac{x^3+8}{x^2-2x+4} &= \frac{(x+2)(x^2-2x+4)}{x^2-2x+4} \\ &= x+2 \end{aligned}$$

25. $\dfrac{x^2+2x+1}{x^2+4x+3} = \dfrac{(x+1)(x+1)}{(x+1)(x+3)} = \dfrac{x+1}{x+3}$

27. $\dfrac{3m-6n}{3n-6m} = \dfrac{3(m-2n)}{3(n-2m)} = \dfrac{m-2n}{n-2m}$

29. $\dfrac{4x^2+24x+32}{16x^2+8x-48} = \dfrac{4(x^2+6x+8)}{8(2x^2+x-6)} = \dfrac{4(x+4)(x+2)}{8(2x-3)(x+2)} = \dfrac{x+4}{2(2x-3)}$

31. $$\begin{aligned} \frac{3x^2-3y^2}{x^2+2y+2x+yx} = \frac{3(x^2-y^2)}{x^2+2x+yx+2y} = \frac{3(x^2-y^2)}{x(x+2)+y(x+2)} &= \frac{3(x+y)(x-y)}{(x+2)(x+y)} \\ &= \frac{3(x-y)}{x+2} \end{aligned}$$

33. $$\begin{aligned} \frac{x-y}{x^3-y^3-x+y} = \frac{x-y}{(x^3-y^3)-(x-y)} &= \frac{x-y}{(x-y)(x^2+xy+y^2)-1(x-y)} \\ &= \frac{x-y}{(x-y)(x^2+xy+y^2-1)} \\ &= \frac{1}{x^2+xy+y^2-1} \end{aligned}$$

35. $\dfrac{x^2y^2}{cd} \cdot \dfrac{c^{-2}d^2}{x} = x^{2-1}y^2c^{-2-1}d^{2-1} = xy^2c^{-3}d = \dfrac{xy^2d}{c^3}$

37. $\dfrac{-x^2y^{-2}}{x^{-1}y^{-3}} \div \dfrac{x^{-3}y^2}{x^4y^{-1}} = \dfrac{-x^2y^{-2}}{x^{-1}y^{-3}} \cdot \dfrac{x^4y^{-1}}{x^{-3}y^2} = \dfrac{-x^6y^{-3}}{x^{-4}y^{-1}} = -x^{6-(-4)}y^{-3-(-1)} = -x^{10}y^{-2} = -\dfrac{x^{10}}{y^2}$

39. $\dfrac{x^2+2x+1}{x} \cdot \dfrac{x^2-x}{x^2-1} = \dfrac{(x+1)(x+1)}{x} \cdot \dfrac{x(x-1)}{(x+1)(x-1)} = x+1$

41. $\dfrac{2x^2-x-3}{x^2-1} \cdot \dfrac{x^2+x-2}{2x^2+x-6} = \dfrac{(2x-3)(x+1)}{(x+1)(x-1)} \cdot \dfrac{(x+2)(x-1)}{(2x-3)(x+2)} = 1$

43. $\dfrac{x^2-16}{x^2-25} \div \dfrac{x+4}{x-5} = \dfrac{x^2-16}{x^2-25} \cdot \dfrac{x-5}{x+4} = \dfrac{(x+4)(x-4)}{(x+5)(x-5)} \cdot \dfrac{x-5}{x+4} = \dfrac{x-4}{x+5}$

45. $\dfrac{a^2+2a-35}{12x} \div \dfrac{ax-3x}{a^2+4a-21} = \dfrac{a^2+2a-35}{12x} \cdot \dfrac{a^2+4a-21}{ax-3x}$

$= \dfrac{(a+7)(a-5)}{12x} \cdot \dfrac{(a+7)(a-3)}{x(a-3)} = \dfrac{(a+7)^2(a-5)}{12x^2}$

47. $\dfrac{3t^2-t-2}{6t^2-5t-6} \cdot \dfrac{4t^2-9}{2t^2+5t+3} = \dfrac{(3t+2)(t-1)}{(3t+2)(2t-3)} \cdot \dfrac{(2t+3)(2t-3)}{(2t+3)(t+1)} = \dfrac{t-1}{t+1}$

49. $\dfrac{3n^2+5n-2}{12n^2-13n+3} \div \dfrac{n^2+3n+2}{4n^2+5n-6} = \dfrac{3n^2+5n-2}{12n^2-13n+3} \cdot \dfrac{4n^2+5n-6}{n^2+3n+2}$

$= \dfrac{(3n-1)(n+2)}{(3n-1)(4n-3)} \cdot \dfrac{(4n-3)(n+2)}{(n+2)(n+1)} = \dfrac{n+2}{n+1}$

51. $(2x^2-15x+25) \div \dfrac{2x^2-3x-5}{x+1} = \dfrac{2x^2-15x+25}{1} \cdot \dfrac{x+1}{2x^2-3x-5}$

$= \dfrac{(2x-5)(x-5)}{1} \cdot \dfrac{x+1}{(2x-5)(x+1)} = x-5$

53. $\dfrac{x^3+y^3}{x^3-y^3} \div \dfrac{x^2-xy+y^2}{x^2+xy+y^2} = \dfrac{x^3+y^3}{x^3-y^3} \cdot \dfrac{x^2+xy+y^2}{x^2-xy+y^2} = \dfrac{(x+y)(x^2-xy+y^2)}{(x-y)(x^2+xy+y^2)} \cdot \dfrac{x^2+xy+y^2}{x^2-xy+y^2}$

$= \dfrac{x+y}{x-y}$

55. $\dfrac{m^2-n^2}{2x^2+3x-2} \cdot \dfrac{2x^2+5x-3}{n^2-m^2} = \dfrac{(m+n)(m-n)}{(2x-1)(x+2)} \cdot \dfrac{(2x-1)(x+3)}{(n+m)(n-m)} = -\dfrac{x+3}{x+2}$

57. $\dfrac{ax+ay+bx+by}{x^3-27} \cdot \dfrac{x^2+3x+9}{xc+xd+yc+yd} = \dfrac{a(x+y)+b(x+y)}{(x-3)(x^2+3x+9)} \cdot \dfrac{x^2+3x+9}{x(c+d)+y(c+d)}$

$= \dfrac{(x+y)(a+b)}{(x-3)(x^2+3x+9)} \cdot \dfrac{x^2+3x+9}{(c+d)(x+y)}$

$= \dfrac{a+b}{(x-3)(c+d)}$

59. $\dfrac{x^2-x-6}{x^2-4}\cdot\dfrac{x^2-x-2}{9-x^2}=\dfrac{(x-3)(x+2)}{(x+2)(x-2)}\cdot\dfrac{(x-2)(x+1)}{(3+x)(3-x)}=-\dfrac{x+1}{x+3}$

61. $\dfrac{2x^2+3xy+y^2}{y^2-x^2}\div\dfrac{6x^2+5xy+y^2}{2x^2-xy-y^2}=\dfrac{2x^2+3xy+y^2}{y^2-x^2}\cdot\dfrac{2x^2-xy-y^2}{6x^2+5xy+y^2}$

$=\dfrac{(2x+y)(x+y)}{(y+x)(y-x)}\cdot\dfrac{(2x+y)(x-y)}{(2x+y)(3x+y)}=-\dfrac{2x+y}{3x+y}$

63. $\dfrac{3x^2y^2}{6x^3y}\cdot\dfrac{-4x^7y^{-2}}{18x^{-2}y}\div\dfrac{36x}{18y^{-2}}=\dfrac{3x^2y^2}{6x^3y}\cdot\dfrac{-4x^7y^{-2}}{18x^{-2}y}\cdot\dfrac{18y^{-2}}{36x}=\dfrac{-2^3\cdot3^3x^9y^{-2}}{2^4\cdot3^5x^2y^2}=-\dfrac{x^7}{18y^4}$

65. $(4x+12)\cdot\dfrac{x^2}{2x-6}\div\dfrac{2}{x-3}=\dfrac{4x+12}{1}\cdot\dfrac{x^2}{2x-6}\cdot\dfrac{x-3}{2}=\dfrac{4(x+3)}{1}\cdot\dfrac{x^2}{2(x-3)}\cdot\dfrac{x-3}{2}$

$=x^2(x+3)$

67. $\dfrac{2x^2-2x-4}{x^2+2x-8}\cdot\dfrac{3x^2+15x}{x+1}\div\dfrac{4x^2-100}{x^2-x-20}=\dfrac{2x^2-2x-4}{x^2+2x-8}\cdot\dfrac{3x^2+15x}{x+1}\cdot\dfrac{x^2-x-20}{4x^2-100}$

$=\dfrac{2(x-2)(x+1)}{(x+4)(x-2)}\cdot\dfrac{3x(x+5)}{x+1}\cdot\dfrac{(x-5)(x+4)}{4(x+5)(x-5)}$

$=\dfrac{3x}{2}$

69. $\dfrac{2x^2+5x-3}{x^2+2x-3}\div\left(\dfrac{x^2+2x-35}{x^2-6x+5}\div\dfrac{x^2-9x+14}{2x^2-5x+2}\right)$

$=\dfrac{2x^2+5x-3}{x^2+2x-3}\div\left(\dfrac{x^2+2x-35}{x^2-6x+5}\cdot\dfrac{2x^2-5x+2}{x^2-9x+14}\right)$

$=\dfrac{2x^2+5x-3}{x^2+2x-3}\div\left(\dfrac{(x+7)(x-5)}{(x-5)(x-1)}\cdot\dfrac{(2x-1)(x-2)}{(x-7)(x-2)}\right)$

$=\dfrac{2x^2+5x-3}{x^2+2x-3}\div\dfrac{(x+7)(2x-1)}{(x-1)(x-7)}=\dfrac{(2x-1)(x+3)}{(x+3)(x-1)}\cdot\dfrac{(x-1)(x-7)}{(x+7)(2x-1)}=\dfrac{x-7}{x+7}$

71. $\dfrac{x^2-x-12}{x^2+x-2}\div\dfrac{x^2-6x+8}{x^2-3x-10}\cdot\dfrac{x^2-3x+2}{x^2-2x-15}=\dfrac{x^2-x-12}{x^2+x-2}\cdot\dfrac{x^2-3x-10}{x^2-6x+8}\cdot\dfrac{x^2-3x+2}{x^2-2x-15}$

$=\dfrac{(x-4)(x+3)}{(x+2)(x-1)}\cdot\dfrac{(x-5)(x+2)}{(x-4)(x-2)}\cdot\dfrac{(x-2)(x-1)}{(x-5)(x+3)}=1$

73. $\dfrac{3}{a+b}-\dfrac{a}{a+b}=\dfrac{3-a}{a+b}$

75. $\dfrac{3x}{2x+2}+\dfrac{x+4}{2x+2}=\dfrac{4x+4}{2x+2}=\dfrac{4(x+1)}{2(x+1)}=2$

77. $\dfrac{5x}{x+1}+\dfrac{3}{x+1}-\dfrac{2x}{x+1}=\dfrac{3x+3}{x+1}=\dfrac{3(x+1)}{x+1}=3$

79. $\dfrac{3(x^2+x)}{x^2-5x+6}+\dfrac{-3(x^2-x)}{x^2-5x+6}=\dfrac{3x^2+3x}{x^2-5x+6}+\dfrac{-3x^2+3x}{x^2-5x+6}=\dfrac{6x}{(x-3)(x-2)}$

81. $$\frac{a}{2}+\frac{2a}{5}=\frac{a\cdot 5}{2\cdot 5}+\frac{2a\cdot 2}{5\cdot 2}=\frac{5a}{10}+\frac{4a}{10}=\frac{9a}{10}$$

83. $$\frac{3}{4x}+\frac{2}{3x}=\frac{3\cdot 3}{4x\cdot 3}+\frac{2\cdot 4}{3x\cdot 4}=\frac{9}{12x}+\frac{8}{12x}=\frac{17}{12x}$$

85. $$\frac{a+b}{3}+\frac{a-b}{7}=\frac{(a+b)7}{3(7)}+\frac{(a-b)3}{7(3)}=\frac{7a+7b}{21}+\frac{3a-3b}{21}=\frac{10a+4b}{21}$$

87. $$\frac{3}{x+2}+\frac{5}{x-4}=\frac{3(x-4)}{(x+2)(x-4)}+\frac{5(x+2)}{(x-4)(x+2)}=\frac{3x-12}{(x+2)(x-4)}+\frac{5x+10}{(x+2)(x-4)}=\frac{8x-2}{(x+2)(x-4)}=\frac{2(4x-1)}{(x+2)(x-4)}$$

89. $$\frac{x+2}{x+5}-\frac{x-3}{x+7}=\frac{(x+2)(x+7)}{(x+5)(x+7)}-\frac{(x-3)(x+5)}{(x+7)(x+5)}=\frac{x^2+9x+14}{(x+5)(x+7)}-\frac{x^2+2x-15}{(x+5)(x+7)}=\frac{7x+29}{(x+5)(x+7)}$$

91. $$x+\frac{1}{x}=\frac{x}{1}+\frac{1}{x}=\frac{x(x)}{1(x)}+\frac{1}{x}=\frac{x^2}{x}+\frac{1}{x}=\frac{x^2+1}{x}$$

93. $$\frac{x+8}{x-3}-\frac{x-14}{3-x}=\frac{x+8}{x-3}-\frac{-x+14}{x-3}=\frac{2x-6}{x-3}=\frac{2(x-3)}{x-3}=2$$

95. $$\begin{aligned}\frac{x}{x^2+5x+6}+\frac{x}{x^2-4}&=\frac{x}{(x+2)(x+3)}+\frac{x}{(x+2)(x-2)}\\&=\frac{x(x-2)}{(x+2)(x+3)(x-2)}+\frac{x(x+3)}{(x+2)(x-2)(x+3)}\\&=\frac{x^2-2x}{(x+2)(x+3)(x-2)}+\frac{x^2+3x}{(x+2)(x+3)(x-2)}\\&=\frac{2x^2+x}{(x+2)(x+3)(x-2)}\end{aligned}$$

97. $$\begin{aligned}\frac{8}{x^2-9}+\frac{2}{x-3}-\frac{6}{x}&=\frac{8}{(x+3)(x-3)}+\frac{2}{x-3}-\frac{6}{x}\\&=\frac{8x}{x(x+3)(x-3)}+\frac{2x(x+3)}{x(x+3)(x-3)}-\frac{6(x+3)(x-3)}{x(x+3)(x-3)}\\&=\frac{8x}{x(x+3)(x-3)}+\frac{2x^2+6x}{x(x+3)(x-3)}-\frac{6(x^2-9)}{x(x+3)(x-3)}\\&=\frac{8x}{x(x+3)(x-3)}+\frac{2x^2+6x}{x(x+3)(x-3)}-\frac{6x^2-54}{x(x+3)(x-3)}\\&=\frac{-4x^2+14x+54}{x(x+3)(x-3)}\end{aligned}$$

99. $$1+x-\frac{x}{x-5}=\frac{x+1}{1}-\frac{x}{x-5}=\frac{(x+1)(x-5)}{1(x-5)}-\frac{x}{x-5}$$
$$=\frac{x^2-4x-5}{x-5}-\frac{x}{x-5}=\frac{x^2-5x-5}{x-5}$$

101. $$\frac{3}{x+1}-\frac{2}{x-1}+\frac{x+3}{x^2-1}=\frac{3}{x+1}-\frac{2}{x-1}+\frac{x+3}{(x+1)(x-1)}$$
$$=\frac{3(x-1)}{(x+1)(x-1)}-\frac{2(x+1)}{(x+1)(x-1)}+\frac{x+3}{(x+1)(x-1)}$$
$$=\frac{3x-3}{(x+1)(x-1)}-\frac{2x+2}{(x+1)(x-1)}+\frac{x+3}{(x+1)(x-1)}$$
$$=\frac{2x-2}{(x+1)(x-1)}=\frac{2(x-1)}{(x+1)(x-1)}=\frac{2}{x+1}$$

103. $$\frac{x-2}{x^2-3x}+\frac{2x-1}{x^2+3x}-\frac{2}{x^2-9}=\frac{x-2}{x(x-3)}+\frac{2x-1}{x(x+3)}-\frac{2}{(x+3)(x-3)}$$
$$=\frac{(x-2)(x+3)}{x(x-3)(x+3)}+\frac{(2x-1)(x-3)}{x(x-3)(x+3)}-\frac{2x}{x(x-3)(x+3)}$$
$$=\frac{x^2+x-6}{x(x-3)(x+3)}+\frac{2x^2-7x+3}{x(x-3)(x+3)}-\frac{2x}{x(x-3)(x+3)}$$
$$=\frac{3x^2-8x-3}{x(x-3)(x+3)}=\frac{(3x+1)(x-3)}{x(x-3)(x+3)}=\frac{3x+1}{x(x+3)}$$

105. $$\frac{\frac{4x}{y}}{\frac{6xz}{y^2}}=\frac{4x}{y}\div\frac{6xz}{y^2}=\frac{4x}{y}\cdot\frac{y^2}{6xz}=\frac{2y}{3z}$$

107. $$\frac{\frac{x-y}{xy}}{\frac{y-x}{x}}=\frac{x-y}{xy}\div\frac{y-x}{x}=\frac{x-y}{xy}\cdot\frac{x}{y-x}=\frac{x-y}{xy}\cdot\frac{x}{-(x-y)}=-\frac{1}{y}$$

109. $$\frac{\frac{1}{a}+\frac{1}{b}}{\frac{1}{a}}=\frac{\left(\frac{1}{a}+\frac{1}{b}\right)\cdot ab}{\frac{1}{a}\cdot ab}=\frac{\frac{1}{a}\cdot ab+\frac{1}{b}\cdot ab}{b}=\frac{b+a}{b}$$

111. $$\frac{\frac{y}{x}-\frac{x}{y}}{\frac{1}{x}+\frac{1}{y}}=\frac{\left(\frac{y}{x}-\frac{x}{y}\right)\cdot xy}{\left(\frac{1}{x}+\frac{1}{y}\right)\cdot xy}=\frac{\frac{y}{x}\cdot xy-\frac{x}{y}\cdot xy}{\frac{1}{x}\cdot xy+\frac{1}{y}\cdot xy}=\frac{y^2-x^2}{y+x}=\frac{(y+x)(y-x)}{y+x}=y-x$$

113. $$\frac{\frac{1}{a}-\frac{1}{b}}{\frac{a}{b}-\frac{b}{a}}=\frac{\left(\frac{1}{a}-\frac{1}{b}\right)\cdot ab}{\left(\frac{a}{b}-\frac{b}{a}\right)\cdot ab}=\frac{\frac{1}{a}\cdot ab-\frac{1}{b}\cdot ab}{\frac{a}{b}\cdot ab-\frac{b}{a}\cdot ab}=\frac{b-a}{a^2-b^2}=\frac{b-a}{(a+b)(a-b)}=-\frac{1}{a+b}$$

115. $$\frac{1+\frac{6}{x}+\frac{8}{x^2}}{1+\frac{1}{x}-\frac{12}{x^2}} = \frac{\left(1+\frac{6}{x}+\frac{8}{x^2}\right)\cdot x^2}{\left(1+\frac{1}{x}-\frac{12}{x^2}\right)\cdot x^2} = \frac{x^2+6x+8}{x^2+x-12} = \frac{(x+4)(x+2)}{(x+4)(x-3)} = \frac{x+2}{x-3}$$

117. $$\frac{\frac{1}{a+1}+1}{\frac{3}{a-1}+1} = \frac{\left(\frac{1}{a+1}+1\right)(a+1)(a-1)}{\left(\frac{3}{a-1}+1\right)(a+1)(a-1)} = \frac{1(a-1)+1(a+1)(a-1)}{3(a+1)+1(a+1)(a-1)} = \frac{a-1+a^2-1}{3a+3+a^2-1}$$
$$= \frac{a^2+a-2}{a^2+3a+2}$$
$$= \frac{(a+2)(a-1)}{(a+2)(a+1)}$$
$$= \frac{a-1}{a+1}$$

119. $$\frac{x^{-1}+y^{-1}}{x^{-1}-y^{-1}} = \frac{\frac{1}{x}+\frac{1}{y}}{\frac{1}{x}-\frac{1}{y}} = \frac{\left(\frac{1}{x}+\frac{1}{y}\right)(xy)}{\left(\frac{1}{x}-\frac{1}{y}\right)(xy)} = \frac{y+x}{y-x}$$

121. $$\frac{x+y}{x^{-1}+y^{-1}} = \frac{x+y}{\frac{1}{x}+\frac{1}{y}} = \frac{(x+y)(xy)}{\left(\frac{1}{x}+\frac{1}{y}\right)(xy)} = \frac{(x+y)xy}{y+x} = xy$$

123. $$\frac{x-y^{-2}}{y-x^{-2}} = \frac{x-\frac{1}{y^2}}{y-\frac{1}{x^2}} = \frac{\left(x-\frac{1}{y^2}\right)(x^2y^2)}{\left(y-\frac{1}{x^2}\right)(x^2y^2)} = \frac{x^3y^2-x^2}{x^2y^3-y^2} = \frac{x^2(xy^2-1)}{y^2(x^2y-1)}$$

125. $$\frac{1+\frac{a}{b}}{1-\frac{a}{1-\frac{a}{b}}} = \frac{1+\frac{a}{b}}{1-\frac{a(b)}{(1-\frac{a}{b})(b)}} = \frac{1+\frac{a}{b}}{1-\frac{ab}{b-a}} = \frac{(1+\frac{a}{b})(b)(b-a)}{\left(1-\frac{ab}{b-a}\right)(b)(b-a)} = \frac{b(b-a)+a(b-a)}{b(b-a)-ab(b)}$$
$$= \frac{b^2-ab+ab-a^2}{b^2-ab-ab^2}$$
$$= \frac{b^2-a^2}{b(b-a-ab)}$$
$$= \frac{(b+a)(b-a)}{b(b-a-ab)}$$

127. $$a+\frac{a}{1+\frac{a}{a+1}} = a+\frac{a(a+1)}{\left(1+\frac{a}{a+1}\right)(a+1)} = a+\frac{a(a+1)}{a+1+a} = a+\frac{a^2+a}{2a+1} = \frac{a(2a+1)}{2a+1}+\frac{a^2+a}{2a+1}$$
$$= \frac{2a^2+a+a^2+a}{2a+1}$$
$$= \frac{3a^2+2a}{2a+1}$$

129. Answers may vary.

131. Answers may vary.

133. $\dfrac{a-3b}{2b-a} = \dfrac{-(3b-a)}{-(a-2b)} = \dfrac{3b-a}{a-2b}$.
The two answers are the same.

135. You can divide out the 4's in parts a and d.

Exercise 7.6 (page 491)

1. $f(1) = 3(1)^2 + 2(1) - 1 = 4$

3. $f(2a) = 3(2a)^2 + 2(2a) - 1$
$= 12a^2 + 4a - 1$

5. $2(x^2 + 4x - 1) + 3(2x^2 - 2x + 2) = 2x^2 + 8x - 2 + 6x^2 - 6x + 6 = 8x^2 + 2x + 4$

7. $P(r)$

9.
$$\begin{array}{r|rrr} 1 & 1 & 1 & -2 \\ & & 1 & 2 \\ \hline & 1 & 2 & 0 \end{array} \Rightarrow \boxed{x+2}$$

11.
$$\begin{array}{r|rrr} 4 & 1 & -7 & 12 \\ & & 4 & -12 \\ \hline & 1 & -3 & 0 \end{array} \Rightarrow \boxed{x-3}$$

13.
$$\begin{array}{r|rrr} -4 & 1 & 6 & 8 \\ & & -4 & -8 \\ \hline & 1 & 2 & 0 \end{array} \Rightarrow \boxed{x+2}$$

15.
$$\begin{array}{r|rrr} -2 & 1 & -5 & 14 \\ & & -2 & 14 \\ \hline & 1 & -7 & 28 \end{array} \Rightarrow \boxed{x-7+\frac{28}{x+2}}$$

17.
$$\begin{array}{r|rrrr} 3 & 3 & -10 & 5 & -6 \\ & & 9 & -3 & 6 \\ \hline & 3 & -1 & 2 & 0 \end{array}$$
$$\Rightarrow \boxed{3x^2 - x + 2}$$

19.
$$\begin{array}{r|rrrr} 2 & 2 & 0 & -5 & -6 \\ & & 4 & 8 & 6 \\ \hline & 2 & 4 & 3 & 0 \end{array}$$
$$\Rightarrow \boxed{2x^2 + 4x + 3}$$

21.
$$\begin{array}{r|rrrr} -1 & 6 & 5 & 0 & 4 \\ & & -6 & 1 & -1 \\ \hline & 6 & -1 & 1 & 3 \end{array}$$
$$\Rightarrow \boxed{6x^2 - x + 1 + \frac{3}{x+1}}$$

23.
$$\begin{array}{r|rrr} 0.2 & 7.2 & -2.1 & 0.5 \\ & & 1.44 & -0.132 \\ \hline & 7.2 & -0.66 & 0.368 \end{array}$$
$$\Rightarrow \boxed{7.2x - 0.66 + \frac{0.368}{x-0.2}}$$

25.
$$\begin{array}{r|rrr} -1.7 & 2.7 & 1.0 & -5.2 \\ & & -4.59 & 6.103 \\ \hline & 2.7 & -3.59 & 0.903 \end{array}$$
$$\Rightarrow \boxed{2.7x - 3.59 + \frac{0.903}{x+1.7}}$$

27.
$$\begin{array}{r|rrrr} -57 & 9 & 0 & 0 & -25 \\ & & -513 & 29241 & -1666737 \\ \hline & 9 & -513 & 29241 & -1666762 \end{array}$$
$$\Rightarrow \boxed{9x^2 - 513x + 29{,}241 - \frac{1{,}666{,}762}{x+57}}$$

29. $P(1) = 2(1)^3 - 4(1)^2 + 2(1) - 1 = \boxed{-1}$

$$\begin{array}{r|rrrr} 1 & 2 & -4 & 2 & -1 \\ & & 2 & -2 & 0 \\ \hline & 2 & -2 & 0 & \boxed{-1} \end{array}$$

31. $P(-2) = 2(-2)^3 - 4(-2)^2 + 2(-2) - 1 = \boxed{-37}$

$$\begin{array}{r|rrrr} -2 & 2 & -4 & 2 & -1 \\ & & -4 & 16 & -36 \\ \hline & 2 & -8 & 18 & \boxed{-37} \end{array}$$

33. $P(3) = 2(3)^3 - 4(3)^2 + 2(3) - 1 = \boxed{23}$

$$\begin{array}{r|rrrr} 3 & 2 & -4 & 2 & -1 \\ & & 6 & 6 & 24 \\ \hline & 2 & 2 & 8 & \boxed{23} \end{array}$$

35. $P(0) = 2(0)^3 - 4(0)^2 + 2(0) - 1 = \boxed{-1}$

$$\begin{array}{r|rrrr} 0 & 2 & -4 & 2 & -1 \\ & & 0 & 0 & 0 \\ \hline & 2 & -4 & 2 & \boxed{-1} \end{array}$$

37. $Q(-1) = (-1)^4 - 3(-1)^3 + 2(-1)^2 + (-1) - 3 = \boxed{2}$

$$\begin{array}{r|rrrrr} -1 & 1 & -3 & 2 & 1 & -3 \\ & & -1 & 4 & -6 & 5 \\ \hline & 1 & -4 & 6 & -5 & \boxed{2} \end{array}$$

39. $Q(2) = (2)^4 - 3(2)^3 + 2(2)^2 + (2) - 3 = \boxed{-1}$

$$\begin{array}{r|rrrrr} 2 & 1 & -3 & 2 & 1 & -3 \\ & & 2 & -2 & 0 & 2 \\ \hline & 1 & -1 & 0 & 1 & \boxed{-1} \end{array}$$

41. $Q(3) = (3)^4 - 3(3)^3 + 2(3)^2 + (3) - 3 = \boxed{18}$

$$\begin{array}{r|rrrrr} 3 & 1 & -3 & 2 & 1 & -3 \\ & & 3 & 0 & 6 & 21 \\ \hline & 1 & 0 & 2 & 7 & \boxed{18} \end{array}$$

43. $Q(-3) = (-3)^4 - 3(-3)^3 + 2(-3)^2 + (-3) - 3 = \boxed{174}$

$$\begin{array}{r|rrrrr} -3 & 1 & -3 & 2 & 1 & -3 \\ & & -3 & 18 & -60 & 177 \\ \hline & 1 & -6 & 20 & -59 & \boxed{174} \end{array}$$

45.

$$\begin{array}{r|rrrr} 2 & 1 & -4 & 1 & -2 \\ & & 2 & -4 & -6 \\ \hline & 1 & -2 & -3 & \boxed{-8} \end{array}$$

47.

$$\begin{array}{r|rrrr} 3 & 2 & 0 & 1 & 2 \\ & & 6 & 18 & 57 \\ \hline & 2 & 6 & 19 & \boxed{59} \end{array}$$

49.

$$\begin{array}{r|rrrrr} -2 & 1 & -2 & 1 & -3 & 2 \\ & & -2 & 8 & -18 & 42 \\ \hline & 1 & -4 & 9 & -21 & \boxed{44} \end{array}$$

51.

$$\begin{array}{r|rrrrrr} -\frac{1}{2} & 3 & 0 & 0 & 0 & 0 & 1 \\ & & -\frac{3}{2} & \frac{3}{4} & -\frac{3}{8} & \frac{3}{16} & -\frac{3}{32} \\ \hline & 3 & -\frac{3}{2} & \frac{3}{4} & -\frac{3}{8} & \frac{3}{16} & \boxed{\frac{29}{32}} \end{array}$$

53. $\begin{array}{r|rrrr} 3 & 1 & -3 & 5 & -15 \\ & & 3 & 0 & 15 \\ \hline & 1 & 0 & 5 & \boxed{0} \end{array}$ $\Rightarrow$ factor

55. $\begin{array}{r|rrr} -2 & 3 & -7 & 4 \\ & & -6 & 26 \\ \hline & 3 & -13 & \boxed{30} \end{array}$ $\Rightarrow$ not a factor

57. $\begin{array}{r|rrrrrrr} 2 & 1 & 0 & 0 & 0 & 0 & 0 & 0 \\ & & 2 & 4 & 8 & 16 & 32 & 64 \\ \hline & 1 & 2 & 4 & 8 & 16 & 32 & \boxed{64} \end{array}$

59. **Answers may vary.**

61. remainder $= P(1) = 1^{100} - 1^{99} + 1^{98} - 1^{97} + \cdots + 1^2 - 1 + 1 = 1$

Chapter 7 Summary (page 494)

1.
$$\begin{aligned} 4(y-1) &= 28 \\ 4y - 4 &= 28 \\ 4y &= 32 \\ y &= 8 \end{aligned}$$

2.
$$\begin{aligned} 3(x+7) &= 42 \\ 3x + 21 &= 42 \\ 3x &= 21 \\ x &= 7 \end{aligned}$$

3.
$$\begin{aligned} 13(x-9) - 2 &= 7x - 5 \\ 13x - 117 - 2 &= 7x - 5 \\ 13x - 119 &= 7x - 5 \\ 13x - 7x - 119 &= 7x - 7x - 5 \\ 6x - 119 + 119 &= -5 + 119 \\ 6x &= 114 \\ \frac{6x}{6} &= \frac{114}{6} \\ x &= 19 \end{aligned}$$

4.
$$\begin{aligned} \frac{8(x-5)}{3} &= 2(x-4) \\ 3 \cdot \frac{8(x-5)}{3} &= 3 \cdot 2(x-4) \\ 8(x-5) &= 6(x-4) \\ 8x - 40 &= 6x - 24 \\ 8x - 6x - 40 &= 6x - 6x - 24 \\ 2x - 40 + 40 &= -24 + 40 \\ 2x &= 16 \\ \frac{2x}{2} &= \frac{16}{2} \\ x &= 8 \end{aligned}$$

5.
$$\begin{aligned} 2x + 4 &= 2(x+3) - 2 \\ 2x + 4 &= 2x + 6 - 2 \\ 2x + 4 &= 2x + 4 \end{aligned}$$
identity

6.
$$\begin{aligned} (3x-2) - x &= 2(x-4) \\ 3x - 2 - x &= 2x - 8 \\ 2x - 2 &= 2x - 8 \\ 2x - 2x - 2 &= 2x - 2x - 8 \\ -2 &\neq -8 \end{aligned}$$
contradiction

7.

$$V = \frac{1}{3}\pi r^2 h$$
$$3V = 3\left(\frac{1}{3}\pi r^2 h\right)$$
$$3V = \pi r^2 h$$
$$\frac{3V}{\pi r^2} = \frac{\pi r^2 h}{\pi r^2}$$
$$\frac{3V}{\pi r^2} = h, \text{ or } h = \frac{3V}{\pi r^2}$$

8.

$$V = \frac{1}{6}ab(x+y)$$
$$6V = 6 \cdot \frac{1}{6}ab(x+y)$$
$$6V = ab(x+y)$$
$$\frac{6V}{ab} = \frac{ab(x+y)}{ab}$$
$$\frac{6V}{ab} = x+y$$
$$\frac{6V}{ab} - y = x + y - y$$
$$\frac{6V}{ab} - y = x, \text{ or } x = \frac{6V}{ab} - y$$

9. Let $x =$ the length of the first piece.
Then $3x =$ the length of the other piece.

$\boxed{\text{Sum of lengths}} = \boxed{\text{Total length}}$

$$x + 3x = 20$$
$$4x = 20$$
$$x = 5$$

He should cut the board 5 feet from one end.

10. Let $w =$ the width of the rectangle. Then $w + 4 =$ the length of the rectangle.

$\boxed{\text{Perimeter}} = 28$

$$w + w + w + 4 + w + 4 = 28$$
$$4w + 8 = 28$$
$$4w = 20$$
$$w = 5$$

The dimensions are 5 meters by 9 meters, for an area of 45 m^2.

11.

$$\frac{1}{3}y - 2 \geq \frac{1}{2}y + 2$$
$$6\left(\frac{1}{3}y - 2\right) \geq 6\left(\frac{1}{2}y + 2\right)$$
$$2y - 12 + 12 \geq 3y + 12 + 12$$
$$2y - 3y \geq 3y - 3y + 24$$
$$-y \geq 24$$
$$y \leq -24$$

solution set: $(-\infty, -24]$

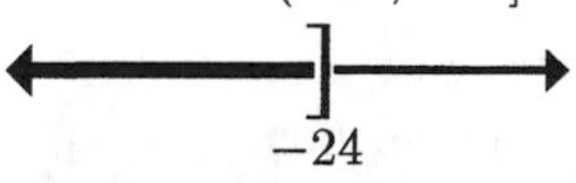

12.

$$\frac{7}{4}(x+3) < \frac{3}{8}(x-3)$$
$$8 \cdot \frac{7}{4}(x+3) < 8 \cdot \frac{3}{8}(x-3)$$
$$14(x+3) < 3(x-3)$$
$$14x + 42 < 3x - 9$$
$$14x < 3x - 51$$
$$11x < -51$$
$$x < -\tfrac{51}{11}$$

solution set: $\left(-\infty, -\frac{51}{11}\right)$

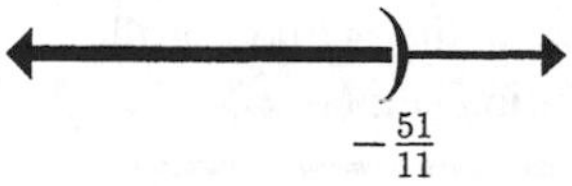

13.

$$3 < 3x + 4 < 10$$
$$3 - 4 < 3x + 4 - 4 < 10 - 4$$
$$-1 < 3x < 6$$
$$-\tfrac{1}{3} < \tfrac{3x}{3} < \tfrac{6}{3}$$

$-\frac{1}{3} < x < 2 \Rightarrow$ solution set: $\left(-\frac{1}{3}, 2\right) \Rightarrow$

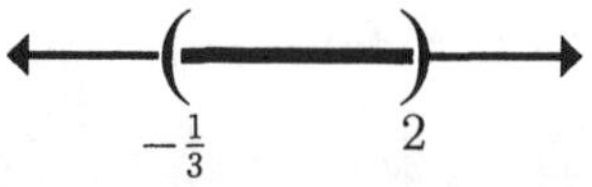

14.
$$4x > 3x + 2 > x - 3$$
$$\begin{array}{ccc} 4x > 3x + 2 & \text{and} & 3x + 2 > x - 3 \\ 4x - 3x > 3x - 3x + 2 & & 3x + 2 - 2 > x - 3 - 2 \\ x > 2 & & 3x - x > x - x - 5 \\ x > 2 & & 2x > -5 \\ & & x > -\frac{5}{2} \end{array}$$

If $x > 2$ **and** $x > -\frac{5}{2}$, $x > 2$. The solution set is $(2, \infty)$.

15.
$$|3x + 1| = 10$$
$$\begin{array}{ccc} 3x + 1 = 10 & \textbf{or} & 3x + 1 = -10 \\ 3x = 9 & & 3x = -11 \\ x = 3 & & x = -\frac{11}{3} \end{array}$$

16.
$$\left|\tfrac{3}{2}x - 4\right| = 9$$
$$\begin{array}{ccc} \frac{3}{2}x - 4 = 9 & \textbf{or} & \frac{3}{2}x - 4 = -9 \\ \frac{3}{2}x = 13 & & \frac{3}{2}x = -5 \\ x = \frac{26}{3} & & x = -\frac{10}{3} \end{array}$$

17.
$$|3x + 2| = |2x - 3|$$
$$\begin{array}{ccc} 3x + 2 = 2x - 3 & \textbf{or} & 3x + 2 = -(2x - 3) \\ x = -5 & & 3x + 2 = -2x + 3 \\ & & 5x = 1 \\ & & x = \frac{1}{5} \end{array}$$

18.
$$|5x - 4| = |4x - 5|$$
$$\begin{array}{ccc} 5x - 4 = 4x - 5 & \textbf{or} & 5x - 4 = -(4x - 5) \\ x = -1 & & 5x - 4 = -4x + 5 \\ & & 9x = 9 \\ & & x = 1 \end{array}$$

19.
$$|2x + 7| < 3$$
$$-3 < 2x + 7 < 3$$
$$-3 - 7 < 2x + 7 - 7 < 3 - 7$$
$$-10 < 2x < -4$$
$$-5 < x < -2$$

solution set: $(-5, -2)$

20.
$$|3x - 8| \geq 4$$
$$\begin{array}{ccc} 3x - 8 \leq -4 & \textbf{or} & 3x - 8 \geq 4 \\ 3x \leq 4 & & 3x \geq 12 \\ x \leq \frac{4}{3} & & x \geq 4 \end{array}$$

solution set: $(-\infty, \frac{4}{3}] \cup [4, \infty)$

21. $\left|\frac{3}{2}x - 14\right| \geq 0$

Since an absolute value is always at least 0, this inequality is true for all real #s. Solution set: $(-\infty, \infty)$

22. $\left|\frac{2}{3}x + 14\right| < 0$

Since an absolute value can never be negative, there is no solution.

23. $4x + 8 = 4(x + 2)$

24. $5x^2y^3 - 10xy^2 = 5xy^2(xy - 2)$

25. $-8x^2y^3z^4 - 12x^4y^3z^2 = -4x^2y^3z^2(2z^2 + 3x^2)$

26. $12a^6b^4c^2 + 15a^2b^4c^6 = 3a^2b^4c^2(4a^4 + 5c^4)$

27. $xy + 2y + 4x + 8 = y(x+2) + 4(x+2)$
$= (x+2)(y+4)$

28. $ac + bc + 3a + 3b = c(a+b) + 3(a+b)$
$= (a+b)(c+3)$

29. $x^{2n} + x^n = x^n(x^{2n-n} + 1) = x^n(x^n + 1)$

30. $y^{2n} - y^{4n} = y^{2n}(1 - y^{4n-2n}) = y^{2n}(1 - y^{2n})$

31. $x^4 + 4y + 4x^2 + x^2y = x^4 + x^2y + 4x^2 + 4y = x^2(x^2+y) + 4(x^2+y) = (x^2+y)(x^2+4)$

32. $a^5 + b^2c + a^2c + a^3b^2 = a^5 + a^3b^2 + a^2c + b^2c = a^3(a^2+b^2) + c(a^2+b^2) = (a^2+b^2)(a^3+c)$

33. $z^2 - 16 = z^2 - 4^2 = (z+4)(z-4)$

34. $y^2 - 121 = y^2 - 11^2 = (y+11)(y-11)$

35. $2x^4 - 98 = 2(x^4 - 49) = 2(x^2+7)(x^2-7)$

36. $3x^6 - 300x^2 = 3x^2(x^4 - 100)$
$= 3x^2(x^2+10)(x^2-10)$

37. $y^2 + 21y + 20 = (y+20)(y+1)$

38. $z^2 - 11z + 30 = (z-5)(z-6)$

39. $-x^2 - 3x + 28 = -(x^2 + 3x - 28)$
$= -(x+7)(x-4)$

40. $-y^2 + 5y + 24 = -(y^2 - 5y - 24)$
$= -(y-8)(y+3)$

41. $y^3 + y^2 - 2y = y(y^2 + y - 2)$
$= y(y+2)(y-1)$

42. $2a^4 + 4a^3 - 6a^2 = 2a^2(a^2 + 2a - 3)$
$= 2a^2(a+3)(a-1)$

43. $15x^2 - 57xy - 12y^2 = 3(5x^2 - 19xy - 4y^2) = 3(5x+y)(x-4y)$

44. $30x^2 + 65xy + 10y^2 = 5(6x^2 + 13xy + 2y^2) = 5(6x+y)(x+2y)$

45. $x^2 + 4x + 4 - 4p^4 = (x^2 + 4x + 4) - 4p^4 = (x+2)^2 - (2p^2)^2$
$= (x + 2 + 2p^2)(x + 2 - 2p^2)$

46. $y^2 + 3y + 2 + 2x + xy = (y^2 + 3y + 2) + 2x + xy = (y+1)(y+2) + x(2+y)$
$= (y+2)(y+1+x)$

47. $x^3 + 343 = x^3 + 7^3$
$= (x+7)(x^2 - 7x + 49)$

48. $a^3 - 125 = a^3 - 5^3$
$= (a-5)(a^2 + 5a + 25)$

49. $8y^3 - 512 = 8(y^3 - 64) = 8(y-4)(y^2 + 4y + 16)$

50. $4x^3y + 108yz^3 = 4y(x^3 + 27z^3) = 4y(x+3z)(x^2 - 3xz + 9z^2)$

51. $\dfrac{248x^2y}{576xy^2} = \dfrac{8 \cdot 31x^2y}{8 \cdot 72xy^2} = \dfrac{31x}{72y}$

52. $\dfrac{x^2 - 49}{x^2 + 14x + 49} = \dfrac{(x+7)(x-7)}{(x+7)(x+7)} = \dfrac{x-7}{x+7}$

53. $\dfrac{x^2 + 4x + 4}{x^2 - x - 6} \cdot \dfrac{x^2 - 9}{x^2 + 5x + 6} = \dfrac{(x+2)(x+2)}{(x+2)(x-3)} \cdot \dfrac{(x+3)(x-3)}{(x+2)(x+3)} = 1$

54. $$\frac{x^3-64}{x^2+4x+16} \div \frac{x^2-16}{x+4} = \frac{x^3-64}{x^2+4x+16} \cdot \frac{x+4}{x^2-16} = \frac{(x-4)(x^2+4x+16)}{x^2+4x+16} \cdot \frac{x+4}{(x+4)(x-4)} = 1$$

55. $$\frac{5y}{x-y} - \frac{3}{x-y} = \frac{5y-3}{x-y}$$

56. $$\frac{3x-1}{x^2+2} + \frac{3(x-2)}{x^2+2} = \frac{3x-1+3x-6}{x^2+2} = \frac{6x-7}{x^2+2}$$

57. $$\frac{3}{x+2} + \frac{2}{x+3} = \frac{3(x+3)}{(x+2)(x+3)} + \frac{2(x+2)}{(x+2)(x+3)} = \frac{3x+9+2x+4}{(x+2)(x+3)} = \frac{5x+13}{(x+2)(x+3)}$$

58. $$\frac{4x}{x-4} - \frac{3}{x+3} = \frac{4x(x+3)}{(x-4)(x+3)} - \frac{3(x-4)}{(x-4)(x+3)} = \frac{4x^2+12x-3x+12}{(x-4)(x+3)} = \frac{4x^2+9x+12}{(x-4)(x+3)}$$

59. $$\frac{x^2+3x+2}{x^2-x-6} \cdot \frac{3x^2-3x}{x^2-3x-4} \div \frac{x^2+3x+2}{x^2-2x-8} = \frac{x^2+3x+2}{x^2-x-6} \cdot \frac{3x^2-3x}{x^2-3x-4} \cdot \frac{x^2-2x-8}{x^2+3x+2}$$
$$= \frac{(x+2)(x+1)}{(x-3)(x+2)} \cdot \frac{3x(x-1)}{(x-4)(x+1)} \cdot \frac{(x-4)(x+2)}{(x+2)(x+1)}$$
$$= \frac{3x(x-1)}{(x-3)(x+1)}$$

60. $$\frac{x^2-x-6}{x^2-3x-10} \div \frac{x^2-x}{x^2-5x} \cdot \frac{x^2-4x+3}{x^2-6x+9} = \frac{x^2-x-6}{x^2-3x-10} \cdot \frac{x^2-5x}{x^2-x} \cdot \frac{x^2-4x+3}{x^2-6x+9}$$
$$= \frac{(x-3)(x+2)}{(x-5)(x+2)} \cdot \frac{x(x-5)}{x(x-1)} \cdot \frac{(x-3)(x-1)}{(x-3)(x-3)} = 1$$

61. $$\frac{2x}{x+1} + \frac{3x}{x+2} + \frac{4x}{x^2+3x+2} = \frac{2x}{x+1} + \frac{3x}{x+2} + \frac{4x}{(x+2)(x+1)}$$
$$= \frac{2x(x+2)}{(x+1)(x+2)} + \frac{3x(x+1)}{(x+1)(x+2)} + \frac{4x}{(x+2)(x+1)}$$
$$= \frac{2x^2+4x+3x^2+3x+4x}{(x+1)(x+2)} = \frac{5x^2+11x}{(x+1)(x+2)}$$

62. $$\frac{5x}{x-3} + \frac{5}{x^2-5x+6} + \frac{x+3}{x-2} = \frac{5x}{x-3} + \frac{5}{(x-3)(x-2)} + \frac{x+3}{x-2}$$
$$= \frac{5x(x-2)}{(x-3)(x-2)} + \frac{5}{(x-3)(x-2)} + \frac{(x+3)(x-3)}{(x-3)(x-2)}$$
$$= \frac{5x^2-10x+5+x^2-9}{(x-3)(x-2)}$$
$$= \frac{6x^2-10x-4}{(x-3)(x-2)} = \frac{2(3x+1)(x-2)}{(x-3)(x-2)} = \frac{2(3x+1)}{x-3}$$

63. $$\frac{3(x+2)}{x^2-1}-\frac{2}{x+1}+\frac{4(x+3)}{x^2-2x+1}=\frac{3(x+2)}{(x+1)(x-1)}-\frac{2}{(x+1)}+\frac{4(x+3)}{(x-1)(x-1)}$$
$$=\frac{3(x+2)(x-1)}{(x+1)(x-1)(x-1)}-\frac{2(x-1)(x-1)}{(x+1)(x-1)(x-1)}+\frac{4(x+3)(x+1)}{(x+1)(x-1)(x-1)}$$
$$=\frac{3(x^2+x-2)-2(x^2-2x+1)+4(x^2+4x+3)}{(x+1)(x-1)(x-1)}$$
$$=\frac{3x^2+3x-6-2x^2+4x-2+4x^2+16x+12}{(x+1)(x-1)(x-1)}=\frac{5x^2+23x+4}{(x+1)(x-1)(x-1)}$$

64. $$\frac{x}{x^2+4x+4}+\frac{2x}{x^2-4}-\frac{x^2-4}{x-2}$$
$$=\frac{x}{(x+2)(x+2)}+\frac{2x}{(x+2)(x-2)}-\frac{x^2-4}{x-2}$$
$$=\frac{x(x-2)}{(x+2)(x+2)(x-2)}+\frac{2x(x+2)}{(x+2)(x-2)(x+2)}-\frac{(x^2-4)(x+2)(x+2)}{(x-2)(x+2)(x+2)}$$
$$=\frac{x^2-2x+2x^2+4x-(x^2-4)(x^2+4x+4)}{(x+2)(x+2)(x-2)}$$
$$=\frac{3x^2+2x-(x^4+4x^3-16x-16)}{(x+2)(x+2)(x-2)}$$
$$=\frac{3x^2+2x-x^4-4x^3+16x+16}{(x+2)(x+2)(x-2)}=\frac{-x^4-4x^3+3x^2+18x+16}{(x+2)(x+2)(x-2)}$$

65. $$\frac{\frac{3}{x}-\frac{2}{y}}{xy}=\frac{\left(\frac{3}{x}-\frac{2}{y}\right)xy}{xy(xy)}=\frac{3y-2x}{x^2y^2}$$

66. $$\frac{\frac{1}{x}+\frac{2}{y}}{\frac{2}{x}-\frac{1}{y}}=\frac{\left(\frac{1}{x}+\frac{2}{y}\right)xy}{\left(\frac{2}{x}-\frac{1}{y}\right)xy}=\frac{y+2x}{2y-x}$$

67. $$\frac{2x+3+\frac{1}{x}}{x+2+\frac{1}{x}}=\frac{\left(2x+3+\frac{1}{x}\right)x}{\left(x+2+\frac{1}{x}\right)x}=\frac{2x^2+3x+1}{x^2+2x+1}=\frac{(2x+1)(x+1)}{(x+1)(x+1)}=\frac{2x+1}{x+1}$$

68. $$\frac{x^{-1}-y^{-1}}{x^{-1}+y^{-1}}=\frac{\frac{1}{x}-\frac{1}{y}}{\frac{1}{x}+\frac{1}{y}}=\frac{\left(\frac{1}{x}-\frac{1}{y}\right)xy}{\left(\frac{1}{x}+\frac{1}{y}\right)xy}=\frac{y-x}{y+x}$$

69.

2	3	2	−7	2
		6	16	18
	3	8	9	20

70.

−2	2	−4	−14	3
		−4	16	−4
	2	−8	2	−1

71.

$$\begin{array}{r|rrrr} 5 & 1 & -3 & -8 & -10 \\ & & 5 & 10 & 10 \\ \hline & 1 & 2 & 2 & \boxed{0} \end{array}$$

factor

72.

$$\begin{array}{r|rrrr} -5 & 1 & 4 & -5 & 5 \\ & & -5 & 5 & 0 \\ \hline & 1 & -1 & 0 & \boxed{5} \end{array}$$

not a factor

Chapter 7 Test (page 498)

1.

$$\begin{aligned} 9(x+4)+4 &= 4(x-5) \\ 9x+36+4 &= 4x-20 \\ 9x+40 &= 4x-20 \\ 9x-4x+40-40 &= 4x-4x-20-40 \\ 5x &= -60 \\ \frac{5x}{5} &= \frac{-60}{5} \\ x &= -12 \end{aligned}$$

2.

$$\begin{aligned} \frac{y-1}{5}+2 &= \frac{2y-3}{3} \\ 15\left(\frac{y-1}{5}+2\right) &= 15\left(\frac{2y-3}{3}\right) \\ 3(y-1)+30 &= 10y-15 \\ 3y-3+30 &= 10y-15 \\ 3y+27 &= 10y-15 \\ 3y-10y+27-27 &= 10y-10y-15-27 \\ -7y &= -42 \\ y &= 6 \end{aligned}$$

3.

$$\begin{aligned} P &= L+\frac{s}{f}i \\ P-L &= \frac{s}{f}i \\ f(P-L) &= f\cdot\frac{s}{f}i \\ f(P-L) &= si \\ \frac{f(P-L)}{s} &= \frac{si}{s} \\ \frac{f(P-L)}{s} &= i\text{, or } i=\frac{f(P-L)}{s} \end{aligned}$$

4.

$$\begin{aligned} n &= \frac{360}{180-a} \\ (180-a)n &= (180-a)\cdot\frac{360}{180-a} \\ 180n-an &= 360 \\ 180n-360 &= an \\ \frac{180n-360}{n} &= \frac{an}{n} \\ \frac{180n-360}{n} &= a\text{, or } a=\frac{180n-360}{n} \end{aligned}$$

5. Let $x =$ the length of the shortest piece. Then $2x$ and $6x$ are the other lengths.

$$\begin{aligned} \boxed{\text{Total length}} &= 20 \\ x+2x+6x &= 20 \\ 9x &= 20 \\ x &= \frac{20}{9}\text{ feet} \end{aligned}$$

$$\text{Longest} = 6\left(\frac{20}{9}\right) = \frac{\not{3}\cdot 2}{1}\cdot\frac{20}{\not{3}\cdot 3} = \frac{40}{3} = 13\frac{1}{3}\text{ ft}$$

The longest piece is $13\frac{1}{3}$ ft. long.

6. Let $w =$ the width of the rectangle. Then $w+5 =$ the length of the rectangle.

$$\begin{aligned} \boxed{\text{Perimeter}} &= 26 \\ w+w+w+5+w+5 &= 26 \\ 4w+10 &= 26 \\ 4w &= 16 \\ w &= 4 \end{aligned}$$

The dimensions are 4 cm by 9 cm, for an area of 36 cm^2.

CHAPTER 7 TEST

7.
$$-2(2x+3) \geq 14$$
$$-4x-6+6 \geq 14+6$$
$$-4x \geq 20$$
$$\frac{-4x}{-4} \leq \frac{20}{-4}$$
$$x \leq -5$$
solution set: $(-\infty, -5]$

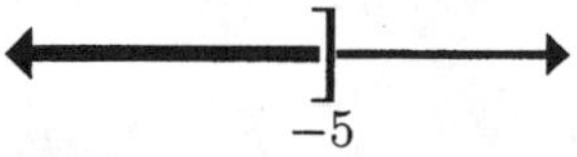

8.
$$-2 < \frac{x-4}{3} < 4$$
$$-6 < x-4 < 12$$
$$-6+4 < x-4+4 \leq 12+4$$
$$-2 < x < 16$$
solution set: $(-2, 16)$

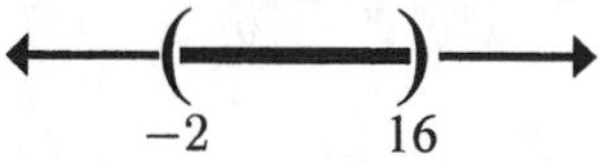

9.
$$|2x+3| = 11$$
$2x+3=11$ **or** $2x+3=-11$

$2x=8$ $\quad$ $2x=-14$

$x=4$ $\quad$ $x=-7$

10.
$$|3x+4| = |x+12|$$
$3x+4=x+12$ **or** $3x+4=-(x+12)$

$2x=8$ $\quad$ $3x+4=-x-12$

$x=4$ $\quad$ $4x=-16$

$x=-4$

11.
$$|x+3| \leq 4$$
$$-4 \leq x+3 \leq 4$$
$$-4-3 \leq x+3-3 \leq 4-3$$
$$-7 \leq x \leq 1$$
solution set: $[-7, 1]$

12.
$$|2x-4| > 22$$
$2x-4 < -22$ **or** $2x-4 > 22$

$2x < -18$ $\quad$ $2x > 26$

$x < -9$ $\quad$ $x > 13$

solution set: $(-\infty, -9) \cup (13, \infty)$

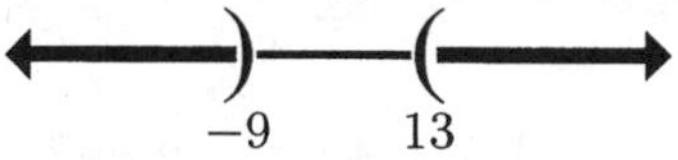

13. $3xy^2 + 6x^2y = 3xy(y+2x)$

14. $12a^3b^2c - 3a^2b^2c^2 + 6abc^3 = 3abc(4a^2b - abc + 2c^2)$

15. $ax - xy + ay - y^2 = x(a-y) + y(a-y) = (a-y)(x+y)$

16. $ax + ay + bx + by - cx - cy = a(x+y) + b(x+y) - c(x+y) = (x+y)(a+b-c)$

17. $x^2 - 49 = x^2 - 7^2 = (x+7)(x-7)$

18. $2x^2 - 32 = 2(x^2-16) = 2(x+4)(x-4)$

19. $4y^4 - 64 = 4(y^4-16) = 4(y^2+4)(y^2-4) = 4(y^2+4)(y+2)(y-2)$

20. $b^3 + 125 = (b+5)(b^2-5b+25)$

21. $b^3 - 27 = (b-3)(b^2+3b+9)$

22. $3u^3 - 24 = 3(u^3-8)$
$= 3(u-2)(u^2+2u+4)$

23. $x^2 + 8x + 15 = (x+3)(x+5)$

24. $6b^2 + b - 2 = (3b+2)(2b-1)$

25. $6u^2 + 9u - 6 = 3(2u^2+3u-2)$
$= 3(2u-1)(u+2)$

26. $x^2+6x+9-y^2=(x^2+6x+9)-y^2=(x+3)^2-y^2=(x+3+y)(x+3-y)$

27. $\dfrac{-12x^2y^3z^2}{18x^3y^4z^2}=-\dfrac{2}{3xy}$

28. $\dfrac{2x^2+7x+3}{4x+12}=\dfrac{(2x+1)(x+3)}{4(x+3)}=\dfrac{2x+1}{4}$

29. $\dfrac{x^2y^{-2}}{x^3z^2}\cdot\dfrac{x^2z^4}{y^2z}=\dfrac{x^4y^{-2}z^4}{x^3y^2z^3}=\dfrac{xz}{y^4}$

30. $\dfrac{u^2+5u+6}{u^2-4}\cdot\dfrac{u^2-5u+6}{u^2-9}=\dfrac{(u+2)(u+3)}{(u+2)(u-2)}\cdot\dfrac{(u-2)(u-3)}{(u+3)(u-3)}=1$

31. $\dfrac{x^3+y^3}{4}\div\dfrac{x^2-xy+y^2}{2x+2y}=\dfrac{x^3+y^3}{4}\cdot\dfrac{2x+2y}{x^2-xy+y^2}=\dfrac{(x+y)(x^2-xy+y^2)}{4}\cdot\dfrac{2(x+y)}{x^2-xy+y^2}$

$$=\frac{(x+y)^2}{2}$$

32. $\dfrac{x+2}{x+1}-\dfrac{x+1}{x+2}=\dfrac{(x+2)(x+2)}{(x+1)(x+2)}-\dfrac{(x+1)(x+1)}{(x+1)(x+2)}=\dfrac{x^2+4x+4-(x^2+2x+1)}{(x+1)(x+2)}$

$$=\frac{x^2+4x+4-x^2-2x-1}{(x+1)(x+2)}$$

$$=\frac{2x+3}{(x+1)(x+2)}$$

33. $\dfrac{\frac{2u^2w^3}{v^2}}{\frac{4uw^4}{uv}}=\dfrac{2u^2w^3}{v^2}\div\dfrac{4uw^4}{uv}=\dfrac{2u^2w^3}{v^2}\cdot\dfrac{uv}{4uw^4}=\dfrac{2u^3w^3v}{4uv^2w^4}=\dfrac{u^2}{2vw}$

34. $\dfrac{\frac{x}{y}+\frac{1}{2}}{\frac{x}{2}-\frac{1}{y}}=\dfrac{\left(\frac{x}{y}+\frac{1}{2}\right)2y}{\left(\frac{x}{2}-\frac{1}{y}\right)2y}=\dfrac{2x+y}{xy-2}$

35.

$$\begin{array}{r|l}
 & x^2 - 5x + 10 \\
x+1 & x^3-4x^2+5x+3 \\
 & \underline{x^3+x^2} \\
 & \quad -5x^2+5x \\
 & \quad \underline{-5x^2-5x} \\
 & \qquad 10x+3 \\
 & \qquad \underline{10x+10} \\
 & \qquad\qquad -7
\end{array}$$

36.

$$\begin{array}{r|rrrr}
2 & 4 & 3 & 2 & -1 \\
 & & 8 & 22 & 48 \\ \hline
 & 4 & 11 & 24 & \boxed{47}
\end{array}$$

Exercise 8.1 (page 506)

1. 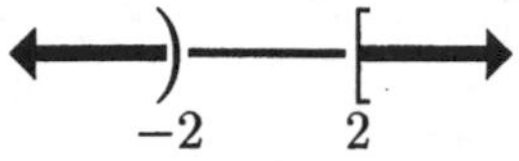

3. $x^2 - x = x(x-1)$

5. $x^3 - 1 = (x-1)(x^2+x+1)$

7. origin

9. y-coordinate

11. x-axis

13. horizontal

15-21.

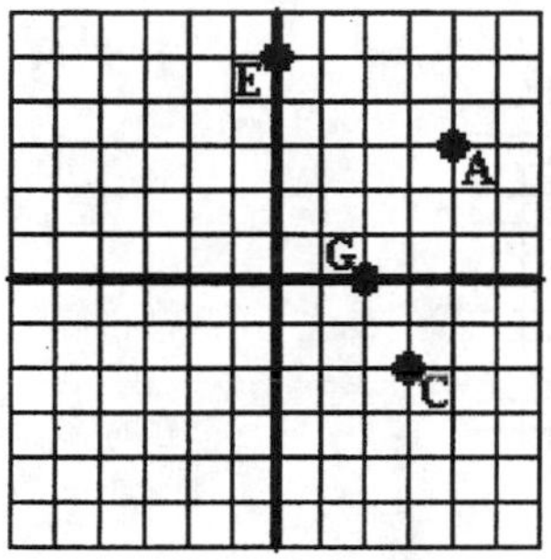

23. $(2,4)$

25. $(-2,-1)$

27. $(4,0)$

29. $(0,0)$

31. $x + y = 4$

x	y
-1	5
0	4
2	2

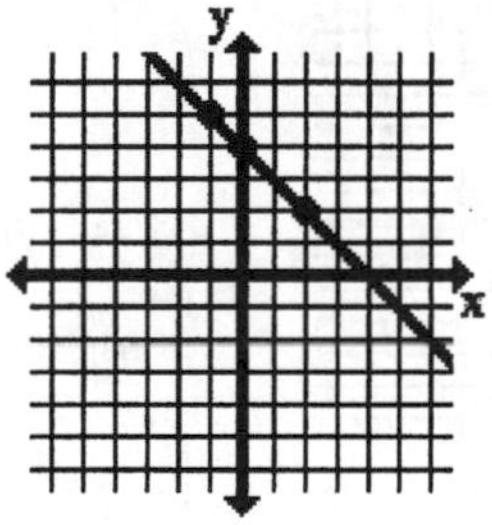

33. $2x - y = 3$

x	y
-1	-5
0	-3
3	3

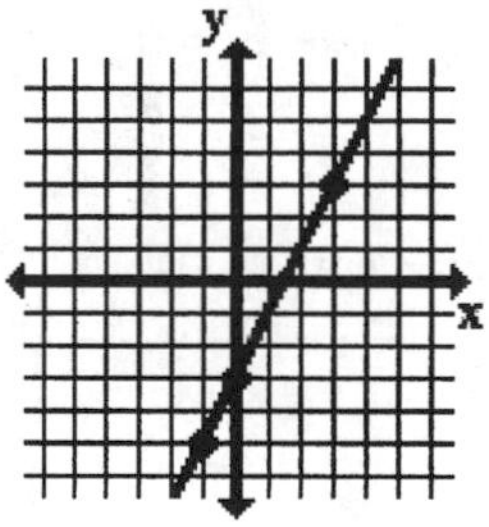

35.

$$
\begin{aligned}
3x + 4y &= 12 \\
3x + 4(0) &= 12 \\
3x &= 12 \\
x &= 4
\end{aligned}
$$

x-intercept: $(4,0)$

$$
\begin{aligned}
3x + 4y &= 12 \\
3(0) + 4y &= 12 \\
4y &= 12 \\
y &= 3
\end{aligned}
$$

y-intercept: $(0,3)$

$3x + 4y = 12$

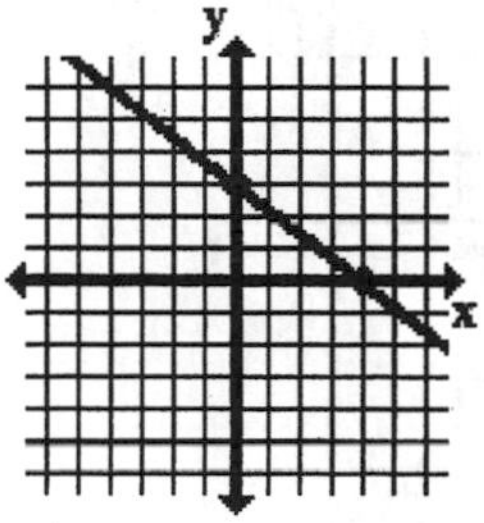

37.

$y = -3x + 2$
$0 = -3x + 2$
$3x = 2$
$x = \frac{2}{3}$
x-intercept: $\left(\frac{2}{3}, 0\right)$

$y = -3x + 2$
$y = -3(0) + 2$
$y = 2$
y-intercept: $(0, 2)$

$y = -3x + 2$

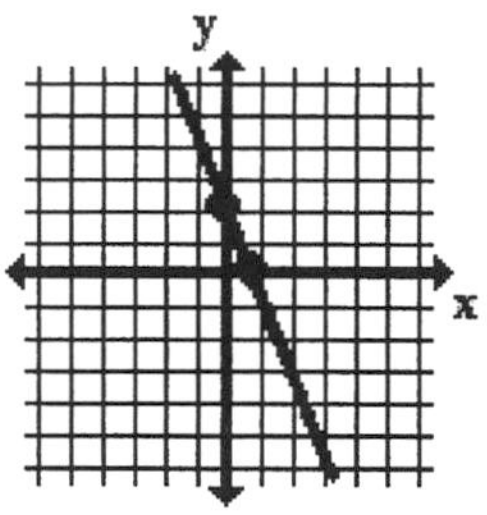

39.

$3y = 6x - 9$
$3(0) = 6x - 9$
$0 = 6x - 9$
$-6x = -9$
$x = \frac{-9}{-6} = \frac{3}{2}$
x-intercept: $\left(\frac{3}{2}, 0\right)$

$3y = 6x - 9$
$3y = 6(0) - 9$
$3y = 0 - 9$
$3y = -9$
$y = \frac{-9}{3} = -3$
y-intercept: $(0, -3)$

$3y = 6x - 9$

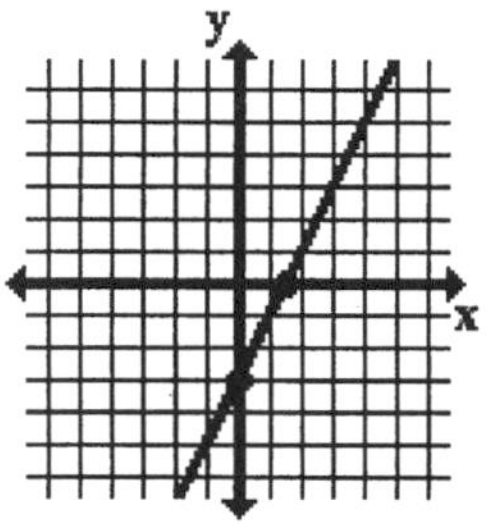

41.

$3x + 4y - 8 = 0$
$3x + 4(0) - 8 = 0$
$3x - 8 = 0$
$3x = 8$
$x = \frac{8}{3}$
x-intercept: $\left(\frac{8}{3}, 0\right)$

$3x + 4y - 8 = 0$
$3(0) + 4y - 8 = 0$
$4y - 8 = 0$
$4y = 8$
$y = 2$
y-intercept: $(0, 2)$

$3x + 4y - 8 = 0$

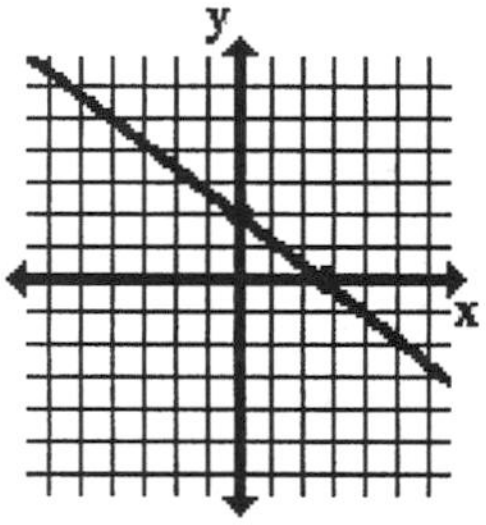

43. $x = 3$
vertical line with x-coordinate of 3

$x = 3$

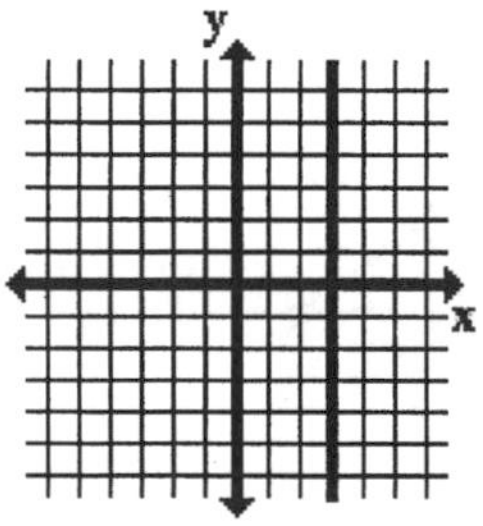

45. $-3y + 2 = 5$
$-3y = 3$
$y = -1$
horizontal line with y-coordinate of -1

$y = -1$

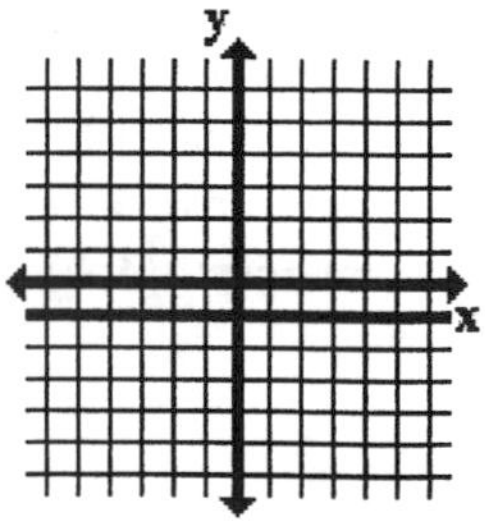

47. $x = \frac{x_1+x_2}{2} = \frac{0+6}{2} = \frac{6}{2} = 3$
$y = \frac{y_1+y_2}{2} = \frac{0+8}{2} = \frac{8}{2} = 4$
midpoint: $(3, 4)$

49. $x = \frac{x_1+x_2}{2} = \frac{6+12}{2} = \frac{18}{2} = 9$
$y = \frac{y_1+y_2}{2} = \frac{8+16}{2} = \frac{24}{2} = 12$
midpoint: $(9, 12)$

51. $x = \frac{x_1+x_2}{2} = \frac{2+5}{2} = \frac{7}{2}$
$y = \frac{y_1+y_2}{2} = \frac{4+8}{2} = \frac{12}{2} = 6$
midpoint: $\left(\frac{7}{2}, 6\right)$

53. $x = \frac{x_1+x_2}{2} = \frac{-2+3}{2} = \frac{1}{2}$
$y = \frac{y_1+y_2}{2} = \frac{-8+4}{2} = \frac{-4}{2} = -2$
midpoint: $\left(\frac{1}{2}, -2\right)$

55. $x = \frac{x_1+x_2}{2} = \frac{-3+(-5)}{2} = \frac{-8}{2} = -4$
$y = \frac{y_1+y_2}{2} = \frac{5+(-5)}{2} = \frac{0}{2} = 0$
midpoint: $(-4, 0)$

57. $x = \frac{x_1+x_2}{2} = \frac{a+4a}{2} = \frac{5a}{2}$
$y = \frac{y_1+y_2}{2} = \frac{b+3b}{2} = \frac{4b}{2} = 2b$
midpoint: $\left(\frac{5a}{2}, 2b\right)$

59. $x = \frac{x_1+x_2}{2} = \frac{(a-b)+(a+b)}{2} = \frac{2a}{2} = a$
$y = \frac{y_1+y_2}{2} = \frac{b+3b}{2} = \frac{4b}{2} = 2b$
midpoint: $(a, 2b)$

61.
$x = \frac{x_1+x_2}{2}$ $\quad$ $y = \frac{y_1+y_2}{2}$
$-2 = \frac{-8+x_2}{2}$ $\quad$ $3 = \frac{5+y_2}{2}$
$-4 = -8 + x_2$ $\quad$ $6 = 5 + y_2$
$4 = x_2$ $\quad$ $1 = y_2$
Q has coordinates $(4, 1)$.

63. To find the value of the house after 5 years, let $x = 5$:
$$\begin{aligned} y &= 7500x + 125{,}000 \\ &= 7500(5) + 125{,}000 \\ &= 37{,}500 + 125{,}000 \\ &= 162{,}500 \end{aligned}$$
It will be worth \$162,500 after 5 years.

To find the value of the house after 10 years, let $x = 10$:
$$\begin{aligned} y &= 7500x + 125{,}000 \\ &= 7500(10) + 125{,}000 \\ &= 75{,}000 + 125{,}000 \\ &= 200{,}000 \end{aligned}$$
It will be worth \$200,000 after 10 years.

65. To find the number of TVs sold, let $p = 150$:
$$\begin{aligned} p &= -\frac{1}{10}q + 170 \\ 150 &= -\frac{1}{10}q + 170 \\ \frac{1}{10}q &= 20 \\ q &= 200 \end{aligned}$$
200 TVs will be sold at a price of \$150.

67. Let $V = 60$ and find v.
$$\begin{aligned} V &= \frac{nv}{N} \\ 60 &= \frac{12v}{20} \\ 1200 &= 12v \\ 100 &= v \end{aligned}$$
The smaller gear has a speed of 100 revolutions per minute.

69. **Answers may vary.**

71. If the line $y = ax + b$ passes through only quadrants I and II, then $a = 0$ and $b > 0$.

Exercise 8.2 (page 517)

1. $\left(x^3y^2\right)^3 = x^9y^6$

3. $\left(x^{-3}y^2\right)^{-4} = x^{12}y^{-8} = \frac{x^{12}}{y^8}$

5. $\left(\frac{3x^2y^3}{8}\right)^0 = 1$

7. y; x

9. $\frac{y_2 - y_1}{x_2 - x_1}$

11. run

13. vertical

15. Parallel

17. $m = \frac{\Delta y}{\Delta x} = \frac{5-(-3)}{2-(-2)} = \frac{8}{4} = 2$

19. $m = \frac{\Delta y}{\Delta x} = \frac{9-0}{3-0} = \frac{9}{3} = 3$

21. $m = \frac{\Delta y}{\Delta x} = \frac{1-8}{6-(-1)} = \frac{-7}{7} = -1$

23. $m = \frac{\Delta y}{\Delta x} = \frac{2-(-1)}{-6-3} = \frac{3}{-9} = -\frac{1}{3}$

25. $m = \frac{\Delta y}{\Delta x} = \frac{5-5}{-9-7} = \frac{0}{-16} = 0$

27. $m = \frac{\Delta y}{\Delta x} = \frac{-2-(-5)}{-7-(-7)} = \frac{3}{0}$: undefined

29. $m = \frac{\Delta y}{\Delta x} = \frac{2.5-3.7}{3.7-2.5} = \frac{-1.2}{1.2} = -1$

31. Find two points on the line:

Let $x = 0$:	Let $y = 0$:
$3x + 2y = 12$	$3x + 2y = 12$
$3(0) + 2y = 12$	$3x + 2(0) = 12$
$2y = 12$	$3x = 12$
$y = 6$	$x = 4$
$(0, 6)$	$(4, 0)$

$m = \frac{\Delta y}{\Delta x} = \frac{6-0}{0-4} = \frac{6}{-4} = -\frac{3}{2}$

33. Find two points on the line:

Let $x = 0$:	Let $y = 0$:
$3x = 4y - 2$	$3x = 4y - 2$
$3(0) = 4y - 2$	$3x = 4(0) - 2$
$0 = 4y - 2$	$3x = -2$
$2 = 4y$	$x = -\frac{2}{3}$
$\frac{1}{2} = y$	$\left(-\frac{2}{3}, 0\right)$
$\left(0, \frac{1}{2}\right)$	

$m = \frac{\Delta y}{\Delta x} = \frac{\frac{1}{2} - 0}{0 - \left(-\frac{2}{3}\right)} = \frac{\frac{1}{2}}{\frac{2}{3}} = \frac{3}{4}$

35. Find two points on the line:

Let $x = 0$:	Let $y = 0$:
$y = \frac{x-4}{2}$	$y = \frac{x-4}{2}$
$y = \frac{0-4}{2}$	$0 = \frac{x-4}{2}$
$y = \frac{-4}{2}$	$0 = x - 4$
$y = -2$	$4 = x$
$(0, -2)$	$(4, 0)$

$m = \frac{\Delta y}{\Delta x} = \frac{-2-0}{0-4} = \frac{-2}{-4} = \frac{1}{2}$

37. $4y = 3(y + 2)$

$4y = 3y + 6$

$y = 6$

horizontal line $\Rightarrow m = 0$

39. negative

41. positive

43. undefined

45. $m_1 \neq m_2 \Rightarrow$ not parallel

$m_1 \cdot m_2 = 3\left(-\frac{1}{3}\right) = -1 \Rightarrow$ perpendicular

47. $m_1 \neq m_2 \Rightarrow$ not parallel

$m_1 \cdot m_2 = 4(0.25) = 1 \Rightarrow$ not perpendicular

49. $m_1 = m_2 \Rightarrow$ parallel

51. $m_{\overline{PQ}} = \frac{\Delta y}{\Delta x} = \frac{2-4}{4-3} = \frac{-2}{1} = -2$

same slope $\Rightarrow$ parallel

53. $m_{\overline{PQ}} = \frac{\Delta y}{\Delta x} = \frac{5-1}{6-(-2)} = \frac{4}{8} = \frac{1}{2}$

opposite reciprocal slope $\Rightarrow$ perpendicular

55. $m_{\overline{PQ}} = \frac{\Delta y}{\Delta x} = \frac{6-4}{6-5} = \frac{2}{1} = 2$

neither parallel nor perpendicular

57. $m_{\overline{PQ}} = \frac{\Delta y}{\Delta x} = \frac{8-4}{4-(-2)} = \frac{4}{6} = \frac{2}{3}$

$m_{\overline{PR}} = \frac{\Delta y}{\Delta x} = \frac{12-4}{8-(-2)} = \frac{8}{10} = \frac{4}{5}$

Since they do not have the same slope, the three points are not on the same line.

59. $m_{\overline{PQ}} = \frac{\Delta y}{\Delta x} = \frac{0-10}{-6-(-4)} = \frac{-10}{-2} = 5$

$m_{\overline{PR}} = \frac{\Delta y}{\Delta x} = \frac{5-10}{-1-(-4)} = \frac{-5}{3} = -\frac{5}{3}$

Since they do not have the same slope, the three points are not on the same line.

61. $m_{\overline{PQ}} = \frac{\Delta y}{\Delta x} = \frac{8-4}{0-(-2)} = \frac{4}{2} = 2$

$m_{\overline{PR}} = \frac{\Delta y}{\Delta x} = \frac{12-4}{2-(-2)} = \frac{8}{4} = 2$

Since they have the same slope and a common point, the three points are on the same line.

63. On the x-axis, all y-coordinates are 0. The equation is $y = 0$, and the slope is 0.

65. Call the points $A(-3,4)$, $B(4,1)$ and $C(-1,-1)$. Compute these slopes:

$$m_{\overline{AB}} = \frac{\Delta y}{\Delta x} = \frac{1-4}{4-(-3)} = \frac{-3}{7} = -\frac{3}{7}$$

$$m_{\overline{AC}} = \frac{\Delta y}{\Delta x} = \frac{-1-4}{-1-(-3)} = \frac{-5}{2} = -\frac{5}{2}$$

$$m_{\overline{BC}} = \frac{\Delta y}{\Delta x} = \frac{-1-1}{-1-4} = \frac{-2}{-5} = \frac{2}{5}$$

Since $\overline{AC}$ and $\overline{BC}$ are perpendicular, it is a right triangle.

67. Call the points $A(a,0)$, $B(0,a)$, $C(-a,0)$ and $D(0,-a)$. Compute these slopes:

$$m_{\overline{AB}} = \frac{\Delta y}{\Delta x} = \frac{a-0}{0-a} = \frac{a}{-a} = -1;\ m_{\overline{BC}} = \frac{\Delta y}{\Delta x} = \frac{0-a}{-a-0} = \frac{-a}{-a} = 1$$

$$m_{\overline{CD}} = \frac{\Delta y}{\Delta x} = \frac{-a-0}{0-(-a)} = \frac{-a}{a} = -1;\ m_{\overline{DA}} = \frac{\Delta y}{\Delta x} = \frac{0-(-a)}{a-0} = \frac{a}{a} = 1$$

Thus, $\overline{AB} \perp \overline{BC}$, $\overline{BC} \perp \overline{CD}$, $\overline{CD} \perp \overline{DA}$, and $\overline{DA} \perp \overline{AB}$.

69. Call the points $A(0,0)$, $B(0,a)$, $C(b,c)$ and $D(b,a+c)$. Compute these slopes:

$$m_{\overline{AB}} = \frac{\Delta y}{\Delta x} = \frac{0-a}{0-0} = \frac{-a}{0} \Rightarrow \text{undefined}$$

$$m_{\overline{BD}} = \frac{\Delta y}{\Delta x} = \frac{a-(a+c)}{0-b} = \frac{a-a-c}{-b} = \frac{-c}{-b} = \frac{c}{b}$$

$$m_{\overline{DC}} = \frac{\Delta y}{\Delta x} = \frac{c-(a+c)}{b-b} = \frac{c-a-c}{0} = \frac{-a}{0} \Rightarrow \text{undefined}$$

$$m_{\overline{CA}} = \frac{\Delta y}{\Delta x} = \frac{c-0}{b-0} = \frac{c}{b}$$

Thus, $\overline{AB} \parallel \overline{DC}$ and $\overline{BD} \parallel \overline{CA}$ and the figure is a parallelogram.

71. $\text{slope} = \frac{\text{rise}}{\text{run}} = \frac{32 \text{ ft}}{1 \text{ mi}} = \frac{32 \text{ ft}}{5280 \text{ ft}} = \frac{1}{165}$

73. $\text{slope} = \frac{\text{rise}}{\text{run}} = \frac{18 \text{ ft}}{5 \text{ ft}} = \frac{18}{5}$

75. $m = \frac{\Delta y}{\Delta x} = \frac{1.4}{50} = \frac{14}{500} = \frac{7}{250} \Rightarrow \frac{7}{250}$ degree per year

77. Let x represent the number of years, and let y represent the enrollment. Then we know two points on the line: $(1,8)$ and $(5,20)$. Find the slope:

$$m = \frac{\Delta y}{\Delta x} = \frac{20-8}{5-1} = \frac{12}{4} = 3 \text{ students per year.}$$

79. Let x represent the number of years, and let y represent the sales. Then we know two points on the line: $(1,85000)$ and $(3,125000)$. Find the slope:

$$m = \frac{\Delta y}{\Delta x} = \frac{125000-85000}{3-1} = \frac{40000}{2} = \$20{,}000 \text{ per year.}$$

81. Answers may vary.

83. Find two points on the line:

Let $x = 0$:

$$\begin{aligned} Ax + By &= C \\ A(0) + By &= C \\ By &= C \\ y &= \tfrac{C}{B} \end{aligned}$$

$\left(0, \frac{C}{B}\right)$ is on the line.

Let $y = 0$:

$$\begin{aligned} Ax + By &= C \\ Ax + B(0) &= C \\ Ax &= C \\ x &= \tfrac{C}{A} \end{aligned}$$

$\left(\frac{C}{A}, 0\right)$ is on the line.

$$m = \frac{\Delta y}{\Delta x} = \frac{\frac{C}{B} - 0}{0 - \frac{C}{A}} = \frac{\frac{C}{B}}{-\frac{C}{A}} = -\frac{A}{B}$$

85. If the three points are on a line, then all slopes must be equal:

$$m = \frac{\Delta y}{\Delta x} = \frac{7-10}{5-7} = \frac{-3}{-2} = \frac{3}{2} \qquad m = \frac{\Delta y}{\Delta x} = \frac{a-7}{3-5} = \frac{a-7}{-2}$$

$$\begin{aligned} \frac{a-7}{-2} &= \frac{3}{2} \\ -2 \cdot \frac{a-7}{-2} &= -2 \cdot \frac{3}{2} \\ a - 7 &= -3 \\ a &= 4 \end{aligned}$$

Exercise 8.3 (page 530)

1.
$$3(x+2)+x=5x$$
$$3x+6+x=5x$$
$$4x+6=5x$$
$$6=x$$

3.
$$\frac{5(2-x)}{3}-1=x+5$$
$$\frac{10-5x}{3}=x+6$$
$$3\cdot\frac{10-5x}{3}=3(x+6)$$
$$10-5x=3x+18$$
$$10=8x+18$$
$$-8=8x$$
$$-1=x$$

5. We want the alloy to contain 25% gold. Let $x=$ the amount of copper added to the alloy.

[Gold at start] + [Gold added] = [Gold at end]

$$20+0=0.25(60+x)$$
$$20=15+0.25x$$
$$5=0.25x$$
$$20=x$$

20 oz of copper should be added to the alloy.

7. $y-y_1=m(x-x_1)$

9. $Ax+By=C$

11. perpendicular

13.
$$y-y_1=m(x-x_1)$$
$$y-7=5(x-0)$$
$$y-7=5x$$
$$-5x+y=7$$
$$5x-y=-7$$

15.
$$y-y_1=m(x-x_1)$$
$$y-0=-3(x-2)$$
$$y=-3x+6$$
$$3x+y=6$$

17. Notice that the line goes through $P(2,5)$ and the point $(-1,3)$. Find the slope:
$$m=\frac{\Delta y}{\Delta x}=\frac{5-3}{2-(-1)}=\frac{2}{3}$$
Use point-slope form to find the equation:
$$y-y_1=m(x-x_1)$$
$$y-5=\frac{2}{3}(x-2)$$
$$3(y-5)=3\cdot\frac{2}{3}(x-2)$$
$$3y-15=2x-4$$
$$-2x+3y=11$$
$$2x-3y=-11$$

19. Find the slope of the line:
$$m=\frac{\Delta y}{\Delta x}=\frac{4-0}{4-0}=\frac{4}{4}=1$$
Use point-slope form to find the equation:
$$y-y_1=m(x-x_1)$$
$$y-0=1(x-0)$$
$$y=x$$

21. Find the slope of the line:

$$m = \frac{\Delta y}{\Delta x} = \frac{4-(-3)}{3-0} = \frac{7}{3}$$

Use point-slope form to find the equation:

$$y - y_1 = m(x - x_1)$$
$$y - 4 = \tfrac{7}{3}(x-3)$$
$$y - 4 = \tfrac{7}{3}x - 7$$
$$y = \tfrac{7}{3}x - 3$$

23. Find the slope of the line:

$$m = \frac{\Delta y}{\Delta x} = \frac{4-(-3)}{-2-2} = \frac{7}{-4} = -\frac{7}{4}$$

Use point-slope form to find the equation:

$$y - y_1 = m(x - x_1)$$
$$y - 4 = -\tfrac{7}{4}(x-(-2))$$
$$y - 4 = -\tfrac{7}{4}(x+2)$$
$$y - 4 = -\tfrac{7}{4}x - \tfrac{7}{2}$$
$$y = -\tfrac{7}{4}x - \tfrac{7}{2} + 4$$
$$y = -\tfrac{7}{4}x + \tfrac{1}{2}$$

25.
$$y = mx + b$$
$$y = 3x + 17$$

27.
$$y = mx + b$$
$$5 = -7(7) + b$$
$$5 = -49 + b$$
$$54 = b$$
$$y = -7x + 54$$

29.
$$y = mx + b$$
$$-4 = 0(2) + b$$
$$-4 = b$$
$$y = 0x + (-4)$$
$$y = -4$$

31. Find the slope of the line:

$$m = \frac{\Delta y}{\Delta x} = \frac{8-10}{6-2} = \frac{-2}{4} = -\frac{1}{2}$$

$$y = mx + b$$
$$8 = -\frac{1}{2}(6) + b$$
$$8 = -3 + b$$
$$11 = b$$
$$y = -\frac{1}{2}x + 11$$

33.
$$y + 1 = x$$
$$y = x - 1$$
$$m = 1;\ b = -1 \Rightarrow (0, -1)$$

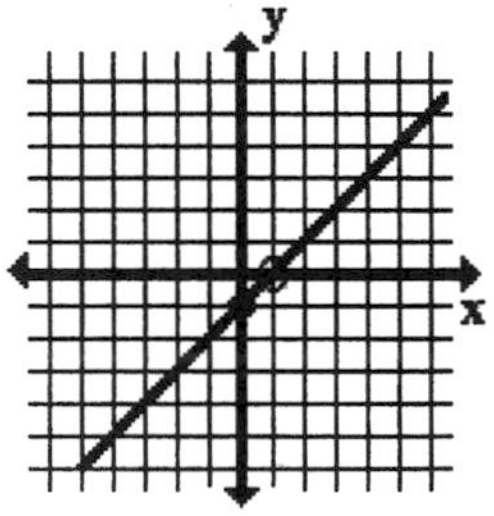

35.
$$x = \tfrac{3}{2}y - 3$$
$$2x = 3y - 6$$
$$-3y = -2x + 6$$
$$y = \tfrac{2}{3}x + 2$$
$$m = \tfrac{2}{3};\ b = 2 \Rightarrow (0, 2)$$

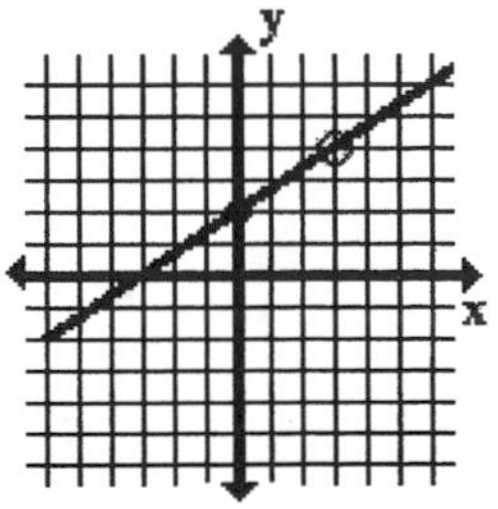

37.
$$3(y-4) = -2(x-3)$$
$$3y - 12 = -2x + 6$$
$$3y = -2x + 18$$
$$y = -\frac{2}{3}x + 6$$
$$m = -\tfrac{2}{3};\ b = 6 \Rightarrow (0, 6)$$

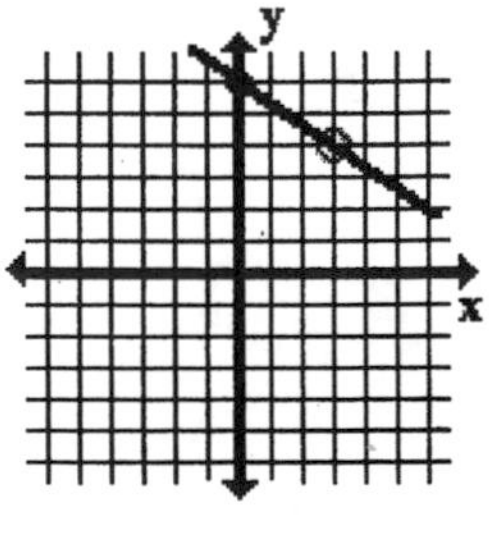

39. $3x - 2y = 8$
$$-2y = -3x + 8$$
$$y = \tfrac{3}{2}x - 4$$
$$m = \tfrac{3}{2};\ b = -4 \Rightarrow (0, -4)$$

41. $-2(x + 3y) = 5$
$$-2x - 6y = 5$$
$$-6y = 2x + 5$$
$$y = -\tfrac{1}{3}x - \tfrac{5}{6}$$
$$m = -\tfrac{1}{3};\ b = -\tfrac{5}{6} \Rightarrow \left(0, -\tfrac{5}{6}\right)$$

43. $x = \dfrac{2y - 4}{7}$
$$7x = 2y - 4$$
$$-2y = -7x - 4$$
$$y = \tfrac{7}{2}x + 2$$
$$m = \tfrac{7}{2};\ b = 2 \Rightarrow (0, 2)$$

45. $y = 3x + 4 \quad y = 3x - 7$
$$m = 3 \qquad m = 3$$
parallel

47. $x + y = 2 \qquad y = x + 5$
$$y = -x + 2 \qquad m = 1$$
$$m = -1$$
perpendicular

49. $y = 3x + 7 \qquad 2y = 6x - 9$
$$m = 3 \qquad y = 3x - \frac{9}{2}$$
$$m = 3$$
parallel

51. $x = 3y + 4 \qquad y = -3x + 7$
$$-3y = -x + 4 \qquad m = -3$$
$$y = \tfrac{1}{3}x - \tfrac{4}{3}$$
$$m = \tfrac{1}{3}$$
perpendicular

53. $y = 3 \qquad x = 4$
horizontal line   vertical line
perpendicular

55. $x = \dfrac{y - 2}{3} \qquad 3(y - 3) + x = 0$
$$3x = y - 2 \qquad 3y - 9 + x = 0$$
$$-y = -3x - 2 \qquad 3y = -x + 9$$
$$y = 3x + 2 \qquad y = -\tfrac{1}{3}x + 3$$
$$m = 3 \qquad m = -\tfrac{1}{3}$$
perpendicular

57. Find the slope of the given line:
$$y = 4x - 7 \Rightarrow m = 4$$
Use the parallel slope.
$$y - y_1 = m(x - x_1)$$
$$y - 0 = 4(x - 0)$$
$$y = 4x$$

59. Find the slope of the given line:
$$4x - y = 7$$
$$-y = -4x + 7$$
$$y = 4x - 7 \Rightarrow m = 4$$
Use the parallel slope.
$$y - y_1 = m(x - x_1)$$
$$y - 5 = 4(x - 2)$$
$$y - 5 = 4x - 8$$
$$y = 4x - 3$$

61. Find the slope of the given line:

$$x = \frac{5}{4}y - 2$$
$$4x = 5y - 8$$
$$-5y = -4x - 8$$
$$y = \tfrac{4}{5}x + \tfrac{8}{5} \Rightarrow m = \tfrac{4}{5}$$

Use the parallel slope.

$$y - y_1 = m(x - x_1)$$
$$y - (-2) = \tfrac{4}{5}(x - 4)$$
$$y + 2 = \tfrac{4}{5}x - \tfrac{16}{5}$$
$$y = \tfrac{4}{5}x - \tfrac{16}{5} - 2$$
$$y = \tfrac{4}{5}x - \tfrac{26}{5}$$

63. Find the slope of the given line:

$$y = 4x - 7 \Rightarrow m = 4$$

Use the perpendicular slope.

$$y - y_1 = m(x - x_1)$$
$$y - 0 = -\tfrac{1}{4}(x - 0)$$
$$y = -\tfrac{1}{4}x$$

65. Find the slope of the given line:

$$4x - y = 7$$
$$-y = -4x + 7$$
$$y = 4x - 7 \Rightarrow m = 4$$

Use the perpendicular slope.

$$y - y_1 = m(x - x_1)$$
$$y - 5 = -\tfrac{1}{4}(x - 2)$$
$$y - 5 = -\tfrac{1}{4}x + \tfrac{1}{2}$$
$$y = -\tfrac{1}{4}x + \tfrac{1}{2} + 5$$
$$y = -\tfrac{1}{4}x + \tfrac{11}{2}$$

67. Find the slope of the given line:

$$x = \tfrac{5}{4}y - 2$$
$$4x = 5y - 8$$
$$-5y = -4x - 8$$
$$y = \tfrac{4}{5}x + \tfrac{8}{5} \Rightarrow m = \tfrac{4}{5}$$

Use the perpendicular slope.

$$y - y_1 = m(x - x_1)$$
$$y - (-2) = -\tfrac{5}{4}(x - 4)$$
$$y + 2 = -\tfrac{5}{4}x + 5$$
$$y = -\tfrac{5}{4}x + 3$$

69. Find the slope of the 1st line. $m = -\frac{A}{B} = -\frac{4}{5}$ Find the slope of the 2nd line. $m = -\frac{A}{B} = -\frac{5}{-4} = \frac{5}{4}$ perpendicular

71. Find the slope of the 1st line. $m = -\frac{A}{B} = -\frac{2}{3}$ Find the slope of the 2nd line. $m = -\frac{A}{B} = -\frac{6}{9} = -\frac{2}{3}$ parallel

73. The line $y = 3$ is horizontal, so any perpendicular line will be vertical.
Find the midpoint of the described segment:

$$x = \frac{x_1 + x_2}{2} = \frac{2 + (-6)}{2} = \frac{-4}{2} = -2; \quad y = \frac{y_1 + y_2}{2} = \frac{4 + 10}{2} = \frac{14}{2} = 7$$

The vertical line through the point $(-2, 7)$ is $x = -2$.

75. The line $x = 3$ is vertical, so any parallel line will be vertical.
Find the midpoint of the described segment:

$$x = \frac{x_1 + x_2}{2} = \frac{2 + 8}{2} = \frac{10}{2} = 5; \quad y = \frac{y_1 + y_2}{2} = \frac{-4 + 12}{2} = \frac{8}{2} = 4$$

The vertical line through the point $(5, 4)$ is $x = 5$.

77. $Ax + By = C$

$$By = -Ax + C$$
$$\frac{By}{B} = \frac{-Ax + C}{B}$$
$$y = -\frac{A}{B}x + \frac{C}{B} \Rightarrow m = -\frac{A}{B};\ b = \frac{C}{B}$$

79. Let x represent the number of years since the truck was purchased, and let y represent the value of the truck. Then we know two points on the depreciation line: $(0, 19984)$ and $(8, 1600)$. Find the slope and then use point-slope form to find the depreciation equation.

$$m = \frac{\Delta y}{\Delta x} = \frac{19984 - 1600}{0 - 8} = \frac{18384}{-8} = -2298$$

$$y - y_1 = m(x - x_1)$$
$$y - 19984 = -2298(x - 0)$$
$$y = -2298x + 19984$$

81. Let x represent the time in years after 1987, and let y represent the value of the painting. Then we know two points on the appreciation line: (0, 36225000) and (20, 72450000). Find the slope and then use point-slope form to find the appreciation equation.

$$m = \frac{\Delta y}{\Delta x} = \frac{72450000 - 36225000}{20 - 0} = \frac{36225000}{20} = 1811250$$
$$y - y_1 = m(x - x_1)$$
$$y - 36225000 = 1811250(x - 0)$$
$$y = 1{,}811{,}250x + 36{,}225{,}000$$

83. Let x represent the number of years since the painting was purchased, and let y represent the value of the painting. Then we know two points on the appreciation line: $(0, 250000)$ and $(5, 500000)$. Find the slope and then use point-slope form to find the appreciation equation.

$$m = \frac{\Delta y}{\Delta x} = \frac{500000 - 250000}{5 - 0} = \frac{250000}{5} = 50000$$

$$y - y_1 = m(x - x_1)$$
$$y - 250000 = 50000(x - 0)$$
$$y = 50{,}000x + 250{,}000$$

85. Let x represent the number of years since the TV was purchased, and let y represent the value of the TV. Then we know two points on the depreciation line: $(0, 1750)$ and $(3, 800)$. Find the slope and then use point-slope form to find the depreciation equation.

$$m = \frac{\Delta y}{\Delta x} = \frac{1750 - 800}{0 - 3} = \frac{950}{-3} = -\frac{950}{3}$$

$$y - y_1 = m(x - x_1)$$
$$y - 1750 = -\tfrac{950}{3}(x - 0)$$
$$y = -\tfrac{950}{3}x + 1750$$

87. Let x represent the number of years since the copy machine was purchased, and let y represent its value. Since the value decreases \$180 per year, the slope of the depreciation line is -180. Since the copier is worth \$1750 new, the y-intercept is 1750. Thus the depreciation equation can be found:

$$y = mx + b$$
$$y = -180x + 1750$$

Now let $x = 7$: $y = -180x + 1750 = -180(7) + 1750 = -1260 + 1750 = \$490$

89. Let x represent the number of years since the house was purchased, and let y represent its value. Since the value increases \$4000 per year, the slope of the depreciation line is 4000. We also know a point on the line: $(2, 122000)$. Use point-slope form to find the appreciation equation.

$$\begin{aligned} y - y_1 &= m(x - x_1) \\ y - 122000 &= 4000(x - 2) \\ y - 122000 &= 4000x - 8000 \\ y &= 4000x + 114000 \end{aligned}$$

Substitute $x = 10$ and find its value:

$$\begin{aligned} y = 4000x + 114000 &= 4000(10) + 114000 \\ &= 40000 + 114000 = \$154{,}000 \end{aligned}$$

91. The relationship between number of copies and the total cost is linear. Let x represent the number of copies (in hundreds), and let y represent the total cost. The slope of the line will be the charge per one hundred copies, or 15. We also know a point on the line: $(3, 75)$. Use point-slope form to find the equation of the line:

$$\begin{aligned} y - y_1 &= m(x - x_1) \\ y - 75 &= 15(x - 3) \\ y - 75 &= 15x - 45 \\ y &= 15x + 30 \end{aligned}$$

Let $x = 10$ and find the cost:

$$y = 15x + 30 = 15(10) + 30 = 150 + 30 = \$180$$

93-99. Answers may vary.

101. To pass through II and IV, the slope must be negative. To pass through II and not III, the y-intercept must be positive. Thus $a < 0$ and $b > 0$.

Exercise 8.4 (page 540)

1.

$$\begin{aligned} \frac{y+2}{2} &= 4(y+2) \\ 2 \cdot \frac{y+2}{2} &= 2 \cdot 4(y+2) \\ y + 2 &= 8(y+2) \\ y + 2 &= 8y + 16 \\ -7y &= 14 \\ y &= -2 \end{aligned}$$

3.

$$\begin{aligned} \frac{2a}{3} + \frac{1}{2} &= \frac{6a-1}{6} \\ 6\left(\frac{2a}{3} + \frac{1}{2}\right) &= 6 \cdot \frac{6a-1}{6} \\ 2(2a) + 3 &= 6a - 1 \\ 4a + 3 &= 6a - 1 \\ -2a &= -4 \\ a &= 2 \end{aligned}$$

5. input

7. x

9. function; input; output

11. range

13. 0

15. $y = mx + b$

17. $y = 2x + 3$ **is a function**, since each value of x corresponds to exactly one value of y.

19. $y = 2x^2$ **is a function**, since each value of x corresponds to exactly one value of y.

21. $y = 3 + 7x^2$ **is a function**, since each value of x corresponds to exactly one value of y.

23. $x = |y|$ **is not a function**, since $x = 2$ corresponds to both $y = 2$ and $y = -2$.

SECTION 8.4

25. $f(x) = 3x$
$f(3) = 3(3) = 9$
$f(-1) = 3(-1) = -3$

27. $f(x) = 2x - 3$
$f(3) = 2(3) - 3 = 3$
$f(-1) = 2(-1) - 3 = -5$

29. $f(x) = 7 + 5x$
$f(3) = 7 + 5(3) = 22$
$f(-1) = 7 + 5(-1) = 2$

31. $f(x) = 9 - 2x$
$f(3) = 9 - 2(3) = 3$
$f(-1) = 9 - 2(-1) = 11$

33. $f(x) = x^2$
$f(2) = 2^2 = 4$
$f(3) = 3^2 = 9$

35. $f(x) = x^3 - 1$
$f(2) = 2^3 - 1 = 8 - 1 = 7$
$f(3) = 3^3 - 1 = 27 - 1 = 26$

37. $f(x) = (x+1)^2$
$f(2) = (2+1)^2 = 3^2 = 9$
$f(3) = (3+1)^2 = 4^2 = 16$

39. $f(x) = 2x^2 - x$
$f(2) = 2(2)^2 - 2 = 2(4) - 2 = 8 - 2 = 6$
$f(3) = 2(3)^2 - 3 = 2(9) - 3 = 18 - 3 = 15$

41. $f(x) = |x| + 2$
$f(2) = |2| + 2 = 2 + 2 = 4$
$f(-2) = |-2| + 2 = 2 + 2 = 4$

43. $f(x) = x^2 - 2$
$f(2) = 2^2 - 2 = 4 - 2 = 2$
$f(-2) = (-2)^2 - 2 = 4 - 2 = 2$

45. $f(x) = \dfrac{1}{x+3}$
$f(2) = \dfrac{1}{2+3} = \dfrac{1}{5}$
$f(-2) = \dfrac{1}{-2+3} = \dfrac{1}{1} = 1$

47. $f(x) = \dfrac{x}{x-3}$
$f(2) = \dfrac{2}{2-3} = \dfrac{2}{-1} = -2$
$f(-2) = \dfrac{-2}{-2-3} = \dfrac{-2}{-5} = \dfrac{2}{5}$

49. $g(x) = 2x$
$g(w) = 2w$
$g(w+1) = 2(w+1) = 2w + 2$

51. $g(x) = 3x - 5$
$g(w) = 3w - 5$
$g(w+1) = 3(w+1) - 5 = 3w + 3 - 5 = 3w - 2$

53. $f(3) + f(2) = 2(3) + 1 + 2(2) + 1 = 6 + 1 + 4 + 1 = 12$

55. $f(b) - f(a) = 2b + 1 - (2a + 1) = 2b + 1 - 2a - 1 = 2b - 2a$

57. $f(b) - 1 = 2b + 1 - 1 = 2b$

59. $f(0) + f(-\frac{1}{2}) = 2(0) + 1 + 2(-\frac{1}{2}) + 1 = 0 + 1 - 1 + 1 = 1$

61. domain $= \{-2, 4, 6\}$; range $= \{3, 5, 7\}$

63. To ensure that the denominator is not 0, x cannot equal 4. The domain is $(-\infty, 4) \cup (4, \infty)$. Because the numerator will never be 0, the range is $(-\infty, 0) \cup (0, \infty)$.

65. not a function

67. The domain is the set of all x-coordinates on the graph. The domain is $(-\infty, \infty)$.
The range is the set of all y-coordinates on the graph. The range is $(-\infty, \infty)$.
Since each vertical line passes through the graph at most once, it is the graph of a function.

69. $f(x) = 2x - 1$
$\mathrm{D} = (-\infty, \infty)$
$\mathrm{R} = (-\infty, \infty)$

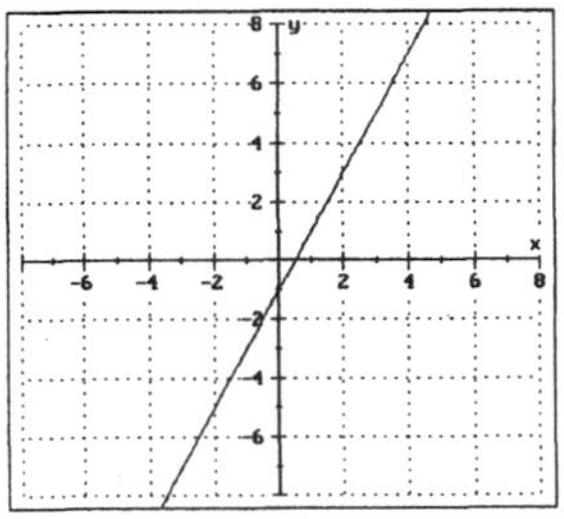

71. $2x - 3y = 6$
$\mathrm{D} = (-\infty, \infty)$
$\mathrm{R} = (-\infty, \infty)$

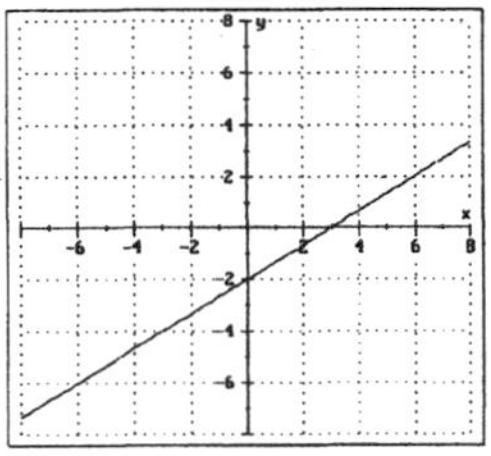

73. $y = 3x^2 + 2 \Rightarrow$ This is not a linear function because the exponent on x is 2.

75. $x = 3y - 4 \Rightarrow y = \frac{1}{3}x + \frac{4}{3} \Rightarrow$ This is a linear function $\left(m = \frac{1}{3}, b = \frac{4}{3}\right)$.

77. Let $t = 3$ and find s.

$$\begin{aligned} s = f(t) &= -16t^2 + 256t \\ &= -16(3)^2 + 256(3) \\ &= -16(9) + 768 \\ &= -144 + 768 \\ &= 624 \end{aligned}$$

The bullet will have a height of 624 ft.

79. Let $h = 16$ and find t.

$$\begin{aligned} h &= -16t^2 + 32t \\ 16 &= -16t^2 + 32t \\ 16t^2 - 32t + 16 &= 0 \\ 16(t^2 - 2t + 1) &= 0 \\ 16(t-1)(t-1) &= 0 \end{aligned}$$

$$t - 1 = 0 \quad \textbf{or} \quad t - 1 = 0$$
$$t = 1 \qquad\qquad t = 1$$

It will take 1 second.

81. Let $C = 25$ and find F.

$$\begin{aligned} F(C) &= \frac{9}{5}C + 32 \\ &= \frac{9}{5}(25) + 32 \\ &= 45 + 32 \\ &= 77 \end{aligned}$$

The temperature is $77°F$.

83. Set $R(x)$ equal to $C(x)$ and solve for x.

$$\begin{aligned} R(x) &= C(x) \\ 120x &= 57.50x + 12000 \\ 62.50x &= 12000 \\ x &= \frac{12000}{62.50} \\ x &= 192 \end{aligned}$$

The company must sell 192 recorders.

85. a. $I(h) = 0.10h + 50$

b. $I(115) = 0.10(115) + 50$
$= 11.5 + 50 = \$61.50$

87. **Answers may vary.**

89. $f(x) + g(x) = (2x + 1) + x^2 = 2x + 1 + x^2$; $g(x) + f(x) = x^2 + (2x + 1) = x^2 + 2x + 1$. They are equal.

Exercise 8.5 (page 554)

1. 41, 43, 47

3. $a \cdot b = b \cdot a$

5. 1

7. squaring

9. absolute value

11. horizontal

13. 2; down

15. 4; to the left

17. rational

19. $f(x) = x^2 - 3$

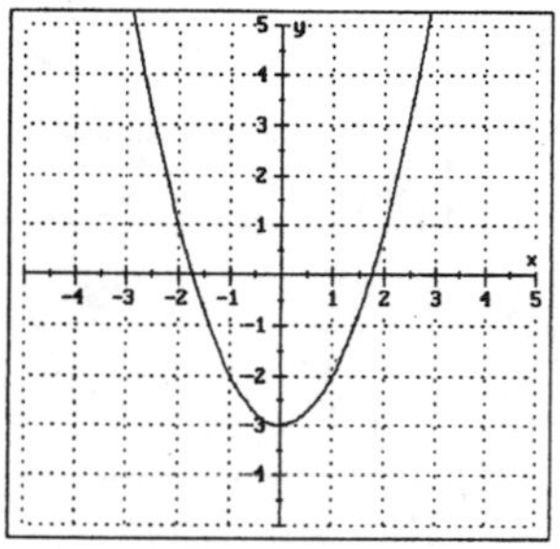

21. $f(x) = (x - 1)^3$

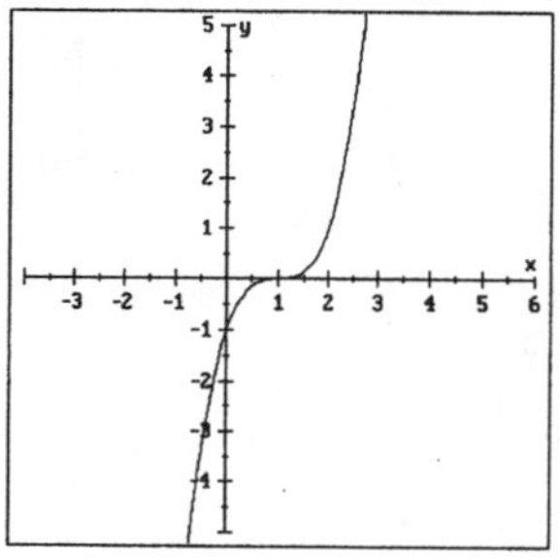

23. $f(x) = |x| - 2$

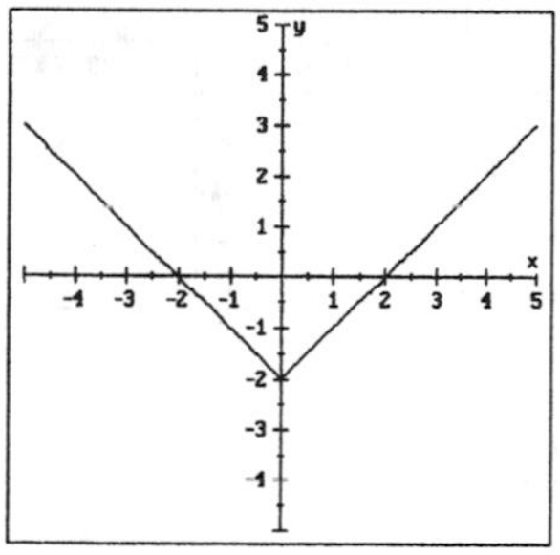

25. $f(x) = |x - 1|$

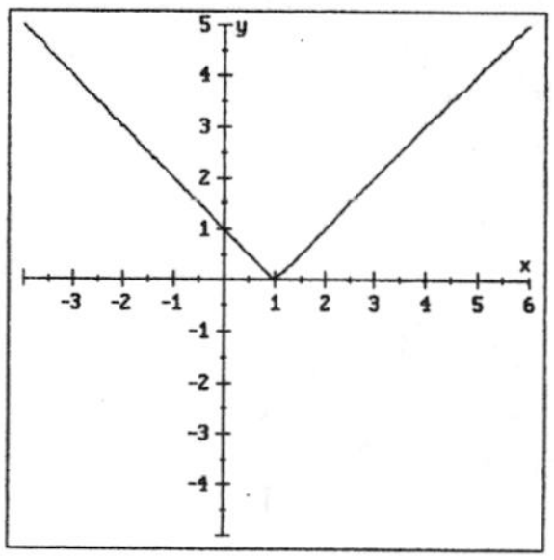

27. $f(x) = x^2 + 8$

x: $[-4, 4]$, y: $[-4, 4]$

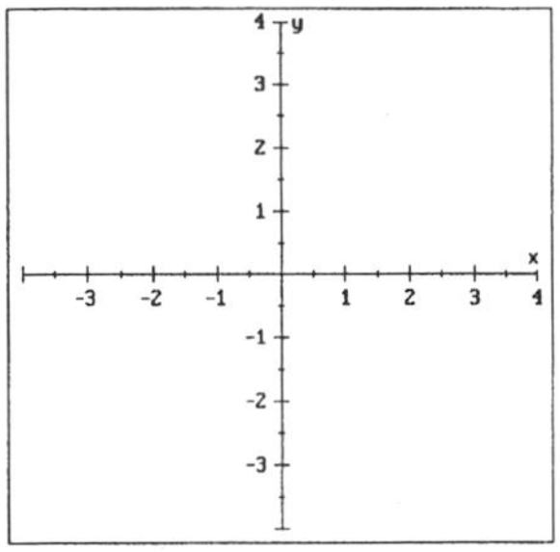

x: $[-7, 7]$, y: $[-2, 12]$

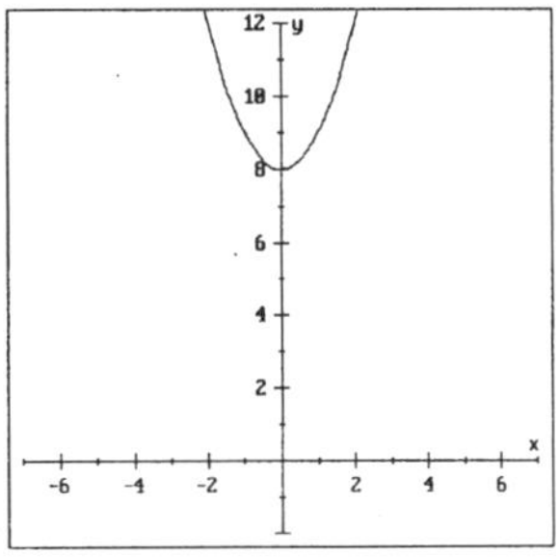

29. $f(x) = |x + 5|$

x: $[-4, 4]$, y: $[-4, 4]$

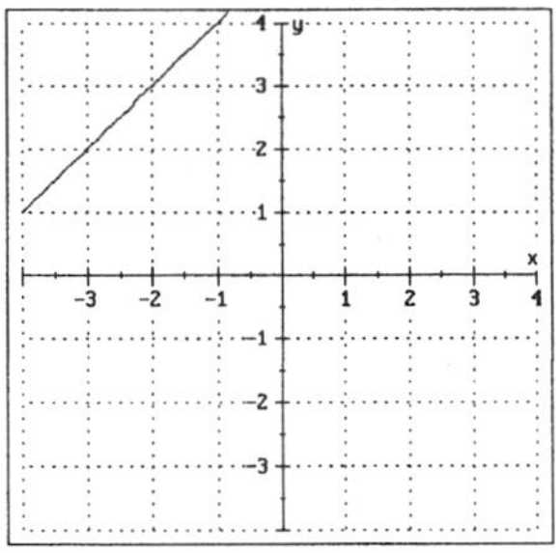

x: $[-12, 2]$, y: $[-4, 10]$

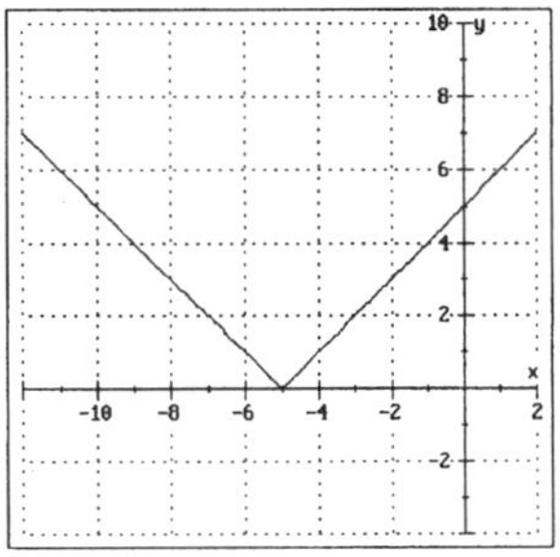

31. $f(x) = (x - 6)^2$

x: $[-4, 4]$, y: $[-4, 4]$

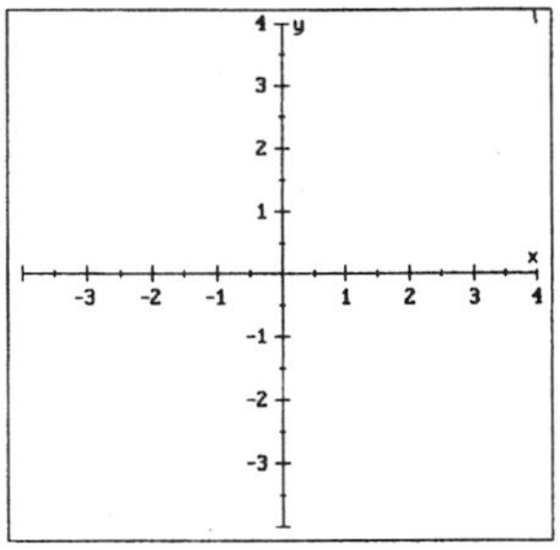

x: $[-2, 12]$, y: $[-2, 12]$

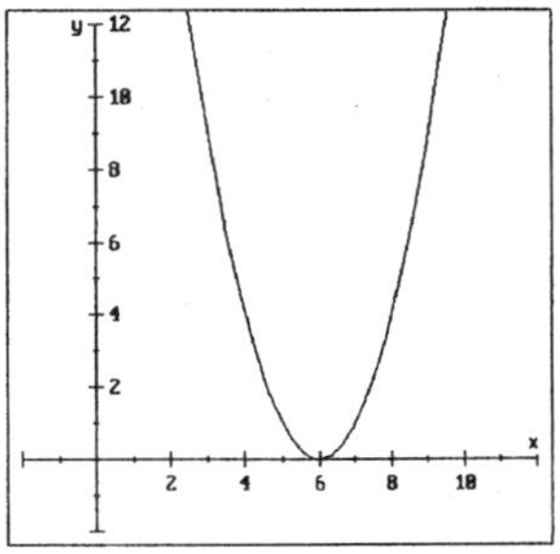

33. $f(x) = x^3 + 8$

x: $[-4, 4]$, y: $[-4, 4]$

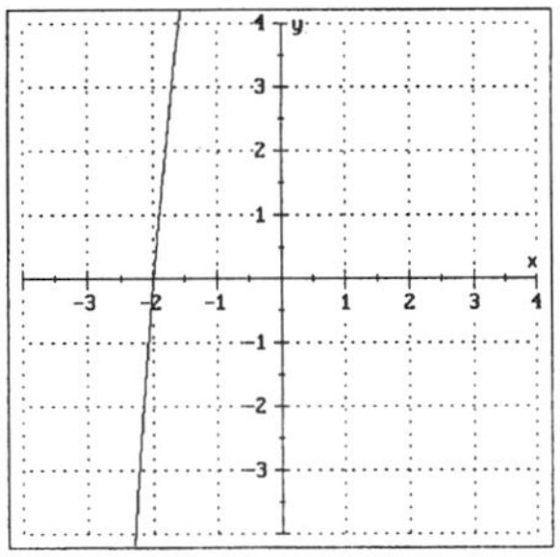

x: $[-10, 10]$, y: $[-4, 16]$

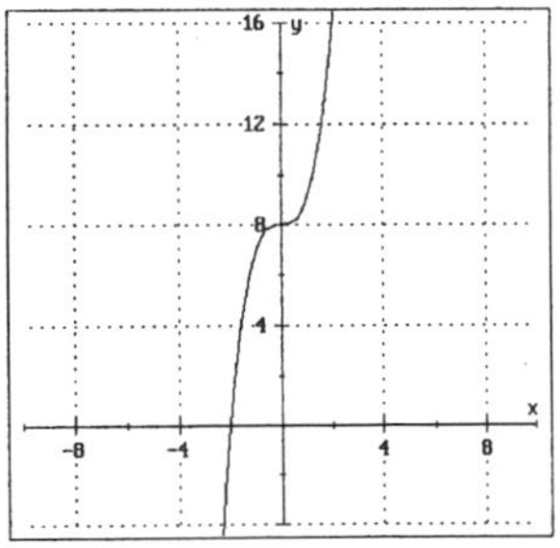

35. $f(x) = x^2 - 5$

Shift $f(x) = x^2$ D 5.

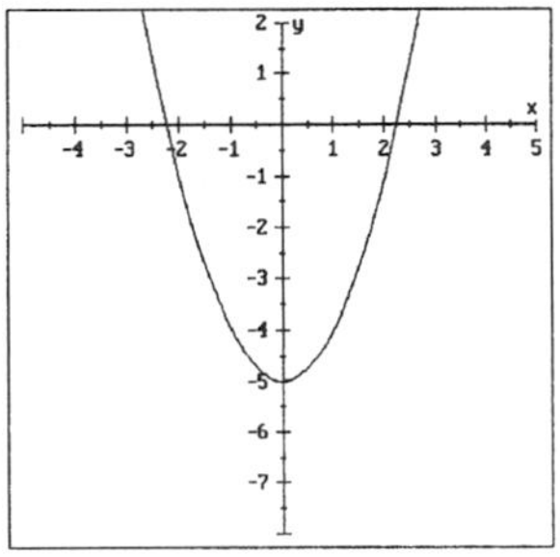

37. $f(x) = (x - 1)^3$

Shift $f(x) = x^3$ R 1.

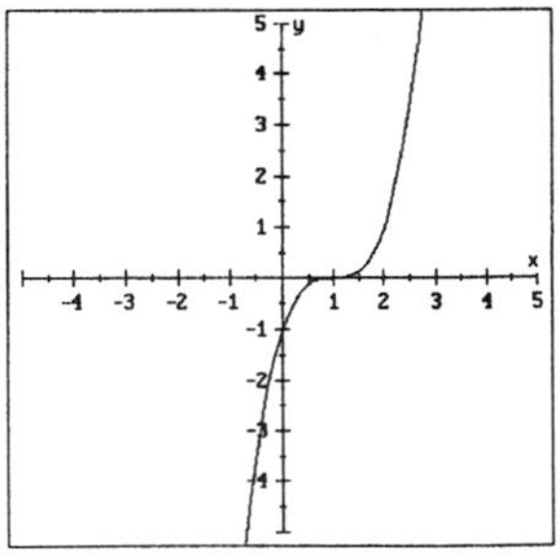

39. $f(x) = |x - 2| - 1$

Shift $f(x) = |x|$ D 1, R 2.

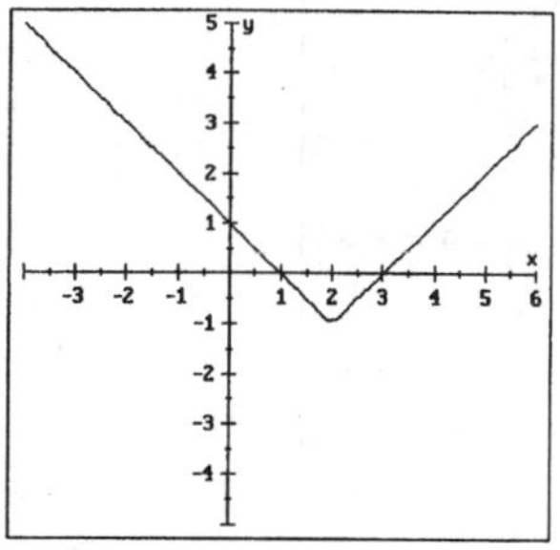

41. $f(x) = (x+1)^3 - 2$

Shift $f(x) = x^3$ D 2, L 1.

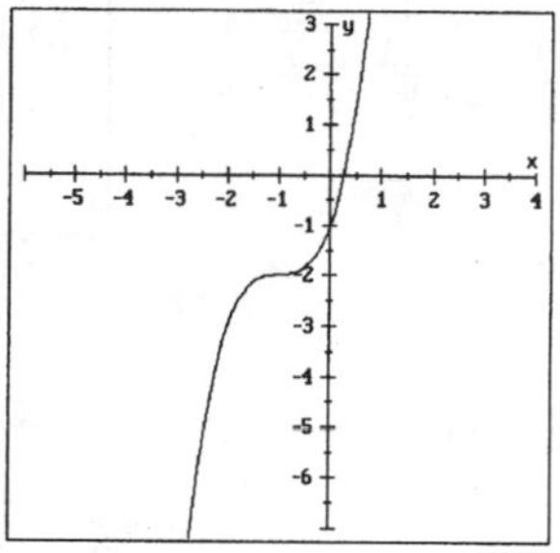

43. $f(x) = -x^2$

Reflect $f(x) = x^2$ about x-axis.

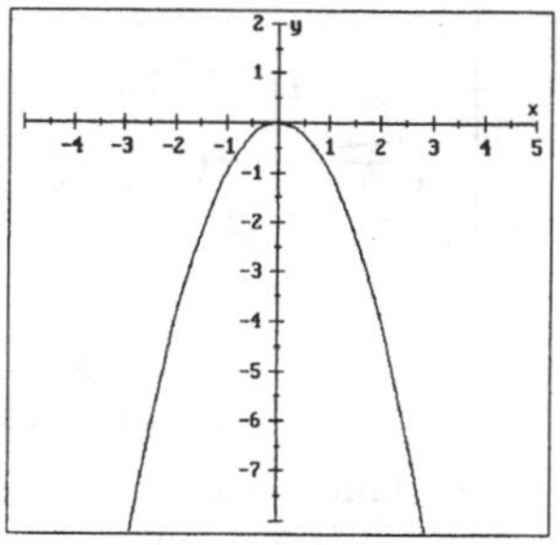

45. $f(x) = -(x-2)^2 - 3$

Reflect $f(x) = x^2$ about x-axis, shift D 3, R 2.

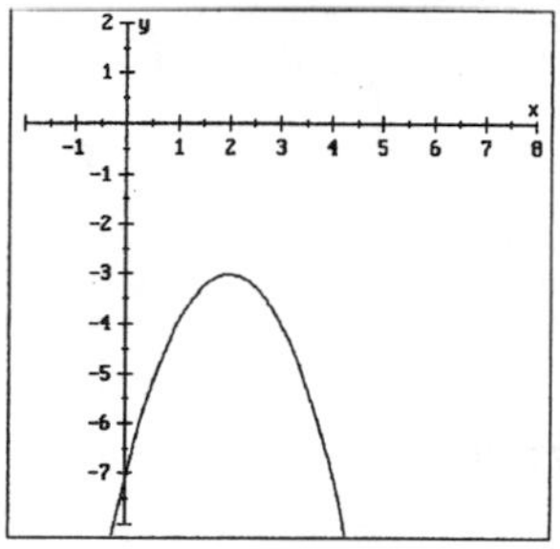

47.
$$t = f(r) = \frac{600}{r}$$
$$f(30) = \frac{600}{30}$$
$$= 20$$
$t = 20$ hours

49.
$$t = f(r) = \frac{600}{r}$$
$$f(50) = \frac{600}{50}$$
$$= 12$$
$t = 12$ hours

51.
$$C = f(p) = \frac{50{,}000p}{100 - p}$$
$$f(10) = \frac{50{,}000(10)}{100 - 10}$$
$$= \frac{500{,}000}{90}$$
$$= 5555.56$$
$C = \$5555.56$

53.
$$C = f(p) = \frac{50{,}000p}{100 - p}$$
$$f(50) = \frac{50{,}000(50)}{100 - 50}$$
$$= \frac{2{,}500{,}000}{50}$$
$$= 50{,}000.00$$
$C = \$50{,}000$

55. $f(x) = \dfrac{x}{x-2}$

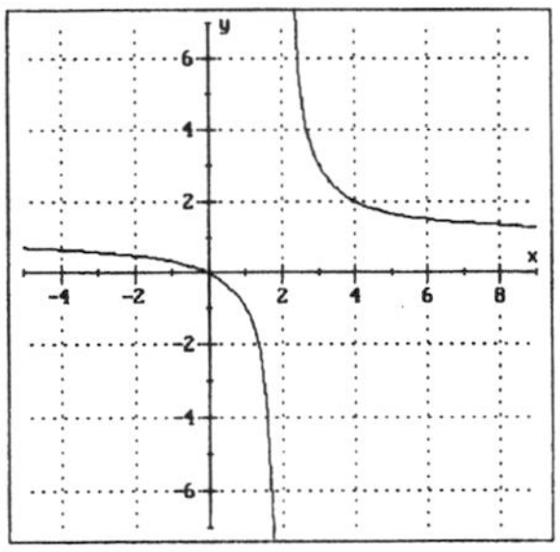

domain: $(-\infty, 2) \cup (2, \infty)$
range: $(-\infty, 1) \cup (1, \infty)$

57. $f(x) = \dfrac{x+1}{x^2-4}$

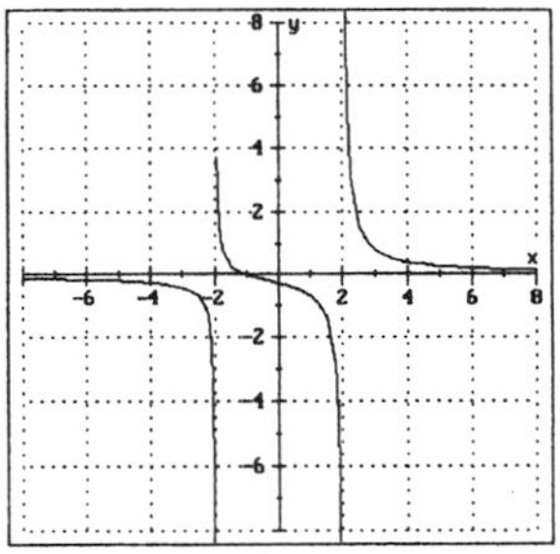

domain: $(-\infty, -2) \cup (-2, 2) \cup (2, \infty)$
range: $(-\infty, \infty)$

59. $c = f(x) = 1.25x + 700$

61. Refer to **#59**. $c = f(500) = 1.25(500) + 700 = \$1325$

63. Refer to **#59**. The mean cost $= \overline{c} = f(x) = \dfrac{1.25x + 700}{x}$.

$$f(1000) = \frac{1.25(1000) + 700}{1000} = \frac{1950}{1000} = \$1.95$$

65. $c = f(n) = 0.09n + 7.50$

67. Refer to **#65**. $c = f(775) = 0.09(775) + 7.50 = 69.75 + 7.50 = \$77.25$

69. Refer to **#65**. The mean cost $= \overline{c} = f(x) = \dfrac{0.09n + 7.50}{n}$.

$$f(1000) = \frac{0.09(1000) + 7.50}{1000} = \frac{97.50}{1000} = \$0.0975 = 9.75¢$$

71. $c = f(x) = 350x + 5000$

73. Refer to **#71**. $c = f(120) = 350(120) + 5000 = \$47{,}000$

75. **Answers may vary.**

Exercise 8.6 (page 563)

1. $(x^2x^3)^2 = (x^2)^2(x^3)^2 = x^4x^6 = x^{10}$

3. $\dfrac{b^0 - 2b^0}{b^0} = \dfrac{1-2}{1} = \dfrac{-1}{1} = -1$

5. $35{,}000 = 3.5 \times 10^4$

7. $2.5 \times 10^{-3} = 0.0025$

9. proportion

11. direct

13. rational

SECTION 8.6

15. joint

17. direct

19. neither

21.
$$\frac{x}{5} = \frac{15}{25}$$
$$25x = 75$$
$$x = 3$$

23.
$$\frac{r-2}{3} = \frac{r}{5}$$
$$5(r-2) = 3r$$
$$5r - 10 = 3r$$
$$2r = 10$$
$$r = 5$$

25.
$$\frac{3}{n} = \frac{2}{n+1}$$
$$3(n+1) = 2n$$
$$3n + 3 = 2n$$
$$n = -3$$

27.
$$\frac{5}{5z+3} = \frac{2z}{2z^2+6}$$
$$5(2z^2+6) = 2z(5z+3)$$
$$10z^2 + 30 = 10z^2 + 6z$$
$$30 = 6z$$
$$5 = z$$

29.
$$\frac{2}{c} = \frac{c-3}{2}$$
$$4 = c^2 - 3c$$
$$0 = c^2 - 3c - 4$$
$$0 = (c+1)(c-4)$$
$$c + 1 = 0 \quad \textbf{or} \quad c - 4 = 0$$
$$c = -1 \qquad\qquad c = 4$$

31.
$$\frac{2}{3x} = \frac{6x}{36}$$
$$72 = 18x^2$$
$$0 = 18x^2 - 72$$
$$0 = 18(x+2)(x-2)$$
$$x + 2 = 0 \quad \textbf{or} \quad x - 2 = 0$$
$$x = -2 \qquad\qquad x = 2$$

33.
$$\frac{2(x+3)}{3} = \frac{4(x-4)}{5}$$
$$10(x+3) = 12(x-4)$$
$$10x + 30 = 12x - 48$$
$$-2x = -78$$
$$x = 39$$

35.
$$\frac{1}{x+3} = \frac{-2x}{x+5}$$
$$x + 5 = -2x^2 - 6x$$
$$2x^2 + 7x + 5 = 0$$
$$(2x+5)(x+1)$$
$$2x + 5 = 0 \quad \textbf{or} \quad x + 1 = 0$$
$$x = -\tfrac{5}{2} \qquad\qquad x = -1$$

37. $A = kp^2$

39. $v = \dfrac{k}{r^3}$

41. $B = kmn$

43. $P = \dfrac{ka^2}{j^3}$

45. L varies jointly with m and n.

47. E varies jointly with a and the square of b.

49. X varies directly with x^2 and inversely with y^2.

51. R varies directly with L and inversely with d^2.

53.
$A = kr^2$
$A = \pi r^2$
$A = \pi(6 \text{ in.})^2$
$\boxed{A = 36\pi \text{ in.}^2}$

55.
$d = kg$
$288 = k(12)$
$24 = k$

$d = 24g$
$d = 24(18)$
$\boxed{d = 432 \text{ mi}}$

57.
$t = \frac{k}{n}$
$10 = \frac{k}{25}$
$250 = k$

$t = \frac{250}{n}$
$t = \frac{250}{10}$
$\boxed{t = 25 \text{ days}}$

59.
$V = \frac{k}{P}$
$20 = \frac{k}{6}$
$120 = k$

$V = \frac{120}{P}$
$V = \frac{120}{10}$
$\boxed{V = 12 \text{ in.}^3}$

61.
$f = \frac{k}{l}$
$256 = \frac{k}{2}$
$512 = k$

$f = \frac{512}{l}$
$f = \frac{512}{6}$
$\boxed{f = 85\frac{1}{3}}$

63.
$V_1 = klwh$
$V_2 = k(2l)(3w)(2h) = 12klwh = 12V_1$
The volume is multiplied by 12.

65.
$g = khr^2$
$g = 23.5hr^2$
$g = 23.5(20)(7.5)^2$
$\boxed{g = 26{,}437.5 \text{ gallons}}$

67.
$V = kC$
$6 = k(2)$
$3 = k$
The resistance is 3 ohms.

69.
$D = \frac{k}{wd^3}$
$1.1 = \frac{k}{4(4)^3}$
$1.1 = \frac{k}{256}$
$281.6 = k$

$D = \frac{281.6}{wd^3}$
$D = \frac{281.6}{2(8)^3}$
$D = \frac{281.6}{1024}$
$\boxed{D = 0.275 \text{ in.}}$

71.
$P = \frac{kT}{V}$
$1 = \frac{k(273)}{1}$
$\frac{1}{273} = k$

$P = \frac{\frac{1}{273}T}{V}$
$1 = \frac{\frac{1}{273}T}{2}$
$2 = \frac{1}{273}T$
$\boxed{546 \text{ K} = T}$

73. **Answers may vary.**

75. **Answers may vary.**

77. This is not direct variation. For this to be direct variation, one temperature would have to be a constant multiple of the other.

79. No. Answers may vary.

Chapter 8 Summary (page 568)

1. $x + y = 4$

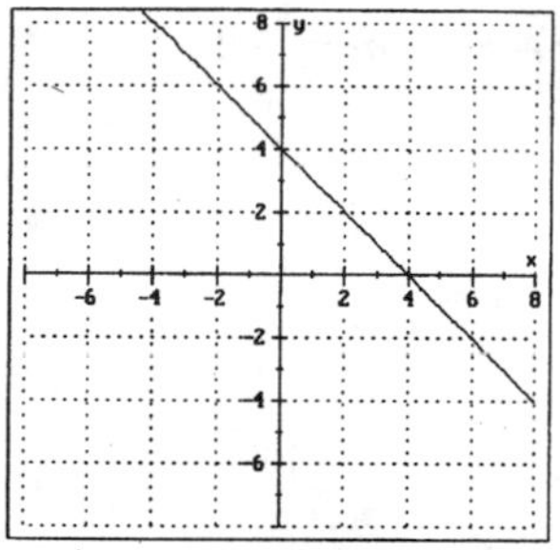

2. $2x - y = 8$

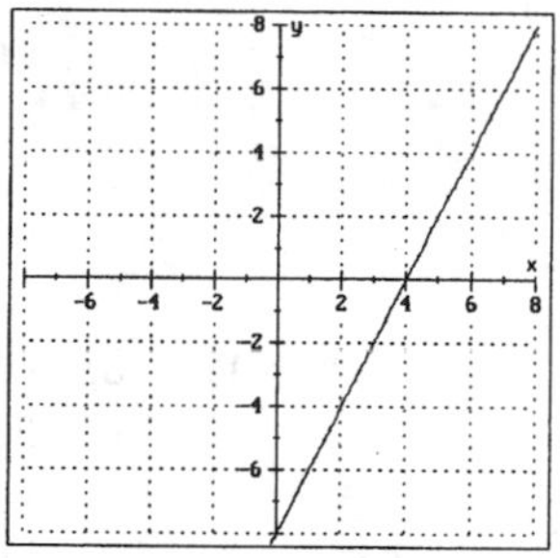

3. $y = 3x + 4$

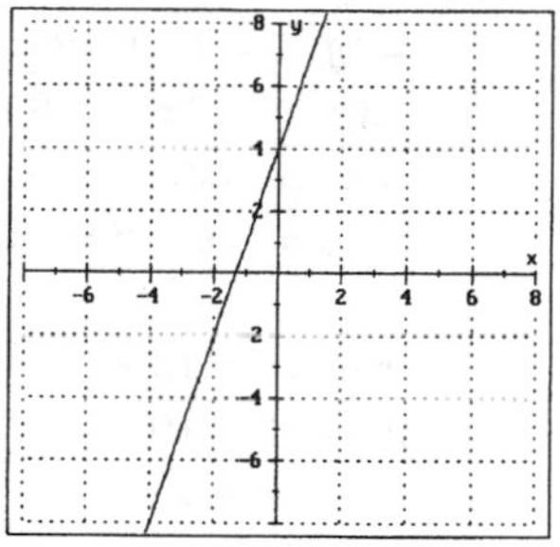

4. $x = 4 - 2y$

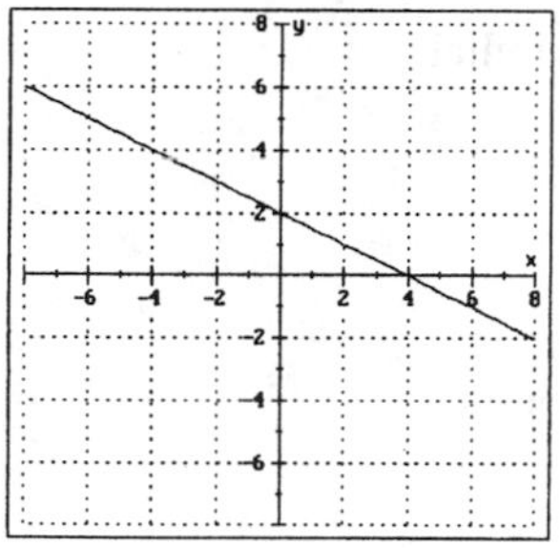

5. $y = 4$

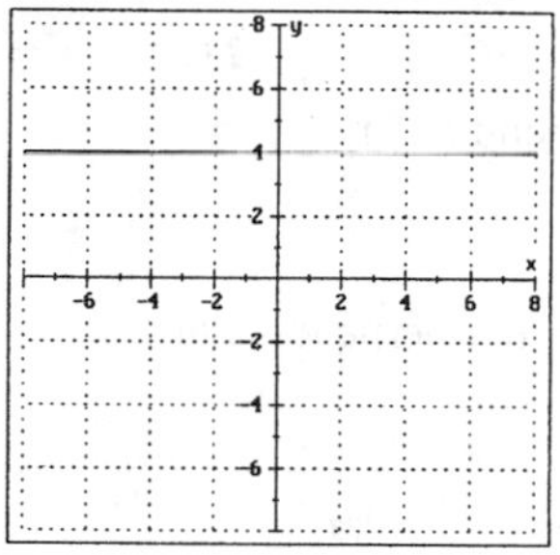

6. $x = -2$

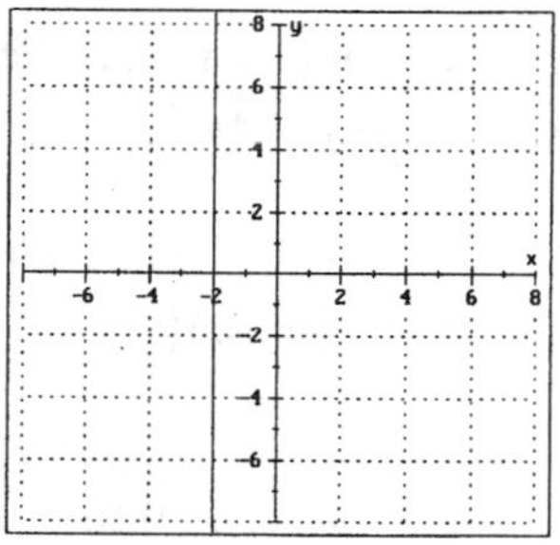

7. $2(x + 3) = x + 2$
$x = -4$

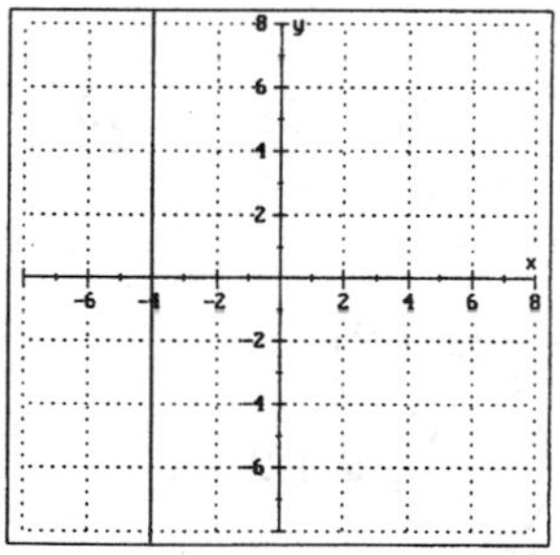

8. $3y = 2(y - 1)$
$y = -2$

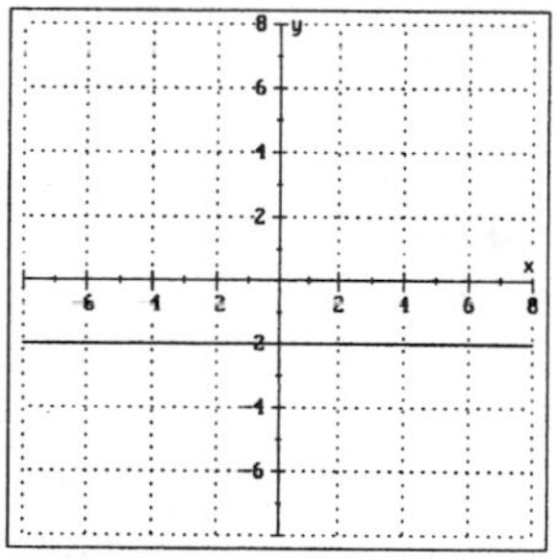

9. $x = \frac{x_1+x_2}{2} = \frac{-3+6}{2} = \frac{3}{2}$; $y = \frac{y_1+y_2}{2} = \frac{5+11}{2} = \frac{16}{2} = 8$; midpoint: $\left(\frac{3}{2}, 8\right)$

10. $m = \dfrac{\Delta y}{\Delta x} = \dfrac{5-8}{2-5} = \dfrac{-3}{-3} = 1$

11. $m = \dfrac{\Delta y}{\Delta x} = \dfrac{-2-12}{-3-6} = \dfrac{-14}{-9} = \dfrac{14}{9}$

12. $m = \dfrac{\Delta y}{\Delta x} = \dfrac{4-(-6)}{-3-(-5)} = \dfrac{10}{2} = 5$

13. $m = \dfrac{\Delta y}{\Delta x} = \dfrac{-4-(-9)}{5-(-6)} = \dfrac{5}{11}$

14. $m = \frac{\Delta y}{\Delta x} = \frac{4-4}{8-(-2)} = \frac{0}{10} = 0$

15. $m = \frac{\Delta y}{\Delta x} = \frac{8-(-4)}{-5-(-5)} = \frac{12}{0} \Rightarrow$ undefined

16.
$$\begin{aligned} 2x - 3y &= 18 \\ -3y &= -2x + 18 \\ \frac{-3y}{-3} &= \frac{-2x}{-3} + \frac{18}{-3} \\ y &= \tfrac{2}{3}x - 6 \Rightarrow m = \tfrac{2}{3} \end{aligned}$$

17.
$$\begin{aligned} 2x + y &= 8 \\ y &= -2x + 8 \Rightarrow m = -2 \end{aligned}$$

18.
$$\begin{aligned} -2(x-3) &= 10 \\ -2x + 6 &= 10 \\ -2x &= 4 \\ x &= -2 \end{aligned}$$
The slope is undefined (vertical).

19.
$$\begin{aligned} 3y + 1 &= 7 \\ 3y &= 6 \\ y &= 2 \end{aligned}$$
$m = 0$ (horizontal)

20. $m_1 \neq m_2 \Rightarrow$ not parallel
$m_1 \cdot m_2 = 4\left(-\frac{1}{4}\right) = -1 \Rightarrow$ perpendicular

21. $m_1 = m_2 \Rightarrow$ parallel

22. $m_1 \neq m_2 \Rightarrow$ not parallel
$m_1 \cdot m_2 = 0.5\left(-\frac{1}{2}\right) = -0.25 \Rightarrow$ not perpendicular

23. $m_1 \neq m_2 \Rightarrow$ not parallel
$m_1 \cdot m_2 = 5(-0.2) = -1 \Rightarrow$ perpendicular

24. Let $y =$ sales and let $x =$ year of business.
$m = \frac{\Delta y}{\Delta x} = \frac{y_2 - y_1}{x_1 - x_1} = \frac{130{,}000 - 65{,}000}{4-1} = \frac{65{,}000}{3} \approx \$21{,}666.67$ per year

25.
$$\begin{aligned} y - y_1 &= m(x - x_1) \\ y - 5 &= 3(x+8) \\ y - 5 &= 3x + 24 \\ -3x + y &= 29 \\ 3x - y &= -29 \end{aligned}$$

26. $m = \frac{\Delta y}{\Delta x} = \frac{4-(-9)}{-2-6} = \frac{13}{-8} = -\frac{13}{8}$
$$\begin{aligned} y - y_1 &= m(x - x_1) \\ y - 4 &= -\frac{13}{8}(x+2) \\ 8(y-4) &= 8\left(-\frac{13}{8}\right)(x+2) \\ 8y - 32 &= -13(x+2) \\ 8y - 32 &= -13x - 26 \\ 13x + 8y &= 6 \end{aligned}$$

27. Find the slope of the given line:

$$3x - 2y = 7$$
$$-2y = -3x + 7$$
$$y = \tfrac{3}{2}x - \tfrac{7}{2} \Rightarrow m = \tfrac{3}{2}$$

Use the parallel slope:

$$y - y_1 = m(x - x_1)$$
$$y + 5 = \tfrac{3}{2}(x + 3)$$
$$2(y + 5) = 2 \cdot \tfrac{3}{2}(x + 3)$$
$$2y + 10 = 3(x + 3)$$
$$2y + 10 = 3x + 9$$
$$-3x + 2y = -1 \Rightarrow 3x - 2y = 1$$

28. Find the slope of the given line:

$$3x - 2y = 7$$
$$-2y = -3x + 7$$
$$y = \tfrac{3}{2}x - \tfrac{7}{2} \Rightarrow m = \tfrac{3}{2}$$

Use the perpendicular slope:

$$y - y_1 = m(x - x_1)$$
$$y + 5 = -\tfrac{2}{3}(x + 3)$$
$$3(y + 5) = 3\left(-\tfrac{2}{3}\right)(x + 3)$$
$$3y + 15 = -2(x + 3)$$
$$3y + 15 = -2x - 6$$
$$2x + 3y = -21$$

29. $y = \frac{2}{3}x + 1;\ m = \frac{2}{3};\ b = 1$

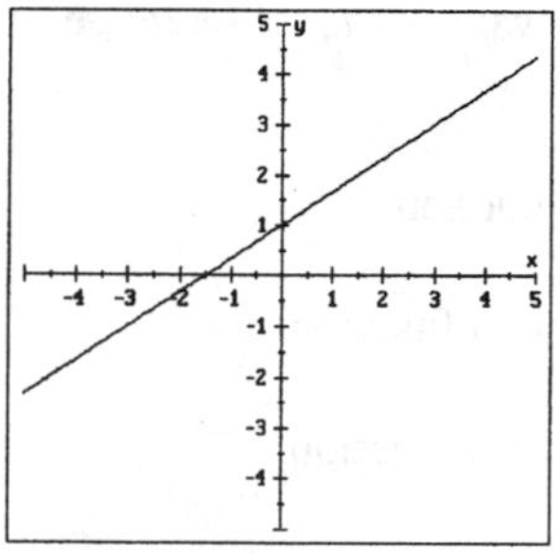

30. $2x + 3y = 8 \Rightarrow m = -\dfrac{A}{B} = -\dfrac{2}{3}$

$3x - 2y = 10 \Rightarrow m = -\dfrac{A}{B} = -\dfrac{3}{-2} = \dfrac{3}{2}$

perpendicular

31. Let x represent the number of years since the copy machine was purchased, and let y represent its value. We know two points on the line: $(0, 8700)$ and $(5, 100)$.

Find the slope:

$$m = \frac{\Delta y}{\Delta x} = \frac{8700 - 100}{0 - 5} = \frac{8600}{-5} = -1720$$

Find the equation of the line:

$$y - y_1 = m(x - x_1)$$
$$y - 8700 = -1720(x - 0)$$
$$y = -1720x + 8700$$

32. $y = 6x - 4$ **is a function**, since each value of x corresponds to exactly one value of y.

33. $y = 4 - x$ **is a function**, since each value of x corresponds to exactly one value of y.

34. $y^2 = x$ **is not a function**, since $x = 9$ corresponds to both $y = 3$ and $y = -3$.

35. $|y| = x^2$ **is not a function**, since $x = 2$ corresponds to both $y = 4$ and $y = -4$.

36. $f(x) = 3x + 2$
$f(-3) = 3(-3) + 2 = -9 + 2 = -7$

37. $g(x) = x^2 - 4$
$g(8) = 8^2 - 4 = 64 - 4 = 60$

38. $g(x) = x^2 - 4$
$g(-2) = (-2)^2 - 4 = 4 - 4 = 0$

39. $f(x) = 3x + 2$
$f(5) = 3(5) + 2 = 15 + 2 = 17$

40. Since any value can be substituted for x, the domain is $(-\infty, \infty)$. The range is also $(-\infty, \infty)$.

41. Since any value can be substituted for x, the domain is $(-\infty, \infty)$. The range is also $(-\infty, \infty)$.

42. Since any value can be substituted for x, the domain is $(-\infty, \infty)$. Since $x^2 \geq 0$, $x^2 + 1$ must be greater than or equal to 1. Thus the range is $[1, \infty)$.

43. Since $x = 2$ makes the denominator equal to 0, the domain is $(-\infty, 2) \cup (2, \infty)$. Since the numerator will never equal 0, the fraction will never equal 0. The range is $(-\infty, 0) \cup (0, \infty)$.

44. Since $x = 3$ makes the denominator equal to 0, the domain is $(-\infty, 3) \cup (3, \infty)$. Since the numerator will never equal 0, the fraction will never equal 0. The range is $(-\infty, 0) \cup (0, \infty)$.

45. x can be any real number, so the domain is $(-\infty, \infty)$. y always has a value of 7, so the range is $\{7\}$.

46. Since each vertical line passes through the graph at most once, it is a function.

47. Since a vertical line can pass through the graph more than once, it is not a function.

48. Since a vertical line can pass through the graph more than once, it is not a function.

49. Since each vertical line passes through the graph at most once, it is a function.

50. $f(x) = x^2 - 3$: Shift $f(x) = x^2$ D 3.

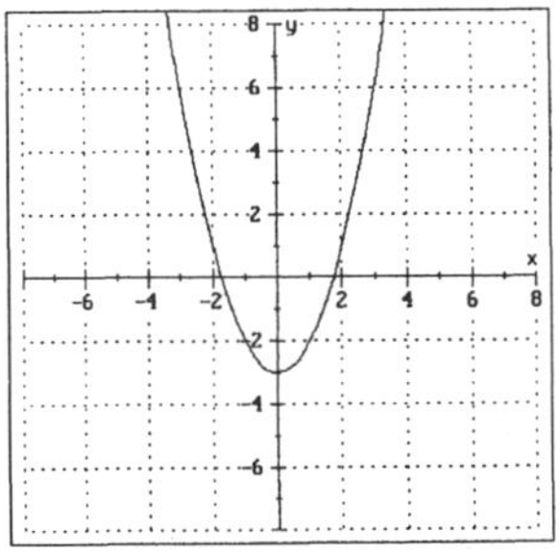

51. $f(x) = |x| - 4$: Shift $f(x) = |x|$ D 4.

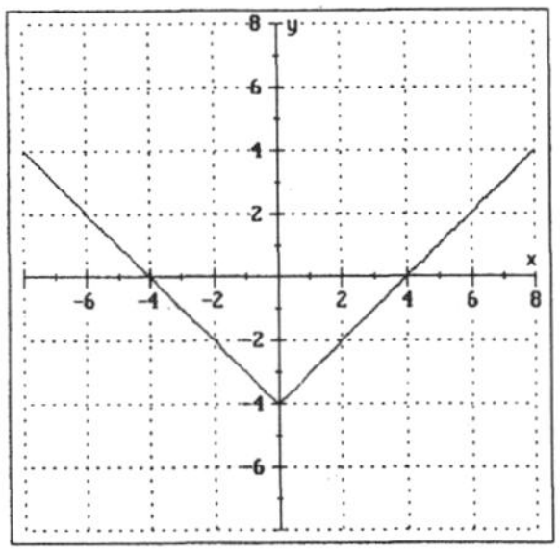

52. $f(x) = (x-2)^3$: Shift $f(x) = x^3$ R 2.

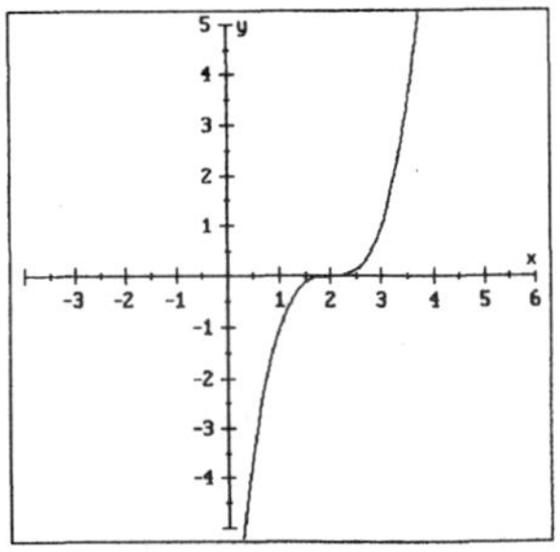

53. $f(x) = (x+4)^2 - 3$: Shift $f(x) = x^2$ D 3 and L 4.

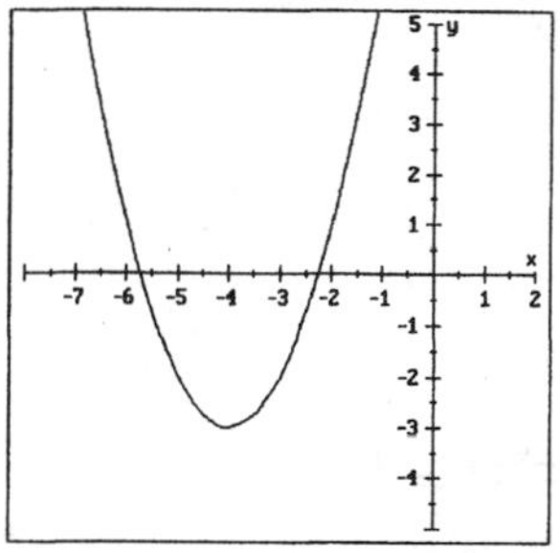

54. $f(x) = -x^3 - 2$: Reflect $f(x) = x^3$ about the x-axis and shift D 2.

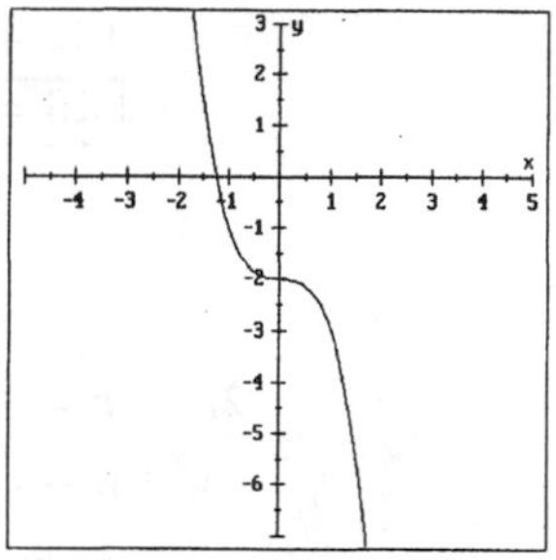

55. $f(x) = -|x-1| + 2$ Reflect $f(x) = |x|$ about the x-axis and shift U 2.

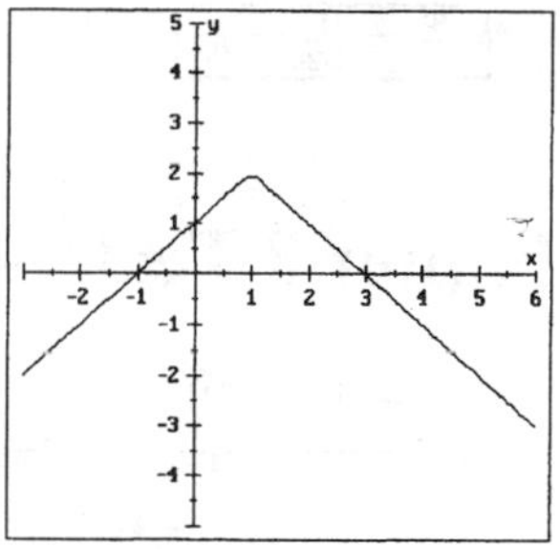

56-61. Compare your graphs to the graphs in numbers **50-55.**

62. $f(x) = \frac{2}{x-2}$

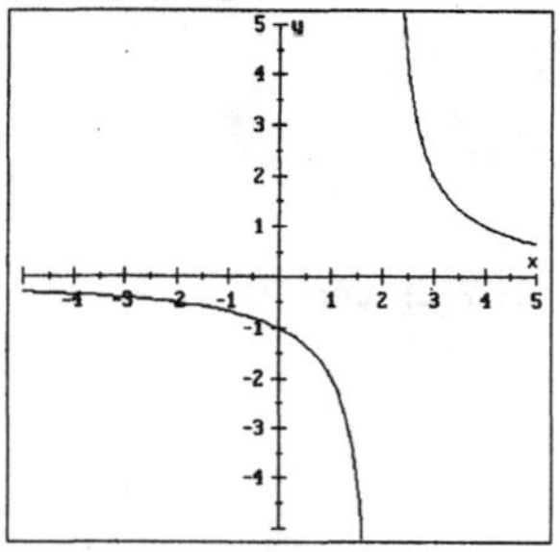

domain $= (-\infty, 2) \cup (2, \infty)$
range $= (-\infty, 0) \cup (0, \infty)$

63. $f(x) = \frac{x}{x+3}$

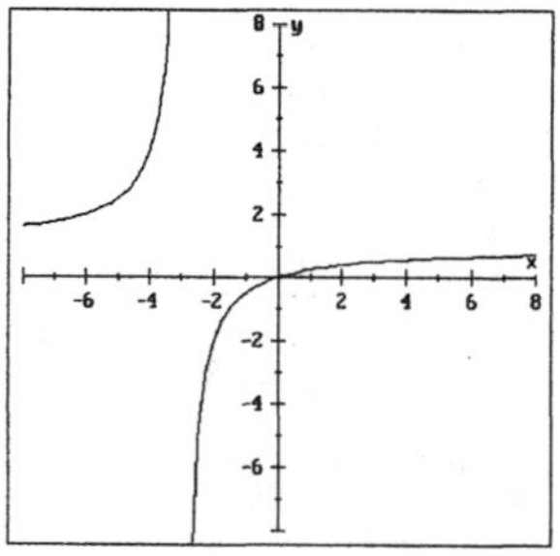

domain $= (-\infty, -3) \cup (-3, \infty)$
range $= (-\infty, 1) \cup (1, \infty)$

64.
$$\begin{aligned} \frac{x+1}{8} &= \frac{4x-2}{23} \\ 23(x+1) &= 8(4x-2) \\ 23x + 23 &= 32x - 16 \\ -9x &= -39 \\ x &= \tfrac{-39}{-9} = \tfrac{13}{3} \end{aligned}$$

65.
$$\begin{aligned} \frac{1}{x+6} &= \frac{x+10}{12} \\ 12 &= (x+6)(x+10) \\ 12 &= x^2 + 16x + 60 \\ 0 &= x^2 + 16x + 48 \\ 0 &= (x+12)(x+4) \end{aligned}$$
$x + 12 = 0$ **or** $x + 4 = 0$
$x = -12$ $\quad$ $x = -4$

66.
$y = kx$ $\quad$ $y = \frac{1}{6}x$
$2 = k(12)$ $\quad$ $y = \frac{1}{6}(12)$
$\frac{1}{6} = k$ $\quad$ $\boxed{y = 2}$

67.
$y = \dfrac{k}{x}$ $\quad$ $y = \dfrac{72}{x}$
$3 = \dfrac{k}{24}$ $\quad$ $y = \dfrac{72}{12}$
$72 = k$ $\quad$ $\boxed{y = 6}$

68.
$$y = kxz$$
$$3 = k(24)(4)$$
$$3 = 96k$$
$$\boxed{\tfrac{3}{96} = \tfrac{1}{32} = k}$$

69.
$$y = \frac{kt}{x}$$
$$64 = \frac{k(8)}{2}$$
$$128 = 8k$$
$$\boxed{16 = k}$$

Chapter 8 Test (page 573)

1. $2x - 5y = 10$

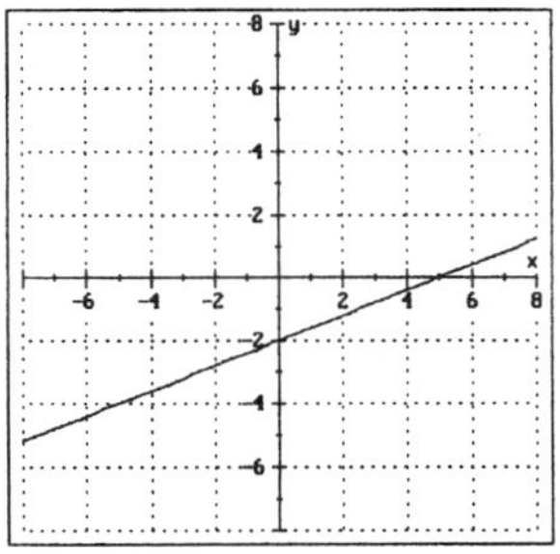

2. $x = \frac{x_1+x_2}{2} = \frac{-3+4}{2} = \frac{1}{2}$
$y = \frac{y_1+y_2}{2} = \frac{3+(-2)}{2} = \frac{1}{2}$
midpoint: $\left(\frac{1}{2}, \frac{1}{2}\right)$

3.
$$y = \frac{x-3}{5}$$
$$0 = \frac{x-3}{5}$$
$$5(0) = 5 \cdot \frac{x-3}{5}$$
$$0 = x - 3$$
$$3 = x$$
x-intercept: $(3, 0)$

$$y = \frac{x-3}{5}$$
$$y = \frac{0-3}{5}$$
$$y = -\frac{3}{5}$$
y-intercept: $\left(0, -\frac{3}{5}\right)$

4. $x - 7 = 0 \Rightarrow x = 7 \Rightarrow$ vertical line

5. $m = \frac{\Delta y}{\Delta x} = \frac{4-8}{-2-6} = \frac{-4}{-8} = \frac{1}{2}$

6.
$$2x - 3y = 8$$
$$-3y = -2x + 8$$
$$y = \frac{2}{3}x - \frac{8}{3} \Rightarrow m = \frac{2}{3}$$

7. The graph of $x = 12$ is a vertical line, so the slope is undefined.

8. The graph of $y = 12$ is a horizontal line, so the slope is 0.

9.
$$y - y_1 = m(x - x_1)$$
$$y + 5 = \frac{2}{3}(x - 4)$$
$$y + 5 = \frac{2}{3}x - \frac{8}{3}$$
$$y = \frac{2}{3}x - \frac{8}{3} - 5$$
$$y = \frac{2}{3}x - \frac{23}{3}$$

10. $m = \frac{\Delta y}{\Delta x} = \frac{6 - (-10)}{-2 - (-4)} = \frac{16}{2} = 8$
$$y - y_1 = m(x - x_1)$$
$$y - 6 = 8(x + 2)$$
$$y - 6 = 8x + 16$$
$$-8x + y = 22$$
$$8x - y = -22$$

11.
$$-2(x - 3) = 3(2y + 5)$$
$$-2x + 6 = 6y + 15$$
$$6y + 15 = -2x + 6$$
$$6y = -2x - 9$$
$$y = \tfrac{-2}{6}x - \tfrac{9}{6}$$
$$y = -\tfrac{1}{3}x - \tfrac{3}{2}$$
$$m = -\tfrac{1}{3}, b = -\tfrac{3}{2} \Rightarrow \left(0, -\tfrac{3}{2}\right)$$

12.
$$4x - y = 12$$
$$-y = -4x + 12$$
$$y = 4x - 12 \Rightarrow m = 4$$
$$y = \tfrac{1}{4}x + 3 \Rightarrow m = \tfrac{1}{4}$$
neither parallel nor perpendicular

13.
$$y = -\tfrac{2}{3}x + 4 \Rightarrow m = -\tfrac{2}{3}$$
$$2y = 3x - 3$$
$$y = \tfrac{3}{2}x - \tfrac{3}{2} \Rightarrow m = \tfrac{3}{2}$$
perpendicular

14. $y = \tfrac{3}{2}x - 7 \Rightarrow m = \tfrac{3}{2}$

Use the parallel slope:
$$y - y_1 = m(x - x_1)$$
$$y - 0 = \tfrac{3}{2}(x - 0)$$
$$y = \tfrac{3}{2}x$$

15. $y = -\tfrac{2}{3}x - 7 \Rightarrow m = -\tfrac{2}{3}$

Use the perpendicular slope:
$$y - y_1 = m(x - x_1)$$
$$y - 6 = \tfrac{3}{2}(x + 3)$$
$$y - 6 = \tfrac{3}{2}x + \tfrac{9}{2}$$
$$y = \tfrac{3}{2}x + \tfrac{21}{2}$$

16. $|y| = x$ **is not a function,** since $x = 2$ corresponds to both $y = 2$ and $y = -2$.

17. Domain: $(-\infty, \infty)$; Range: $[0, \infty)$

18. Domain: $(-\infty, \infty)$; Range: $(-\infty, \infty)$

19.
$$f(x) = 3x + 1$$
$$f(3) = 3(3) + 1 = 9 + 1 = 10$$

20.
$$g(x) = x^2 - 2$$
$$g(0) = 0^2 - 2 = 0 - 2 = -2$$

21.
$$f(x) = 3x + 1$$
$$f(a) = 3a + 1$$

22.
$$g(x) = x^2 - 2$$
$$g(-x) = (-x)^2 - 2 = x^2 - 2$$

23. function

24. not a function

25. $f(x) = x^2 - 1$: Shift $f(x) = x^2$ D 1.

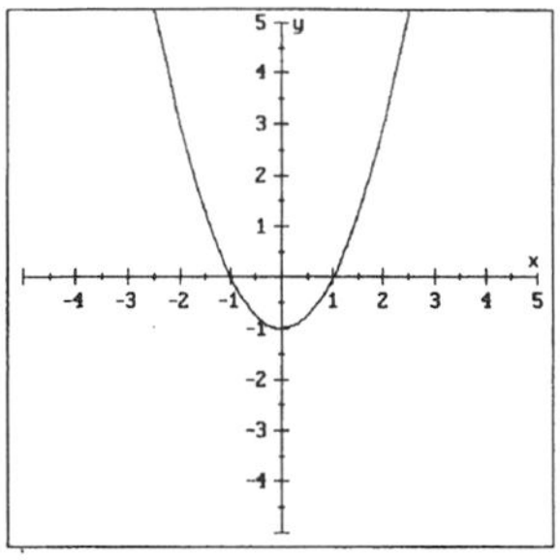

26. $f(x) = -|x + 2|$: Reflect $f(x) = |x|$ about the x-axis and shift L 2.

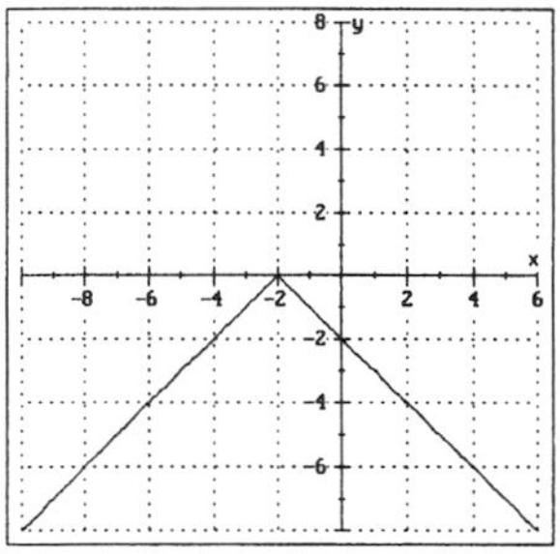

27.

$$\frac{3}{x-2} = \frac{x+3}{2x}$$
$$6x = (x+3)(x-2)$$
$$6x = x^2 + x - 6$$
$$0 = x^2 - 5x - 6$$
$$0 = (x-6)(x+1)$$
$$x - 6 = 0 \quad \textbf{or} \quad x + 1 = 0$$
$$x = 6 \qquad\qquad x = -1$$

28.

$$y = kx \qquad y = \tfrac{2}{15}x$$
$$4 = k(30) \qquad y = \tfrac{2}{15}(9)$$
$$\tfrac{4}{30} = k \qquad y = \tfrac{18}{15}$$
$$\tfrac{2}{15} = k \qquad \boxed{y = \tfrac{6}{5}}$$

29.

$$V = \frac{k}{t} \qquad V = \frac{1100}{t}$$
$$55 = \frac{k}{20} \qquad 75 = \frac{1100}{t}$$
$$1100 = k \qquad 75t = 1100$$
$$\boxed{t = \tfrac{44}{3}}$$

30. function

Cumulative Review Exercises (page 575)

1. natural numbers: $1, 2, 6, 7$

2. whole numbers: $0, 1, 2, 6, 7$

3. rational numbers: $-2, 0, 1, 2, \frac{13}{12}, 6, 7$

4. irrational numbers: $\sqrt{5}, \pi$

5. negative numbers: -2

6. real numbers: $-2, 0, 1, 2, \frac{13}{12}, 6, 7, \sqrt{5}, \pi$

7. prime numbers: $2, 7$

8. composite numbers: 6

9. even numbers: $-2, 0, 2, 6$

10. odd numbers: $1, 7$

11. $-2 < x \le 5 \Rightarrow$ 

CUMULATIVE REVIEW EXERCISES

12. $[-5, 0) \cup [3, 6] \Rightarrow$ 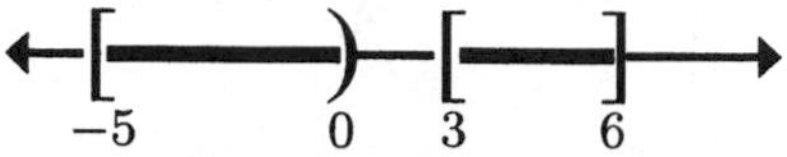

13. $-|5| + |-3| = -5 + 3 = -2$

14. $\dfrac{|-5| + |-3|}{-|4|} = \dfrac{5+3}{-4} = \dfrac{8}{-4} = -2$

15. $2 + 4 \cdot 5 = 2 + 20 = 22$

16. $\dfrac{8-4}{2-4} = \dfrac{4}{-2} = -2$

17. $20 \div (-10 \div 2) = 20 \div (-5) = -4$

18. $\dfrac{6 + 3(6+4)}{2(3-9)} = \dfrac{6 + 3(10)}{2(-6)} = \dfrac{6+30}{-12} = \dfrac{36}{-12} = -3$

19. $-x - 2y = -2 - 2(-3) = -2 + 6 = 4$

20. $\dfrac{x^2 - y^2}{2x + y} = \dfrac{2^2 - (-3)^2}{2(2) + (-3)} = \dfrac{4-9}{4-3} = \dfrac{-5}{1} = -5$

21. associative property of addition

22. distributive property

23. commutative property of addition

24. associative property of multiplication

25. $(x^2y^3)^4 = (x^2)^4(y^3)^4 = x^8y^{12}$

26. $\dfrac{c^4c^8}{(c^5)^2} = \dfrac{c^{12}}{c^{10}} = c^2$

27. $\left(-\dfrac{a^3b^{-2}}{ab}\right)^{-1} = \left(-\dfrac{a^2}{b^3}\right)^{-1} = -\dfrac{b^3}{a^2}$

28. $\left(\dfrac{-3a^3b^{-2}}{6a^{-2}b^3}\right)^0 = 1$

29. $0.00000497 = 4.97 \times 10^{-6}$

30. $9.32 \times 10^8 = 932{,}000{,}000$

Exercise 9.1 (page 587)

1. $\dfrac{x^2+7x+12}{x^2-16} = \dfrac{(x+4)(x+3)}{(x+4)(x-4)} = \dfrac{x+3}{x-4}$

3. $\dfrac{x^2-x-6}{x^2-2x-3} \cdot \dfrac{x^2-1}{x^2+x-2} = \dfrac{(x-3)(x+2)}{(x-3)(x+1)} \cdot \dfrac{(x+1)(x-1)}{(x+2)(x-1)} = 1$

5. $\dfrac{3}{m+1} + \dfrac{3m}{m-1} = \dfrac{3(m-1)}{(m+1)(m-1)} + \dfrac{3m(m+1)}{(m+1)(m-1)} = \dfrac{3m^2+6m-3}{(m+1)(m-1)} = \dfrac{3(m^2+2m-1)}{(m+1)(m-1)}$

7. $(5x^2)^2$

9. positive

11. 3; up

13. x

15. odd

17. 0

19. $3x^2$

21. a^2+b^3

23. $\sqrt{121} = \sqrt{11^2} = 11$

25. $-\sqrt{64} = -\sqrt{8^2} = -8$

27. $\sqrt{\frac{1}{9}} = \sqrt{\left(\frac{1}{3}\right)^2} = \frac{1}{3}$

29. $-\sqrt{\frac{25}{49}} = -\sqrt{\left(\frac{5}{7}\right)^2} = -\frac{5}{7}$

31. $\sqrt{-25}$: not a real number

33. $\sqrt{0.16} = 0.4$

35. $\sqrt{(-4)^2} = \sqrt{16} = 4$

37. $\sqrt{-36}$: not a real number

39. $\sqrt{12} \approx 3.4641$

41. $\sqrt{679.25} \approx 26.0624$

43. $\sqrt{4x^2} = \sqrt{(2x)^2} = |2x| = 2|x|$

45. $\sqrt{9a^4} = \sqrt{(3a^2)^2} = |3a^2| = 3a^2$

47. $\sqrt{(t+5)^2} = |t+5|$

49. $\sqrt{(-5b)^2} = |-5b| = 5|b|$

51. $\sqrt{a^2+6a+9} = \sqrt{(a+3)^2} = |a+3|$

53. $\sqrt{t^2+24t+144} = \sqrt{(t+12)^2} = |t+12|$

55. $\sqrt[3]{1} = \sqrt[3]{1^3} = 1$

57. $\sqrt[3]{-125} = \sqrt[3]{(-5)^3} = -5$

59. $\sqrt[3]{-\frac{8}{27}} = \sqrt[3]{\left(-\frac{2}{3}\right)^3} = -\frac{2}{3}$

61. $\sqrt[3]{0.064} = 0.4$

63. $\sqrt[3]{8a^3} = \sqrt[3]{(2a)^3} = 2a$

65. $\sqrt[3]{-1000p^3q^3} = \sqrt[3]{(-10pq)^3} = -10pq$

67. $\sqrt[3]{-\frac{1}{8}m^6n^3} = \sqrt[3]{\left(-\frac{1}{2}m^2n\right)^3} = -\frac{1}{2}m^2n$

69. $\sqrt[3]{0.008z^9} = \sqrt[3]{(0.2z^3)^3} = 0.2z^3$

71. $\sqrt[4]{81} = \sqrt[4]{3^4} = 3$

73. $-\sqrt[5]{243} = -\sqrt[5]{3^5} = -3$

75. $\sqrt[5]{-32} = \sqrt[5]{(-2)^5} = -2$

77. $\sqrt[4]{\frac{16}{625}} = \sqrt[4]{\left(\frac{2}{5}\right)^4} = \frac{2}{5}$

79. $-\sqrt[5]{-\frac{1}{32}} = -\sqrt[5]{\left(-\frac{1}{2}\right)^5} = -\left(-\frac{1}{2}\right) = \frac{1}{2}$

81. $\sqrt[4]{-256} \Rightarrow$ not a real number

83. $\sqrt[4]{16x^4} = \sqrt[4]{(2x)^4} = |2x| = 2|x|$

85. $\sqrt[3]{8a^3} = \sqrt[3]{(2a)^3} = 2a$

87. $\sqrt[4]{\frac{1}{16}x^4} = \sqrt[4]{\left(\frac{1}{2}x\right)^4} = \left|\frac{1}{2}x\right| = \frac{1}{2}|x|$

89. $\sqrt[4]{x^{12}} = \sqrt[4]{(x^3)^4} = |x^3|$

91. $\sqrt[5]{-x^5} = \sqrt[5]{(-x)^5} = -x$

93. $\sqrt[3]{-27a^6} = \sqrt[3]{(-3a^2)^3} = -3a^2$

95. $\sqrt[25]{(x+2)^{25}} = x+2$

97. $\sqrt[8]{0.00000001x^{16}y^8} = \sqrt[8]{(0.1x^2y)^8} = |0.1x^2y| = 0.1x^2|y|$

99. $f(4) = \sqrt{4-4} = \sqrt{0} = 0$

101. $f(20) = \sqrt{20-4} = \sqrt{16} = 4$

103. $g(9) = \sqrt{9-8} = \sqrt{1} = 1$

105. $g(8.25) = \sqrt{8.25-8} = \sqrt{0.25} = 0.5$

107. $f(4) = \sqrt{4^2+1} = \sqrt{17} \approx 4.1231$

109. $f(2.35) = \sqrt{(2.35)^2+1} = \sqrt{6.5225} \approx 2.5539$

111. $f(x) = \sqrt{x+4}$; Shift $y = \sqrt{x}$ left 4.

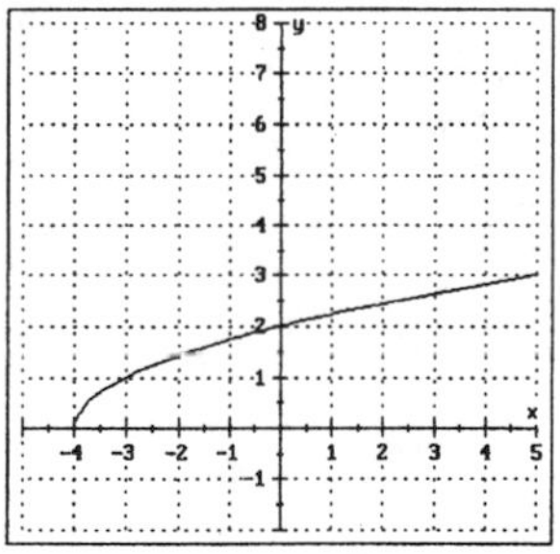

$D = [-4, \infty)$; $R = [0, \infty)$

113. $f(x) = -\sqrt{x} - 3$; Reflect $y = \sqrt{x}$ about the x-axis and shift down 3.

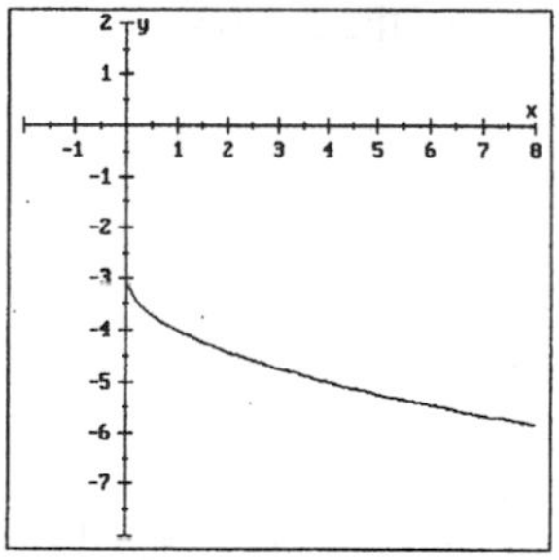

$D = [0, \infty)$; $R = (-\infty, -3]$

115. mean $= \dfrac{2+5+5+6+7}{5} = \dfrac{25}{5} = 5$

Original term	Mean	Difference (term−mean)	Square of difference
2	5	−3	9
5	5	0	0
5	5	0	0
6	5	1	1
7	5	2	4

st. dev. $= \sqrt{\dfrac{9+0+0+1+4}{5}} \approx 1.67$

117. $s_{\bar{x}} = \dfrac{s}{\sqrt{N}} = \dfrac{65}{\sqrt{30}} \approx 11.8673$

119. $r = \sqrt{\frac{A}{\pi}} = \sqrt{\frac{9\pi}{\pi}} = \sqrt{9} = 3$ units

121. $t = \frac{\sqrt{s}}{4} = \frac{\sqrt{256}}{4} = \frac{16}{4} = 4$ seconds

123. $I = \sqrt{\frac{P}{18}} = \sqrt{\frac{980}{18}} \approx \sqrt{54.44} \approx 7.4$ amps

125. **Answers may vary.**

127. $\sqrt{x^2 - 4x + 4} = \sqrt{(x-2)^2} = |x-2|$.
$|x-2| = x - 2$ when $x - 2 \geq 0$, or $x \geq 2$.

Exercise 9.2 (page 594)

1. $(4x+2)(3x-5) = 12x^2 - 14x - 10$

3. $(5t+4s)(3t-2s) = 15t^2 + 2ts - 8s^2$

5. hypotenuse

7. $a^2 + b^2 = c^2$

9. distance

11.
$$\begin{aligned} c^2 &= a^2 + b^2 \\ c^2 &= 6^2 + 8^2 \\ c^2 &= 36 + 64 \\ c^2 &= 100 \\ c &= \sqrt{100} \\ c &= 10 \text{ ft} \end{aligned}$$

13.
$$\begin{aligned} c^2 &= a^2 + b^2 \\ 82^2 &= a^2 + 18^2 \\ 6724 &= a^2 + 324 \\ 6400 &= a^2 \\ \sqrt{6400} &= a \\ 80 \text{ m} &= a \end{aligned}$$

15.
$$\begin{aligned} c^2 &= a^2 + b^2 \\ 50^2 &= 14^2 + b^2 \\ 2500 &= 196 + b^2 \\ 2304 &= b^2 \\ \sqrt{2304} &= b \\ 48 \text{ in.} &= b \end{aligned}$$

17. Let $x =$ the length of the diagonal.
$$\begin{aligned} 7^2 + 7^2 &= x^2 \\ 49 + 49 &= x^2 \\ 98 &= x^2 \\ \sqrt{98} &= x \\ 9.9 \text{ cm} &\approx x \end{aligned}$$

19. $d = \sqrt{(x_2 - x_1)^2 + (y_2 - y_1)^2} = \sqrt{(0-3)^2 + [0-(-4)]^2} = \sqrt{(-3)^2 + (4)^2} = \sqrt{9+16}$
$= \sqrt{25} = 5$

21. $d = \sqrt{(x_2 - x_1)^2 + (y_2 - y_1)^2} = \sqrt{(2-5)^2 + (4-8)^2} = \sqrt{(-3)^2 + (-4)^2} = \sqrt{9+16}$
$= \sqrt{25} = 5$

23. $d = \sqrt{(x_2 - x_1)^2 + (y_2 - y_1)^2} = \sqrt{(-2-3)^2 + (-8-4)^2} = \sqrt{(-5)^2 + (-12)^2} = \sqrt{25+144}$
$= \sqrt{169} = 13$

25. $d = \sqrt{(x_2 - x_1)^2 + (y_2 - y_1)^2} = \sqrt{(6-12)^2 + (8-16)^2} = \sqrt{(-6)^2 + (-8)^2} = \sqrt{36+64}$
$= \sqrt{100} = 10$

27. $d = \sqrt{(x_2 - x_1)^2 + (y_2 - y_1)^2} = \sqrt{[-3 - (-5)]^2 + [5 - (-5)]^2} = \sqrt{2^2 + 10^2} = \sqrt{4 + 100}$
$= \sqrt{104} \approx 10.2$

29. Let the points be represented by $A(5,1)$, $B(7,0)$ and $C(3,0)$. Find the length of $\overline{AB}$ and $\overline{AC}$:
$\overline{AB}: \sqrt{(5-7)^2 + (1-0)^2} = \sqrt{(-2)^2 + 1^2} = \sqrt{5}$
$\overline{AC}: \sqrt{(5-3)^2 + (1-0)^2} = \sqrt{(2)^2 + 1^2} = \sqrt{5}$
Since $\overline{AB}$ and $\overline{AC}$ have the same length, $(5,1)$ is equidistant from $(7,0)$ and $(3,0)$.

31. Let the points be represented by $A(-2,4)$, $B(2,8)$ and $C(6,4)$. Find the length of each side:
$\overline{AB}: \sqrt{(-2-2)^2 + (4-8)^2} = \sqrt{(-4)^2 + (-4)^2} = \sqrt{32}$
$\overline{AC}: \sqrt{(-2-6)^2 + (4-4)^2} = \sqrt{(-8)^2 + 0^2} = \sqrt{64}$
$\overline{BC}: \sqrt{(2-6)^2 + (8-4)^2} = \sqrt{(-4)^2 + 4^2} = \sqrt{32}$
Since $\overline{AB}$ and $\overline{BC}$ have the same length, the triangle is isosceles.

33. $d^2 = 5^2 + 12^2$
$d^2 = 25 + 144$
$d^2 = 169$
$d = 13$ ft

35. Let $x =$ distance to 2nd
$x^2 = 90^2 + 90^2$
$x^2 = 8100 + 8100$
$x^2 = 16200$
$x \approx 127$ ft

37. Refer to the diagram provided. The 3rd baseman is at B, so $\overline{BC}$ has a length of 10 ft. Let $\overline{AC}$ and $\overline{AB}$ both have a length x.

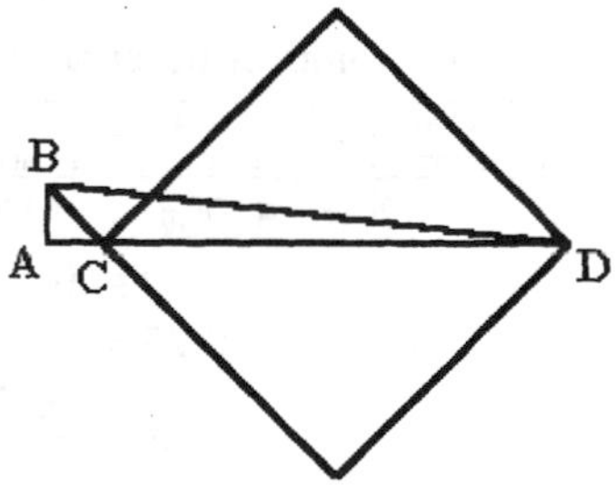

$x^2 + x^2 = 10^2$
$2x^2 = 100$
$x^2 = 50$
$x = \sqrt{50} \approx 7.1$ ft

From **#35**, the length of $\overline{CD} \approx 127.3$ ft, so $\overline{AD}$ has a length of about $127.3 + 7.1 = 134.4$ ft. Let $y =$ the length of $\overline{BD}$.
$y^2 = (7.1)^2 + (134.4)^2$
$y^2 = 18113.77$
$y = \sqrt{18113.77} \approx 135$ ft

39. $d = \sqrt{a^2 + b^2 + c^2} = \sqrt{12^2 + 24^2 + 17^2} = \sqrt{1009} = 31.76 \Rightarrow$ The racket will not fit.

41. $d = \sqrt{a^2 + b^2 + c^2} = \sqrt{21^2 + 21^2 + 21^2} = \sqrt{1323} = 36.4 \Rightarrow$ The femur will fit.

43. Let $x =$ direct distance from A to D.
$x^2 = 52^2 + (105+60)^2$
$x^2 = 52^2 + 165^2$
$x^2 = 29929$
$x = 173$ yd

45. Let $x =$ half the length of the stretched wire.
$x^2 = 20^2 + 1^2$
$x^2 = 401$
$x = 20.025$ ft
The stretched wire has a length of 40.05 ft.
It has been stretched by 0.05 ft.

47. $A = 6\sqrt[3]{V^2} = 6\sqrt[3]{8^2} = 6\sqrt[3]{64}$
$= 6(4) = 24 \text{ cm}^2$

49. **Answers may vary.**

51. $I = \frac{703w}{h^2} = \frac{703(104)}{(54.1)^2} = \frac{73112}{2926.81} \approx 25$

Exercise 9.3 (page 604)

1.
$5x - 4 < 11$
$5x < 15$
$x < 3$

3.
$\frac{4}{5}(r-3) > \frac{2}{3}(r+2)$
$15 \cdot \frac{4}{5}(r-3) > 15 \cdot \frac{2}{3}(r+2)$
$12(r-3) > 10(r+2)$
$2r > 56$
$r > 28$

5. Let $x =$ pints of water added (0% alcohol).
[Alcohol at start] + [Alcohol added] = [Alcohol at end]
$0.20(5) + 0(x) = 0.15(5+x)$
$1 + 0 = 0.75 + 0.15x$
$0.25 = 0.15x$
$x = \frac{0.25}{0.15} = \frac{5}{3} \Rightarrow 1\frac{2}{3}$ pints of water should be added.

7. $a \cdot a \cdot a \cdot a$

9. a^{mn}

11. $\frac{a^n}{b^n}$

13. $\frac{1}{a^n}$; 0

15. $\left(\frac{b}{a}\right)^n$

17. $|x|$

19. $7^{1/3} = \sqrt[3]{7}$

21. $8^{1/5} = \sqrt[5]{8}$

23. $(3x)^{1/4} = \sqrt[4]{3x}$

25. $\left(\frac{1}{2}x^3y\right)^{1/4} = \sqrt[4]{\frac{1}{2}x^3y}$

27. $(4a^2b^3)^{1/5} = \sqrt[5]{4a^2b^3}$

29. $(x^2+y^2)^{1/2} = \sqrt{x^2+y^2}$

31. $\sqrt{11} = 11^{1/2}$

33. $\sqrt[4]{3a} = (3a)^{1/4}$

35. $3\sqrt[5]{a} = 3a^{1/5}$

37. $\sqrt[6]{\frac{1}{7}abc} = \left(\frac{1}{7}abc\right)^{1/6}$

39. $\sqrt[5]{\frac{1}{2}mn} = \left(\frac{1}{2}mn\right)^{1/5}$

41. $\sqrt[3]{a^2-b^2}=(a^2-b^2)^{1/3}$

43. $4^{1/2}=\sqrt{4}=2$

45. $27^{1/3}=\sqrt[3]{27}=3$

47. $16^{1/4}=\sqrt[4]{16}=2$

49. $32^{1/5}=\sqrt[5]{32}=2$

51. $\left(\frac{1}{4}\right)^{1/2}=\sqrt{\frac{1}{4}}=\frac{1}{2}$

53. $\left(\frac{1}{8}\right)^{1/3}=\sqrt[3]{\frac{1}{8}}=\frac{1}{2}$

55. $-16^{1/4}=-\sqrt[4]{16}=-2$

57. $(-27)^{1/3}=\sqrt[3]{-27}=-3$

59. $(-64)^{1/2}=\sqrt{-64}$; not a real number

61. $0^{1/3}=\sqrt[3]{0}=0$r

63. $(25y^2)^{1/2}=\left[(5y)^2\right]^{1/2}=|5y|=5|y|$

65. $(16x^4)^{1/4}=\left[(2x)^4\right]^{1/4}=|2x|=2|x|$

67. $(243x^5)^{1/5}=\left[(3x)^5\right]^{1/5}=3x$

69. $(-64x^8)^{1/4}\Rightarrow$ not a real number

71. $36^{3/2}=\left(36^{1/2}\right)^3=6^3=216$

73. $81^{3/4}=\left(81^{1/4}\right)^3=3^3=27$

75. $144^{3/2}=\left(144^{1/2}\right)^3=12^3=1728$

77. $\left(\frac{1}{8}\right)^{2/3}=\left[\left(\frac{1}{8}\right)^{1/3}\right]^2=\left(\frac{1}{2}\right)^2=\frac{1}{4}$

79. $\left(25x^4\right)^{3/2}=\left[\left(25x^4\right)^{1/2}\right]^3=\left(5x^2\right)^3=125x^6$

81. $\left(\frac{8x^3}{27}\right)^{2/3}=\left[\left(\frac{8x^3}{27}\right)^{1/3}\right]^2=\left(\frac{2x}{3}\right)^2=\frac{4x^2}{9}$

Problems 83-85 are to be solved using a calculator. The keystrokes needed to solve each problem using a TI-83 graphing calculator appear in each solution. There may be other solutions. Keystrokes for other calculators may be slightly different.

83. [1] [5] [^] [(] [1] [÷] [3] [)] [ENTER] {2.4662...} $\Rightarrow$ 2.47

85. [1] [.] [0] [4] [5] [^] [(] [1] [÷] [5] [)] [ENTER] {1.0088...} $\Rightarrow$ 1.01

87. $4^{-1/2}=\frac{1}{4^{1/2}}=\frac{1}{2}$

89. $4^{-3/2}=\frac{1}{4^{3/2}}=\frac{1}{\left(4^{1/2}\right)^3}=\frac{1}{2^3}=\frac{1}{8}$

91. $(16x^2)^{-3/2}=\frac{1}{(16x^2)^{3/2}}=\frac{1}{\left[(16x^2)^{1/2}\right]^3}=\frac{1}{(4x)^3}=\frac{1}{64x^3}$

93. $\left(-27y^3\right)^{-2/3}=\frac{1}{(-27y^3)^{2/3}}=\frac{1}{\left[(-27y^3)^{1/3}\right]^2}=\frac{1}{(-3y)^2}=\frac{1}{9y^2}$

95. $\left(-32p^5\right)^{-2/5}=\frac{1}{(-32p^5)^{2/5}}=\frac{1}{\left[(-32p^5)^{1/5}\right]^2}=\frac{1}{(-2p)^2}=\frac{1}{4p^2}$

97. $\left(\frac{1}{4}\right)^{-3/2} = \left(\frac{4}{1}\right)^{3/2} = 4^{3/2} = \left(4^{1/2}\right)^3 = 2^3 = 8$

99. $\left(\frac{27}{8}\right)^{-4/3} = \left(\frac{8}{27}\right)^{4/3} = \left[\left(\frac{8}{27}\right)^{1/3}\right]^4 = \left(\frac{2}{3}\right)^4 = \frac{16}{81}$

101. $\left(-\frac{8x^3}{27}\right)^{-1/3} = \left(-\frac{27}{8x^3}\right)^{1/3} = -\frac{3}{2x}$

Problems 103-105 are to be solved using a calculator. The keystrokes needed to solve each problem using a TI-83 graphing calculator appear in each solution. There may be other solutions. Keystrokes for other calculators may be slightly different.

103. [1] [7] [^] [(] [(-)] [1] [÷] [2] [)] [ENTER] $\{0.2425...\} \Rightarrow 0.24$

105. [(] [(-)] [.] [2] [5] [)] [^] [(] [(-)] [1] [÷] [5] [)] [ENTER] $\{-1.3195...\} \Rightarrow -1.32$

107. $5^{4/9}5^{4/9} = 5^{4/9+4/9} = 5^{8/9}$

109. $\left(4^{1/5}\right)^3 = 4^{(1/5)\cdot 3} = 4^{3/5}$

111. $\frac{9^{4/5}}{9^{3/5}} = 9^{4/5-3/5} = 9^{1/5}$

113. $\frac{7^{1/2}}{7^0} = 7^{1/2-0} = 7^{1/2}$

115. $6^{-2/3}6^{-4/3} = 6^{-6/3} = 6^{-2} = \frac{1}{36}$

117. $\frac{2^{5/6}2^{1/3}}{2^{1/2}} = \frac{2^{7/6}}{2^{1/2}} = 2^{4/6} = 2^{2/3}$

119. $a^{2/3}a^{1/3} = a^{3/3} = a^1 = a$

121. $\left(a^{2/3}\right)^{1/3} = a^{(2/3)(1/3)} = a^{2/9}$

123. $\left(a^{1/2}b^{1/3}\right)^{3/2} = a^{3/4}b^{1/2}$

125. $\left(mn^{-2/3}\right)^{-3/5} = m^{-3/5}n^{2/5} = \frac{n^{2/5}}{m^{3/5}}$

127. $\frac{\left(4x^3y\right)^{1/2}}{(9xy)^{1/2}} = \frac{2x^{3/2}y^{1/2}}{3x^{1/2}y^{1/2}} = \frac{2x}{3}$

129. $\left(27x^{-3}\right)^{-1/3} = (27)^{-1/3}x = \frac{1}{3}x$

131. $y^{1/3}\left(y^{2/3} + y^{5/3}\right) = y^{3/3} + y^{6/3} = y + y^2$

133. $x^{3/5}\left(x^{7/5} - x^{2/5} + 1\right) = x^{10/5} - x^{5/5} + x^{3/5} = x^2 - x + x^{3/5}$

135. $\left(x^{1/2} + 2\right)\left(x^{1/2} - 2\right) = x^{2/2} - 2x^{1/2} + 2x^{1/2} - 4 = x - 4$

137. $\left(x^{2/3} - x\right)\left(x^{2/3} + x\right) = x^{4/3} + x^{5/3} - x^{5/3} - x^2 = x^{4/3} - x^2$

139. $\left(x^{2/3} + y^{2/3}\right)^2 = \left(x^{2/3} + y^{2/3}\right)\left(x^{2/3} + y^{2/3}\right) = x^{4/3} + x^{2/3}y^{2/3} + x^{2/3}y^{2/3} + y^{4/3}$
$= x^{4/3} + 2x^{2/3}y^{2/3} + y^{4/3}$

141. $\left(a^{3/2}-b^{3/2}\right)^2=\left(a^{3/2}-b^{3/2}\right)\left(a^{3/2}-b^{3/2}\right)=a^{6/2}-a^{3/2}b^{3/2}-a^{3/2}b^{3/2}+b^{6/2}$
$=a^3-2a^{3/2}b^{3/2}+b^3$

143. $\sqrt[6]{p^3}=\left(p^3\right)^{1/6}=p^{3/6}=p^{1/2}=\sqrt{p}$

145. $\sqrt[4]{25b^2}=\left(5^2b^2\right)^{1/4}=5^{1/2}b^{1/2}=\sqrt{5b}$

147. Answers may vary.

149. $16^{2/4}=2^2=4; 16^{1/2}=4$

Exercise 9.4 (page 613)

1. $3x^2y^3(-5x^3y^{-4})=-15x^5y^{-1}=\dfrac{-15x^5}{y}$

3. $(3t+2)^2=(3t+2)(3t+2)$
$=9t^2+12t+4$

5.
$$\begin{array}{r|l} & 3p+\quad 4+\frac{-5}{2p-5} \\ 2p-5 & 6p^2-\ 7p-\ 25 \\ & \underline{6p^2-15p} \\ & \qquad 8p-\ 25 \\ & \qquad \underline{8p-\ 20} \\ & \qquad\qquad -5 \end{array}$$

7. $\sqrt[n]{a}\sqrt[n]{b}$

9. $\sqrt{6}\sqrt{6}=\sqrt{36}=6$

11. $\sqrt{t}\sqrt{t}=\sqrt{t^2}=t$

13. $\sqrt[3]{5x^2}\sqrt[3]{25x}=\sqrt[3]{125x^3}=5x$

15. $\dfrac{\sqrt{500}}{\sqrt{5}}=\sqrt{\dfrac{500}{5}}=\sqrt{100}=10$

17. $\dfrac{\sqrt{98x^3}}{\sqrt{2x}}=\sqrt{\dfrac{98x^3}{2x}}=\sqrt{49x^2}=7x$

19. $\dfrac{\sqrt{180ab^4}}{\sqrt{5ab^2}}=\sqrt{\dfrac{180ab^4}{5ab^2}}=\sqrt{36b^2}=6b$

21. $\dfrac{\sqrt[3]{48}}{\sqrt[3]{6}}=\sqrt[3]{\dfrac{48}{6}}=\sqrt[3]{8}=2$

23. $\dfrac{\sqrt[3]{189a^4}}{\sqrt[3]{7a}}=\sqrt[3]{\dfrac{189a^4}{7a}}=\sqrt[3]{27a^3}=3a$

25. $\sqrt{20}=\sqrt{4\cdot 5}=\sqrt{4}\sqrt{5}=2\sqrt{5}$

27. $-\sqrt{200}=-\sqrt{100\cdot 2}=-\sqrt{100}\sqrt{2}$
$=-10\sqrt{2}$

29. $\sqrt[3]{80}=\sqrt[3]{8\cdot 10}=\sqrt[3]{8}\sqrt[3]{10}=2\sqrt[3]{10}$

31. $\sqrt[3]{-81}=\sqrt[3]{-27\cdot 3}=\sqrt[3]{-27}\sqrt[3]{3}=-3\sqrt[3]{3}$

33. $\sqrt[4]{32}=\sqrt[4]{16\cdot 2}=\sqrt[4]{16}\sqrt[4]{2}=2\sqrt[4]{2}$

35. $\sqrt[5]{96}=\sqrt[5]{32\cdot 3}=\sqrt[5]{32}\sqrt[5]{3}=2\sqrt[5]{3}$

37. $\sqrt{\dfrac{7}{9}}=\dfrac{\sqrt{7}}{\sqrt{9}}=\dfrac{\sqrt{7}}{3}$

39. $\sqrt[3]{\dfrac{7}{64}}=\dfrac{\sqrt[3]{7}}{\sqrt[3]{64}}=\dfrac{\sqrt[3]{7}}{4}$

41. $\sqrt[4]{\dfrac{3}{10{,}000}}=\dfrac{\sqrt[4]{3}}{\sqrt[4]{10{,}000}}=\dfrac{\sqrt[4]{3}}{10}$

SECTION 9.4

43. $\sqrt[5]{\frac{3}{32}} = \frac{\sqrt[5]{3}}{\sqrt[5]{32}} = \frac{\sqrt[5]{3}}{2}$

45. $\sqrt{50x^2} = \sqrt{25x^2 \cdot 2} = \sqrt{25x^2}\sqrt{2} = 5x\sqrt{2}$

47. $\sqrt{32b} = \sqrt{16 \cdot 2b} = \sqrt{16}\sqrt{2b} = 4\sqrt{2b}$

49. $-\sqrt{112a^3} = -\sqrt{16a^2 \cdot 7a} = -\sqrt{16a^2}\sqrt{7a} = -4a\sqrt{7a}$

51. $\sqrt{175a^2b^3} = \sqrt{25a^2b^2 \cdot 7b} = \sqrt{25a^2b^2}\sqrt{7b} = 5ab\sqrt{7b}$

53. $-\sqrt{300xy} = -\sqrt{100 \cdot 3xy} = -\sqrt{100}\sqrt{3xy} = -10\sqrt{3xy}$

55. $\sqrt[3]{-54x^6} = \sqrt[3]{-27x^6 \cdot 2} = \sqrt[3]{-27x^6}\sqrt[3]{2}$
$= -3x^2\sqrt[3]{2}$

57. $\sqrt[3]{16x^{12}y^3} = \sqrt[3]{8x^{12}y^3 \cdot 2} = \sqrt[3]{8x^{12}y^3}\sqrt[3]{2}$
$= 2x^4y\sqrt[3]{2}$

59. $\sqrt[4]{32x^{12}y^4} = \sqrt[4]{16x^{12}y^4 \cdot 2} = \sqrt[4]{16x^{12}y^4}\sqrt[4]{2}$
$= 2x^3y\sqrt[4]{2}$

61. $\sqrt{\frac{z^2}{16x^2}} = \frac{\sqrt{z^2}}{\sqrt{16x^2}} = \frac{z}{4x}$

63. $\sqrt[4]{\frac{5x}{16z^4}} = \frac{\sqrt[4]{5x}}{\sqrt[4]{16z^4}} = \frac{\sqrt[4]{5x}}{2z}$

65. $4\sqrt{2x} + 6\sqrt{2x} = 10\sqrt{2x}$

67. $8\sqrt[5]{7a^2} - 7\sqrt[5]{7a^2} = \sqrt[5]{7a^2}$

69. $\sqrt{3} + \sqrt{27} = \sqrt{3} + \sqrt{9}\sqrt{3}$
$= \sqrt{3} + 3\sqrt{3} = 4\sqrt{3}$

71. $\sqrt{2} - \sqrt{8} = \sqrt{2} - \sqrt{4}\sqrt{2} = \sqrt{2} - 2\sqrt{2} = -\sqrt{2}$

73. $\sqrt{98} - \sqrt{50} = \sqrt{49}\sqrt{2} - \sqrt{25}\sqrt{2} = 7\sqrt{2} - 5\sqrt{2} = 2\sqrt{2}$

75. $3\sqrt{24} + \sqrt{54} = 3\sqrt{4}\sqrt{6} + \sqrt{9}\sqrt{6} = 3(2)\sqrt{6} + 3\sqrt{6} = 6\sqrt{6} + 3\sqrt{6} = 9\sqrt{6}$

77. $\sqrt[3]{24} + \sqrt[3]{3} = \sqrt[3]{8}\sqrt[3]{3} + \sqrt[3]{3} = 2\sqrt[3]{3} + \sqrt[3]{3} = 3\sqrt[3]{3}$

79. $\sqrt[3]{32} - \sqrt[3]{108} = \sqrt[3]{8}\sqrt[3]{4} - \sqrt[3]{27}\sqrt[3]{4} = 2\sqrt[3]{4} - 3\sqrt[3]{4} = -\sqrt[3]{4}$

81. $2\sqrt[3]{125} - 5\sqrt[3]{64} = 2(5) - 5(4) = 10 - 20 = -10$

83. $14\sqrt[4]{32} - 15\sqrt[4]{162} = 14\sqrt[4]{16}\sqrt[4]{2} - 15\sqrt[4]{81}\sqrt[4]{2} = 14(2)\sqrt[4]{2} - 15(3)\sqrt[4]{2} = 28\sqrt[4]{2} - 45\sqrt[4]{2}$
$= -17\sqrt[4]{2}$

85. $3\sqrt[4]{512} + 2\sqrt[4]{32} = 3\sqrt[4]{256}\sqrt[4]{2} + 2\sqrt[4]{16}\sqrt[4]{2} = 3(4)\sqrt[4]{2} + 2(2)\sqrt[4]{2} = 12\sqrt[4]{2} + 4\sqrt[4]{2} = 16\sqrt[4]{2}$

87. $\sqrt{98} - \sqrt{50} - \sqrt{72} = \sqrt{49}\sqrt{2} - \sqrt{25}\sqrt{2} - \sqrt{36}\sqrt{2} = 7\sqrt{2} - 5\sqrt{2} - 6\sqrt{2} = -4\sqrt{2}$

89. $\sqrt{18}+\sqrt{300}-\sqrt{243}=\sqrt{9}\sqrt{2}+\sqrt{100}\sqrt{3}-\sqrt{81}\sqrt{3}=3\sqrt{2}+10\sqrt{3}-9\sqrt{3}=3\sqrt{2}+\sqrt{3}$

91. $2\sqrt[3]{16}-\sqrt[3]{54}-3\sqrt[3]{128}=2\sqrt[3]{8}\sqrt[3]{2}-\sqrt[3]{27}\sqrt[3]{2}-3\sqrt[3]{64}\sqrt[3]{2}=2(2)\sqrt[3]{2}-3\sqrt[3]{2}-3(4)\sqrt[3]{2}$
$=4\sqrt[3]{2}-3\sqrt[3]{2}-12\sqrt[3]{2}=-11\sqrt[3]{2}$

93. $\sqrt{25y^2z}-\sqrt{16y^2z}=\sqrt{25y^2}\sqrt{z}-\sqrt{16y^2}\sqrt{z}=5y\sqrt{z}-4y\sqrt{z}=y\sqrt{z}$

95. $\sqrt{36xy^2}+\sqrt{49xy^2}=\sqrt{36y^2}\sqrt{x}+\sqrt{49y^2}\sqrt{x}=6y\sqrt{x}+7y\sqrt{x}=13y\sqrt{x}$

97. $2\sqrt[3]{64a}+2\sqrt[3]{8a}=2\sqrt[3]{64}\sqrt[3]{a}+2\sqrt[3]{8}\sqrt[3]{a}=2(4)\sqrt[3]{a}+2(2)\sqrt[3]{a}=8\sqrt[3]{a}+4\sqrt[3]{a}=12\sqrt[3]{a}$

99. $\sqrt{y^5}-\sqrt{9y^5}-\sqrt{25y^5}=\sqrt{y^4}\sqrt{y}-\sqrt{9y^4}\sqrt{y}-\sqrt{25y^4}\sqrt{y}=y^2\sqrt{y}-3y^2\sqrt{y}-5y^2\sqrt{y}$
$=-7y^2\sqrt{y}$

101. $\sqrt[5]{x^6y^2}+\sqrt[5]{32x^6y^2}+\sqrt[5]{x^6y^2}=\sqrt[5]{x^5}\sqrt[5]{xy^2}+\sqrt[5]{32x^5}\sqrt[5]{xy^2}+\sqrt[5]{x^5}\sqrt[5]{xy^2}$
$=x\sqrt[5]{xy^2}+2x\sqrt[5]{xy^2}+x\sqrt[5]{xy^2}=4x\sqrt[5]{xy^2}$

103. $\sqrt{x^2+2x+1}+\sqrt{x^2+2x+1}=\sqrt{(x+1)^2}+\sqrt{(x+1)^2}=x+1+x+1=2x+2$

105. $x=2.00$;
$h=2\sqrt{2}\approx 2.83$

107. $h=2(5)=10.00$;
$x=5\sqrt{3}\approx 8.66$

109. $x=\frac{9.37}{2}\approx 4.69$;
$y\approx 4.69\sqrt{3}\approx 8.11$

111. $x=y=\frac{17.12}{\sqrt{2}}\approx 12.11$

113. $x=5\sqrt{3}\approx 8.66$ mm
$h=2x=2\left(5\sqrt{3}\right)=10\sqrt{3}\approx 17.32$ mm

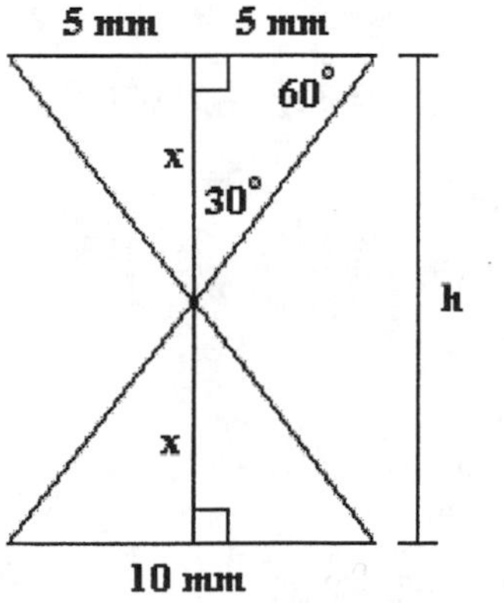

115. Answers may vary.

117. $\sqrt{a+b}=\sqrt{a}+\sqrt{b}$ if either a or b equals 0 and the other is nonnegative.

Exercise 9.5 (page 622)

1. $$\begin{aligned}\frac{2}{3-a} &= 1\\ 2 &= 3-a\\ a &= 1\end{aligned}$$

3. $$\begin{aligned}\frac{8}{b-2}+\frac{3}{2-b} &= -\frac{1}{b}\\ \frac{8}{b-2}+\frac{-3}{b-2} &= -\frac{1}{b}\\ \frac{5}{b-2} &= \frac{-1}{b}\\ 5b &= -b+2\\ 6b &= 2\\ b &= \frac{2}{6}=\frac{1}{3}\end{aligned}$$

5. 2; $\sqrt{7}$; $\sqrt{5}$

7. FOIL

9. conjugate

11. $\sqrt{2}\sqrt{8}=\sqrt{16}=4$

13. $\sqrt{5}\sqrt{10}=\sqrt{50}=\sqrt{25}\sqrt{2}=5\sqrt{2}$

15. $2\sqrt{3}\sqrt{6}=2\sqrt{18}=2\sqrt{9}\sqrt{2}=2(3)\sqrt{2}$
$=6\sqrt{2}$

17. $\sqrt[3]{5}\sqrt[3]{25}=\sqrt[3]{125}=5$

19. $\left(3\sqrt[3]{9}\right)\left(2\sqrt[3]{3}\right)=6\sqrt[3]{27}=6(3)=18$

21. $\sqrt[3]{2}\sqrt[3]{12}=\sqrt[3]{24}=\sqrt[3]{8}\sqrt[3]{3}=2\sqrt[3]{3}$

23. $\sqrt{ab^3}\sqrt{ab}=\sqrt{a^2b^4}=ab^2$

25. $\sqrt{5ab}\sqrt{5a}=\sqrt{25a^2b}=\sqrt{25a^2}\sqrt{b}$
$=5a\sqrt{b}$

27. $\sqrt[3]{5r^2s}\sqrt[3]{2r}=\sqrt[3]{10r^3s}=\sqrt[3]{r^3}\sqrt[3]{10s}$
$=r\sqrt[3]{10s}$

29. $\sqrt[3]{a^5b}\sqrt[3]{16ab^5}=\sqrt[3]{16a^6b^6}=\sqrt[3]{8a^6b^6}\sqrt[3]{2}$
$=2a^2b^2\sqrt[3]{2}$

31. $\sqrt{x(x+3)}\sqrt{x^3(x+3)}=\sqrt{x^4(x+3)^2}=x^2(x+3)$

33. $\sqrt[3]{6x^2(y+z)^2}\sqrt[3]{18x(y+z)}=\sqrt[3]{108x^3(y+z)^3}=\sqrt[3]{27x^3(y+z)^3}\sqrt[3]{4}=3x(y+z)\sqrt[3]{4}$

35. $3\sqrt{5}\left(4-\sqrt{5}\right)=12\sqrt{5}-3\sqrt{25}=12\sqrt{5}-3(5)=12\sqrt{5}-15$

37. $3\sqrt{2}\left(4\sqrt{3}+2\sqrt{7}\right)=12\sqrt{6}+6\sqrt{14}$

39. $-2\sqrt{5x}\left(4\sqrt{2x}-3\sqrt{3}\right)=-8\sqrt{10x^2}+6\sqrt{15x}=-8x\sqrt{10}+6\sqrt{15x}$

41. $\left(\sqrt{2}+1\right)\left(\sqrt{2}-3\right)=\sqrt{4}-3\sqrt{2}+\sqrt{2}-3=2-2\sqrt{2}-3=-1-2\sqrt{2}$

43. $(4\sqrt{x}+3)(2\sqrt{x}-5)=8\sqrt{x^2}-20\sqrt{x}+6\sqrt{x}-15=8x-14\sqrt{x}-15$

SECTION 9.5

45. $\left(\sqrt{5z}+\sqrt{3}\right)\left(\sqrt{5z}+\sqrt{3}\right)=\sqrt{25z^2}+\sqrt{15z}+\sqrt{15z}+\sqrt{9}=5z+2\sqrt{15z}+3$

47. $\left(\sqrt{3x}-\sqrt{2y}\right)\left(\sqrt{3x}+\sqrt{2y}\right)=\sqrt{9x^2}+\sqrt{6xy}-\sqrt{6xy}-\sqrt{4y^2}=3x-2y$

49. $\left(2\sqrt{3a}-\sqrt{b}\right)\left(\sqrt{3a}+3\sqrt{b}\right)=2\sqrt{9a^2}+6\sqrt{3ab}-\sqrt{3ab}-3\sqrt{b^2}=6a+5\sqrt{3ab}-3b$

51. $\left(3\sqrt{2r}-2\right)^2=\left(3\sqrt{2r}-2\right)\left(3\sqrt{2r}-2\right)=9\sqrt{4r^2}-6\sqrt{2r}-6\sqrt{2r}+4=18r-12\sqrt{2r}+4$

53.
$$\begin{aligned}-2\left(\sqrt{3x}+\sqrt{3}\right)^2&=-2\left(\sqrt{3x}+\sqrt{3}\right)\left(\sqrt{3x}+\sqrt{3}\right)\\&=-2\left(\sqrt{9x^2}+\sqrt{9x}+\sqrt{9x}+\sqrt{9}\right)\\&=-2(3x+3\sqrt{x}+3\sqrt{x}+3)=-2(3x+6\sqrt{x}+3)=-6x-12\sqrt{x}-6\end{aligned}$$

55. $\sqrt{\frac{1}{7}}=\frac{\sqrt{1}}{\sqrt{7}}=\frac{1\sqrt{7}}{\sqrt{7}\sqrt{7}}=\frac{\sqrt{7}}{7}$

57. $\sqrt{\frac{2}{3}}=\frac{\sqrt{2}}{\sqrt{3}}=\frac{\sqrt{2}\sqrt{3}}{\sqrt{3}\sqrt{3}}=\frac{\sqrt{6}}{3}$

59. $\frac{\sqrt{5}}{\sqrt{8}}=\frac{\sqrt{5}\sqrt{2}}{\sqrt{8}\sqrt{2}}=\frac{\sqrt{10}}{\sqrt{16}}=\frac{\sqrt{10}}{4}$

61. $\frac{\sqrt{8}}{\sqrt{2}}=\sqrt{\frac{8}{2}}=\sqrt{4}=2$

63. $\frac{1}{\sqrt[3]{2}}=\frac{1\sqrt[3]{4}}{\sqrt[3]{2}\sqrt[3]{4}}=\frac{\sqrt[3]{4}}{\sqrt[3]{8}}=\frac{\sqrt[3]{4}}{2}$

65. $\frac{3}{\sqrt[3]{9}}=\frac{3\sqrt[3]{3}}{\sqrt[3]{9}\sqrt[3]{3}}=\frac{3\sqrt[3]{3}}{\sqrt[3]{27}}=\frac{3\sqrt[3]{3}}{3}=\sqrt[3]{3}$

67. $\frac{\sqrt[3]{2}}{\sqrt[3]{9}}=\frac{\sqrt[3]{2}\sqrt[3]{3}}{\sqrt[3]{9}\sqrt[3]{3}}=\frac{\sqrt[3]{6}}{\sqrt[3]{27}}=\frac{\sqrt[3]{6}}{3}$

69. $\frac{\sqrt{8x^2y}}{\sqrt{xy}}=\sqrt{\frac{8x^2y}{xy}}=\sqrt{8x}=2\sqrt{2x}$

71. $\frac{\sqrt{10xy^2}}{\sqrt{2xy^3}}=\sqrt{\frac{10xy^2}{2xy^3}}=\sqrt{\frac{5}{y}}=\frac{\sqrt{5}\sqrt{y}}{\sqrt{y}\sqrt{y}}=\frac{\sqrt{5y}}{y}$

73. $\frac{\sqrt[3]{4a^2}}{\sqrt[3]{2ab}}=\sqrt[3]{\frac{4a^2}{2ab}}=\sqrt[3]{\frac{2a}{b}}=\frac{\sqrt[3]{2a}\sqrt[3]{b^2}}{\sqrt[3]{b}\sqrt[3]{b^2}}=\frac{\sqrt[3]{2ab^2}}{b}$

75. $\frac{1}{\sqrt[4]{4}}=\frac{1\sqrt[4]{4}}{\sqrt[4]{4}\sqrt[4]{4}}=\frac{\sqrt[4]{4}}{\sqrt[4]{16}}=\frac{\sqrt[4]{4}}{2}$

77. $\frac{1}{\sqrt[5]{16}}=\frac{1\sqrt[5]{2}}{\sqrt[5]{16}\sqrt[5]{2}}=\frac{\sqrt[5]{2}}{\sqrt[5]{32}}=\frac{\sqrt[5]{2}}{2}$

79. $\frac{1}{\sqrt{2}-1}=\frac{1\left(\sqrt{2}+1\right)}{\left(\sqrt{2}-1\right)\left(\sqrt{2}+1\right)}=\frac{\sqrt{2}+1}{\sqrt{4}-1}=\frac{\sqrt{2}+1}{2-1}=\frac{\sqrt{2}+1}{1}=\sqrt{2}+1$

81. $$\frac{\sqrt{2}}{\sqrt{5}+3}=\frac{\sqrt{2}\left(\sqrt{5}-3\right)}{\left(\sqrt{5}+3\right)\left(\sqrt{5}-3\right)}=\frac{\sqrt{2}\left(\sqrt{5}-3\right)}{\sqrt{25}-9}=\frac{\sqrt{2}\left(\sqrt{5}-3\right)}{5-9}=\frac{\sqrt{2}\left(\sqrt{5}-3\right)}{-4}$$
$$=\frac{\sqrt{10}-3\sqrt{2}}{-4}$$
$$=\frac{3\sqrt{2}-\sqrt{10}}{4}$$

83. $$\frac{\sqrt{3}+1}{\sqrt{3}-1}=\frac{\left(\sqrt{3}+1\right)\left(\sqrt{3}+1\right)}{\left(\sqrt{3}-1\right)\left(\sqrt{3}+1\right)}=\frac{\sqrt{9}+\sqrt{3}+\sqrt{3}+1}{\sqrt{9}-1}=\frac{4+2\sqrt{3}}{2}=\frac{2\left(2+\sqrt{3}\right)}{2}$$
$$=2+\sqrt{3}$$

85. $$\frac{\sqrt{7}-\sqrt{2}}{\sqrt{2}+\sqrt{7}}=\frac{\left(\sqrt{7}-\sqrt{2}\right)\left(\sqrt{2}-\sqrt{7}\right)}{\left(\sqrt{2}+\sqrt{7}\right)\left(\sqrt{2}-\sqrt{7}\right)}=\frac{\sqrt{14}-7-2+\sqrt{14}}{\sqrt{4}-\sqrt{49}}=\frac{2\sqrt{14}-9}{-5}=\frac{9-2\sqrt{14}}{5}$$

87. $$\frac{2}{\sqrt{x}+1}=\frac{2(\sqrt{x}-1)}{(\sqrt{x}+1)(\sqrt{x}-1)}=\frac{2(\sqrt{x}-1)}{\sqrt{x^2}-1}=\frac{2(\sqrt{x}-1)}{x-1}$$

89. $$\frac{x}{\sqrt{x}-4}=\frac{x(\sqrt{x}+4)}{(\sqrt{x}-4)(\sqrt{x}+4)}=\frac{x(\sqrt{x}+4)}{\sqrt{x^2}-16}=\frac{x(\sqrt{x}+4)}{x-16}$$

91. $$\frac{2z-1}{\sqrt{2z}-1}=\frac{(2z-1)\left(\sqrt{2z}+1\right)}{\left(\sqrt{2z}-1\right)\left(\sqrt{2z}+1\right)}=\frac{(2z-1)\left(\sqrt{2z}+1\right)}{\sqrt{4z^2}-1}=\frac{(2z-1)\left(\sqrt{2z}+1\right)}{2z-1}$$
$$=\sqrt{2z}+1$$

93. $$\frac{\sqrt{x}-\sqrt{y}}{\sqrt{x}+\sqrt{y}}=\frac{(\sqrt{x}-\sqrt{y})(\sqrt{x}-\sqrt{y})}{(\sqrt{x}+\sqrt{y})(\sqrt{x}-\sqrt{y})}=\frac{\sqrt{x^2}-\sqrt{xy}-\sqrt{xy}+\sqrt{y^2}}{\sqrt{x^2}-\sqrt{y^2}}=\frac{x-2\sqrt{xy}+y}{x-y}$$

95. $$\frac{\sqrt{3}+1}{2}=\frac{\left(\sqrt{3}+1\right)\left(\sqrt{3}-1\right)}{2\left(\sqrt{3}-1\right)}=\frac{\sqrt{9}-1}{2\left(\sqrt{3}-1\right)}=\frac{2}{2\left(\sqrt{3}-1\right)}=\frac{1}{\sqrt{3}-1}$$

97. $$\frac{\sqrt{x}+3}{x}=\frac{(\sqrt{x}+3)(\sqrt{x}-3)}{x(\sqrt{x}-3)}=\frac{\sqrt{x^2}-9}{x(\sqrt{x}-3)}=\frac{x-9}{x(\sqrt{x}-3)}$$

99. $$\frac{\sqrt{x}+\sqrt{y}}{\sqrt{x}}=\frac{(\sqrt{x}+\sqrt{y})(\sqrt{x}-\sqrt{y})}{\sqrt{x}(\sqrt{x}-\sqrt{y})}=\frac{\sqrt{x^2}-\sqrt{y^2}}{\sqrt{x^2}-\sqrt{xy}}=\frac{x-y}{x-\sqrt{xy}}$$

101. $r = \sqrt{\frac{A}{\pi}} = \frac{\sqrt{A}}{\sqrt{\pi}} = \frac{\sqrt{A}\sqrt{\pi}}{\sqrt{\pi}\sqrt{\pi}} = \frac{\sqrt{A\pi}}{\pi}$

103. If the area of the aperture is again cut in half, the area will equal $9\pi/4\ \text{cm}^2$.

$$\begin{aligned} A &= \pi r^2 \\ \frac{9\pi}{4} &= \pi\left(\frac{d}{2}\right)^2 \\ \frac{9\pi}{4} &= \frac{\pi d^2}{4} \\ 36\pi &= 4\pi d^2 \\ \frac{36\pi}{4\pi} &= d^2 \\ 9 &= d^2 \\ 3 &= d \end{aligned}$$

$$\begin{aligned} f\text{-number} &= \frac{f}{d} \\ &= \frac{12}{3} \\ &= 4 \end{aligned}$$

$\boxed{f/4}$

105. Answers may vary.

107. $\frac{\sqrt{x}-3}{4} = \frac{(\sqrt{x}-3)(\sqrt{x}+3)}{4(\sqrt{x}+3)} = \frac{x-9}{4(\sqrt{x}+3)}$

Exercise 9.6 (page 631)

1. $f(0) = 3(0)^2 - 4(0) + 2 = 2$

3. $f(2) = 3(2)^2 - 4(2) + 2 = 6$

5. $x^n = y^n$

7. square

9. extraneous

11.

$$\begin{aligned} \sqrt{5x-6} &= 2 \\ \left(\sqrt{5x-6}\right)^2 &= 2^2 \\ 5x-6 &= 4 \\ 5x &= 10 \\ x &= 2 \end{aligned}$$

The answer checks.

13.

$$\begin{aligned} \sqrt{6x+1}+2 &= 7 \\ \sqrt{6x+1} &= 5 \\ \left(\sqrt{6x+1}\right)^2 &= 5^2 \\ 6x+1 &= 25 \\ 6x &= 24 \\ x &= 4 \end{aligned}$$

The answer checks.

15.

$$\begin{aligned} 2\sqrt{4x+1} &= \sqrt{x+4} \\ \left(2\sqrt{4x+1}\right)^2 &= \left(\sqrt{x+4}\right)^2 \\ 4(4x+1) &= x+4 \\ 16x+4 &= x+4 \\ 15x &= 0 \\ x &= 0 \end{aligned}$$

The answer checks.

17.

$$\begin{aligned} \sqrt[3]{7n-1} &= 3 \\ \left(\sqrt[3]{7n-1}\right)^3 &= 3^3 \\ 7n-1 &= 27 \\ 7n &= 28 \\ n &= 4 \end{aligned}$$

The answer checks.

19.

$$\sqrt[4]{10p+1} = \sqrt[4]{11p-7}$$
$$\left(\sqrt[4]{10p+1}\right)^4 = \left(\sqrt[4]{11p-7}\right)^4$$
$$10p+1 = 11p-7$$
$$-p = -8$$
$$p = 8$$

The answer checks.

21.

$$x = \frac{\sqrt{12x-5}}{2}$$
$$2x = \sqrt{12x-5}$$
$$(2x)^2 = \left(\sqrt{12x-5}\right)^2$$
$$4x^2 = 12x-5$$
$$4x^2 - 12x + 5 = 0$$
$$(2x-1)(2x-5)$$
$$2x-1=0 \quad \textbf{or} \quad 2x-5=0$$
$$2x = 1 \qquad 2x = 5$$
$$x = \tfrac{1}{2} \qquad x = \tfrac{5}{2}$$

Both answers check.

23.

$$\sqrt{x+2} = \sqrt{4-x}$$
$$\left(\sqrt{x+2}\right)^2 = \left(\sqrt{4-x}\right)^2$$
$$x+2 = 4-x$$
$$2x = 2$$
$$x = 1$$

The answer checks.

25.

$$2\sqrt{x} = \sqrt{5x-16}$$
$$\left(2\sqrt{x}\right)^2 = \left(\sqrt{5x-16}\right)^2$$
$$4x = 5x-16$$
$$-x = -16$$
$$x = 16$$

The answer checks.

27.

$$r-9 = \sqrt{2r-3}$$
$$(r-9)^2 = \left(\sqrt{2r-3}\right)^2$$
$$r^2 - 18r + 81 = 2r-3$$
$$r^2 - 20r + 84 = 0$$
$$(r-14)(r-6) = 0$$
$$r-14=0 \quad \textbf{or} \quad r-6=0$$
$$r = 14 \qquad r = 6$$

solution | not a solution

29.

$$\sqrt{-5x+24} = 6-x$$
$$\left(\sqrt{-5x+24}\right)^2 = (6-x)^2$$
$$-5x+24 = 36-12x+x^2$$
$$0 = x^2 - 7x + 12$$
$$0 = (x-3)(x-4)$$
$$x-3=0 \quad \textbf{or} \quad x-4=0$$
$$x = 3 \qquad x = 4$$

solution | solution

31.

$$\sqrt{y+2} = 4-y$$
$$\left(\sqrt{y+2}\right) = (4-y)^2$$
$$y+2 = 16-8y+y^2$$
$$0 = y^2 - 9y + 14$$
$$0 = (y-2)(y-7)$$
$$y-2=0 \quad \textbf{or} \quad y-7=0$$
$$y = 2 \qquad y = 7$$

solution | not a solution

33.

$$\sqrt{x}\sqrt{x+16} = 15$$
$$\left(\sqrt{x}\sqrt{x+16}\right)^2 = 15^2$$
$$x(x+16) = 225$$
$$x^2 + 16x - 225 = 0$$
$$(x-9)(x+25) = 0$$
$$x-9=0 \quad \textbf{or} \quad x+25=0$$
$$x = 9 \qquad x = -25$$

solution | not a solution

35.
$$\sqrt[3]{x^3-7}=x-1$$
$$\left(\sqrt[3]{x^3-7}\right)^3=(x-1)^3$$
$$x^3-7=x^3-3x^2+3x-1$$
$$3x^2-3x-6=0$$
$$3(x-2)(x+1)=0$$
$$x-2=0 \quad \textbf{or} \quad x+1=0$$
$$x=2 \qquad\qquad x=-1$$
solution   solution

37.
$$\sqrt[4]{x^4+4x^2-4}=-x$$
$$\left(\sqrt[4]{x^4+4x^2-4}\right)^4=(-x)^4$$
$$x^4+4x^2-4=x^4$$
$$4x^2-4=0$$
$$4(x+1)(x-1)=0$$
$$x+1=0 \quad \textbf{or} \quad x-1=0$$
$$x=-1 \qquad\qquad x=1$$
solution   not a solution

39.
$$\sqrt[4]{12t+4}+2=0$$
$$\sqrt[4]{12t+4}=-2$$
$$\left(\sqrt[4]{12t+4}\right)^4=(-2)^4$$
$$12t+4=16$$
$$12t=12$$
$$t=1$$
The answer does not check. ⇒ no solution

41.
$$\sqrt{2y+1}=1-2\sqrt{y}$$
$$\left(\sqrt{2y+1}\right)^2=\left(1-2\sqrt{y}\right)^2$$
$$2y+1=1-4\sqrt{y}+4y$$
$$4\sqrt{y}=2y$$
$$\left(4\sqrt{y}\right)^2=(2y)^2$$
$$16y=4y^2$$
$$0=4y^2-16y$$
$$0=4y(y-4)$$
$$4y=0 \quad \textbf{or} \quad y-4=0$$
$$y=0 \qquad\qquad y=4$$
solution   not a solution

43.
$$\sqrt{y+7}+3=\sqrt{y+4}$$
$$\left(\sqrt{y+7}+3\right)^2=\left(\sqrt{y+4}\right)^2$$
$$y+7+6\sqrt{y+7}+9=y+4$$
$$6\sqrt{y+7}=-12$$
$$\left(6\sqrt{y+7}\right)^2=(-12)^2$$
$$36(y+7)=144$$
$$y+7=4$$
$$y=-3$$
The answer does not check. ⇒ no solution

45.
$$\sqrt{v}+\sqrt{3}=\sqrt{v+3}$$
$$\left(\sqrt{v}+\sqrt{3}\right)^2=\left(\sqrt{v+3}\right)^2$$
$$v+2\sqrt{3v}+3=v+3$$
$$2\sqrt{3v}=0$$
$$\left(2\sqrt{3v}\right)^2=0^2$$
$$12v=0$$
$$v=0$$
The answer checks.

47.
$$2+\sqrt{u}=\sqrt{2u+7}$$
$$\left(2+\sqrt{u}\right)^2=\left(\sqrt{2u+7}\right)^2$$
$$4+4\sqrt{u}+u=2u+7$$
$$4\sqrt{u}=u+3$$
$$\left(4\sqrt{u}\right)^2=(u+3)^2$$
$$16u=u^2+6u+9$$
$$0=u^2-10u+9$$
$$0=(u-9)(u-1)$$
$u-9=0$ **or** $u-1=0$

$u=9$ $\quad$ $u=1$

solution $\quad$ solution

49.
$$\sqrt{6t+1}-3\sqrt{t}=-1$$
$$\sqrt{6t+1}=3\sqrt{t}-1$$
$$\left(\sqrt{6t+1}\right)^2=\left(3\sqrt{t}-1\right)^2$$
$$6t+1=9t-6\sqrt{t}+1$$
$$6\sqrt{t}=3t$$
$$\left(6\sqrt{t}\right)^2=(3t)^2$$
$$36t=9t^2$$
$$0=9t^2-36t$$
$$0=9t(t-4)$$
$9t=0$ **or** $t-4=0$

$t=0$ $\quad$ $t=4$

not a solution $\quad$ solution

51.
$$\sqrt{2x+5}+\sqrt{x+2}=5$$
$$\sqrt{2x+5}=5-\sqrt{x+2}$$
$$\left(\sqrt{2x+5}\right)^2=\left(5-\sqrt{x+2}\right)^2$$
$$2x+5=25-10\sqrt{x+2}+x+2$$
$$10\sqrt{x+2}=-x+22$$
$$\left(10\sqrt{x+2}\right)^2=(-x+22)^2$$
$$100(x+2)=x^2-44x+484$$
$$0=x^2-144x+284$$
$$0=(x-142)(x-2)$$
$x-142=0$ **or** $x-2=0$

$x=142$ $\quad$ $x=2$

not a solution $\quad$ solution

53.
$$\sqrt{z-1}+\sqrt{z+2}=3$$
$$\sqrt{z-1}=3-\sqrt{z+2}$$
$$\left(\sqrt{z-1}\right)^2=\left(3-\sqrt{z+2}\right)^2$$
$$z-1=9-6\sqrt{z+2}+z+2$$
$$6\sqrt{z+2}=12$$
$$\sqrt{z+2}=2$$
$$\left(\sqrt{z+2}\right)^2=2^2$$
$$z+2=4$$
$z=2$: The answer checks.

55.
$$\begin{aligned}
\sqrt{x-5}-\sqrt{x+3}&=4\\
\sqrt{x-5}&=\sqrt{x+3}+4\\
\left(\sqrt{x-5}\right)^2&=\left(\sqrt{x+3}+4\right)^2\\
x-5&=x+3+8\sqrt{x+3}+16\\
-24&=8\sqrt{x+3}\\
-3&=\sqrt{x+3}\\
(-3)^2&=\left(\sqrt{x+3}\right)^2\\
9&=x+3\\
6&=x
\end{aligned}$$

The answer does not check. ⇒ no solution

57.
$$\begin{aligned}
\sqrt{x+1}+\sqrt{3x}&=\sqrt{5x+1}\\
\left(\sqrt{x+1}+\sqrt{3x}\right)^2&=\left(\sqrt{5x+1}\right)^2\\
x+1+2\sqrt{3x(x+1)}+3x&=5x+1\\
2\sqrt{3x^2+3x}&=x\\
\left(2\sqrt{3x^2+3x}\right)^2&=x^2\\
12x^2+12x&=x^2\\
11x^2+12x&=0\\
x(11x+12)&=0
\end{aligned}$$

$x=0$ **or** $11x+12=0$

$x=-\frac{12}{11}$

solution   not a solution

59.
$$\begin{aligned}
\sqrt{\sqrt{a}+\sqrt{a+8}}&=2\\
\left(\sqrt{\sqrt{a}+\sqrt{a+8}}\right)^2&=2^2\\
\sqrt{a}+\sqrt{a+8}&=4\\
\left(\sqrt{a}+\sqrt{a+8}\right)^2&=4^2\\
a+2\sqrt{a(a+8)}+a+8&=16\\
2\sqrt{a^2+8a}&=-2a+8\\
\left(2\sqrt{a^2+8a}\right)^2&=(-2a+8)^2\\
4a^2+32a&=4a^2-32a+64\\
64a&=64
\end{aligned}$$

$a=1$: The answer checks.

61.

$$\frac{6}{\sqrt{x+5}} = \sqrt{x}$$
$$\left(\frac{6}{\sqrt{x+5}}\right)^2 = \left(\sqrt{x}\right)^2$$
$$\frac{36}{x+5} = x$$
$$36 = x^2 + 5x$$
$$0 = x^2 + 5x - 36$$
$$0 = (x+9)(x-4)$$

$x + 9 = 0$ **or** $x - 4 = 0$

$x = -9$ $x = 4$

not a solution   solution

63.

$$\sqrt{x+2} + \sqrt{2x-3} = \sqrt{11-x}$$
$$\left(\sqrt{x+2} + \sqrt{2x-3}\right)^2 = \left(\sqrt{11-x}\right)^2$$
$$x + 2 + 2\sqrt{(x+2)(2x-3)} + 2x - 3 = 11 - x$$
$$2\sqrt{2x^2 + x - 6} = -4x + 12$$
$$\left(2\sqrt{2x^2 + x - 6}\right)^2 = (-4x + 12)^2$$
$$8x^2 + 4x - 24 = 16x^2 - 96x + 144$$
$$0 = 8x^2 - 100x + 168$$
$$0 = 4(2x - 21)(x - 2)$$

$2x - 21 = 0$ **or** $x - 2 = 0$

$x = \frac{21}{2}$ $x = 2$

not a solution   solution

65.

$$v = \sqrt{2gh}$$
$$v^2 = \left(\sqrt{2gh}\right)^2$$
$$v^2 = 2gh$$
$$\frac{v^2}{2g} = \frac{2gh}{2g}$$
$$\frac{v^2}{2g} = h, \text{ or } h = \frac{v^2}{2g}$$

67.

$$T = 2\pi\sqrt{\frac{l}{32}}$$
$$T^2 = \left(2\pi\sqrt{\frac{l}{32}}\right)^2$$
$$T^2 = 4\pi^2 \cdot \frac{l}{32}$$
$$32\left(T^2\right) = 32\left(4\pi^2 \cdot \frac{l}{32}\right)$$
$$32T^2 = 4\pi^2 l$$
$$\frac{32T^2}{4\pi^2} = \frac{4\pi^2 l}{4\pi^2}$$
$$\frac{8T^2}{\pi^2} = l, \text{ or } l = \frac{8T^2}{\pi^2}$$

69.
$$r = \sqrt[3]{\frac{A}{P}} - 1$$
$$r + 1 = \sqrt[3]{\frac{A}{P}}$$
$$(r+1)^3 = \left(\sqrt[3]{\frac{A}{P}}\right)^3$$
$$(r+1)^3 = \frac{A}{P}$$
$$P(r+1)^3 = P\left(\frac{A}{P}\right)$$
$$P(r+1)^3 = A, \text{ or } A = P(r+1)^3$$

71.
$$L_A = L_B\sqrt{1 - \frac{v^2}{c^2}}$$
$$L_A^2 = \left(L_B\sqrt{1 - \frac{v^2}{c^2}}\right)^2$$
$$L_A^2 = L_B^2\left(1 - \frac{v^2}{c^2}\right)$$
$$\frac{L_A^2}{L_B^2} = 1 - \frac{v^2}{c^2}$$
$$\frac{v^2}{c^2} = 1 - \frac{L_A^2}{L_B^2}$$
$$c^2\left(\frac{v^2}{c^2}\right) = c^2\left(1 - \frac{L_A^2}{L_B^2}\right)$$
$$v^2 = c^2\left(1 - \frac{L_A^2}{L_B^2}\right)$$

73.
$$s = 1.45\sqrt{r}$$
$$65 = 1.45\sqrt{r}$$
$$65^2 = \left(1.45\sqrt{r}\right)^2$$
$$4225 = 2.1025r$$
$$2010 \text{ ft} \approx r$$

75.
$$v = \sqrt[3]{\frac{P}{0.02}}$$
$$v = \sqrt[3]{\frac{500}{0.02}}$$
$$v = \sqrt[3]{25000}$$
$$v \approx 29 \text{ mph}$$

77. $r = 1 - \sqrt[n]{\frac{T}{C}} = 1 - \sqrt[5]{\frac{9000}{22000}} \approx 1 - \sqrt[5]{0.40909} \approx 1 - 0.836 \approx 0.164 \approx 16\%$

79. Graph $y = \sqrt{5x}$ and $y = \sqrt{100 - 3x^2}$:

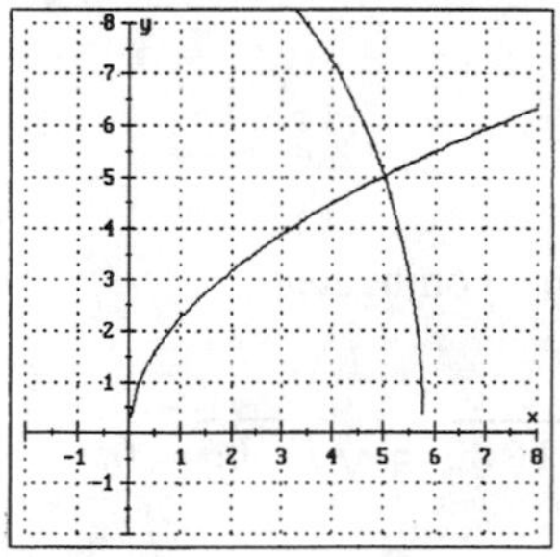

equilibrium price: $x = \$5$

81.
$$r = \sqrt[4]{\frac{8kl}{\pi R}}$$
$$r^4 = \left(\sqrt[4]{\frac{8kl}{\pi R}}\right)^4$$
$$r^4 = \frac{8kl}{\pi R}$$
$$\pi R r^4 = 8kl$$
$$R = \frac{8kl}{\pi r^4}$$

83. Answers may vary.

85.
$$\sqrt[3]{2x} = \sqrt{x}$$
$$\left(\sqrt[3]{2x}\right)^2 = \left(\sqrt{x}\right)^2$$
$$\left[\left(\sqrt[3]{2x}\right)^2\right]^3 = x^3$$
$$4x^2 = x^3$$
$$0 = x^3 - 4x^2$$
$$0 = x^2(x-4)$$
$x = 0$ or $x = 4$

Exercise 9.7 (page 642)

1. $$\frac{x^2-x-6}{9-x^2}\cdot\frac{x^2+x-6}{x^2-4} = \frac{(x-3)(x+2)}{(3+x)(3-x)}\cdot\frac{(x+3)(x-2)}{(x+2)(x-2)} = -1$$

3. Let $w =$ the speed of the wind.

	Rate	Time	Dist.
Downwind	$200+w$	$\frac{330}{200+w}$	330
Upwind	$200-w$	$\frac{330}{200-w}$	330

$$\frac{330}{200+w} + \frac{330}{200-w} = \frac{10}{3}$$
$$\left(\frac{330}{200+w} + \frac{330}{200-w}\right)3(200+w)(200-w) = \frac{10}{3}\cdot 3(200+w)(200-w)$$
$$330(3)(200-w) + 330(3)(200+w) = 10(40{,}000 - w^2)$$
$$198{,}000 - 990w + 198{,}000 + 990w = 400{,}000 - 10w^2$$
$$10w^2 = 4000$$
$$w^2 = 400$$
$$w = 20 \text{ mph}$$

5. imaginary

7. -1

9. 1

11. $\frac{\sqrt{a}}{\sqrt{b}}$

13. 5; 7

15. conjugates

17. $\sqrt{-9} = \sqrt{-1\cdot 9} = \sqrt{i^2\cdot 3^2} = 3i$

19. $\sqrt{-36} = \sqrt{-1\cdot 36} = \sqrt{i^2\cdot 6^2} = 6i$

21. $\sqrt{-7} = \sqrt{-1\cdot 7} = \sqrt{i^2\cdot 7} = i\sqrt{7}$

23. $3+7i \stackrel{?}{=} \sqrt{9} + (5+2)i$
$3+7i \stackrel{?}{=} 3+7i$
They are equal.

25. $8+5i \stackrel{?}{=} 2^3+\sqrt{25}i^3$
$8+5i \stackrel{?}{=} 8+5(-i)$
$8+5i \stackrel{?}{=} 8-5i$
They are not equal.

27. $\sqrt{4}+\sqrt{-4} \stackrel{?}{=} 2-2i$
$2+\sqrt{-1\cdot 4} \stackrel{?}{=} 2-2i$
$2+2i = 2-2i$
They are not equal.

29. $(3+4i)+(5-6i) = 3+4i+5-6i = 8-2i$

31. $(7-3i)-(4+2i) = 7-3i-4-2i = 3-5i$

33. $(8+5i)+(7+2i) = 8+5i+7+2i = 15+7i$

35. $(1+i)-2i+(5-7i) = 1+i-2i+5-7i = 6-8i$

37. $(5+3i)-(3-5i)+\sqrt{-1} = 5+3i-3+5i+i = 2+9i$

39. $(-8-\sqrt{3}i)-(7-3\sqrt{3}i) = -8-\sqrt{3}i-7+3\sqrt{3}i = -15+2\sqrt{3}i$

41. $3i(2-i) = 6i-3i^2 = 6i-3(-1) = 6i+3 = 3+6i$

43. $-5i(5-5i) = -25i+25i^2 = -25i+25(-1) = -25i-25 = -25-25i$

45. $(2+i)(3-i) = 6-2i+3i-i^2 = 6+i-(-1) = 6+i+1 = 7+i$

47. $(2-4i)(3+2i) = 6+4i-12i-8i^2 = 6-8i-8(-1) = 6-8i+8 = 14-8i$

49. $(2+\sqrt{2}i)(3-\sqrt{2}i) = 6-2\sqrt{2}i+3\sqrt{2}i-2i^2 = 6+\sqrt{2}i-2(-1) = 8+\sqrt{2}i$

51. $(8-\sqrt{-1})(-2-\sqrt{-16}) = (8-i)(-2-4i) = -16-32i+2i+4i^2 = -16-30i-4 = -20-30i$

53. $(2+i)^2 = (2+i)(2+i) = 4+2i+2i+i^2 = 4+4i-1 = 3+4i$

55. $(2+3i)^2 = (2+3i)(2+3i) = 4+6i+6i+9i^2 = 4+12i-9 = -5+12i$

57. $i(5+i)(3-2i) = i(15-10i+3i-2i^2) = i(15-7i+2) = i(17-7i) = 17i-7i^2 = 7+17i$

59. $(2+i)(2-i)(1+i) = (4-2i+2i-i^2)(1+i) = 5(1+i) = 5+5i$

61. $(3+i)[(3-2i)+(2+i)] = (3+i)(5-i) = 15-3i+5i-i^2 = 16+2i$

63. $\frac{1}{i} = \frac{1i^3}{ii^3} = \frac{i^3}{i^4} = \frac{i^3}{1} = i^3 = -i = 0-i$

65. $\frac{4}{5i^3} = \frac{4i}{5i^3i} = \frac{4i}{5i^4} = \frac{4i}{5(1)} = \frac{4}{5}i = 0+\frac{4}{5}i$

67. $\frac{3i}{8\sqrt{-9}} = \frac{3i}{8(3i)} = \frac{1}{8} = \frac{1}{8}+0i$

69. $\frac{-3}{5i^5} = \frac{-3i^3}{5i^5i^3} = \frac{-3(-i)}{5i^8} = \frac{3i}{5} = 0+\frac{3}{5}i$

71. $\dfrac{5}{2-i} = \dfrac{5(2+i)}{(2-i)(2+i)} = \dfrac{5(2+i)}{4-i^2} = \dfrac{5(2+i)}{5} = 2+i$

73. $\dfrac{13i}{5+i} = \dfrac{13i(5-i)}{(5+i)(5-i)} = \dfrac{65i-13i^2}{25-i^2} = \dfrac{13+65i}{26} = \dfrac{13}{26} + \dfrac{65}{26}i = \dfrac{1}{2} + \dfrac{5}{2}i$

75. $\dfrac{-12}{7-\sqrt{-1}} = \dfrac{-12}{7-i} = \dfrac{-12(7+i)}{(7-i)(7+i)} = \dfrac{-84-12i}{49-i^2} = \dfrac{-84-12i}{50} = \dfrac{-84}{50} - \dfrac{12}{50}i = -\dfrac{42}{25} - \dfrac{6}{25}i$

77. $\dfrac{5i}{6+2i} = \dfrac{5i(6-2i)}{(6+2i)(6-2i)} = \dfrac{30i-10i^2}{36-4i^2} = \dfrac{10+30i}{40} = \dfrac{10}{40} + \dfrac{30}{40}i = \dfrac{1}{4} + \dfrac{3}{4}i$

79. $\dfrac{3-2i}{3+2i} = \dfrac{(3-2i)(3-2i)}{(3+2i)(3-2i)} = \dfrac{9-6i-6i+4i^2}{9-4i^2} = \dfrac{5-12i}{13} = \dfrac{5}{13} - \dfrac{12}{13}i$

81. $\dfrac{3+2i}{3+i} = \dfrac{(3+2i)(3-i)}{(3+i)(3-i)} = \dfrac{9-3i+6i-2i^2}{9-i^2} = \dfrac{11+3i}{10} = \dfrac{11}{10} + \dfrac{3}{10}i$

83. $\dfrac{\sqrt{5}-\sqrt{3}i}{\sqrt{5}+\sqrt{3}i} = \dfrac{\left(\sqrt{5}-\sqrt{3}i\right)\left(\sqrt{5}-\sqrt{3}i\right)}{\left(\sqrt{5}+\sqrt{3}i\right)\left(\sqrt{5}-\sqrt{3}i\right)} = \dfrac{5-\sqrt{15}i-\sqrt{15}i+3i^2}{5-3i^2} = \dfrac{2-2\sqrt{15}i}{8}$

$$= \frac{2}{8} - \frac{2\sqrt{15}}{8}i$$
$$= \frac{1}{4} - \frac{\sqrt{15}}{4}i$$

85. $\left(\dfrac{i}{3+2i}\right)^2 = \dfrac{i^2}{(3+2i)^2} = \dfrac{-1}{(3+2i)(3+2i)} = \dfrac{-1}{9+12i+4i^2} = \dfrac{-1}{5+12i} = \dfrac{-1(5-12i)}{(5+12i)(5-12i)}$

$$= \frac{-5+12i}{25-144i^2}$$
$$= -\frac{5}{169} + \frac{12}{169}i$$

87. $\dfrac{i(3-i)}{3+i} = \dfrac{(3i-i^2)(3-i)}{(3+i)(3-i)} = \dfrac{(1+3i)(3-i)}{9-i^2} = \dfrac{3-i+9i-3i^2}{10} = \dfrac{6+8i}{10} = \dfrac{3}{5} + \dfrac{4}{5}i$

89. $\dfrac{(2-5i)-(5-2i)}{5-i} = \dfrac{2-5i-5+2i}{5-i} = \dfrac{-3-3i}{5-i} = \dfrac{(-3-3i)(5+i)}{(5-i)(5+i)} = \dfrac{-15-3i-15i-3i^2}{25-i^2}$

$$= \frac{-12-18i}{26}$$
$$= \frac{-12}{26} - \frac{18}{26}i$$
$$= -\frac{6}{13} - \frac{9}{13}i$$

91. $i^{21} = i^{20}i^1 = (i^4)^5 i = 1^5 i = i$

93. $i^{27} = i^{24}i^3 = (i^4)^6 i^3 = 1^6 i^3 = i^3 = -i$

95. $i^{100} = (i^4)^{25} = 1^{25} = 1$

97. $i^{97} = i^{96}i^1 = (i^4)^{24} i = 1^{24} i = i$

99. $|6 + 8i| = \sqrt{6^2 + 8^2} = \sqrt{36 + 64} = \sqrt{100} = 10$

101. $|12 - 5i| = \sqrt{12^2 + (-5)^2} = \sqrt{144 + 25} = \sqrt{169} = 13$

103. $|5 + 7i| = \sqrt{5^2 + 7^2} = \sqrt{25 + 49} = \sqrt{74}$

105. $\left|\frac{3}{5} - \frac{4}{5}i\right| = \sqrt{\left(\frac{3}{5}\right)^2 + \left(\frac{4}{5}\right)^2} = \sqrt{\frac{9}{25} + \frac{16}{25}} = \sqrt{\frac{25}{25}} = \sqrt{1} = 1$

107.
$$\begin{aligned} x^2 - 2x + 26 &= 0 \\ (1 - 5i)^2 - 2(1 - 5i) + 26 &= 0 \\ (1 - 5i)(1 - 5i) - 2 + 10i + 26 &= 0 \\ 1 - 10i + 25i^2 + 24 + 10i &= 0 \\ 1 - 10i - 25 + 24 + 10i &= 0 \\ 0 &= 0 \end{aligned}$$

109.
$$\begin{aligned} x^4 - 3x^2 - 4 &= 0 \\ i^4 - 3i^2 - 4 &= 0 \\ 1 - 3(-1) - 4 &= 0 \\ 1 + 3 - 4 &= 0 \\ 0 &= 0 \end{aligned}$$

111. $V = IR = (2 - 3i)(2 + i) = 4 + 2i - 6i - 3i^2 = 4 - 4i - 3(-1) = 4 - 4i + 3 = 7 - 4i$ volts

113. $Z = \frac{V}{I} = \frac{1.7 + 0.5i}{0.5i} = \frac{(1.7 + 0.5i)i}{(0.5i)i} = \frac{1.7i + 0.5i^2}{0.5i^2} = \frac{-0.5 + 1.7i}{-0.5} = 1 - 3.4i$

115. **Answers may vary.**

117. $\frac{3 - i}{2} = \frac{(3 - i)(3 + i)}{2(3 + i)} = \frac{9 - i^2}{2(3 + i)} = \frac{10}{2(3 + i)} = \frac{5}{3 + i}$

Chapter 9 Summary (page 645)

1. $\sqrt{49} = \sqrt{7^2} = 7$

2. $-\sqrt{121} = -\sqrt{11^2} = -11$

3. $-\sqrt{36} = -\sqrt{6^2} = -6$

4. $\sqrt{225} = \sqrt{15^2} = 15$

5. $\sqrt[3]{-27} = \sqrt[3]{(-3)^3} = -3$

6. $-\sqrt[3]{216} = -\sqrt[3]{6^3} = -6$

7. $\sqrt[4]{625} = \sqrt[4]{5^4} = 5$

8. $\sqrt[5]{-32} = \sqrt[5]{(-2)^5} = -2$

9. $\sqrt{25x^2} = \sqrt{5^2x^2} = |5x| = 5|x|$

10. $\sqrt{x^2 + 4x + 4} = \sqrt{(x + 2)^2} = |x + 2|$

11. $\sqrt[3]{27a^6b^3} = 3a^2b$

12. $\sqrt[4]{256x^8y^4} = |4x^2y| = 4x^2|y|$

13. $y = f(x) = \sqrt{x+2}$; Shift $y = \sqrt{x}$ left 2.

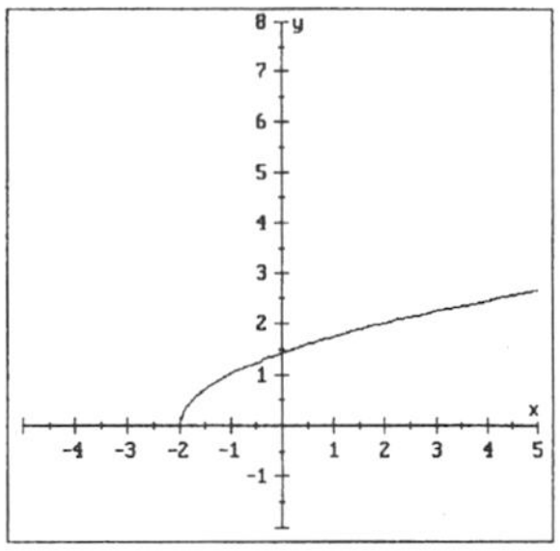

14. $y = f(x) = -\sqrt{x-1}$; Reflect $y = \sqrt{x}$ about the x-axis and shift right 1.

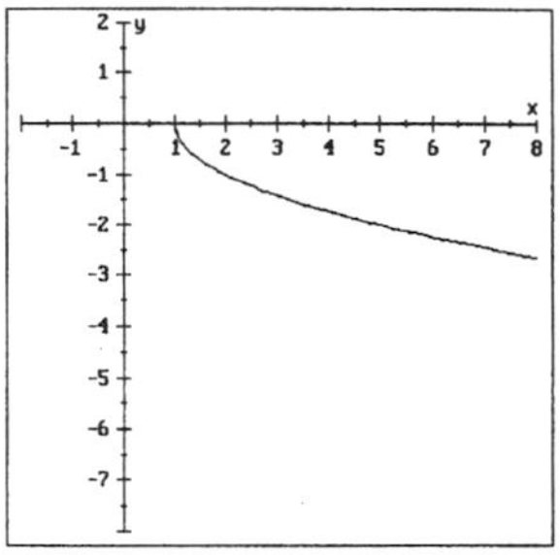

15. $y = f(x) = -\sqrt{x} + 2$; Reflect $y = \sqrt{x}$ about the x-axis and shift up 2.

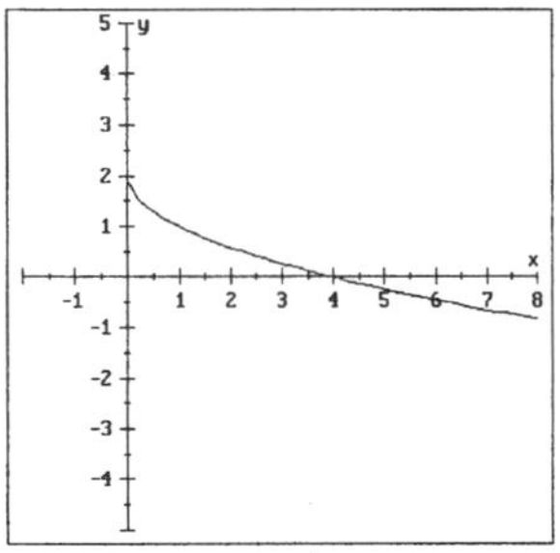

16. $y = f(x) = -\sqrt[3]{x} + 3$; Reflect $y = \sqrt[3]{x}$ about the x-axis and shift up 3.

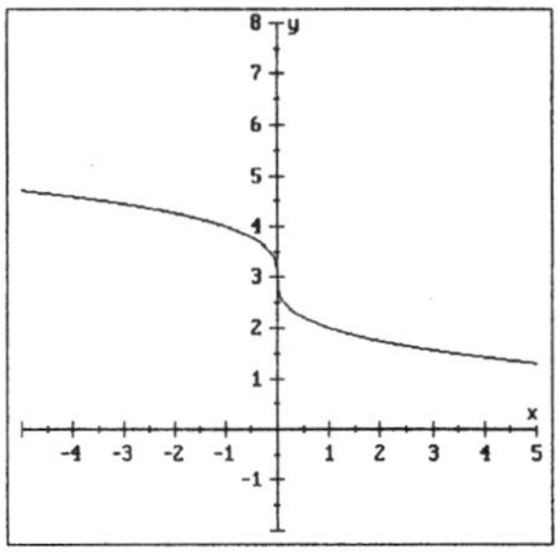

17. mean $= \frac{4+8+12+16+20}{5} = \frac{60}{5} = 12$

18.

Original term	Mean	Difference (term−mean)	Square of difference
4	12	−8	64
8	12	−4	16
12	12	0	0
16	12	4	16
20	12	8	64

$$\begin{aligned} s &= \sqrt{\frac{64+16+0+16+64}{5}} \\ &= \sqrt{\frac{160}{5}} \\ &= \sqrt{32} \\ &\approx 5.7 \end{aligned}$$

19. $d = 1.4\sqrt{h}$

$d = 1.4\sqrt{4.7} \approx 3.0$ miles

20.

$$\begin{aligned} d &= 1.4\sqrt{h} \\ 4 &= 1.4\sqrt{h} \\ 16 &= 1.96h \\ 8.2 \text{ ft} &= h \end{aligned}$$

21. Let $x =$ one-half of d.

$$125^2 = x^2 + 117^2$$
$$15625 = x^2 + 13689$$
$$1936 = x^2$$
$$44 = x$$
$$d = 2x = 2(44) = 88 \text{ yd}$$

22. Let $x =$ one-half of d.

$$8900^2 = x^2 + 3900^2$$
$$79{,}210{,}000 = x^2 + 15{,}210{,}000$$
$$64{,}000{,}000 = x^2$$
$$8000 = x$$
$$d = 2x = 2(8000) = 16{,}000 \text{ yd}$$

23. $d = \sqrt{(x_2 - x_1)^2 + (y_2 - y_1)^2} = \sqrt{(0-5)^2 + [0-(-12)]^2} = \sqrt{(-5)^2 + 12^2} = \sqrt{169} = 13$

24. $d = \sqrt{(x_2 - x_1)^2 + (y_2 - y_1)^2} = \sqrt{(-4-(-2))^2 + (6-8)^2} = \sqrt{(-2)^2 + (-2)^2} = \sqrt{8} \approx 2.83$

25. $25^{1/2} = 5$

26. $-36^{1/2} = -6$

27. $9^{3/2} = \left(9^{1/2}\right)^3 = 3^3 = 27$

28. $16^{3/2} = \left(16^{1/2}\right)^3 = 4^3 = 64$

29. $(-8)^{1/3} = -2$

30. $-8^{2/3} = -(8^{1/3})^2 = -2^2 = -4$

31. $8^{-2/3} = \dfrac{1}{8^{2/3}} = \dfrac{1}{(8^{1/3})^2} = \dfrac{1}{2^2} = \dfrac{1}{4}$

32. $8^{-1/3} = \dfrac{1}{8^{1/3}} = \dfrac{1}{2}$

33. $-49^{5/2} = -(49^{1/2})^5 = -7^5 = -16{,}807$

34. $\dfrac{1}{25^{5/2}} = \dfrac{1}{(25^{1/2})^5} = \dfrac{1}{5^5} = \dfrac{1}{3125}$

35. $\left(\dfrac{1}{4}\right)^{-3/2} = 4^{3/2} = (4^{1/2})^3 = 2^3 = 8$

36. $\left(\dfrac{4}{9}\right)^{-3/2} = \left(\dfrac{9}{4}\right)^{3/2} = \left[\left(\dfrac{9}{4}\right)^{1/2}\right]^3 = \left(\dfrac{3}{2}\right)^3 = \dfrac{27}{8}$

37. $(27x^3y)^{1/3} = 3xy^{1/3}$

38. $(81x^4y^2)^{1/4} = 3xy^{1/2}$

39. $(25x^3y^4)^{3/2} = 125x^{9/2}y^6$

40. $(8u^2v^3)^{-2/3} = \dfrac{1}{4u^{4/3}v^2}$

41. $5^{1/4}5^{1/2} = 5^{1/4+1/2} = 5^{3/4}$

42. $a^{3/7}a^{2/7} = a^{3/7+2/7} = a^{5/7}$

43. $u^{1/2}\left(u^{1/2} - u^{-1/2}\right) = u^{2/2} - u^0 = u - 1$

44. $v^{2/3}\left(v^{1/3} + v^{4/3}\right) = v^{3/3} + v^{6/3} = v + v^2$

45.
$$\left(x^{1/2} + y^{1/2}\right)^2 = \left(x^{1/2} + y^{1/2}\right)\left(x^{1/2} + y^{1/2}\right) = x^{2/2} + x^{1/2}y^{1/2} + x^{1/2}y^{1/2} + y^{2/2}$$
$$= x + 2x^{1/2}y^{1/2} + y$$

46. $\left(a^{2/3} + b^{2/3}\right)\left(a^{2/3} - b^{2/3}\right) = a^{4/3} - a^{2/3}b^{2/3} + a^{2/3}b^{2/3} - b^{4/3} = a^{4/3} - b^{4/3}$

47. $\sqrt[6]{5^2} = 5^{2/6} = 5^{1/3} = \sqrt[3]{5}$

48. $\sqrt[8]{x^4} = x^{4/8} = x^{1/2} = \sqrt{x}$

49. $\sqrt[9]{27a^3b^6} = (3^3a^3b^6)^{1/9} = 3^{3/9}a^{3/9}b^{6/9} = 3^{1/3}a^{1/3}b^{2/3} = \sqrt[3]{3ab^2}$

50. $\sqrt[4]{25a^2b^2} = 5^{2/4}a^{2/4}b^{2/4} = 5^{1/2}a^{1/2}b^{1/2} = \sqrt{5ab}$

51. $\sqrt{240} = \sqrt{16 \cdot 15} = \sqrt{16}\sqrt{15} = 4\sqrt{15}$

52. $\sqrt[3]{54} = \sqrt[3]{27 \cdot 2} = \sqrt[3]{27}\sqrt[3]{2} = 3\sqrt[3]{2}$

53. $\sqrt[4]{32} = \sqrt[4]{16 \cdot 2} = \sqrt[4]{16}\sqrt[4]{2} = 2\sqrt[4]{2}$

54. $\sqrt[5]{96} = \sqrt[5]{32 \cdot 3} = \sqrt[5]{32}\sqrt[5]{3} = 2\sqrt[5]{3}$

55. $\sqrt{8x^3} = \sqrt{4x^2}\sqrt{2x} = 2x\sqrt{2x}$

56. $\sqrt{18x^4y^3} = \sqrt{9x^4y^2}\sqrt{2y} = 3x^2y\sqrt{2y}$

57. $\sqrt[3]{16x^5y^4} = \sqrt[3]{8x^3y^3}\sqrt[3]{2x^2y} = 2xy\sqrt[3]{2x^2y}$

58. $\sqrt[3]{54x^7y^3} = \sqrt[3]{27x^6y^3}\sqrt[3]{2x} = 3x^2y\sqrt[3]{2x}$

59. $\dfrac{\sqrt{32x^3}}{\sqrt{2x}} = \sqrt{\dfrac{32x^3}{2x}} = \sqrt{16x^2} = 4x$

60. $\dfrac{\sqrt[3]{16x^5}}{\sqrt[3]{2x^2}} = \sqrt[3]{\dfrac{16x^5}{2x^2}} = \sqrt[3]{8x^3} = 2x$

61. $\sqrt[3]{\dfrac{2a^2b}{27x^3}} = \dfrac{\sqrt[3]{2a^2b}}{3x}$

62. $\sqrt{\dfrac{17xy}{64a^4}} = \dfrac{\sqrt{17xy}}{8a^2}$

63. $\sqrt{2} + \sqrt{8} = \sqrt{2} + \sqrt{4}\sqrt{2} = \sqrt{2} + 2\sqrt{2}$
$= 3\sqrt{2}$

64. $\sqrt{20} - \sqrt{5} = \sqrt{4}\sqrt{5} - \sqrt{5} = 2\sqrt{5} - \sqrt{5}$
$= \sqrt{5}$

65. $2\sqrt[3]{3} - \sqrt[3]{24} = 2\sqrt[3]{3} - \sqrt[3]{8}\sqrt[3]{3} = 2\sqrt[3]{3} - 2\sqrt[3]{3} = 0$

66. $\sqrt[4]{32} + 2\sqrt[4]{162} = \sqrt[4]{16}\sqrt[4]{2} + 2\sqrt[4]{81}\sqrt[4]{2} = 2\sqrt[4]{2} + 2(3)\sqrt[4]{2} = 2\sqrt[4]{2} + 6\sqrt[4]{2} = 8\sqrt[4]{2}$

67.
$$\begin{aligned} 2x\sqrt{8} + 2\sqrt{200x^2} + \sqrt{50x^2} &= 2x\sqrt{4}\sqrt{2} + 2\sqrt{100x^2}\sqrt{2} + \sqrt{25x^2}\sqrt{2} \\ &= 2x(2)\sqrt{2} + 2(10x)\sqrt{2} + 5x\sqrt{2} \\ &= 4x\sqrt{2} + 20x\sqrt{2} + 5x\sqrt{2} = 29x\sqrt{2} \end{aligned}$$

68.
$$\begin{aligned} 3\sqrt{27a^3} - 2a\sqrt{3a} + 5\sqrt{75a^3} &= 3\sqrt{9a^2}\sqrt{3a} - 2a\sqrt{3a} + 5\sqrt{25a^2}\sqrt{3a} \\ &= 3(3a)\sqrt{3a} - 2a\sqrt{3a} + 5(5a)\sqrt{3a} \\ &= 9a\sqrt{3a} - 2a\sqrt{3a} + 25a\sqrt{3a} = 32a\sqrt{3a} \end{aligned}$$

69.
$$\begin{aligned} \sqrt[3]{54} - 3\sqrt[3]{16} + 4\sqrt[3]{128} &= \sqrt[3]{27}\sqrt[3]{2} - 3\sqrt[3]{8}\sqrt[3]{2} + 4\sqrt[3]{64}\sqrt[3]{2} \\ &= 3\sqrt[3]{2} - 3(2)\sqrt[3]{2} + 4(4)\sqrt[3]{2} = 3\sqrt[3]{2} - 6\sqrt[3]{2} + 16\sqrt[3]{2} = 13\sqrt[3]{2} \end{aligned}$$

70.
$$\begin{aligned} 2\sqrt[4]{32x^5} + 4\sqrt[4]{162x^5} - 5x\sqrt[4]{512x} &= 2\sqrt[4]{16x^4}\sqrt[4]{2x} + 4\sqrt[4]{81x^4}\sqrt[4]{2x} - 5x\sqrt[4]{256}\sqrt[4]{2x} \\ &= 2(2x)\sqrt[4]{2x} + 4(3x)\sqrt[4]{2x} - 5x(4)\sqrt[4]{2x} \\ &= 4x\sqrt[4]{2x} + 12x\sqrt[4]{2x} - 20x\sqrt[4]{2x} = -4x\sqrt[4]{2x} \end{aligned}$$

71. hypotenuse $= 7\sqrt{2}$ m

72. shorter leg $= \dfrac{1}{2}\left(12\sqrt{3}\right) = 6\sqrt{3}$ cm
longer leg $= \sqrt{3}\left(6\sqrt{3}\right) = 6(3) = 18$ cm

73. $x = 5\sqrt{2} \approx 7.07$ in.

74. $x = \sqrt{3}\left(\frac{1}{2} \cdot 10\right) = 5\sqrt{3} \approx 8.66$ cm

75. $\left(2\sqrt{5}\right)\left(3\sqrt{2}\right) = 6\sqrt{10}$

76. $2\sqrt{6}\sqrt{216} = 2\sqrt{6}\sqrt{36}\sqrt{6} = 2(6)(6) = 72$

77. $\sqrt{9x}\sqrt{x} = \sqrt{9x^2} = 3x$

78. $\sqrt[3]{3}\sqrt[3]{9} = \sqrt[3]{27} = 3$

79. $-\sqrt[3]{2x^2}\sqrt[3]{4x} = -\sqrt[3]{8x^3}$
$= -2x$

80. $-\sqrt[4]{256x^5y^{11}}\sqrt[4]{625x^9y^3} = -\sqrt[4]{256x^4y^8}\sqrt[4]{xy^3}\sqrt[4]{625x^8}\sqrt[4]{xy^3} = -4xy^2(5x^2)\sqrt[4]{x^2y^6}$
$= -20x^3y^2\sqrt[4]{y^4}\sqrt[4]{x^2y^2}$
$= -20x^3y^3\sqrt[4]{x^2y^2} = -20x^3y^3\sqrt{xy}$

81. $\sqrt{2}\left(\sqrt{8} - 3\right) = \sqrt{16} - 3\sqrt{2} = 4 - 3\sqrt{2}$

82. $\sqrt{2}\left(\sqrt{2} + 3\right) = \sqrt{4} + 3\sqrt{2} = 2 + 3\sqrt{2}$

83. $\sqrt{5}\left(\sqrt{2} - 1\right) = \sqrt{10} - \sqrt{5}$

84. $\sqrt{3}\left(\sqrt{3} + \sqrt{2}\right) = \sqrt{9} + \sqrt{6} = 3 + \sqrt{6}$

85. $\left(\sqrt{2} + 1\right)\left(\sqrt{2} - 1\right) = \sqrt{4} - \sqrt{2} + \sqrt{2} - 1 = 1$

86. $\left(\sqrt{3} + \sqrt{2}\right)\left(\sqrt{3} + \sqrt{2}\right) = \sqrt{9} + \sqrt{6} + \sqrt{6} + \sqrt{4} = 5 + 2\sqrt{6}$

87. $(\sqrt{x} + \sqrt{y})(\sqrt{x} - \sqrt{y}) = \sqrt{x^2} - \sqrt{xy} + \sqrt{xy} - \sqrt{y^2} = x - y$

88. $(2\sqrt{u} + 3)(3\sqrt{u} - 4) = 6\sqrt{u^2} - 8\sqrt{u} + 9\sqrt{u} - 12 = 6u + \sqrt{u} - 12$

89. $\dfrac{1}{\sqrt{3}} = \dfrac{1\sqrt{3}}{\sqrt{3}\sqrt{3}} = \dfrac{\sqrt{3}}{3}$

90. $\dfrac{\sqrt{3}}{\sqrt{5}} = \dfrac{\sqrt{3}\sqrt{5}}{\sqrt{5}\sqrt{5}} = \dfrac{\sqrt{15}}{5}$

91. $\dfrac{x}{\sqrt{xy}} = \dfrac{x\sqrt{xy}}{\sqrt{xy}\sqrt{xy}} = \dfrac{x\sqrt{xy}}{xy} = \dfrac{\sqrt{xy}}{y}$

92. $\dfrac{\sqrt[3]{uv}}{\sqrt[3]{u^5v^7}} = \dfrac{\sqrt[3]{uv}\sqrt[3]{uv^2}}{\sqrt[3]{u^5v^7}\sqrt[3]{uv^2}} = \dfrac{\sqrt[3]{u^2v^3}}{\sqrt[3]{u^6v^9}} = \dfrac{v\sqrt[3]{u^2}}{u^2v^3} = \dfrac{\sqrt[3]{u^2}}{u^2v^2}$

93. $\dfrac{2}{\sqrt{2} - 1} = \dfrac{2\left(\sqrt{2} + 1\right)}{\left(\sqrt{2} - 1\right)\left(\sqrt{2} + 1\right)} = \dfrac{2\left(\sqrt{2} + 1\right)}{\sqrt{4} - 1} = \dfrac{2\left(\sqrt{2} + 1\right)}{2 - 1} = \dfrac{2\left(\sqrt{2} + 1\right)}{1} = 2\left(\sqrt{2} + 1\right)$

94. $\dfrac{\sqrt{2}}{\sqrt{3}-1}=\dfrac{\sqrt{2}\left(\sqrt{3}+1\right)}{\left(\sqrt{3}-1\right)\left(\sqrt{3}+1\right)}=\dfrac{\sqrt{2}\left(\sqrt{3}+1\right)}{\sqrt{9}-1}=\dfrac{\sqrt{2}\left(\sqrt{3}+1\right)}{3-1}=\dfrac{\sqrt{2}\left(\sqrt{3}+1\right)}{2}$

$=\dfrac{\sqrt{6}+\sqrt{2}}{2}$

95. $\dfrac{2x-32}{\sqrt{x}+4}=\dfrac{(2x-32)(\sqrt{x}-4)}{(\sqrt{x}+4)(\sqrt{x}-4)}=\dfrac{2(x-16)(\sqrt{x}-4)}{\sqrt{x^2}-16}=\dfrac{2(x-16)(\sqrt{x}-4)}{x-16}=2(\sqrt{x}-4)$

96. $\dfrac{\sqrt{a}+1}{\sqrt{a}-1}=\dfrac{(\sqrt{a}+1)(\sqrt{a}+1)}{(\sqrt{a}-1)(\sqrt{a}+1)}=\dfrac{\sqrt{a^2}+\sqrt{a}+\sqrt{a}+1}{\sqrt{a^2}-1}=\dfrac{a+2\sqrt{a}+1}{a-1}$

97. $\dfrac{\sqrt{3}}{5}=\dfrac{\sqrt{3}\sqrt{3}}{5\sqrt{3}}=\dfrac{3}{5\sqrt{3}}$

98. $\dfrac{\sqrt[3]{9}}{3}=\dfrac{\sqrt[3]{9}\sqrt[3]{3}}{3\sqrt[3]{3}}=\dfrac{\sqrt[3]{27}}{3\sqrt[3]{3}}=\dfrac{3}{3\sqrt[3]{3}}=\dfrac{1}{\sqrt[3]{3}}$

99. $\dfrac{3-\sqrt{x}}{2}=\dfrac{(3-\sqrt{x})(3+\sqrt{x})}{2(3+\sqrt{x})}=\dfrac{9-\sqrt{x^2}}{2(3+\sqrt{x})}=\dfrac{9-x}{2(3+\sqrt{x})}$

100. $\dfrac{\sqrt{a}-\sqrt{b}}{\sqrt{a}}=\dfrac{\left(\sqrt{a}-\sqrt{b}\right)\left(\sqrt{a}+\sqrt{b}\right)}{\sqrt{a}\left(\sqrt{a}+\sqrt{b}\right)}=\dfrac{\sqrt{a^2}-\sqrt{b^2}}{\sqrt{a^2}+\sqrt{ab}}=\dfrac{a-b}{a+\sqrt{ab}}$

101.
$$\begin{aligned}\sqrt{y+3}&=\sqrt{2y-19}\\ \left(\sqrt{y+3}\right)^2&=\left(\sqrt{2y-19}\right)^2\\ y+3&=2y-19\\ -y&=-22\\ y&=22\end{aligned}$$
The answer checks.

102.
$$\begin{aligned}u&=\sqrt{25u-144}\\ u^2&=\left(\sqrt{25u-144}\right)^2\\ u^2&=25u-144\end{aligned}$$
$u^2-25u+144=0$

$(u-9)(u-16)=0$

$u=9$ or $u=16$; Both answers check.

103.
$$\begin{aligned}r&=\sqrt{12r-27}\\ r^2&=\left(\sqrt{12r-27}\right)^2\\ r^2&=12r-27\end{aligned}$$
$r^2-12r+27=0$

$(r-9)(r-3)=0$

$r=9$ or $r=3$; Both answers check.

104.
$$\begin{aligned}\sqrt{z+1}+\sqrt{z}&=2\\ \sqrt{z+1}&=2-\sqrt{z}\\ \left(\sqrt{z+1}\right)^2&=\left(2-\sqrt{z}\right)^2\\ z+1&=4-4\sqrt{z}+z\\ 4\sqrt{z}&=3\\ \left(4\sqrt{z}\right)^2&=3^2\\ 16z&=9\\ z&=\tfrac{9}{16}\end{aligned}$$
; The answer checks.

105. $\sqrt{2x+5}-\sqrt{2x}=1$
$$\sqrt{2x+5}=1+\sqrt{2x}$$
$$\left(\sqrt{2x+5}\right)^2=\left(1+\sqrt{2x}\right)^2$$
$$2x+5=1+2\sqrt{2x}+2x$$
$$4=2\sqrt{2x}$$
$$4^2=\left(2\sqrt{2x}\right)^2$$
$$16=8x$$
$2=x$; The answer checks.

106. $\sqrt[3]{x^3+8}=x+2$
$$\left(\sqrt[3]{x^3+8}\right)^3=(x+2)^3$$
$$x^3+8=x^3+6x^2+12x+8$$
$$0=6x^2+12x$$
$$0=6x(x+2)$$
$x=0$ or $x=-2$; Both answers check.

107. $(5+4i)+(7-12i)=5+4i+7-12i=12-8i$

108. $(-6-40i)-(-8+28i)=-6-40i+8-28i=2-68i$

109. $(-32+\sqrt{-144})-(64+\sqrt{-81})=-32+\sqrt{144i^2}-64-\sqrt{81i^2}=-32+12i-64-9i$
$=-96+3i$

110. $(-8+\sqrt{-8})+(6-\sqrt{-32})=-8+\sqrt{4i^2}\sqrt{2}+6-\sqrt{16i^2}\sqrt{2}$
$=-8+2i\sqrt{2}+6-4i\sqrt{2}=-2-2\sqrt{2}i$

111. $(2-7i)(-3+4i)=-6+8i+21i-28i^2=-6+29i+28=22+29i$

112. $(-5+6i)(2+i)=-10-5i+12i+6i^2=-10+7i-6=-16+7i$

113. $(5-\sqrt{-27})(-6+\sqrt{-12})=(5-3i\sqrt{3})(-6+2i\sqrt{3})=-30+10i\sqrt{3}+18i\sqrt{3}-6i^2(3)$
$=-30+28i\sqrt{3}+18=-12+28\sqrt{3}i$

114. $(2+\sqrt{-128})(3-\sqrt{-98})=(2+8i\sqrt{2})(3-7i\sqrt{2})=6-14i\sqrt{2}+24i\sqrt{2}-56i^2(2)$
$=6+10i\sqrt{2}+112=118+10\sqrt{2}i$

115. $\dfrac{3}{4i}=\dfrac{3}{4i}\cdot\dfrac{i}{i}=\dfrac{3i}{4i^2}=\dfrac{3i}{-4}=-\dfrac{3}{4}i=0-\dfrac{3}{4}i$

116. $\dfrac{-2}{5i^3}=\dfrac{-2}{5i^3}\cdot\dfrac{i}{i}=\dfrac{-2i}{5i^4}=\dfrac{-2i}{5}=0-\dfrac{2}{5}i$

117. $\dfrac{6}{2+i}=\dfrac{6}{2+i}\cdot\dfrac{2-i}{2-i}=\dfrac{6(2-i)}{4-i^2}=\dfrac{6(2-i)}{5}=\dfrac{12-6i}{5}=\dfrac{12}{5}-\dfrac{6}{5}i$

118. $\dfrac{7}{3-i}=\dfrac{7}{3-i}\cdot\dfrac{3+i}{3+i}=\dfrac{7(3+i)}{9-i^2}=\dfrac{7(3+i)}{10}=\dfrac{21+7i}{10}=\dfrac{21}{10}+\dfrac{7}{10}i$

119. $\dfrac{4+i}{4-i}=\dfrac{4+i}{4-i}\cdot\dfrac{4+i}{4+i}=\dfrac{16+8i+i^2}{16-i^2}=\dfrac{15+8i}{17}=\dfrac{15}{17}+\dfrac{8}{17}i$

120. $\dfrac{3-i}{3+i} = \dfrac{3-i}{3+i} \cdot \dfrac{3-i}{3-i} = \dfrac{9-6i+i^2}{9-i^2} = \dfrac{8-6i}{10} = \dfrac{8}{10} - \dfrac{6}{10}i = \dfrac{4}{5} - \dfrac{3}{5}i$

121. $\dfrac{3}{5+\sqrt{-4}} = \dfrac{3}{5+2i} = \dfrac{3}{5+2i} \cdot \dfrac{5-2i}{5-2i} = \dfrac{3(5-2i)}{25-4i^2} = \dfrac{3(5-2i)}{29} = \dfrac{15-6i}{29} = \dfrac{15}{29} - \dfrac{6}{29}i$

122. $\dfrac{2}{3-\sqrt{-9}} = \dfrac{2}{3-3i} = \dfrac{2}{3-3i} \cdot \dfrac{3+3i}{3+3i} = \dfrac{2(3+3i)}{9-9i^2} = \dfrac{2(3+3i)}{18} = \dfrac{3+3i}{9} = \dfrac{3}{9} + \dfrac{3}{9}i = \dfrac{1}{3} + \dfrac{1}{3}i$

123. $|9+12i| = \sqrt{9^2+12^2} = \sqrt{81+144} = \sqrt{225} = 15 = 15+0i$

124. $|24-10i| = \sqrt{24^2+(-10)^2} = \sqrt{576+100} = \sqrt{676} = 26 = 26+0i$

125. $i^{12} = (i^4)^3 = 1^3 = 1$

126. $i^{583} = i^{580}i^3 = (i^4)^{145}i^3 = 1^{145}(-i) = -i$

Chapter 9 Test (page 651)

1. $\sqrt{49} = \sqrt{7^2} = 7$

2. $\sqrt[3]{64} = \sqrt[3]{4^3} = 4$

3. $\sqrt{4x^2} = \sqrt{(2x)^2} = |2x| = 2|x|$

4. $\sqrt[3]{8x^3} = \sqrt[3]{(2x)^3} = 2x$

5. $f(x) = \sqrt{x-2}$; Shift $y = \sqrt{x}$ right 2.

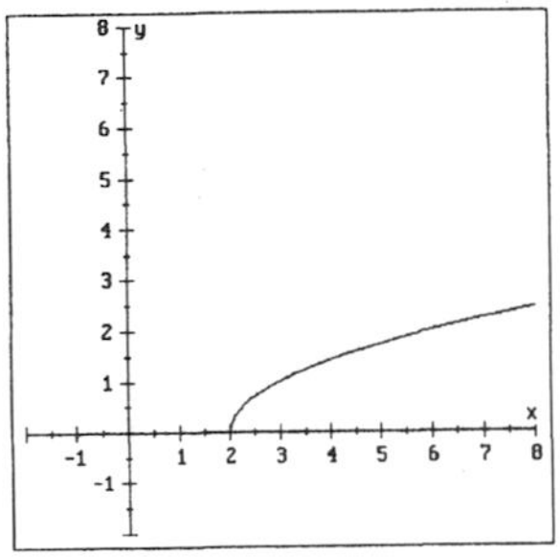

domain $= [2, \infty)$, range $= [0, \infty)$

6. $f(x) = \sqrt[3]{x} + 3$; Shift $y = \sqrt[3]{x}$ up 3.

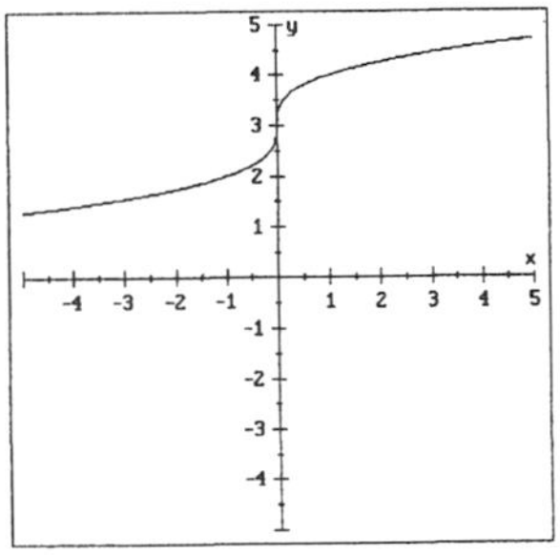

domain $= (-\infty, \infty)$, range $= (-\infty, \infty)$

7.
$$\begin{aligned} 53^2 &= 45^2 + h^2 \\ 2809 &= 2025 + h^2 \\ 784 &= h^2 \\ 28 \text{ in.} &= h \end{aligned}$$

8.
$$\begin{aligned} 2^2 &= \left(\frac{w}{2}\right)^2 + (1.9)^2 \\ 4 &= \frac{w^2}{4} + 3.61 \\ 0.39 &= \frac{w^2}{4} \\ 1.56 &= w^2 \\ 1.25 \text{ meters} &= w \end{aligned}$$

9. $d = \sqrt{(x_2-x_1)^2 + (y_2-y_1)^2} = \sqrt{(6-0)^2+(8-0)^2} = \sqrt{6^2+8^2} = \sqrt{100} = 10$

CHAPTER 9 TEST

10. $d = \sqrt{(x_2 - x_1)^2 + (y_2 - y_1)^2} = \sqrt{(-2-22)^2 + (5-12)^2} = \sqrt{(-24)^2 + (-7)^2} = \sqrt{625} = 25$

11. $16^{1/4} = 2$

12. $27^{2/3} = \left(27^{1/3}\right)^2 = 3^2 = 9$

13. $36^{-3/2} = \dfrac{1}{36^{3/2}} = \dfrac{1}{\left(36^{1/2}\right)^3} = \dfrac{1}{6^3} = \dfrac{1}{216}$

14. $\left(-\dfrac{8}{27}\right)^{-2/3} = \left(-\dfrac{27}{8}\right)^{2/3} = \left[\left(-\dfrac{27}{8}\right)^{1/3}\right]^2 = \left(-\dfrac{3}{2}\right)^2 = \dfrac{9}{4}$

15. $\dfrac{2^{5/3}2^{1/6}}{2^{1/2}} = \dfrac{2^{10/6}2^{1/6}}{2^{3/6}} = 2^{10/6+1/6-3/6} = 2^{8/6} = 2^{4/3}$

16. $\dfrac{(8x^3y)^{1/2}(8xy^5)^{1/2}}{(x^3y^6)^{1/3}} = \dfrac{8^{1/2}x^{3/2}y^{1/2}8^{1/2}x^{1/2}y^{5/2}}{x^{3/3}y^{6/3}} = \dfrac{8^{2/2}x^{4/2}y^{6/2}}{xy^2} = \dfrac{8x^2y^3}{xy^2} = 8xy$

17. $\sqrt{48} = \sqrt{16}\sqrt{3} = 4\sqrt{3}$

18. $\sqrt{250x^3y^5} = \sqrt{25x^2y^4}\sqrt{10xy} = 5xy^2\sqrt{10xy}$

19. $\dfrac{\sqrt[3]{24x^{15}y^4}}{\sqrt[3]{y}} = \sqrt[3]{\dfrac{24x^{15}y^4}{y}} = \sqrt[3]{24x^{15}y^3} = \sqrt[3]{8x^{15}y^3}\sqrt[3]{3} = 2x^5y\sqrt[3]{3}$

20. $\sqrt{\dfrac{3a^5}{48a^7}} = \sqrt{\dfrac{1}{16a^2}} = \dfrac{1}{4a}$

21. $\sqrt{12x^2} = \sqrt{4x^2}\sqrt{3} = 2|x|\sqrt{3}$

22. $\sqrt{8x^6} = \sqrt{4x^6}\sqrt{2} = 2|x^3|\sqrt{2}$

23. $\sqrt[3]{81x^3} = \sqrt[3]{27x^3}\sqrt[3]{3} = 3x\sqrt[3]{3}$

24. $\sqrt{18x^4y^9} = \sqrt{9x^4y^8}\sqrt{2y} = |3x^2y^4|\sqrt{2y} = 3x^2y^4\sqrt{2y}$

25. $\sqrt{12} - \sqrt{27} = \sqrt{4}\sqrt{3} - \sqrt{9}\sqrt{3} = 2\sqrt{3} - 3\sqrt{3} = -\sqrt{3}$

26. $2\sqrt[3]{40} - \sqrt[3]{5000} + 4\sqrt[3]{625} = 2\sqrt[3]{8}\sqrt[3]{5} - \sqrt[3]{1000}\sqrt[3]{5} + 4\sqrt[3]{125}\sqrt[3]{5} = 2(2)\sqrt[3]{5} - 10\sqrt[3]{5} + 4(5)\sqrt[3]{5}$
$= 4\sqrt[3]{5} - 10\sqrt[3]{5} + 20\sqrt[3]{5}$
$= 14\sqrt[3]{5}$

27. $2\sqrt{48y^5} - 3y\sqrt{12y^3} = 2\sqrt{16y^4}\sqrt{3y} - 3y\sqrt{4y^2}\sqrt{3y} = 2(4y^2)\sqrt{3y} - 3y(2y)\sqrt{3y}$
$= 8y^2\sqrt{3y} - 6y^2\sqrt{3y} = 2y^2\sqrt{3y}$

28. $\sqrt[4]{768z^5} + z\sqrt[4]{48z} = \sqrt[4]{256z^4}\sqrt[4]{3z} + z\sqrt[4]{16}\sqrt[4]{3z} = 4z\sqrt[4]{3z} + 2z\sqrt[4]{3z} = 6z\sqrt[4]{3z}$

29. $-2\sqrt{xy}(3\sqrt{x} + \sqrt{xy^3}) = -6\sqrt{x^2y} - 2\sqrt{x^2y^4} = -6x\sqrt{y} - 2xy^2$

30. $\left(3\sqrt{2}+\sqrt{3}\right)\left(2\sqrt{2}-3\sqrt{3}\right)=6\sqrt{4}-9\sqrt{6}+2\sqrt{6}-3\sqrt{9}=12-7\sqrt{6}-9=3-7\sqrt{6}$

31. $\dfrac{1}{\sqrt{5}}=\dfrac{1\sqrt{5}}{\sqrt{5}\sqrt{5}}=\dfrac{\sqrt{5}}{5}$

32. $\dfrac{3t-1}{\sqrt{3t}-1}=\dfrac{(3t-1)\left(\sqrt{3t}+1\right)}{\left(\sqrt{3t}-1\right)\left(\sqrt{3t}+1\right)}=\dfrac{(3t-1)\left(\sqrt{3t}+1\right)}{\sqrt{9t^2}-1}=\dfrac{(3t-1)\left(\sqrt{3t}+1\right)}{3t-1}=\sqrt{3t}+1$

33. $\dfrac{\sqrt{3}}{\sqrt{7}}=\dfrac{\sqrt{3}\sqrt{3}}{\sqrt{7}\sqrt{3}}=\dfrac{3}{\sqrt{21}}$

34. $\dfrac{\sqrt{a}+\sqrt{b}}{\sqrt{a}-\sqrt{b}}=\dfrac{\left(\sqrt{a}+\sqrt{b}\right)\left(\sqrt{a}-\sqrt{b}\right)}{\left(\sqrt{a}-\sqrt{b}\right)\left(\sqrt{a}-\sqrt{b}\right)}=\dfrac{\sqrt{a^2}-\sqrt{b^2}}{\sqrt{a^2}-\sqrt{ab}-\sqrt{ab}+\sqrt{b^2}}=\dfrac{a-b}{a-2\sqrt{ab}+b}$

35.
$$\begin{aligned}\sqrt[3]{6n+4}-4&=0\\ \sqrt[3]{6n+4}&=4\\ \left(\sqrt[3]{6n+4}\right)^3&=4^3\\ 6n+4&=64\\ 6n&=60\\ n&=10\end{aligned}$$
The answer checks.

36.
$$\begin{aligned}1-\sqrt{u}&=\sqrt{u-3}\\ \left(1-\sqrt{u}\right)^2&=\left(\sqrt{u-3}\right)^2\\ 1-2\sqrt{u}+u&=u-3\\ 4&=2\sqrt{u}\\ 4^2&=\left(2\sqrt{u}\right)^2\\ 16&=4u\\ 4&=u\end{aligned}$$
The answer does not check. ⇒ no solution

37. $(2+4i)+(-3+7i)=2+4i-3+7i=-1+11i$

38. $(3-\sqrt{-9})-(-1+\sqrt{-16})=3-3i+1-4i=4-7i$

39. $2i(3-4i)=6i-8i^2=6i+8=8+6i$

40. $(3+2i)(-4-i)=-12-3i-8i-2i^2=-12-11i+2=-10-11i$

41. $\dfrac{1}{i\sqrt{2}}=\dfrac{1}{i\sqrt{2}}\cdot\dfrac{i\sqrt{2}}{i\sqrt{2}}=\dfrac{i\sqrt{2}}{2i^2}=-\dfrac{\sqrt{2}i}{2}=0-\dfrac{\sqrt{2}}{2}i$

42. $\dfrac{2+i}{3-i}=\dfrac{2+i}{3-i}\cdot\dfrac{3+i}{3+i}=\dfrac{6+5i+i^2}{9-i^2}=\dfrac{5+5i}{10}=\dfrac{1}{2}+\dfrac{1}{2}i$

Exercise 10.1 (page 663)

1.
$$\begin{aligned} \frac{t+9}{2}+\frac{t+2}{5} &= \frac{8}{5}+4t \\ 5(t+9)+2(t+2) &= 2(8)+40t \\ 5t+45+2t+4 &= 16+40t \\ 33 &= 33t \\ 1 &= t \end{aligned}$$

3.
$$\begin{aligned} 3(t-3)+3t &\le 2(t+1)+t+1 \\ 3t-9+3t &\le 2t+2+t+1 \\ 3t &\le 12 \\ t &\le 4 \end{aligned}$$

5. $x=\sqrt{c};\ x=-\sqrt{c}$

7. positive or negative

9.
$$\begin{aligned} &6x^2+12x=0 \\ &6x(x+2)=0 \\ &6x=0 \quad \textbf{or} \quad x+2=0 \\ &x=0 \qquad\qquad x=-2 \end{aligned}$$

11.
$$\begin{aligned} &2y^2-50=0 \\ &2(y+5)(y-5)=0 \\ &y+5=0 \quad \textbf{or} \quad y-5=0 \\ &y=-5 \qquad\qquad y=5 \end{aligned}$$

13.
$$\begin{aligned} &r^2+6r+8=0 \\ &(r+2)(r+4)=0 \\ &r+2=0 \quad \textbf{or} \quad r+4=0 \\ &r=-2 \qquad\qquad r=-4 \end{aligned}$$

15.
$$\begin{aligned} &7x-6=x^2 \\ &0=x^2-7x+6 \\ &0=(x-6)(x-1) \\ &x-6=0 \quad \textbf{or} \quad x-1=0 \\ &x=6 \qquad\qquad x=1 \end{aligned}$$

17.
$$\begin{aligned} &2z^2-5z+2=0 \\ &(2z-1)(z-2)=0 \\ &2z-1=0 \quad \textbf{or} \quad z-2=0 \\ &z=\tfrac{1}{2} \qquad\qquad z=2 \end{aligned}$$

19.
$$\begin{aligned} &6s^2+11s-10=0 \\ &(2s+5)(3s-2)=0 \\ &2s+5=0 \quad \textbf{or} \quad 3s-2=0 \\ &s=-\tfrac{5}{2} \qquad\qquad s=\tfrac{2}{3} \end{aligned}$$

21. $x^2=36$

$x=\pm\sqrt{36}=\pm 6$

23. $z^2=5$

$z=\pm\sqrt{5}$

25.
$$\begin{aligned} 3x^2-16 &= 0 \\ 3x^2 &= 16 \\ x^2 &= \frac{16}{3} \\ x &= \pm\sqrt{\frac{16}{3}} \\ &= \pm\frac{4}{\sqrt{3}} \\ &= \pm\frac{4\sqrt{3}}{3} \end{aligned}$$

27.
$$\begin{aligned} (y+1)^2 &= 1 \\ y+1 &= \pm\sqrt{1} \\ y+1 &= \pm 1 \\ y &= -1\pm 1 \\ y=0 &\text{ or } y=-2 \end{aligned}$$

29.
$$\begin{aligned} (s-7)^2-9 &= 0 \\ (s-7)^2 &= 9 \\ s-7 &= \pm\sqrt{9} \\ s-7 &= \pm 3 \\ s &= 7\pm 3 \\ s=10 &\text{ or } s=4 \end{aligned}$$

31.
$$\begin{aligned}(x+5)^2-3&=0\\(x+5)^2&=3\\x+5&=\pm\sqrt{3}\\x&=-5\pm\sqrt{3}\end{aligned}$$

33.
$$\begin{aligned}(x-2)^2-5&=0\\(x-2)^2&=5\\x-2&=\pm\sqrt{5}\\x&=2\pm\sqrt{5}\end{aligned}$$

35.
$$\begin{aligned}p^2+16&=0\\p^2&=-16\\p&=\pm\sqrt{-16}\\p&=\pm 4i\end{aligned}$$

37.
$$\begin{aligned}4m^2+81&=0\\4m^2&=-81\\m^2&=-\frac{81}{4}\\m&=\pm\sqrt{-\frac{81}{4}}\\m&=\pm\frac{9}{2}i\end{aligned}$$

39.
$$\begin{aligned}2d^2&=3h\\d^2&=\frac{3h}{2}\\d&=\sqrt{\frac{3h}{2}}\\d&=\frac{\sqrt{3h}}{\sqrt{2}}\cdot\frac{\sqrt{2}}{\sqrt{2}}\\d&=\frac{\sqrt{6h}}{2}\end{aligned}$$

41.
$$\begin{aligned}E&=mc^2\\\frac{E}{m}&=c^2\\\sqrt{\frac{E}{m}}&=c\\\frac{\sqrt{E}}{\sqrt{m}}\cdot\frac{\sqrt{m}}{\sqrt{m}}&=c\\\frac{\sqrt{Em}}{m}&=c\end{aligned}$$

43.
$$\begin{aligned}x^2+2x-8&=0\\x^2+2x&=8\\x^2+2x+1&=8+1\\(x+1)^2&=9\\x+1&=\pm 3\\x&=-1\pm 3\end{aligned}$$
$x=2$ or $x=-4$

45.
$$\begin{aligned}x^2-6x+8&=0\\x^2-6x&=-8\\x^2-6x+9&=-8+9\\(x-3)^2&=1\\x-3&=\pm 1\\x&=3\pm 1\end{aligned}$$
$x=4$ or $x=2$

47.
$$\begin{aligned}x^2+5x+4&=0\\x^2+5x&=-4\\x^2+5x+\frac{25}{4}&=-4+\frac{25}{4}\\\left(x+\frac{5}{2}\right)^2&=\frac{9}{4}\\x+\frac{5}{2}&=\pm\frac{3}{2}\\x&=-\frac{5}{2}\pm\frac{3}{2}\end{aligned}$$
$x=-\frac{2}{2}=-1$
or $x=-\frac{8}{2}=-4$

49.
$$\begin{aligned}x+1&=2x^2\\-2x^2+x&=-1\\x^2-\frac{1}{2}x&=\frac{1}{2}\\x^2-\frac{1}{2}x+\frac{1}{16}&=\frac{1}{2}+\frac{1}{16}\\\left(x-\frac{1}{4}\right)^2&=\frac{9}{16}\\x-\frac{1}{4}&=\pm\frac{3}{4}\\x&=\frac{1}{4}\pm\frac{3}{4}\end{aligned}$$
$x=\frac{4}{4}=1$ or $x=-\frac{2}{4}=-\frac{1}{2}$

51.
$$\begin{aligned}6x^2+11x+3&=0\\6x^2+11x&=-3\\x^2+\frac{11}{6}x&=-\frac{1}{2}\\x^2+\frac{11}{6}x+\frac{121}{144}&=-\frac{1}{2}+\frac{121}{144}\\\left(x+\frac{11}{12}\right)^2&=\frac{49}{144}\\x+\frac{11}{12}&=\pm\frac{7}{12}\\x&=-\frac{11}{12}\pm\frac{7}{12}\end{aligned}$$
$x=-\frac{4}{12}=-\frac{1}{3}$ or $x=-\frac{18}{12}=-\frac{3}{2}$

53.
$$\begin{aligned}
9-6r&=8r^2\\
-8r^2-6r&=-9\\
r^2+\frac{3}{4}r&=\frac{9}{8}\\
r^2+\frac{3}{4}r+\frac{9}{64}&=\frac{9}{8}+\frac{9}{64}\\
\left(r+\frac{3}{8}\right)^2&=\frac{81}{64}\\
r+\frac{3}{8}&=\pm\frac{9}{8}\\
r&=-\frac{3}{8}\pm\frac{9}{8}
\end{aligned}$$
$r=\frac{6}{8}=\frac{3}{4}$ or $r=-\frac{12}{8}=-\frac{3}{2}$

55.
$$\begin{aligned}
\frac{7x+1}{5}&=-x^2\\
7x+1&=-5x^2\\
5x^2+7x&=-1\\
x^2+\frac{7}{5}x&=-\frac{1}{5}\\
x^2+\frac{7}{5}x+\frac{49}{100}&=-\frac{1}{5}+\frac{49}{100}\\
\left(x+\frac{7}{10}\right)^2&=\frac{29}{100}\\
x+\frac{7}{10}&=\pm\frac{\sqrt{29}}{10}\\
x&=-\frac{7}{10}\pm\frac{\sqrt{29}}{10}
\end{aligned}$$

57.
$$\begin{aligned}
p^2+2p+2&=0\\
p^2+2p&=-2\\
p^2+2p+1&=-2+1\\
(p+1)^2&=-1\\
p+1&=\pm\sqrt{-1}\\
p+1&=\pm i\\
p&=-1\pm i
\end{aligned}$$

59.
$$\begin{aligned}
y^2+8y+18&=0\\
y^2+8y&=-18\\
y^2+8y+16&=-18+16\\
(y+4)^2&=-2\\
y+4&=\pm\sqrt{-2}\\
y&=-4\pm i\sqrt{2}
\end{aligned}$$

61.
$$\begin{aligned}
3m^2-2m+3&=0\\
m^2-\frac{2}{3}m&=-1\\
m^2-\frac{2}{3}m+\frac{1}{9}&=-1+\frac{1}{9}\\
\left(m-\frac{1}{3}\right)^2&=-\frac{8}{9}\\
m-\frac{1}{3}&=\pm\sqrt{-\frac{8}{9}}\\
m&=\frac{1}{3}\pm\frac{\sqrt{8}}{3}i\\
m&=\frac{1}{3}\pm\frac{2\sqrt{2}}{3}i
\end{aligned}$$

63.
$$\begin{aligned}
f(x)&=0\\
2x^2+x-5&=0\\
x^2+\frac{1}{2}x&=\frac{5}{2}\\
x^2+\frac{1}{2}x+\frac{1}{16}&=\frac{5}{2}+\frac{1}{16}\\
\left(x+\frac{1}{4}\right)^2&=\frac{41}{16}\\
x+\frac{1}{4}&=\pm\sqrt{\frac{41}{16}}\\
x+\frac{1}{4}&=\pm\frac{\sqrt{41}}{4}\\
x&=-\frac{1}{4}\pm\frac{\sqrt{41}}{4}
\end{aligned}$$

65.
$$\begin{aligned} f(x) &= 0 \\ x^2 + x - 3 &= 0 \\ x^2 + x &= 3 \\ x^2 + x + \frac{1}{4} &= 3 + \frac{1}{4} \\ \left(x + \frac{1}{2}\right)^2 &= \frac{13}{4} \\ x + \frac{1}{2} &= \pm\sqrt{\frac{13}{4}} \\ x &= -\frac{1}{2} \pm \frac{\sqrt{13}}{2} \end{aligned}$$

67.
$$\begin{aligned} s &= 16t^2 \\ 256 &= 16t^2 \\ 16 &= t^2 \\ \pm\sqrt{16} &= t \\ \pm 4 &= t \end{aligned}$$
$t = 4$ is the only answer that makes sense, so it will take 4 seconds.

69.
$$\begin{aligned} s^2 &= 10.5l \\ s^2 &= 10.5(500) \\ s^2 &= 5250 \\ s &= \pm\sqrt{5250} \\ s &\approx \pm 72.5 \end{aligned}$$
s must be positive, so the speed was about 72.5 mph.

71.
$$\begin{aligned} A &= P(1+r)^t \\ 9193.60 &= 8500(1+r)^2 \\ \frac{9193.60}{8500} &= (1+r)^2 \\ 1.0816 &= (1+r)^2 \\ \pm\sqrt{1.0816} &= \sqrt{(1+r)^2} \\ \pm 1.04 &= 1 + r \\ -1 \pm 1.04 &= r \end{aligned}$$
$r = 0.04$ or $r = -2.04$; r must be positive, so $r = 0.04$, or 4%.

73.
$$\begin{aligned} A &= 100 \\ lw &= 100 \\ (1.9x)(x) &= 100 \\ 1.9x^2 &= 100 \\ x^2 &= \frac{100}{1.9} \\ x &= \pm\sqrt{\frac{100}{1.9}} \approx \pm 7.25 \end{aligned}$$
x must be positive, so $x = 7\frac{1}{4}$ ft.
$1.9x \approx 1.9(7.25) \approx 13.75 = 13\frac{3}{4}$ ft.

75. **Answers may vary.**

77. $\left(\frac{1}{2}\sqrt{3}\right)^2 = \left(\frac{\sqrt{3}}{2}\right)^2 = \frac{3}{4}$

Exercise 10.2 (page 670)

1.
$$\begin{aligned} Ax + By &= C \\ By &= -Ax + C \\ B &= \frac{-Ax + C}{y} \end{aligned}$$

3. $\sqrt{24} = \sqrt{4}\sqrt{6} = 2\sqrt{6}$

5. $\frac{3}{\sqrt{3}} = \frac{3}{\sqrt{3}} \cdot \frac{\sqrt{3}}{\sqrt{3}} = \frac{3\sqrt{3}}{3} = \sqrt{3}$

7. 3; −2; 6

9.
$$x^2 + 3x + 2 = 0$$
$$a = 1, b = 3, c = 2$$
$$x = \frac{-b \pm \sqrt{b^2 - 4ac}}{2a} = \frac{-3 \pm \sqrt{3^2 - 4(1)(2)}}{2(1)} = \frac{-3 \pm \sqrt{9-8}}{2} = \frac{-3 \pm \sqrt{1}}{2} = \frac{-3 \pm 1}{2}$$
$x = -\frac{2}{2} = -1$ or $x = -\frac{4}{2} = -2$

11.
$$x^2 - 2x - 15 = 0$$
$$a = 1, b = -2, c = -15$$
$$x = \frac{-b \pm \sqrt{b^2 - 4ac}}{2a} = \frac{2 \pm \sqrt{(-2)^2 - 4(1)(-15)}}{2(1)} = \frac{2 \pm \sqrt{4+60}}{2} = \frac{2 \pm \sqrt{64}}{2} = \frac{2 \pm 8}{2}$$
$x = \frac{10}{2} = 5$ or $x = -\frac{6}{2} = -3$

13.
$$x^2 + 12x = -36$$
$$x^2 + 12x + 36 = 0$$
$$a = 1, b = 12, c = 36$$
$$x = \frac{-b \pm \sqrt{b^2 - 4ac}}{2a} = \frac{-12 \pm \sqrt{12^2 - 4(1)(36)}}{2(1)} = \frac{-12 \pm \sqrt{144 - 144}}{2} = \frac{-12 \pm \sqrt{0}}{2} = \frac{-12 \pm 0}{2}$$
$x = -\frac{12}{2} = -6$ or $x = -\frac{12}{2} = -6$

15.
$$2x^2 - x - 3 = 0$$
$$a = 2, b = -1, c = -3$$
$$x = \frac{-b \pm \sqrt{b^2 - 4ac}}{2a} = \frac{1 \pm \sqrt{(-1)^2 - 4(2)(-3)}}{2(2)} = \frac{1 \pm \sqrt{1+24}}{4} = \frac{1 \pm \sqrt{25}}{4} = \frac{1 \pm 5}{4}$$
$x = \frac{6}{4} = \frac{3}{2}$ or $x = -\frac{4}{4} = -1$

17.
$$6x^2 - x - 1 = 0$$
$$a = 6, b = -1, c = -1$$
$$x = \frac{-b \pm \sqrt{b^2 - 4ac}}{2a} = \frac{1 \pm \sqrt{(-1)^2 - 4(6)(-1)}}{2(6)} = \frac{1 \pm \sqrt{1+24}}{12} = \frac{1 \pm \sqrt{25}}{12} = \frac{1 \pm 5}{12}$$
$x = \frac{6}{12} = \frac{1}{2}$ or $x = -\frac{4}{12} = -\frac{1}{3}$

19.
$$15x^2 - 14x = 8$$
$$15x^2 - 14x - 8 = 0$$
$$a = 15, b = -14, c = -8$$
$$x = \frac{-b \pm \sqrt{b^2 - 4ac}}{2a} = \frac{14 \pm \sqrt{(-14)^2 - 4(15)(-8)}}{2(15)} = \frac{14 \pm \sqrt{196 + 480}}{30} = \frac{14 \pm \sqrt{676}}{30} = \frac{14 \pm 26}{30}$$
$x = \frac{40}{30} = \frac{4}{3}$ or $x = -\frac{12}{30} = -\frac{2}{5}$

21.
$$8u = -4u^2 - 3$$
$$4u^2 + 8u + 3 = 0$$
$$a = 4, b = 8, c = 3$$
$$u = \frac{-b \pm \sqrt{b^2 - 4ac}}{2a}$$
$$= \frac{-8 \pm \sqrt{8^2 - 4(4)(3)}}{2(4)}$$
$$= \frac{-8 \pm \sqrt{64 - 48}}{8}$$
$$= \frac{-8 \pm \sqrt{16}}{8} = \frac{-8 \pm 4}{8}$$
$u = \frac{-4}{8} = -\frac{1}{2}$ or $u = \frac{-12}{8} = -\frac{3}{2}$

23.
$$16y^2 + 8y - 3 = 0$$
$$a = 16, b = 8, c = -3$$
$$y = \frac{-b \pm \sqrt{b^2 - 4ac}}{2a}$$
$$= \frac{-8 \pm \sqrt{8^2 - 4(16)(-3)}}{2(16)}$$
$$= \frac{-8 \pm \sqrt{64 + 192}}{32}$$
$$= \frac{-8 \pm \sqrt{256}}{32} = \frac{-8 \pm 16}{32}$$
$y = \frac{8}{32} = \frac{1}{4}$ or $y = \frac{-24}{32} = -\frac{3}{4}$

25.
$$5x^2 + 5x + 1 = 0$$
$$a = 5, b = 5, c = 1$$
$$x = \frac{-b \pm \sqrt{b^2 - 4ac}}{2a}$$
$$= \frac{-5 \pm \sqrt{5^2 - 4(5)(1)}}{2(5)}$$
$$= \frac{-5 \pm \sqrt{25 - 20}}{10}$$
$$= \frac{-5 \pm \sqrt{5}}{10}$$
$$= \frac{-5}{10} \pm \frac{\sqrt{5}}{10} = -\frac{1}{2} \pm \frac{\sqrt{5}}{10}$$

27.
$$\frac{x^2}{2} + \frac{5}{2}x = -1$$
$$x^2 + 5x = -2$$
$$x^2 + 5x + 2 = 0$$
$$a = 1, b = 5, c = 2$$
$$x = \frac{-b \pm \sqrt{b^2 - 4ac}}{2a}$$
$$= \frac{-5 \pm \sqrt{5^2 - 4(1)(2)}}{2(1)}$$
$$= \frac{-5 \pm \sqrt{25 - 8}}{2}$$
$$= \frac{-5 \pm \sqrt{17}}{2} = -\frac{5}{2} \pm \frac{\sqrt{17}}{2}$$

29.
$$2x^2 - 1 = 3x$$
$$2x^2 - 3x - 1 = 0$$
$$a = 2, b = -3, c = -1$$
$$x = \frac{-b \pm \sqrt{b^2 - 4ac}}{2a}$$
$$= \frac{3 \pm \sqrt{(-3)^2 - 4(2)(-1)}}{2(2)}$$
$$= \frac{3 \pm \sqrt{9 + 8}}{4}$$
$$= \frac{3 \pm \sqrt{17}}{4} = \frac{3}{4} \pm \frac{\sqrt{17}}{4}$$

31.
$$x^2 + 2x + 2 = 0$$
$$a = 1, b = 2, c = 2$$
$$x = \frac{-b \pm \sqrt{b^2 - 4ac}}{2a}$$
$$= \frac{-2 \pm \sqrt{2^2 - 4(1)(2)}}{2(1)}$$
$$= \frac{-2 \pm \sqrt{4 - 8}}{2}$$
$$= \frac{-2 \pm \sqrt{-4}}{2}$$
$$= \frac{-2 \pm \sqrt{-1 \cdot 4}}{2}$$
$$= \frac{-2 \pm 2i}{2} = \frac{-2}{2} \pm \frac{2i}{2} = -1 \pm i$$

33.
$$2x^2 + x + 1 = 0$$
$$a = 2, b = 1, c = 1$$
$$\begin{aligned} x &= \frac{-b \pm \sqrt{b^2 - 4ac}}{2a} \\ &= \frac{-1 \pm \sqrt{1^2 - 4(2)(1)}}{2(2)} \\ &= \frac{-1 \pm \sqrt{1-8}}{4} \\ &= \frac{-1 \pm \sqrt{-7}}{4} \\ &= \frac{-1 \pm \sqrt{-1 \cdot 7}}{4} \\ &= \frac{-1 \pm i\sqrt{7}}{4} = -\frac{1}{4} \pm \frac{\sqrt{7}}{4}i \end{aligned}$$

35.
$$3x^2 - 4x = -2$$
$$3x^2 - 4x + 2 = 0$$
$$a = 3, b = -4, c = 2$$
$$\begin{aligned} x &= \frac{-b \pm \sqrt{b^2 - 4ac}}{2a} \\ &= \frac{4 \pm \sqrt{(-4)^2 - 4(3)(2)}}{2(3)} \\ &= \frac{4 \pm \sqrt{16 - 24}}{6} \\ &= \frac{4 \pm \sqrt{-8}}{6} \\ &= \frac{4 \pm \sqrt{-1 \cdot 4 \cdot 2}}{6} \\ &= \frac{4 \pm 2i\sqrt{2}}{6} \\ &= \frac{4}{6} \pm \frac{2\sqrt{2}}{6}i = \frac{2}{3} \pm \frac{\sqrt{2}}{3}i \end{aligned}$$

37.
$$3x^2 - 2x = -3$$
$$3x^2 - 2x + 3 = 0$$
$$a = 3, b = -2, c = 3$$
$$\begin{aligned} x &= \frac{-b \pm \sqrt{b^2 - 4ac}}{2a} \\ &= \frac{2 \pm \sqrt{(-2)^2 - 4(3)(3)}}{2(3)} \\ &= \frac{2 \pm \sqrt{4 - 36}}{6} \\ &= \frac{2 \pm \sqrt{-32}}{6} \\ &= \frac{2 \pm \sqrt{-1 \cdot 16 \cdot 2}}{6} \\ &= \frac{2 \pm 4i\sqrt{2}}{6} \\ &= \frac{2}{6} \pm \frac{4\sqrt{2}}{6}i = \frac{1}{3} \pm \frac{2\sqrt{2}}{3}i \end{aligned}$$

39.
$$f(x) = 0$$
$$4x^2 + 4x - 19 = 0$$
$$a = 4, b = 4, c = -19$$
$$\begin{aligned} x &= \frac{-b \pm \sqrt{b^2 - 4ac}}{2a} \\ &= \frac{-4 \pm \sqrt{4^2 - 4(4)(-19)}}{2(4)} \\ &= \frac{-4 \pm \sqrt{16 + 304}}{8} \\ &= \frac{-4 \pm \sqrt{320}}{8} \\ &= -\frac{4}{8} \pm \frac{8\sqrt{5}}{8} = -\frac{1}{2} \pm \sqrt{5} \end{aligned}$$

41.
$$\begin{aligned} f(x) &= 0 \\ 3x^2 + 2x + 2 &= 0 \\ a = 3, b = 2, c &= 2 \\ x &= \frac{-b \pm \sqrt{b^2 - 4ac}}{2a} \\ &= \frac{-2 \pm \sqrt{2^2 - 4(3)(2)}}{2(3)} \\ &= \frac{-2 \pm \sqrt{4 - 24}}{6} \\ &= \frac{-2 \pm \sqrt{-20}}{6} \\ &= -\frac{2}{6} \pm \frac{2i\sqrt{5}}{6} = -\frac{1}{3} \pm \frac{\sqrt{5}}{3}i \end{aligned}$$

43.
$$\begin{aligned} &0.7x^2 - 3.5x - 25 = 0 \\ &a = 0.7, b = -3.5, c = -25 \\ x &= \frac{-b \pm \sqrt{b^2 - 4ac}}{2a} \\ &= \frac{3.5 \pm \sqrt{(-3.5)^2 - 4(0.7)(-25)}}{2(0.7)} \\ &= \frac{3.5 \pm \sqrt{12.25 + 70}}{1.4} \\ &= \frac{3.5 \pm \sqrt{82.25}}{1.4} \\ &= \frac{3.5 \pm 9.069}{1.4} \end{aligned}$$
$x = 8.98$ or $x = -3.98$

45.
$$\begin{aligned} C &= \frac{N^2 - N}{2} \\ 2C &= N^2 - N \\ N^2 - N - 2C &= 0 \\ a = 1, b = -1, c &= -2C \\ N &= \frac{-b \pm \sqrt{b^2 - 4ac}}{2a} \\ &= \frac{1 \pm \sqrt{(-1)^2 - 4(1)(-2C)}}{2(1)} \\ &= \frac{1 \pm \sqrt{1 + 8C}}{2} \end{aligned}$$

47. Let x and $x + 2$ represent the integers.
$$\begin{aligned} x(x + 2) &= 288 \\ x^2 + 2x - 288 &= 0 \\ (x + 18)(x - 16) &= 0 \end{aligned}$$
$x = -18$ or $x = 16$

Since the integers are positive, they must be 16 and 18.

49. Let x and $x + 1$ represent the integers.
$$\begin{aligned} x^2 + (x + 1)^2 &= 85 \\ x^2 + x^2 + 2x + 1 &= 85 \\ 2x^2 + 2x - 84 &= 0 \\ 2(x + 7)(x - 6) &= 0 \end{aligned}$$
$x = -7$ or $x = 6$

Since the integers are positive, they must be 6 and 7.

51.
$$\begin{aligned} (x - 3)(x - 5) &= 0 \\ x^2 - 8x + 15 &= 0 \end{aligned}$$

53.
$$\begin{aligned} (x - 2)(x - 3)(x + 4) &= 0 \\ (x^2 - 5x + 6)(x + 4) &= 0 \\ x^3 - x^2 - 14x + 24 &= 0 \end{aligned}$$

55.
$$\begin{aligned} \text{length} \cdot \text{width} &= \text{Area} \\ (x + 4)x &= 96 \\ x^2 + 4x - 96 &= 0 \\ (x + 12)(x - 8) &= 0 \end{aligned}$$
$x = -12$ or $x = 8$

Since the width is positive, the dimensions are 8 ft by 12 ft.

57. Let $s =$ the length of a side.

$$\text{Area} = \text{perimeter}$$
$$s^2 = 4s$$
$$s^2 - 4s = 0$$
$$s(s-4) = 0$$
$$s = 0 \text{ or } s = 4$$

Since the length cannot be 0, the length of a side is 4 units.

59. Let b represent the base.
Then $3b + 5$ represents the height.

$$\tfrac{1}{2}\text{base} \cdot \text{height} = \text{Area}$$
$$\tfrac{1}{2}b(3b+5) = 6$$
$$b(3b+5) = 12$$
$$3b^2 + 5b - 12 = 0$$
$$(3b-4)(b+3) = 0$$
$$b = \tfrac{4}{3} \text{ or } b = -3$$

Since the base is positive, it must be $\frac{4}{3}$ cm.

61. Let $r =$ the slower rate. Then $r + 20 =$ the faster rate.

	Rate	Time	Dist.
Slower	r	$\frac{150}{r}$	150
Faster	$r+20$	$\frac{150}{r+20}$	150

$$\boxed{\text{Faster time}} + 2 = \boxed{\text{Slower time}}$$
$$\frac{150}{r+20} + 2 = \frac{150}{r}$$
$$\left(\frac{150}{r+20} + 2\right)(r)(r+20) = \frac{150}{r} \cdot r(r+20)$$
$$150r + 2r(r+20) = 150(r+20)$$
$$2r^2 + 40r - 3000 = 0$$
$$2(r+50)(r-30) = 0$$

$r = -50$ or $r = 30$ $\quad$ $r = 30$ is the only answer that makes sense.
Her original speed was 30 mph.

63. Let $x =$ the number of 10¢ increases. Then the ticket price will be $4 + 0.10x$, while the projected attendance will be $300 - 5x$, for total receipts of $(4 + 0.10x)(300 - 5x)$.

$$\text{Total} = 1248$$
$$(4 + 0.10x)(300 - 5x) = 1248$$
$$1200 + 10x - 0.5x^2 = 1248$$
$$-0.5x^2 + 10x - 48 = 0$$
$$x^2 - 20x + 96 = 0$$
$$(x-12)(x-8) = 0$$

$x = 12$ or $x = 8 \Rightarrow 4 + 0.10(12) = 5.20;\ 4 + 0.10(8) = 4.80$

The ticket price would be either \$5.20 or \$4.80.

65. Let $x =$ the number of additional subscribers. Then the profit per subscriber will be $20 + 0.01x$, for a total profit of $(20 + 0.01x)(3000 + x)$.

$$\begin{aligned}\text{Total profit} &= 120000\\(20+0.01x)(3000+x) &= 120000\\60{,}000+50x+0.01x^2 &= 120000\\0.01x^2+50x-60{,}000 &= 0\\x^2+5000x-6{,}000{,}000 &= 0\\(x+6000)(x-1000) &= 0\end{aligned}$$

$x = -6000$ (impossible) or $x = 1000 \Rightarrow$ The total number of subscribers would be 4000.

67. Let $w =$ the constant width.

$$\begin{aligned}\text{Frame area} &= \text{Picture area}\\(12+2w)(10+2w)-12(10) &= 12(10)\\120+44w+4w^2-240 &= 0\\4w^2+44w-120 &= 0\\4(w^2+11w-30) &= 0\\w^2+11w-30 &= 0 \Rightarrow a=1, b=11, c=-30\end{aligned}$$

$$w = \frac{-11 \pm \sqrt{11^2 - 4(1)(-30)}}{2(1)}$$

$$= \frac{-11 \pm \sqrt{121+120}}{2} = \frac{-11 \pm \sqrt{241}}{2} = \frac{-11 \pm 15.52}{2} = \frac{4.52}{2} \text{ or } \frac{-26.52}{2} \text{ (impossible)}$$

$w = 2.26$ in.

69.

$$\begin{aligned}P &= 0.03x^2 - 1.37x + 82.51\\75 &= 0.03x^2 - 1.37x + 82.51\\0 &= 0.03x^2 - 1.37x + 7.51\end{aligned}$$

$a = 0.03$, $b = -1.37$, $c = 7.51$

$$x = \frac{-b \pm \sqrt{b^2 - 4ac}}{2a}$$

$$= \frac{1.37 \pm \sqrt{(-1.37)^2 - 4(0.03)(7.51)}}{2(0.03)} = \frac{1.37 \pm \sqrt{0.9757}}{0.06} \approx 39.3 \text{ or } 6.4$$

Since $x \leq 30$, the only answer that works is 6.4. The model indicates the desired result happened in early 1976.

71. Let $[H^+]$ (and then $[A^-]$) $= x$ and $[HA] = 0.1 - x$.

$$\frac{[H^+][A^-]}{[HA]} = 4 \times 10^{-4}$$

$$\frac{x^2}{0.1 - x} = 4 \times 10^{-4}$$

$$x^2 = 4 \times 10^{-5} - (4 \times 10^{-4})x$$

$$x^2 + (4 \times 10^{-4})x - 4 \times 10^{-5} = 0$$

$$x = \frac{-b \pm \sqrt{b^2 - 4ac}}{2a} = \frac{-4 \times 10^{-4} \pm \sqrt{(4 \times 10^{-4})^2 - 4(1)(-4 \times 10^{-5})}}{2(1)}$$

$$\approx \frac{-4 \times 10^{-4} \pm 0.012655}{2} \approx \frac{0.012255}{2} \text{ or } -\frac{0.013055}{2} \text{ (impossible)}$$

The concentration is about $0.00613 \text{ M} = 6.13 \times 10^{-3}$ M.

73. **Answers may vary.**

75. $x^2 + 2\sqrt{2}x - 6 = 0$

$a = 1, b = 2\sqrt{2}, c = -6$

$$x = \frac{-b \pm \sqrt{b^2 - 4ac}}{2a}$$

$$= \frac{-2\sqrt{2} \pm \sqrt{\left(2\sqrt{2}\right)^2 - 4(1)(-6)}}{2(1)}$$

$$= \frac{-2\sqrt{2} \pm \sqrt{8 + 24}}{2}$$

$$= \frac{-2\sqrt{2} \pm \sqrt{32}}{2} = \frac{-2\sqrt{2} \pm 4\sqrt{2}}{2}$$

$x = \frac{2\sqrt{2}}{2} = \sqrt{2}$ or $x = \frac{-6\sqrt{2}}{2} = -3\sqrt{2}$

77. $x^2 - 3ix - 2 = 0$

$a = 1, b = -3i, c = -2$

$$x = \frac{-b \pm \sqrt{b^2 - 4ac}}{2a}$$

$$= \frac{3i \pm \sqrt{(-3i)^2 - 4(1)(-2)}}{2(1)}$$

$$= \frac{3i \pm \sqrt{9i^2 + 8}}{2}$$

$$= \frac{3i \pm \sqrt{-1}}{2} = \frac{3i \pm i}{2}$$

$x = \frac{4i}{2} = 2i$ or $x = \frac{2i}{2} = i$

Exercise 10.3 (page 680)

1.

$$\frac{1}{4} + \frac{1}{t} = \frac{1}{2t}$$

$$\left(\frac{1}{4} + \frac{1}{t}\right)4t = \frac{1}{2t} \cdot 4t$$

$$t + 4 = 2$$

$$t = -2$$

3. $m = \frac{\Delta y}{\Delta x} = \frac{-4 - 5}{-2 - 3} = \frac{-9}{-5} = \frac{9}{5}$

5. $b^2 - 4ac$

7. rational; unequal

9. $4x^2 - 4x + 1 = 0; a = 4, b = -4, c = 1$

$$b^2 - 4ac = (-4)^2 - 4(4)(1) = 16 - 16 = 0$$

The solutions are rational and equal.

11. $5x^2 + x + 2 = 0; a = 5, b = 1, c = 2$

$$b^2 - 4ac = 1^2 - 4(5)(2) = 1 - 40 = -39$$

The solutions are complex conjugates.

13. $2x^2 = 4x - 1$

$2x^2 - 4x + 1 = 0; a = 2, b = -4, c = 1$

$$b^2 - 4ac = (-4)^2 - 4(2)(1) = 16 - 8 = 8$$

The solutions are irrational and unequal.

15. $x(2x - 3) = 20$

$2x^2 - 3x - 20 = 0; a = 2, b = -3, c = -20$

$$b^2 - 4ac = (-3)^2 - 4(2)(-20) = 9 + 160 = 169$$

The solutions are rational and unequal.

17. $x^2 + kx + 9 = 0; a = 1, b = k, c = 9$

Set the discriminant equal to 0:

$$\begin{aligned} b^2 - 4ac &= 0 \\ k^2 - 4(1)(9) &= 0 \\ k^2 - 36 &= 0 \\ k^2 &= 36 \\ k &= \pm 6 \end{aligned}$$

19. $9x^2 + 4 = -kx$

$9x^2 + kx + 4 = 0; a = 9, b = k, c = 4$

Set the discriminant equal to 0:

$$\begin{aligned} b^2 - 4ac &= 0 \\ k^2 - 4(9)(4) &= 0 \\ k^2 - 144 &= 0 \\ k^2 &= 144 \\ k &= \pm 12 \end{aligned}$$

21. $(k - 1)x^2 + (k - 1)x + 1 = 0$

$a = k - 1, b = k - 1, c = 1$

Set the discriminant equal to 0:

$$\begin{aligned} b^2 - 4ac &= 0 \\ (k - 1)^2 - 4(k - 1)(1) &= 0 \\ k^2 - 2k + 1 - 4k + 4 &= 0 \\ k^2 - 6k + 5 &= 0 \\ (k - 5)(k - 1) &= 0 \end{aligned}$$

$k - 5 = 0$ **or** $k - 1 = 0$

$k = 5$ $\quad$ $k = 1$

doesn't work

23. $(k + 4)x^2 + 2kx + 9 = 0$

$a = k + 4, b = 2k, c = 9$

Set the discriminant equal to 0:

$$\begin{aligned} b^2 - 4ac &= 0 \\ (2k)^2 - 4(k + 4)(9) &= 0 \\ 4k^2 - 36(k + 4) &= 0 \\ 4k^2 - 36k - 144 &= 0 \\ 4(k^2 - 9k - 36) &= 0 \\ 4(k - 12)(k + 3) &= 0 \end{aligned}$$

$k - 12 = 0$ **or** $k + 3 = 0$

$k = 12$ $\quad$ $k = -3$

25. $1492x^2 + 1776x - 1984 = 0$

$a = 1492, b = 1776, c = -1984$

$$\begin{aligned} b^2 - 4ac &= (1776)^2 - 4(1492)(-1984) \\ &= 3{,}154{,}176 + 11{,}840{,}512 \\ &= 14{,}994{,}688 \end{aligned}$$

The solutions are real numbers.

27. $3x^2 + 4x = k$

$3x^2 + 4x - k = 0$

$a = 3, b = 4, c = -k$

Set the discriminant less than 0:

$$\begin{aligned} b^2 - 4ac &< 0 \\ 4^2 - 4(3)(-k) &< 0 \\ 16 + 12k &< 0 \\ 12k &< -16 \\ k &< -\frac{16}{12}, \text{ or } k < -\frac{4}{3} \end{aligned}$$

29.
$$\begin{aligned} x^4 - 17x^2 + 16 &= 0 \\ (x^2 - 16)(x^2 - 1) &= 0 \end{aligned}$$
$$x^2 - 16 = 0 \quad \textbf{or} \quad x^2 - 1 = 0$$
$$x^2 = 16 \qquad x^2 = 1$$
$$x = \pm 4 \qquad x = \pm 1$$

31.
$$\begin{aligned} x^4 - 3x^2 &= -2 \\ x^4 - 3x^2 + 2 &= 0 \\ (x^2 - 2)(x^2 - 1) &= 0 \end{aligned}$$
$$x^2 - 2 = 0 \quad \textbf{or} \quad x^2 - 1 = 0$$
$$x^2 = 2 \qquad x^2 = 1$$
$$x = \pm\sqrt{2} \qquad x = \pm 1$$

33.
$$\begin{aligned} x^4 &= 6x^2 - 5 \\ x^4 - 6x^2 + 5 &= 0 \\ (x^2 - 5)(x^2 - 1) &= 0 \end{aligned}$$
$$x^2 - 5 = 0 \quad \textbf{or} \quad x^2 - 1 = 0$$
$$x^2 = 5 \qquad x^2 = 1$$
$$x = \pm\sqrt{5} \qquad x = \pm 1$$

35.
$$\begin{aligned} 2x^4 - 10x^2 &= -8 \\ 2x^4 - 10x^2 + 8 &= 0 \\ 2(x^2 - 4)(x^2 - 1) &= 0 \end{aligned}$$
$$x^2 - 4 = 0 \quad \textbf{or} \quad x^2 - 1 = 0$$
$$x^2 = 4 \qquad x^2 = 1$$
$$x = \pm 2 \qquad x = \pm 1$$

37.
$$\begin{aligned} 2x^4 + 24 &= 26x^2 \\ 2x^4 - 26x^2 + 24 &= 0 \\ 2(x^2 - 12)(x^2 - 1) &= 0 \end{aligned}$$
$$x^2 - 12 = 0 \quad \textbf{or} \quad x^2 - 1 = 0$$
$$x^2 = 12 \qquad x^2 = 1$$
$$x = \pm\sqrt{12} \qquad x = \pm 1$$
$$= \pm 2\sqrt{3}$$

39.
$$\begin{aligned} t^4 + 3t^2 &= 28 \\ t^4 + 3t^2 - 28 &= 0 \\ (t^2 + 7)(t^2 - 4) &= -0 \end{aligned}$$
$$t^2 + 7 = 0 \quad \textbf{or} \quad t^2 - 4 = 0$$
$$t^2 = -7 \qquad t^2 = 4$$
$$t = \pm\sqrt{-7} \qquad t = \pm 2$$
$$= \pm i\sqrt{7}$$

41.
$$\begin{aligned} x - 6\sqrt{x} + 8 &= 0 \\ (\sqrt{x} - 2)(\sqrt{x} - 4) &= 0 \end{aligned}$$
$$\sqrt{x} - 2 = 0 \quad \textbf{or} \quad \sqrt{x} - 4 = 0$$
$$\sqrt{x} = 2 \qquad \sqrt{x} = 4$$
$$\left(\sqrt{x}\right)^2 = 2^2 \qquad \left(\sqrt{x}\right)^2 = 4^2$$
$$x = 4 \qquad x = 16$$
Solution Solution

43.
$$\begin{aligned} 2x - \sqrt{x} &= 3 \\ 2x - \sqrt{x} - 3 &= 0 \\ (2\sqrt{x} - 3)(\sqrt{x} + 1) &= 0 \end{aligned}$$
$$2\sqrt{x} - 3 = 0 \quad \textbf{or} \quad \sqrt{x} + 1 = 0$$
$$2\sqrt{x} = 3 \qquad \sqrt{x} = -1$$
$$\left(2\sqrt{x}\right)^2 = 3^2 \qquad \left(\sqrt{x}\right)^2 = (-1)^2$$
$$4x = 9 \qquad x = 1$$
$$x = \tfrac{9}{4}$$
Solution Does not check

45.
$$\begin{aligned} 2x + x^{1/2} - 3 &= 0 \\ \left(2x^{1/2} + 3\right)\left(x^{1/2} - 1\right) &= 0 \end{aligned}$$
$$2x^{1/2} + 3 = 0 \quad \textbf{or} \quad x^{1/2} - 1 = 0$$
$$2x^{1/2} = -3 \qquad x^{1/2} = 1$$
$$x^{1/2} = -\tfrac{3}{2} \qquad \left(x^{1/2}\right)^2 = 1^2$$
$$\left(x^{1/2}\right)^2 = \left(-\tfrac{3}{2}\right)^2 \qquad x = 1$$
$$x = \tfrac{9}{4}$$
Does not check Solution

47.
$$\begin{aligned} 3x + 5x^{1/2} + 2 &= 0 \\ \left(3x^{1/2} + 2\right)\left(x^{1/2} + 1\right) &= 0 \end{aligned}$$
$$3x^{1/2} + 2 = 0 \quad \textbf{or} \quad x^{1/2} + 1 = 0$$
$$3x^{1/2} = -2 \qquad x^{1/2} = -1$$
$$x^{1/2} = -\tfrac{2}{3} \qquad \left(x^{1/2}\right)^2 = (-1)^2$$
$$\left(x^{1/2}\right)^2 = \left(-\tfrac{2}{3}\right)^2 \qquad x = 1$$
$$x = \tfrac{4}{9}$$
Does not check Does not check

49.

$$x^{2/3} + 5x^{1/3} + 6 = 0$$
$$\left(x^{1/3} + 2\right)\left(x^{1/3} + 3\right) = 0$$

$x^{1/3} + 2 = 0$ **or** $x^{1/3} + 3 = 0$

$x^{1/3} = -2$ $\quad$ $x^{1/3} = -3$

$\left(x^{1/3}\right)^3 = (-2)^3$ $\quad$ $\left(x^{1/3}\right)^3 = (-3)^3$

$x = -8$ $\quad$ $x = -27$

Solution $\quad$ Solution

51.

$$x^{2/3} - 2x^{1/3} - 3 = 0$$
$$\left(x^{1/3} - 3\right)\left(x^{1/3} + 1\right) = 0$$

$x^{1/3} - 3 = 0$ **or** $x^{1/3} + 1 = 0$

$x^{1/3} = 3$ $\quad$ $x^{1/3} = -1$

$\left(x^{1/3}\right)^3 = (3)^3$ $\quad$ $\left(x^{1/3}\right)^3 = (-1)^3$

$x = 27$ $\quad$ $x = -1$

Solution $\quad$ Solution

53.

$$x + 5 + \frac{4}{x} = 0$$
$$x\left(x + 5 + \frac{4}{x}\right) = x(0)$$
$$x^2 + 5x + 4 = 0$$
$$(x + 4)(x + 1) = 0$$

$x + 4 = 0$ **or** $x + 1 = 0$

$x = -4$ $\quad$ $x = -1$

55.

$$x + 1 = \frac{20}{x}$$
$$x + 1 - \frac{20}{x} = 0$$
$$x\left(x + 1 - \frac{20}{x}\right) = x(0)$$
$$x^2 + x - 20 = 0$$
$$(x + 5)(x - 4) = 0$$

$x + 5 = 0$ **or** $x - 4 = 0$

$x = -5$ $\quad$ $x = 4$

57.

$$\frac{1}{x-1} + \frac{3}{x+1} = 2$$
$$\left(\frac{1}{x-1} + \frac{3}{x+1}\right)(x-1)(x+1) = 2(x+1)(x-1)$$
$$1(x+1) + 3(x-1) = 2(x^2 - 1)$$
$$x + 1 + 3x - 3 = 2x^2 - 2$$
$$0 = 2x^2 - 4x$$
$$0 = 2x(x - 2)$$

$2x = 0$ **or** $x - 2 = 0$

$x = 0$ $\quad$ $x = 2$

59.

$$\frac{1}{x+2} + \frac{24}{x+3} = 13$$
$$\left(\frac{1}{x+2} + \frac{24}{x+3}\right)(x+2)(x+3) = 13(x+2)(x+3)$$
$$1(x+3) + 24(x+2) = 13(x^2 + 5x + 6)$$
$$x + 3 + 24x + 48 = 13x^2 + 65x + 78$$
$$0 = 13x^2 + 40x + 27$$
$$0 = (13x + 27)(x + 1)$$

$13x + 27 = 0$ **or** $x + 1 = 0$

$13x = -27$ $\quad$ $x = -1$

$x = -\frac{27}{13}$

61.
$$\begin{aligned}
x+\frac{2}{x-2}&=0\\
\left(x+\frac{2}{x-2}\right)(x-2)&=0(x-2)\\
x(x-2)+2&=0\\
x^2-2x+2&=0\\
x^2-2x&=-2\\
x^2-2x+1&=-2+1\\
(x-1)^2&=-1\\
x-1&=\pm\sqrt{-1}\\
x&=1\pm i
\end{aligned}$$

63.
$$\begin{aligned}
x^{-4}-2x^{-2}+1&=0\\
(x^{-2}-1)(x^{-2}-1)&=0
\end{aligned}$$
$$\begin{aligned}
x^{-2}-1&=0 &\quad \textbf{or} \quad x^{-2}-1&=0\\
x^{-2}&=1 & x^{-2}&=1\\
\frac{1}{x^2}&=1 & \frac{1}{x^2}&=1\\
1&=x^2 & 1&=x^2\\
\pm 1&=x & \pm 1&=x
\end{aligned}$$

65.
$$\begin{aligned}
8a^{-2}-10a^{-1}-3&=0\\
(2a^{-1}-3)(4a^{-1}+1)&=0
\end{aligned}$$
$$\begin{aligned}
2a^{-1}-3&=0 &\quad \textbf{or} \quad 4a^{-1}+1&=0\\
2a^{-1}&=3 & 4a^{-1}&=-1\\
\frac{2}{a}&=3 & \frac{4}{a}&=-1\\
2&=3a & 4&=-a\\
\tfrac{2}{3}&=a & -4&=a
\end{aligned}$$

67. Let $y=m+1$.
$$\begin{aligned}
8(m+1)^{-2}-30(m+1)^{-1}+7&=0\\
8y^{-2}-30y^{-1}+7&=0\\
(4y^{-1}-1)(2y^{-1}-7)&=0
\end{aligned}$$
$$\begin{aligned}
4y^{-1}-1&=0 &\quad \textbf{or} \quad 2y^{-1}-7&=0\\
4y^{-1}&=1 & 2y^{-1}&=7\\
\frac{4}{y}&=1 & \frac{2}{y}&=7\\
4&=y & 2&=7y\\
& & \tfrac{2}{7}&-y\\
4&=m+1 & \tfrac{2}{7}&=m+1\\
3&=m & -\tfrac{5}{7}&=m
\end{aligned}$$

69.
$$\begin{aligned}
x^2+y^2&=r^2\\
x^2&=r^2-y^2\\
x&=\pm\sqrt{r^2-y^2}
\end{aligned}$$

71.
$$\begin{aligned}
I&=\frac{k}{d^2}\\
Id^2&=k\\
d^2&=\frac{k}{I}\\
d&=\pm\sqrt{\frac{k}{I}}=\pm\frac{\sqrt{kI}}{I}
\end{aligned}$$

73. $xy^2 + 3xy + 7 = 0;\ a = x, b = 3x, c = 7$

$$y = \frac{-b \pm \sqrt{b^2 - 4ac}}{2a} = \frac{-3x \pm \sqrt{(3x)^2 - 4(x)(7)}}{2x} = \frac{-3x \pm \sqrt{9x^2 - 28x}}{2x}$$

75. $\sigma = \sqrt{\frac{\Sigma x^2}{N} - \mu^2}$

$$\sigma^2 = \frac{\Sigma x^2}{N} - \mu^2$$

$$\mu^2 = \frac{\Sigma x^2}{N} - \sigma^2$$

77. $12x^2 - 5x - 2 = 0;\ a = 12, b = -5, c = -2$

$(4x + 1)(3x - 2) = 0$

$4x + 1 = 0$ **or** $3x - 2 = 0$

$4x = -1 \qquad 3x = 2$

$x = -\frac{1}{4} \qquad x = \frac{2}{3}$

$$-\frac{b}{a} = -\frac{-5}{12} = \frac{5}{12} \qquad -\frac{1}{4} + \frac{2}{3} = -\frac{3}{12} + \frac{8}{12} = \frac{5}{12}$$

$$\frac{c}{a} = \frac{-2}{12} = -\frac{1}{6} \qquad \left(-\frac{1}{4}\right)\left(\frac{2}{3}\right) = -\frac{1}{6}$$

79. $2x^2 + 5x + 1 = 0;\ a = 2, b = 5, c = 1;\ -\frac{b}{a} = -\frac{5}{2};\ \frac{c}{a} = \frac{1}{2}$

$$x = \frac{-b \pm \sqrt{b^2 - 4ac}}{2a} = \frac{-5 \pm \sqrt{5^2 - 4(2)(1)}}{2(2)} = \frac{-5 \pm \sqrt{17}}{4} = -\frac{5}{4} \pm \frac{\sqrt{17}}{4}$$

$$\frac{-5 + \sqrt{17}}{4} + \frac{-5 - \sqrt{17}}{4} = \frac{-10}{4} = -\frac{5}{2}$$

$$\left(\frac{-5 + \sqrt{17}}{4}\right)\left(\frac{-5 - \sqrt{17}}{4}\right) = \frac{25 + 5\sqrt{17} - 5\sqrt{17} - 17}{16} = \frac{8}{16} = \frac{1}{2}$$

81. $3x^2 - 2x + 4 = 0;\ a = 3, b = -2, c = 4;\ -\frac{b}{a} = -\frac{-2}{3} = \frac{2}{3};\ \frac{c}{a} = \frac{4}{3}$

$$x = \frac{-b \pm \sqrt{b^2 - 4ac}}{2a} = \frac{-(-2) \pm \sqrt{(-2)^2 - 4(3)(4)}}{2(3)} = \frac{2 \pm \sqrt{-44}}{6} = \frac{1}{3} \pm \frac{i\sqrt{11}}{3}$$

$$\frac{1 + i\sqrt{11}}{3} + \frac{1 - i\sqrt{11}}{3} = \frac{2}{3}$$

$$\left(\frac{1 + i\sqrt{11}}{3}\right)\left(\frac{1 - i\sqrt{11}}{3}\right) = \frac{1 - i\sqrt{11} + i\sqrt{11} - 11i^2}{9} = \frac{1 + 11}{9} = \frac{12}{9} = \frac{4}{3}$$

83. $x^2 + 2x + 5 = 0;\ a = 1, b = 2, c = 5;\ -\frac{b}{a} = -\frac{2}{1} = -2;\ \frac{c}{a} = \frac{5}{1} = 5$

$$x = \frac{-b \pm \sqrt{b^2 - 4ac}}{2a} = \frac{-2 \pm \sqrt{2^2 - 4(1)(5)}}{2(1)} = \frac{-2 \pm \sqrt{-16}}{2} = \frac{-2 \pm 4i}{2} = -1 \pm 2i$$

$(-1 + 2i) + (-1 - 2i) = -2;\ (-1 + 2i)(-1 - 2i) = 1 + 2i - 2i - 4i^2 = 1 + 4 = 5$

85. **Answers may vary.**

87. No

Exercise 10.4 (page 693)

1. $3x + 5 = 5x - 15$
$-2x = -20$
$x = 10$

3. Let t = the time of the second train.
Then $t + 3$ = the time of the first train.

	Rate	Time	Dist.
First	30	$t+3$	$30(t+3)$
Second	55	t	$55t$

1st distance = 2nd distance
$30(t+3) = 55t$
$30t + 90 = 55t$
$-25t = -90$
$t = \frac{-90}{-25} = \frac{18}{5} = 3\frac{3}{5}$ hours

5. $f(x) = ax^2 + bx + c;$
$a \neq 0$

7. vertex

9. upward

11. to the right

13. upward

15. $f(x) = x^2$
vertex: $(0,0)$; opens U

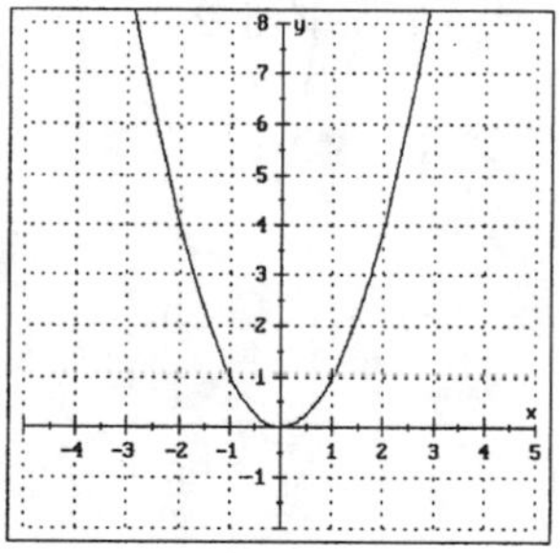

17. $f(x) = x^2 + 2$
vertex: $(0,2)$; opens U

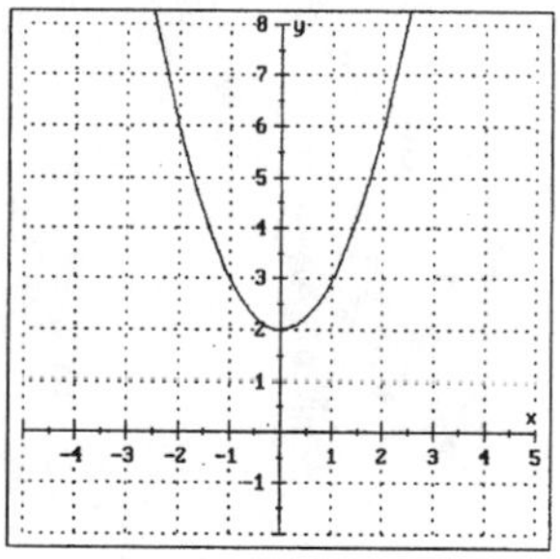

19. $f(x) = -(x-2)^2$
vertex: $(2,0)$; opens D

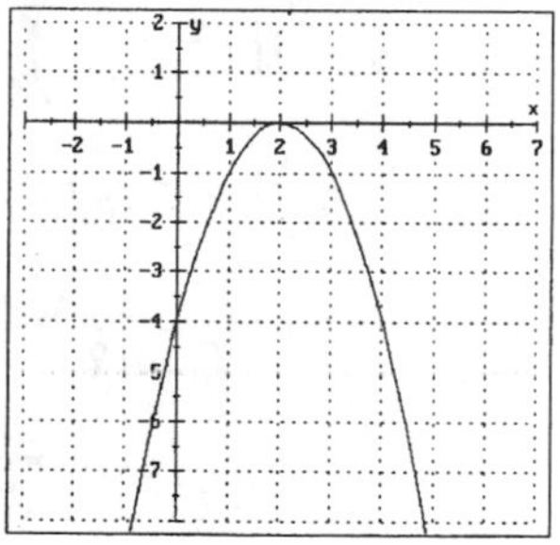

21. $f(x) = (x-3)^2 + 2$
vertex: $(3,2)$; opens U

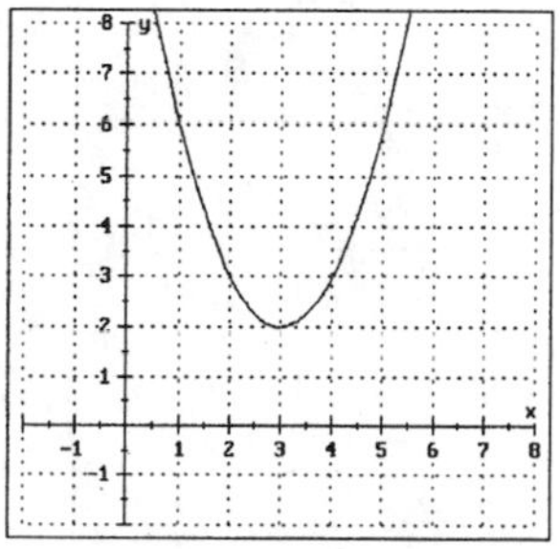

23. $f(x) = x^2 + x - 6$
$f(x) = \left(x + \frac{1}{2}\right)^2 - \frac{25}{4}$
vertex: $\left(-\frac{1}{2}, -\frac{25}{4}\right)$; opens U

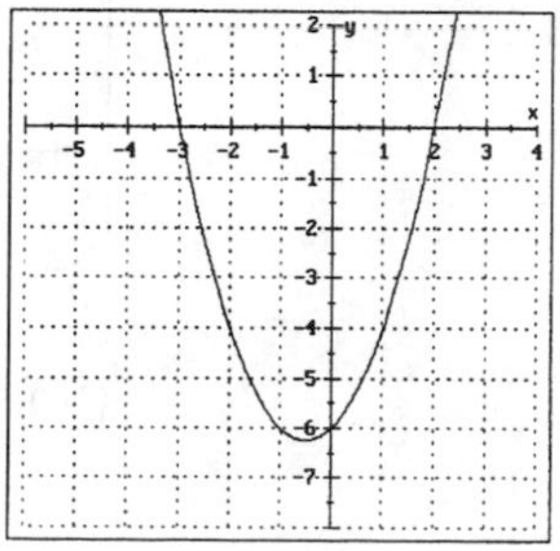

25. $f(x) = -2x^2 + 4x + 1$
$f(x) = -2(x-1)^2 + 3$
vertex: $(1,3)$; opens D

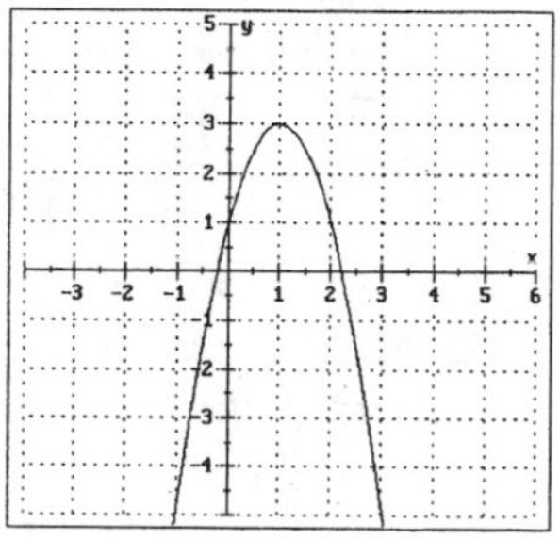

27. $f(x) = 3x^2 - 12x + 10$
$f(x) = 3(x-2)^2 - 2$
vertex: $(2, -2)$; opens U

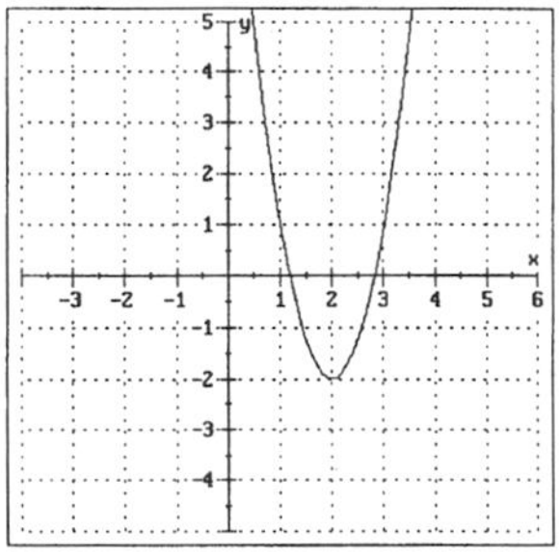

29. $y = (x-1)^2 + 2; V(1, 2)$; axis: $x = 1$

31. $y = 2(x+3)^2 - 4$;
$V(-3, -4)$; axis: $x = -3$

33. $y = -3x^2 \Rightarrow y = -3(x-0)^2 + 0$
$V(0, 0)$; axis: $x = 0$

35. $y = 2x^2 - 4x$
$y = 2(x^2 - 2x)$
$y = 2(x^2 - 2x + 1) - 2$
$y = 2(x-1)^2 - 2$
$V(1, -2)$; axis: $x = 1$

37. $y = -4x^2 + 16x + 5$
$y = -4(x^2 - 4x) + 5$
$y = -4(x^2 - 4x + 4) + 5 + 16$
$y = -4(x-2)^2 + 21$
$V(2, 21)$; axis: $x = 2$

39. $y - 7 = 6x^2 - 5x$

$$y = 6\left(x^2 - \frac{5}{6}x\right) + 7$$

$$y = 6\left(x^2 - \frac{5}{6}x + \frac{25}{144}\right) + 7 - \frac{25}{24}$$

$$y = 6\left(x - \frac{5}{12}\right)^2 + \frac{143}{24}$$

$V\left(\frac{5}{12}, \frac{143}{24}\right)$; axis: $x = \frac{5}{12}$

41. $y - 2 = (x-5)^2$
$y = (x-5)^2 + 2 \Rightarrow V(5, 2)$

43. $y = 2x^2 - x + 1$

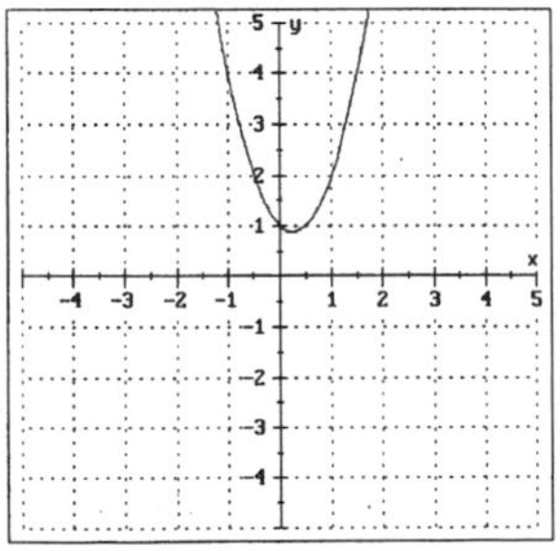

$V(0.25, 0.88)$

45. $y = 7 + x - x^2$

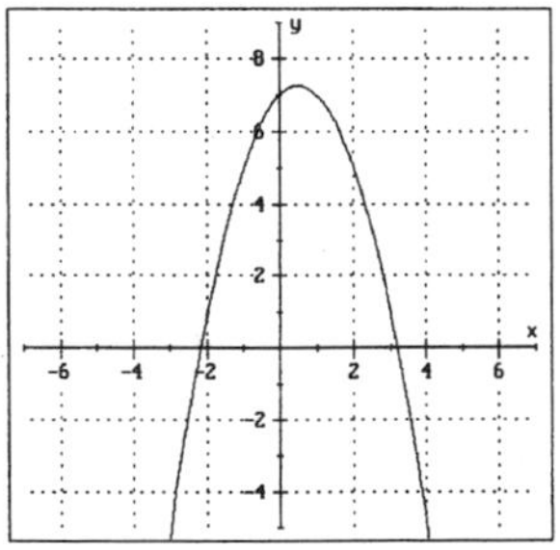

$V(0.5, 7.25)$

47. $y = x^2 + x - 6$

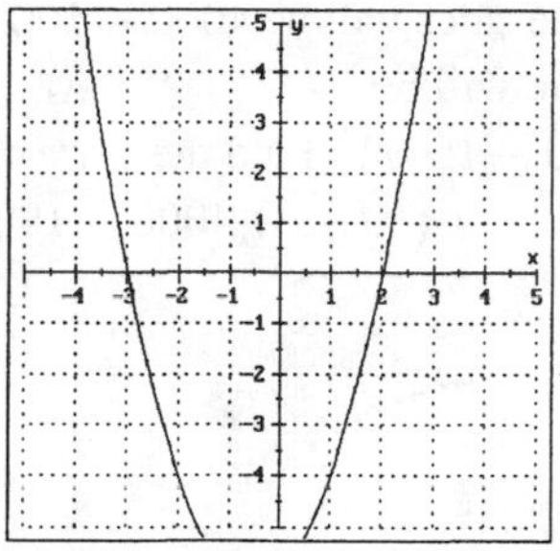

solution set: $\{2, -3\}$

49. $y = 0.5x^2 - 0.7x - 3$

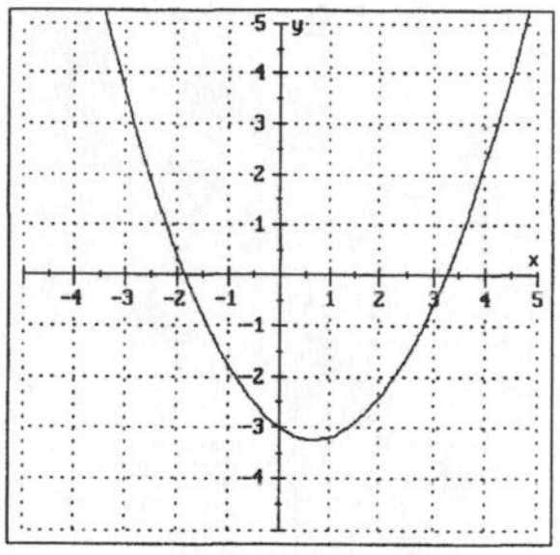

solution set: $\{-1.85, 3.25\}$

51. Since the graph of the height equation is a parabola, the max. height occurs at the vertex.

$s = 48t - 16t^2$

$s = -16(t^2 - 3t)$

$s = -16\left(t^2 - 3t + \frac{9}{4}\right) + 36$

$s = -16\left(t - \frac{3}{2}\right)^2 + 36$

$V\left(\frac{3}{2}, 36\right) \Rightarrow$ max. height $= 36$ ft

To find the time it takes for the ball to return to earth, set $s = 0$ and solve for t.

$s = 48t - 16t^2$

$0 = 48t - 16t^2$

$0 = 16t(3 - t)$

$t = 0$ or $t = 3$

The ball returns to earth after 3 seconds.
(0 seconds is when the ball is originally thrown.)

53. Let $w =$ the width of the rectangle.
Then $100 - w =$ the length.

$A = w(100 - w)$

$A = -w^2 + 100w$

$A = -(w^2 - 100w + 2500) + 2500$

$A = -(w - 50)^2 + 2500$

dim: 50 ft by 50 ft; area $= 2500$ ft^2

55. Let $w =$ the width of the rectangle.
Then $150 - w =$ the length.

$A = w(150 - w)$

$A = -w^2 + 150w$

$A = -(w^2 - 150w + 5625) + 5625$

$A = -(w - 75)^2 + 5625$

dim: 75 ft by 75 ft; area $= 5625$ ft^2

57. Graph $y = H = 3.3x^2 - 59.4x + 281.3$ and find the y-coordinate of the vertex:

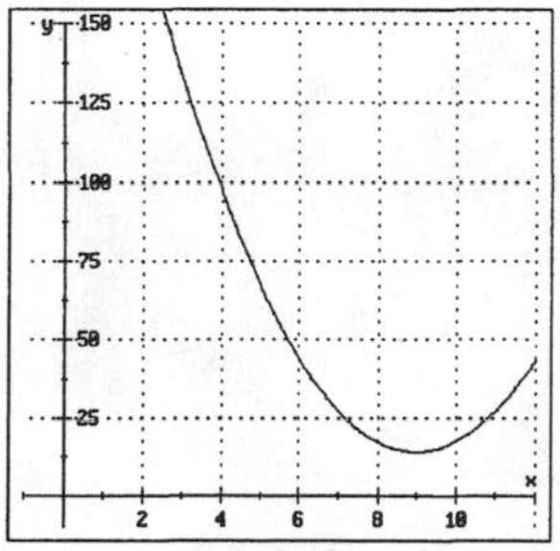

The minimum level was 14 feet.

59. Graph $y = R = -\frac{x^2}{1000} + 10x$ and find the x-coordinate of the vertex:

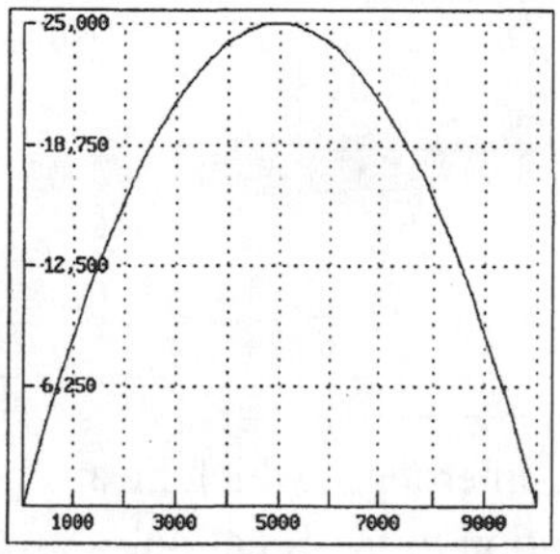

5000 stereos should be sold.

61. Graph $y = R = -\frac{x^2}{728} + 9x$ and find the x- and y- coordinates of the vertex:

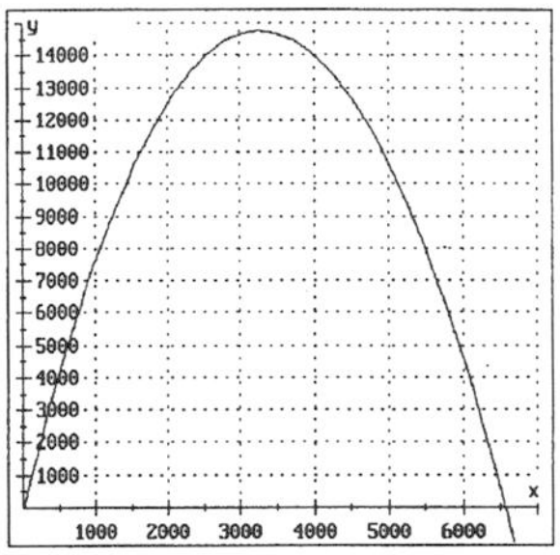

Max. revenue = \$14,742; # radios = 3276

63. Let $x =$ the number of \$1 increases to the price. Then the sales will be $4000 - 100x$, and the revenue will be $(30 + x)(4000 - 100x)$. Find the vertex of the parabola $y = (30 + x)(4000 - 100x)$.

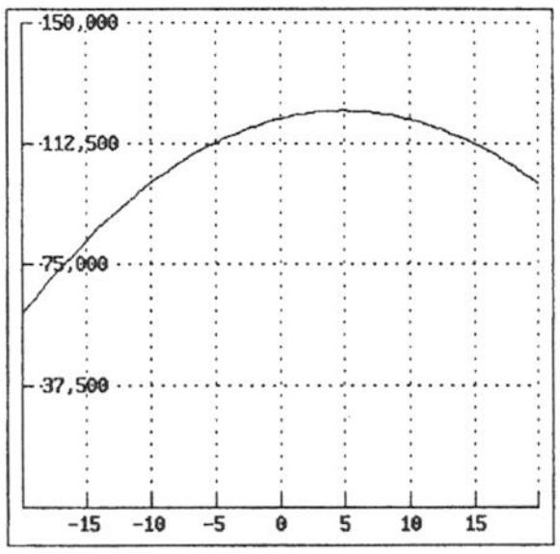

The price should increase \$5, to a total of \$35.

65. Graph $y = 50x(1 - x)$ and find the x-coordinate(s) when $y = 9.375$.

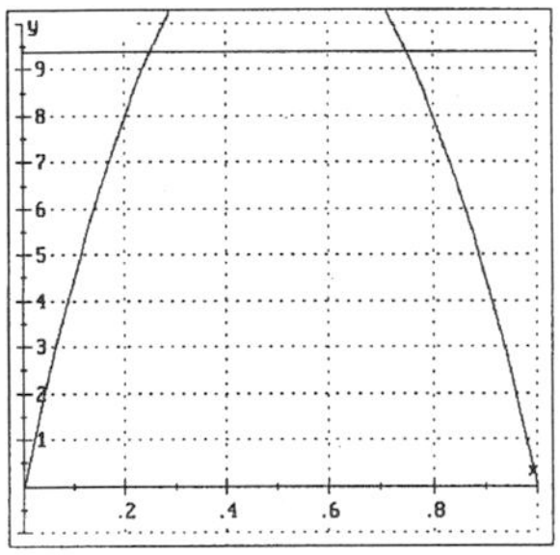

$p = 0.25$ or $p = 0.75$

67. **Answers may vary.**

69. Graph $y = x^2 + x + 1$ to find x-intercept(s):

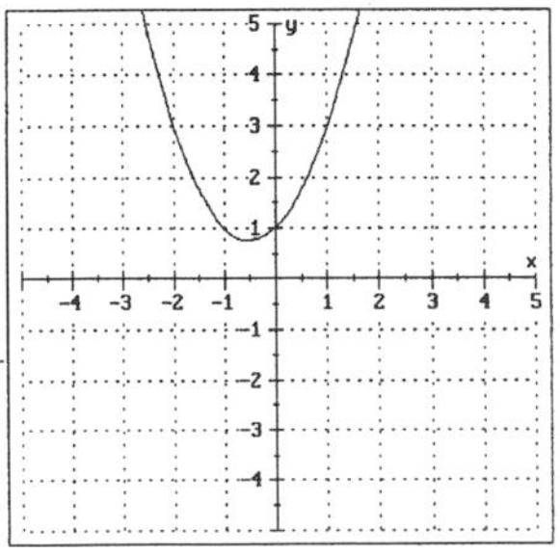

There are no x-intercepts, which means there is no solution to the equation.

Exercise 10.5 (page 703)

1. $y = kx$

3. $t = kxy$

5. $y = 3x - 4$
$m = 3$

7. greater

9. undefined

11. $x^2 - 5x + 4 < 0$
$(x - 4)(x - 1) < 0$
$x - 4$ – – – – – – – 0 ++++
$x - 1$ – – – 0+++++++++++
←—(——)—→
1 4
solution set: $(1, 4)$

13. $x^2 - 8x + 15 > 0$
$(x - 5)(x - 3) > 0$
$x - 5$ – – – – – – – – –0++++
$x - 3$ – – – 0+++++++++++
←—)——(—→
3 5
solution set: $(-\infty, 3) \cup (5, \infty)$

15. $x^2 + x - 12 \le 0$
$(x - 3)(x + 4) \le 0$
$x - 3$ – – – – – – – – – 0++++
$x + 4$ – – – 0+++++++++++
←— [——]—→
−4 3
solution set: $[-4, 3]$

17. $x^2 + 2x \ge 15$
$x^2 + 2x - 15 \ge 0$
$(x - 3)(x + 5) \ge 0$
$x - 3$ – – – – – – – – –0++++
$x + 5$ – – – 0+++++++++++
←—]——[—→
−5 3
solution set: $(-\infty, -5] \cup [3, \infty)$

19. $x^2 + 8x < -16$
$x^2 + 8x + 16 < 0$
$(x + 4)(x + 4) < 0$
$x + 4$ – – – – 0+++++++
$x + 4$ – – – – 0+++++++
←————→
−4
Since the product is never negative, there is no solution.

21. $x^2 \ge 9$
$x^2 - 9 \ge 0$
$(x - 3)(x + 3) \ge 0$
$x - 3$ – – – – – – – – –0++++
$x + 3$ – – – 0+++++++++++
←—]——[—→
−3 3
solution set: $(-\infty, -3] \cup [3, \infty)$

23. $2x^2 - 50 < 0$
$2(x - 5)(x + 5) < 0$
$x - 5$ – – – – – – – – – 0++++
$x + 5$ – – – 0+++++++ ++++
←— (——)—→
−5 5
solution set: $(-5, 5)$

25.

$$\frac{1}{x} < 2$$
$$\frac{1}{x} - 2 < 0$$
$$\frac{1}{x} - \frac{2x}{x} < 0$$
$$\frac{1-2x}{x} < 0$$

$1 - 2x$ +++++++++ 0 − − −
x − − − 0++++++++++++

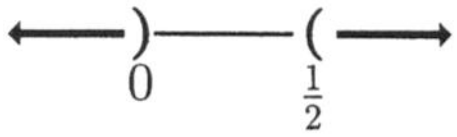

solution set: $(-\infty, 0) \cup \left(\frac{1}{2}, \infty\right)$

27.

$$\frac{4}{x} \geq 2$$
$$\frac{4}{x} - 2 \geq 0$$
$$\frac{4}{x} - \frac{2x}{x} \geq 0$$
$$\frac{4-2x}{x} \geq 0$$

$4 - 2x$ ++++++++++0 − − −
x − − − 0 ++++++++++++

solution set: $(0, 2]$

29.

$$-\frac{5}{x} < 3$$
$$-\frac{5}{x} - 3 < 0$$
$$\frac{-5}{x} - \frac{3x}{x} < 0$$
$$\frac{-5-3x}{x} < 0$$

$-5 - 3x$ +++ 0 − − − − − − − −
x − − − − − − − − − 0++++

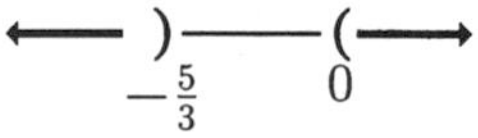

solution set: $\left(-\infty, -\frac{5}{3}\right) \cup (0, \infty)$

31.

$$\frac{x^2 - x - 12}{x - 1} < 0$$
$$\frac{(x-4)(x+3)}{x-1} < 0$$

$x - 4$ − − − − − − − − − − − 0++++
$x - 1$ − − − − − − − − 0+++ +++++
$x + 3$ − − − 0++++++ ++++ +++++

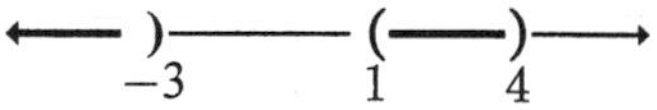

solution set: $(-\infty, -3) \cup (1, 4)$

33.

$$\frac{x^2 + x - 20}{x + 2} \geq 0$$
$$\frac{(x-4)(x+5)}{x+2} \geq 0$$

$x - 4$ − − − − − − − − − − − − 0++++
$x + 2$ − − − − − − − − 0++++ +++++
$x + 5$ − − − 0+++++++++++ +++++

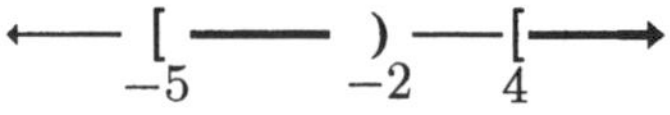

solution set: $[-5, -2) \cup [4, \infty)$

35.

$$\frac{x^2 - 4x + 4}{x + 4} < 0$$
$$\frac{(x-2)(x-2)}{x+4} < 0$$

$x - 2$ − − − − − − − − 0++++
$x - 2$ − − − − − − − − 0++++
$x + 4$ − − − 0++++++ +++++

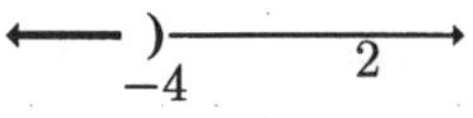

solution set: $(-\infty, -4)$

37.

$$\frac{6x^2-5x+1}{2x+1}>0$$

$$\frac{(2x-1)(3x-1)}{2x+1}>0$$

```
2x − 1   −−−−−−−− −−−− 0++++
3x − 1   −−−−−−−− 0+++++++++
2x + 1   −−− 0+++++++++++++++++
```

$\longleftarrow\ (\longrightarrow)\ \ (\longrightarrow$ at $-\frac{1}{2}$, $\frac{1}{3}$, $\frac{1}{2}$

solution set: $\left(-\frac{1}{2},\frac{1}{3}\right)\cup\left(\frac{1}{2},\infty\right)$

39.

$$\frac{3}{x-2}<\frac{4}{x}$$

$$\frac{3}{x-2}-\frac{4}{x}<0$$

$$\frac{3x}{x(x-2)}-\frac{4(x-2)}{x(x-2)}<0$$

$$\frac{-x+8}{x(x-2)}<0$$

```
−x + 8   +++++++++++++++0−−−
x − 2    −−−−−−− 0+++++++++
x        −−− 0++++ ++++++++++
```

$\longleftarrow(\longrightarrow)\ \ (\longrightarrow$ at 0, 2, 8

solution set: $(0,2)\cup(8,\infty)$

41.

$$\frac{-5}{x+2}\geq\frac{4}{2-x}$$

$$\frac{-5}{x+2}-\frac{4}{2-x}\geq 0$$

$$\frac{-5(2-x)}{(x+2)(2-x)}-\frac{4(x+2)}{(x+2)(2-x)}\geq 0$$

$$\frac{x-18}{(x+2)(2-x)}\geq 0$$

```
x − 18   −−−−−−−−− −−−− 0++++
2 − x    ++++++++++++ 0−−−−−−
x + 2    −−− 0+++++++++++++++++
```

$\longleftarrow)\ \ (\longrightarrow]\longrightarrow$ at -2, 2, 18

solution set: $(-\infty,-2)\cup(2,18]$

43.

$$\frac{7}{x-3}\geq\frac{2}{x+4}$$

$$\frac{7}{x-3}-\frac{2}{x+4}\geq 0$$

$$\frac{7(x+4)}{(x-3)(x+4)}-\frac{2(x-3)}{(x-3)(x+4)}\geq 0$$

$$\frac{5x+34}{(x-3)(x+4)}\geq 0$$

```
x − 3     −−−−−−−− −−−− 0++++
x + 4     −−−−−−−− 0++++ +++++
5x + 34   −−− 0++++++ +++++ +++++
```

$\longleftarrow\ [\longrightarrow)\ \ (\longrightarrow$ at $-\frac{34}{5}$, -4, 3

solution set: $\left[-\frac{34}{5},-4\right)\cup(3,\infty)$

45.

$$\frac{x}{x+4}\leq\frac{1}{x+1}$$

$$\frac{x}{x+4}-\frac{1}{x+1}\leq 0$$

$$\frac{x(x+1)}{(x+4)(x+1)}-\frac{1(x+4)}{(x+4)(x+1)}\leq 0$$

$$\frac{x^2-4}{(x+4)(x+1)}\leq 0$$

$$\frac{(x+2)(x-2)}{(x+4)(x+1)}\leq 0$$

```
x − 2   −−−−−−−−−− −−−−0+++
x + 1   −−−−−−−−−− 0++++ ++++
x + 2   −−−−−−0++++ +++++ ++++
x + 4   −− 0+++++++++ +++++ ++++
```

$\longleftarrow(\longrightarrow]\ \ (\longrightarrow]\longrightarrow$ at -4, -2, -1, 2

solution set: $(-4,-2]\cup(-1,2]$

47.

$$\frac{x}{x+16} > \frac{1}{x+1}$$
$$\frac{x}{x+16} - \frac{1}{x+1} > 0$$
$$\frac{x(x+1)}{(x+16)(x+1)} - \frac{1(x+16)}{(x+16)(x+1)} > 0$$
$$\frac{x^2-16}{(x+16)(x+1)} > 0$$
$$\frac{(x+4)(x-4)}{(x+16)(x+1)} > 0$$

$x-4$ ------- ---- ----0+++
$x+1$ ------- ---- 0++++ ++++
$x+4$ ------- 0++++ +++++ ++++
$x+16$ --- 0++++ +++++ +++++ ++++
⟵) — (—) —(⟶
−16 −4 −1 4

solution set: $(-\infty, -16) \cup (-4, -1) \cup (4, \infty)$

49.

$$(x+2)^2 > 0$$
$$(x+2)(x+2) > 0$$

$x+2$ -----0+++++++
$x+2$ -----0+++++++
⟵) (⟶
−2

solution set: $(-\infty, -2) \cup (-2, \infty)$

51. $x^2 - 2x - 3 < 0$

Graph $y = x^2 - 2x - 3$ and find the x-coordinates of points below the x-axis.

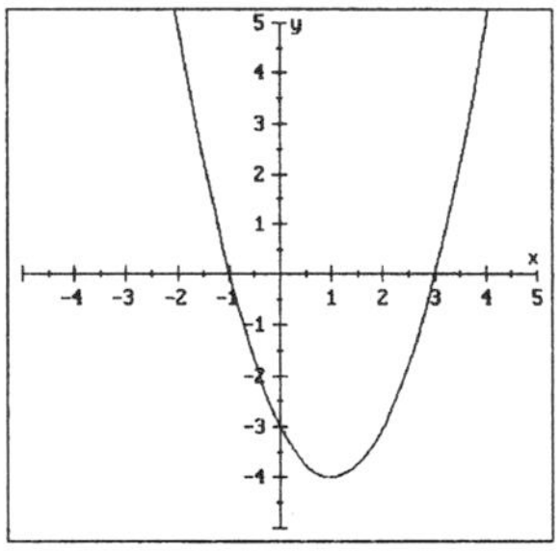

$(-1, 3)$

53. $\frac{x+3}{x-2} > 0$

Graph $y = (x+3)/(x-2)$ and find the x-coordinates of points above the x-axis.

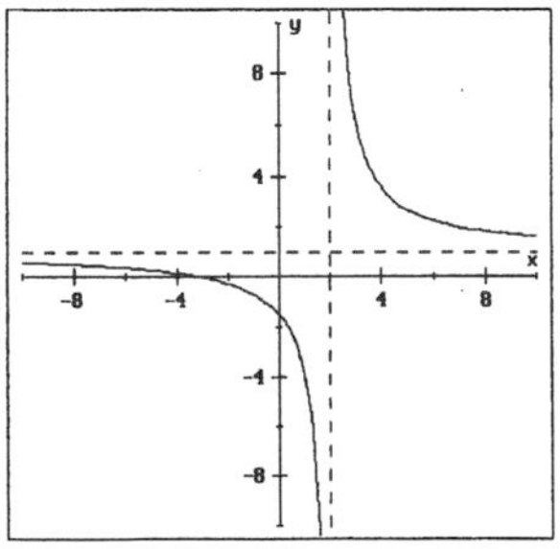

$(-\infty, -3) \cup (2, \infty)$

55. $y < x^2 + 1$

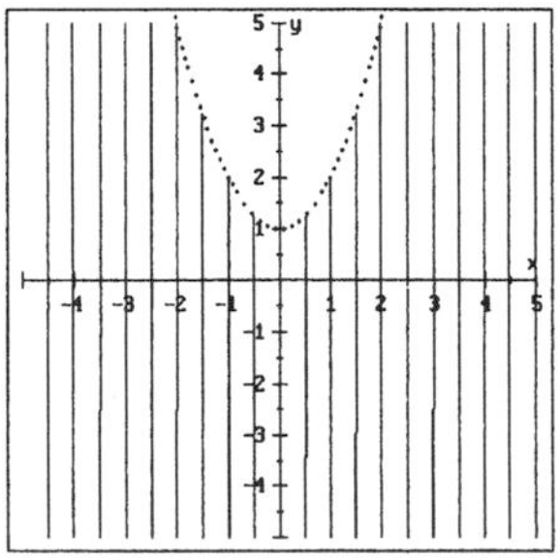

57. $y \le x^2 + 5x + 6$

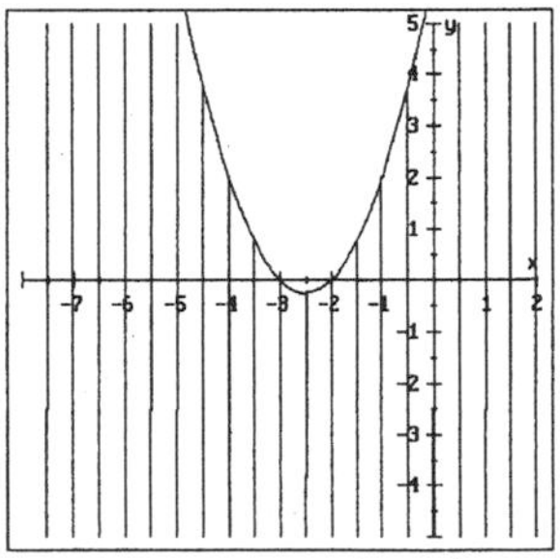

59. $y \geq (x-1)^2$

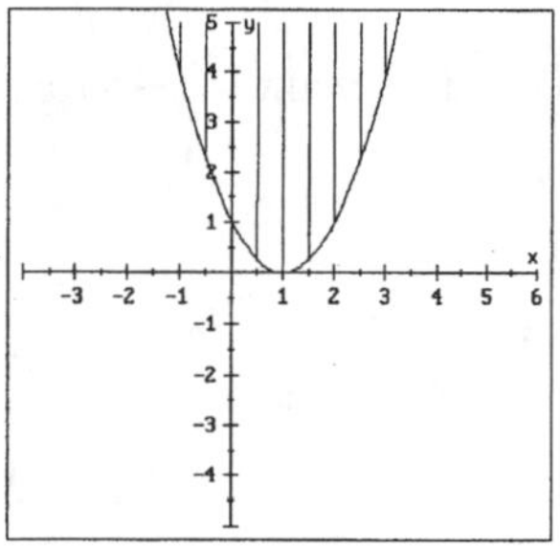

61. $-x^2 - y + 6 > -x$
$-x^2 + x + 6 > y$

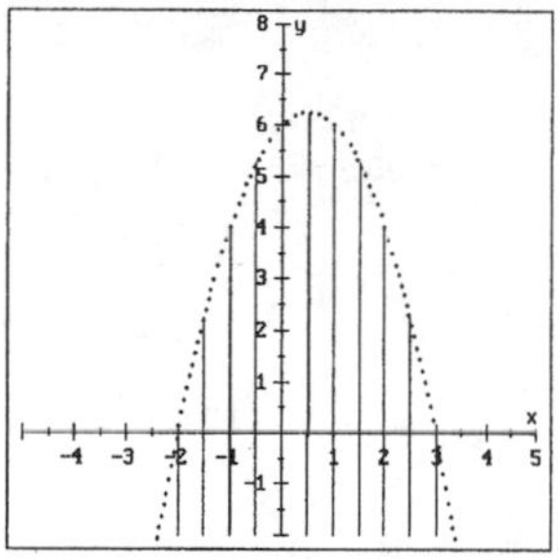

63. $y < |x+4|$

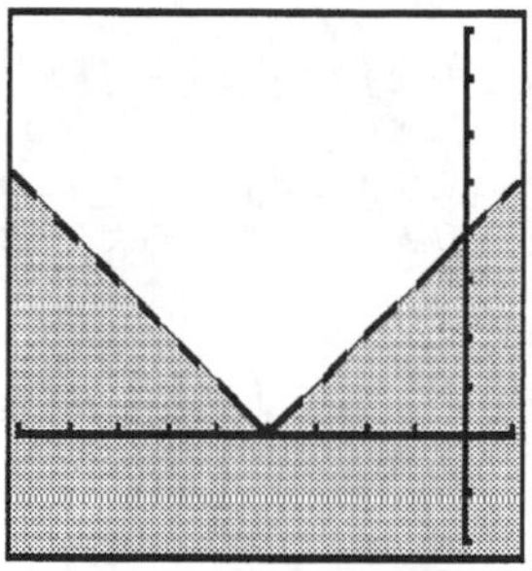

65. $y \leq -|x| + 2$

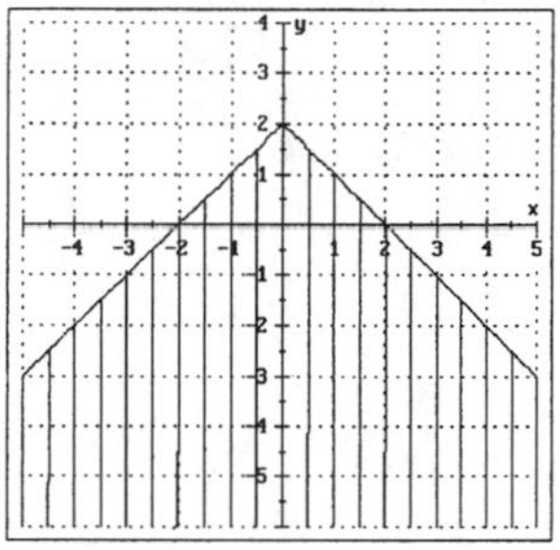

67. **Answers may vary.**

69. It will be positive if 4, 2 or 0 factors are negative.

Exercise 10.6 (page 711)

1. $\dfrac{3x^2+x-14}{4-x^2} = \dfrac{(3x+7)(x-2)}{(2+x)(2-x)} = -\dfrac{3x+7}{x+2}$

3. $\dfrac{8+2x-x^2}{12+x-3x^2} \div \dfrac{3x^2+5x-2}{3x-1} = \dfrac{x^2-2x-8}{3x^2-x-12} \cdot \dfrac{3x-1}{3x^2+5x-2}$

$= \dfrac{(x-4)(x+2)}{3x^2-x-12} \cdot \dfrac{3x-1}{(3x-1)(x+2)} = \dfrac{x-4}{3x^2-x-12}$

5. $f(x) + g(x)$

7. $f(x)g(x)$

9. domain

11. $f(x)$

13. $f + g = f(x) + g(x) = 3x + 4x = 7x$
domain $= (-\infty, \infty)$

15. $f \cdot g = f(x) \cdot g(x) = 3x \cdot 4x = 12x^2$
domain $= (-\infty, \infty)$

17. $g - f = g(x) - f(x) = 4x - 3x = x$
domain $= (-\infty, \infty)$

19. $g/f = \dfrac{g(x)}{f(x)} = \dfrac{4x}{3x} = \dfrac{4}{3}$ (for $x \neq 0$)
domain $= (-\infty, 0) \cup (0, \infty)$

21. $f + g = f(x) + g(x) = 2x + 1 + x - 3 = 3x - 2$; domain $= (-\infty, \infty)$

23. $f \cdot g = f(x) \cdot g(x) = (2x+1)(x-3) = 2x^2 - 5x - 3$; domain $= (-\infty, \infty)$

25. $g - f = g(x) - f(x) = (x-3) - (2x+1) = x - 3 - 2x - 1 = -x - 4$; domain $= (-\infty, \infty)$

27. $g/f = \dfrac{g(x)}{f(x)} = \dfrac{x-3}{2x+1}$; domain $= \left(-\infty, -\dfrac{1}{2}\right) \cup \left(-\dfrac{1}{2}, \infty\right)$

29. $f - g = f(x) - g(x) = (3x-2) - (2x^2+1) = 3x - 2 - 2x^2 - 1 = -2x^2 + 3x - 3$
domain $= (-\infty, \infty)$

31. $f/g = \dfrac{f(x)}{g(x)} = \dfrac{3x-2}{2x^2+1}$; domain $= (-\infty, \infty)$

33. $f - g = f(x) - g(x) = (x^2-1) - (x^2-4) = x^2 - 1 - x^2 + 4 = 3$; domain $= (-\infty, \infty)$

35. $g/f = \dfrac{g(x)}{f(x)} = \dfrac{x^2-4}{x^2-1} = \dfrac{(x+2)(x-2)}{(x+1)(x-1)}$; domain $= (-\infty, -1) \cup (-1, 1) \cup (1, \infty)$

37. $(f \circ g)(2) = f(g(2)) = f(2^2 - 1) = f(3) = 2(3) + 1 = 7$

39. $(g \circ f)(-3) = g(f(-3)) = g(2(-3) + 1) = g(-5) = (-5)^2 - 1 = 24$

41. $(f \circ g)(0) = f(g(0)) = f(0^2 - 1) = f(-1) = 2(-1) + 1 = -1$

43. $(f \circ g)\left(\dfrac{1}{2}\right) = f\left(g\left(\dfrac{1}{2}\right)\right) = f\left(\left(\dfrac{1}{2}\right)^2 - 1\right) = f\left(-\dfrac{3}{4}\right) = 2\left(-\dfrac{3}{4}\right) + 1 = -\dfrac{3}{2} + 1 = -\dfrac{1}{2}$

45. $(f \circ g)(x) = f(g(x)) = f(x^2 - 1) = 2(x^2 - 1) + 1 = 2x^2 - 2 + 1 = 2x^2 - 1$

47. $(g \circ f)(2x) = g(f(2x)) = g(2(2x) + 1) = g(4x + 1) = (4x+1)^2 - 1 = 16x^2 + 8x + 1 - 1$
$= 16x^2 + 8x$

49. $(f \circ g)(4) = f(g(4)) = f(4^2 + 4) = f(20) = 3(20) - 2 = 58$

51. $(g \circ f)(-3) = g(f(-3)) = g(3(-3) - 2) = g(-11) = (-11)^2 + (-11) = 110$

53. $(g \circ f)(0) = g(f(0)) = g(3(0) - 2) = g(-2) = (-2)^2 + (-2) = 2$

55. $(g \circ f)(x) = g(f(x)) = g(3x - 2) = (3x-2)^2 + 3x - 2 = 9x^2 - 12x + 4 + 3x - 2$
$= 9x^2 - 9x + 2$

57. $\dfrac{f(x+h) - f(x)}{h} = \dfrac{2(x+h) + 3 - (2x+3)}{h} = \dfrac{2x + 2h + 3 - 2x - 3}{h} = \dfrac{2h}{h} = 2$

59. $\dfrac{f(x+h) - f(x)}{h} = \dfrac{(x+h)^2 - x^2}{h} = \dfrac{x^2 + 2xh + h^2 - x^2}{h} = \dfrac{2xh + h^2}{h} = 2x + h$

SECTION 10.6

61. $$\frac{f(x+h)-f(x)}{h}=\frac{2(x+h)^2-1-(2x^2-1)}{h}=\frac{2x^2+4xh+2h^2-1-2x^2+1}{h}=\frac{4xh+2h^2}{h}=4x+2h$$

63. $$\frac{f(x+h)-f(x)}{h}=\frac{(x+h)^2+(x+h)-(x^2+x)}{h}=\frac{x^2+2xh+h^2+x+h-x^2-x}{h}=\frac{2xh+h^2+h}{h}=2x+h+1$$

65. $$\frac{f(x+h)-f(x)}{h}=\frac{(x+h)^2+3(x+h)-4-(x^2+3x-4)}{h}=\frac{x^2+2xh+h^2+3x+3h-4-x^2-3x+4}{h}=\frac{2xh+h^2+3h}{h}=2x+h+3$$

67. $$\frac{f(x+h)-f(x)}{h}=\frac{2(x+h)^2+3(x+h)-7-(2x^2+3x-7)}{h}=\frac{2x^2+4xh+2h^2+3x+3h-7-2x^2-3x+7}{h}=\frac{4xh+2h^2+3h}{h}=4x+2h+3$$

69. $$\frac{f(x)-f(a)}{x-a}=\frac{(2x+3)-(2a+3)}{x-a}=\frac{2x+3-2a-3}{x-a}=\frac{2x-2a}{x-a}=\frac{2(x-a)}{x-a}=2$$

71. $$\frac{f(x)-f(a)}{x-a}=\frac{x^2-a^2}{x-a}=\frac{(x+a)(x-a)}{x-a}=x+a$$

73. $$\frac{f(x)-f(a)}{x-a}=\frac{(2x^2-1)-(2a^2-1)}{x-a}=\frac{2x^2-1-2a^2+1}{x-a}=\frac{2x^2-2a^2}{x-a}=\frac{2(x+a)(x-a)}{x-a}=2(x+a)=2x+2a$$

75. $$\frac{f(x)-f(a)}{x-a}=\frac{(x^2+x)-(a^2+a)}{x-a}=\frac{x^2+x-a^2-a}{x-a}=\frac{x^2-a^2+x-a}{x-a}=\frac{(x+a)(x-a)+1(x-a)}{x-a}=\frac{(x-a)(x+a+1)}{x-a}=x+a+1$$

77. $$\frac{f(x)-f(a)}{x-a}=\frac{(x^2+3x-4)-(a^2+3a-4)}{x-a}=\frac{x^2+3x-4-a^2-3a+4}{x-a}$$
$$=\frac{x^2-a^2+3x-3a}{x-a}$$
$$=\frac{(x+a)(x-a)+3(x-a)}{x-a}$$
$$=\frac{(x-a)(x+a+3)}{x-a}=x+a+3$$

79. $$\frac{f(x)-f(a)}{x-a}=\frac{(2x^2+3x-7)-(2a^2+3a-7)}{x-a}=\frac{2x^2+3x-7-2a^2-3a+7}{x-a}$$
$$=\frac{2x^2-2a^2+3x-3a}{x-a}$$
$$=\frac{2(x+a)(x-a)+3(x-a)}{x-a}$$
$$=\frac{(x-a)(2(x+a)+3)}{x-a}$$
$$=2(x+a)+3=2x+2a+3$$

81. $(f\circ g)(x)=f(g(x))=f(2x-5)=(2x-5)+1=2x-4$
$(g\circ f)(x)=g(f(x))=g(x+1)=2(x+1)-5=2x+2-5=2x-3$

83. $f(a)=a^2+2a-3;\ f(h)=h^2+2h-3\Rightarrow f(a)+f(h)=a^2+h^2+2a+2h-6$
$f(a+h)=(a+h)^2+2(a+h)-3=a^2+2ah+h^2+2a+2h-3$
$=a^2+h^2+2ah+2a+2h-3$

85. $$\frac{f(x+h)-f(x)}{h}=\frac{(x+h)^3-1-(x^3-1)}{h}=\frac{x^3+3x^2h+3xh^2+h^3-1-x^3+1}{h}$$
$$=\frac{3x^2h+3xh^2+h^3}{h}$$
$$=\frac{h(3x^2+3xh+h^2)}{h}=3x^2+3xh+h^2$$

87. $F(t)=2700-200t;\ C(F)=\frac{5}{9}(F-32)$
$C(F(t))=C(2700-200t)=\frac{5}{9}(2700-200t-32)=\frac{5}{9}(2668-200t)$

89. **Answers may vary.**

91. It is associative. Examples will vary.

Exercise 10.7 (page 720)

1. $3-\sqrt{-64}=3-\sqrt{64i^2}=3-8i$

3. $(3+4i)(2-3i)=6-9i+8i-12i^2=6-i-12(-1)=6-i+12=18-i$

5. $|6-8i|=\sqrt{6^2+(-8)^2}=\sqrt{36+64}=\sqrt{100}=10$

7. one-to-one **9.** 2 **11.** x

13. Each input has a different output. one-to-one

15. The inputs $x = 2$ and $x = -2$ have the same output. not one-to-one

17. one-to-one **19.** one-to-one **21.** not one-to-one **23.** one-to-one

25. inverse $= \{(2,3),(1,2),(0,1)\}$. Since each x-coordinate is paired with only one y-coordinate, the inverse relation **is a function.**

27. inverse $= \{(2,1),(3,2),(3,1),(5,1)\}$. Since $x = 3$ is paired with more than one y-coordinate, the inverse relation **is not a function.**

29. inverse $= \{(1,1),(4,2),(9,3),(16,4)\}$. Since each x-coordinate is paired with only one y-coordinate, the inverse relation **is a function.**

31.

$$\begin{aligned} f(x) &= 3x+1 \\ y &= 3x+1 \\ x &= 3y+1 \\ x-1 &= 3y \\ \frac{x-1}{3} &= y \\ \frac{1}{3}x - \frac{1}{3} &= y \\ f^{-1}(x) &= \tfrac{1}{3}x - \tfrac{1}{3} \end{aligned}$$

$f \circ f^{-1}$

$$\begin{aligned} f[f^{-1}(x)] &= f\left(\tfrac{1}{3}x - \tfrac{1}{3}\right) \\ &= 3 \cdot \left(\tfrac{1}{3}x - \tfrac{1}{3}\right) + 1 \\ &= x - 1 + 1 \\ &= x \end{aligned}$$

$f^{-1} \circ f$

$$\begin{aligned} f^{-1}[f(x)] &= f^{-1}(3x+1) \\ &= \tfrac{1}{3}(3x+1) - \tfrac{1}{3} \\ &= x + \tfrac{1}{3} - \tfrac{1}{3} \\ &= x \end{aligned}$$

33.

$$\begin{aligned} x+4 &= 5y \\ y = f(x) &= \frac{x+4}{5} \\ x &= \frac{y+4}{5} \\ 5x &= y+4 \\ 5x-4 &= y \\ f^{-1}(x) &= 5x-4 \end{aligned}$$

$f \circ f^{-1}$

$$\begin{aligned} f[f^{-1}(x)] &= f(5x-4) \\ &= \frac{(5x-4)+4}{5} \\ &= \frac{5x}{5} \\ &= x \end{aligned}$$

$f^{-1} \circ f$

$$\begin{aligned} f^{-1}[f(x)] &= f^{-1}\left(\frac{x+4}{5}\right) \\ &= 5 \cdot \left(\frac{x+4}{5}\right) - 4 \\ &= x + 4 - 4 \\ &= x \end{aligned}$$

35.

$$\begin{aligned} f(x) &= \frac{x-4}{5} \\ y &= \frac{x-4}{5} \\ x &= \frac{y-4}{5} \\ 5x &= y-4 \\ 5x+4 &= y \\ f^{-1}(x) &= 5x+4 \end{aligned}$$

$f \circ f^{-1}$

$$\begin{aligned} f[f^{-1}(x)] &= f(5x+4) \\ &= \frac{(5x+4)-4}{5} \\ &= \frac{5x}{5} \\ &= x \end{aligned}$$

$f^{-1} \circ f$

$$\begin{aligned} f^{-1}[f(x)] &= f^{-1}\left(\frac{x-4}{5}\right) \\ &= 5 \cdot \left(\frac{x-4}{5}\right) + 4 \\ &= x - 4 + 4 \\ &= x \end{aligned}$$

37.

$$\begin{aligned} 4x - 5y &= 20 \\ 5y &= 4x - 20 \\ y = f(x) &= \tfrac{4}{5}x - 4 \\ x &= \tfrac{4}{5}y - 4 \\ x + 4 &= \tfrac{4}{5}y \\ \tfrac{5}{4}(x+4) &= y \\ f^{-1}(x) &= \tfrac{5}{4}x + 5 \end{aligned}$$

$f \circ f^{-1}$

$$\begin{aligned} f\left[f^{-1}(x)\right] &= f\left(\tfrac{5}{4}x + 5\right) \\ &= \tfrac{4}{5}\left(\tfrac{5}{4}x + 5\right) - 4 \\ &= x + 4 - 4 \\ &= x \end{aligned}$$

$f^{-1} \circ f$

$$\begin{aligned} f^{-1}[f(x)] &= f^{-1}\left(\tfrac{4}{5}x - 4\right) \\ &= \tfrac{5}{4}\left(\tfrac{4}{5}x - 4\right) + 5 \\ &= x - 5 + 5 \\ &= x \end{aligned}$$

39.

$$\begin{aligned} y &= 4x + 3 \\ x &= 4y + 3 \\ x - 3 &= 4y \\ \frac{x-3}{4} &= y \end{aligned}$$

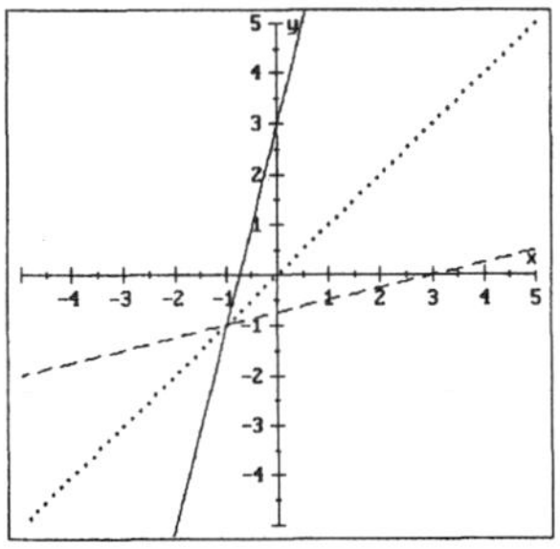

41.

$$\begin{aligned} x &= \frac{y-2}{3} \\ y &= \frac{x-2}{3} \end{aligned}$$

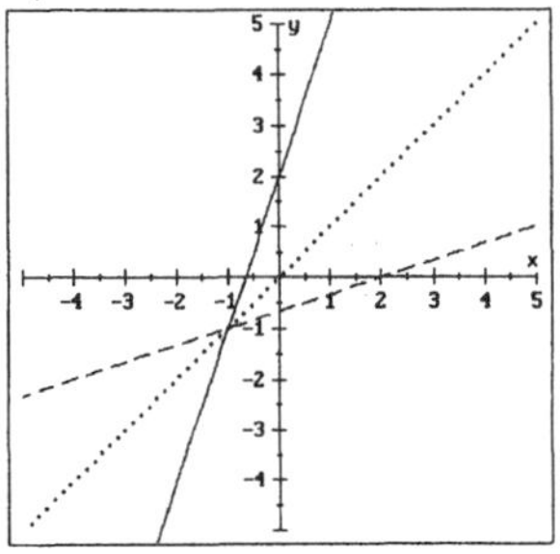

43.

$$\begin{aligned} 3x - y &= 5 \\ 3y - x &= 5 \\ 3y &= x + 5 \\ y &= \frac{x+5}{3} \end{aligned}$$

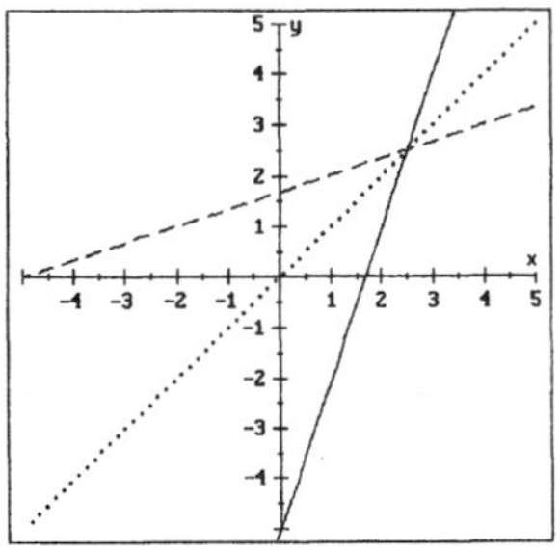

45.

$$\begin{aligned} 3(x + y) &= 2x + 4 \\ 3(y + x) &= 2y + 4 \\ 3y + 3x &= 2y + 4 \\ y &= 4 - 3x \end{aligned}$$

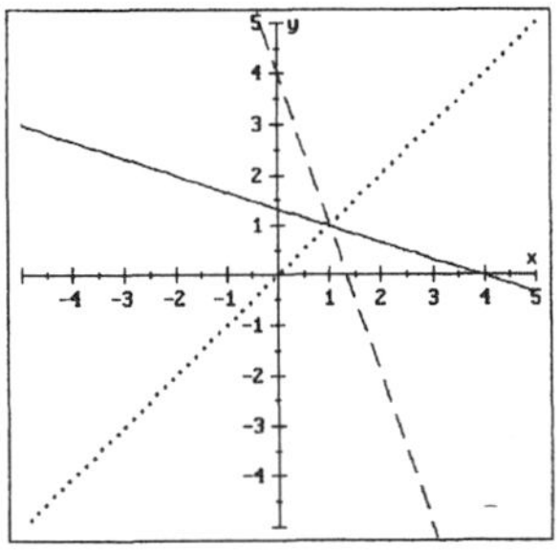

47.

$$\begin{aligned} y &= x^2 + 4 \\ x &= y^2 + 4 \\ x - 4 &= y^2 \\ \pm\sqrt{x-4} &= y \end{aligned}$$

The relation **is not** a function.

49.

$$\begin{aligned} y &= x^3 \\ x &= y^3 \\ \sqrt[3]{x} &= y \end{aligned}$$

The relation **is** a function.

51. $y = |x|$
$x = |y|$
The relation **is not** a function.

53.
$$y = 2x^3 - 3$$
$$x = 2y^3 - 3$$
$$x + 3 = 2y^3$$
$$\frac{x+3}{2} = y^3$$
$$\sqrt[3]{\frac{x+3}{2}} = \sqrt[3]{y^3}, \text{ or } y = f^{-1}(x) = \sqrt[3]{\frac{x+3}{2}}$$

55. $y = x^2 + 1$
inverse: $x = y^2 + 1$

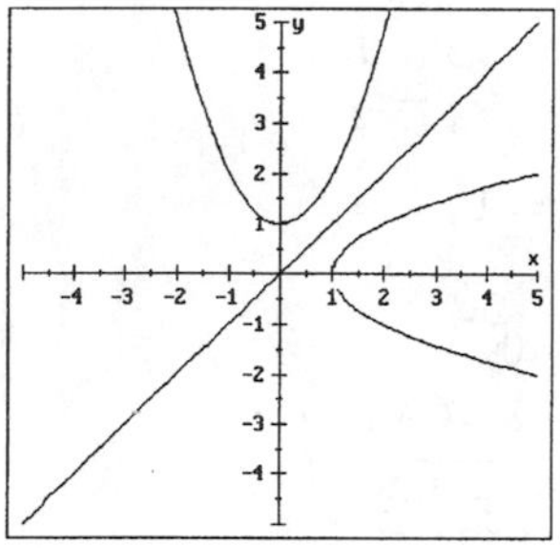

57. $y = \sqrt{x}$
inverse: $x = \sqrt{y}$

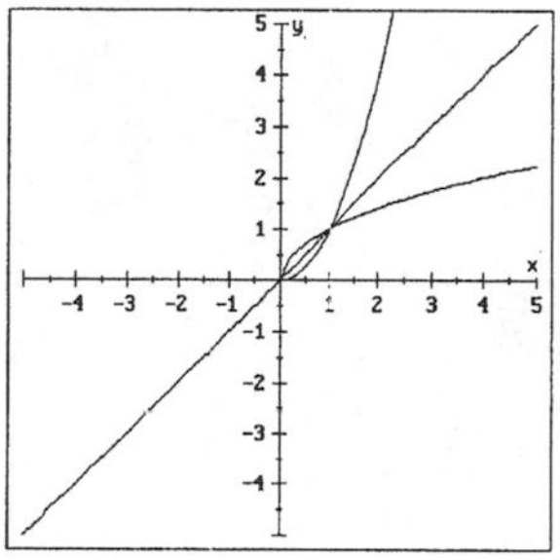

59. **Answers may vary.**

61.
$$y = \frac{x+1}{x-1}$$
$$x = \frac{y+1}{y-1}$$
$$x(y-1) = y + 1$$
$$xy - x = y + 1$$
$$xy - y = x + 1$$
$$y(x-1) = x + 1$$
$$y = \frac{x+1}{x-1}$$

Chapter 10 Summary (page 724)

1.
$$12x^2 + x - 6 = 0$$
$$(4x+3)(3x-2) = 0$$
$$4x + 3 = 0 \quad \text{or} \quad 3x - 2 = 0$$
$$x = -\tfrac{3}{4} \qquad\qquad x = \tfrac{2}{3}$$

2.
$$6x^2 + 17x + 5 = 0$$
$$(2x+5)(3x+1) = 0$$
$$2x + 5 = 0 \quad \text{or} \quad 3x + 1 = 0$$
$$x = -\tfrac{5}{2} \qquad\qquad x = -\tfrac{1}{3}$$

3.
$$15x^2 + 2x - 8 = 0$$
$$(3x-2)(5x+4) = 0$$
$$3x - 2 = 0 \quad \text{or} \quad 5x + 4 = 0$$
$$x = \tfrac{2}{3} \qquad\qquad x = -\tfrac{4}{5}$$

4.
$$(x+2)^2 = 36$$
$$x + 2 = \pm\sqrt{36}$$
$$x + 2 = \pm 6$$
$$x = -2 \pm 6$$
$$x = 4 \quad \text{or} \quad x = -8$$

5.

$$\begin{aligned} x^2+6x+8&=0\\ x^2+6x&=-8\\ x^2+6x+9&=-8+9\\ (x+3)^2&=1\\ x+3&=\pm 1\\ x&=-3\pm 1 \end{aligned}$$

$x=-2$ or $x=-4$

6.

$$\begin{aligned} 2x^2-9x+7&=0\\ x^2-\frac{9}{2}x+\frac{7}{2}&=0\\ x^2-\frac{9}{2}x&=-\frac{7}{2}\\ x^2-\frac{9}{2}x+\frac{81}{16}&=-\frac{56}{16}+\frac{81}{16}\\ \left(x-\frac{9}{4}\right)^2&=\frac{25}{16}\\ x-\frac{9}{4}&=\pm\frac{5}{4}\\ x&=\frac{9}{4}\pm\frac{5}{4} \end{aligned}$$

$x=\frac{7}{2}$ or $x=1$

7.

$$\begin{aligned} 2x^2-x-5&=0\\ x^2-\frac{1}{2}x-\frac{5}{2}&=0\\ x^2-\frac{1}{2}x&=\frac{5}{2}\\ x^2-\frac{1}{2}x+\frac{1}{16}&=\frac{5}{2}+\frac{1}{16}\\ \left(x-\frac{1}{4}\right)^2&=\frac{41}{16}\\ x-\frac{1}{4}&=\pm\frac{\sqrt{41}}{4}\\ x&=\frac{1}{4}\pm\frac{\sqrt{41}}{4} \end{aligned}$$

8.

$x^2-8x-9=0$

$a=1, b=-8, c=-9$

$$\begin{aligned} x&=\frac{-b\pm\sqrt{b^2-4ac}}{2a}\\ &=\frac{-(-8)\pm\sqrt{(-8)^2-4(1)(-9)}}{2(1)}\\ &=\frac{8\pm\sqrt{64+36}}{2}\\ &=\frac{8\pm\sqrt{100}}{2}=\frac{8\pm 10}{2} \end{aligned}$$

$x=\frac{18}{2}=9$ or $x=\frac{-2}{2}=-1$

9.

$x^2-10x=0$

$a=1, b=-10, c=0$

$$\begin{aligned} x&=\frac{-b\pm\sqrt{b^2-4ac}}{2a}\\ &=\frac{-(-10)\pm\sqrt{(-10)^2-4(1)(0)}}{2(1)}\\ &=\frac{10\pm\sqrt{100+0}}{2}\\ &=\frac{10\pm\sqrt{100}}{2}=\frac{10\pm 10}{2} \end{aligned}$$

$x=\frac{20}{2}=10$ or $x=\frac{0}{2}=0$

10.

$2x^2+13x-7=0$

$a=2, b=13, c=-7$

$$\begin{aligned} x&=\frac{-b\pm\sqrt{b^2-4ac}}{2a}\\ &=\frac{-(13)\pm\sqrt{13^2-4(2)(-7)}}{2(2)}\\ &=\frac{-13\pm\sqrt{169+56}}{4}\\ &=\frac{-13\pm\sqrt{225}}{4}=\frac{-13\pm 15}{4} \end{aligned}$$

$x=\frac{2}{4}=\frac{1}{2}$ or $x=\frac{-28}{4}=-7$

11. $3x^2 + 20x - 7 = 0$

$a = 3, b = 20, c = -7$

$$x = \frac{-b \pm \sqrt{b^2 - 4ac}}{2a}$$
$$= \frac{-20 \pm \sqrt{(20)^2 - 4(3)(-7)}}{2(3)}$$
$$= \frac{-20 \pm \sqrt{400 + 84}}{6}$$
$$= \frac{-20 \pm \sqrt{484}}{6} = \frac{-20 \pm 22}{6}$$

$x = \frac{2}{6} = \frac{1}{3}$ or $x = \frac{-42}{6} = -7$

12. $2x^2 - x - 2 = 0$

$a = 2, b = -1, c = -2$

$$x = \frac{-b \pm \sqrt{b^2 - 4ac}}{2a}$$
$$= \frac{-(-1) \pm \sqrt{(-1)^2 - 4(2)(-2)}}{2(2)}$$
$$= \frac{1 \pm \sqrt{1 + 16}}{4}$$
$$= \frac{1 \pm \sqrt{17}}{4} = \frac{1}{4} \pm \frac{\sqrt{17}}{4}$$

13. $x^2 + x + 2 = 0$

$a = 1, b = 1, c = 2$

$$x = \frac{-b \pm \sqrt{b^2 - 4ac}}{2a}$$
$$= \frac{-1 \pm \sqrt{1^2 - 4(1)(2)}}{2(1)}$$
$$= \frac{-1 \pm \sqrt{1 - 8}}{2}$$
$$= \frac{-1 \pm \sqrt{-7}}{2} = -\frac{1}{2} \pm \frac{\sqrt{7}}{2}i$$

14. Let w represent the original width.
Then $w + 2$ represents the original length.
The new dimensions are then $2w$ and
$2(w + 2) = 2w + 4$.

Old Area + 72 = New Area

$$w(w + 2) + 72 = 2w(2w + 4)$$
$$w^2 + 2w + 72 = 4w^2 + 8w$$
$$0 = 3w^2 + 6w - 72$$
$$0 = 3(w + 6)(w - 4)$$

$w = -6$ or $w = 4$

Since the width is positive, the dimensions are 4 cm by 6 cm.

15. Let w represent the original width.
Then $w + 1$ represents the original length.
The new dimensions are then $2w$ and
$3(w + 1) = 3w + 3$.

Old Area + 30 = New Area

$$w(w + 1) + 30 = 2w(3w + 3)$$
$$w^2 + w + 30 = 6w^2 + 6w$$
$$0 = 5w^2 + 5w - 30$$
$$0 = 5(w + 3)(w - 2)$$

$w = -3$ or $w = 2$

Since the width is positive, the dimensions are 2 ft by 3 ft.

16. When the rocket hits the ground, $h = 0$:

$h = 112t - 16t^2$

$0 = 112t - 16t^2$

$0 = 16t(7 - t)$

$t = 0$ or $t = 7$

It hits the ground after 7 seconds.

17. The maximum height occurs at the vertex:

$$h = 112t - 16t^2$$
$$h = -16t^2 + 112t$$
$$h = -16(t^2 - 7t)$$
$$h = -16\left(t^2 - 7t + \frac{49}{4}\right) + 196$$
$$h = -16\left(t^2 - 7t + \frac{49}{4}\right) + 196$$
$$h = -16\left(t - \frac{7}{2}\right)^2 + 196$$

Vertex: $\left(\frac{7}{2}, 196\right) \Rightarrow$ max. height = 196 ft

18. $3x^2 + 4x - 3 = 0$

$a = 3, b = 4, c = -3$

$$b^2 - 4ac = 4^2 - 4(3)(-3) = 16 + 36 = 52$$

irrational unequal solutions

19. $4x^2 - 5x + 7 = 0$

$a = 4, b = -5, c = 7$

$$b^2 - 4ac = (-5)^2 - 4(4)(7) = 25 - 112 = -87$$

complex conjugate solutions

20. $(k-8)x^2 + (k+16)x = -49$

$(k-8)x^2 + (k+16)x + 49 = 0$

$a = k - 8, b = k + 16, c = 49$

Set the discriminant equal to 0:

$$b^2 - 4ac = 0$$
$$(k+16)^2 - 4(k-8)(49) = 0$$
$$k^2 + 32k + 256 - 196k + 1568 = 0$$
$$k^2 - 164k + 1824 = 0$$
$$(k-12)(k-152) = 0$$

$k - 12 = 0$ **or** $k - 152 = 0$

$k = 12$ $\quad$ $k = 152$

21. $3x^2 + 4x = k + 1$

$3x^2 + 4x - k - 1 = 0$

$a = 3, b = 4, c = -k - 1$

Set the discriminant ≥ 0:

$$b^2 - 4ac \geq 0$$
$$4^2 - 4(3)(-k-1) \geq 0$$
$$16 + 12k + 12 \geq 0$$
$$12k \geq -28$$
$$k \geq -\frac{28}{12}$$
$$k \geq -\frac{7}{3}$$

22. $x - 13x^{1/2} + 12 = 0$

$\left(x^{1/2} - 12\right)\left(x^{1/2} - 1\right) = 0$

$x^{1/2} - 12 = 0$ **or** $x^{1/2} - 1 = 0$

$x^{1/2} = 12$ $\quad$ $x^{1/2} = 1$

$\left(x^{1/2}\right)^2 = (12)^2$ $\quad$ $\left(x^{1/2}\right)^2 = 1^2$

$x = 144$ $\quad$ $x = 1$

Solution. $\quad$ Solution.

23. $a^{2/3} + a^{1/3} - 6 = 0$

$\left(a^{1/3} - 2\right)\left(a^{1/3} + 3\right) = 0$

$a^{1/3} - 2 = 0$ **or** $a^{1/3} + 3 = 0$

$a^{1/3} = 2$ $\quad$ $a^{1/3} = -3$

$\left(a^{1/3}\right)^3 = (2)^3$ $\quad$ $\left(a^{1/3}\right)^3 = (-3)^3$

$a = 8$ $\quad$ $a = -27$

Solution. $\quad$ Solution.

24.

$$\frac{1}{x+1} - \frac{1}{x} = -\frac{1}{x+1}$$
$$\left(\frac{1}{x+1} - \frac{1}{x}\right)(x)(x+1) = -\frac{1}{x+1}(x)(x+1)$$
$$1(x) - 1(x+1) = -x$$
$$x - x - 1 = -x$$
$$-1 = -x$$
$$1 = x$$

25.
$$\frac{6}{x+2}+\frac{6}{x+1}=5$$
$$\left(\frac{6}{x+2}+\frac{6}{x+1}\right)(x+2)(x+1)=5(x+2)(x+1)$$
$$6(x+1)+6(x+2)=5(x^2+3x+2)$$
$$6x+6+6x+12=5x^2+15x+10$$
$$0=5x^2+3x-8$$
$$0=(5x+8)(x-1)\qquad 5x+8=0 \quad \textbf{or} \quad x-1=0$$
$$x=-\tfrac{8}{5} \qquad\qquad x=1$$

26. $3x^2-14x+3=0$

sum $=-\dfrac{b}{a}=-\dfrac{-14}{3}=\dfrac{14}{3}$

27. $3x^2-14x+3=0$

product $=\dfrac{c}{a}=\dfrac{3}{3}=1$

28. $y=2x^2-3$

$y=2(x-0)^2-3$

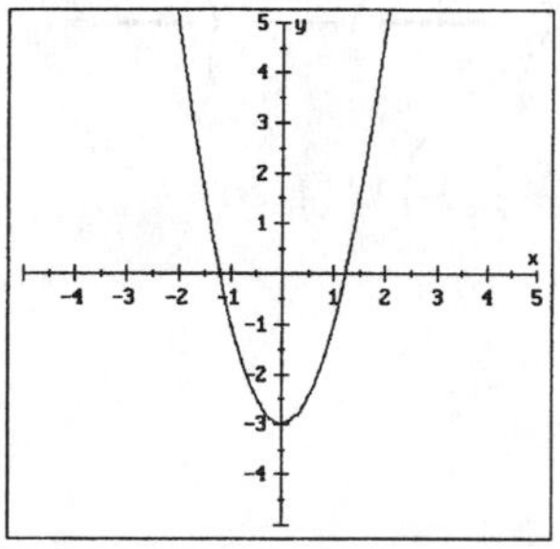

vertex: $(0,-3)$

29. $y=-2x^2-1$

$y=-2(x-0)^2-1$

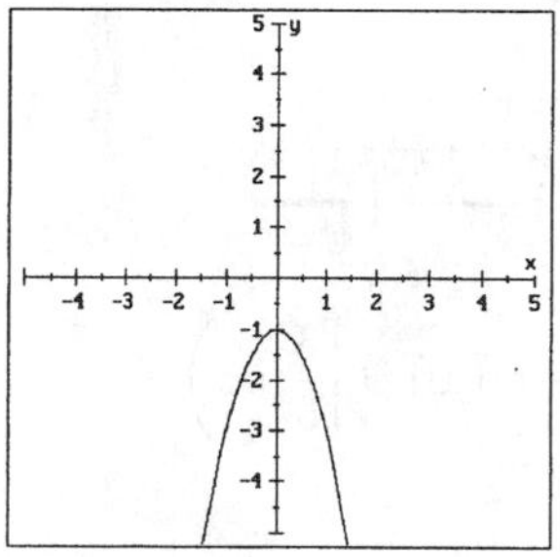

vertex: $(0,-1)$

30. $y=-4(x-2)^2+1$

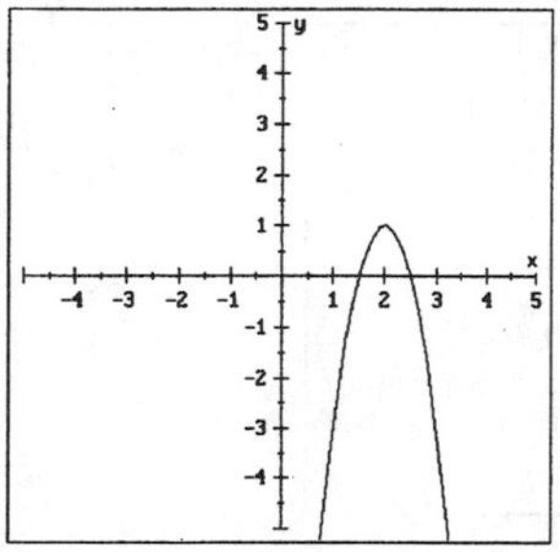

vertex: $(2,1)$

31.
$$\begin{aligned} y &= 5x^2+10x-1 \\ &= 5(x^2+2x)-1 \\ &= 5(x^2+2x+1)-1-5 \\ &= 5(x+1)^2-6 \end{aligned}$$

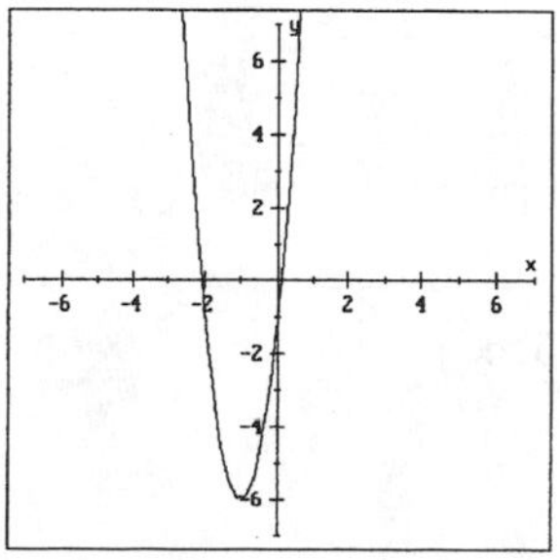

vertex: $(-1,-6)$

32. $y = 3x^2 - 12x - 5 = 3(x^2 - 4x) - 5 = 3(x^2 - 4x + 4) - 5 - 12 = 3(x-2)^2 - 17$
vertex: $(2, -17)$

33. $x^2 + 2x - 35 > 0$

$(x+7)(x-5) > 0$

$x - 5$ − − − − − − − − − 0 ++++
$x + 7$ − − − 0++++++ ++++

⟵) ——— (⟶
-7 5

solution set: $(-\infty, -7) \cup (5, \infty)$

34. $x^2 + 7x - 18 < 0$

$(x-2)(x+9) < 0$

$x - 2$ − − − − − − − − − 0++++
$x + 9$ − − − 0++++++++++

⟵ (———) ⟶
-9 2

solution set: $(-9, 2)$

35. $\frac{3}{x} \le 5$

$\frac{3}{x} - 5 \le 0$

$\frac{3}{x} - \frac{5x}{x} \le 0$

$\frac{3 - 5x}{x} \le 0$

$3 - 5x$ +++++++++ 0 − − −
x − − − 0++++++++++

⟵) ——— [⟶
0 $\frac{3}{5}$

solution set: $(-\infty, 0) \cup \left[\frac{3}{5}, \infty\right)$

36. $\frac{2x^2 - x - 28}{x - 1} > 0$

$\frac{(2x+7)(x-4)}{x-1} > 0$

$x - 4$ − − − − − − − − − − − − 0++++
$x - 1$ − − − − − − − − − 0++++++++++
$2x + 7$ − − − 0+++++++++++++++++

⟵ (———) —— (⟶
$-\frac{7}{2}$ 1 4

solution set: $\left(-\frac{7}{2}, 1\right) \cup (4, \infty)$

37. $x^2 + 2x - 35 > 0$
Graph $y = x^2 + 2x - 35$ and find the x-coordinates of points above the x-axis.

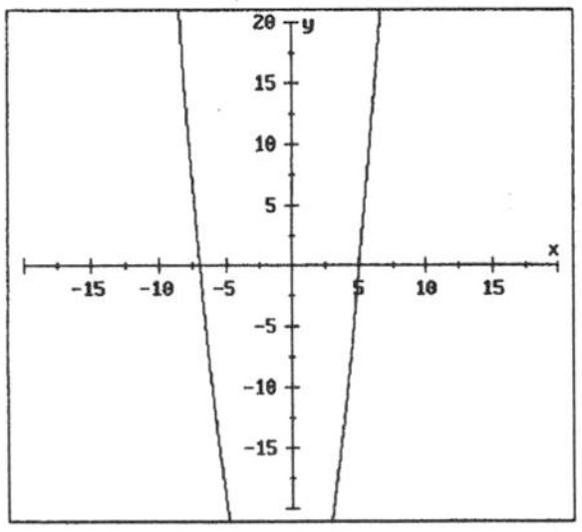

$(-\infty, -7) \cup (5, \infty)$

38. $x^2 + 7x - 18 < 0$
Graph $y = x^2 + 7x - 18$ and find the x-coordinates of points below the x-axis.

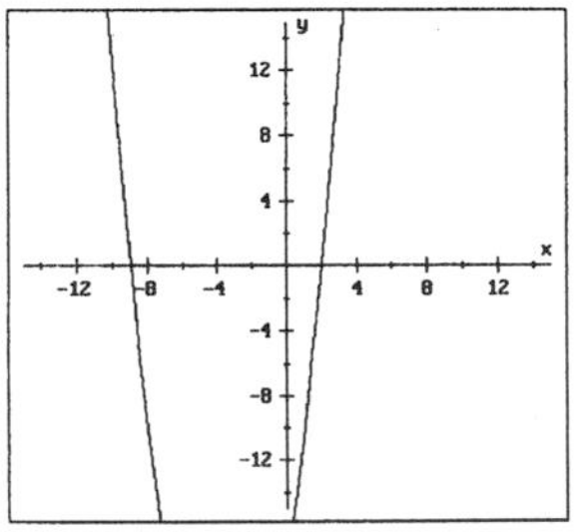

$(-9, 2)$

39. $\frac{3}{x} \leq 5 \Rightarrow \frac{3}{x} - 5 \leq 0$
Graph $y = (3/x) - 5$
and find the x-coordinates
of points below or on the x-axis.

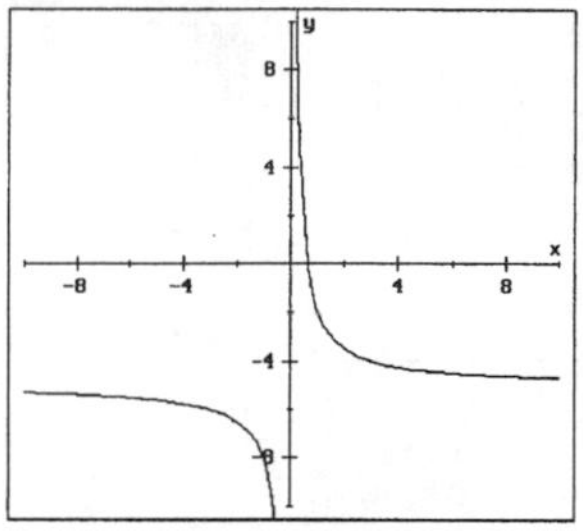

$(-\infty, 0) \cup \left[\frac{3}{5}, \infty\right)$

40. $\frac{2x^2-x-28}{x-1} > 0$
Graph $y = (2x^2 - x - 28)/(x-1)$
and find the x-coordinates
of points above the x-axis.

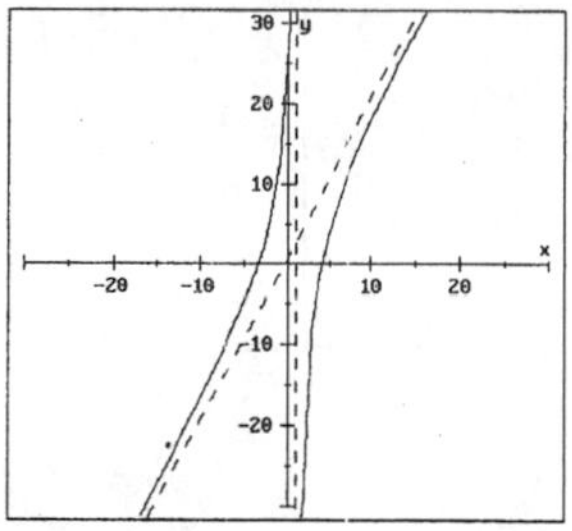

$\left(-\frac{7}{2}, 1\right) \cup (4, \infty)$

41. $y < \frac{1}{2}x^2 - 1$

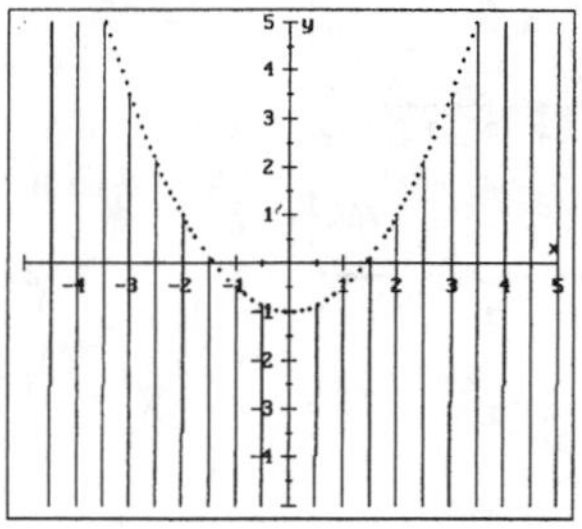

42. $y \geq -|x|$

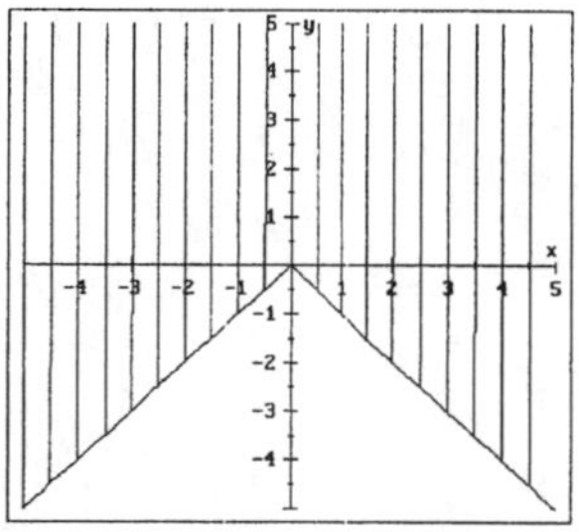

43. $f + g = f(x) + g(x) = 2x + x + 1$
$= 3x + 1$

44. $f - g = f(x) - g(x) = 2x - (x + 1)$
$= x - 1$

45. $f \cdot g = f(x)g(x) = 2x(x+1) = 2x^2 + 2x$

46. $f/g = \frac{f(x)}{g(x)} = \frac{2x}{x+1}$

47. $(f \circ g)(2) = f(g(2)) = f(2+1)$
$= f(3) = 2(3) = 6$

48. $(g \circ f)(-1) = g(f(-1)) = g(2(-1))$
$= g(-2)$
$= -2 + 1 = -1$

49. $(f \circ g)(x) = f(g(x)) = f(x+1) = 2(x+1)$

50. $(g \circ f)(x) = g(f(x)) = g(2x) = 2x + 1$

51. $f(x) = 2(x-3)$

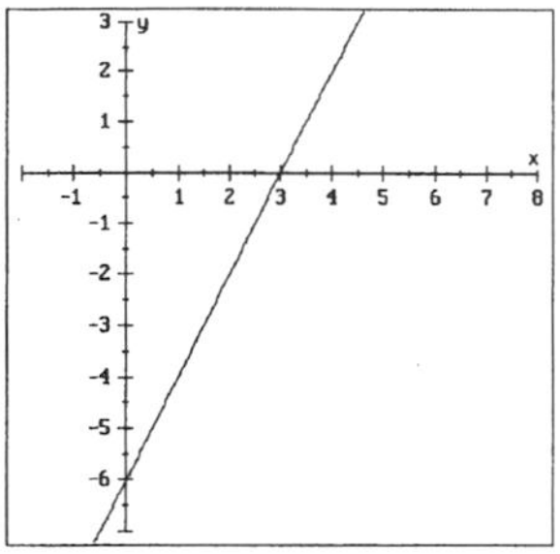

one-to-one

52. $f(x) = x(2x-3)$

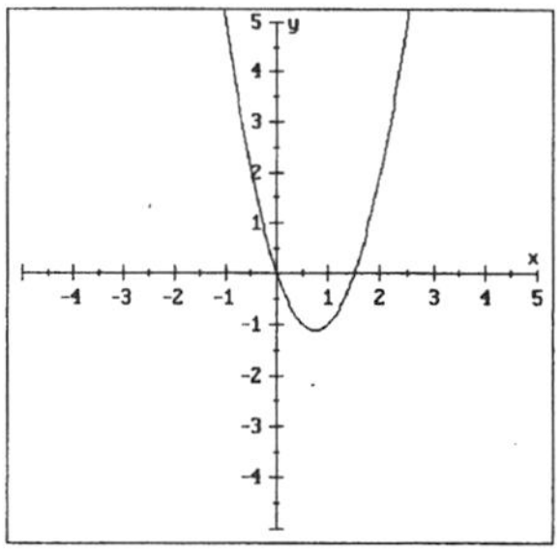

not one-to-one

53. $f(x) = -3(x-2)^2 + 5$

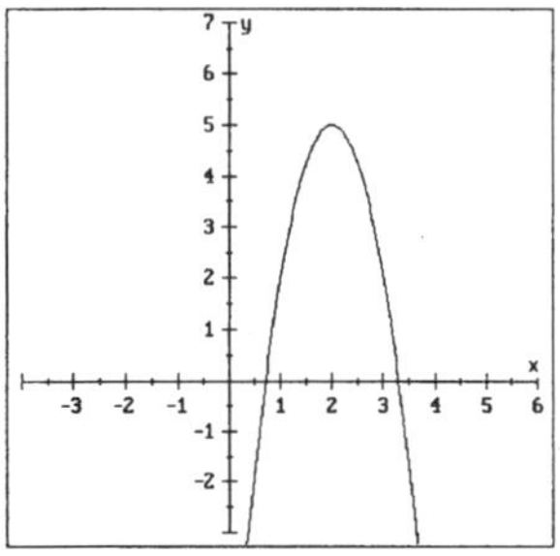

not one-to-one

54. $f(x) = |x|$

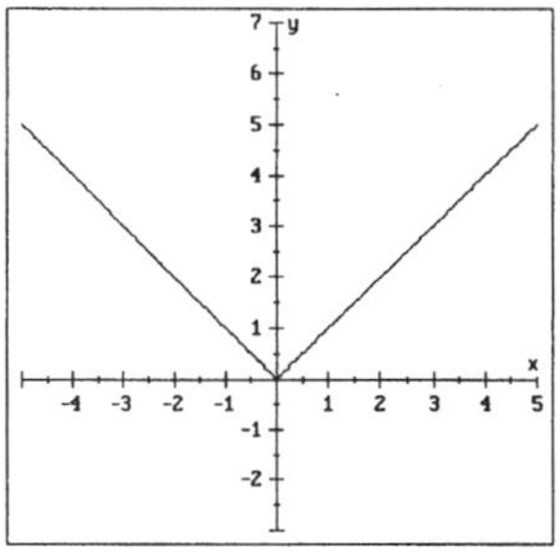

not one-to-one

55.
$$y = 6x - 3$$
$$x = 6y - 3$$
$$x + 3 = 6y$$
$$\tfrac{x+3}{6} = y, \text{ or } y = f^{-1}(x) = \tfrac{x+3}{6}$$

56.
$$y = 4x + 5$$
$$x = 4y + 5$$
$$x - 5 = 4y$$
$$\tfrac{x-5}{4} = y, \text{ or } y = f^{-1}(x) = \tfrac{x-5}{4}$$

57.
$$y = 2x^2 - 1$$
$$x = 2y^2 - 1$$
$$x + 1 = 2y^2$$
$$\tfrac{x+1}{2} = y^2$$
$$\sqrt{\tfrac{x+1}{2}} = y, \text{ or } y = f^{-1}(x) = \sqrt{\tfrac{x+1}{2}}$$

58.
$$y = |x|$$
$$x = |y|$$

Chapter 10 Test (page 728)

1.
$$x^2 + 3x - 18 = 0$$
$$(x+6)(x-3) = 0$$
$$x + 6 = 0 \quad \textbf{or} \quad x - 3 = 0$$
$$x = -6 \qquad\qquad x = 3$$

2.
$$x(6x + 19) = -15$$
$$6x^2 + 19x + 15 = 0$$
$$(2x+3)(3x+5) = 0$$
$$2x + 3 = 0 \quad \textbf{or} \quad 3x + 5 = 0$$
$$x = -\tfrac{3}{2} \qquad\qquad x = -\tfrac{5}{3}$$

3. $\left(\frac{1}{2}\cdot 24\right)^2 = 12^2 = 144$

4. $\left(\frac{1}{2}\cdot(-50)\right)^2 = (-25)^2 = 625$

5.
$$\begin{aligned} x^2+4x+1&=0\\ x^2+4x&=-1\\ x^2+4x+4&=-1+4\\ (x+2)^2&=3\\ x+2&=\pm\sqrt{3}\\ x&=-2\pm\sqrt{3} \end{aligned}$$

6.
$$\begin{aligned} x^2-5x-3&=0\\ x^2-5x&=3\\ x^2-5x+\frac{25}{4}&=3+\frac{25}{4}\\ \left(x-\frac{5}{2}\right)^2&=\frac{37}{4}\\ x-\frac{5}{2}&=\pm\sqrt{\frac{37}{4}}\\ x&=\frac{5}{2}\pm\frac{\sqrt{37}}{2} \end{aligned}$$

7.
$$\begin{aligned} &2x^2+5x+1=0\\ &a=2, b=5, c=1\\ x&=\frac{-b\pm\sqrt{b^2-4ac}}{2a}\\ &=\frac{-5\pm\sqrt{5^2-4(2)(1)}}{2(2)}\\ &=\frac{-5\pm\sqrt{25-8}}{4}\\ &=\frac{-5\pm\sqrt{17}}{4}=-\frac{5}{4}\pm\frac{\sqrt{17}}{4} \end{aligned}$$

8.
$$\begin{aligned} &x^2-x+3=0\\ &a=1, b=-1, c=3\\ x&=\frac{-b\pm\sqrt{b^2-4ac}}{2a}\\ &=\frac{-(-1)\pm\sqrt{(-1)^2-4(1)(3)}}{2(1)}\\ &=\frac{1\pm\sqrt{1-12}}{2}\\ &=\frac{1\pm\sqrt{-11}}{2}=\frac{1}{2}\pm\frac{\sqrt{11}}{2}i \end{aligned}$$

9.
$$\begin{aligned} &3x^2+5x+17=0\\ &a=3, b=5, c=17\\ &b^2-4ac=5^2-4(3)(17)\\ &\qquad=25-208=-183 \end{aligned}$$
nonreal solutions

10.
$4x^2-2kx+k-1=0$

$a=4, b=-2k, c=k-1$

Set the discriminant equal to 0:
$$\begin{aligned} b^2-4ac&=0\\ (-2k)^2-4(4)(k-1)&=0\\ 4k^2-16k+16&=0\\ 4(k-2)(k-2)&=0 \end{aligned}$$
$k-2=0$ **or** $k-2=0$

$k=2$ $\qquad$ $k=2$

11. Let $x =$ the length of the shorter leg.
Then $x+14 =$ the other length.
$$\begin{aligned} x^2+(x+14)^2&=26^2\\ x^2+x^2+28x+196&=676\\ 2x^2+28x-480&=0\\ 2(x+24)(x-10)&=0 \end{aligned}$$
$x=-24$ or $x=10$

The shorter leg is 10 inches long.

12.
$$\begin{aligned} 2y-3y^{1/2}+1&=0\\ \left(2y^{1/2}-1\right)\left(y^{1/2}-1\right)&=0 \end{aligned}$$
$2y^{1/2}-1=0$ **or** $y^{1/2}-1=0$

$2y^{1/2}=1$ $\qquad$ $y^{1/2}=1$

$\left(y^{1/2}\right)^2=\left(\frac{1}{2}\right)^2$ $\qquad$ $\left(y^{1/2}\right)^2=1^2$

$y=\frac{1}{4}$ $\qquad$ $y=1$

Solution $\qquad$ Solution

13. $y = \frac{1}{2}x^2 - 4 = \frac{1}{2}(x-0)^2 - 4$

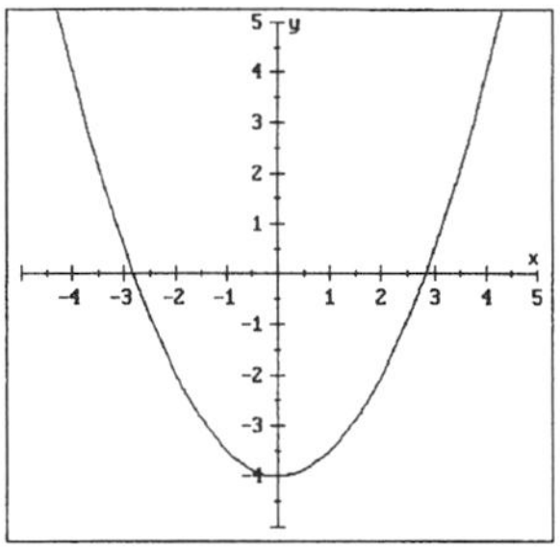

vertex: $(0, -4)$

14. $y = -2x^2 + 8x - 7$
$= -2(x^2 - 4x) - 7$
$= -2(x^2 - 4x + 4) - 7 + 8$
$= -2(x-2)^2 + 1$
Vertex: $(2, 1)$

15. $y \le -x^2 + 3$

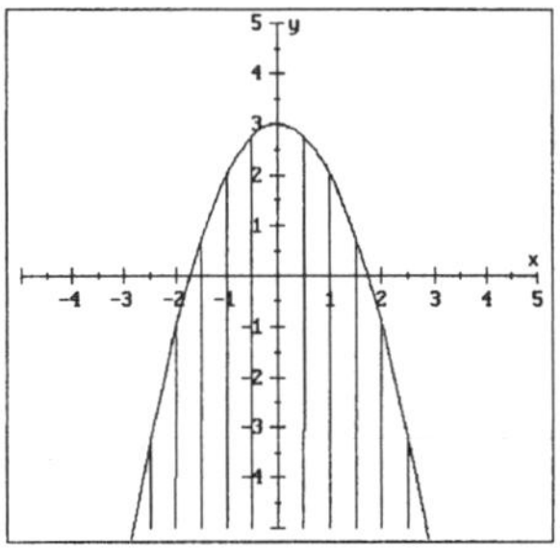

16. $x^2 - 2x - 8 > 0$
$(x+2)(x-4) > 0$

```
x − 4   ---------0 ++++
x + 2   ---  0+++++++ ++++
      ←——— ) ————( ———→
          −2      4
```

solution set: $(-\infty, -2) \cup (4, \infty)$

17. $\frac{x-2}{x+3} \le 0$

```
x − 2   --------0  ++++
x + 3   ---0+++++++ ++++
     ←——— (————] ———→
         −3     2
```

solution set: $(-3, 2]$

18. $g + f = g(x) + f(x) = x - 1 + 4x$
$= 5x - 1$

19. $f - g = f(x) - g(x) = 4x - (x-1)$
$= 3x + 1$

20. $g \cdot f = g(x)f(x) = (x-1)4x = 4x^2 - 4x$

21. $g/f = \frac{g(x)}{f(x)} = \frac{x-1}{4x}$

22. $(g \circ f)(1) = g(f(1)) = g(4(1))$
$= g(4) = 4 - 1 = 3$

23. $(f \circ g)(0) = f(g(0)) = f(0-1)$
$= f(-1) = 4(-1) = -4$

24. $(f \circ g)(-1) = f(g(-1)) = f(-1-1) = f(-2) = 4(-2) = -8$

25. $(g \circ f)(-2) = g(f(-2)) = g(4(-2)) = g(-8) = -8 - 1 = -9$

26. $(f \circ g)(x) = f(g(x)) = f(x-1) = 4(x-1)$

27. $(g \circ f)(x) = g(f(x)) = g(4x) = 4x - 1$

28.
$$\begin{aligned} 3x + 2y &= 12 \\ 3y + 2x &= 12 \\ 3y &= -2x + 12 \\ y &= \frac{-2x+12}{3} \end{aligned}$$

29.
$$\begin{aligned} y &= 3x^2 + 4 \\ x &= 3y^2 + 4 \\ x - 4 &= 3y^2 \\ \frac{x-4}{3} &= y^2 \\ -\sqrt{\frac{x-4}{3}} &= y \end{aligned}$$

Cumulative Review Exercises (page 729)

1.
$y = f(x) = 2x^2 - 3$
domain $= (-\infty, \infty)$
range $= [-3, \infty)$

2.
$y = f(x) = -|x - 4|$
domain $= (-\infty, \infty)$
range $= (-\infty, 0]$

3.
$$\begin{aligned} y - y_1 &= m(x - x_1) \\ y + 4 &= 3(x + 2) \\ y &= 3x + 2 \end{aligned}$$

4.
$$\begin{aligned} 2x + 3y &= 6 \\ 3y &= -2x + 6 \\ y &= -\frac{2}{3}x + 2 \\ y - y_1 &= m(x - x_1) \\ y + 2 &= -\frac{2}{3}(x - 0) \\ y &= -\frac{2}{3}x - 2 \end{aligned}$$

5. $(2a^2 + 4a - 7) - 2(3a^2 - 4a) = 2a^2 + 4a - 7 - 6a^2 + 8a = -4a^2 + 12a - 7$

6. $(3x + 2)(2x - 3) = 6x^2 - 9x + 4x - 6 = 6x^2 - 5x - 6$

7. $x^4 - 16y^4 = (x^2 + 4y^2)(x^2 - 4y^2) = (x^2 + 4y^2)(x + 2y)(x - 2y)$

8. $15x^2 - 2x - 8 = (5x - 4)(3x + 2)$

9.
$$\begin{aligned} x^2 - 5x - 6 &= 0 \\ (x - 6)(x + 1) &= 0 \end{aligned}$$
$x - 6 = 0$ **or** $x + 1 = 0$
$x = 6$ $\quad\quad$ $x = -1$

10.
$$\begin{aligned} 6a^3 - 2a &= a^2 \\ 6a^3 - a^2 - 2a &= 0 \\ a(6a^2 - a - 2) &= 0 \\ a(3a - 2)(2a + 1) &= 0 \end{aligned}$$
$a = 0$ **or** $3a - 2 = 0$ **or** $2a + 1 = 0$
$a = \frac{2}{3}$ $\quad\quad$ $a = -\frac{1}{2}$

11. $\sqrt{25x^4} = 5x^2$

12. $\sqrt{48t^3} = \sqrt{16t^2}\sqrt{3t} = 4t\sqrt{3t}$

13. $\sqrt[3]{-27x^3} = -3x$

14. $\sqrt[3]{\dfrac{128x^4}{2x}} = \sqrt[3]{64x^3} = 4x$

15. $8^{-1/3} = \dfrac{1}{8^{1/3}} = \dfrac{1}{2}$

16. $64^{2/3} = \left(64^{1/3}\right)^2 = 4^2 = 16$

17. $\dfrac{y^{2/3}y^{5/3}}{y^{1/3}} = \dfrac{y^{7/3}}{y^{1/3}} = y^{6/3} = y^2$

18. $\dfrac{x^{5/3}x^{1/2}}{x^{3/4}} = \dfrac{x^{13/6}}{x^{3/4}} = x^{17/12}$

19. $f(x) = \sqrt{x-2}$; Shift $y = \sqrt{x}$ right 2.

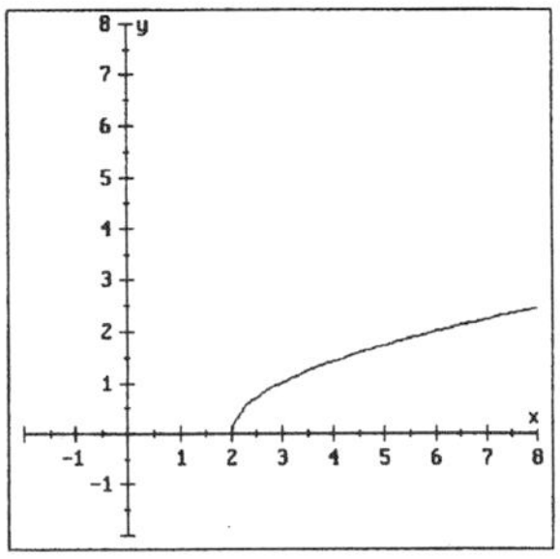

$D = [2, \infty), R = [0, \infty)$

20. $f(x) = -\sqrt{x+2}$; Reflect $y = \sqrt{x}$ about the x-axis and shift left 2.

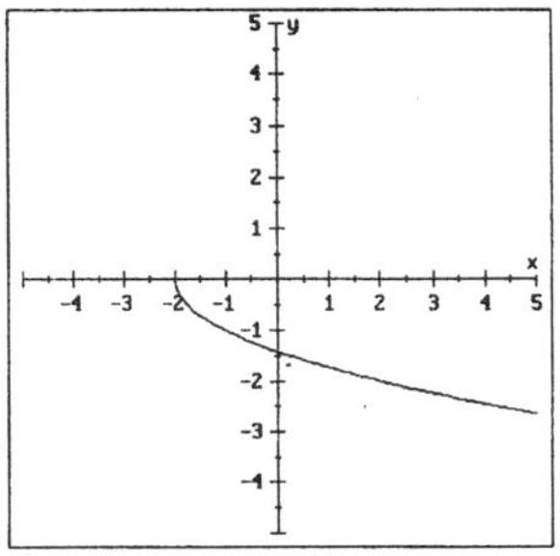

$D = [-2, \infty), R = (-\infty, 0]$

21. $\left(x^{2/3} - x^{1/3}\right)\left(x^{2/3} + x^{1/3}\right) = x^{4/3} + x^{3/3} - x^{3/3} - x^{2/3} = x^{4/3} - x^{2/3}$

22. $\left(x^{-1/2} + x^{1/2}\right)^2 = \left(x^{-1/2} + x^{1/2}\right)\left(x^{-1/2} + x^{1/2}\right) = x^{-2/2} + x^0 + x^0 + x^{2/2} = x + 2 + \frac{1}{x}$

23. $\sqrt{50} - \sqrt{8} + \sqrt{32} = \sqrt{25}\sqrt{2} - \sqrt{4}\sqrt{2} + \sqrt{16}\sqrt{2} = 5\sqrt{2} - 2\sqrt{2} + 4\sqrt{2} = 7\sqrt{2}$

24.
$$\begin{aligned} -3\sqrt[4]{32} - 2\sqrt[4]{162} + 5\sqrt[4]{48} &= -3\sqrt[4]{16}\sqrt[4]{2} - 2\sqrt[4]{81}\sqrt[4]{2} + 5\sqrt[4]{16}\sqrt[4]{3} \\ &= -3(2)\sqrt[4]{2} - 2(3)\sqrt[4]{2} + 5(2)\sqrt[4]{3} \\ &= -6\sqrt[4]{2} - 6\sqrt[4]{2} + 10\sqrt[4]{3} = -12\sqrt[4]{2} + 10\sqrt[4]{3} \end{aligned}$$

25. $3\sqrt{2}(2\sqrt{3} - 4\sqrt{12}) = 6\sqrt{6} - 12\sqrt{24} = 6\sqrt{6} - 12\sqrt{4}\sqrt{6} = 6\sqrt{6} - 24\sqrt{6} = -18\sqrt{6}$

26. $\dfrac{5}{\sqrt[3]{x}} = \dfrac{5}{\sqrt[3]{x}} \cdot \dfrac{\sqrt[3]{x^2}}{\sqrt[3]{x^2}} = \dfrac{5\sqrt[3]{x^2}}{\sqrt[3]{x^3}} = \dfrac{5\sqrt[3]{x^2}}{x}$

27. $\dfrac{\sqrt{x}+2}{\sqrt{x}-1} = \dfrac{\sqrt{x}+2}{\sqrt{x}-1} \cdot \dfrac{\sqrt{x}+1}{\sqrt{x}+1} = \dfrac{x + 3\sqrt{x} + 2}{x - 1}$

28. $\sqrt[6]{x^3y^3} = (x^3y^3)^{1/6} = x^{3/6}y^{3/6} = x^{1/2}y^{1/2} = \sqrt{xy}$

29.
$$5\sqrt{x+2} = x+8$$
$$\left(5\sqrt{x+2}\right)^2 = (x+8)^2$$
$$25(x+2) = x^2+16x+64$$
$$25x+50 = x^2+16x+64$$
$$0 = x^2-9x+14$$
$$0 = (x-7)(x-2)$$
$x = 7$ or $x = 2$ (Both check.)

30.
$$\sqrt{x}+\sqrt{x+2} = 2$$
$$\sqrt{x} = 2-\sqrt{x+2}$$
$$\left(\sqrt{x}\right)^2 = \left(2-\sqrt{x+2}\right)^2$$
$$x = 4-4\sqrt{x+2}+x+2$$
$$4\sqrt{x+2} = 6$$
$$\left(4\sqrt{x+2}\right)^2 = 6^2$$
$$16(x+2) = 36$$
$$16x+32 = 36$$
$$16x = 4$$
$$x = \frac{4}{16} = \frac{1}{4}$$

31. hypotenuse $= 3\sqrt{2}$ in.

32. hypotenuse $= 2 \cdot \dfrac{3}{\sqrt{3}} = \dfrac{6\sqrt{3}}{3} = 2\sqrt{3}$ in.

33. $d = \sqrt{(-2-4)^2+(6-14)^2} = \sqrt{(-6)^2+(-8)^2} = \sqrt{36+64} = \sqrt{100} = 10$

34. $\left(\frac{1}{2}\cdot 6\right)^2 = 3^2 = 9$

35.
$$2x^2+x-3 = 0$$
$$x^2+\frac{1}{2}x-\frac{3}{2} = 0$$
$$x^2+\frac{1}{2}x = \frac{3}{2}$$
$$x^2+\frac{1}{2}x+\frac{1}{16} = \frac{3}{2}+\frac{1}{16}$$
$$\left(x+\frac{1}{4}\right)^2 = \frac{25}{16}$$
$$x+\frac{1}{4} = \pm\frac{5}{4}$$
$$x = -\frac{1}{4}\pm\frac{5}{4}$$
$$x = \frac{4}{4} = 1 \quad \text{or} \quad x = -\frac{6}{4} = -\frac{3}{2}$$

36.
$$3x^2+4x-1 = 0$$
$$a = 3, b = 4, c = -1$$
$$x = \frac{-b\pm\sqrt{b^2-4ac}}{2a}$$
$$= \frac{-4\pm\sqrt{4^2-4(3)(-1)}}{2(3)}$$
$$= \frac{-4\pm\sqrt{16+12}}{6}$$
$$= \frac{-4\pm\sqrt{28}}{6}$$
$$= \frac{-4\pm 2\sqrt{7}}{6} = -\frac{2}{3}\pm\frac{\sqrt{7}}{3}$$

37. $y = \frac{1}{2}x^2 + 5 = \frac{1}{2}(x-0)^2 + 5$

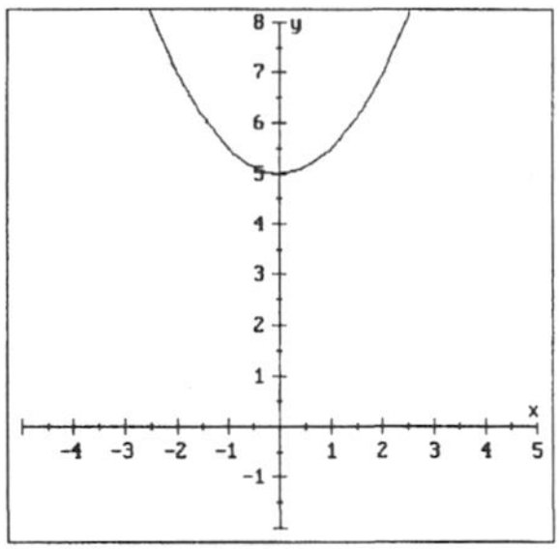

vertex: $(0, 5)$

38. $y \le -x^2 + 3$

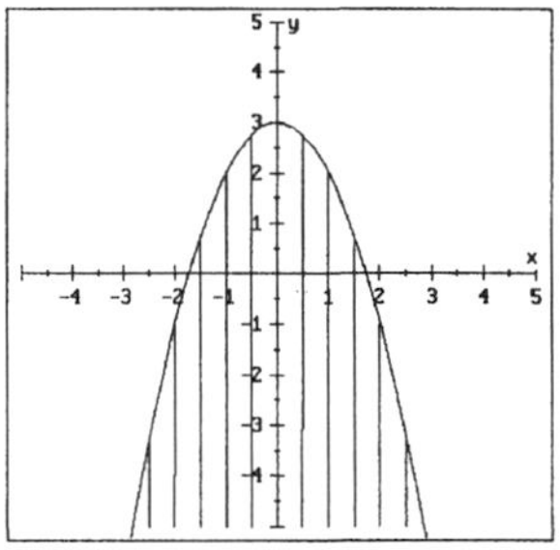

vertex: $(0, 3)$

39. $(3+5i) + (4-3i) = 3 + 5i + 4 - 3i = 7 + 2i$

40. $(7-4i) - (12+3i) = 7 - 4i - 12 - 3i = -5 - 7i$

41. $(2-3i)(2+3i) = 4 + 6i - 6i - 9i^2 = 4 + 9 = 13 = 13 + 0i$

42. $(3+i)(3-3i) = 9 - 9i + 3i - 3i^2 = 9 - 6i + 3 = 12 - 6i$

43. $(3-2i) - (4+i)^2 = 3 - 2i - (16 + 8i + i^2) = 3 - 2i - (15 + 8i) = 3 - 2i - 15 - 8i$
$= -12 - 10i$

44. $\frac{5}{3-i} = \frac{5}{3-i} \cdot \frac{3+i}{3+i} = \frac{5(3+i)}{9-i^2} = \frac{5(3+i)}{10} = \frac{3+i}{2} = \frac{3}{2} + \frac{1}{2}i$

45. $|3+2i| = \sqrt{3^2 + 2^2} = \sqrt{9+4} = \sqrt{13}$

46. $|5-6i| = \sqrt{5^2 + (-6)^2} = \sqrt{25+36} = \sqrt{61}$

47.

$$2x^2 + 4x = k$$
$$2x^2 + 4x - k = 0$$
$$a = 2, b = 4, c = -k$$

Set the discriminant equal to 0:

$$b^2 - 4ac = 0$$
$$4^2 - 4(2)(-k) = 0$$
$$16 + 8k = 0$$
$$8k = -16$$
$$k = -2$$

48.

$$a - 7a^{1/2} + 12 = 0$$
$$\left(a^{1/2} - 3\right)\left(a^{1/2} - 4\right) = 0$$

$a^{1/2} - 3 = 0$ **or** $a^{1/2} - 4 = 0$

$a^{1/2} = 3$	$a^{1/2} = 4$
$\left(a^{1/2}\right)^2 = 3^2$	$\left(a^{1/2}\right)^2 = 4^2$
$a = 9$	$a = 16$
Solution	Solution

49. $x^2 - x - 6 > 0$

$(x+2)(x-3) > 0$

$x - 3$ ‒ ‒ ‒ ‒ ‒ ‒ ‒ ‒ ‒ 0 ++++
$x + 2$ ‒ ‒ ‒ 0+++++++ ++++

⟵) —— (⟶
-2 3

solution set: $(-\infty, -2) \cup (3, \infty)$

50. $x^2 - x - 6 \leq 0$

$(x+2)(x-3) \leq 0$

$x - 3$ ‒ ‒ ‒ ‒ ‒ ‒ ‒ ‒ 0 ++++
$x + 2$ ‒ ‒ ‒ 0+++++++ ++++

⟵ [——] ⟶
-2 3

solution set: $[-2, 3]$

51. $f(-1) = 3(-1)^2 + 2 = 3(1) + 2 = 3 + 2 = 5$

52. $(g \circ f)(2) = g(f(2)) = g(3(2)^2 + 2) = g(14) = 2(14) - 1 = 27$

53. $(f \circ g)(x) = f(g(x)) = f(2x - 1) = 3(2x-1)^2 + 2 = 3(4x^2 - 4x + 1) + 2 = 12x^2 - 12x + 5$

54. $(g \circ f)(x) = g(f(x)) = g(3x^2 + 2) = 2(3x^2 + 2) - 1 = 6x^2 + 3$

55.
$$\begin{aligned} y &= 3x + 2 \\ x &= 3y + 2 \\ x - 2 &= 3y \\ \frac{x-2}{3} &= y, \text{ or } y = f^{-1}(x) = \frac{x-2}{3} \end{aligned}$$

56.
$$\begin{aligned} y &= x^3 + 4 \\ x &= y^3 + 4 \\ x - 4 &= y^3 \\ \sqrt[3]{x-4} &= y, \text{ or } y = f^{-1}(x) = \sqrt[3]{x-4} \end{aligned}$$

Exercise 11.1 (page 742)

1.
$$\begin{aligned} 3x + 2x - 20 &= 180 \\ 5x &= 200 \\ x &= 40 \end{aligned}$$

3. $m(\angle 2) = 3x = 3(40) = 120°$

5. exponential

7. $(0, \infty)$

9. increasing

11. $P\left(1 + \frac{r}{k}\right)^{kt}$

Problems 13-15 are to be solved using a calculator. The keystrokes needed to solve each problem using a TI-83 graphing calculator appear in each solution. There may be other solutions. Keystrokes for other calculators may be slightly different.

13. $2^{\sqrt{2}} \Rightarrow$ [2] [^] [√] [2] [ENTER]
$\{2.6651\}$

15. $5^{\sqrt{5}} \Rightarrow$ [5] [^] [√] [5] [ENTER]
$\{36.5548\}$

17. $\left(2^{\sqrt{3}}\right)^{\sqrt{3}} = 2^{(\sqrt{3})(\sqrt{3})} = 2^3 = 8$

19. $7^{\sqrt{3}} 7^{\sqrt{12}} = 7^{\sqrt{3}+\sqrt{12}} = 7^{\sqrt{3}+2\sqrt{3}} = 7^{3\sqrt{3}}$

21. $y = f(x) = 3^x$
through $(0, 1)$ and $(1, 3)$

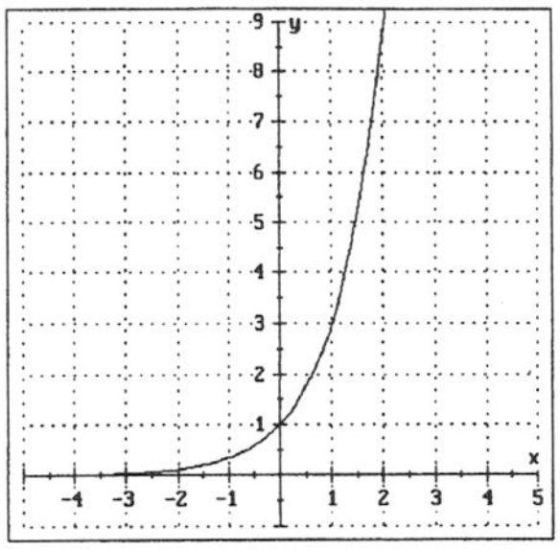

23. $y = f(x) = \left(\frac{1}{3}\right)^x$
through $(0, 1)$ and $\left(1, \frac{1}{3}\right)$

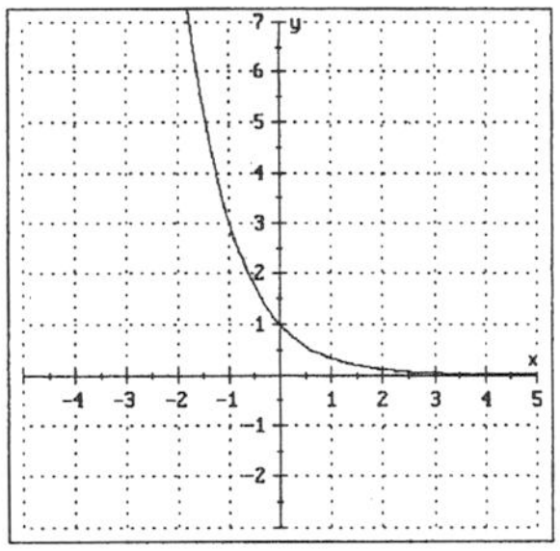

25. $f(x) = 3^x - 2$
Shift $y = 3^x$ down 2.

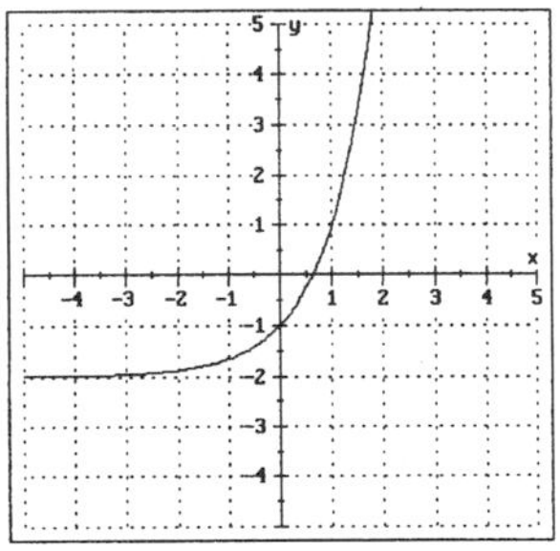

27. $f(x) = 3^{x-1}$
Shift $y = 3^x$ right 1.

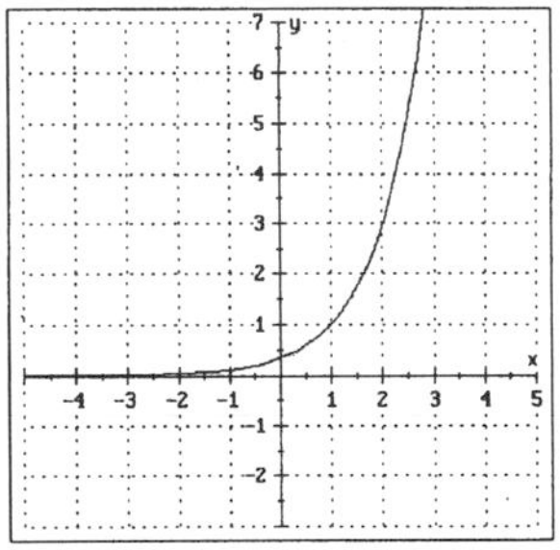

29. $y = b^x$
$\frac{1}{2} = b^1$
$\frac{1}{2} = b$

31. $y = b^x$
$2 = b^0$
no such value of b

33. $y = b^x$
$2 = b^1$
$2 = b$

35. $y = b^x$
$9 = b^2$
$3 = b$

37. $y = f(x) = \frac{1}{2}\left(3^{x/2}\right)$

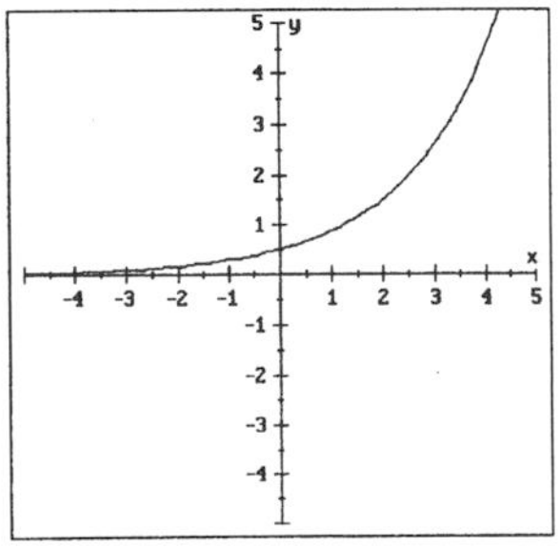

increasing

39. $y = f(x) = 2\left(3^{-x/2}\right)$

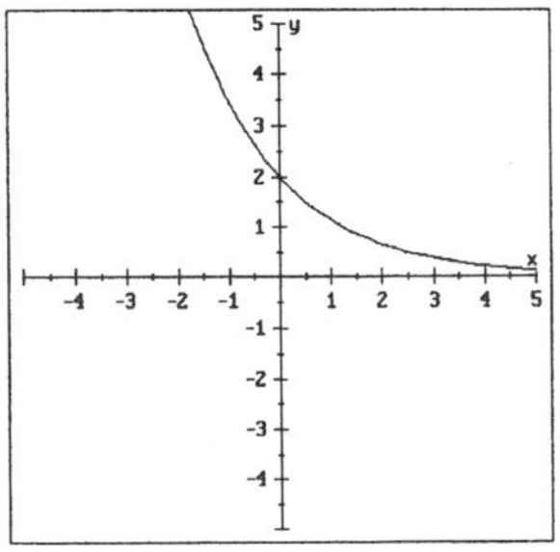

decreasing

41. $S(n) = 5.74(1.39)^n = 5.74(1.39)^5$
≈ 29.8
There were about 29.8 million users in 1995.

43. $A = A_0\left(\frac{2}{3}\right)^t$
$A = A_0\left(\frac{2}{3}\right)^5$
$A = \frac{32}{243}A_0$

45. $C = (3 \times 10^{-4})(0.7)^t = (3 \times 10^{-4})(0.7)^5 \approx 5.0421 \times 10^{-5}$ coulombs

47. $A = P\left(1 + \frac{r}{k}\right)^{kt}$
$= 10{,}000\left(1 + \frac{0.08}{4}\right)^{4(10)}$
$= 10{,}000(1.02)^{40} \approx \$22{,}080.40$

49. $A = P\left(1 + \frac{r}{k}\right)^{kt}$
$= 1000\left(1 + \frac{0.05}{4}\right)^{4(5)}$
$= 1000(1.0125)^{20}$
$\approx \$1282.040$

$A = P\left(1 + \frac{r}{k}\right)^{kt}$
$= 1000\left(1 + \frac{0.055}{4}\right)^{4(5)}$
$= 1000(1.01375)^{20}$
$\approx \$1314.07$

difference $= \$1314.07 - \$1282.04$
$= \$32.03$

51. $A = P\left(1 + \frac{r}{k}\right)^{kt}$
$= 1\left(1 + \frac{0.05}{1}\right)^{1(300)}$
$= 1(1.05)^{300}$
$\approx \$2{,}273{,}996.13$

53. $A = P\left(1 + \frac{r}{k}\right)^{kt}$
$= 4700\left(1 + \frac{-0.25}{1}\right)^{1(5)}$
$= 4700(0.75)^5$
$\approx \$1115.33$

55. **Answers may vary.**

57. If the base were 0, then the function would not be defined for $x = 0 \Rightarrow y = 0^0$.

Exercise 11.2 (page 750)

1. $\sqrt{240x^5} = \sqrt{16x^4}\sqrt{15x} = 4x^2\sqrt{15x}$

3. $4\sqrt{48y^3} - 3y\sqrt{12y} = 4\sqrt{16y^2}\sqrt{3y} - 3y\sqrt{4}\sqrt{3y} = 4(4y)\sqrt{3y} - 3y(2)\sqrt{3y}$
$= 16y\sqrt{3y} - 6y\sqrt{3y} = 10y\sqrt{3y}$

5. 2.72

7. increasing

9. $A = Pe^{rt}$

11. $y = f(x) = e^x + 1$
Shift $y = e^x$ up 1.

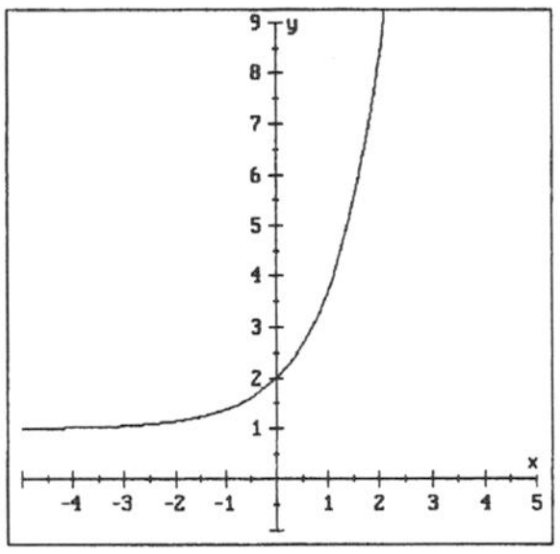

13. $y = f(x) = e^{(x+3)}$
Shift $y = e^x$ left 3.

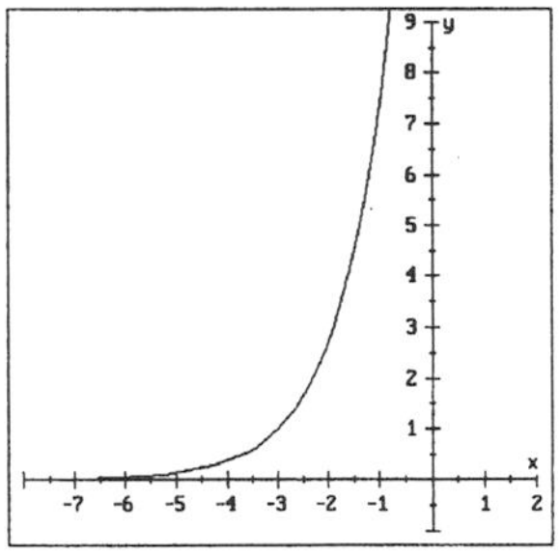

15. $y = f(x) = -e^x$; Reflect $y = e^x$ about the x-axis.

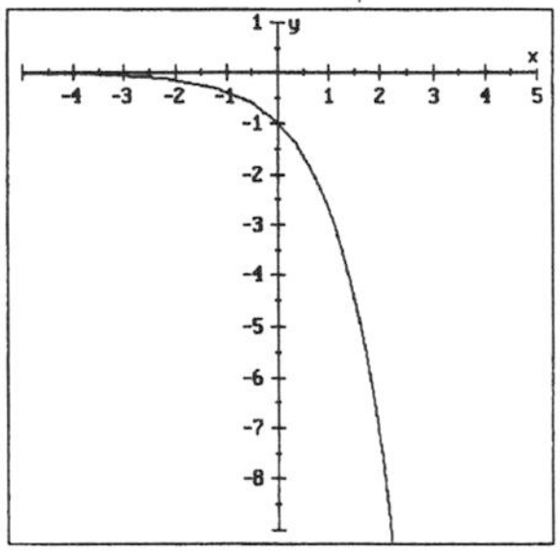

17. $y = f(x) = 2e^x$; Stretch $y = e^x$ vertically by a factor of 2.

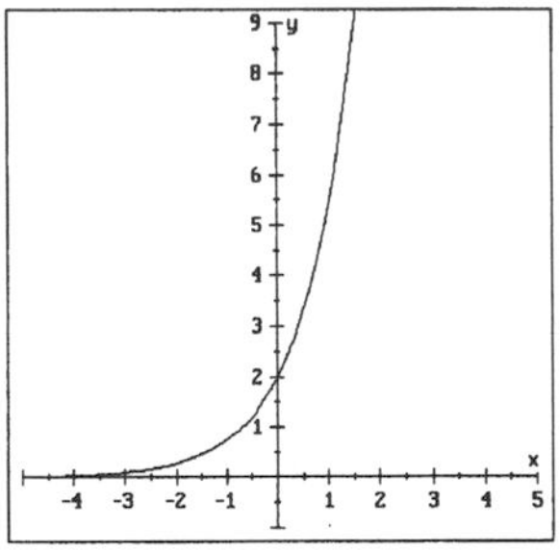

19. The graph should be increasing. The graph could not look like this.

21. The graph should go through the point $(0, 1)$. The graph could not look like this.

23.
$$\begin{aligned} A &= Pe^{rt} \\ &= 5000e^{0.06(12)} \\ &= 5000e^{0.72} \\ &\approx \$10{,}272.17 \end{aligned}$$

25.
$$\begin{aligned} A &= Pe^{rt} \\ 12000 &= Pe^{0.07(9)} \\ 12000 &= Pe^{0.63} \\ 12000 &\approx P(1.8776106) \\ P &\approx \frac{12000}{1.8776106} \\ P &\approx \$6{,}391.10 \end{aligned}$$

27.
$$\begin{aligned} A &= Pe^{rt} \\ &= 5000e^{0.085(5)} \\ &= 5000e^{0.425} \\ &\approx \$7{,}647.95 \quad \text{(continuous)} \end{aligned}$$

$$\begin{aligned} A &= P\left(1 + \frac{r}{k}\right)^{kt} \\ &= 5000\left(1 + \frac{0.085}{1}\right)^{1(5)} \\ &= 5000(1.085)^5 \\ &\approx \$7{,}518.28 \quad \text{(annual)} \end{aligned}$$

29. $A = Pe^{rt}$
$= 6e^{0.019(30)}$
$= 6e^{0.57}$
$\approx$ 10.6 billion people

31. $A = Pe^{rt}$
$= 6e^{0.019(50)}$
$= 6e^{0.95}$
$\approx 6(2.6)$
It will increase by a factor of about 2.6.

33. $A = 8000e^{-0.008t}$
$= 8000e^{-0.008(20)}$
$= 8000e^{-0.16}$
≈ 6817

35. $A = A_0e^{-0.087t}$
$= 50e^{-0.087(30)}$
$= 50e^{-2.61}$
$\approx$ 3.68 grams

37. $A = A_0e^{-0.00000693t}$
$= 2500e^{-0.00000693(100)}$
$= 2500e^{-0.000693}$
$\approx$ 2498.27 grams

39. $P = 0.3\left(1 - e^{-0.05t}\right)$
$= 0.3\left(1 - e^{-0.05(15)}\right)$
$= 0.3\left(1 - e^{-0.75}\right)$
$\approx 0.3(1 - 0.47237)$
$\approx 0.3(0.52763)$
≈ 0.16

41. $x = 0.08\left(1 - e^{-0.1t}\right)$
$= 0.08\left(1 - e^{-0.1(0)}\right)$
$= 0.08\left(1 - e^{0}\right)$
$= 0.08(1 - 1)$
$= 0.08(0) = 0$

43. $v = 50\left(1 - e^{-0.2t}\right)$
$= 50\left(1 - e^{-0.2(20)}\right)$
$= 50\left(1 - e^{-4}\right)$
$\approx 50(1 - 0.01832)$
$\approx 50(0.98168) \approx 49$ mps

45. $A = Pe^{rt}$
$= 4570e^{-0.06(6.5)}$
$\approx \$3094.15$

47. $y = 1000e^{0.02x}$
$y = 31x + 2000$

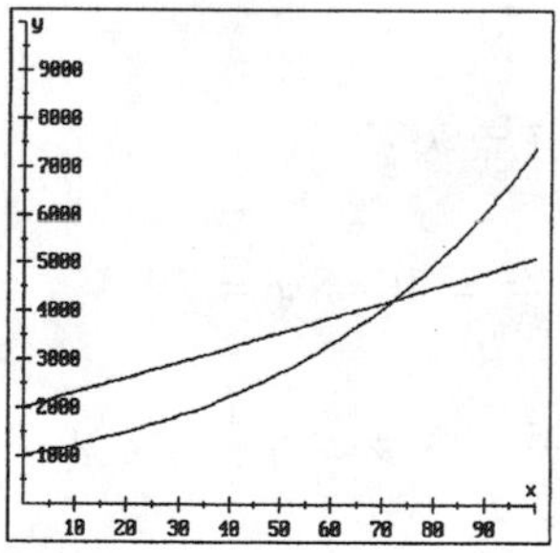

about 72 years

49. **Answers may vary.**

51. $e \approx 2.7182$

$$1 + 1 + \frac{1}{2} + \frac{1}{2 \cdot 3} + \frac{1}{2 \cdot 3 \cdot 4} + \frac{1}{2 \cdot 3 \cdot 4 \cdot 5} \approx 2.7167$$

53.

$$\begin{aligned} e^{t+5} &= ke^{t} \\ e^{t} \cdot e^{5} &= ke^{t} \\ e^{5}e^{t} &= ke^{t} \\ k &= e^{5} \end{aligned}$$

Exercise 11.3 (page 760)

1.
$$\begin{aligned} \sqrt[3]{6x+4} &= 4 \\ 6x+4 &= 4^3 \\ 6x+4 &= 64 \\ 6x &= 60 \\ x &= 10 \end{aligned}$$

3.
$$\begin{aligned} \sqrt{a+1}-1 &= 3a \\ \sqrt{a+1} &= 3a+1 \\ a+1 &= (3a+1)^2 \\ a+1 &= 9a^2+6a+1 \\ 0 &= 9a^2+5a \\ 0 &= a(9a+5) \end{aligned}$$
$a=0$ or $a=-\frac{5}{9}$ ($-\frac{5}{9}$ does not check.)

5. $x = b^y$

7. range

9. inverse

11. exponent

13. $(b,1)$; $(1,0)$

15. $20 \log \frac{E_O}{E_I}$

17. $\log_3 27 = 3 \Rightarrow 3^3 = 27$

19. $\log_{1/2} \frac{1}{4} = 2 \Rightarrow \left(\frac{1}{2}\right)^2 = \frac{1}{4}$

21. $\log_4 \frac{1}{64} = -3 \Rightarrow 4^{-3} = \frac{1}{64}$

23. $\log_{1/2} \frac{1}{8} = 3 \Rightarrow \left(\frac{1}{2}\right)^3 = \frac{1}{8}$

25. $6^2 = 36 \Rightarrow \log_6 36 = 2$

27. $5^{-2} = \frac{1}{25} \Rightarrow \log_5 \frac{1}{25} = -2$

29. $\left(\frac{1}{2}\right)^{-5} = 32 \Rightarrow \log_{1/2} 32 = -5$

31. $x^y = z \Rightarrow \log_x z = y$

33. $\log_2 16 = x \Rightarrow 2^x = 16 \Rightarrow x = 4$

35. $\log_4 16 = x \Rightarrow 4^x = 16 \Rightarrow x = 2$

37. $\log_{1/2} \frac{1}{8} = x \Rightarrow \left(\frac{1}{2}\right)^x = \frac{1}{8} \Rightarrow x = 3$

39. $\log_9 3 = x \Rightarrow 9^x = 3 \Rightarrow x = \frac{1}{2}$

41. $\log_{1/2} 8 = x \Rightarrow \left(\frac{1}{2}\right)^x = 8 \Rightarrow x = -3$

43. $\log_7 x = 2 \Rightarrow 7^2 = x \Rightarrow x = 49$

45. $\log_6 x = 1 \Rightarrow 6^1 = x \Rightarrow x = 6$

47. $\log_{25} x = \frac{1}{2} \Rightarrow 25^{1/2} = x \Rightarrow x = 5$

49. $\log_5 x = -2 \Rightarrow 5^{-2} = x \Rightarrow x = \frac{1}{25}$

51. $\log_{36} x = -\frac{1}{2} \Rightarrow 36^{-1/2} = x \Rightarrow x = \frac{1}{6}$

53. $\log_{100} \frac{1}{1000} = x \Rightarrow 100^x = \frac{1}{1000} \Rightarrow x = -\frac{3}{2}$

55. $\log_{27} 9 = x \Rightarrow 27^x = 9 \Rightarrow x = \dfrac{2}{3}$

57. $\log_x 5^3 = 3 \Rightarrow x^3 = 5^3 \Rightarrow x = 5$

59. $\log_x \dfrac{9}{4} = 2 \Rightarrow x^2 = \dfrac{9}{4} \Rightarrow x = \dfrac{3}{2}$

61. $\log_x \dfrac{1}{64} = -3 \Rightarrow x^{-3} = \dfrac{1}{64} \Rightarrow x = 4$

63. $\log_{2\sqrt{2}} x = 2 \Rightarrow (2\sqrt{2})^2 = x \Rightarrow x = 8$

65. $2^{\log_2 4} = x \Rightarrow x = 4$

67. $x^{\log_4 6} = 6 \Rightarrow x = 4$

69. $\log 10^3 = x \Rightarrow 10^x = 10^3 \Rightarrow x = 3$

71. $10^{\log x} = 100 \Rightarrow \log x = 2 \Rightarrow x = 100$

73. $\log 8.25 \approx 0.9165$

75. $\log 0.00867 \approx -2.0620$

77. $\log y = 1.4023 \Rightarrow y = 25.25$

79. $\log y = 4.24 \Rightarrow y = 17{,}378.01$

81. $\log y = -3.71 \Rightarrow y = 0.00$

83. $\log y = \log 8 \Rightarrow \log y = 0.9030 \Rightarrow y = 8$

85. $y = f(x) = \log_3 x$
through $(1, 0)$ and $(3, 1)$

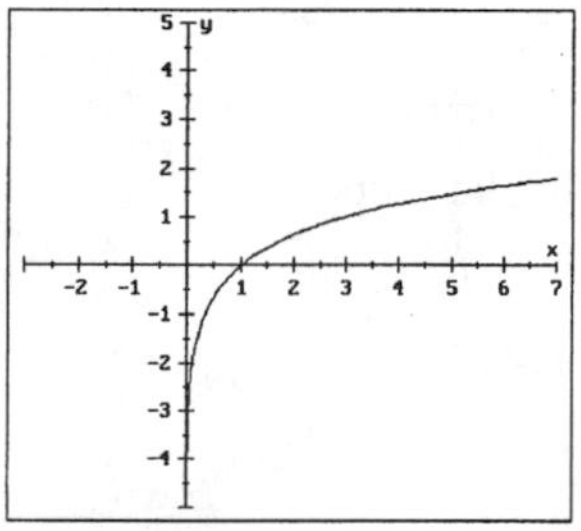

increasing

87. $y = f(x) = \log_{1/2} x$
through $(1, 0)$ and $\left(\frac{1}{2}, 1\right)$

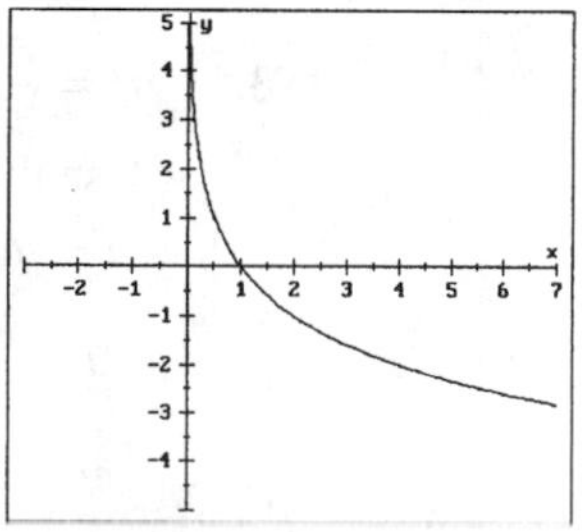

decreasing

89. $y = f(x) = 3 + \log_3 x$
Shift $y = \log_3 x$ up 3.

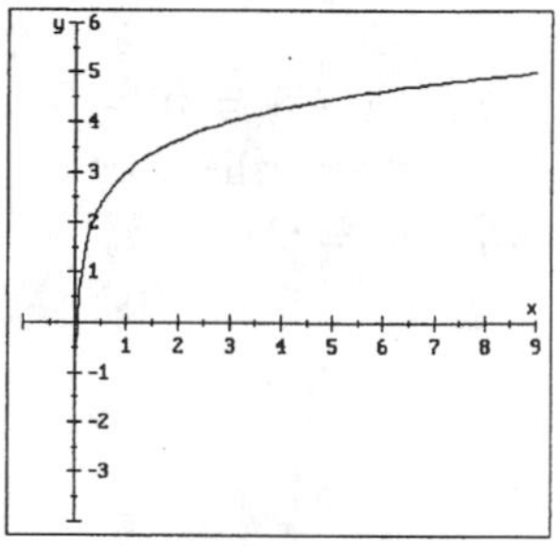

91. $y = f(x) = \log_{1/2}(x - 2)$
Shift $y = \log_{1/2} x$ right 2.

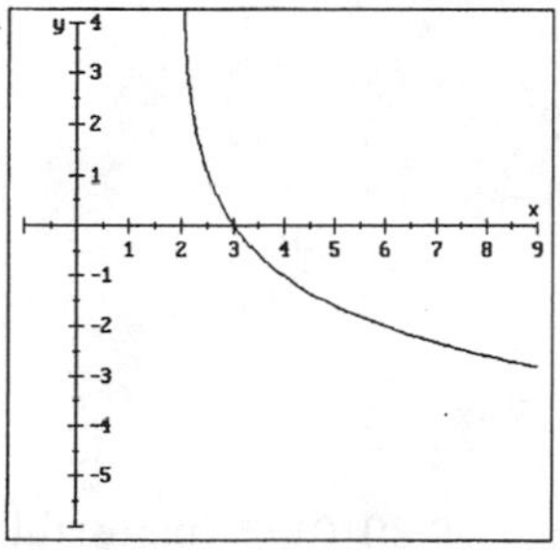

93. $y = f(x) = 2^x$
$y = g(x) = \log_2 x$

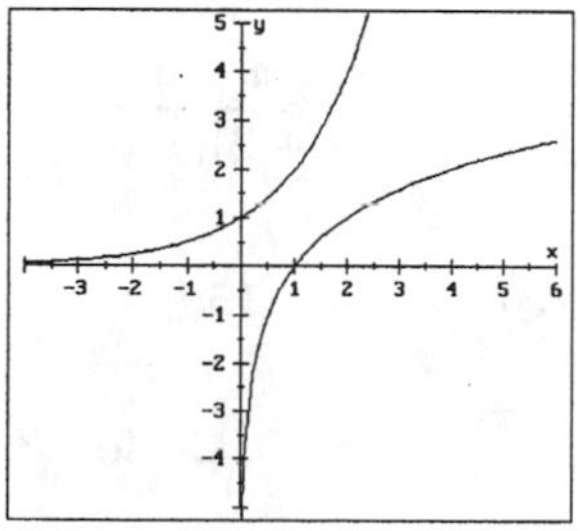

95. $y = f(x) = \left(\dfrac{1}{4}\right)^x$
$y = g(x) = \log_{1/4} x$

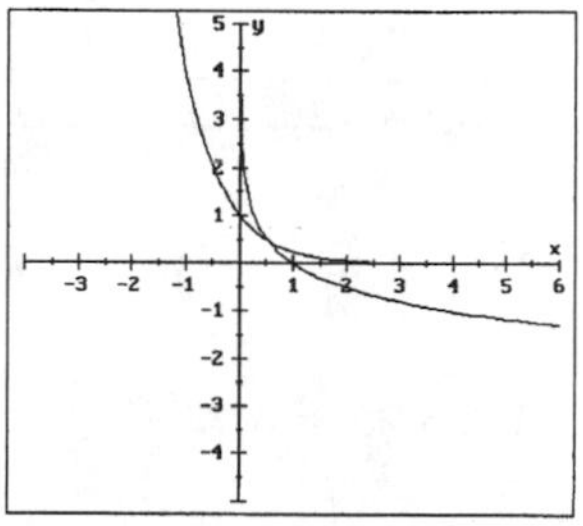

97. $\log_b 2 = 0 \Rightarrow b^0 = 2$
No such value exists.

99. $\log_b 9 = 2 \Rightarrow b^2 = 9 \Rightarrow b = 3$

101. $\text{dB gain} = 20 \log \frac{E_O}{E_I} = 20 \log \frac{20}{0.71} = 20 \log 28.169 \approx 29.0 \text{ dB}$

103. $\text{dB gain} = 20 \log \frac{E_O}{E_I} = 20 \log \frac{30}{0.1} = 20 \log 300 \approx 49.5 \text{ dB}$

105. $R = \log \frac{A}{P} = \log \frac{5000}{0.2} = \log 25{,}000 \approx 4.4$

107. $R = \log \frac{A}{P} = \log \frac{2500}{0.25} = \log 10000 = 4$

109. $n = \frac{\log V - \log C}{\log\left(1 - \frac{2}{N}\right)} = \frac{\log 2000 - \log 17000}{\log\left(1 - \frac{2}{5}\right)} \approx \frac{-0.929419}{-0.221849} \approx 4.2$ years old

111. $n = \frac{\log\left[\frac{Ar}{P} + 1\right]}{\log(1+r)} = \frac{\log\left[\frac{20{,}000(0.12)}{1000} + 1\right]}{\log(1 + 0.12)} = \frac{\log 3.4}{\log 1.12} \approx 10.8$ years

113-117. Answers may vary.

Exercise 11.4 (page 767)

1. $y = 5x + 8 \Rightarrow m = 5$
Use the parallel slope:
$$y - y_1 = m(x - x_1)$$
$$y - 0 = 5(x - 0)$$
$$y = 5x$$

3. $y = \frac{2}{3}x - 12 \Rightarrow m = \frac{2}{3}$
Use the perpendicular slope:
$$y - y_1 = m(x - x_1)$$
$$y - 2 = -\frac{3}{2}(x - 3)$$
$$y - 2 = -\frac{3}{2}x + \frac{9}{2}$$
$$y = -\frac{3}{2}x + \frac{13}{2}$$

5. $x = 5$

7. $\frac{2x+3}{4x^2-9} = \frac{2x+3}{(2x+3)(2x-3)} = \frac{1}{2x-3}$

9. $\frac{x^2+3x+2}{3x+12} \cdot \frac{x+4}{x^2-4} = \frac{(x+2)(x+1)}{3(x+4)} \cdot \frac{x+4}{(x+2)(x-2)} = \frac{x+1}{3(x-2)}$

11. $\log_e x$

13. $(-\infty, \infty)$

15. 10

17. $\frac{\ln 2}{r}$

19. $\ln 25.25 \approx 3.2288$

21. $\ln 9.89 \approx 2.2915$

23. $\log(\ln 2) \approx \log(0.6931) \approx -0.1592$

25. $\ln(\log 0.5) = \ln(-0.3010) \Rightarrow$ impossible

27. $\ln y = 2.3015 \Rightarrow y = 9.9892$

29. $\ln y = 3.17 \Rightarrow y = 23.8075$

31. $\ln y = -4.72 \Rightarrow y = 0.0089$

33. $\log y = \ln 6 \Rightarrow \log y \approx 1.7918 \Rightarrow y \approx 61.9098$
(The answer will vary if rounding is used on the calculator.)

35. The graph must be increasing. The graph could not look like this.

37. The graph must go through $(1, 0)$. The graph could not look like this.

39. $y = -\ln x$

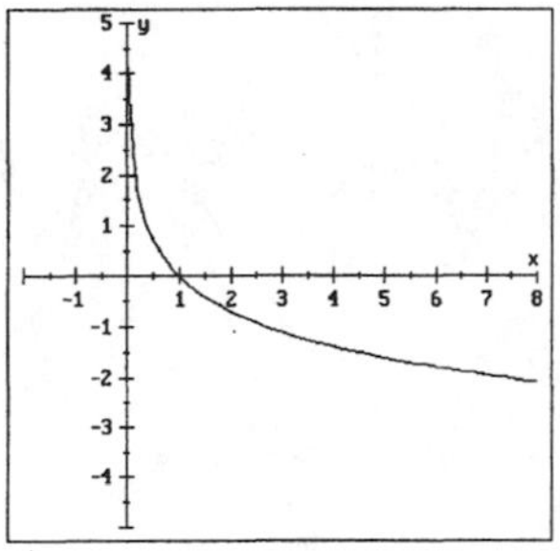

41. $y = \ln(-x)$

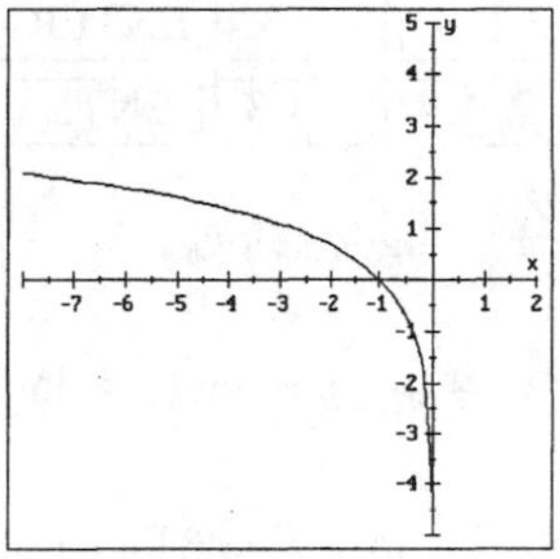

43. $t = \dfrac{\ln 2}{r} = \dfrac{\ln 2}{0.12} \approx 5.8$ years

45. $t = \dfrac{\ln 3}{r} = \dfrac{\ln 3}{0.12} \approx 9.2$ years

47. $t = -\dfrac{1}{0.9} \ln \dfrac{50 - T_r}{200 - T_r} = -\dfrac{1}{0.9} \ln \dfrac{50 - 38}{200 - 38} = -\dfrac{1}{0.9} \ln \dfrac{12}{162} \approx -\dfrac{1}{0.9}(-2.6027) \approx 2.9$ hours

49. **Answers may vary.**

51. $P = P_0 e^{rt} = P_0 e^{r \frac{\ln 3}{r}} = P_0 e^{\ln 3} = 3P_0$

53. Let $t = \dfrac{\ln 5}{r} \Rightarrow P = P_0 e^{rt} = P_0 e^{r \frac{\ln 5}{r}} = P_0 e^{\ln 5} = 5P_0$

Exercise 11.5 (page 777)

1. $m = \dfrac{\Delta y}{\Delta x} = \dfrac{3 - (-4)}{-2 - 4} = \dfrac{7}{-6} = -\dfrac{7}{6}$

3. $x = \frac{-2+4}{2} = \frac{2}{2} = 1$
$y = \frac{3+(-4)}{2} = \frac{-1}{2} = -\frac{1}{2}$
midpoint: $\left(1, -\frac{1}{2}\right)$

5. 0

7. M; N

9. x; y

11. x

13. $\neq$

15. 0

17. 7

19. 10

21. 1

23. 0

25. 7

27. 10

29. 1

Problems 31-35 are to be solved using a calculator. The keystrokes needed to solve each problem using a TI-83 graphing calculator appear in each solution. There may be other solutions. Keystrokes for other calculators may be slightly different.

31. [log] [2] [.] [5] [×] [3] [.] [7] [ENTER] $\{0.96614\}$
[log] [2] [.] [5] [)] [+] [log] [3] [.] [7] [)] [ENTER] $\{0.96614\}$

33. [ln] [2] [.] [2] [5] [^] [4] [ENTER] $\{3.24372\}$
[4] [ln] [2] [.] [2] [5] [ENTER] $\{3.24372\}$

35. [log] [√] [2] [4] [.] [3] [ENTER] $\{0.69280\}$
[.] [5] [log] [2] [4] [.] [3] [ENTER] $\{0.69280\}$

37. $\log_b xyz = \log_b x + \log_b y + \log_b z$

39. $\log_b \dfrac{2x}{y} = \log_b 2x - \log_b y = \log_b 2 + \log_b x - \log_b y$

41. $\log_b x^3y^2 = \log_b x^3 + \log_b y^2 = 3\log_b x + 2\log_b y$

43. $\log_b (xy)^{1/2} = \dfrac{1}{2}\log_b xy = \dfrac{1}{2}(\log_b x + \log_b y) = \dfrac{1}{2}\log_b x + \dfrac{1}{2}\log_b y$

45. $\log_b x\sqrt{z} = \log_b xz^{1/2} = \log_b x + \log_b z^{1/2} = \log_b x + \dfrac{1}{2}\log_b z$

47.
$$\begin{aligned}\log_b \frac{\sqrt[3]{x}}{\sqrt[4]{yz}} = \log_b \frac{x^{1/3}}{(yz)^{1/4}} = \log_b x^{1/3} - \log_b (yz)^{1/4} &= \frac{1}{3}\log_b x - \frac{1}{4}\log_b yz \\ &= \frac{1}{3}\log_b x - \frac{1}{4}(\log_b y + \log_b z) \\ &= \frac{1}{3}\log_b x - \frac{1}{4}\log_b y - \frac{1}{4}\log_b z\end{aligned}$$

49. $\log_b (x+1) - \log_b x = \log_b \dfrac{x+1}{x}$

51. $2\log_b x + \dfrac{1}{2}\log_b y = \log_b x^2 + \log_b y^{1/2} = \log_b x^2y^{1/2}$

53. $-3\log_b x - 2\log_b y + \dfrac{1}{2}\log_b z = \log_b x^{-3} + \log_b y^{-2} + \log_b z^{1/2} = \log_b x^{-3}y^{-2}z^{1/2} = \log_b \dfrac{z^{1/2}}{x^3y^2}$

55. $\log_b \left(\dfrac{x}{z} + x\right) - \log_b \left(\dfrac{y}{z} + y\right) = \log_b \dfrac{\frac{x}{z} + x}{\frac{y}{z} + y} = \log_b \dfrac{x + xz}{y + yz} = \log_b \dfrac{x(1+z)}{y(1+z)} = \log_b \dfrac{x}{y}$

57. $\log_b 0 = 1 \Rightarrow b^1 = 0 \Rightarrow b = 0 \Rightarrow$ FALSE $(b \neq 0)$

59. $\log_b xy = (\log_b x)(\log_b y) \Rightarrow$ FALSE $(\log_b xy = \log_b x + \log_b y)$

61. $\log_7 7^7 = 7 \Rightarrow 7^7 = 7^7 \Rightarrow$ TRUE

63. $\dfrac{\log_b A}{\log_b B} = \log_b A - \log_b B \Rightarrow$ FALSE $\left(\log_b \dfrac{A}{B} = \log_b A - \log_b B\right)$

65. $3\log_b \sqrt[3]{a} = 3\log_b a^{1/3} = \dfrac{1}{3}\cdot 3\log_b a = \log_b a \Rightarrow$ TRUE

67. $\log_b \dfrac{1}{a} = \log_b 1 - \log_b a = 0 - \log_b a = -\log_b a \Rightarrow$ TRUE

69. $\log 28 = \log 4\cdot 7 = \log 4 + \log 7 = 0.6021 + 0.8451 = 1.4472$

71. $\log 2.25 = \log \dfrac{9}{4} = \log 9 - \log 4 = 0.9542 - 0.6021 = 0.3521$

73. $\log \dfrac{63}{4} = \log \dfrac{7\cdot 9}{4} = \log 7 + \log 9 - \log 4 = 0.8451 + 0.9542 - 0.6021 = 1.1972$

75. $\log 252 = \log 4\cdot 7\cdot 9 = \log 4 + \log 7 + \log 9 = 0.6021 + 0.8451 + 0.9542 = 2.4014$

77. $\log 112 = \log 4^2 \cdot 7 = \log 4^2 + \log 7 = 2\log 4 + \log 7 = 2(0.6021) + 0.8451 = 2.0493$

79. $$\log \frac{144}{49} = \log \frac{16\cdot 9}{7^2} = \log \frac{4^2\cdot 9}{7^2} = \log 4^2 + \log 9 - \log 7^2 = 2\log 4 + \log 9 - 2\log 7$$
$$= 2(0.6021) + 0.9542 - 2(0.8451)$$
$$= 0.4682$$

81. $\log_3 7 = \dfrac{\log 7}{\log 3} \approx 1.7712$

83. $\log_{1/3} 3 = \dfrac{\log 3}{\log \frac{1}{3}} \approx -1.0000$

85. $\log_3 8 = \dfrac{\log 8}{\log 3} \approx 1.8928$

87. $\log_{\sqrt{2}} \sqrt{5} = \dfrac{\log \sqrt{5}}{\log \sqrt{2}} \approx 2.3219$

89. $\text{pH} = -\log[\text{H}^+] = -\log(1.7\times 10^{-5}) \approx 4.77$

91. low pH:

$$\text{pH} = -\log[\text{H}^+]$$
$$6.8 = -\log[\text{H}^+]$$
$$-6.8 = \log[\text{H}^+]$$
$$[\text{H}^+] = 1.5849\times 10^{-7}$$

high pH:

$$\text{pH} = -\log[\text{H}^+]$$
$$7.6 = -\log[\text{H}^+]$$
$$-7.6 = \log[\text{H}^+]$$
$$[\text{H}^+] = 2.5119\times 10^{-8}$$

93. $$k\ln 2I = k(\ln 2 + \ln I)$$
$$= k\ln 2 + k\ln I$$
$$= k\ln 2 + L$$

The loudness increases by $k\ln 2$.

95. $$L = 3k\ln I = k\cdot 3\ln I$$
$$= k\ln I^3$$

The intensity must be cubed.

97. **Answers may vary.**

99. $\ln(e^x) = \log_e(e^x) = x$

101. Let $\log_{b^2} x = y$. Then

$$\begin{aligned} (b^2)^y &= x \\ b^{2y} &= x \\ (b^{2y})^{1/2} &= x^{1/2} \\ b^y &= x^{1/2} \\ \log_b x^{1/2} &= y \\ \frac{1}{2}\log_b x &= y \end{aligned}$$

Exercise 11.6 (page 788)

1.
$$\begin{aligned} 5x^2 - 25x &= 0 \\ 5x(x-5) &= 0 \\ 5x = 0 \quad &\textbf{or} \quad x - 5 = 0 \\ x = 0 \quad &\quad\quad x = 5 \end{aligned}$$

3.
$$\begin{aligned} 3p^2 + 10p &= 8 \\ 3p^2 + 10p - 8 &= 0 \\ (3p-2)(p+4) &= 0 \\ 3p - 2 = 0 \quad &\textbf{or} \quad p + 4 = 0 \\ p = \tfrac{2}{3} \quad &\quad\quad p = -4 \end{aligned}$$

5. exponential

7. $A_0 2^{-t/h}$

9.
$$\begin{aligned} 4^x &= 5 \\ \log 4^x &= \log 5 \\ x \log 4 &= \log 5 \\ x &= \frac{\log 5}{\log 4} \\ x &\approx 1.1610 \end{aligned}$$

11.
$$\begin{aligned} e^t &= 50 \\ \ln e^t &= \ln 50 \\ t \ln e &= \ln 50 \\ t &= \ln 50 \\ t &\approx 3.9120 \end{aligned}$$

13.
$$\begin{aligned} 5 &= 2.1(1.04)^t \\ \frac{5}{2.1} &= (1.04)^t \\ \log \frac{5}{2.1} &= \log (1.04)^t \\ \log \frac{5}{2.1} &= t \log 1.04 \\ \frac{\log \frac{5}{2.1}}{\log 1.04} &= t \\ 22.1184 &\approx t \end{aligned}$$

15.
$$\begin{aligned} 13^{x-1} &= 2 \\ \log 13^{x-1} &= \log 2 \\ (x-1) \log 13 &= \log 2 \\ x - 1 &= \frac{\log 2}{\log 13} \\ x &= \frac{\log 2}{\log 13} + 1 \\ x &\approx 1.2702 \end{aligned}$$

17.
$$\begin{aligned} 2^{x+1} &= 3^x \\ \log 2^{x+1} &= \log 3^x \\ (x+1)\log 2 &= x\log 3 \\ x\log 2 + \log 2 &= x\log 3 \\ \log 2 &= x\log 3 - x\log 2 \\ \log 2 &= x(\log 3 - \log 2) \\ \frac{\log 2}{\log 3 - \log 2} &= x \\ 1.7095 &\approx x \end{aligned}$$

19.
$$\begin{aligned} 2^x &= 3^x \\ \log 2^x &= \log 3^x \\ x\log 2 &= x\log 3 \\ 0 &= x\log 3 - x\log 2 \\ 0 &= x(\log 3 - \log 2) \\ \frac{0}{\log 3 - \log 2} &= x \\ 0 &= x \end{aligned}$$

21.
$$\begin{aligned} 7^{x^2} &= 10 \\ \log 7^{x^2} &= \log 10 \\ x^2\log 7 &= \log 10 \\ x^2 &= \frac{\log 10}{\log 7} \\ x^2 &\approx 1.1833 \\ x &\approx \pm 1.0878 \end{aligned}$$

23.
$$\begin{aligned} 8^{x^2} &= 9^x \\ \log 8^{x^2} &= \log 9^x \\ x^2\log 8 &= x\log 9 \\ x^2\log 8 - x\log 9 &= 0 \\ x(x\log 8 - \log 9) &= 0 \\ x = 0 \quad \textbf{or} \quad x\log 8 - \log 9 &= 0 \\ x\log 8 &= \log 9 \\ x &= \frac{\log 9}{\log 8} \\ x &\approx 1.0566 \end{aligned}$$

25.
$$\begin{aligned} 2^{x^2-2x} &= 8 \\ 2^{x^2-2x} &= 2^3 \\ x^2 - 2x &= 3 \\ x^2 - 2x - 3 &= 0 \\ (x+1)(x-3) &= 0 \\ x+1 = 0 \quad &\textbf{or} \quad x - 3 = 0 \\ x = -1 \quad & \quad\quad x = 3 \end{aligned}$$

27.
$$\begin{aligned} 3^{x^2+4x} &= \frac{1}{81} \\ 3^{x^2+4x} &= 3^{-4} \\ x^2 + 4x &= -4 \\ x^2 + 4x + 4 &= 0 \\ (x+2)(x+2) &= 0 \\ x+2 = 0 \quad &\textbf{or} \quad x+2 = 0 \\ x = -2 \quad & \quad\quad x = -2 \end{aligned}$$

29.
$$\begin{aligned} 4^{x+2} - 4^x &= 15 \\ 4^x 4^2 - 4^x &= 15 \\ 16\cdot 4^x - 4^x &= 15 \\ 15\cdot 4^x &= 15 \\ 4^x &= 1 \\ x &= 0 \end{aligned}$$

31.
$$\begin{aligned} 2(3^x) &= 6^{2x} \\ \log 2(3^x) &= \log 6^{2x} \\ \log 2 + \log 3^x &= 2x\log 6 \\ \log 2 + x\log 3 &= 2x\log 6 \\ \log 2 &= 2x\log 6 - x\log 3 \\ \log 2 &= x(2\log 6 - \log 3) \\ \frac{\log 2}{2\log 6 - \log 3} &= x \\ 0.2789 &\approx x \end{aligned}$$

33. $2^{x+1} = 7 \Rightarrow$ Graph $y = 2^{x+1}$ and $y = 7$.

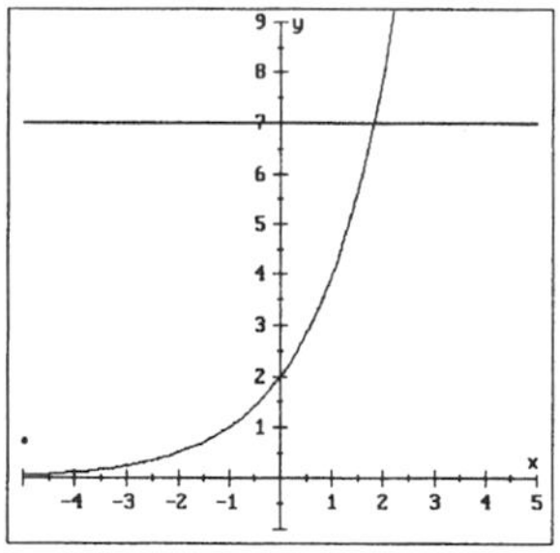

$x \approx 1.8$

35. $2^{x^2-2x} - 8 = 0 \Rightarrow$ Graph $y = 2^{x^2-2x} - 8$ and find any x-intercept(s).

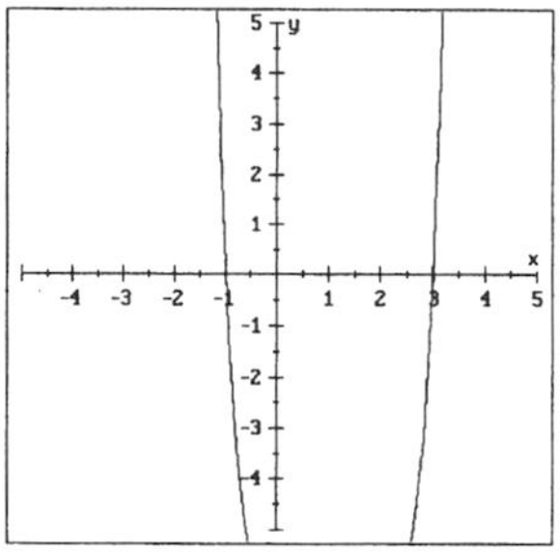

$x = 3$ or $x = -1$

37.

$$\begin{aligned} \log 2x &= \log 4 \\ 2x &= 4 \\ x &= 2 \end{aligned}$$

39.

$$\begin{aligned} \log(3x+1) &= \log(x+7) \\ 3x+1 &= x+7 \\ 2x &= 6 \\ x &= 3 \end{aligned}$$

41.

$$\begin{aligned} \log(3-2x) - \log(x+24) &= 0 \\ \log(3-2x) &= \log(x+24) \\ 3-2x &= x+24 \\ -21 &= 3x \\ -7 &= x \end{aligned}$$

43.

$$\begin{aligned} \log \frac{4x+1}{2x+9} &= 0 \\ 10^0 &= \frac{4x+1}{2x+9} \\ 1 &= \frac{4x+1}{2x+9} \\ 2x+9 &= 4x+1 \\ 8 &= 2x \\ 4 &= x \end{aligned}$$

45.

$$\begin{aligned} \log x^2 &= 2 \\ 10^2 &= x^2 \\ 100 &= x^2 \\ \pm 10 &= x \end{aligned}$$

47.

$$\begin{aligned} \log x + \log(x-48) &= 2 \\ \log x(x-48) &= 2 \\ 10^2 &= x(x-48) \\ 0 &= x^2 - 48x - 100 \\ 0 &= (x-50)(x+2) \end{aligned}$$

$x - 50 = 0$ **or** $x + 2 = 0$

$x = 50$ $\qquad$ $x = -2$

Does not check.

49.

$$\begin{aligned} \log x + \log(x-15) &= 2 \\ \log x(x-15) &= 2 \\ 10^2 &= x(x-15) \\ 0 &= x^2 - 15x - 100 \\ 0 &= (x-20)(x+5) \end{aligned}$$

$x - 20 = 0$ **or** $x + 5 = 0$

$x = 20$ $\quad$ $x = -5$

Does not check.

51.

$$\begin{aligned} \log(x+90) &= 3 - \log x \\ \log x + \log(x+90) &= 3 \\ \log x(x+90) &= 3 \\ 10^3 &= x(x+90) \\ 0 &= x^2 + 90x - 1000 \\ 0 &= (x-10)(x+100) \end{aligned}$$

$x - 10 = 0$ **or** $x + 100 = 0$

$x = 10$ $\quad$ $x = -100$

Does not check.

53.

$$\begin{aligned} \log(x-6) - \log(x-2) &= \log \frac{5}{x} \\ \log \frac{x-6}{x-2} &= \log \frac{5}{x} \\ \frac{x-6}{x-2} &= \frac{5}{x} \\ x(x-6) &= 5(x-2) \\ x^2 - 6x &= 5x - 10 \\ x^2 - 11x + 10 &= 0 \\ (x-10)(x-1) &= 0 \end{aligned}$$

$x - 10 = 0$ **or** $x - 1 = 0$

$x = 10$ $\quad$ $x = 1$

Does not check.

55.

$$\begin{aligned} \log x^2 &= (\log x)^2 \\ 2 \log x &= (\log x)^2 \\ 0 &= (\log x)^2 - 2 \log x \\ 0 &= \log x (\log x - 2) \end{aligned}$$

$\log x = 0$ **or** $\log x - 2 = 0$

$x = 1$ $\quad$ $\log x = 2$

$x = 100$

57.

$$\begin{aligned} \frac{\log(3x-4)}{\log x} &= 2 \\ \log(3x-4) &= 2 \log x \\ \log(3x-4) &= \log x^2 \\ 3x - 4 &= x^2 \\ 0 &= x^2 - 3x + 4 \end{aligned}$$

$b^2 - 4ac = (-3)^2 - 4(1)(4) = -7 \Rightarrow$

solutions are nonreal $\Rightarrow$ no solution

59.

$$\begin{aligned} \frac{\log(5x+6)}{2} &= \log x \\ \log(5x+6) &= 2 \log x \\ \log(5x+6) &= \log x^2 \\ 5x + 6 &= x^2 \\ 0 &= x^2 - 5x - 6 \\ 0 &= (x-6)(x+1) \end{aligned}$$

$x - 6 = 0$ **or** $x + 1 = 0$

$x = 6$ $\quad$ $x = -1$

Does not check.

61.
$$\log_3 x = \log_3 \left(\frac{1}{x}\right) + 4$$
$$\log_3 x = \log_3 \left(\frac{1}{x}\right) + \log_3 81$$
$$\log_3 x = \log_3 \left(\frac{81}{x}\right)$$
$$x = \frac{81}{x}$$
$$x^2 = 81$$
$x = 9$ (-9 does not check.)

63.
$$2 \log_2 x = 3 + \log_2 (x - 2)$$
$$\log_2 x^2 - \log_2 (x - 2) = 3$$
$$\log_2 \frac{x^2}{x - 2} = 3$$
$$\frac{x^2}{x - 2} = 8$$
$$x^2 = 8(x - 2)$$
$$x^2 = 8x - 16$$
$$x^2 - 8x + 16 = 0$$
$$(x - 4)(x - 4) = 0$$
$x - 4 = 0$ **or** $x - 4 = 0$
$x = 4$ $\quad$ $x = 4$

65.
$$\log (7y + 1) = 2 \log (y + 3) - \log 2$$
$$\log (7y + 1) = \log (y + 3)^2 - \log 2$$
$$\log (7y + 1) = \log \frac{y^2 + 6y + 9}{2}$$
$$7y + 1 = \frac{y^2 + 6y + 9}{2}$$
$$2(7y + 1) = y^2 + 6y + 9$$
$$14y + 2 = y^2 + 6y + 9$$
$$0 = y^2 - 8y + 7$$
$$0 = (y - 7)(y - 1)$$
$y - 7 = 0$ **or** $y - 1 = 0$
$y = 7$ $\quad$ $y = 1$

67. $\log x + \log (x - 15) = 2$
Graph $y = \log x + \log (x - 15)$ and $y = 2$.

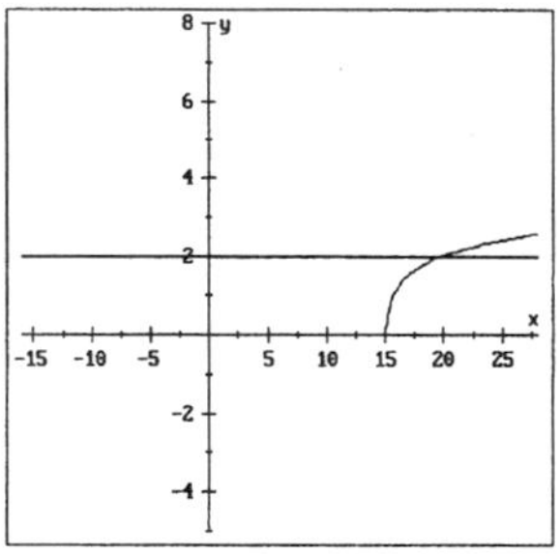

$x = 20$

69. $\ln (2x + 5) - \ln 3 = \ln (x - 1) \Rightarrow$ Graph $y = \ln (2x + 5) - \ln 3$ and $y = \ln (x - 1)$.

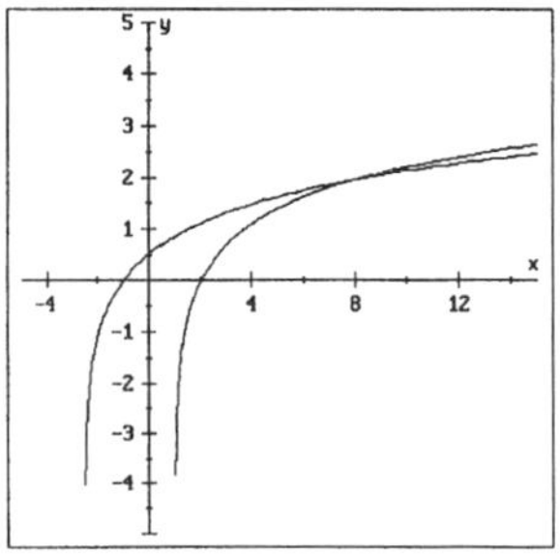

$x = 8$

71.
$$A = A_0 e^{-0.013t}$$
$$0.5A_0 = A_0 e^{-0.013t}$$
$$0.5 = e^{-0.013t}$$
$$\ln 0.5 = \ln e^{-0.013t}$$
$$\ln 0.5 = -0.013t$$
$$\frac{\ln 0.5}{-0.013} = t$$
$$53 \text{ days} \approx t$$

73.

$$\begin{aligned} A &= A_0 2^{-t/h} \\ 0.80A_0 &= A_0 2^{-2/h} \\ 0.80 &= 2^{-2/h} \\ \log 0.80 &= \log 2^{-2/h} \\ \log 0.80 &= -\frac{2}{h}\log 2 \\ h\log 0.80 &= -2\log 2 \\ h &= \frac{-2\log 2}{\log 0.80} \\ h &\approx 6.2 \text{ years} \end{aligned}$$

75.

$$\begin{aligned} A &= A_0 2^{-t/h} \\ 0.20A_0 &= A_0 2^{-t/18.4} \\ 0.20 &= 2^{-t/18.4} \\ \log 0.20 &= \log 2^{-t/18.4} \\ \log 0.20 &= -\frac{t}{18.4}\log 2 \\ \frac{\log 0.20}{-\log 2} &= \frac{t}{18.4} \\ -18.4\frac{\log 0.20}{\log 2} &= t \\ 42.7 \text{ days} &\approx t \end{aligned}$$

77.

$$\begin{aligned} A &= A_0 2^{-t/h} \\ 0.60A_0 &= A_0 2^{-t/5700} \\ 0.60 &= 2^{-t/5700} \\ \log 0.60 &= \log 2^{-t/5700} \\ \log 0.60 &= -\frac{t}{5700}\log 2 \\ \frac{\log 0.60}{-\log 2} &= \frac{t}{5700} \\ -5700\frac{\log 0.60}{\log 2} &= t \\ 4200 \text{ years} &\approx t \end{aligned}$$

79.

$$\begin{aligned} A &= P\left(1+\frac{r}{k}\right)^{kt} \\ 800 &= 500\left(1+\frac{0.085}{2}\right)^{2t} \\ 1.6 &= (1.0425)^{2t} \\ \log 1.6 &= \log (1.0425)^{2t} \\ \log 1.6 &= 2t\log(1.0425) \\ \frac{\log 1.6}{2\log 1.0425} &= t \\ 5.6 \text{ years} &\approx t \end{aligned}$$

81.

$$\begin{aligned} A &= P\left(1+\frac{r}{k}\right)^{kt} \\ 2100 &= 1300\left(1+\frac{0.09}{4}\right)^{4t} \\ \frac{2100}{1300} &= (1.0225)^{4t} \\ \log\frac{2100}{1300} &= \log(1.0225)^{4t} \\ \log\frac{21}{13} &= 4t\log(1.0225) \\ \frac{\log\frac{21}{13}}{4\log 1.0225} &= t \\ 5.4 \text{ years} &\approx t \end{aligned}$$

83.

$$\begin{aligned} \text{doubling time} = t &= \frac{\ln 2}{r} = \frac{100\ln 2}{100r} \\ &\approx \frac{70}{100r} \\ &= \frac{70}{r\text{, written as a \%}} \end{aligned}$$

85.
$$\begin{aligned} P &= P_0e^{kt} \\ 2P_0 &= P_0e^{5k} \\ 2 &= e^{5k} \\ \ln 2 &= \ln e^{5k} \\ \ln 2 &= 5k \\ \frac{\ln 2}{5} &= k \end{aligned}$$

$$\begin{aligned} P &= P_0e^{kt} \\ 1{,}000{,}000 &= 30{,}000e^{\frac{\ln 2}{5}t} \\ 33.333 &\approx e^{\frac{\ln 2}{5}t} \\ \ln 33.333 &\approx \ln e^{\frac{\ln 2}{5}t} \\ \ln 33.333 &\approx \frac{\ln 2}{5}t \\ \frac{5 \ln 33.333}{\ln 2} &\approx t \\ 25.3 \text{ years} &\approx t \end{aligned}$$

87.
$$\begin{aligned} P &= P_0e^{kt} \\ 2P_0 &= P_0e^{24k} \\ 2 &= e^{24k} \\ \ln 2 &= \ln e^{24k} \\ \ln 2 &= 24k \\ \frac{\ln 2}{24} &= k \end{aligned}$$

$$\begin{aligned} P &= P_0e^{kt} \\ P &= P_0e^{\frac{\ln 2}{24}(36)} \\ P &= P_0e^{\frac{3\ln 2}{2}} \\ P &= P_0(2.828) \end{aligned}$$

It will be about 2.828 times larger.

89. $n = \dfrac{1}{\log 2}\left(\log\dfrac{B}{b}\right) = \dfrac{1}{\log 2}\log\dfrac{5 \times 10^6}{500} = \dfrac{1}{\log 2}\log 10{,}000 \approx 13.3$ generations

91. **Answers may vary.**

93. Since the logarithm of a negative number is not defined (as a real number), the values $x - 3$ and $x^2 + 2$ must be nonnegative. Since $x^2 + 2$ is always greater than 0, the only restriction is that $x - 3 > 0$, or $x > 3$. Thus, x cannot be a solution if $x \le 3$.

Chapter 11 Summary (page 792)

1. $5^{\sqrt{2}} \cdot 5^{\sqrt{2}} = 5^{\sqrt{2}+\sqrt{2}} = 5^{2\sqrt{2}}$

2. $\left(2^{\sqrt{5}}\right)^{\sqrt{2}} = 2^{(\sqrt{5})(\sqrt{2})} = 2^{\sqrt{10}}$

3. $y = 3^x$; through $(0, 1)$ and $(1, 3)$

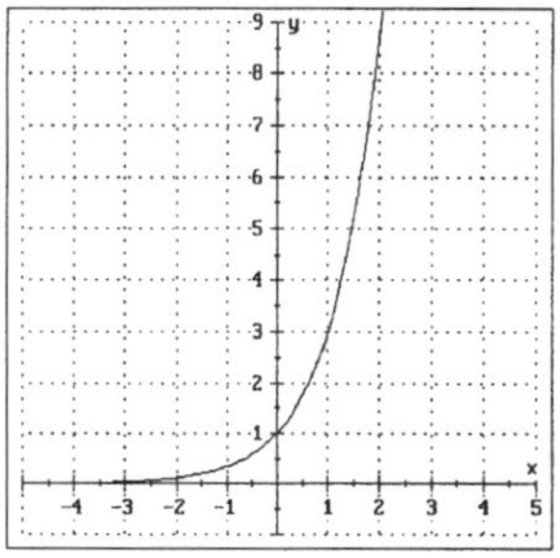

4. $y = \left(\frac{1}{3}\right)^x$; through $(0, 1)$ and $\left(1, \frac{1}{3}\right)$

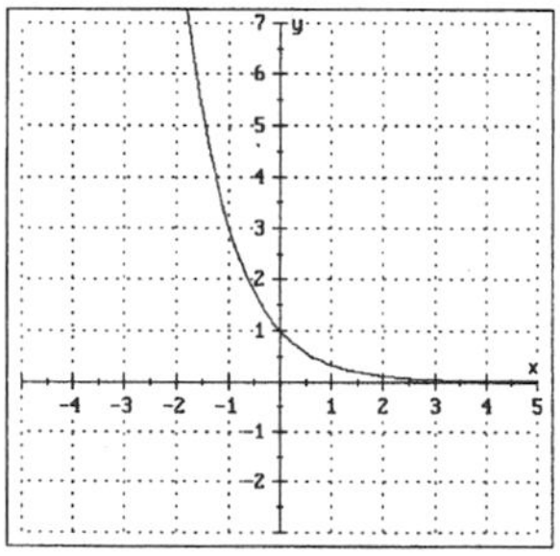

5. The graph will go through $(0, 1)$ and $(1, 6)$, so $x = 1$ and $y = 6$.

6. domain $= (-\infty, \infty)$; range $= (0, \infty)$

7. $y = f(x) = \left(\frac{1}{2}\right)^x - 2$
Shift $y = \left(\frac{1}{2}\right)^x$ down 2.

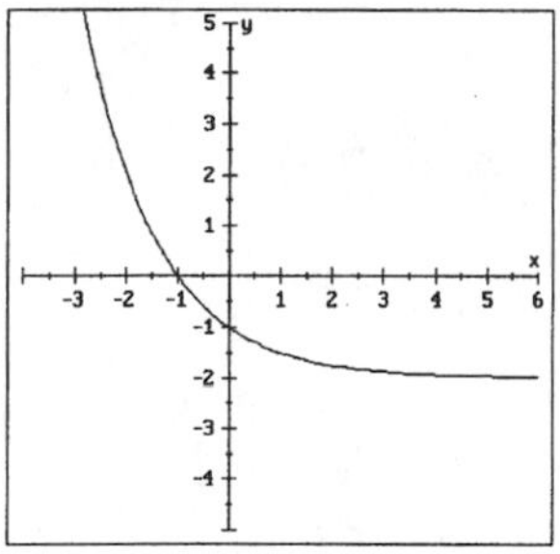

8. $y = f(x) = \left(\frac{1}{2}\right)^{x+2}$
Shift $y = \left(\frac{1}{2}\right)^x$ left 2.

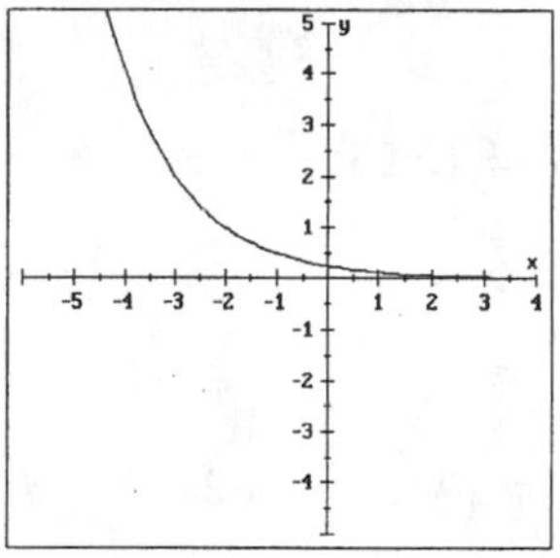

9. $A = P\left(1 + \frac{r}{k}\right)^{kt} = 10500\left(1 + \frac{0.09}{4}\right)^{4 \cdot 60} = 10500(1.0225)^{240} \approx \$2{,}189{,}703.45$

10. $A = Pe^{rt} = 10500e^{0.09(60)} = 10500e^{5.4} \approx \$2{,}324{,}767.37$

11. $y = f(x) = e^x + 1$; Shift $y = e^x$ up 1.

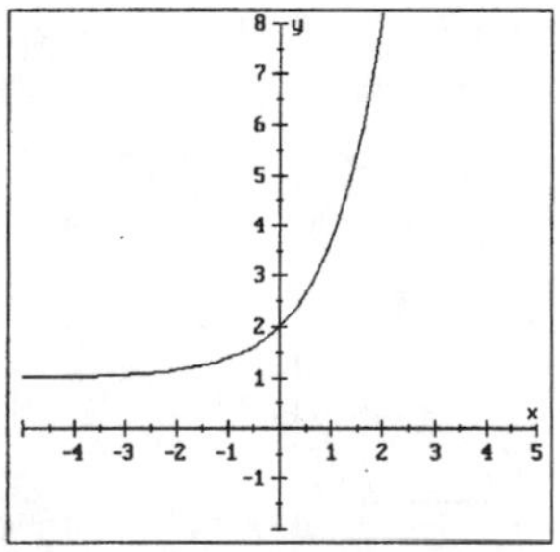

12. $y = f(x) = e^{x-3}$; Shift $y = e^x$ right 3.

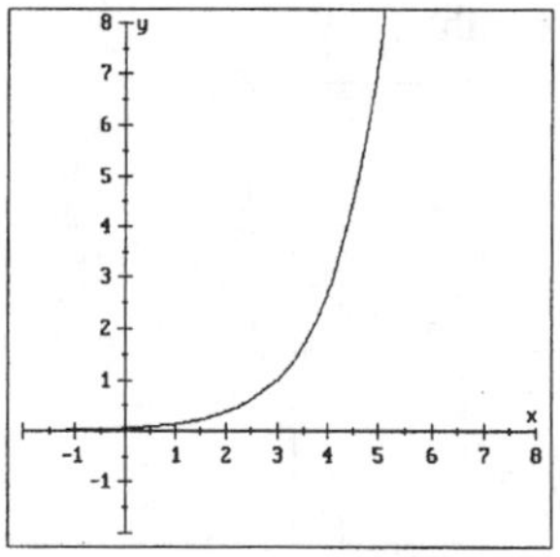

13. $P = P_0e^{kt} = 275{,}000{,}000e^{0.015(50)}$
$\approx 582{,}000{,}000$

14. $A = A_0e^{-0.0244t} = 50e^{-0.0244(20)} = 50e^{-0.488}$
≈ 30.69 g

15. domain $= (0, \infty)$; range $= (-\infty, \infty)$

16. **Answers will vary.**

17. $\log_3 9 = 2$

18. $\log_9 \frac{1}{3} = -\frac{1}{2}$

19. $\log_\pi 1 = 0$

20. $\log_5 0.04 = \log_5 \frac{1}{25} = -2$

21. $\log_a \sqrt{a} = \log_a a^{1/2} = \frac{1}{2}$

22. $\log_a \sqrt[3]{a} = \log_a a^{1/3} = \frac{1}{3}$

23. $\log_2 x = 5 \Rightarrow 2^5 = x \Rightarrow x = 32$

24. $\log_{\sqrt{3}} x = 4 \Rightarrow \left(\sqrt{3}\right)^4 = x \Rightarrow x = 9$

25. $\log_{\sqrt{3}} x = 6 \Rightarrow \left(\sqrt{3}\right)^6 = x \Rightarrow x = 27$

26. $\log_{0.1} 10 = x \Rightarrow (0.1)^x = 10$
$\left(\frac{1}{10}\right)^x = 10 \Rightarrow x = -1$

27. $\log_x 2 = -\frac{1}{3} \Rightarrow x^{-1/3} = 2$

$\left(x^{-1/3}\right)^{-3} = 2^{-3} \Rightarrow x = \frac{1}{8}$

28. $\log_x 32 = 5 \Rightarrow x^5 = 32 \Rightarrow x = 2$

29. $\log_{0.25} x = -1 \Rightarrow (0.25)^{-1} = x$

$\left(\frac{1}{4}\right)^{-1} = x \Rightarrow x = 4$

30. $\log_{0.125} x = -\frac{1}{3} \Rightarrow (0.125)^{-1/3} = x$

$\left(\frac{1}{8}\right)^{-1/3} = x \Rightarrow x = 2$

31. $\log_{\sqrt{2}} 32 = x \Rightarrow \left(\sqrt{2}\right)^x = 32$

$\left(2^{1/2}\right)^x = 2^5 \Rightarrow \frac{1}{2}x = 5 \Rightarrow x = 10$

32. $\log_{\sqrt{5}} x = -4 \Rightarrow \left(\sqrt{5}\right)^{-4} = x$

$\left(5^{1/2}\right)^{-4} = x \Rightarrow 5^{-2} = x \Rightarrow x = \frac{1}{25}$

33. $\log_{\sqrt{3}} 9\sqrt{3} = x \Rightarrow \left(\sqrt{3}\right)^x = 9\sqrt{3}$

$\left(3^{1/2}\right)^x = 3^{5/2} \Rightarrow x = 5$

34. $\log_{\sqrt{5}} 5\sqrt{5} = x \Rightarrow \left(\sqrt{5}\right)^x = 5\sqrt{5}$

$\left(5^{1/2}\right)^x = 5^{3/2} \Rightarrow x = 3$

35. $y = f(x) = \log(x-2)$
Shift $y = \log x$ right 2.

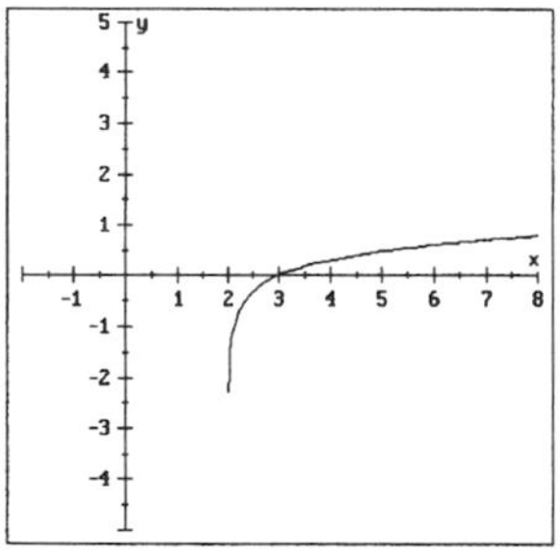

36. $y = f(x) = 3 + \log x$
Shift $y = \log x$ up 3.

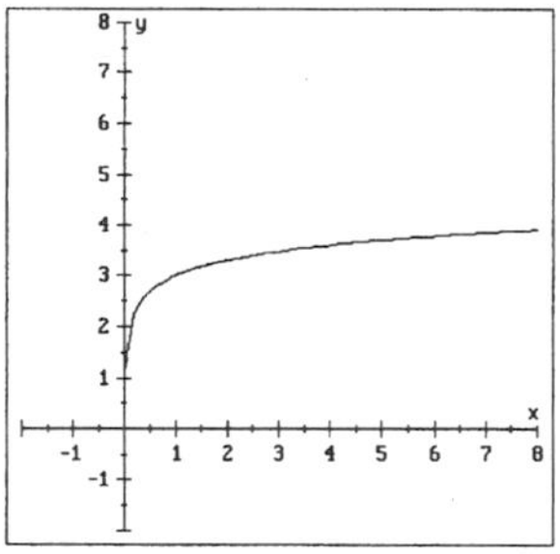

37. $y = 4^x$
$y = \log_4 x$

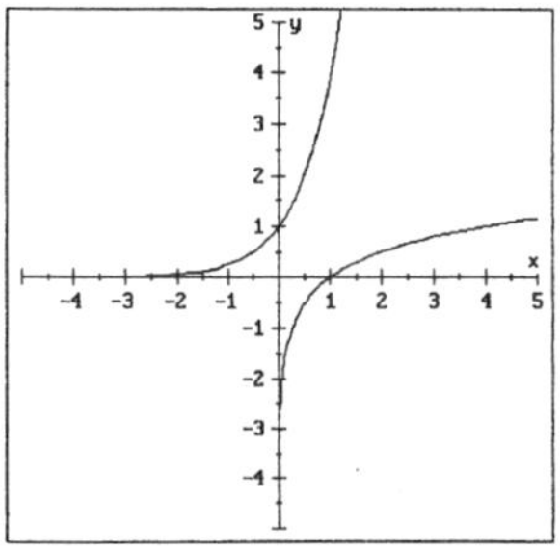

38. $y = \left(\frac{1}{3}\right)^x$
$y = \log_{1/3} x$

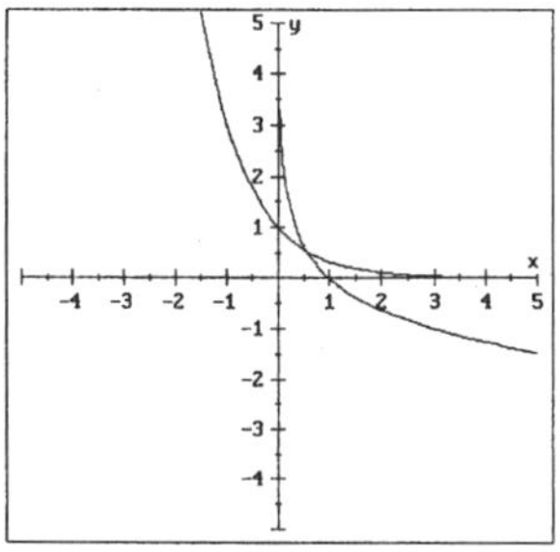

39. dB gain $= 20 \log \frac{E_O}{E_I} = 20 \log \frac{18}{0.04} = 20 \log 450 \approx 53$ dB

40. $R = \log \frac{A}{P} = \log \frac{7500}{0.3} = \log 25{,}000 \approx 4.4$

41. $\ln 452 \approx 6.1137$

42. $\ln(\log 7.85) \approx \ln 0.8949 \approx -0.1111$

43. $\ln x = 2.336 \Rightarrow x = 10.3398$

44. $\ln x = \log 8.8 \Rightarrow x = 2.5715$

45. $y = f(x) = 1 + \ln x$
Shift $y = \ln x$ up 1.

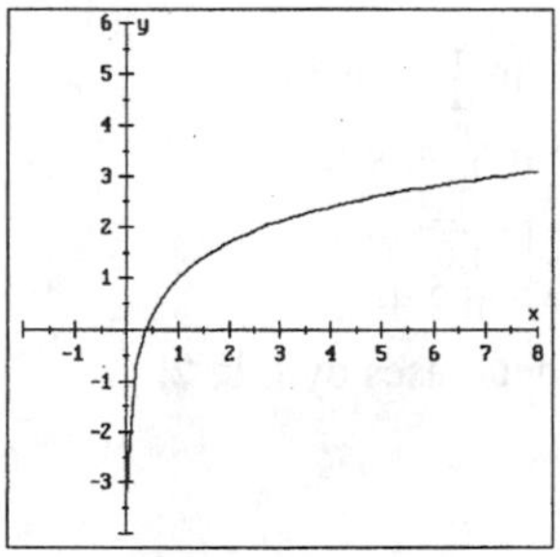

46. $y = f(x) = \ln(x+1)$
Shift $y = \ln x$ left 1.

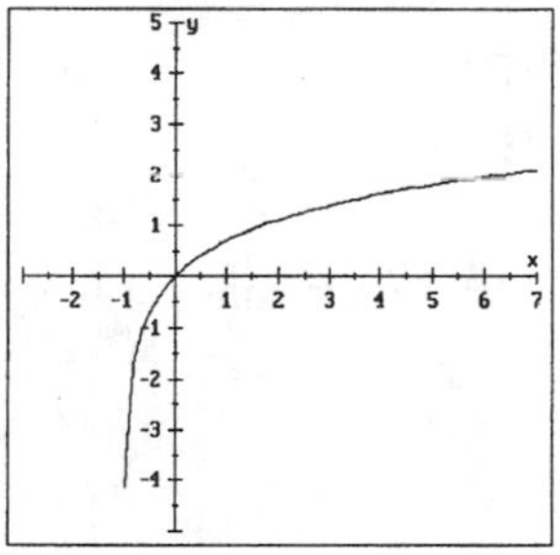

47. $t = \dfrac{\ln 2}{r} = \dfrac{\ln 2}{0.03} \approx 23$ years

48. $\log_7 1 = 0$

49. $\log_7 7 = 1$

50. $\log_7 7^3 = 3$

51. $7^{\log_7 4} = 4$

52. $\ln e^4 = 4$

53. $\ln 1 = 0$

54. $10^{\log_{10} 7} = 7$

55. $e^{\ln 3} = 3$

56. $\log_b b^4 = 4$

57. $\ln e^9 = 9$

58. $\log_b \dfrac{x^2y^3}{z^4} = \log_b x^2 + \log_b y^3 - \log_b z^4 = 2\log_b x + 3\log_b y - 4\log_b z$

59. $\log_b \sqrt{\dfrac{x}{yz^2}} = \log_b \left(\dfrac{x}{yz^2}\right)^{1/2} = \dfrac{1}{2}\log_b \dfrac{x}{yz^2} = \dfrac{1}{2}(\log_b x - \log_b y - \log_b z^2)$
$= \dfrac{1}{2}(\log_b x - \log_b y - 2\log_b z)$

60. $3\log_b x - 5\log_b y + 7\log_b z = \log_b x^3 - \log_b y^5 + \log_b z^7 = \log_b \dfrac{x^3z^7}{y^5}$

61. $\dfrac{1}{2}\log_b x + 3\log_b y - 7\log_b z = \log_b x^{1/2} + \log_b y^3 - \log_b z^7 = \log_b \dfrac{y^3\sqrt{x}}{z^7}$

62. $\log abc = \log a + \log b + \log c = 0.6 + 0.36 + 2.4 = 3.36$

63. $\log a^2b = \log a^2 + \log b = 2\log a + \log b = 2(0.6) + 0.36 = 1.56$

64. $\log \dfrac{ac}{b} = \log a + \log c - \log b = 0.6 + 2.4 - 0.36 = 2.64$

65. $\log \frac{a^2}{c^3b^2} = \log a^2 - \log c^3 - \log b^2 = 2\log a - 3\log c - 2\log b$
$= 2(0.6) - 3(2.4) - 2(0.36) = -6.72$

66. $\log_5 17 = \frac{\log 17}{\log 5} \approx 1.7604$

67.
$$\begin{aligned} \text{pH} &= -\log\,[\text{H}^+] \\ 3.1 &= -\log\,[\text{H}^+] \\ -3.1 &= \log\,[\text{H}^+] \\ [\text{H}^+] &= 7.94 \times 10^{-4} \text{ gram-ions per liter} \end{aligned}$$

68.
$$\begin{aligned} k\ln\left(\tfrac{1}{2}I\right) &= k\left(\ln\tfrac{1}{2} + \ln I\right) \\ &= k\ln\tfrac{1}{2} + k\ln I \\ &= k\ln 2^{-1} + L \\ &= -k\ln 2 + L \end{aligned}$$
The loudness decreases by $k\ln 2$.

69.
$$\begin{aligned} 3^x &= 7 \\ \log 3^x &= \log 7 \\ x\log 3 &= \log 7 \\ x &= \frac{\log 7}{\log 3} \approx 1.7712 \end{aligned}$$

70.
$$\begin{aligned} 5^{x+2} &= 625 \\ 5^{x+2} &= 5^4 \\ x+2 &= 4 \\ x &= 2 \end{aligned}$$

71.
$$\begin{aligned} 25 &= 5.5(1.05)^t \\ \frac{25}{5.5} &= (1.05)^t \\ \log\frac{25}{5.5} &= \log\,(1.05)^t \\ \log\frac{25}{5.5} &= t\log 1.05 \\ \frac{\log\frac{25}{5.5}}{\log 1.05} &= t \\ 31.0335 &\approx t \end{aligned}$$

72.
$$\begin{aligned} 4^{2t-1} &= 64 \\ 4^{2t-1} &= 4^3 \\ 2t-1 &= 3 \\ 2t &= 4 \\ t &= 2 \end{aligned}$$

73.
$$\begin{aligned} 2^x &= 3^{x-1} \\ \log 2^x &= \log 3^{x-1} \\ x\log 2 &= (x-1)\log 3 \\ x\log 2 &= x\log 3 - \log 3 \\ \log 3 &= x\log 3 - x\log 2 \\ \log 3 &= x(\log 3 - \log 2) \\ \frac{\log 3}{\log 3 - \log 2} &= x \\ 2.7095 &\approx x \end{aligned}$$

74.
$$\begin{aligned} 2^{x^2+4x} &= \frac{1}{8} \\ 2^{x^2+4x} &= 2^{-3} \\ x^2+4x &= -3 \\ x^2+4x+3 &= 0 \\ (x+3)(x+1) &= 0 \end{aligned}$$
$x+3=0$ **or** $x+1=0$
$x=-3$ $\qquad x=-1$

75.
$$\begin{aligned} \log x + \log(29 - x) &= 2 \\ \log x(29 - x) &= 2 \\ 10^2 &= x(29 - x) \\ 100 &= 29x - x^2 \\ x^2 - 29x + 100 &= 0 \\ (x - 25)(x - 4) &= 0 \end{aligned}$$
$x - 25 = 0$ **or** $x - 4 = 0$

$x = 25$ $\quad$ $x = 4$

76.
$$\begin{aligned} \log_2 x + \log_2(x - 2) &= 3 \\ \log_2 x(x - 2) &= 3 \\ x(x - 2) &= 2^3 \\ x^2 - 2x &= 8 \\ x^2 - 2x - 8 &= 0 \\ (x - 4)(x + 2) &= 0 \end{aligned}$$
$x - 4 = 0$ **or** $x + 2 = 0$

$x = 4$ $\quad$ $x = -2$

Does not check.

77.
$$\begin{aligned} \log_2(x + 2) + \log_2(x - 1) &= 2 \\ \log_2(x + 2)(x - 1) &= 2 \\ (x + 2)(x - 1) &= 2^2 \\ x^2 + x - 2 &= 4 \\ x^2 + x - 6 &= 0 \\ (x - 2)(x + 3) &= 0 \end{aligned}$$
$x - 2 = 0$ **or** $x + 3 = 0$

$x = 2$ $\quad$ $x = -3$

Does not check.

78.
$$\begin{aligned} \frac{\log(7x - 12)}{\log x} &= 2 \\ \log(7x - 12) &= 2\log x \\ \log(7x - 12) &= \log x^2 \\ 7x - 12 &= x^2 \\ 0 &= x^2 - 7x + 12 \\ 0 &= (x - 4)(x - 3) \end{aligned}$$
$x - 4 = 0$ **or** $x - 3 = 0$

$x = 4$ $\quad$ $x = 3$

79.
$$\begin{aligned} \log x + \log(x - 5) &= \log 6 \\ \log x(x - 5) &= \log 6 \\ x(x - 5) &= 6 \\ x^2 - 5x - 6 &= 0 \\ (x - 6)(x + 1) &= 0 \end{aligned}$$
$x - 6 = 0$ **or** $x + 1 = 0$

$x = 6$ $\quad$ $x = -1$

Does not check.

80.
$$\begin{aligned} \log 3 - \log(x - 1) &= -1 \\ \log\frac{3}{x - 1} &= -1 \\ \frac{3}{x - 1} &= 10^{-1} \\ \frac{3}{x - 1} &= \frac{1}{10} \\ 30 &= x - 1 \\ 31 &= x \end{aligned}$$

81.
$$\begin{aligned} e^{x \ln 2} &= 9 \\ e^{\ln 2^x} &= 9 \\ 2^x &= 9 \\ \ln 2^x &= \ln 9 \\ x \ln 2 &= \ln 9 \\ x &= \frac{\ln 9}{\ln 2} \approx 3.1699 \end{aligned}$$

82.
$$\begin{aligned} \ln x &= \ln(x - 1) \\ x &= x - 1 \\ 0 &= -1 \end{aligned}$$
There is no solution.

83.
$$\begin{aligned}\ln x &= \ln(x-1)+1\\ \ln x-\ln(x-1) &= 1\\ \ln\frac{x}{x-1} &= 1\\ \frac{x}{x-1} &= e^1\\ x &= e(x-1)\\ x &= ex-e\\ e &= ex-x\\ e &= x(e-1)\\ \frac{e}{e-1} &= x\\ 1.5820 &\approx x\end{aligned}$$

84.
$$\begin{aligned}\ln x &= \log_{10} x\\ \ln x &= \frac{\ln x}{\ln 10}\\ \ln x \ln 10 &= \ln x\\ \ln x \ln 10-\ln x &= 0\\ \ln x(\ln 10-1) &= 0\\ \ln x &= 0\\ x &= 1\end{aligned}$$

85.
$$\begin{aligned}A &= A_0 2^{-t/h}\\ \frac{2}{3}A_0 &= A_0 2^{-t/5700}\\ \frac{2}{3} &= 2^{-t/5700}\\ \log\frac{2}{3} &= \log 2^{-t/5700}\\ \log\frac{2}{3} &= -\frac{t}{5700}\log 2\\ \frac{\log\frac{2}{3}}{-\log 2} &= \frac{t}{5700}\\ -5700\frac{\log\frac{2}{3}}{\log 2} &= t\\ 3300\text{ years} &\approx t\end{aligned}$$

Chapter 11 Test (page 797)

1. $f(x)=2^x+1$; Shift $y=2^x$ up 1.

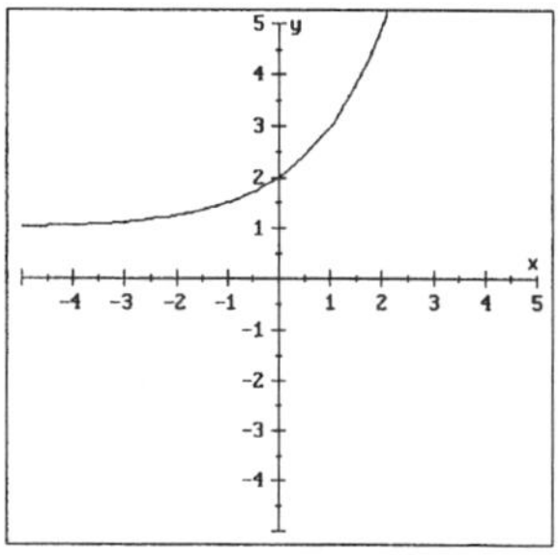

2. $f(x)=2^{-x}$; Reflect $y=2^x$ about y-axis.

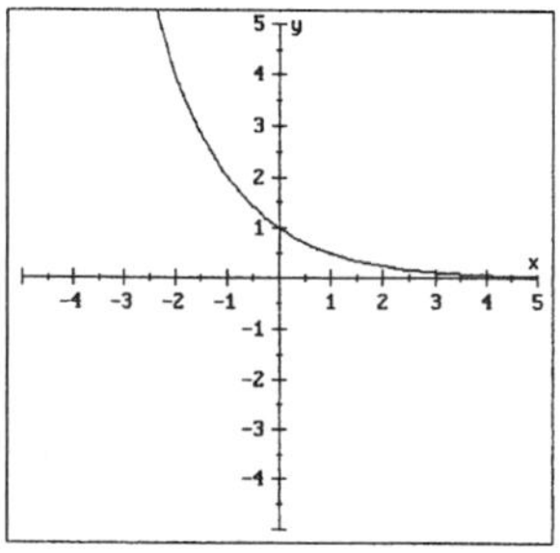

3. $A=A_0(2)^{-t}=3(2)^{-6}=\frac{3}{2^6}=\frac{3}{64}$ gram

4. $A = A_0\left(1+\frac{r}{k}\right)^{kt} = 1000\left(1+\frac{0.06}{2}\right)^{2(1)} = 1000(1.03)^2 \approx \$1060.90$

5. $f(x) = e^x$

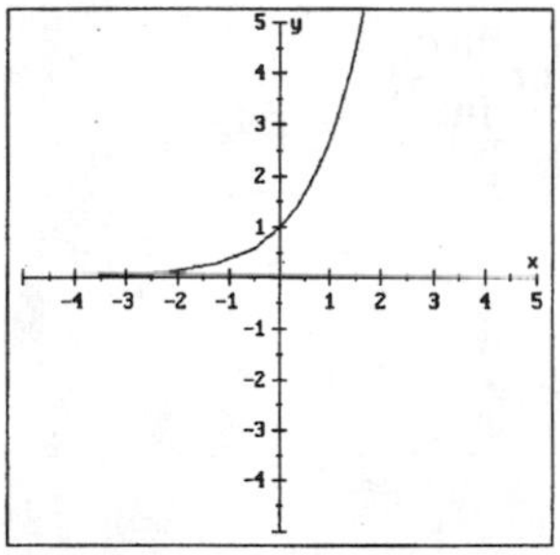

6. $A = A_0e^{rt} = 2000e^{(0.08)10}$
$= 2000e^{0.8}$
$\approx \$4451.08$

7. $\log_4 16 = x \Rightarrow 4^x = 16 \Rightarrow x = 2$

8. $\log_x 81 = 4 \Rightarrow x^4 = 81 \Rightarrow x = 3$

9. $\log_3 x = -3 \Rightarrow 3^{-3} = x \Rightarrow x = \frac{1}{27}$

10. $\log_x 100 = 2 \Rightarrow x^2 = 100 \Rightarrow x = 10$

11. $\log_{3/2}\frac{9}{4} = x \Rightarrow \left(\frac{3}{2}\right)^x = \frac{9}{4} \Rightarrow x = 2$

12. $\log_{2/3} x = -3 \Rightarrow \left(\frac{2}{3}\right)^{-3} = x \Rightarrow x = \frac{27}{8}$

13. $f(x) = -\log_3 x$

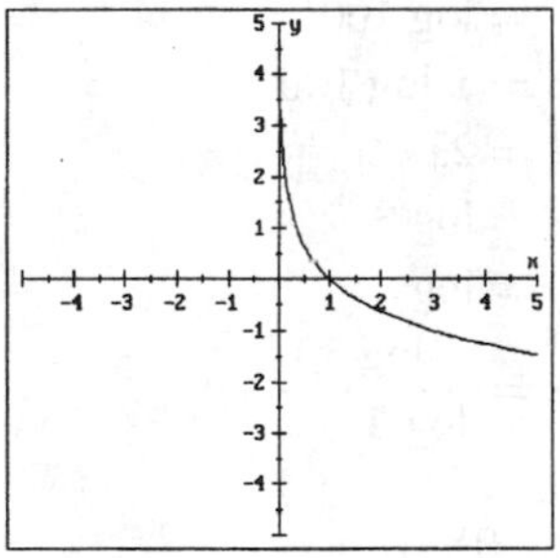

14. $f(x) = \ln x$

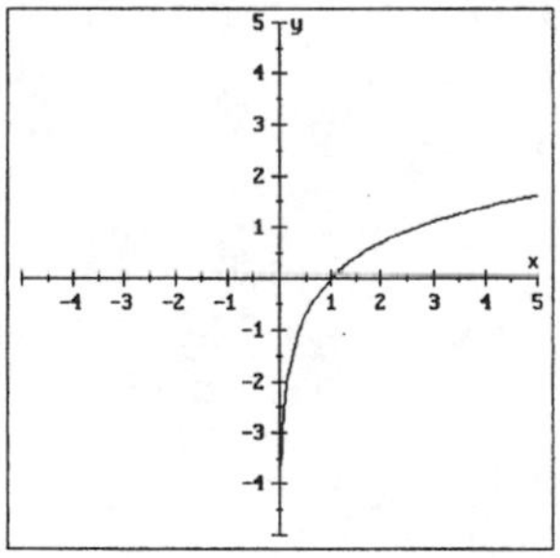

15. $\log a^2bc^3 = \log a^2 + \log b + \log c^3 = 2\log a + \log b + 3\log c$

16. $\ln\sqrt{\frac{a}{b^2c}} = \ln\left(\frac{a}{b^2c}\right)^{1/2} = \frac{1}{2}\ln\frac{a}{b^2c} = \frac{1}{2}(\ln a - \ln b^2 - \ln c) = \frac{1}{2}\ln a - \ln b - \frac{1}{2}\ln c$

17. $\frac{1}{2}\log(a+2) + \log b - 3\log c = \log(a+2)^{1/2} + \log b - \log c^3 = \log\frac{b\sqrt{a+2}}{c^3}$

18. $\frac{1}{3}(\log a - 2\log b) - \log c = \frac{1}{3}(\log a - \log b^2) - \log c = \frac{1}{3}\log\frac{a}{b^2} - \log c = \log\sqrt[3]{\frac{a}{b^2}} - \log c$

$= \log\frac{\sqrt[3]{a}}{c\sqrt[3]{b^2}}$

19. $\log 24 = \log 8 \cdot 3 = \log 2^3 \cdot 3 = \log 2^3 + \log 3 = 3 \log 2 + \log 3 = 3(0.3010) + 0.4771 = 1.3801$

20. $\log \frac{8}{3} = \log \frac{2^3}{3} = \log 2^3 - \log 3 = 3 \log 2 - \log 3 = 3(0.3010) - 0.4771 = 0.4259$

21. $\log_7 3 = \frac{\log 3}{\log 7} \text{ or } \frac{\ln 3}{\ln 7}$

22. $\log_\pi e = \frac{\log e}{\log \pi} \text{ or } \frac{\ln e}{\ln \pi}$

23. $\log_a ab = \log_a a + \log_a b = 1 + \log_a b \Rightarrow$ TRUE

24. $\frac{\log a}{\log b} = \log a - \log b \Rightarrow$ FALSE $\left(\log \frac{a}{b} = \log a - \log b\right)$

25. $\log a^{-3} = -3 \log a \neq \frac{1}{3 \log a} \Rightarrow$ FALSE

26. $\ln(-x) = -\ln x \Rightarrow$ FALSE (This implies one of the logarithms is negative, which is impossible.)

27. $\text{pH} = -\log [\text{H}^+] = -\log(3.7 \times 10^{-7}) \approx 6.4$

28. $\text{dB gain} = 20 \log \frac{E_O}{E_I} = 20 \log \frac{60}{0.3} = 20 \log 200 \approx 46$

29.
$$\begin{aligned} 5^x &= 3 \\ \log 5^x &= \log 3 \\ x \log 5 &= \log 3 \\ x &= \frac{\log 3}{\log 5} \end{aligned}$$

30.
$$\begin{aligned} 3^{x-1} &= 100^x \\ \log 3^{x-1} &= \log 100^x \\ (x-1)\log 3 &= x \log 100 \\ x \log 3 - \log 3 &= 2x \\ x \log 3 - 2x &= \log 3 \\ x(\log 3 - 2) &= \log 3 \\ x &= \frac{\log 3}{\log 3 - 2} \end{aligned}$$

31.
$$\begin{aligned} \log(5x+2) &= \log(2x+5) \\ 5x + 2 &= 2x + 5 \\ 3x &= 3 \\ x &= 1 \end{aligned}$$

32.
$$\begin{aligned} \log x + \log(x-9) &= 1 \\ \log x(x-9) &= 1 \\ x(x-9) &= 10 \\ x^2 - 9x - 10 &= 0 \\ (x-10)(x+1) &= 0 \end{aligned}$$
$$x - 10 = 0 \quad \textbf{or} \quad x + 1 = 0$$
$$x = 10 \qquad\qquad x = -1$$

Does not check.

Exercise 12.1 (page 810)

1.
$$|3x-4|=11$$
$$3x-4=11 \quad \textbf{or} \quad 3x-4=-11$$
$$3x=15 \qquad\qquad 3x=-7$$
$$x=5 \qquad\qquad x=-\tfrac{7}{3}$$

3.
$$|3x+4|=|5x-2|$$
$$3x+4=5x-2 \quad \textbf{or} \quad 3x+4=-(5x-2)$$
$$-2x=-6 \qquad\qquad 3x+4=-5x+2$$
$$x=3 \qquad\qquad 8x=-2$$
$$x=-\tfrac{1}{4}$$

5. circle; plane

7. $r^2<0$

9. parabola; $(3,2)$; right

11. $x^2+y^2=9$

C $(0,0)$; $r=\sqrt{9}=3$

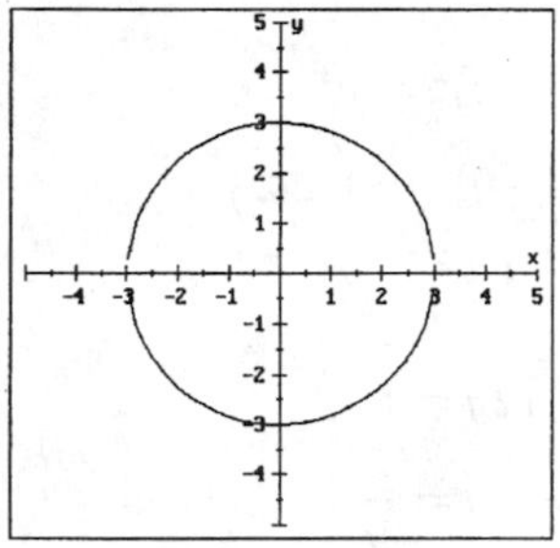

13. $(x-2)^2+y^2=9$

C $(2,0)$; $r=\sqrt{9}=3$

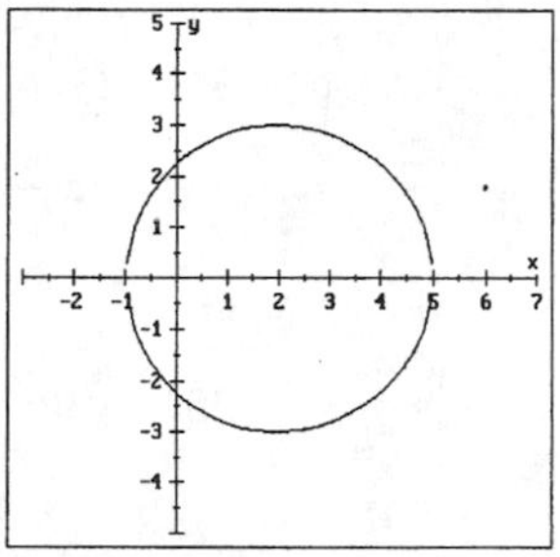

15. $(x-2)^2+(y-4)^2=4$

C $(2,4)$; $r=\sqrt{4}=2$

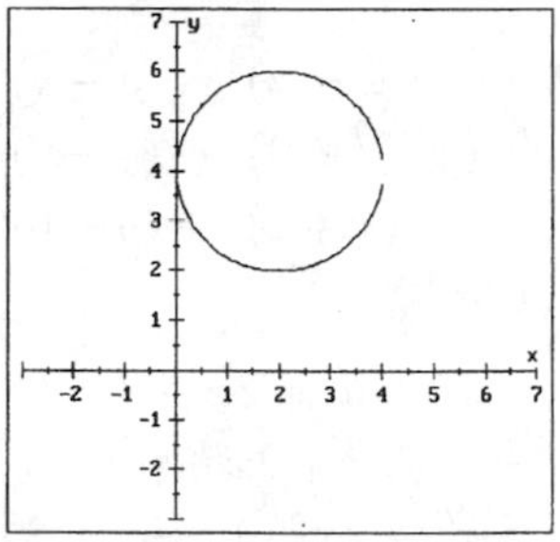

17. $(x+3)^2+(y-1)^2=16$

C $(-3,1)$; $r=\sqrt{16}=4$

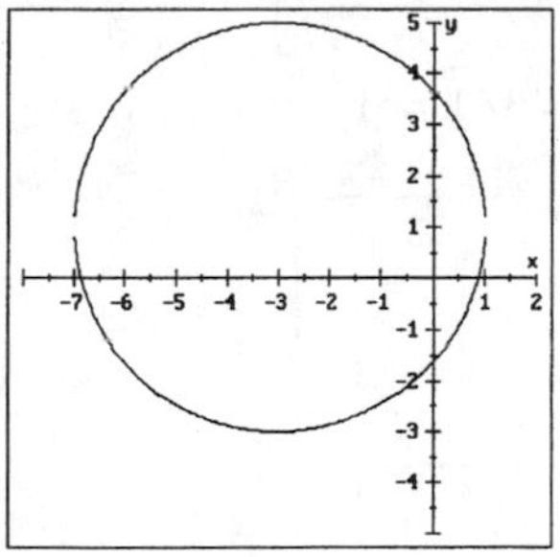

19. $x^2+(y+3)^2=1$

C $(0,-3)$; $r=\sqrt{1}=1$

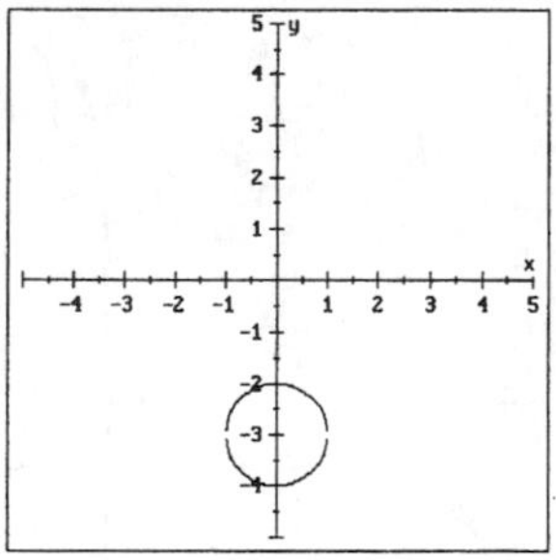

21.
$$3x^2+3y^2=16$$
$$3y^2=16-3x^2$$
$$y^2=\frac{16-3x^2}{3}$$
$$y=\pm\sqrt{\frac{16-3x^2}{3}}$$

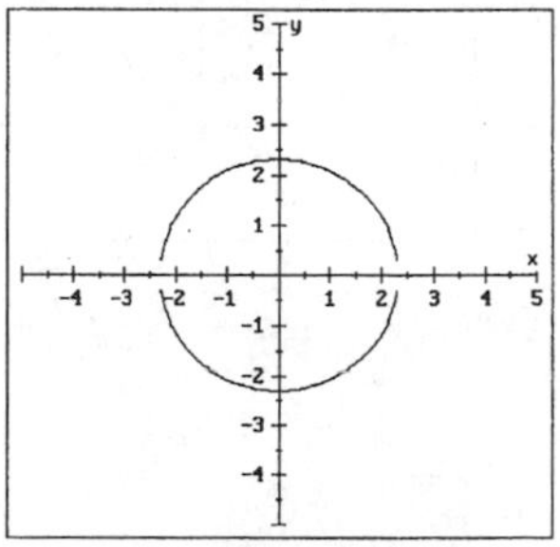

23. $(x+1)^2 + y^2 = 16$

$$y^2 = 16 - (x+1)^2$$

$$y = \pm\sqrt{16-(x+1)^2}$$

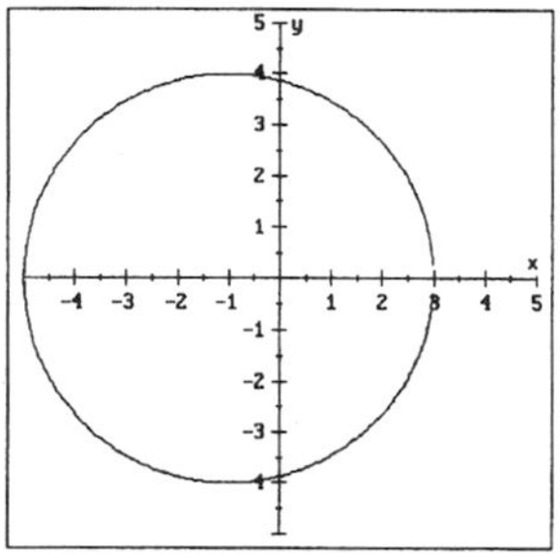

25. $(x-h)^2 + (y-k)^2 = r^2$

$$(x-0)^2 + (y-0)^2 = 1^2$$

$$x^2 + y^2 = 1$$

27. $(x-h)^2 + (y-k)^2 = r^2$

$$(x-6)^2 + (y-8)^2 = 5^2$$

$$(x-6)^2 + (y-8)^2 = 25$$

29. $(x-h)^2 + (y-k)^2 = r^2$

$$(x-(-2))^2 + (y-6)^2 = 12^2$$

$$(x+2)^2 + (y-6)^2 = 144$$

31. $(x-h)^2 + (y-k)^2 = r^2$

$$(x-0)^2 + (y-0)^2 = \left(\sqrt{2}\right)^2$$

$$x^2 + y^2 = 2$$

33. $x^2 + y^2 + 2x - 8 = 0$

$$x^2 + 2x + y^2 = 8$$

$$x^2 + 2x + 1 + y^2 = 8 + 1$$

$$(x+1)^2 + y^2 = 9$$

C $(-1, 0)$; $r = \sqrt{9} = 3$

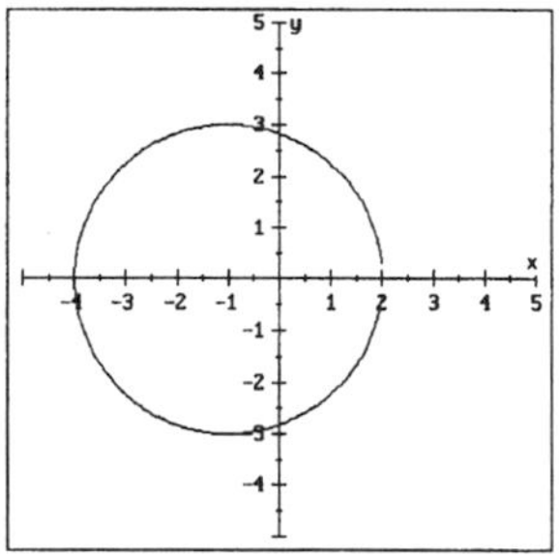

35. $9x^2 + 9y^2 - 12y = 5$

$$x^2 + y^2 - \frac{4}{3}y = \frac{5}{9}$$

$$x^2 + y^2 - \frac{4}{3}y + \frac{4}{9} = \frac{5}{9} + \frac{4}{9}$$

$$x^2 + \left(y - \frac{2}{3}\right)^2 = 1$$

C $\left(0, \frac{2}{3}\right)$; $r = \sqrt{1} = 1$

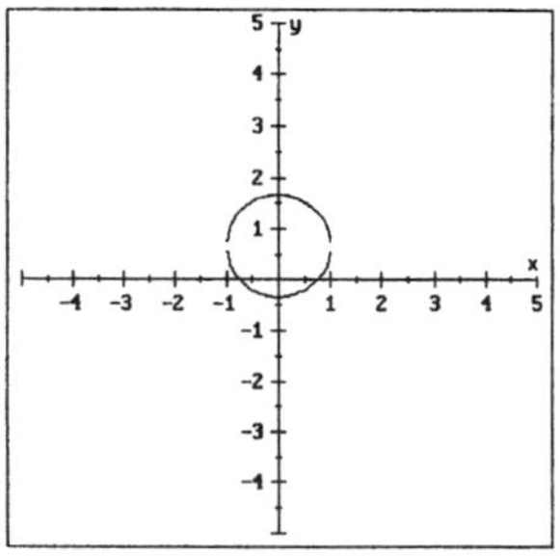

37.
$$x^2 + y^2 - 2x + 4y = -1$$
$$x^2 - 2x + y^2 + 4y = -1$$
$$x^2 - 2x + 1 + y^2 + 4y + 4 = -1 + 1 + 4$$
$$(x-1)^2 + (y+2)^2 = 4$$
C $(1, -2)$; $r = \sqrt{4} = 2$

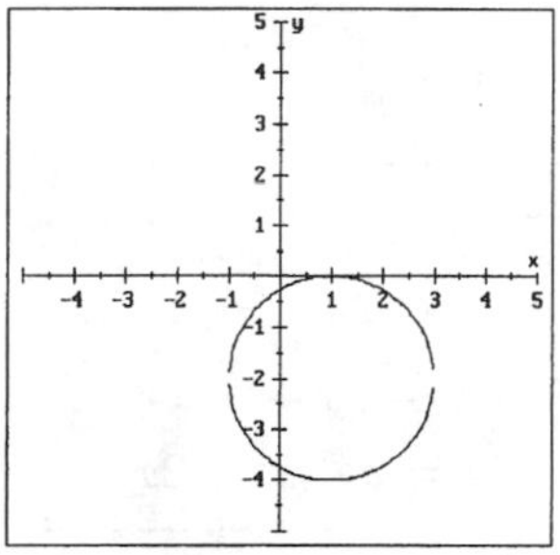

39.
$$x^2 + y^2 + 6x - 4y = -12$$
$$x^2 + 6x + y^2 - 4y = -12$$
$$x^2 + 6x + 9 + y^2 - 4y + 4 = -12 + 9 + 4$$
$$(x+3)^2 + (y-2)^2 = 1$$
C $(-3, 2)$; $r = \sqrt{1} = 1$

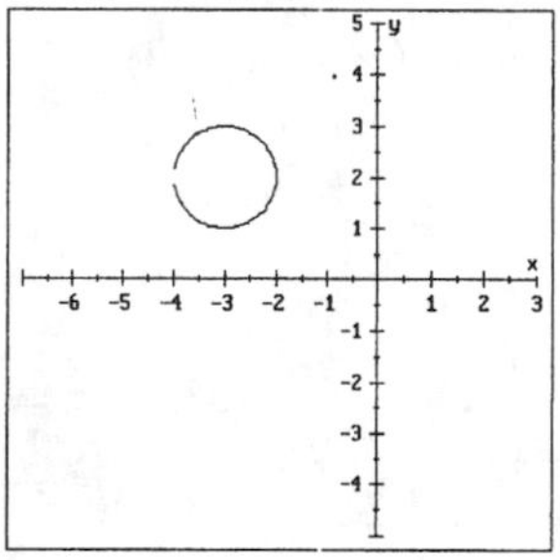

41. $x = y^2$
$$x = (y-0)^2 + 0$$
V $(0, 0)$; opens R

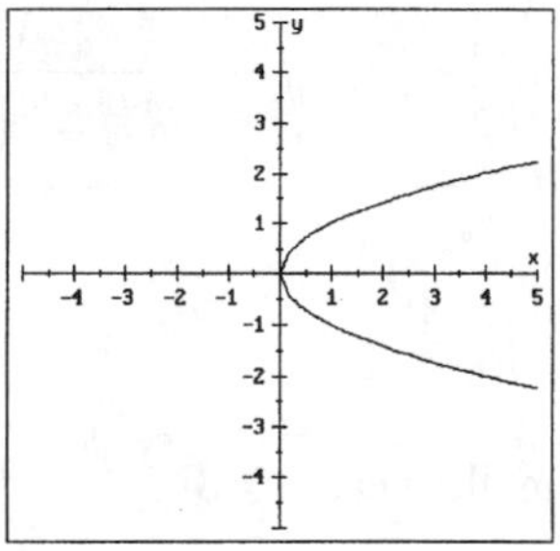

43. $x = -\frac{1}{4}y^2$
$$x = -\frac{1}{4}(y-0)^2 + 0$$
V $(0, 0)$; opens L

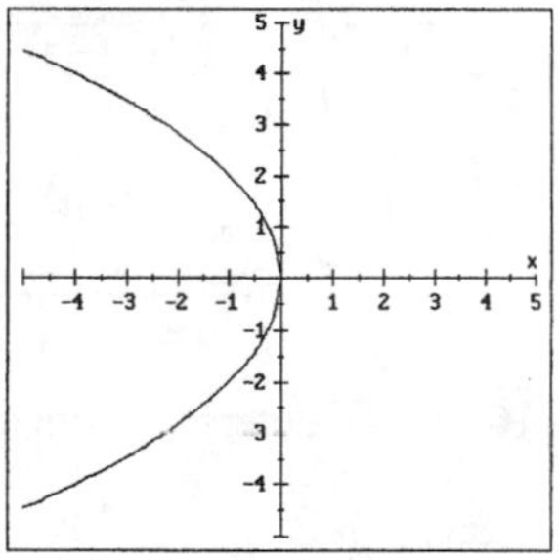

45. $y = x^2 + 4x + 5$
$$y = x^2 + 4x + 4 + 5 - 4$$
$$y = (x+2)^2 + 1$$
V $(-2, 1)$; opens U

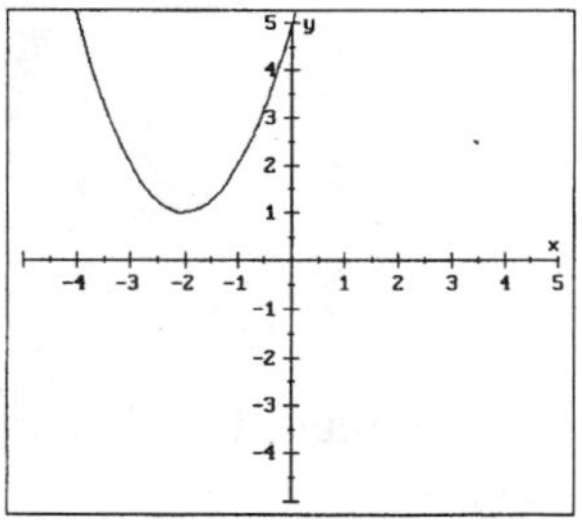

47. $y = -x^2 - x + 1$
$$y = -(x^2 + x) + 1$$
$$y = -\left(x^2 + x + \frac{1}{4}\right) + 1 + \frac{1}{4}$$
$$y = -\left(x + \frac{1}{2}\right)^2 + \frac{5}{4}$$
V $\left(-\frac{1}{2}, \frac{5}{4}\right)$; opens D

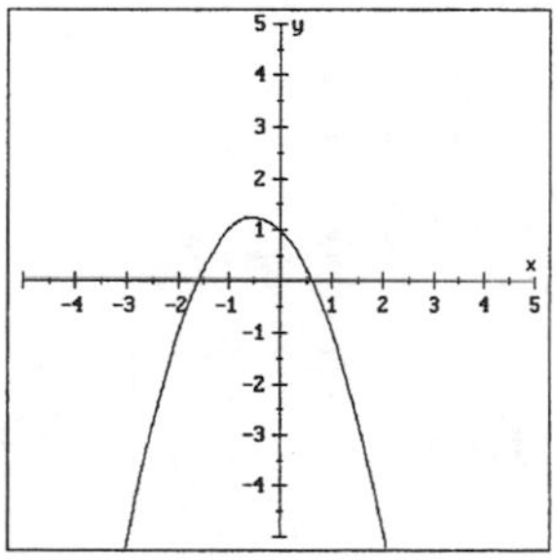

49. $y^2 + 4x - 6y = -1$

$$4x = -y^2 + 6y - 1$$
$$x = -\frac{1}{4}y^2 + \frac{3}{2}y - \frac{1}{4}$$
$$x = -\frac{1}{4}(y^2 - 6y) - \frac{1}{4}$$
$$x = -\frac{1}{4}(y^2 - 6y + 9) - \frac{1}{4} + \frac{9}{4}$$
$$x = -\frac{1}{4}(y-3)^2 + 2$$

V $(2,3)$; opens L

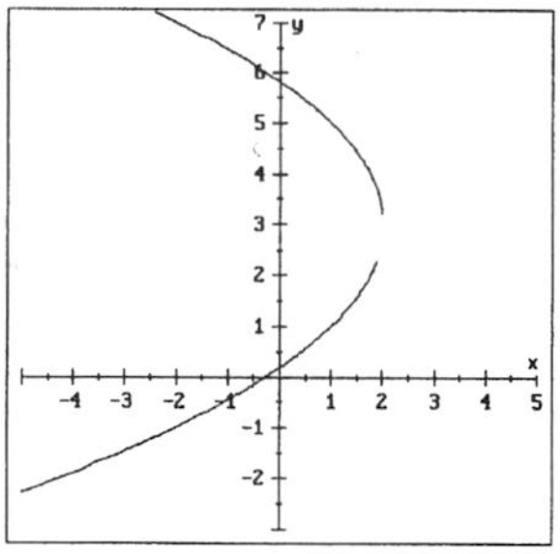

51. $y = 2(x-1)^2 + 3$

V $(1,3)$; opens U

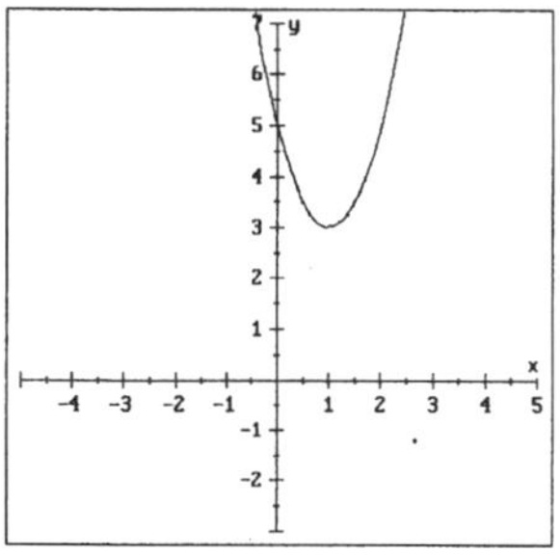

53. $x = 2y^2$

$$y^2 = \frac{x}{2}$$
$$y = \pm\sqrt{\frac{x}{2}}$$

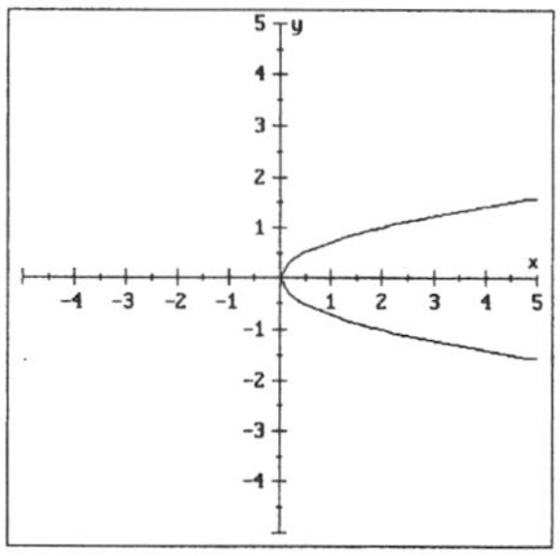

55. $x^2 - 2x + y = 6$

$$y = 6 - x^2 + 2x$$

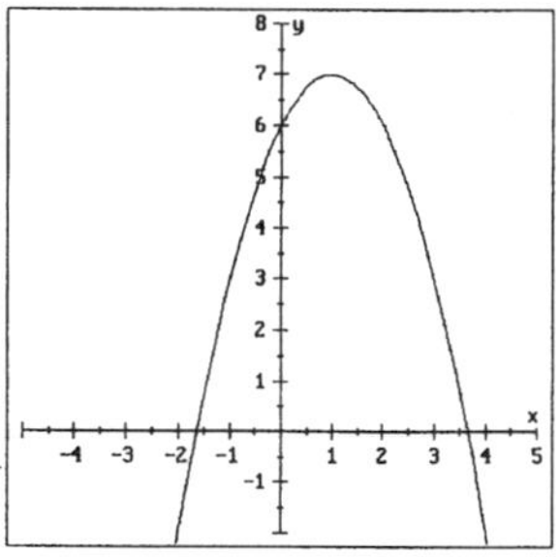

57. The radius of the larger gear is $\sqrt{16} = 4$. Centers: 7 units apart $\Rightarrow$ smaller gear $r = 3$.

$(x-h)^2 + (y-k)^2 = r^2$

$(x-7)^2 + (y-0)^2 = 3^2 \Rightarrow (x-7)^2 + y^2 = 9$

59.

$$x^2 + y^2 - 8x - 20y + 16 = 0$$
$$x^2 - 8x + y^2 - 20y = -16$$
$$x^2 - 8x + 16 + y^2 - 20y + 100 = -16 + 16 + 100$$
$$(x-4)^2 + (y-10)^2 = 100$$

center: $(4,10)$; radius $= 10$

$$x^2 + y^2 + 2x + 4y - 11 = 0$$
$$x^2 + 2x + 1 + y^2 + 4y + 4 = 11 + 1 + 4$$
$$(x+1)^2 + (y+2)^2 = 16$$

center: $(-1,-2)$; radius $= 4$

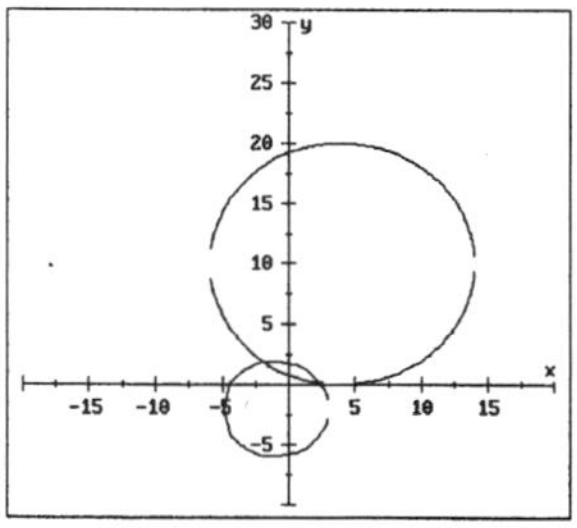

Since the ranges overlap (see graph), they can not be licensed for the same frequency.

61. Set $y = 0$:
$y = 30x - x^2$
$0 = 30x - x^2$
$0 = x(30 - x)$
$x = 0$ or $x = 30$
It lands 30 feet away.

63. Find the vertex:
$2y^2 - 9x = 18$
$-9x = -2y^2 + 18$
$x = \frac{2}{9}y^2 - 2$
$x = \frac{2}{9}(y - 0)^2 - 2$
vertex: $(-2, 0) \Rightarrow$ distance $= 2$ AU

65. **Answers may vary.**

67. **Answers may vary.**

Exercise 12.2 (page 822)

1. $3x^{-2}y^2(4x^2 + 3y^{-2}) = 12x^0y^2 + 9x^{-2}y^0 = 12y^2 + \frac{9}{x^2}$

3. $\frac{x^{-2} + y^{-2}}{x^{-2} - y^{-2}} = \frac{x^{-2} + y^{-2}}{x^{-2} - y^{-2}} \cdot \frac{x^2y^2}{x^2y^2} = \frac{y^2 + x^2}{y^2 - x^2}$

5. ellipse; sum

7. center

9. $(0, 0)$

11. $\frac{x^2}{4} + \frac{y^2}{9} = 1$
C $(0, 0)$; move 2 horiz. and 3 vert.

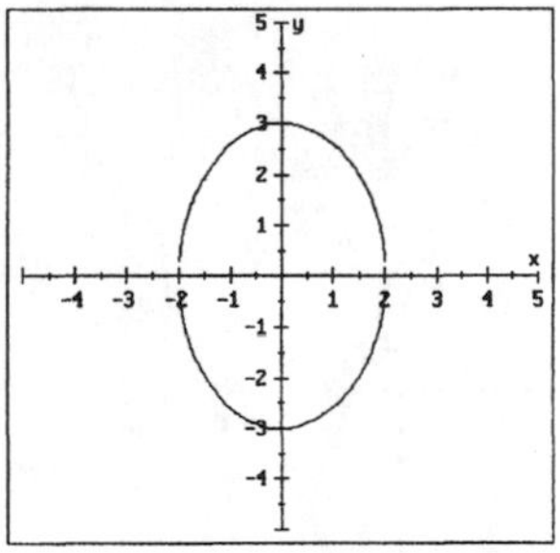

13. $x^2 + 9y^2 = 9$
$\frac{x^2}{9} + \frac{9y^2}{9} = \frac{9}{9}$
$\frac{x^2}{9} + \frac{y^2}{1} = 1$
C $(0, 0)$; move 3 horiz. and 1 vert.

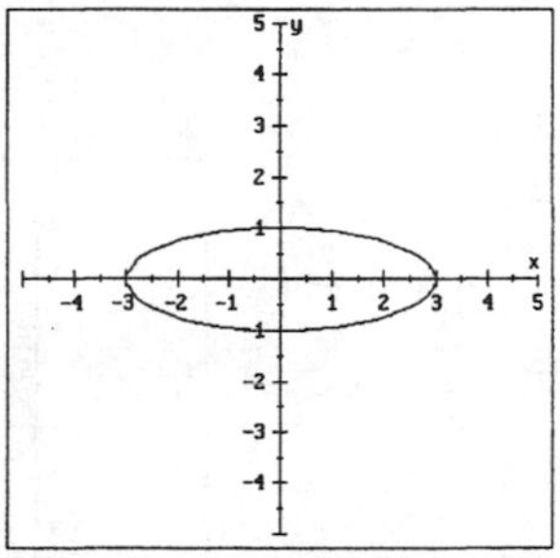

15. $16x^2 + 4y^2 = 64$
$\frac{16x^2}{64} + \frac{4y^2}{64} = \frac{64}{64}$
$\frac{x^2}{4} + \frac{y^2}{16} = 1$
C $(0, 0)$; move 2 horiz. and 4 vert.

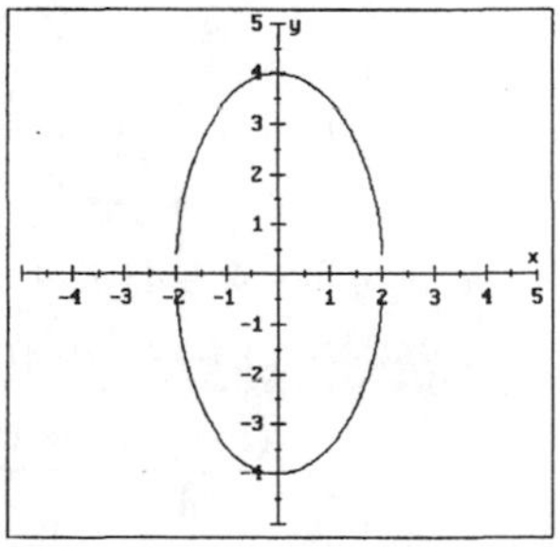

17. $\dfrac{(x-2)^2}{9}+\dfrac{(y-1)^2}{4}=1$

C $(2,1)$; move 3 horiz. and 2 vert.

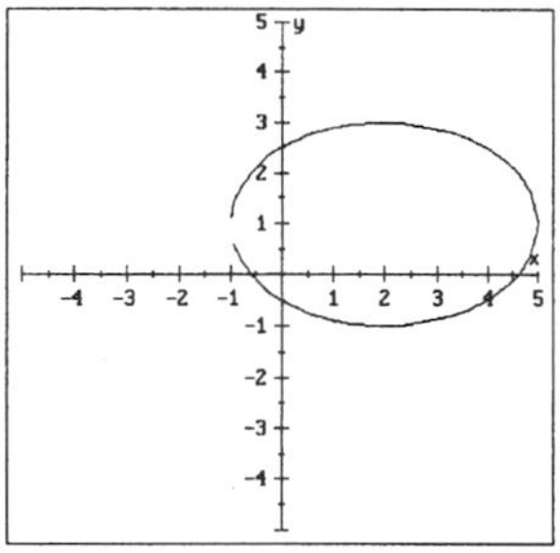

19.
$$(x+1)^2+4(y+2)^2=4$$
$$\frac{(x+1)^2}{4}+\frac{4(y+2)^2}{4}=\frac{4}{4}$$
$$\frac{(x+1)^2}{4}+\frac{(y+2)^2}{1}=1$$

C $(-1,-2)$; move 2 horiz. and 1 vert.

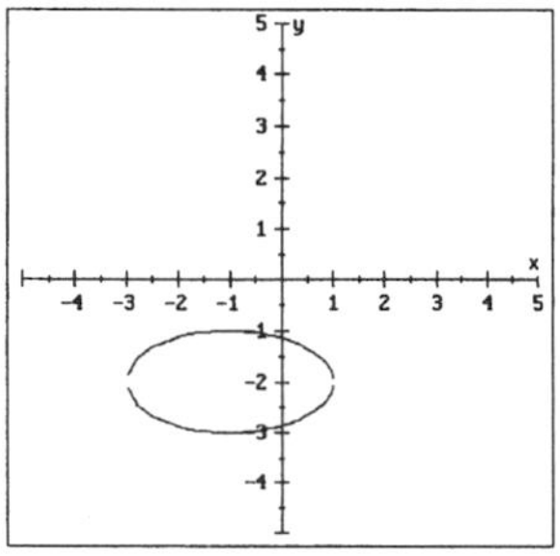

21.
$$\frac{x^2}{9}+\frac{y^2}{4}=1$$
$$\frac{y^2}{4}=1-\frac{x^2}{9}$$
$$y^2=4\left(1-\frac{x^2}{9}\right)$$
$$y=\pm\sqrt{4\left(1-\frac{x^2}{9}\right)}$$

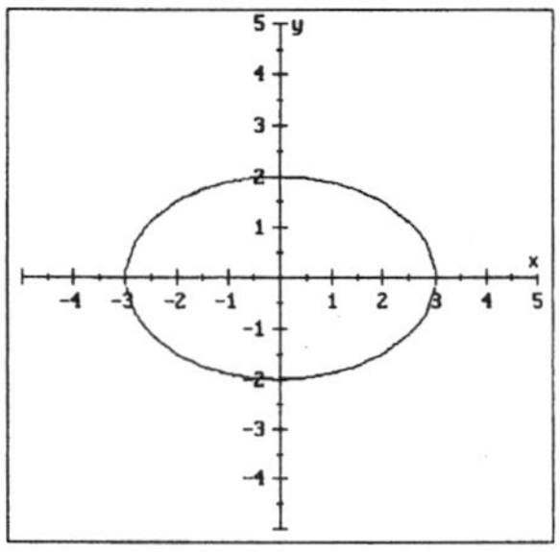

23.
$$\frac{x^2}{4}+\frac{(y-1)^2}{9}=1$$
$$\frac{(y-1)^2}{9}=1-\frac{x^2}{4}$$
$$(y-1)^2=9\left(1-\frac{x^2}{4}\right)$$
$$y-1=\pm\sqrt{9\left(1-\frac{x^2}{4}\right)}$$
$$y=1\pm\sqrt{9\left(1-\frac{x^2}{4}\right)}$$

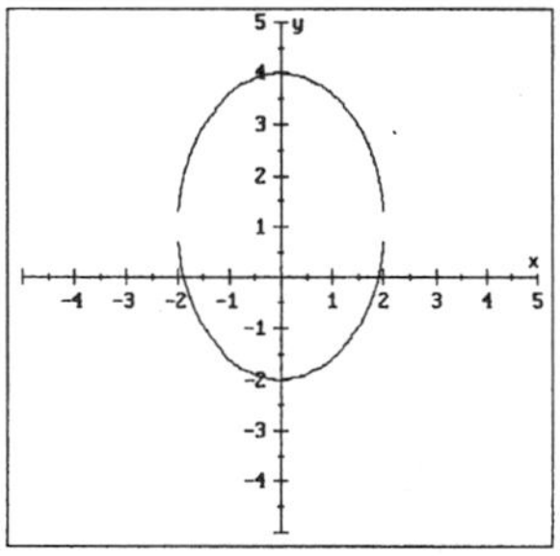

25.
$$x^2+4y^2-4x+8y+4=0$$
$$x^2-4x+4\left(y^2+2y\right)=-4$$
$$x^2-4x+4+4\left(y^2+2y+1\right)=-4+4+4$$
$$(x-2)^2+4(y+1)^2=4$$
$$\frac{(x-2)^2}{4}+\frac{(y+1)^2}{1}=1$$

C $(2,-1)$; move 2 horiz. and 1 vert.

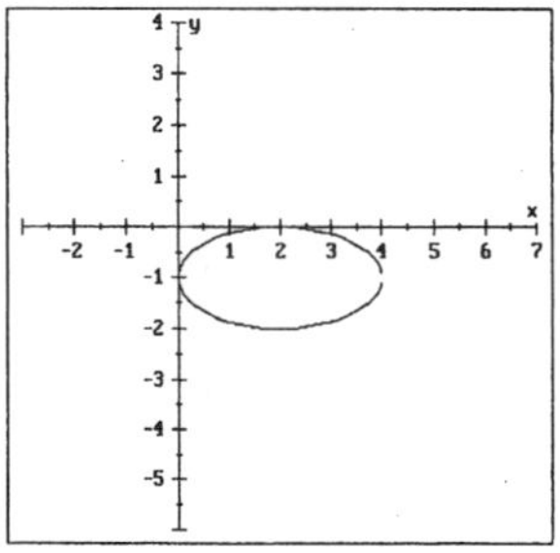

27.

$$9x^2 + 4y^2 - 18x + 16y = 11$$
$$9(x^2 - 2x) + 4(y^2 + 4y) = 11$$
$$9(x^2 - 2x + 1) + 4(y^2 + 4y + 4) = 11 + 9 + 16$$
$$9(x-1)^2 + 4(y+2)^2 = 36$$
$$\frac{(x-1)^2}{4} + \frac{(y+2)^2}{9} = 1$$

C $(1, -2)$; move 2 horiz. and 3 vert.

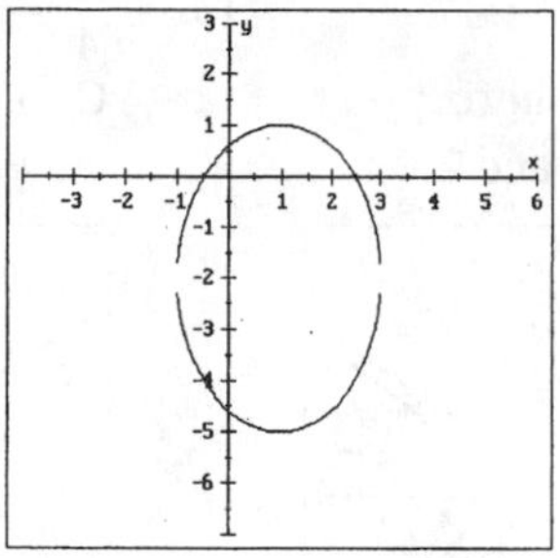

29. $a = 24/2 = 12,\ b = 10/2 = 5$

$$\frac{x^2}{a^2} + \frac{y^2}{b^2} = 1$$
$$\frac{x^2}{12^2} + \frac{y^2}{5^2} = 1$$
$$\frac{x^2}{144} + \frac{y^2}{25} = 1$$

31. Note: $a = 40/2 = 20,\ b = 10$

$$\frac{x^2}{a^2} + \frac{y^2}{b^2} = 1$$
$$\frac{x^2}{400} + \frac{y^2}{100} = 1$$
$$x^2 + 4y^2 = 400$$
$$4y^2 = 400 - x^2$$
$$y^2 = \tfrac{1}{4}(400 - x^2)$$
$$y = \tfrac{1}{2}\sqrt{400 - x^2}$$

33. $9x^2 + 16y^2 = 144$

$$\frac{9x^2}{144} + \frac{16y^2}{144} = \frac{144}{144}$$
$$\frac{x^2}{16} + \frac{y^2}{9} = 1$$

area $= \pi ab = \pi(4)(3)$

$= 12\pi$ square units

35. **Answers may vary.**

37. It is a circle.

Exercise 12.3 (page 832)

1. $-6x^4 + 9x^3 - 6x^2 = -3x^2(2x^2 - 3x + 2)$

3. $15a^2 - 4ab - 4b^2 = (3a - 2b)(5a + 2b)$

5. hyperbola; difference

7. center

9. $(0, 0)$

11. $\dfrac{x^2}{9} - \dfrac{y^2}{4} = 1$

C $(0, 0)$; open horiz.;
move 3 horiz. and 2 vert.

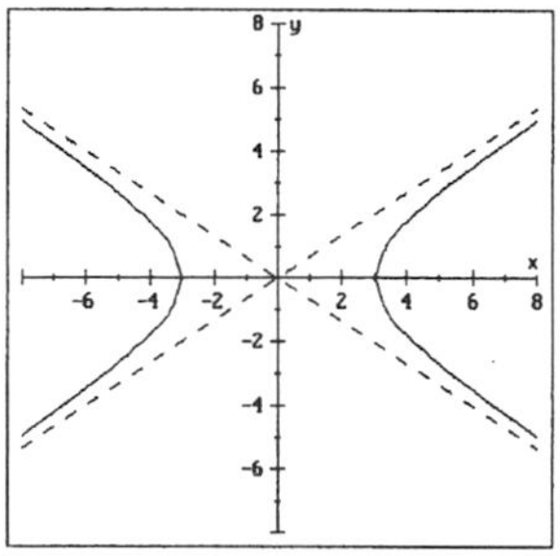

13. $\dfrac{y^2}{4} - \dfrac{x^2}{9} = 1$

C $(0, 0)$; open vert.;
move 3 horiz. and 2 vert.

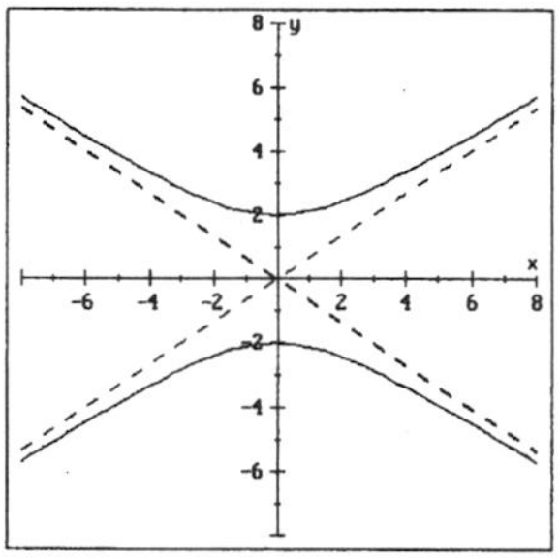

15. $25x^2 - y^2 = 25$

$\dfrac{25x^2}{25} - \dfrac{y^2}{25} = 1$

$\dfrac{x^2}{1} - \dfrac{y^2}{25} = 1$

C $(0, 0)$; open horiz.;
move 1 horiz. and 5 vert.

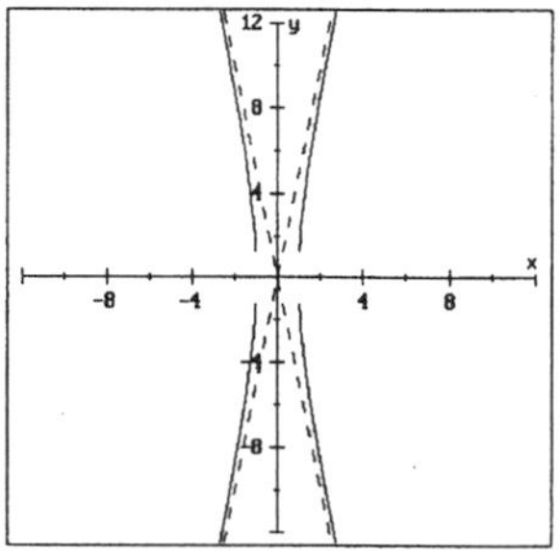

17. $\dfrac{(x-2)^2}{9} - \dfrac{y^2}{16} = 1$

C $(2, 0)$; open horiz.;
move 3 horiz. and 4 vert.

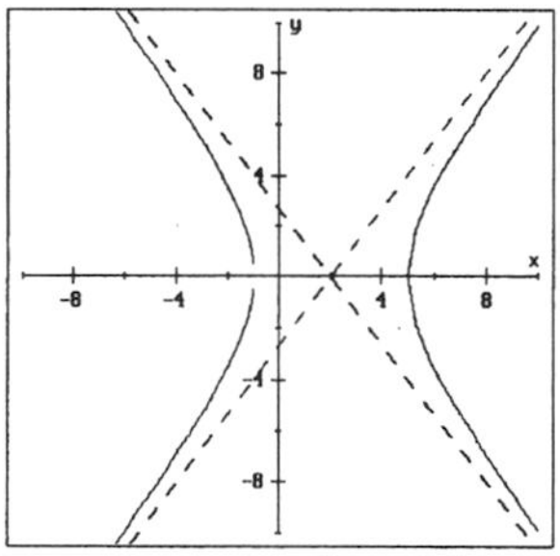

19. $\dfrac{(y+1)^2}{1} - \dfrac{(x-2)^2}{4} = 1$

C $(2, -1)$; open vert.;
move 2 horiz. and 1 vert.

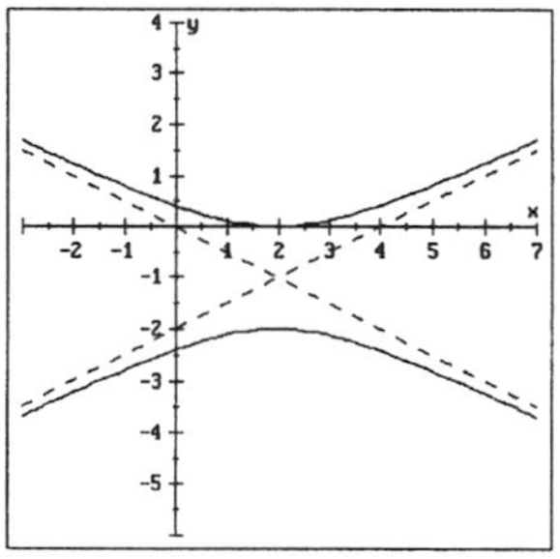

21. $4(x+3)^2 - (y-1)^2 = 4$

$\dfrac{4(x+3)^2}{4} - \dfrac{(y-1)^2}{4} = \dfrac{4}{4}$

$\dfrac{(x+3)^2}{1} - \dfrac{(y-1)^2}{4} = 1$

C $(-3, 1)$; open horiz.;
move 1 horiz. and 2 vert.

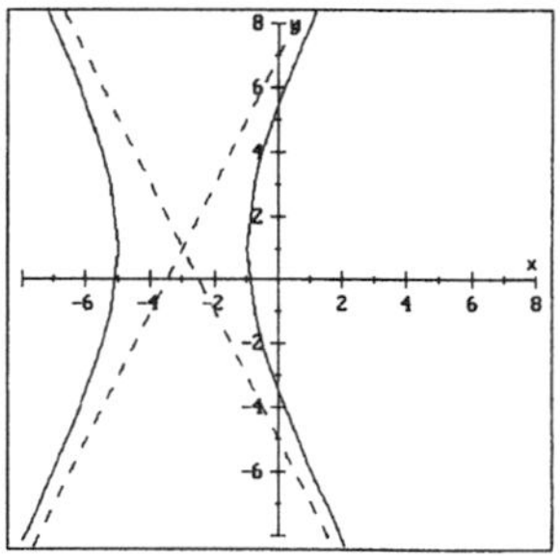

23. $xy = 8$

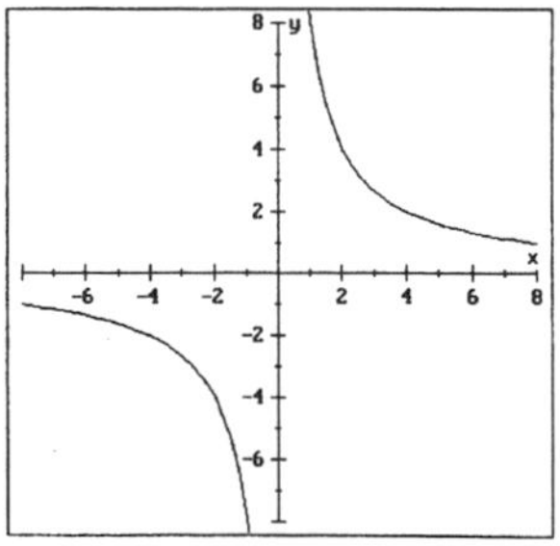

25.
$$\frac{x^2}{9} - \frac{y^2}{4} = 1$$
$$\frac{y^2}{4} = \frac{x^2}{9} - 1$$
$$y^2 = 4\left(\frac{x^2}{9} - 1\right)$$
$$y = \pm\sqrt{4\left(\frac{x^2}{9} - 1\right)}$$

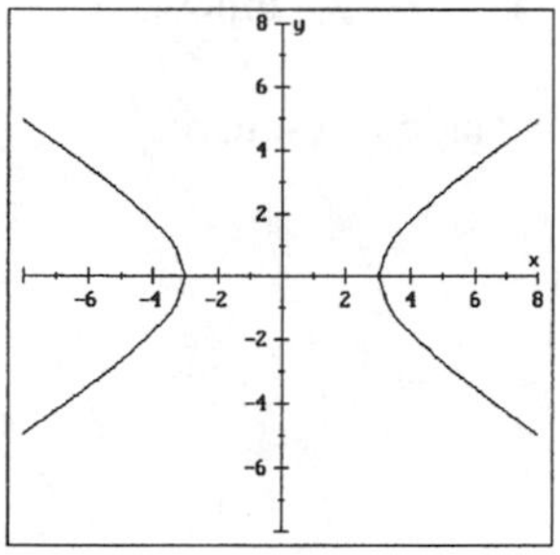

27.
$$\frac{x^2}{4} - \frac{(y-1)^2}{9} = 1$$
$$\frac{(y-1)^2}{9} = \frac{x^2}{4} - 1$$
$$(y-1)^2 = 9\left(\frac{x^2}{4} - 1\right)$$
$$y - 1 = \pm\sqrt{9\left(\frac{x^2}{4} - 1\right)}$$
$$y = 1 \pm \sqrt{9\left(\frac{x^2}{4} - 1\right)}$$

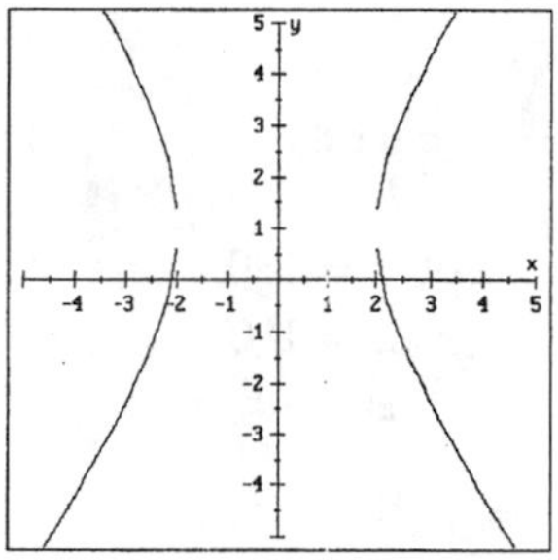

29.
$$4x^2 - y^2 + 8x - 4y = 4$$
$$4x^2 + 8x - y^2 - 4y = 4$$
$$4(x^2 + 2x) - (y^2 + 4y) = 4$$
$$4(x^2 + 2x + 1) - (y^2 + 4y + 4) = 4 + 4 - 4$$
$$4(x+1)^2 - (y+2)^2 = 4$$
$$\frac{(x+1)^2}{1} - \frac{(y+2)^2}{4} = 1$$

C $(-1, -2)$; opens horiz.; move 1 horiz. and 2 vert.

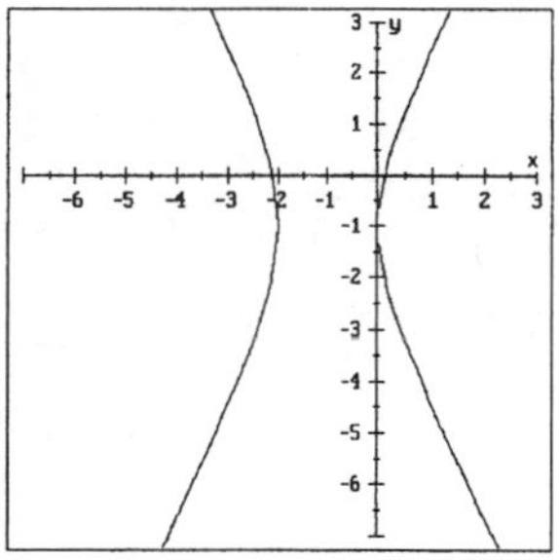

31.
$$4y^2 - x^2 + 8y + 4x = 4$$
$$4y^2 + 8y - x^2 + 4x = 4$$
$$4(y^2 + 2y) - (x^2 - 4x) = 4$$
$$4(y^2 + 2y + 1) - (x^2 - 4x + 4) = 4 + 4 - 4$$
$$4(y+1)^2 - (x-2)^2 = 4$$
$$\frac{(y+1)^2}{1} - \frac{(x-2)^2}{4} = 1$$

C $(2, -1)$; opens vert.; move 2 horiz. and 1 vert.

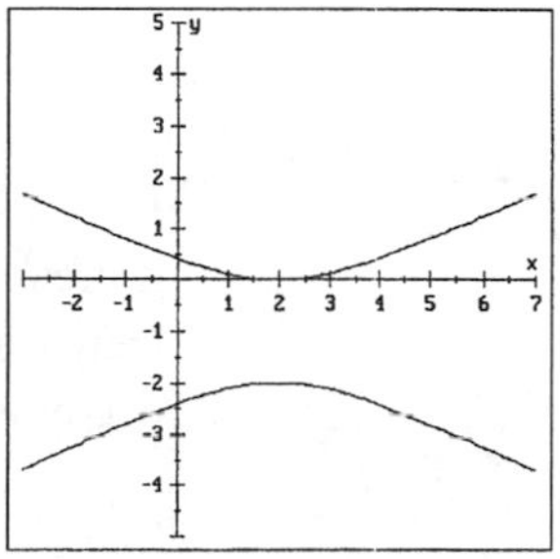

33. $9y^2 - x^2 = 81$

$$\frac{9y^2}{81} - \frac{x^2}{81} = \frac{81}{81}$$

$$\frac{y^2}{9} - \frac{x^2}{81} = 1$$

distance $= \sqrt{9} = 3$ units

35. $y^2 - x^2 = 25$

$$\frac{y^2}{25} - \frac{x^2}{25} = 1$$

vertex: $(0, 5)$

Let $y = 10$:

$$10^2 - x^2 = 25$$
$$-x^2 = -75$$
$$x = \sqrt{75} = 5\sqrt{3}$$

width $= 2(5\sqrt{3}) = 10\sqrt{3}$ miles

37. **Answers may vary.**

39. If $a = b$, the rectangle is a square.

Exercise 12.4 (page 839)

1. $(6x - 10) + (3x + 10) = 180$

$$9x = 180$$
$$x = 20$$

3. domains

5. constant; $f(x)$

7. step

9. increasing on $(-\infty, 0)$, decreasing on $(0, \infty)$

11. decreasing on $(-\infty, 0)$, constant on $(0, 2)$ increasing on $(2, \infty)$

13. $f(x) = \begin{cases} -1 \text{ if } x \leq 0 \\ x \text{ if } x > 0 \end{cases}$

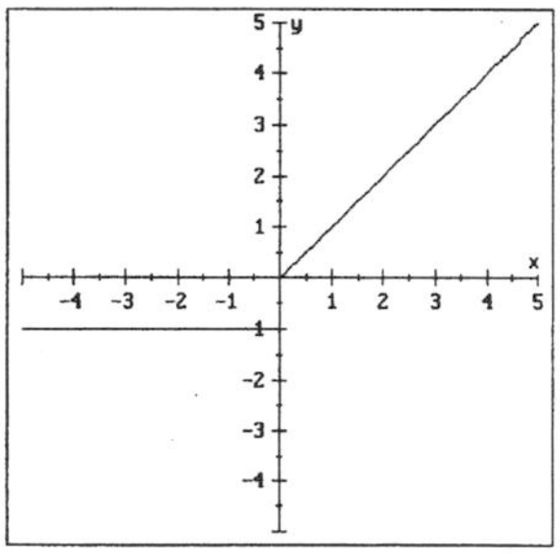

constant on $(-\infty, 0)$
increasing on $(0, \infty)$

15. $f(x) = \begin{cases} -x \text{ if } x \leq 0 \\ x \text{ if } 0 < x < 2 \\ -x \text{ if } x \geq 2 \end{cases}$

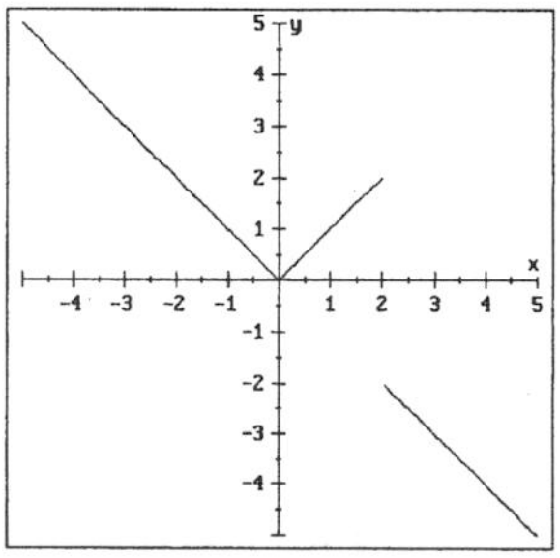

decreasing on $(-\infty, 0)$
increasing on $(0, 2)$
decreasing on $(2, \infty)$

17. $f(x) = -[[x]]$

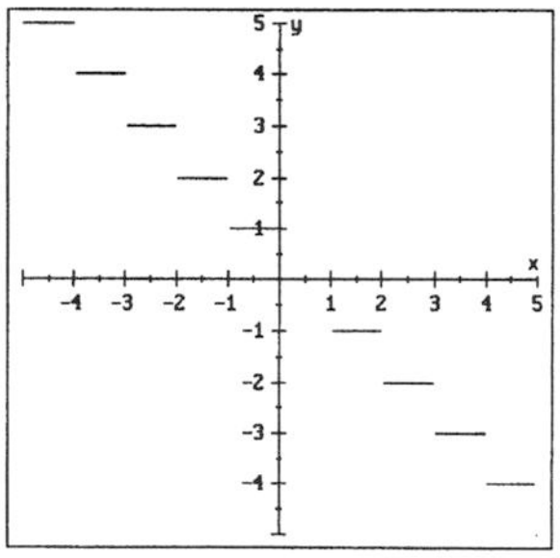

19. $f(x) = 2[[x]]$

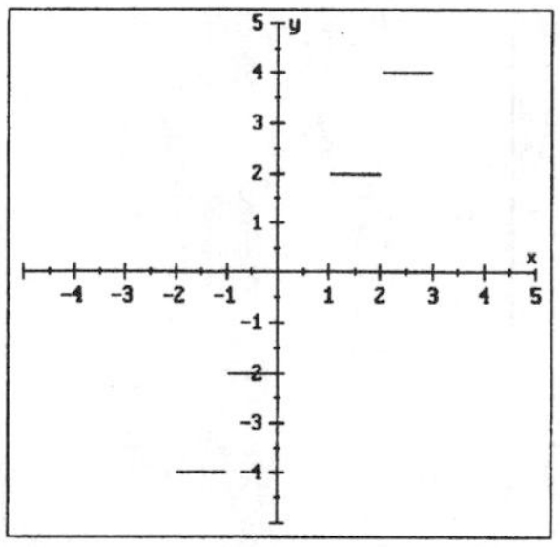

21. $f(x) = \begin{cases} -1 & \text{if } x < 0 \\ 0 & \text{if } x = 0 \\ 1 & \text{if } x > 0 \end{cases}$

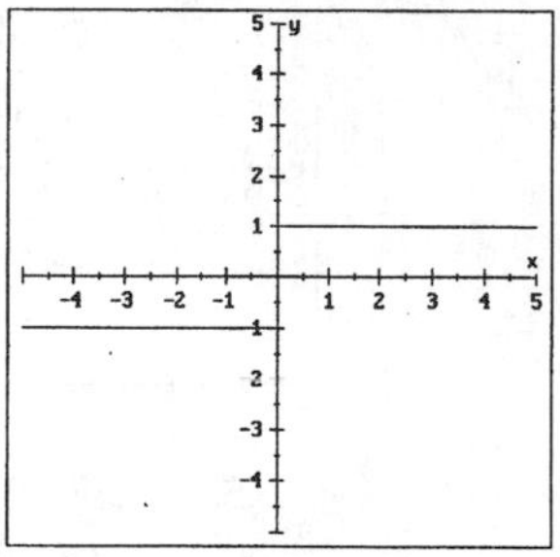

23. Find y when $x = 2.5$.
cost = \$30

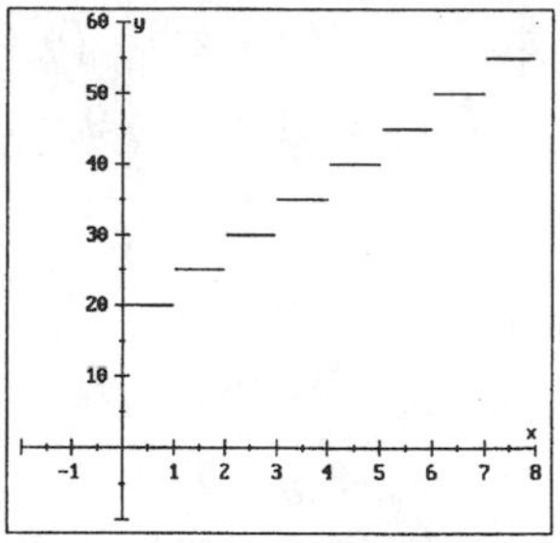

25. After 2 hours, B is cheaper.

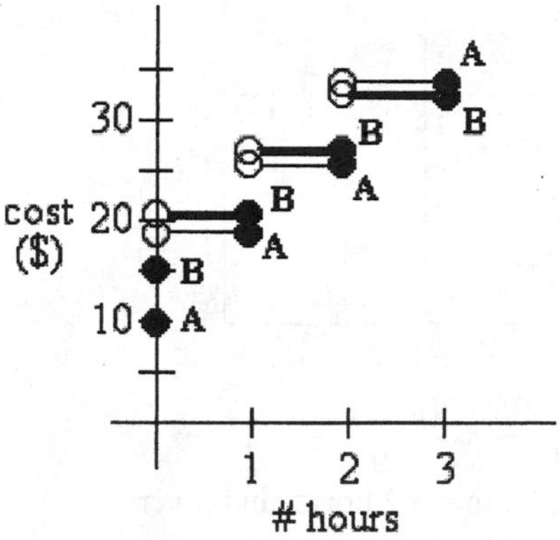

27. **Answers may vary.**

29. $f(x) = \begin{cases} x & \text{if } x < -2 \\ -x & \text{if } x > -2 \end{cases}$

Chapter 12 Summary (page 842)

1. $(x-1)^2 + (y+2)^2 = 9$
C $(1, -2)$; $r = \sqrt{9} = 3$

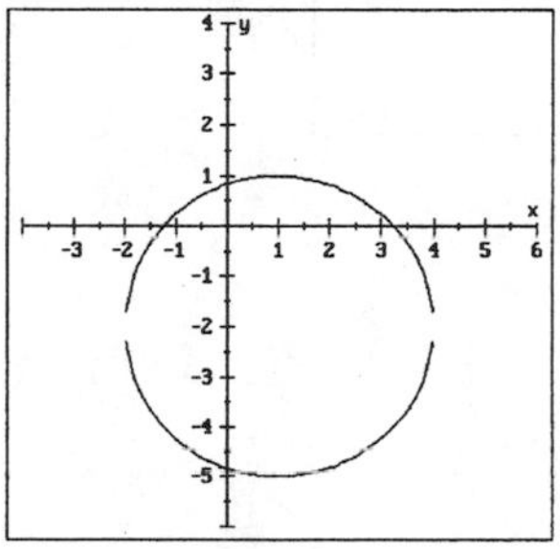

2. $x^2 + y^2 = 16$
C $(0, 0)$; $r = \sqrt{16} = 4$

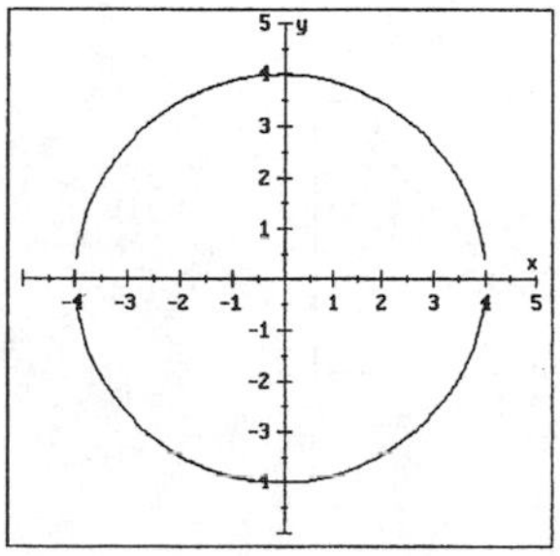

3.

$$x^2 + y^2 + 4x - 2y = 4$$
$$x^2 + 4x + y^2 - 2y = 4$$
$$x^2 + 4x + 4 + y^2 - 2y + 1 = 4 + 4 + 1$$
$$(x+2)^2 + (y-1)^2 = 9$$

C $(-2, 1)$; $r = \sqrt{9} = 3$

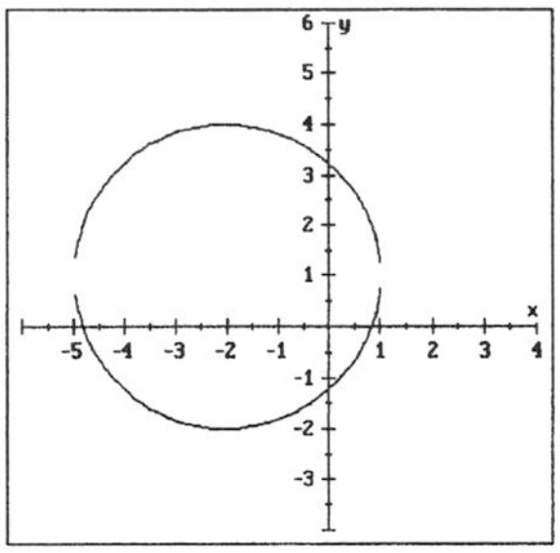

4. $x = -3(y-2)^2 + 5$

V $(5, 2)$; opens L

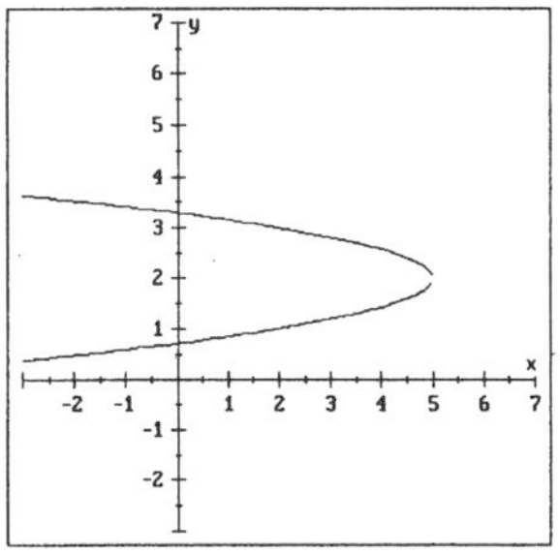

5. $x = 2(y+1)^2 - 2$

V $(-2, -1)$; opens R

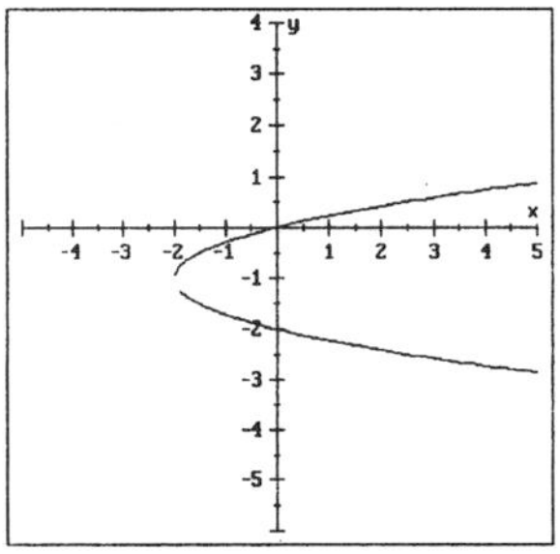

6.

$$9x^2 + 16y^2 = 144$$
$$\frac{9x^2}{144} + \frac{16y^2}{144} = \frac{144}{144}$$
$$\frac{x^2}{16} + \frac{y^2}{9} = 1$$

C $(0, 0)$; move 4 horiz. and 3 vert.

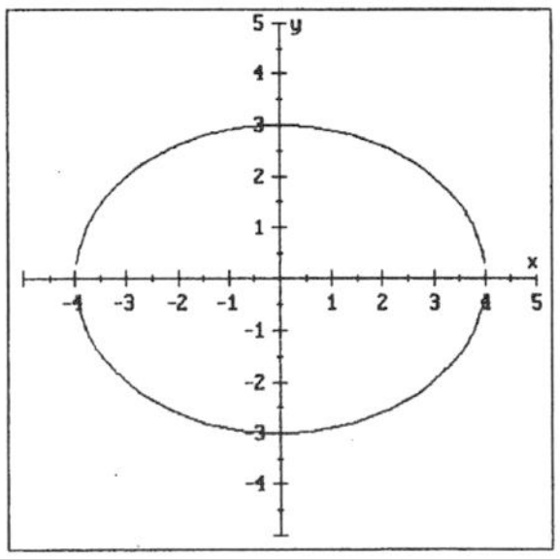

7. $\dfrac{(x-2)^2}{4} + \dfrac{(y-1)^2}{9} = 1$

C $(2, 1)$; move 2 horiz. and 3 vert.

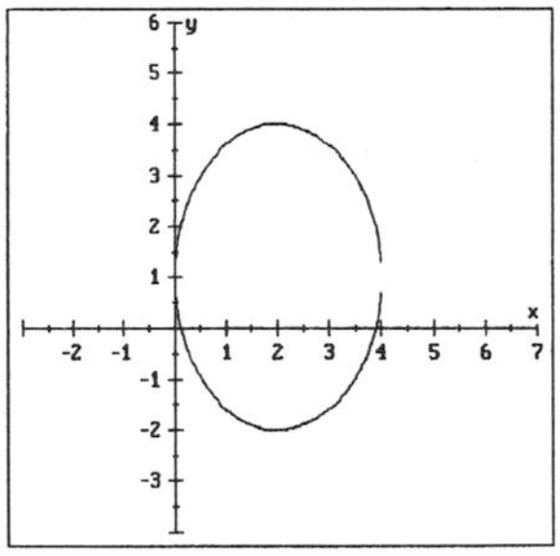

8.

$$4x^2 + 9y^2 + 8x - 18y = 23$$
$$4x^2 + 8x + 9y^2 - 18y = 23$$
$$4(x^2 + 2x) + 9(y^2 - 2y) = 23$$
$$4(x^2 + 2x + 1) + 9(y^2 - 2y + 1) = 23 + 4 + 9$$
$$4(x+1)^2 + 9(y-1)^2 = 36$$
$$\frac{(x+1)^2}{9} + \frac{(y-1)^2}{4} = 1$$

C $(-1, 1)$; move 3 horiz. and 2 vert.

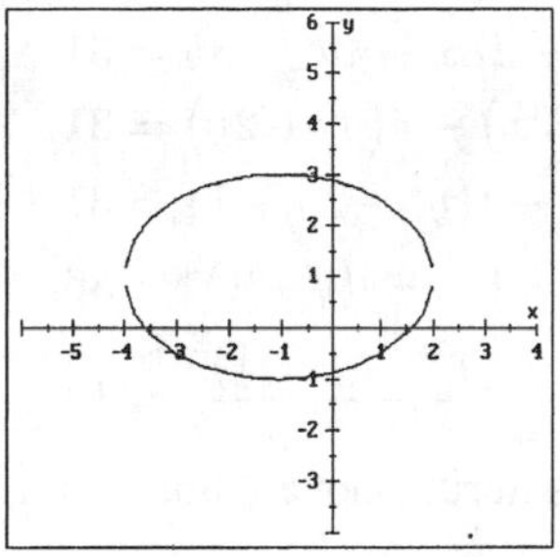

9.

$$9x^2 - y^2 = -9$$
$$\frac{9x^2}{-9} - \frac{y^2}{-9} = \frac{-9}{-9}$$
$$\frac{y^2}{9} - \frac{x^2}{1} = 1$$

C $(0, 0)$; opens vert.; move 1 horiz. and 3 vert.

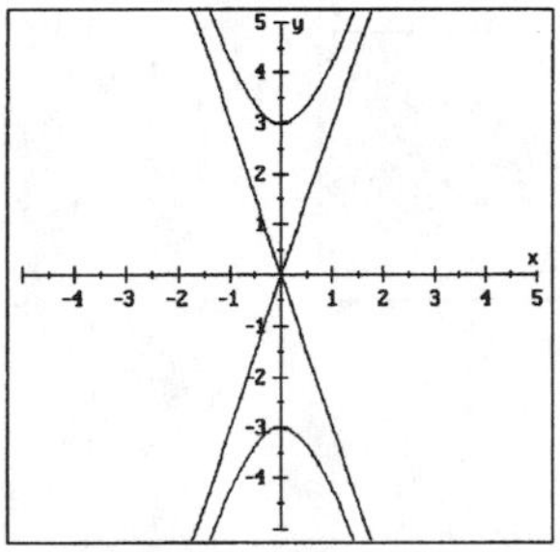

10. $xy = 9$

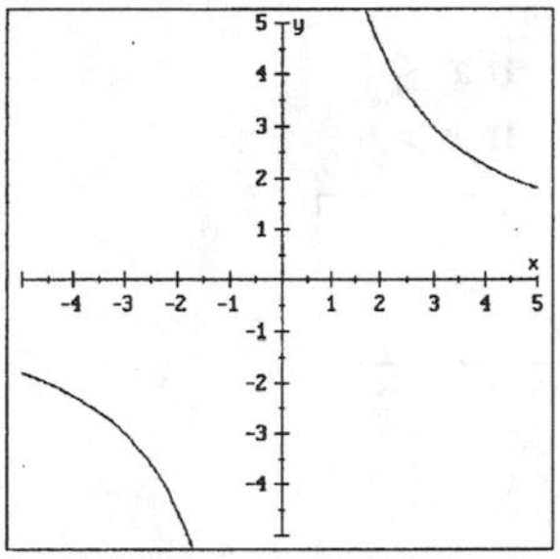

11.

$$4x^2 - 2y^2 + 8x - 8y = 8$$
$$4x^2 + 8x - 2y^2 - 8y = 8$$
$$4(x^2 + 2x) - 2(y^2 + 4y) = 8$$
$$4(x^2 + 2x + 1) - 2(y^2 + 4y + 4) = 8$$
$$4(x+1)^2 - 2(y+2)^2 = 8$$
$$\frac{(x+1)^2}{2} - \frac{(y+2)^2}{4} = 1 \Rightarrow \text{hyperbola}$$

12.

$$\begin{aligned}
9x^2 - 4y^2 - 18x - 8y &= 31 \\
9x^2 - 18x - 4y^2 - 8y &= 31 \\
9(x^2 - 2x) - 4(y^2 + 2y) &= 31 \\
9(x^2 - 2x + 1) - 4(y^2 + 2y + 1) &= 31 + 9 - 4 \\
9(x-1)^2 - 4(y+1)^2 &= 36 \\
\frac{(x-1)^2}{4} - \frac{(y+1)^2}{9} &= 1
\end{aligned}$$

C $(1, -1)$; opens horiz.; move 2 horiz. and 3 vert.

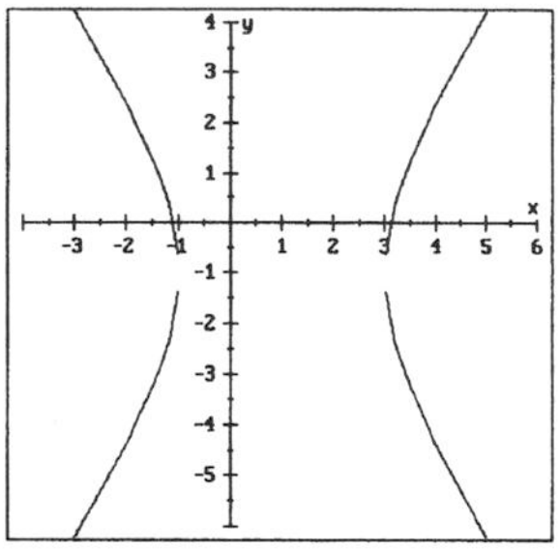

13. increasing on $(-\infty, -2)$; constant on $(-2, 1)$; decreasing on $(1, \infty)$

14. $f(x) = \begin{cases} x & \text{if } x \le 1 \\ -x^2 & \text{if } x > 1 \end{cases}$

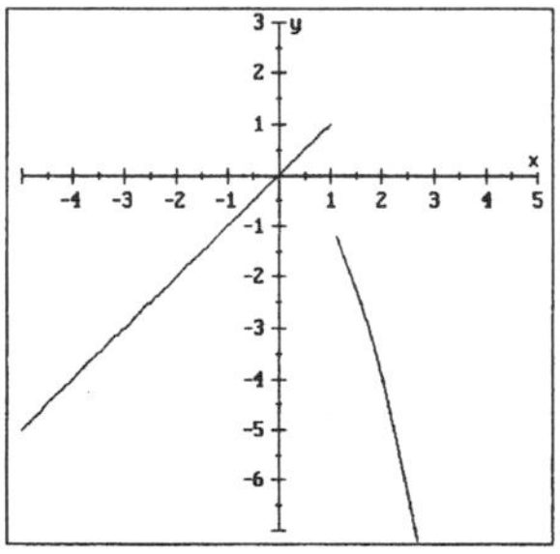

15. $f(x) = 3[[x]]$

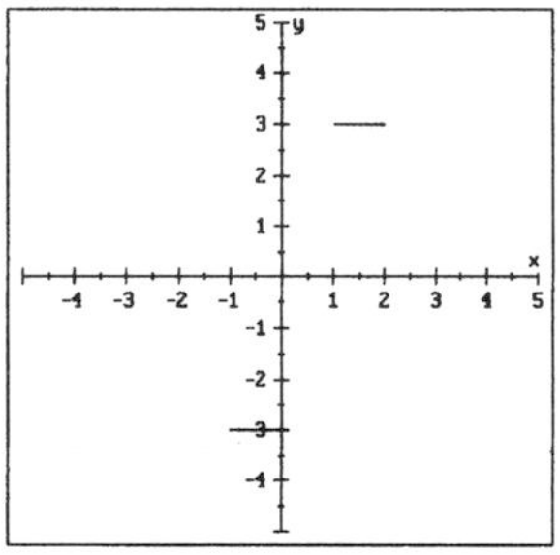

Chapter 12 Test (page 845)

1. $(x-2)^2 + (y+3)^2 = 4$

Center: $(2, -3)$; radius $= 2$

2.

$$\begin{aligned}
x^2 + y^2 + 4x - 6y &= 3 \\
x^2 + 4x + y^2 - 6y &= 3 \\
x^2 + 4x + 4 + y^2 - 6y + 9 &= 3 + 4 + 9 \\
(x+2)^2 + (y-3)^2 &= 16
\end{aligned}$$

Center: $(-2, 3)$; radius $= 4$

3. $(x+1)^2 + (y-2)^2 = 9$

C $(-1, 2)$; $r = \sqrt{9} = 3$

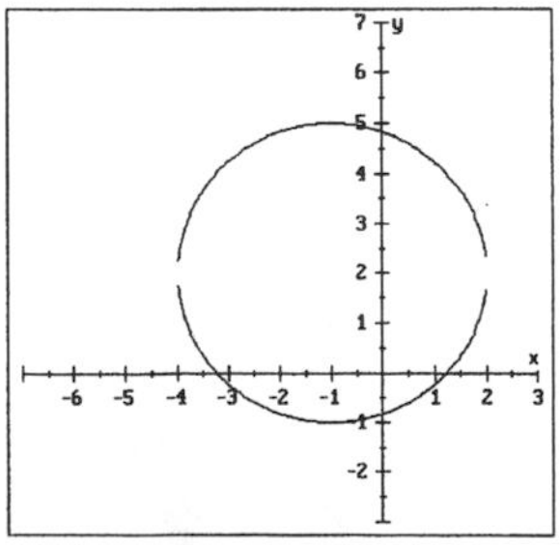

4. $x = (y-2)^2 - 1$

V $(-1, 2)$; opens R

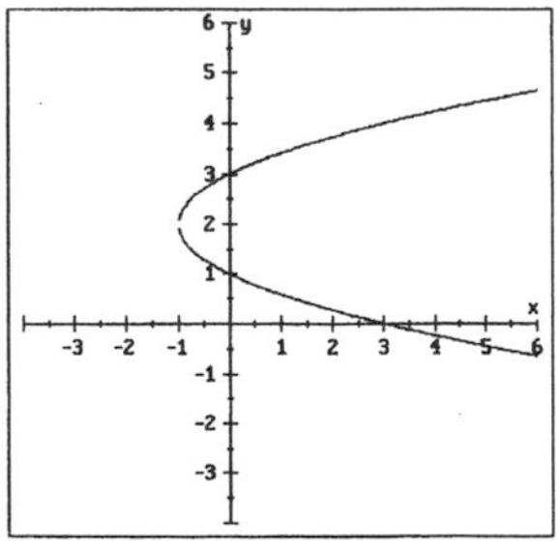

5. $9x^2 + 4y^2 = 36$

$$\frac{9x^2}{36} + \frac{4y^2}{36} = \frac{36}{36}$$

$$\frac{x^2}{4} + \frac{y^2}{9} = 1$$

C $(0,0)$; move 2 horiz. and 3 vert.

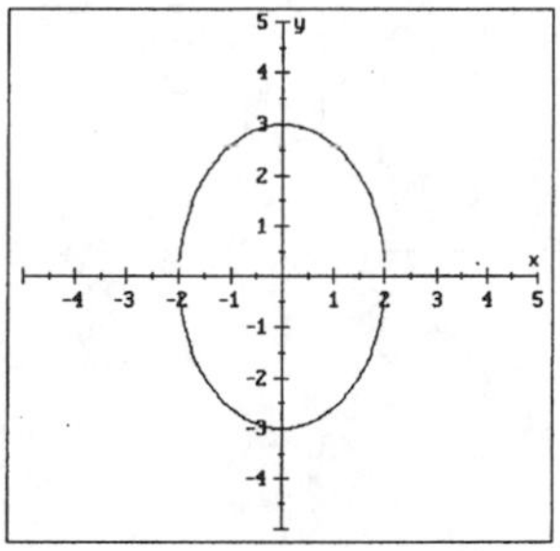

6. $\frac{(x-2)^2}{9} - y^2 = 1$

C $(2,0)$; opens horiz; move 3 horiz and 1 vert

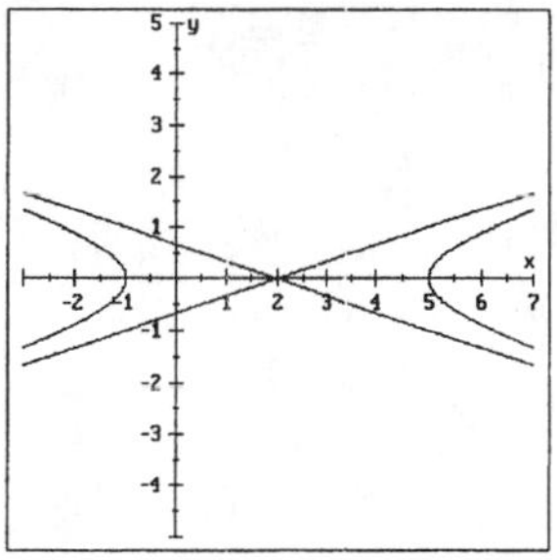

7.

$$4x^2 + y^2 - 24x + 2y = -33$$

$$4(x^2 - 6x) + (y^2 + 2y) = -33$$

$$4(x^2 - 6x + 9) + (y^2 + 2y + 1) = -33 + 36 + 1$$

$$4(x-3)^2 + (y+1)^2 = 4$$

$$\frac{(x-3)^2}{1} + \frac{(y+1)^2}{4} = 1$$

C $(3,-1)$; move 1 horiz. and 2 vert.

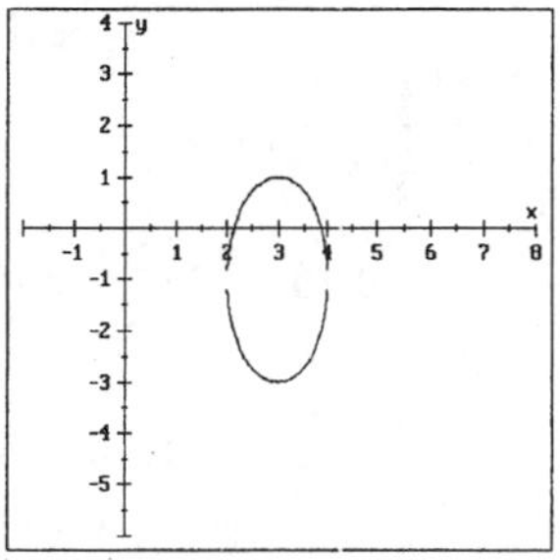

8.

$$x^2 - 9y^2 + 2x + 36y = 44$$

$$x^2 + 2x - 9y^2 + 36y = 44$$

$$(x^2 + 2x) - 9(y^2 - 4y) = 44$$

$$(x^2 + 2x + 1) - 9(y^2 - 4y + 4) = 44 + 1 - 36$$

$$(x+1)^2 - 9(y-2)^2 = 9$$

$$\frac{(x+1)^2}{9} - \frac{(y-2)^2}{1} = 1$$

C $(-1,2)$; opens horiz.; move 3 horiz. and 1 vert.

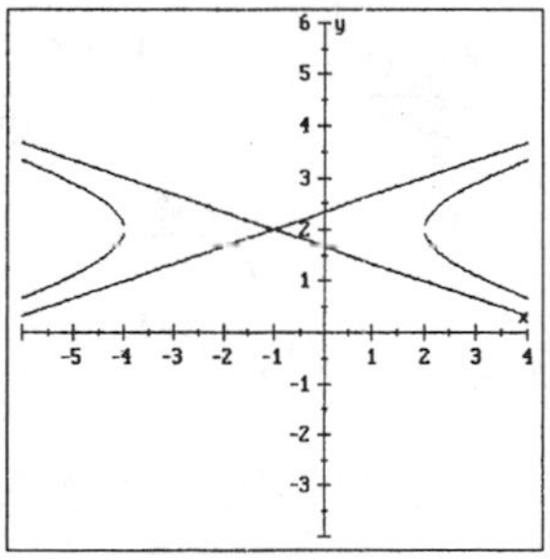

9. increasing: $(-3, 0)$; decreasing: $(0, 3)$

10. $f(x) = \begin{cases} -x^2 & \text{when } x < 0 \\ -x & \text{when } x \geq 0 \end{cases}$

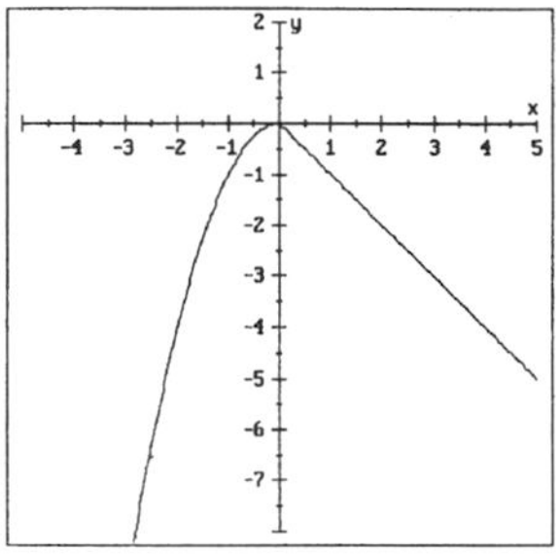

Cumulative Review Exercises (page 846)

1. $(4x - 3y)(3x + y) = 12x^2 + 4xy - 9xy - 3y^2 = 12x^2 - 5xy - 3y^2$

2. $(a^n + 1)(a^n - 3) = a^n a^n - 3a^n + a^n - 3 = a^{2n} - 2a^n - 3$

3. $\dfrac{5a - 10}{a^2 - 4a + 4} = \dfrac{5(a-2)}{(a-2)(a-2)} = \dfrac{5}{a-2}$

4. $\dfrac{a^4 - 5a^2 + 4}{a^2 + 3a + 2} = \dfrac{(a^2-4)(a^2-1)}{(a+2)(a+1)} = \dfrac{(a+2)(a-2)(a+1)(a-1)}{(a+2)(a+1)} = (a-2)(a-1)$
$= a^2 - 3a + 2$

5. $\dfrac{a^2 - a - 6}{a^2 - 4} \div \dfrac{a^2 - 9}{a^2 + a - 6} = \dfrac{a^2 - a - 6}{a^2 - 4} \cdot \dfrac{a^2 + a - 6}{a^2 - 9} = \dfrac{(a-3)(a+2)}{(a+2)(a-2)} \cdot \dfrac{(a+3)(a-2)}{(a+3)(a-3)} = 1$

6.
$$\begin{aligned}
\frac{2}{a-2} + \frac{3}{a+2} - \frac{a-1}{a^2-4} &= \frac{2}{a-2} + \frac{3}{a+2} - \frac{a-1}{(a+2)(a-2)} \\
&= \frac{2(a+2)}{(a-2)(a+2)} + \frac{3(a-2)}{(a+2)(a-2)} - \frac{a-1}{(a+2)(a-2)} \\
&= \frac{2(a+2) + 3(a-2) - (a-1)}{(a+2)(a-2)} \\
&= \frac{2a + 4 + 3a - 6 - a + 1}{(a+2)(a-2)} = \frac{4a-1}{(a+2)(a-2)}
\end{aligned}$$

7.
$$\begin{aligned}
3x - 4y &= 12 \\
-4y &= -3x + 12 \\
y &= \frac{3}{4}x - 3 \\
m &= \frac{3}{4}
\end{aligned}
\qquad
\begin{aligned}
y &= \frac{3}{4}x - 5 \\
m &= \frac{3}{4}
\end{aligned}$$

Parallel

8.
$$\begin{aligned}
y &= 3x + 4 \\
m &= 3
\end{aligned}
\qquad
\begin{aligned}
x &= -3y + 4 \\
3y &= -x + 4 \\
y &= -\frac{1}{3}x + \frac{4}{3} \\
m &= -\frac{1}{3}
\end{aligned}$$

Perpendicular

9.
$$\begin{aligned} y - y_1 &= m(x - x_1) \\ y - 5 &= -2(x - 0) \\ y - 5 &= -2x \\ y &= -2x + 5 \end{aligned}$$

10.
$$m = \frac{y_2 - y_1}{x_2 - x_1} = \frac{4 - (-5)}{-5 - 8} = \frac{9}{-13} = -\frac{9}{13}$$
$$\begin{aligned} y - y_1 &= m(x - x_1) \\ y - (-5) &= -\frac{9}{13}(x - 8) \\ y + 5 &= -\frac{9}{13}x + \frac{72}{13} \\ y &= -\frac{9}{13}x + \frac{7}{13} \end{aligned}$$

11. $2x - 3y < 6$

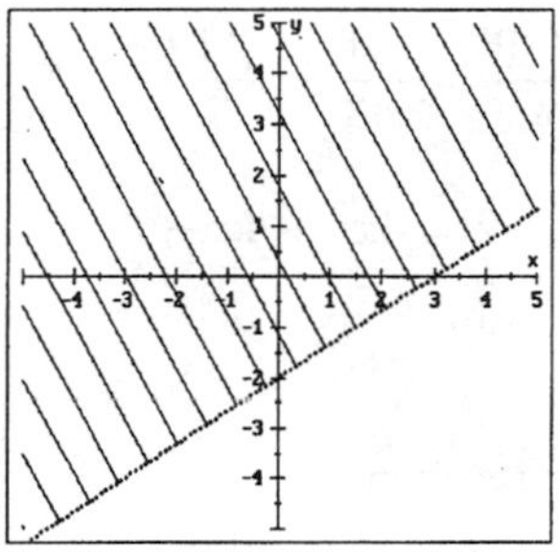

12. $y \geq x^2 - 4$

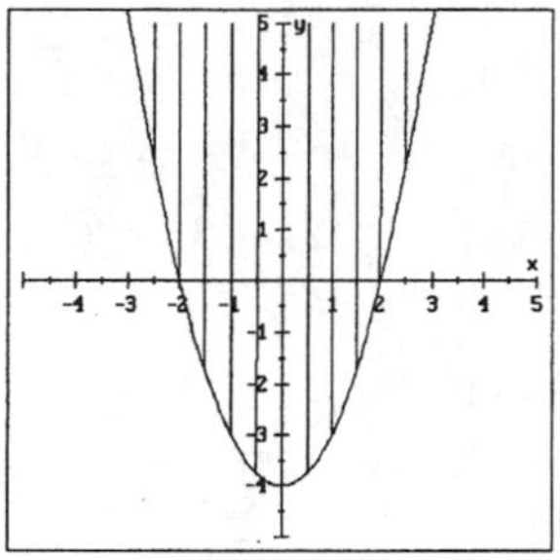

13. $\sqrt{98} + \sqrt{8} - \sqrt{32} = \sqrt{49}\sqrt{2} + \sqrt{4}\sqrt{2} - \sqrt{16}\sqrt{2} = 7\sqrt{2} + 2\sqrt{2} - 4\sqrt{2} = 5\sqrt{2}$

14.
$$\begin{aligned} 12\sqrt[3]{648x^4} + 3\sqrt[3]{81x^4} &= 12\sqrt[3]{216x^3}\sqrt[3]{3x} + 3\sqrt[3]{27x^3}\sqrt[3]{3x} = 12(6x)\sqrt[3]{3x} + 3(3x)\sqrt[3]{3x} \\ &= 72x\sqrt[3]{3x} + 9x\sqrt[3]{3x} = 81x\sqrt[3]{3x} \end{aligned}$$

15.
$$\begin{aligned} \sqrt{3a+1} &= a - 1 \\ \left(\sqrt{3a+1}\right)^2 &= (a-1)^2 \\ 3a + 1 &= a^2 - 2a + 1 \\ 0 &= a^2 - 5a \\ 0 &= a(a-5) \end{aligned}$$

$a = 0$ **or** $a - 5 = 0$

doesn't check $\quad a = 5$

16.
$$\begin{aligned} \sqrt{x+3} - \sqrt{3} &= \sqrt{x} \\ \sqrt{x+3} &= \sqrt{x} + \sqrt{3} \\ \left(\sqrt{x+3}\right)^2 &= \left(\sqrt{x} + \sqrt{3}\right)^2 \\ x + 3 &= x + 2\sqrt{3x} + 3 \\ 0 &= 2\sqrt{3x} \\ 0^2 &= \left(2\sqrt{3x}\right)^2 \\ 0 &= 4(3x) \\ 0 &= 12x \\ 0 &= x \end{aligned}$$

17.

$$6a^2 + 5a - 6 = 0$$
$$(2a + 3)(3a - 2) = 0$$
$$2a + 3 = 0 \quad \textbf{or} \quad 3a - 2 = 0$$
$$a = -\tfrac{3}{2} \qquad\qquad a = \tfrac{2}{3}$$

18.

$$3x^2 + 8x - 1 = 0$$
$$a = 3, b = 8, c = -1$$
$$x = \frac{-b \pm \sqrt{b^2 - 4ac}}{2a}$$
$$= \frac{-8 \pm \sqrt{8^2 - 4(3)(-1)}}{2(3)}$$
$$= \frac{-8 \pm \sqrt{64 + 12}}{6}$$
$$= \frac{-8 \pm \sqrt{76}}{6}$$
$$= -\frac{8}{6} \pm \frac{2\sqrt{19}}{6} = -\frac{4}{3} \pm \frac{\sqrt{19}}{3}$$

19. $(f \circ g)(x) = f(g(x)) = f(2x + 1) = (2x + 1)^2 - 2 = 4x^2 + 4x + 1 - 2 = 4x^2 + 4x - 1$

20.

$$y = 2x^3 - 1$$
$$x = 2y^3 - 1$$
$$x + 1 = 2y^3$$
$$\frac{x + 1}{2} = y^3$$
$$\sqrt[3]{\frac{x + 1}{2}} = y$$
$$y = f^{-1}(x) = \sqrt[3]{\frac{x + 1}{2}}$$

21. $y = \left(\frac{1}{2}\right)^x$

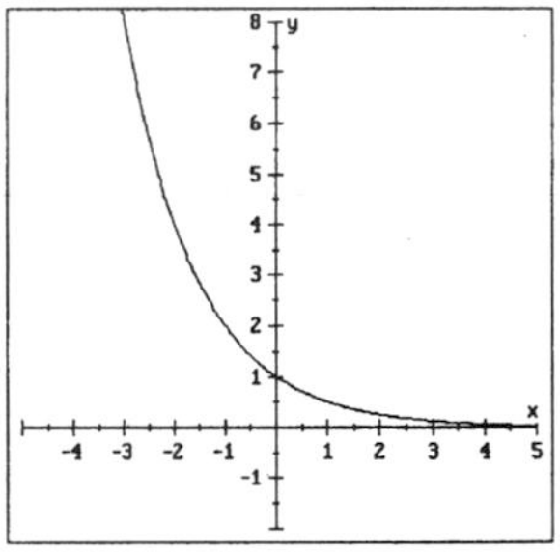

22. $y = \log_2 x \Rightarrow 2^y = x$

23.

$$2^{x+2} = 3^x$$
$$\log 2^{x+2} = \log 3^x$$
$$(x + 2) \log 2 = x \log 3$$
$$x \log 2 + 2 \log 2 = x \log 3$$
$$2 \log 2 = x \log 3 - x \log 2$$
$$2 \log 2 = x(\log 3 - \log 2)$$
$$\frac{2 \log 2}{\log 3 - \log 2} = x$$

24.

$$2 \log 5 + \log x - \log 4 = 2$$
$$\log 5^2 + \log x - \log 4 = 2$$
$$\log \tfrac{25x}{4} = 2$$
$$10^2 = \tfrac{25x}{4}$$
$$400 = 25x$$
$$16 = x$$

25. $x^2 + (y+1)^2 = 9$

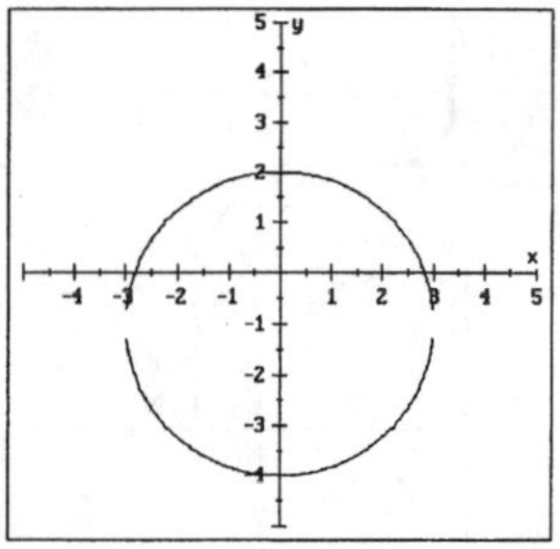

26. $x^2 - 9(y+1)^2 = 9$

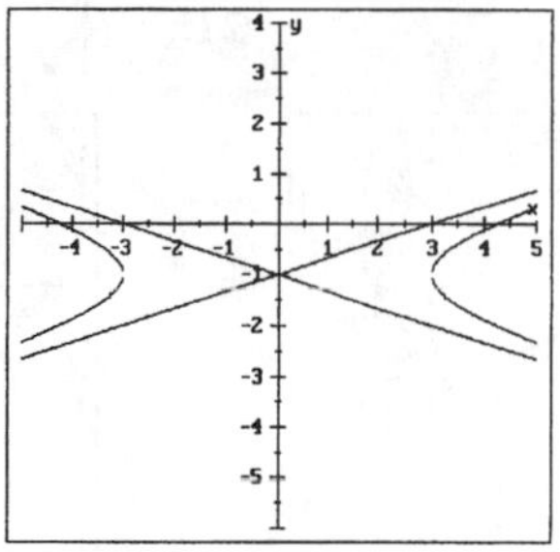

Exercise 13.1 (page 859)

1. $$\left(a^2a^3\right)^2\left(a^4a^2\right)^2 = \left(a^5\right)^2\left(a^6\right)^2 = a^{10}a^{12} = a^{22}$$

3. $$\left(\frac{-3x^3y^4}{x^{-5}y^3}\right)^{-4} = \left(\frac{x^{-5}y^3}{-3x^3y^4}\right)^4 = \left(\frac{1}{-3x^8y}\right)^4 = \frac{1}{81x^{32}y^4}$$

5.
$$A = p + prt$$
$$A - p = prt$$
$$\frac{A-p}{pt} = r, \text{ or } r = \frac{A-p}{pt}$$

7.
$$\frac{1}{r} = \frac{1}{r_1} + \frac{1}{r_2}$$
$$\frac{1}{r} \cdot rr_1r_2 = \left(\frac{1}{r_1} + \frac{1}{r_2}\right)rr_1r_2$$
$$r_1r_2 = rr_2 + rr_1$$
$$r_1r_2 = r(r_2 + r_1)$$
$$\frac{r_1r_2}{r_2 + r_1} = r, \text{ or } r = \frac{r_1r_2}{r_2 + r_1}$$

9. consistent

11. independent

13. $\begin{cases} x - y = 4 \\ 2x + y = 5 \end{cases}$

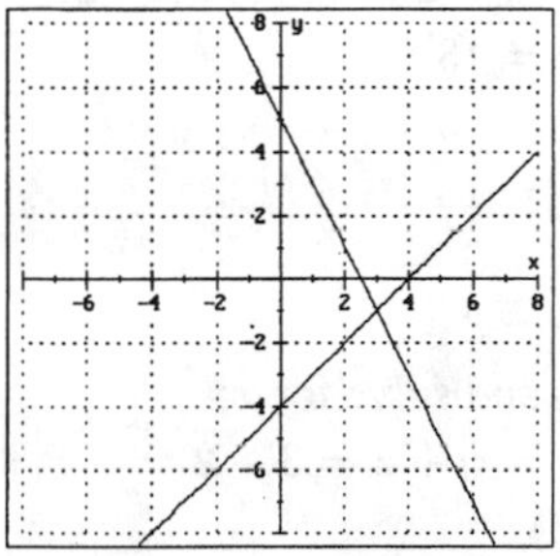

$(3, -1)$ is the solution.

15. $\begin{cases} x = 13 - 4y \\ 3x = 4 + 2y \end{cases}$

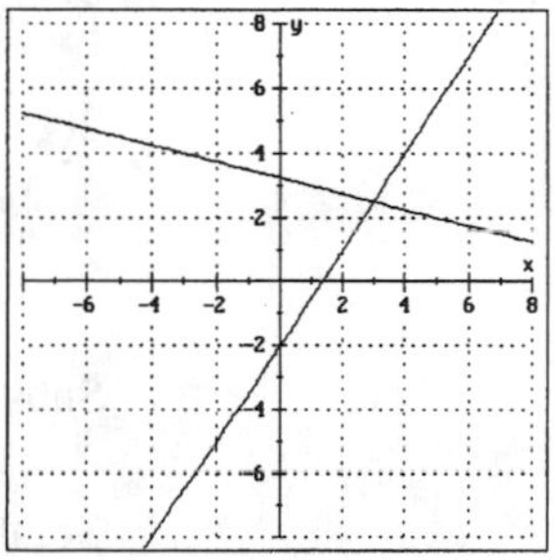

$(3, \frac{5}{2})$ is the solution.

17. $\begin{cases} 2x + 3y = 0 \\ 2x + y - 4 \end{cases}$

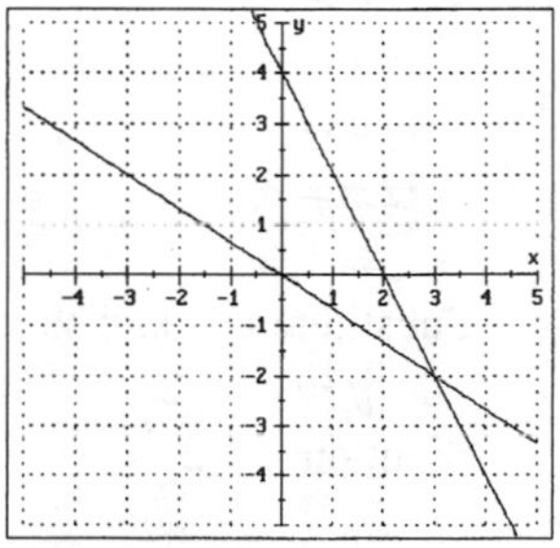

$(3, -2)$ is the solution.

19. $\begin{cases} x = 3 - 2y \\ 2x + 4y = 6 \end{cases}$

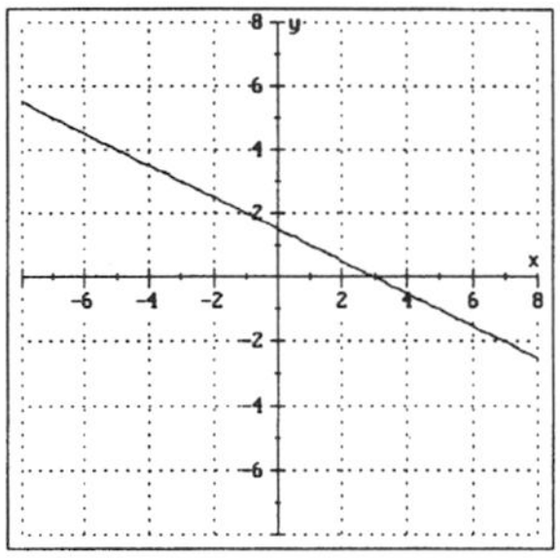

dependent equations

21. $\begin{cases} y = 3 \\ x = 2 \end{cases}$

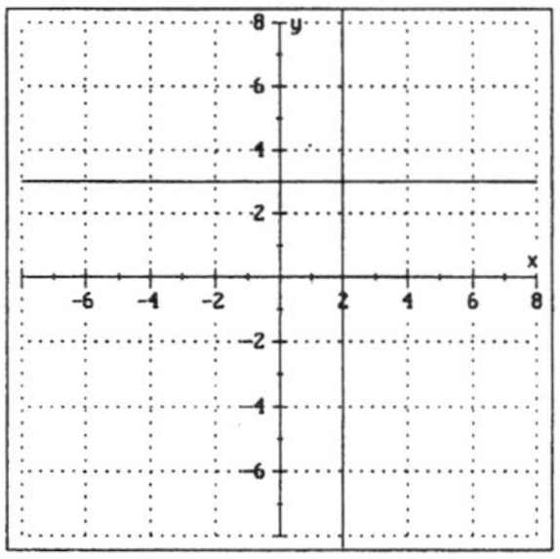

$(2, 3)$ is the solution.

23. $\begin{cases} x = \dfrac{11 - 2y}{3} \\ y = \dfrac{11 - 6x}{4} \end{cases}$

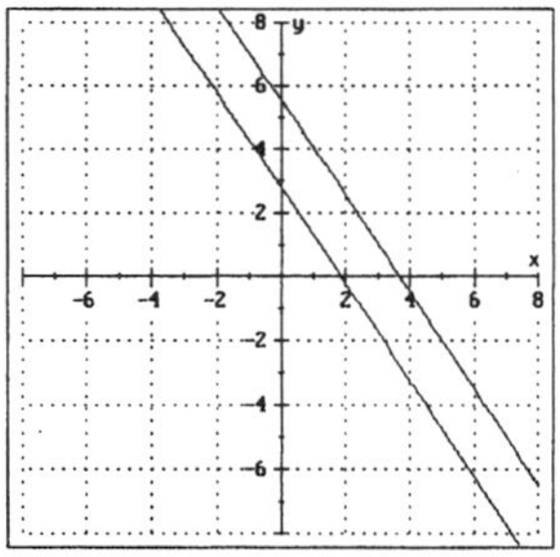

inconsistent system

25. $\begin{cases} \dfrac{5}{2}x + y = \dfrac{1}{2} \\ 2x - \dfrac{3}{2}y = 5 \end{cases}$

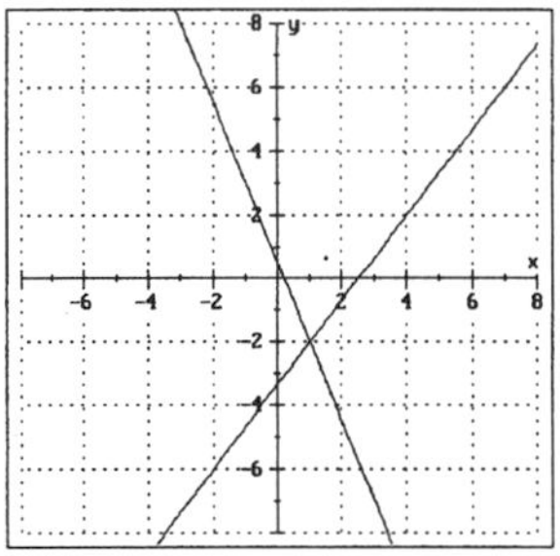

$(1, -2)$ is the solution.

27. $\begin{cases} y = 3.2x - 1.5 \\ y = -2.7x - 3.7 \end{cases}$

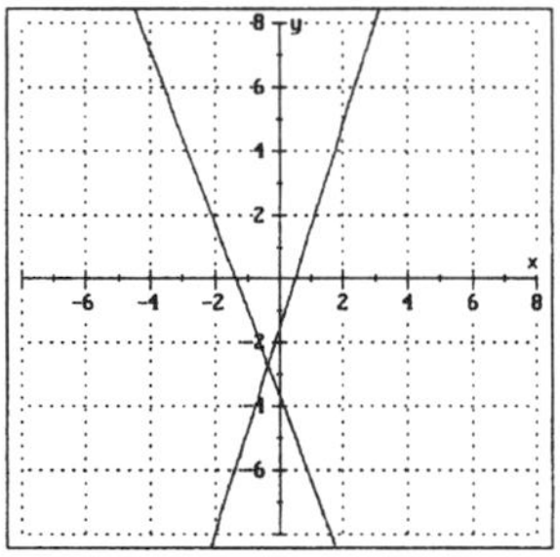

solution: $(-0.37, -2.69)$

29. $\begin{cases} 1.7x + 2.3y = 3.2 \\ y = 0.25x + 8.95 \end{cases}$

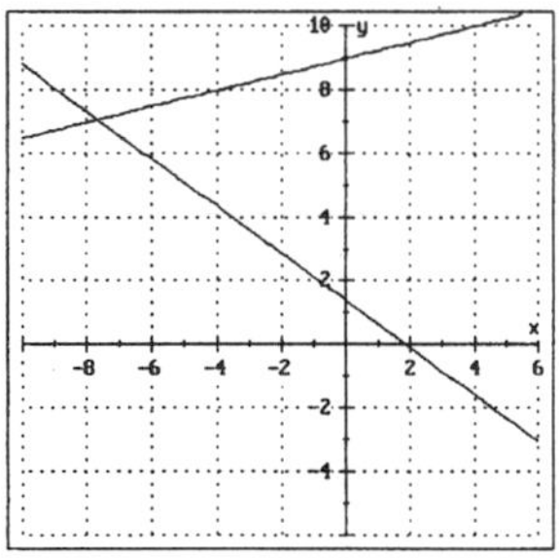

solution: $(-7.64, 7.04)$

31. $\begin{cases} (1) \quad y = x \\ (2) \quad x + y = 4 \end{cases}$

Substitute $y = x$ from (1) into (2):

$$x + \boldsymbol{y} = 4$$
$$x + \boldsymbol{x} = 4$$
$$2x = 4$$
$$x = 2$$

Substitute this and solve for y:

$$y = x = 2$$

Solution: $(2, 2)$

33. $\begin{cases} (1) \quad x - y = 2 \\ (2) \quad 2x + y = 13 \end{cases}$

Substitute $x = y + 2$ from (1) into (2):

$$2\boldsymbol{x} + y = 13$$
$$2(\boldsymbol{y+2}) + y = 13$$
$$2y + 4 + y = 13$$
$$3y = 9$$
$$y = 3$$

Substitute this and solve for x:

$$x = y + 2 = 3 + 2 = 5$$

Solution: $(5, 3)$

35. $\begin{cases} (1) & x + 2y = 6 \\ (2) & 3x - y = -10 \end{cases}$

Substitute $x = -2y + 6$ from (1) into (2):

$$\begin{aligned} 3\boldsymbol{x} - y &= -10 \\ 3(\boldsymbol{-2y+6}) - y &= -10 \\ -6y + 18 - y &= -10 \\ -7y &= -28 \\ y &= 4 \end{aligned}$$

Substitute this and solve for x:

$$\begin{aligned} x &= -2y + 6 \\ x &= -2(4) + 6 \\ x &= -8 + 6 = -2 \end{aligned}$$

Solution: $(-2, 4)$

37. $\begin{cases} (1) & 3x = 2y - 4 \\ (2) & 6x - 4y = -4 \end{cases}$

Substitute $x = \dfrac{2y-4}{3}$ from (1) into (2):

$$\begin{aligned} 6\boldsymbol{x} - 4y &= -4 \\ 6\left(\frac{\boldsymbol{2y-4}}{\boldsymbol{3}}\right) - 4y &= -4 \\ 2(2y-4) - 4y &= -4 \\ 4y - 8 - 4y &= -4 \\ -8 &= -4 \end{aligned}$$

Impossible $\Rightarrow$ no solution

39. $\begin{cases} (1) & 3x - 4y = 9 \\ (2) & x + 2y = 8 \end{cases}$

Substitute $x = -2y + 8$ from (2) into (1):

$$\begin{aligned} 3\boldsymbol{x} - 4y &= 9 \\ 3(\boldsymbol{-2y+8}) - 4y &= 9 \\ -6y + 24 - 4y &= 9 \\ -10y &= -15 \\ y &= \tfrac{3}{2} \end{aligned}$$

Substitute this and solve for x:

$$\begin{aligned} x &= -2y + 8 \\ x &= -2\left(\frac{3}{2}\right) + 8 \\ x &= -3 + 8 = 5 \end{aligned}$$

Solution: $(5, \frac{3}{2})$

41. $\begin{cases} (1) & 2x + 2y = -1 \\ (2) & 3x + 4y = 0 \end{cases}$

Substitute $y = \dfrac{-2x-1}{2}$ from (1) into (2):

$$\begin{aligned} 3x + 4\boldsymbol{y} &= 0 \\ 3x + 4\left(\frac{\boldsymbol{-2x-1}}{\boldsymbol{2}}\right) &= 0 \\ 3x + 2(-2x-1) &= 0 \\ 3x - 4x - 2 &= 0 \\ -x &= 2 \\ x &= -2 \end{aligned}$$

Substitute this and solve for y:

$$\begin{aligned} y &= \frac{-2x-1}{2} \\ y &= \frac{-2(-2)-1}{2} \\ y &= \frac{4-1}{2} = \frac{3}{2} \end{aligned}$$

Solution: $\left(-2, \dfrac{3}{2}\right)$

43.

$$\begin{aligned} x - y &= 3 \\ x + y &= 7 \\ \hline 2x \quad &= 10 \\ x \quad &= 5 \end{aligned}$$

Substitute and solve for y:

$$\begin{aligned} x + y &= 7 \\ 5 + y &= 7 \\ y &= 2 \end{aligned}$$

The solution is $(5, 2)$.

45.

$$\begin{aligned} 2x + y &= -10 \\ 2x - y &= -6 \\ \hline 4x \quad &= -16 \\ x \quad &= -4 \end{aligned}$$

Substitute and solve for y:

$$\begin{aligned} 2x + y &= -10 \\ 2(-4) + y &= -10 \\ -8 + y &= -10 \\ y &= -2 \end{aligned}$$

The solution is $(-4, -2)$.

47.

$$\begin{array}{ll} 2x + 3y = 8 \Rightarrow \times(2) & 4x + 6y = 16 \\ 3x - 2y = -1 \Rightarrow \times(3) & 9x - 6y = -3 \\ & \overline{13x \quad = 13} \\ & x \quad = 1 \end{array}$$

$$\begin{aligned} 2x + 3y &= 8 \\ 2(1) + 3y &= 8 \\ 3y &= 6 \\ y &= 2 \end{aligned}$$

Solution: $\boxed{(1, 2)}$

49.

$$\begin{array}{ll} 4x + 9y = 8 & 4x + 9y = 8 \\ 2x - 6y = -3 \Rightarrow \times(-2) & -4x + 12y = 6 \\ & \overline{21y = 14} \\ & y = \frac{14}{21} = \frac{2}{3} \end{array}$$

$$\begin{aligned} 2x - 6y &= -3 \\ 2x - 6\left(\tfrac{2}{3}\right) &= -3 \\ 2x - 4 &= -3 \\ 2x &= 1 \\ x &= \tfrac{1}{2} \end{aligned}$$

Solution: $\boxed{\left(\frac{1}{2}, \frac{2}{3}\right)}$

51.

$$\begin{array}{lll} 8x - 4y = 16 \Rightarrow & 8x - 4y = 16 \Rightarrow & 8x - 4y = 16 \\ 2x - 4 = y \Rightarrow & 2x - y = 4 \Rightarrow \times(-4) & -8x + 4y = -16 \\ & & \overline{0 = 0} \Rightarrow \boxed{\text{Dependent equations}} \end{array}$$

53.

$$\begin{array}{llll} x = \frac{3}{2}y + 5 \Rightarrow \times(2) & 2x = 3y + 10 \Rightarrow & 2x - 3y = 10 \Rightarrow & 2x - 3y = 10 \\ 2x - 3y = 8 \Rightarrow & 2x - 3y = 8 \Rightarrow & 2x - 3y = 8 \Rightarrow \times(-1) & -2x + 3y = -8 \\ & & & \overline{0 \neq 2} \Rightarrow \boxed{\text{No solution}} \end{array}$$

55.

$$\begin{array}{ll} \frac{x}{2} + \frac{y}{2} = 6 \Rightarrow \times 2 & x + y = 12 \\ \frac{x}{2} - \frac{y}{2} = -2 \Rightarrow \times 2 & x - y = -4 \\ & \overline{2x \quad = 8} \\ & x \quad = 4 \end{array}$$

$$\begin{aligned} x + y &= 12 \\ 4 + y &= 12 \\ y &= 8 \end{aligned}$$

Solution: $\boxed{(4, 8)}$

57.

$$\begin{array}{lll} \frac{3}{4}x + \frac{2}{3}y = 7 \Rightarrow \times 12 & 9x + 8y = 84 \Rightarrow \times 2 & 18x + 16y = 168 \\ \frac{3}{5}x - \frac{1}{2}y = 18 \Rightarrow \times 10 & 6x - 5y = 180 \Rightarrow \times(-3) & -18x + 15y = -540 \\ & & \overline{31y = -372} \\ & & y = -12 \end{array}$$

$$\begin{aligned} 6x - 5y &= 180 \\ 6x - 5(-12) &= 180 \\ 6x &= 120 \\ x &= 20 \end{aligned}$$

Solution: $\boxed{(20, -12)}$

59. $\begin{cases} y < 3x + 2 \\ y < -2x + 3 \end{cases}$

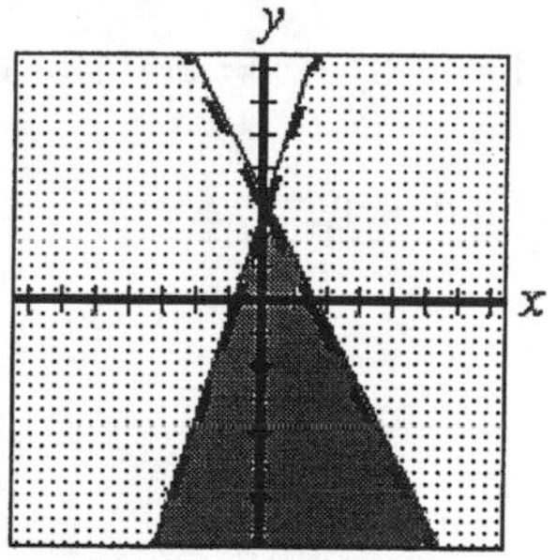

61. $\begin{cases} 3x + 2y > 6 \\ x + 3y \le 2 \end{cases}$

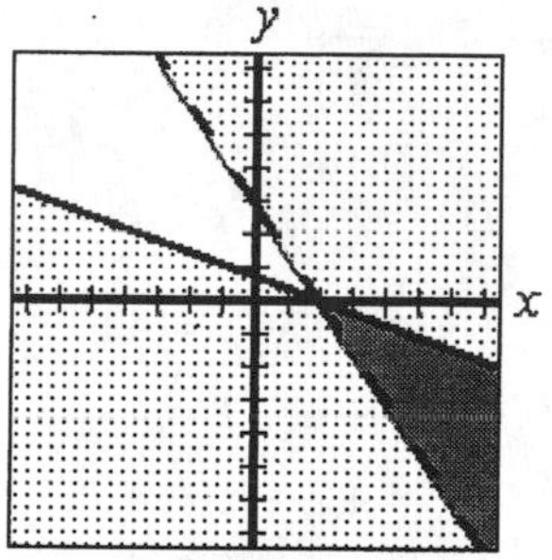

63. **a.** The point (15, 2.0) is on the graph of the cost function, so it costs $2 million to manufacture 15,000 cameras.

b. The point (20, 3.0) is on the graph of the revenue function, so there is a revenue of $3 million when 20,000 cameras are sold.

c. The graphs of the cost function and the revenue function meet at the point (10, 1.5), so the revenue and cost functions are equal for 10,000 cameras.

65. Let $x =$ the cost of the pair of shoes and $y =$ the cost of the sweater.

(1) $x + y = 98$
(2) $y = x + 16$

$$\begin{aligned} x + \boldsymbol{y} &= 98 \\ x + \boldsymbol{x + 16} &= 98 \\ 2x &= 82 \\ x &= 41 \end{aligned} \qquad \begin{aligned} y &= x + 16 \\ y &= 41 + 16 \\ y &= 57 \end{aligned}$$

The sweater cost $57.

67. (1) $R_1 + R_2 = 1375$
(2) $R_1 = R_2 + 125$

$$\begin{aligned} \boldsymbol{R_1} + R_2 &= 1375 \\ \boldsymbol{R_2 + 125} + R_2 &= 1375 \\ 2R_2 &= 1250 \\ R_2 &= 625 \end{aligned} \qquad \begin{aligned} R_1 &= R_2 + 125 \\ R_1 &= 625 + 125 \\ R_1 &= 750 \end{aligned}$$

The resistances are $R_1 = 750$ ohms and $R_2 = 625$ ohms.

69. Let $l =$ the length of the field and $w =$ the width of the field.

$$\begin{aligned} 2w + 2l = 72 &\Rightarrow \times(-1) & -2w - 2l &= -72 \\ \underline{3w + 2l = 88} &\Rightarrow & \underline{3w + 2l} &= \underline{88} \\ & & w &= 16 \end{aligned}$$

$$\begin{aligned} 2w + 2l &= 72 \\ 2(16) + 2l &= 72 \\ 32 + 2l &= 72 \\ 2l &= 40 \\ l &= 20 \end{aligned}$$

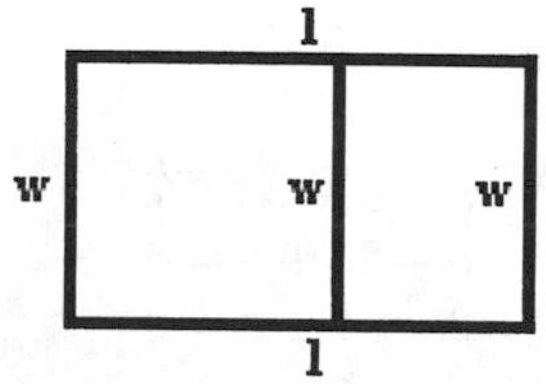

The dimensions of the field are 20 meters by 16 meters.

71-75. Answers may vary.

77. No.

Exercise 13.2 (page 869)

1. $m = \dfrac{\Delta y}{\Delta x} = \dfrac{-4-5}{-2-3} = \dfrac{-9}{-5} = \dfrac{9}{5}$

3. $f(0) = 2(0)^2 + 1 = 2(0) + 1 = 0 + 1 = 1$

5. $f(s) = 2s^2 + 1$

7. plane

9. infinitely

11.

$$\begin{array}{ccc} x - y + z = 2 & 2x + y - z = 4 & 2x - 3y + z = 2 \\ 2 - 1 + 1 \stackrel{?}{=} 2 & 2(2) + 1 - 1 \stackrel{?}{=} 4 & 2(2) - 3(1) + 1 \stackrel{?}{=} 2 \\ 2 = 2 & 4 + 1 - 1 \stackrel{?}{=} 4 & 4 - 3 + 1 \stackrel{?}{=} 2 \\ & 4 = 4 & 2 = 2 \end{array}$$

$(2, 1, 1)$ is a solution to the system.

13.

$$\begin{array}{ll} (1) & x + y + z = 4 \\ (2) & 2x + y - z = 1 \\ (3) & 2x - 3y + z = 1 \end{array} \qquad \begin{array}{ll} (1) & x + y + z = 4 \\ (2) & 2x + y - z = 1 \\ \hline (4) & 3x + 2y \quad = 5 \end{array} \qquad \begin{array}{ll} (2) & 2x + y - z = 1 \\ (3) & 2x - 3y + z = 1 \\ \hline (5) & 4x - 2y \quad = 2 \end{array}$$

$$\begin{array}{ll} (4) & 3x + 2y = 5 \\ (5) & 4x - 2y = 2 \\ \hline & 7x \quad = 7 \\ & x \quad = 1 \end{array} \qquad \begin{array}{r} 3x + 2y = 5 \\ 3(1) + 2y = 5 \\ 3 + 2y = 5 \\ 2y = 2 \\ y = 1 \end{array} \qquad \begin{array}{r} x + y + z = 4 \\ 1 + 1 + z = 4 \\ 2 + z = 4 \\ z = 2 \end{array}$$

$\boxed{\text{The solution is } (1, 1, 2).}$

15.

$$\begin{array}{ll} (1) & 2x + 2y + 3z = 10 \\ (2) & 3x + y - z = 0 \\ (3) & x + y + 2z = 6 \end{array} \qquad \begin{array}{ll} (1) & 2x + 2y + 3z = 10 \\ 3 \cdot (2) & 9x + 3y - 3z = 0 \\ \hline (4) & 11x + 5y \quad = 10 \end{array} \qquad \begin{array}{ll} (3) & x + y + 2z = 6 \\ 2 \cdot (2) & 6x + 2y - 2z = 0 \\ \hline (5) & 7x + 3y \quad = 6 \end{array}$$

$$\begin{array}{l} 11x + 5y = 10 \Rightarrow \times 3 \\ 7x + 3y = 6 \Rightarrow \times (-5) \\ \hline \end{array} \qquad \begin{array}{rcr} 33x + 15y & = & 30 \\ -35x - 15y & = & -30 \\ \hline -2x & = & 0 \\ x & = & 0 \end{array} \qquad \begin{array}{r} 11x + 5y = 10 \\ 11(0) + 5y = 10 \\ 0 + 5y = 10 \\ 5y = 10 \\ y = 2 \end{array} \qquad \begin{array}{r} x + y + 2z = 6 \\ 0 + 2 + 2z = 6 \\ 2 + 2z = 6 \\ 2z = 4 \\ z = 2 \end{array}$$

Solution: $\boxed{(0, 2, 2)}$

17.

$$\begin{array}{ll} (1) & a + b + 2c = 7 \\ (2) & a + 2b + c = 8 \\ (3) & 2a + b + c = 9 \end{array} \qquad \begin{array}{ll} (1) & a + b + 2c = 7 \\ -2 \cdot (2) & -2a - 4b - 2c = -16 \\ \hline (4) & -a - 3b \quad = -9 \end{array} \qquad \begin{array}{ll} (1) & a + b + 2c = 7 \\ -2 \cdot (3) & -4a - 2b - 2c = -18 \\ \hline (5) & -3a - b \quad = -11 \end{array}$$

continued on next page...

17. continued

$$\begin{array}{ll} -a-3b = -9 \Rightarrow \times(-3) & 3a+9b = 27 \\ \underline{-3a - b = -11} \Rightarrow & \underline{-3a - b = -11} \\ & 8b = 16 \\ & b = 2 \end{array}$$

$$\begin{array}{r} -a-3b=-9 \\ -a-3(2)=-9 \\ -a-6=-9 \\ -a=-3 \\ a=3 \end{array} \qquad \begin{array}{r} 2a+b+c=9 \\ 2(3)+2+c=9 \\ 6+2+c=9 \\ 8+c=9 \\ c=1 \end{array}$$

Solution: $\boxed{(3,2,1)}$

19.

$$\begin{array}{ll} (1) & 2x+y-z=1 \\ (2) & x+2y+2z=2 \\ (3) & 4x+5y+3z=3 \end{array} \qquad \begin{array}{ll} (2) & x+2y+2z=2 \\ 2\cdot(1) & \underline{4x+2y-2z=2} \\ (4) & 5x+4y \quad =4 \end{array} \qquad \begin{array}{ll} (3) & 4x+5y+3z=3 \\ 3\cdot(1) & \underline{6x+3y-3z=3} \\ (5) & 10x+8y \quad =6 \end{array}$$

$$\begin{array}{ll} 5x+4y=4 \Rightarrow \times(-2) & -10x-8y=-8 \\ \underline{10x+8y=6} \Rightarrow & \underline{10x+8y=6} \\ & 0=-2 \end{array}$$

Since this equation is always false, there is no solution. The system is inconsistent.

21.

$$\begin{array}{ll} (1) & 4x+3z=4 \\ (2) & 2y-6z=-1 \\ (3) & 8x+4y+3z=9 \end{array} \qquad \begin{array}{ll} (2) & 2y-6z=-1 \\ 2\cdot(1) & \underline{8x \quad +6z=8} \\ (4) & 8x+2y \quad =7 \end{array} \qquad \begin{array}{ll} (2) & 2y-6z=-1 \\ 2\cdot(3) & \underline{16x+8y+6z=18} \\ (5) & 16x+10y \quad =17 \end{array}$$

$$\begin{array}{ll} 8x+2y=7 \Rightarrow \times(-2) & -16x-4y=-14 \\ \underline{16x+10y=17} \Rightarrow & \underline{16x+10y=17} \\ & 6y=3 \\ & y=\frac{1}{2} \end{array}$$

$$\begin{array}{r} 8x+2y=7 \\ 8x+2(\frac{1}{2})=7 \\ 8x+1=7 \\ 8x=6 \\ x=\frac{3}{4} \end{array} \qquad \begin{array}{r} 4x+3z=4 \\ 4(\frac{3}{4})+3z=4 \\ 3+3z=4 \\ 3z=1 \\ z=\frac{1}{3} \end{array}$$

Solution: $\boxed{\left(\frac{3}{4},\frac{1}{2},\frac{1}{3}\right)}$

23.

$$\begin{array}{ll} (1) & 2x+3y+4z=6 \\ (2) & 2x-3y-4z=-4 \\ (3) & 4x+6y+8z=12 \end{array} \qquad \begin{array}{ll} (3) & 4x+6y+8z=12 \\ -2\cdot(1) & \underline{-4x-6y-8z=-12} \\ (4) & 0=0 \end{array}$$

Since equation (4) is always true, the equations are dependent, and there are infinitely many solutions.

25.

$$\begin{array}{lll} x+\frac{1}{3}y+z=13 & \Rightarrow \times 3 & (1)\ 3x+y+3z=39 \\ \frac{1}{2}x-y+\frac{1}{3}z=-2 & \Rightarrow \times 6 & (2)\ 3x-6y+2z=-12 \\ x+\frac{1}{2}y-\frac{1}{3}z=2 & \Rightarrow \times 6 & (3)\ 6x+3y-2z=12 \end{array}$$

$$\begin{array}{ll} (2) & 3x-6y+2z=-12 \\ 6\cdot(1) & \underline{18x+6y+18z=234} \\ (4) & 21x \quad +20z=222 \end{array} \qquad \begin{array}{ll} (2) & 3x-6y+2z=-12 \\ 2\cdot(3) & \underline{12x+6y-4z=24} \\ (5) & 15x \quad -2z=12 \end{array}$$

continued on next page...

25. continued

$$\begin{array}{l} 21x + 20z = 222 \Rightarrow \\ \underline{15x - \ \ 2z = \ \ 12} \Rightarrow \times 10 \end{array} \quad \begin{array}{rl} 21x + 20z &= 222 \\ \underline{150x - 20z} &\underline{= 120} \\ 171x &= 342 \\ x &= 2 \end{array} \quad \begin{array}{r} 15x - 2z = 12 \\ 15(2) - 2z = 12 \\ 30 - 2z = 12 \\ -2z = -18 \\ z = 9 \end{array} \quad \begin{array}{r} 3x + y + 3z = 39 \\ 3(2) + y + 3(9) = 39 \\ 6 + y + 27 = 39 \\ y = 6 \end{array}$$

Solution: $\boxed{(2, 6, 9)}$

27. Let $x =$ the first integer, $y =$ the second integer and $z =$ the third integer.

$$\begin{array}{rcl} x + y + z = 18 & \Rightarrow & (1)\ x + y + z = 18 \\ z = 4y & \Rightarrow & (2)\ -4y + z = 0 \\ y = x + 6 & \Rightarrow & (3)\ -x + y = 6 \end{array} \quad \begin{array}{ll} (1) & x + \ y + z = 18 \\ (3) & \underline{-x + \ y \qquad = \ \ 6} \\ (4) & \qquad 2y + z = 24 \end{array}$$

$$\begin{array}{l} 2y + z = 24 \Rightarrow \times 2 \\ \underline{-4y + z = \ \ 0} \Rightarrow \end{array} \quad \begin{array}{r} 4y + 2z = 48 \\ \underline{-4y + \ \ z = \ \ 0} \\ 3z = 48 \\ z = 16 \end{array} \quad \begin{array}{r} 2y + z = 24 \\ 2y + 16 = 24 \\ 2y = 8 \\ y = 4 \end{array} \quad \begin{array}{r} x + y + z = 18 \\ x + 4 + 16 = 18 \\ x = -2 \end{array}$$

The integers are $-2, 4$ and 16.

29. Let A, B and C represent the measures of the three angles.

$$\begin{array}{rcl} A + B + C = 180 & \Rightarrow & (1)\ A + B + C = 180 \\ A = B + C - 100 & \Rightarrow & (2)\ A - B - C = -100 \\ C = 2B - 40 & \Rightarrow & (3)\ -2B + C = -40 \end{array} \quad \begin{array}{ll} (1) & A + B + C = \ \ 180 \\ (2) & \underline{A - B - C = -100} \\ (4) & 2A \qquad\quad = \ \ 80 \\ & A \qquad\quad\ = \ \ 40 \end{array}$$

$$\begin{array}{r} A + B + C = 180 \\ 40 + B + C = 180 \\ (5)\quad B + C = 140 \end{array} \quad \begin{array}{rr} (-1)\cdot 3 & 2B - C = \ \ 40 \\ \underline{(5)} & \underline{B + C = 140} \\ & 3B \qquad = 180 \\ & B \qquad = \ \ 60 \end{array} \quad \begin{array}{r} B + C = 140 \\ 60 + C = 140 \\ C = 80 \end{array}$$

The angles have measures of 40°, 60° and 80°.

31. Let $A =$ the units of food A, $B =$ the units of food B and $C =$ the units of food C.

$$\begin{array}{lrl} (1) & A + 2B + 2C = 11 & \text{(fat)} \\ (2) & A + B + C = 6 & \text{(carbohydrate)} \\ (3) & 2A + B + 2C = 10 & \text{(protein)} \end{array} \quad \begin{array}{rl} (1) & A + 2B + 2C = \ \ 11 \\ -2\cdot(2) & \underline{-2A - 2B - 2C = -12} \\ (4) & -A \qquad\qquad\quad = \ -1 \\ & A \qquad\qquad\qquad = \ \ \ 1 \end{array}$$

$$\begin{array}{rl} (1) & A + 2B + 2C = \ \ 11 \\ -2\cdot(3) & \underline{-4A - 2B - 4C = -20} \\ (5) & -3A \qquad - 2C = \ -9 \end{array} \quad \begin{array}{r} -3A - 2C = -9 \\ -3(1) - 2C = -9 \\ -3 - 2C = -9 \\ -2C = -6 \\ C = 3 \end{array} \quad \begin{array}{r} A + B + C = 6 \\ 1 + B + 3 = 6 \\ B + 4 = 6 \\ B = 2 \end{array}$$

1 unit of food A, 2 units of food B and 3 units of food C should be used.

33. Let $x =$ the number of \$5 statues, $y =$ the number of \$4 statues and $z =$ the number of \$3 statues.

$$\begin{array}{lrl} (1) & x + y + z = 180 & \text{(total number made)} \\ (2) & 5x + 4y + 3z = 650 & \text{(total cost)} \\ (3) & 20x + 12y + 9z = 2100 & \text{(total revenue)} \end{array} \quad \begin{array}{rl} -3\cdot(1) & -3x - 3y - 3z = -540 \\ (2) & 5x + 4y + 3z = 650 \\ \hline (4) & 2x + y = 110 \end{array}$$

$$\begin{array}{rl} -9\cdot(1) & -9x - 9y - 9z = -1620 \\ (3) & 20x + 12y + 9z = 2100 \\ \hline (5) & 11x + 3y = 480 \end{array} \quad \begin{array}{l} 2x + y = 110 \Rightarrow \times(-3) \\ \underline{11x + 3y = 480 \Rightarrow} \end{array} \quad \begin{array}{l} -6x - 3y = -330 \\ 11x + 3y = 480 \\ \hline 5x = 150 \\ x = 30 \end{array}$$

$$\begin{array}{r} 2x + y = 110 \\ 2(30) + y = 110 \\ 60 + y = 110 \\ y = 50 \end{array} \quad \begin{array}{r} x + y + z = 180 \\ 30 + 50 + z = 180 \\ z = 100 \end{array}$$

30 of the \$5, 50 of the \$4 and 100 of the \$3 statues should be made.

35. Let $x =$ the number of \$5 tickets, $y =$ the number of \$3 tickets and $z =$ the number of \$2 tickets.

$$\begin{array}{lrl} (1) & x + y + z = 750 & \text{(total sold)} \\ (2) & x = 2z \Rightarrow x - 2z = 0 & \text{(twice as many)} \\ (3) & 5x + 3y + 2z = 2625 & \text{(total revenue)} \end{array} \quad \begin{array}{rl} 2\cdot(1) & 2x + 2y + 2z = 1500 \\ (2) & x - 2z = 0 \\ \hline (4) & 3x + 2y = 1500 \end{array}$$

$$\begin{array}{rl} (2) & x - 2z = 0 \\ (3) & 5x + 3y + 2z = 2625 \\ \hline (5) & 6x + 3y = 2625 \end{array} \quad \begin{array}{l} 3x + 2y = 1500 \Rightarrow \times(-2) \\ \underline{6x + 3y = 2625 \Rightarrow} \end{array} \quad \begin{array}{r} -6x - 4y = -3000 \\ 6x + 3y = 2625 \\ \hline -y = -375 \\ y = 375 \end{array}$$

$$\begin{array}{r} 3x + 2y = 1500 \\ 3x + 2(375) = 1500 \\ 3x + 750 = 1500 \\ 3x = 750 \\ x = 250 \end{array} \quad \begin{array}{r} x + y + z = 750 \\ 250 + 375 + z = 750 \\ 625 + z = 750 \\ z = 125 \end{array}$$

250 of the \$5, 375 of the \$3 and 125 of the \$2 tickets were sold.

37. Let $x =$ the number of totem poles, $y =$ the number of bears and $z =$ the number of deer.

$$\begin{array}{lrl} (1) & 2x + 2y + z = 14 & \text{(carving)} \\ (2) & x + 2y + 2z = 15 & \text{(sanding)} \\ (3) & 3x + 2y + 2z = 21 & \text{(painting)} \end{array} \quad \begin{array}{rl} -2\cdot(1) & -4x - 4y - 2z = -28 \\ (2) & x + 2y + 2z = 15 \\ \hline (4) & -3x - 2y = -13 \end{array}$$

$$\begin{array}{rl} -2\cdot(1) & -4x - 4y - 2z = -28 \\ (3) & 3x + 2y + 2z = 21 \\ \hline (5) & -x - 2y = -7 \end{array} \quad \begin{array}{l} -3x - 2y = -13 \Rightarrow \times(-1) \\ \underline{-x - 2y = -7 \Rightarrow} \end{array} \quad \begin{array}{r} 3x + 2y = 13 \\ -x - 2y = -7 \\ \hline 2x = 6 \\ x = 3 \end{array}$$

$$\begin{array}{r} 3x + 2y = 13 \\ 3(3) + 2y = 13 \\ 9 + 2y = 13 \\ 2y = 4 \\ y = 2 \end{array} \quad \begin{array}{r} 2x + 2y + z = 14 \\ 2(3) + 2(2) + z = 14 \\ 6 + 4 + z = 14 \\ z = 4 \end{array}$$

3 totem poles, 2 bears and 4 deer should be made.

39. Let $x =$ the % of nitrogen, $y =$ the % of oxygen and $z =$ the % of other gases.

$$\begin{array}{lll} (1) & x+y+z=100 & \\ (2) & x=3(y+z)+12 \Rightarrow & x-3y-3z=12 \\ (3) & z=y-20 \Rightarrow & -y+z=-20 \end{array}$$

$$\begin{array}{lrcl} (1) & x+y+z & = & 100 \\ (3) & -y+z & = & -20 \\ \hline (4) & x+2z & = & 80 \end{array}$$

$$\begin{array}{lrcl} 3\cdot(1) & 3x+3y+3z & = & 300 \\ (2) & x-3y-3z & = & 12 \\ \hline (5) & 4x & = & 312 \\ & x & = & 78 \end{array}$$

$$\begin{aligned} x+2z&=80 \\ 78+2z&=80 \\ 2z&=2 \\ z&=1 \end{aligned} \qquad \begin{aligned} x+y+z&=100 \\ 78+y+1&=100 \\ y+79&=100 \\ y&=21 \end{aligned}$$

The composition is 78% nitrogen, 21% oxygen, and 1% other gases.

41. Substitute the coordinates of each point for x and y in the equation $y = ax^2 + bx + c$.

$$\begin{aligned} y&=ax^2+bx+c \\ 0&=a(0)^2+b(0)+c \\ 0&=c \end{aligned} \qquad \begin{aligned} y&=ax^2+bx+c \\ -4&=a(2)^2+b(2)+c \\ -4&=4a+2b+c \end{aligned} \qquad \begin{aligned} y&=ax^2+bx+c \\ 0&=a(4)^2+b(4)+c \\ 0&=16a+4b+c \end{aligned}$$

Solve the system of equations formed from the three equations:

$$\begin{array}{lrcl} (1) & c & = & 0 \\ (2) & 4a+2b+c & = & -4 \\ (3) & 16a+4b+c & = & 0 \end{array}$$

$$\begin{array}{rcl} 4a+2b & = & -4 \Rightarrow \times(-2) \\ 16a+4b & = & 0 \Rightarrow \\ \hline \end{array} \qquad \begin{array}{rcl} -8a-4b & = & 8 \\ 16a+4b & = & 0 \\ \hline 8a & = & 8 \\ a & = & 1 \end{array}$$

$$\begin{aligned} 16a+4b&=0 \\ 16(1)+4b&=0 \\ 16+4b&=0 \\ 4b&=-16 \\ b&=-4 \end{aligned}$$

The equation is $y = x^2 - 4x$.

43. Substitute the coordinates of each point for x and y in the equation $x^2 + y^2 + cx + dy + e = 0$.

$$\begin{aligned} x^2+y^2+cx+dy+e&=0 \\ (1)^2+(3)^2+c(1)+d(3)+e&=0 \\ 1+9+c+3d+e&=0 \\ c+3d+e&=-10 \end{aligned} \qquad \begin{aligned} x^2+y^2+cx+dy+e&=0 \\ (3)^2+(1)^2+c(3)+d(1)+e&=0 \\ 9+1+3c+d+e&=0 \\ 3c+d+e&=-10 \end{aligned}$$

$$\begin{aligned} x^2+y^2+cx+dy+e&=0 \\ (1)^2+(-1)^2+c(1)+d(-1)+e&=0 \\ 1+1+c-d+e&=0 \\ c-d+e&=-2 \end{aligned} \qquad \begin{array}{lrcl} (1) & c+3d+e & = & -10 \\ (2) & 3c+d+e & = & -10 \\ (3) & c-d+e & = & -2 \end{array}$$

$$\begin{array}{lrcl} (1) & c+3d+e & = & -10 \\ -1\cdot(2) & -3c-d-e & = & 10 \\ \hline (4) & -2c+2d & = & 0 \end{array} \qquad \begin{array}{lrcl} (1) & c+3d+e & = & -10 \\ -1\cdot(3) & -c+d-e & = & 2 \\ \hline (5) & 4d & = & -8 \\ & d & = & -2 \end{array}$$

$$\begin{aligned} -2c+2d&=0 \\ -2c+2(-2)&=0 \\ -2c-4&=0 \\ -2c&=4 \\ c&=-2 \end{aligned}$$

$$\begin{aligned} c+3d+e&=-10 \\ -2+3(-2)+e&=-10 \\ -2-6+e&=-10 \\ -8+e&=-10 \\ e&=-2 \end{aligned}$$

The equation is $x^2 + y^2 - 2x - 2y - 2 = 0$.

45. Answers may vary.

47.

$$\begin{array}{ll} (1) & x+y+z+w=3 \\ (2) & x-y-z-w=-1 \\ (3) & x+y-z-w=1 \\ (4) & x+y-z+w=3 \end{array} \qquad \begin{array}{rl} (1)\ x+y+z+w= & 3 \\ (2)\ x-y-z-w= & -1 \\ \hline 2x \qquad\qquad = & 2 \\ x \qquad\qquad = & 1 \end{array} \qquad \begin{array}{rl} (1)\ x+\ y+z+w= & 3 \\ (3)\ x+\ y-z-w= & 1 \\ \hline 2x+2y \qquad = & 4 \end{array}$$

$$\begin{array}{rl} (1) & x+\ y+z+\ w=3 \\ (4) & x+\ y-z+\ w=3 \\ \hline & 2x+2y \quad +2w=6 \end{array} \qquad \begin{aligned} 2x+2y&=4 \\ 2(1)+2y&=4 \\ 2y&=2 \\ y&=1 \end{aligned} \qquad \begin{aligned} 2x+2y+2w&=6 \\ 2(1)+2(1)+2w&=6 \\ 4+2w&=6 \\ 2w&=2 \\ w&=1 \end{aligned} \qquad \begin{aligned} x+y+z+w&=3 \\ 1+1+z+1&=3 \\ 3+z&=3 \\ z&=0 \end{aligned}$$

The solution is $(1,1,0,1)$.

Exercise 13.3 (page 879)

Note: The notation $3R_1 + R_3 \Rightarrow R_2$ means to multiply Row #1 of the previous matrix, add that result to Row #3 of the previous matrix, and write the final result in Row #2 of the current matrix.

1. $93{,}000{,}000 = 9.3 \times 10^7$

3. $63 \times 10^3 = 6.3 \times 10^1 \times 10^3 = 6.3 \times 10^4$

5. matrix

7. 3; columns

9. augmented

11. type 1

13. nonzero

15. $\begin{bmatrix} 2 & 1 & 1 \\ 5 & 4 & 1 \end{bmatrix} \overset{R_2+(-R_1)\Rightarrow R_2}{\Rightarrow} \begin{bmatrix} 2 & 1 & 1 \\ 3 & 3 & \boxed{0} \end{bmatrix}$

17. $\begin{bmatrix} 3 & -2 & 1 \\ -1 & 2 & 4 \end{bmatrix} \overset{2R_2\Rightarrow R_2}{\Rightarrow} \begin{bmatrix} 3 & -2 & 1 \\ -2 & 4 & \boxed{8} \end{bmatrix}$

19. $\left[\begin{array}{cc|c} 1 & 1 & 2 \\ 1 & -1 & 0 \end{array}\right] \overset{R_1+(-R_2)\Rightarrow R_2}{\Rightarrow} \left[\begin{array}{cc|c} 1 & 1 & 2 \\ 0 & 2 & 2 \end{array}\right] \overset{\frac{1}{2}R_2\Rightarrow R_2}{\Rightarrow} \left[\begin{array}{cc|c} 1 & 1 & 2 \\ 0 & 1 & 1 \end{array}\right]$

From R_2, $y=1$. From R_1:

$$\begin{aligned} x+y&=2 \\ x+1&=2 \Rightarrow x=1 \end{aligned}$$

The solution is $(1,1)$.

21. $\left[\begin{array}{cc|c} 1 & 2 & -4 \\ 2 & 1 & 1 \end{array}\right] \overset{-2R_1+R_2\Rightarrow R_2}{\Rightarrow} \left[\begin{array}{cc|c} 1 & 2 & -4 \\ 0 & -3 & 9 \end{array}\right] \overset{-\frac{1}{3}R_2\Rightarrow R_2}{\Rightarrow} \left[\begin{array}{cc|c} 1 & 2 & -4 \\ 0 & 1 & -3 \end{array}\right]$

From R_2, $y=-3$. From R_1:

$$\begin{aligned} x+2y&=-4 \\ x+2(-3)&=-4 \\ x-6&=-4 \Rightarrow x=2 \end{aligned}$$

The solution is $(2,-3)$.

23. $\left[\begin{array}{cc|c} 3 & 4 & -12 \\ 9 & -2 & 6 \end{array}\right] \overset{-3R_1+R_2 \Rightarrow R_2}{\Rightarrow} \left[\begin{array}{cc|c} 3 & 4 & -12 \\ 0 & -14 & 42 \end{array}\right] \overset{-\frac{1}{14}R_2 \Rightarrow R_2}{\Rightarrow} \left[\begin{array}{cc|c} 3 & 4 & -12 \\ 0 & 1 & -3 \end{array}\right]$

From R_2, $y = -3$. From R_1: The solution is $(0, -3)$.

$$\begin{aligned} 3x + 4y &= -12 \\ 3x + 4(-3) &= -12 \\ 3x - 12 &= -12 \\ 3x &= 0 \Rightarrow x = 0 \end{aligned}$$

25. $\left[\begin{array}{cc|c} 5 & -2 & 4 \\ 2 & -4 & -8 \end{array}\right] \overset{\frac{1}{2}R_2 \Rightarrow R_2}{\Rightarrow} \left[\begin{array}{cc|c} 5 & -2 & 4 \\ 1 & -2 & -4 \end{array}\right] \overset{R_1+(-5R_2) \Rightarrow R_2}{\Rightarrow} \left[\begin{array}{cc|c} 5 & -2 & 4 \\ 0 & 8 & 24 \end{array}\right] \overset{\frac{1}{8}R_2 \Rightarrow R_2}{\Rightarrow} \left[\begin{array}{cc|c} 5 & -2 & 4 \\ 0 & 1 & 3 \end{array}\right]$

From R_2, $y = 3$. From R_1: The solution is $(2, 3)$.

$$\begin{aligned} 5x - 2y &= 4 \\ 5x - 2(3) &= 4 \\ 5x &= 10 \Rightarrow x = 2 \end{aligned}$$

27. $\begin{cases} 5a = 24 + 2b \\ 5b = 3a + 16 \end{cases} \Rightarrow \begin{cases} 5a - 2b = 24 \\ -3a + 5b = 16 \end{cases}$

$\left[\begin{array}{cc|c} 5 & -2 & 24 \\ -3 & 5 & 16 \end{array}\right] \overset{2R_2+R_1 \Rightarrow R_1}{\Rightarrow} \left[\begin{array}{cc|c} -1 & 8 & 56 \\ -3 & 5 & 16 \end{array}\right] \overset{-R_1 \Rightarrow R_1}{\Rightarrow} \left[\begin{array}{cc|c} 1 & -8 & -56 \\ -3 & 5 & 16 \end{array}\right] \overset{3R_1+R_2 \Rightarrow R_2}{\Rightarrow} \left[\begin{array}{cc|c} 1 & -8 & -56 \\ 0 & -19 & -152 \end{array}\right]$

$\overset{-\frac{1}{19}R_2 \Rightarrow R_2}{\Rightarrow} \left[\begin{array}{cc|c} 1 & -8 & -56 \\ 0 & 1 & 8 \end{array}\right]$

From R_2, $b = 8$. From R_1: The solution is $(8, 8)$.

$$\begin{aligned} a - 8b &= -56 \\ a - 8(8) &= -56 \\ a - 64 &= -56 \Rightarrow a = 8 \end{aligned}$$

29. $\left[\begin{array}{ccc|c} 1 & 1 & 1 & 6 \\ 1 & 2 & 1 & 8 \\ 1 & 1 & 2 & 9 \end{array}\right] \overset{\substack{-R_1+R_2 \Rightarrow R_2 \\ -R_1+R_3 \Rightarrow R_3}}{\Rightarrow} \left[\begin{array}{ccc|c} 1 & 1 & 1 & 6 \\ 0 & 1 & 0 & 2 \\ 0 & 0 & 1 & 3 \end{array}\right]$

From R_3, $z = 3$. From R_2, $y = 2$. From R_1: The solution is $(1, 2, 3)$.

$$\begin{aligned} x + y + z &= 6 \\ x + 2 + 3 &= 6 \\ x + 5 &= 6 \\ x &= 1 \end{aligned}$$

31.
$$\begin{bmatrix} 2 & 1 & 3 & | & 3 \\ -2 & -1 & 1 & | & 5 \\ 4 & -2 & 2 & | & 2 \end{bmatrix} \overset{\substack{R_1+R_2 \Rightarrow R_2 \\ -2R_1+R_3 \Rightarrow R_3}}{\Rightarrow} \begin{bmatrix} 2 & 1 & 3 & | & 3 \\ 0 & 0 & 4 & | & 8 \\ 0 & -4 & -4 & | & -4 \end{bmatrix} \overset{\substack{-\frac{1}{4}R_3 \Rightarrow R_2 \\ \frac{1}{4}R_2 \Rightarrow R_3}}{\Rightarrow} \begin{bmatrix} 2 & 1 & 3 & | & 3 \\ 0 & 1 & 1 & | & 1 \\ 0 & 0 & 1 & | & 2 \end{bmatrix}$$

From R_3, $z = 2$. From R_2:

$$y + z = 1$$
$$y + 2 = 1$$
$$y = -1$$

From R_1:

$$2x + y + 3z = 3$$
$$2x + (-1) + 3(2) = 3$$
$$2x - 1 + 6 = 3$$
$$2x + 5 = 3$$
$$2x = -2$$
$$x = -1$$

The solution is $(-1, -1, 2)$.

33.
$$\begin{bmatrix} 3 & 1 & -3 & | & 5 \\ 1 & -2 & 4 & | & 10 \\ 1 & 1 & 1 & | & 13 \end{bmatrix} \overset{\substack{-3R_2+R_1 \Rightarrow R_2 \\ -3R_3+R_1 \Rightarrow R_3}}{\Rightarrow} \begin{bmatrix} 3 & 1 & -3 & | & 5 \\ 0 & 7 & -15 & | & -25 \\ 0 & -2 & -6 & | & -34 \end{bmatrix} \overset{\frac{2}{7}R_2+R_3 \Rightarrow R_3}{\Rightarrow} \begin{bmatrix} 3 & 1 & -3 & | & 5 \\ 0 & 7 & -15 & | & -25 \\ 0 & 0 & -\frac{72}{7} & | & -\frac{288}{7} \end{bmatrix}$$

From R_3:

$$-\frac{72}{7}c = -\frac{288}{7}$$
$$-\frac{7}{72}\left(-\frac{72}{7}c\right) = -\frac{7}{72}\left(-\frac{288}{7}\right)$$
$$c = 4$$

From R_2:

$$7b - 15c = -25$$
$$7b - 15(4) = -25$$
$$7b = 35$$
$$b = 5$$

From R_1:

$$3a + b - 3c = 5$$
$$3a + 5 - 3(4) = 5$$
$$3a - 7 = 5$$
$$3a = 12$$
$$a = 4$$

The solution is (4, 5, 4).

35.
$$\begin{bmatrix} 3 & -2 & 4 & | & 4 \\ 1 & 1 & 1 & | & 3 \\ 6 & -2 & -3 & | & 10 \end{bmatrix} \overset{\substack{-3R_2+R_1 \Rightarrow R_2 \\ -2R_1+R_3 \Rightarrow R_3}}{\Rightarrow} \begin{bmatrix} 3 & -2 & 4 & | & 4 \\ 0 & -5 & 1 & | & -5 \\ 0 & 2 & -11 & | & 2 \end{bmatrix} \overset{\frac{2}{5}R_2+R_3 \Rightarrow R_3}{\Rightarrow} \begin{bmatrix} 3 & -2 & 4 & | & 4 \\ 0 & -5 & 1 & | & -5 \\ 0 & 0 & -\frac{53}{5} & | & 0 \end{bmatrix}$$

From R_3, $z = 0$. From R_2:

$$-5y + z = -5$$
$$-5y + 0 = -5$$
$$-5y = -5$$
$$y = 1$$

From R_1:

$$3x - 2y + 4z = 4$$
$$3x - 2(1) + 4(0) = 4$$
$$3x - 2 = 4$$
$$3x = 6$$
$$x = 2$$

The solution is $(2, 1, 0)$.

37.

$$\begin{array}{c}\\ \begin{bmatrix}1 & 1 & | & 3\\3 & -1 & | & 1\\2 & 1 & | & 4\end{bmatrix}\end{array} \begin{array}{c} \\ \Rightarrow \end{array} \begin{array}{c}-3R_1+R_2\Rightarrow R_2\\-2R_1+R_3\Rightarrow R_3\\ \begin{bmatrix}1 & 1 & | & 3\\0 & -4 & | & -8\\0 & -1 & | & -2\end{bmatrix}\end{array} \begin{array}{c} \\ \Rightarrow \end{array} \begin{array}{c}-\frac{1}{4}R_2\Rightarrow R_2\\-4R_3+R_2\Rightarrow R_3\\ \begin{bmatrix}1 & 1 & | & 3\\0 & 1 & | & 2\\0 & 0 & | & 0\end{bmatrix}\end{array}$$

From R_2, $y=2$. From R_1: The solution is $(1,2)$.

$$\begin{aligned} x+y&=3\\ x+2&=3\\ x&=1\end{aligned}$$

39.

$$\begin{array}{c}\\ \begin{bmatrix}2 & -1 & | & 4\\1 & 3 & | & 2\\-1 & -4 & | & -2\end{bmatrix}\end{array} \begin{array}{c} \\ \Rightarrow \end{array} \begin{array}{c}-2R_2+R_1\Rightarrow R_2\\2R_3+R_1\Rightarrow R_3\\ \begin{bmatrix}2 & -1 & | & 4\\0 & -7 & | & 0\\0 & -9 & | & 0\end{bmatrix}\end{array} \begin{array}{c} \\ \Rightarrow \end{array} \begin{array}{c}-\frac{1}{7}R_2\Rightarrow R_2\\-\frac{7}{9}R_2+R_3\Rightarrow R_3\\ \begin{bmatrix}2 & -1 & | & 4\\0 & 1 & | & 0\\0 & 0 & | & 0\end{bmatrix}\end{array}$$

From R_2, $y=0$. From R_1: The solution is $(2,0)$.

$$\begin{aligned} 2x-y&=4\\ 2x-0&=4\\ 2x&=4\\ x&=2\end{aligned}$$

41.

$$\begin{array}{c}\\ \begin{bmatrix}2 & 1 & | & 7\\1 & -1 & | & 2\\-1 & 3 & | & -2\end{bmatrix}\end{array} \begin{array}{c} \\ \Rightarrow \end{array} \begin{array}{c}-2R_2+R_1\Rightarrow R_2\\2R_3+R_1\Rightarrow R_3\\ \begin{bmatrix}2 & 1 & | & 7\\0 & 3 & | & 3\\0 & 7 & | & 3\end{bmatrix}\end{array} \begin{array}{c} \\ \Rightarrow \end{array} \begin{array}{c}\frac{1}{3}R_2\Rightarrow R_2\\-\frac{7}{3}R_2+R_3\Rightarrow R_3\\ \begin{bmatrix}2 & 1 & | & 7\\0 & 1 & | & 1\\0 & 0 & | & -4\end{bmatrix}\end{array}$$

R_3: $0x+0y=-4$, or $0=-4$, which is an impossible equation. NO SOLUTION

43.

$$\begin{array}{c}\\ \begin{bmatrix}1 & 3 & | & 7\\1 & 1 & | & 3\\3 & 1 & | & 5\end{bmatrix}\end{array} \begin{array}{c} \\ \Rightarrow \end{array} \begin{array}{c}-R_2+R_1\Rightarrow R_2\\-3R_1+R_3\Rightarrow R_3\\ \begin{bmatrix}1 & 3 & | & 7\\0 & 2 & | & 4\\0 & -8 & | & -16\end{bmatrix}\end{array} \begin{array}{c} \\ \Rightarrow \end{array} \begin{array}{c}\frac{1}{2}R_2\Rightarrow R_2\\4R_2+R_3\Rightarrow R_3\\ \begin{bmatrix}1 & 3 & | & 7\\0 & 1 & | & 2\\0 & 0 & | & 0\end{bmatrix}\end{array}$$

From R_2, $y=2$. From R_1: The solution is $(1,2)$.

$$\begin{aligned} x+3y&=7\\ x+3(2)&=7\\ x+6&=7\Rightarrow x=1\end{aligned}$$

45. $\left[\begin{array}{ccc|c} 1 & 2 & 3 & -2 \\ -1 & -1 & -2 & 4 \end{array}\right] \overset{R_1+R_2\Rightarrow R_2}{\Rightarrow} \left[\begin{array}{ccc|c} 1 & 2 & 3 & -2 \\ 0 & 1 & 1 & 2 \end{array}\right]$

From R_2:

$$y+z=2$$
$$y=2-z$$

From R_1:

$$x+2y+3z=-2$$
$$x+2(2-z)+3z=-2$$
$$x+4-2z+3z=-2$$
$$x+z=-6$$
$$x=-6-z$$

The solution is $(-6-z, 2-z, z)$.

47. $\left[\begin{array}{ccc|c} 1 & -1 & 0 & 1 \\ 0 & 1 & 1 & 1 \\ 1 & 0 & 1 & 2 \end{array}\right] \overset{-R_1+R_3\Rightarrow R_3}{\Rightarrow} \left[\begin{array}{ccc|c} 1 & -1 & 0 & 1 \\ 0 & 1 & 1 & 1 \\ 0 & 1 & 1 & 1 \end{array}\right] \overset{-R_2+R_3\Rightarrow R_3}{\Rightarrow} \left[\begin{array}{ccc|c} 1 & -1 & 0 & 1 \\ 0 & 1 & 1 & 1 \\ 0 & 0 & 0 & 0 \end{array}\right]$

From R_2:

$$y+z=1$$
$$y=1-z$$

From R_1:

$$x-y=1$$
$$x-(1-z)=1$$
$$x-1+z=1$$
$$x+z=2$$
$$x=2-z$$

The solution is $(2-z, 1-z, z)$.

49. Let $x=$ the measure of the first angle and $y=$ the measure of the second angle. Form and solve this system of equations: $\begin{cases} x+y=90 \\ y=x+46 \end{cases} \Rightarrow \begin{cases} x+y=90 \\ -x+y=46 \end{cases}$

$\left[\begin{array}{cc|c} 1 & 1 & 90 \\ -1 & 1 & 46 \end{array}\right] \overset{R_1+R_2\Rightarrow R_2}{\Rightarrow} \left[\begin{array}{cc|c} 1 & 1 & 90 \\ 0 & 2 & 136 \end{array}\right] \overset{\frac{1}{2}R_2\Rightarrow R_2}{\Rightarrow} \left[\begin{array}{cc|c} 1 & 1 & 90 \\ 0 & 1 & 68 \end{array}\right]$

From R_2, $y=68$.

From R_1:

$$x+y=90$$
$$x+68=90$$
$$x=22$$

The angles have measures of 22° and 68°.

51. Let A, B and C represent the measures of the three angles.

$$\begin{cases} A+B+C=180 \\ B=A+25 \\ C=2A-5 \end{cases} \Rightarrow \begin{cases} A+B+C=180 \\ -A+B=25 \\ -2A+C=-5 \end{cases}$$

$\left[\begin{array}{ccc|c} 1 & 1 & 1 & 180 \\ -1 & 1 & 0 & 25 \\ -2 & 0 & 1 & -5 \end{array}\right] \overset{\substack{R_1+R_2\Rightarrow R_2 \\ 2R_1+R_3\Rightarrow R_3}}{\Rightarrow} \left[\begin{array}{ccc|c} 1 & 1 & 1 & 180 \\ 0 & 2 & 1 & 205 \\ 0 & 2 & 3 & 355 \end{array}\right] \overset{-R_2+R_3\Rightarrow R_3}{\Rightarrow} \left[\begin{array}{ccc|c} 1 & 1 & 1 & 180 \\ 0 & 2 & 1 & 205 \\ 0 & 0 & 2 & 150 \end{array}\right] \overset{\frac{1}{2}R_3\Rightarrow R_3}{\Rightarrow} \left[\begin{array}{ccc|c} 1 & 1 & 1 & 180 \\ 0 & 2 & 1 & 205 \\ 0 & 0 & 1 & 75 \end{array}\right]$

continued on next page...

51. continued

From R_3, $C = 75$.

From R_2:
$$\begin{aligned} 2B + C &= 205 \\ 2B + 75 &= 205 \\ 2B &= 130 \\ B &= 65 \end{aligned}$$

From R_1:
$$\begin{aligned} A + B + C &= 180 \\ A + 65 + 75 &= 180 \\ A + 140 &= 180 \\ A &= 40 \end{aligned}$$

The angles have measures of 40°, 65° and 75°.

53. Plug the coordinates of the points into the general equation to form and solve a system of equations.

$$\begin{aligned} y &= ax^2 + bx + c \\ 1 &= a(0)^2 + b(0) + c \\ 1 &= c \end{aligned} \qquad \begin{aligned} y &= ax^2 + bx + c \\ 2 &= a(1)^2 + b(1) + c \\ 2 &= a + b + c \end{aligned} \qquad \begin{aligned} y &= ax^2 + bx + c \\ 4 &= a(-1)^2 + b(-1) + c \\ 4 &= a - b + c \end{aligned}$$

$$\begin{cases} a + b + c = 2 \\ a - b + c = 4 \\ c = 1 \end{cases} \Rightarrow \left[\begin{array}{ccc|c} 1 & 1 & 1 & 2 \\ 1 & -1 & 1 & 4 \\ 0 & 0 & 1 & 1 \end{array}\right] \overset{-R_2 + R_1 \Rightarrow R_2}{\Rightarrow} \left[\begin{array}{ccc|c} 1 & 1 & 1 & 2 \\ 0 & 2 & 0 & -2 \\ 0 & 0 & 1 & 1 \end{array}\right] \overset{\frac{1}{2}R_2 \Rightarrow R_2}{\Rightarrow} \left[\begin{array}{ccc|c} 1 & 1 & 1 & 2 \\ 0 & 1 & 0 & -1 \\ 0 & 0 & 1 & 1 \end{array}\right]$$

From R_3, $c = 1$.

From R_2, $b = -1$. From R_1:
$$\begin{aligned} a + b + c &= 2 \\ a + (-1) + 1 &= 2 \\ a &= 2 \end{aligned}$$

The equation is $y = 2x^2 - x + 1$.

55. Let $x =$ the measure of the first angle and $y =$ the measure of the second angle. Form and solve this system of equations:
$$\begin{cases} x + y = 180 \\ y = x + 28 \end{cases} \Rightarrow \begin{cases} x + y = 180 \\ -x + y = 28 \end{cases}$$

$$\left[\begin{array}{cc|c} 1 & 1 & 180 \\ -1 & 1 & 28 \end{array}\right] \overset{R_1 + R_2 \Rightarrow R_2}{\Rightarrow} \left[\begin{array}{cc|c} 1 & 1 & 180 \\ 0 & 2 & 208 \end{array}\right] \overset{\frac{1}{2}R_2 \Rightarrow R_2}{\Rightarrow} \left[\begin{array}{cc|c} 1 & 1 & 180 \\ 0 & 1 & 104 \end{array}\right]$$

From R_2, $y = 104$.

From R_1:
$$\begin{aligned} x + y &= 180 \\ x + 104 &= 180 \\ x &= 76 \end{aligned}$$

The angles have measures of 76° and 104°.

57. Let $x =$ the number of nickels, $y =$ the number of dimes, and $z =$ the number of quarters.
$$\begin{cases} x + y + z = 64 \\ 5x + 10y + 25z = 600 \\ 10x + 5y + 25z = 500 \end{cases}$$

$$\left[\begin{array}{ccc|c} 1 & 1 & 1 & 64 \\ 5 & 10 & 25 & 600 \\ 10 & 5 & 25 & 500 \end{array}\right] \overset{\substack{-5R_1 + R_2 \Rightarrow R_2 \\ -10R_1 + R_3 \Rightarrow R_3}}{\Rightarrow} \left[\begin{array}{ccc|c} 1 & 1 & 1 & 64 \\ 0 & 5 & 20 & 280 \\ 0 & -5 & 15 & -140 \end{array}\right] \overset{R_2 + R_3 \Rightarrow R_3}{\Rightarrow} \left[\begin{array}{ccc|c} 1 & 1 & 1 & 64 \\ 0 & 5 & 20 & 280 \\ 0 & 0 & 35 & 140 \end{array}\right] \overset{\substack{\frac{1}{5}R_2 \Rightarrow R_2 \\ \frac{1}{35}R_3 \Rightarrow R_3}}{\Rightarrow} \left[\begin{array}{ccc|c} 1 & 1 & 1 & 64 \\ 0 & 1 & 4 & 56 \\ 0 & 0 & 1 & 4 \end{array}\right]$$

continued on next page...

57. continued

From R_3, $z = 4$.

From R_2,
$$\begin{aligned} y + 4z &= 56 \\ y + 4(4) &= 56 \\ y + 16 &= 56 \\ y &= 40 \end{aligned}$$

From R_3,
$$\begin{aligned} x + y + z &= 64 \\ x + 40 + 4 &= 64 \\ x + 44 &= 64 \\ x &= 20 \end{aligned}$$

There are 20 nickels, 40 dimes, and 4 quarters.

59. Answers may vary.

61. The last equation represents the equation $0x + 0y + 0z = k$, or $0 = k$. If $k = 0$, then the system can be solved. However, if $k \neq 0$, the system will have no solution.

Exercise 13.4 (page 889)

1.
$$\begin{aligned} 3(x+2) - (2-x) &= x - 5 \\ 3x + 6 - 2 + x &= x - 5 \\ 4x + 4 &= x - 5 \\ 3x &= -9 \\ x &= -3 \end{aligned}$$

3.
$$\begin{aligned} \frac{5}{3}(5x+6) - 10 &= 0 \\ 3 \cdot \frac{5}{3}(5x+6) - 3 \cdot 10 &= 3 \cdot 0 \\ 5(5x+6) - 30 &= 0 \\ 25x + 30 - 30 &= 0 \\ 25x &= 0 \\ x &= 0 \end{aligned}$$

5. number

7. $\begin{vmatrix} a_2 & c_2 \\ a_3 & c_3 \end{vmatrix}$

9. $\begin{vmatrix} 3 & 4 \\ 2 & -3 \end{vmatrix}$

11.
$$\begin{aligned} \begin{vmatrix} 2 & 3 \\ -2 & 1 \end{vmatrix} &= 2(1) - 3(-2) \\ &= 2 + 6 = 8 \end{aligned}$$

13.
$$\begin{aligned} \begin{vmatrix} -1 & 2 \\ 3 & -4 \end{vmatrix} &= -1(-4) - 2(3) \\ &= 4 - 6 = -2 \end{aligned}$$

15.
$$\begin{vmatrix} x & y \\ y & x \end{vmatrix} = x(x) - y(y) = x^2 - y^2$$

17.
$$\begin{vmatrix} 1 & 0 & 1 \\ 0 & 1 & 0 \\ 1 & 1 & 1 \end{vmatrix} = 1\begin{vmatrix} 1 & 0 \\ 1 & 1 \end{vmatrix} - 0\begin{vmatrix} 0 & 0 \\ 1 & 1 \end{vmatrix} + 1\begin{vmatrix} 0 & 1 \\ 1 & 1 \end{vmatrix} = 1(1) - 0(0) + 1(-1) = 1 - 0 - 1 = 0$$

19.
$$\begin{aligned} \begin{vmatrix} -1 & 2 & 1 \\ 2 & 1 & -3 \\ 1 & 1 & 1 \end{vmatrix} &= -1\begin{vmatrix} 1 & -3 \\ 1 & 1 \end{vmatrix} - 2\begin{vmatrix} 2 & -3 \\ 1 & 1 \end{vmatrix} + 1\begin{vmatrix} 2 & 1 \\ 1 & 1 \end{vmatrix} = -1(4) - 2(5) + 1(1) = -4 - 10 + 1 \\ &= -13 \end{aligned}$$

21.
$$\begin{aligned} \begin{vmatrix} 1 & -2 & 3 \\ -2 & 1 & 1 \\ -3 & -2 & 1 \end{vmatrix} &= 1\begin{vmatrix} 1 & 1 \\ -2 & 1 \end{vmatrix} - (-2)\begin{vmatrix} -2 & 1 \\ -3 & 1 \end{vmatrix} + 3\begin{vmatrix} -2 & 1 \\ -3 & -2 \end{vmatrix} = 1(3) + 2(1) + 3(7) = 3 + 2 + 21 \\ &= 26 \end{aligned}$$

23. $\begin{vmatrix} 1 & 2 & 3 \\ 4 & 5 & 6 \\ 7 & 8 & 9 \end{vmatrix} = 1\begin{vmatrix} 5 & 6 \\ 8 & 9 \end{vmatrix} - 2\begin{vmatrix} 4 & 6 \\ 7 & 9 \end{vmatrix} + 3\begin{vmatrix} 4 & 5 \\ 7 & 8 \end{vmatrix} = 1(-3) - 2(-6) + 3(-3) = -3 + 12 - 9 = 0$

25. $\begin{vmatrix} a & 2a & -a \\ 2 & -1 & 3 \\ 1 & 2 & -3 \end{vmatrix} = a\begin{vmatrix} -1 & 3 \\ 2 & -3 \end{vmatrix} - 2a\begin{vmatrix} 2 & 3 \\ 1 & -3 \end{vmatrix} + (-a)\begin{vmatrix} 2 & -1 \\ 1 & 2 \end{vmatrix} = a(-3) - 2a(-9) - a(5)$

$= -3a + 18a - 5a = 10a$

27. $\begin{vmatrix} 1 & a & b \\ 1 & 2a & 2b \\ 1 & 3a & 3b \end{vmatrix} = 1\begin{vmatrix} 2a & 2b \\ 3a & 3b \end{vmatrix} - a\begin{vmatrix} 1 & 2b \\ 1 & 3b \end{vmatrix} + b\begin{vmatrix} 1 & 2a \\ 1 & 3a \end{vmatrix} = 1(0) - a(b) + b(a) = 0 - ab + ab = 0$

29. $x = \dfrac{\begin{vmatrix} 6 & 1 \\ 2 & -1 \end{vmatrix}}{\begin{vmatrix} 1 & 1 \\ 1 & -1 \end{vmatrix}} = \dfrac{-6-2}{-1-1} = \dfrac{-8}{-2} = 4$; $y = \dfrac{\begin{vmatrix} 1 & 6 \\ 1 & 2 \end{vmatrix}}{\begin{vmatrix} 1 & 1 \\ 1 & -1 \end{vmatrix}} = \dfrac{2-6}{-2} = \dfrac{-4}{-2} = 2$; solution: $(4, 2)$

31. $x = \dfrac{\begin{vmatrix} 1 & 1 \\ -7 & -2 \end{vmatrix}}{\begin{vmatrix} 2 & 1 \\ 1 & -2 \end{vmatrix}} = \dfrac{-2-(-7)}{-4-1} = \dfrac{5}{-5} = -1$; $y = \dfrac{\begin{vmatrix} 2 & 1 \\ 1 & -7 \end{vmatrix}}{\begin{vmatrix} 2 & 1 \\ 1 & -2 \end{vmatrix}} = \dfrac{-14-1}{-5} = \dfrac{-15}{-5} = 3$

solution: $(-1, 3)$

33. $x = \dfrac{\begin{vmatrix} 0 & 3 \\ -4 & -6 \end{vmatrix}}{\begin{vmatrix} 2 & 3 \\ 4 & -6 \end{vmatrix}} = \dfrac{0-(-12)}{-12-12} = \dfrac{12}{-24} = -\dfrac{1}{2}$; $y = \dfrac{\begin{vmatrix} 2 & 0 \\ 4 & -4 \end{vmatrix}}{\begin{vmatrix} 2 & 3 \\ 4 & -6 \end{vmatrix}} = \dfrac{-8-0}{-24} = \dfrac{-8}{-24} = \dfrac{1}{3}$

solution: $\left(-\frac{1}{2}, \frac{1}{3}\right)$

35. $\begin{cases} y = \dfrac{-2x+1}{3} \\ 3x - 2y = 8 \end{cases} \Rightarrow \begin{cases} 3y = -2x + 1 \\ 3x - 2y = 8 \end{cases} \Rightarrow \begin{cases} 2x + 3y = 1 \\ 3x - 2y = 8 \end{cases}$

$x = \dfrac{\begin{vmatrix} 1 & 3 \\ 8 & -2 \end{vmatrix}}{\begin{vmatrix} 2 & 3 \\ 3 & -2 \end{vmatrix}} = \dfrac{-2-24}{-4-9} = \dfrac{-26}{-13} = 2$; $y = \dfrac{\begin{vmatrix} 2 & 1 \\ 3 & 8 \end{vmatrix}}{\begin{vmatrix} 2 & 3 \\ 3 & -2 \end{vmatrix}} = \dfrac{16-3}{-13} = \dfrac{13}{-13} = -1$; solution: $(2, -1)$

37. $\begin{cases} y = \dfrac{11-3x}{2} \\ x = \dfrac{11-4y}{6} \end{cases} \Rightarrow \begin{cases} 2y = 11-3x \\ 6x = 11-4y \end{cases} \Rightarrow \begin{cases} 3x+2y = 11 \\ 6x+4y = 11 \end{cases}$

$$x = \frac{\begin{vmatrix} 11 & 2 \\ 11 & 4 \end{vmatrix}}{\begin{vmatrix} 3 & 2 \\ 6 & 4 \end{vmatrix}} = \frac{44-22}{12-12} = \frac{22}{0} \Rightarrow \text{denominator} = 0, \text{ numerator} \neq 0 \Rightarrow \text{no solution}$$

39. $\begin{cases} x = \dfrac{5y-4}{2} \\ y = \dfrac{3x-1}{5} \end{cases} \Rightarrow \begin{cases} 2x = 5y-4 \\ 5y = 3x-1 \end{cases} \Rightarrow \begin{cases} 2x-5y = -4 \\ -3x+5y = -1 \end{cases}$

$$x = \frac{\begin{vmatrix} -4 & -5 \\ -1 & 5 \end{vmatrix}}{\begin{vmatrix} 2 & -5 \\ -3 & 5 \end{vmatrix}} = \frac{-20-5}{10-15} = \frac{-25}{-5} = 5;\ y = \frac{\begin{vmatrix} 2 & -4 \\ -3 & -1 \end{vmatrix}}{\begin{vmatrix} 2 & -5 \\ -3 & 5 \end{vmatrix}} = \frac{-2-12}{-5} = \frac{-14}{-5} = \frac{14}{5}$$

solution: $(5, \frac{14}{5})$

Note: In the following problems, D stands for the denominator determinant, while N_x, N_y and N_z stand for the numerator determinants for x, y and z, respectively.

41. $D = \begin{vmatrix} 1 & 1 & 1 \\ 1 & 1 & -1 \\ 1 & -1 & 1 \end{vmatrix} = 1\begin{vmatrix} 1 & -1 \\ -1 & 1 \end{vmatrix} - 1\begin{vmatrix} 1 & -1 \\ 1 & 1 \end{vmatrix} + 1\begin{vmatrix} 1 & 1 \\ 1 & -1 \end{vmatrix} = 1(0) - 1(2) + 1(-2) = -4$

$N_x = \begin{vmatrix} 4 & 1 & 1 \\ 0 & 1 & -1 \\ 2 & -1 & 1 \end{vmatrix} = 4\begin{vmatrix} 1 & -1 \\ -1 & 1 \end{vmatrix} - 1\begin{vmatrix} 0 & -1 \\ 2 & 1 \end{vmatrix} + 1\begin{vmatrix} 0 & 1 \\ 2 & -1 \end{vmatrix} = 4(0) - 1(2) + 1(-2) = -4$

$N_y = \begin{vmatrix} 1 & 4 & 1 \\ 1 & 0 & -1 \\ 1 & 2 & 1 \end{vmatrix} = 1\begin{vmatrix} 0 & -1 \\ 2 & 1 \end{vmatrix} - 4\begin{vmatrix} 1 & -1 \\ 1 & 1 \end{vmatrix} + 1\begin{vmatrix} 1 & 0 \\ 1 & 2 \end{vmatrix} = 1(2) - 4(2) + 1(2) = -4$

$N_z = \begin{vmatrix} 1 & 1 & 4 \\ 1 & 1 & 0 \\ 1 & -1 & 2 \end{vmatrix} = 1\begin{vmatrix} 1 & 0 \\ -1 & 2 \end{vmatrix} - 1\begin{vmatrix} 1 & 0 \\ 1 & 2 \end{vmatrix} + 4\begin{vmatrix} 1 & 1 \\ 1 & -1 \end{vmatrix} = 1(2) - 1(2) + 4(-2) = -8$

$x = \dfrac{N_x}{D} = \dfrac{-4}{-4} = 1;\ y = \dfrac{N_y}{D} = \dfrac{-4}{-4} = 1;\ z = \dfrac{N_z}{D} = \dfrac{-8}{-4} = 2 \Rightarrow$ solution: $(1, 1, 2)$

43. $D = \begin{vmatrix} 1 & 1 & 2 \\ 1 & 2 & 1 \\ 2 & 1 & 1 \end{vmatrix} = 1\begin{vmatrix} 2 & 1 \\ 1 & 1 \end{vmatrix} - 1\begin{vmatrix} 1 & 1 \\ 2 & 1 \end{vmatrix} + 2\begin{vmatrix} 1 & 2 \\ 2 & 1 \end{vmatrix} = 1(1) - 1(-1) + 2(-3) = -4$

$N_x = \begin{vmatrix} 7 & 1 & 2 \\ 8 & 2 & 1 \\ 9 & 1 & 1 \end{vmatrix} = 7\begin{vmatrix} 2 & 1 \\ 1 & 1 \end{vmatrix} - 1\begin{vmatrix} 8 & 1 \\ 9 & 1 \end{vmatrix} + 2\begin{vmatrix} 8 & 2 \\ 9 & 1 \end{vmatrix} = 7(1) - 1(-1) + 2(-10) = -12$

continued on next page...

43. continued

$$N_y = \begin{vmatrix} 1 & 7 & 2 \\ 1 & 8 & 1 \\ 2 & 9 & 1 \end{vmatrix} = 1\begin{vmatrix} 8 & 1 \\ 9 & 1 \end{vmatrix} - 7\begin{vmatrix} 1 & 1 \\ 2 & 1 \end{vmatrix} + 2\begin{vmatrix} 1 & 8 \\ 2 & 9 \end{vmatrix} = 1(-1) - 7(-1) + 2(-7) = -8$$

$$N_z = \begin{vmatrix} 1 & 1 & 7 \\ 1 & 2 & 8 \\ 2 & 1 & 9 \end{vmatrix} = 1\begin{vmatrix} 2 & 8 \\ 1 & 9 \end{vmatrix} - 1\begin{vmatrix} 1 & 8 \\ 2 & 9 \end{vmatrix} + 7\begin{vmatrix} 1 & 2 \\ 2 & 1 \end{vmatrix} = 1(10) - 1(-7) + 7(-3) = -4$$

$x = \dfrac{N_x}{D} = \dfrac{-12}{-4} = 3$; $y = \dfrac{N_y}{D} = \dfrac{-8}{-4} = 2$; $z = \dfrac{N_z}{D} = \dfrac{-4}{-4} = 1 \Rightarrow$ solution: $(3, 2, 1)$

45. $$D = \begin{vmatrix} 2 & 1 & -1 \\ 1 & 2 & 2 \\ 4 & 5 & 3 \end{vmatrix} = 2\begin{vmatrix} 2 & 2 \\ 5 & 3 \end{vmatrix} - 1\begin{vmatrix} 1 & 2 \\ 4 & 3 \end{vmatrix} + (-1)\begin{vmatrix} 1 & 2 \\ 4 & 5 \end{vmatrix} = 2(-4) - 1(-5) - 1(-3) = 0$$

$$N_x = \begin{vmatrix} 1 & 1 & -1 \\ 2 & 2 & 2 \\ 3 & 5 & 3 \end{vmatrix} = 1\begin{vmatrix} 2 & 2 \\ 5 & 3 \end{vmatrix} - 1\begin{vmatrix} 2 & 2 \\ 3 & 3 \end{vmatrix} + (-1)\begin{vmatrix} 2 & 2 \\ 3 & 5 \end{vmatrix} = 1(-4) - 1(0) - 1(4) = -8$$

$x = \dfrac{N_x}{D} = \dfrac{-8}{0} \Rightarrow$ denominator $= 0$, numerator $\neq 0 \Rightarrow$ no solution.

47. $$D = \begin{vmatrix} 2 & 1 & 1 \\ 1 & -2 & 3 \\ 1 & 1 & -4 \end{vmatrix} = 2\begin{vmatrix} -2 & 3 \\ 1 & -4 \end{vmatrix} - 1\begin{vmatrix} 1 & 3 \\ 1 & -4 \end{vmatrix} + 1\begin{vmatrix} 1 & -2 \\ 1 & 1 \end{vmatrix} = 2(5) - 1(-7) + 1(3) = 20$$

$$N_x = \begin{vmatrix} 5 & 1 & 1 \\ 10 & -2 & 3 \\ -3 & 1 & -4 \end{vmatrix} = 5\begin{vmatrix} -2 & 3 \\ 1 & -4 \end{vmatrix} - 1\begin{vmatrix} 10 & 3 \\ -3 & -4 \end{vmatrix} + 1\begin{vmatrix} 10 & -2 \\ -3 & 1 \end{vmatrix}$$
$$= 5(5) - 1(-31) + 1(4) = 60$$

$$N_y = \begin{vmatrix} 2 & 5 & 1 \\ 1 & 10 & 3 \\ 1 & -3 & -4 \end{vmatrix} = 2\begin{vmatrix} 10 & 3 \\ -3 & -4 \end{vmatrix} - 5\begin{vmatrix} 1 & 3 \\ 1 & -4 \end{vmatrix} + 1\begin{vmatrix} 1 & 10 \\ 1 & -3 \end{vmatrix}$$
$$= 2(-31) - 5(-7) + 1(-13) = -40$$

$$N_z = \begin{vmatrix} 2 & 1 & 5 \\ 1 & -2 & 10 \\ 1 & 1 & -3 \end{vmatrix} = 2\begin{vmatrix} -2 & 10 \\ 1 & -3 \end{vmatrix} - 1\begin{vmatrix} 1 & 10 \\ 1 & -3 \end{vmatrix} + 5\begin{vmatrix} 1 & -2 \\ 1 & 1 \end{vmatrix} = 2(-4) - 1(-13) + 5(3) = 20$$

$x = \dfrac{N_x}{D} = \dfrac{60}{20} = 3$; $y = \dfrac{N_y}{D} = \dfrac{-40}{20} = -2$; $z = \dfrac{N_z}{D} = \dfrac{20}{20} = 1 \Rightarrow$ solution: $(3, -2, 1)$

49. $$D = \begin{vmatrix} 2 & 3 & 4 \\ 2 & -3 & -4 \\ 4 & 6 & 8 \end{vmatrix} = 2\begin{vmatrix} -3 & -4 \\ 6 & 8 \end{vmatrix} - 3\begin{vmatrix} 2 & -4 \\ 4 & 8 \end{vmatrix} + 4\begin{vmatrix} 2 & -3 \\ 4 & 6 \end{vmatrix} = 2(0) - 3(32) + 4(24) = 0$$

$$N_x = \begin{vmatrix} 6 & 3 & 4 \\ -4 & -3 & -4 \\ 12 & 6 & 8 \end{vmatrix} = 6\begin{vmatrix} -3 & -4 \\ 6 & 8 \end{vmatrix} - 3\begin{vmatrix} -4 & -4 \\ 12 & 8 \end{vmatrix} + 4\begin{vmatrix} -4 & -3 \\ 12 & 6 \end{vmatrix} = 6(0) - 3(16) + 4(12) = 0$$

$x = \dfrac{N_x}{D} = \dfrac{0}{0} \Rightarrow$ denominator $= 0$, numerator $= 0 \Rightarrow$ dependent equations.

51. $\begin{cases} x+y=1 \Rightarrow & x+y=1 \\ \frac{1}{2}y+z=\frac{5}{2} \Rightarrow & y+2z=5 \\ x-z=-3 \Rightarrow & x-z=-3 \end{cases}$

$$D=\begin{vmatrix} 1 & 1 & 0 \\ 0 & 1 & 2 \\ 1 & 0 & -1 \end{vmatrix}=1\begin{vmatrix} 1 & 2 \\ 0 & -1 \end{vmatrix}-1\begin{vmatrix} 0 & 2 \\ 1 & -1 \end{vmatrix}+0\begin{vmatrix} 0 & 1 \\ 1 & 0 \end{vmatrix}=1(-1)-1(-2)+0(-1)=1$$

$$N_x=\begin{vmatrix} 1 & 1 & 0 \\ 5 & 1 & 2 \\ -3 & 0 & -1 \end{vmatrix}=1\begin{vmatrix} 1 & 2 \\ 0 & -1 \end{vmatrix}-1\begin{vmatrix} 5 & 2 \\ -3 & -1 \end{vmatrix}+0\begin{vmatrix} 5 & 1 \\ -3 & 0 \end{vmatrix}=1(-1)-1(1)+0(3)=-2$$

$$N_y=\begin{vmatrix} 1 & 1 & 0 \\ 0 & 5 & 2 \\ 1 & -3 & -1 \end{vmatrix}=1\begin{vmatrix} 5 & 2 \\ -3 & -1 \end{vmatrix}-1\begin{vmatrix} 0 & 2 \\ 1 & -1 \end{vmatrix}+0\begin{vmatrix} 0 & 5 \\ 1 & -3 \end{vmatrix}=1(1)-1(-2)+0(-5)=3$$

$$N_z=\begin{vmatrix} 1 & 1 & 1 \\ 0 & 1 & 5 \\ 1 & 0 & -3 \end{vmatrix}=1\begin{vmatrix} 1 & 5 \\ 0 & -3 \end{vmatrix}-1\begin{vmatrix} 0 & 5 \\ 1 & -3 \end{vmatrix}+1\begin{vmatrix} 0 & 1 \\ 1 & 0 \end{vmatrix}=1(-3)-1(-5)+1(-1)=1$$

$$x=\frac{N_x}{D}=\frac{-2}{1}=-2;\ y=\frac{N_y}{D}=\frac{3}{1}=3;\ z=\frac{N_z}{D}=\frac{1}{1}=1 \Rightarrow \text{solution: } (-2,3,1)$$

53. $\begin{cases} 2x-y+4z+2=0 \Rightarrow & 2x-y+4z=-2 \\ 5x+8y+7z=-8 \Rightarrow & 5x+8y+7z=-8 \\ x+3y+z+3=0 \Rightarrow & x+3y+z=-3 \end{cases}$

$$D=\begin{vmatrix} 2 & -1 & 4 \\ 5 & 8 & 7 \\ 1 & 3 & 1 \end{vmatrix}=2\begin{vmatrix} 8 & 7 \\ 3 & 1 \end{vmatrix}-(-1)\begin{vmatrix} 5 & 7 \\ 1 & 1 \end{vmatrix}+4\begin{vmatrix} 5 & 8 \\ 1 & 3 \end{vmatrix}=2(-13)+1(-2)+4(7)=0$$

$$N_x=\begin{vmatrix} -2 & -1 & 4 \\ -8 & 8 & 7 \\ -3 & 3 & 1 \end{vmatrix}=-2\begin{vmatrix} 8 & 7 \\ 3 & 1 \end{vmatrix}-(-1)\begin{vmatrix} -8 & 7 \\ -3 & 1 \end{vmatrix}+4\begin{vmatrix} -8 & 8 \\ -3 & 3 \end{vmatrix}=-2(-13)+1(13)+4(0)$$

$$=39$$

$$x=\frac{N_x}{D}=\frac{39}{0} \Rightarrow \text{denominator}=0,\ \text{numerator}\neq 0 \Rightarrow \text{no solution.}$$

55. $\begin{vmatrix} x & 1 \\ 3 & 2 \end{vmatrix}=1$

$2x-3=1$

$2x=4$

$x=2$

57. $\begin{vmatrix} x & -2 \\ 3 & 1 \end{vmatrix}=\begin{vmatrix} 4 & 2 \\ x & 3 \end{vmatrix}$

$x-(-6)=12-2x$

$x+6=12-2x$

$3x=6$

$x=2$

59. $\begin{cases} 2x+y=180 \\ y=x+30 \end{cases} \Rightarrow \begin{cases} 2x+y=180 \\ -x+y=30 \end{cases}$

$$x=\frac{\begin{vmatrix} 180 & 1 \\ 30 & 1 \end{vmatrix}}{\begin{vmatrix} 2 & 1 \\ -1 & 1 \end{vmatrix}}=\frac{180-30}{2-(-1)}=\frac{150}{3}=50;\ y=\frac{\begin{vmatrix} 2 & 180 \\ -1 & 30 \end{vmatrix}}{\begin{vmatrix} 2 & 1 \\ -1 & 1 \end{vmatrix}}=\frac{60-(-180)}{3}=\frac{240}{3}=80$$

61. Let $x =$ the amount invested in HiTech, $y =$ the amount invested in SaveTel and $z =$ the amount invested in HiGas. Form and solve the following system of equations:

$$\begin{cases} x+y+z=20000 \Rightarrow & x+y+z=20000 \\ 0.10x+0.05y+0.06z=0.066(20000) \Rightarrow & 10x+5y+6x=132000 \\ y+z=3x \Rightarrow & -3x+y+z=0 \end{cases}$$

$$D = \begin{vmatrix} 1 & 1 & 1 \\ 10 & 5 & 6 \\ -3 & 1 & 1 \end{vmatrix} = 1\begin{vmatrix} 5 & 6 \\ 1 & 1 \end{vmatrix} - 1\begin{vmatrix} 10 & 6 \\ -3 & 1 \end{vmatrix} + 1\begin{vmatrix} 10 & 5 \\ -3 & 1 \end{vmatrix} = 1(-1) - 1(28) + 1(25) = -4$$

$$N_x = \begin{vmatrix} 20000 & 1 & 1 \\ 132000 & 5 & 6 \\ 0 & 1 & 1 \end{vmatrix} = 20000\begin{vmatrix} 5 & 6 \\ 1 & 1 \end{vmatrix} - 1\begin{vmatrix} 132000 & 6 \\ 0 & 1 \end{vmatrix} + 1\begin{vmatrix} 132000 & 5 \\ 0 & 1 \end{vmatrix}$$

$$= 20000(-1) - 1(132000) + 1(132000) = -20000$$

$$N_y = \begin{vmatrix} 1 & 20000 & 1 \\ 10 & 132000 & 6 \\ -3 & 0 & 1 \end{vmatrix} = 1\begin{vmatrix} 132000 & 6 \\ 0 & 1 \end{vmatrix} - 20000\begin{vmatrix} 10 & 6 \\ -3 & 1 \end{vmatrix} + 1\begin{vmatrix} 10 & 132000 \\ -3 & 0 \end{vmatrix}$$

$$= 1(132000) - 20000(28) + 1(396000) = -32000$$

$$N_z = \begin{vmatrix} 1 & 1 & 20000 \\ 10 & 5 & 132000 \\ -3 & 1 & 0 \end{vmatrix} = 1\begin{vmatrix} 5 & 132000 \\ 1 & 0 \end{vmatrix} - 1\begin{vmatrix} 10 & 132000 \\ -3 & 0 \end{vmatrix} + 20000\begin{vmatrix} 10 & 5 \\ -3 & 1 \end{vmatrix}$$

$$= 1(-132000) - 1(396000) + 20000(25) = -28000$$

$$x = \frac{N_x}{D} = \frac{-20000}{-4} = 5000;\ y = \frac{N_y}{D} = \frac{-32000}{-4} = 8000;\ z = \frac{N_z}{D} = \frac{-28000}{-4} = 7000$$

He should invest \$5000 in HiTech, \$8000 in SaveTel and \$7000 in HiGas.

63. $\begin{vmatrix} 2 & -3 & 4 \\ -1 & 2 & 4 \\ 3 & -3 & 1 \end{vmatrix} = -23$

65. $\begin{vmatrix} 2 & 1 & -3 \\ -2 & 2 & 4 \\ 1 & -2 & 2 \end{vmatrix} = 26$

67. **Answers may vary.**

69.
$$\begin{vmatrix} x & y & 1 \\ -2 & 3 & 1 \\ 3 & 5 & 1 \end{vmatrix} = 0$$

$$x\begin{vmatrix} 3 & 1 \\ 5 & 1 \end{vmatrix} - y\begin{vmatrix} -2 & 1 \\ 3 & 1 \end{vmatrix} + 1\begin{vmatrix} -2 & 3 \\ 3 & 5 \end{vmatrix} = 0$$

$$x(3-5) - y(-2-3) + 1(-10-9) = 0$$

$$-2x + 5y - 19 = 0$$

$$2x - 5y = -19$$ Verify that both points satisfy this equation:

71.
$$\begin{vmatrix} 1 & 0 & 2 & 1 \\ 2 & 1 & 1 & 3 \\ 1 & 1 & 1 & 1 \\ 2 & 1 & 1 & 1 \end{vmatrix} = 1\begin{vmatrix} 1 & 1 & 3 \\ 1 & 1 & 1 \\ 1 & 1 & 1 \end{vmatrix} - 0\begin{vmatrix} 2 & 1 & 3 \\ 1 & 1 & 1 \\ 2 & 1 & 1 \end{vmatrix} + 2\begin{vmatrix} 2 & 1 & 3 \\ 1 & 1 & 1 \\ 2 & 1 & 1 \end{vmatrix} - 1\begin{vmatrix} 2 & 1 & 1 \\ 1 & 1 & 1 \\ 2 & 1 & 1 \end{vmatrix}$$

$$= 1(0) - 0(???) + 2(-2) - 1(0) = -4$$

Exercise 13.5 (page 897)

1. $\sqrt{200x^2} - 3\sqrt{98x^2} = \sqrt{100x^2}\sqrt{2} - 3\sqrt{49x^2}\sqrt{2} = 10x\sqrt{2} - 3(7x)\sqrt{2} = -11x\sqrt{2}$

3. $\dfrac{3t\sqrt{2t} - 2\sqrt{2t^3}}{\sqrt{18t} - \sqrt{2t}} = \dfrac{3t\sqrt{2t} - 2\sqrt{t^2}\sqrt{2t}}{\sqrt{9}\sqrt{2t} - \sqrt{2t}} = \dfrac{3t\sqrt{2t} - 2t\sqrt{2t}}{3\sqrt{2t} - \sqrt{2t}} = \dfrac{t\sqrt{2t}}{2\sqrt{2t}} = \dfrac{t}{2}$

5. graphing; substitution

7. $\begin{cases} 8x^2 + 32y^2 = 256 \\ x = 2y \end{cases}$

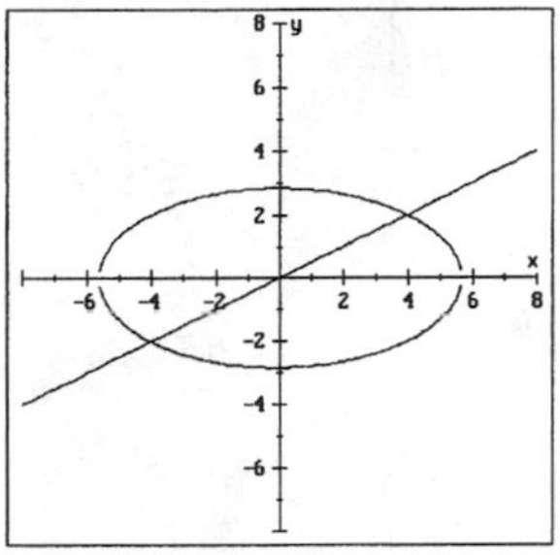

$(-4, -2), (4, 2)$

9. $\begin{cases} x^2 + y^2 = 10 \\ y = 3x^2 \end{cases}$

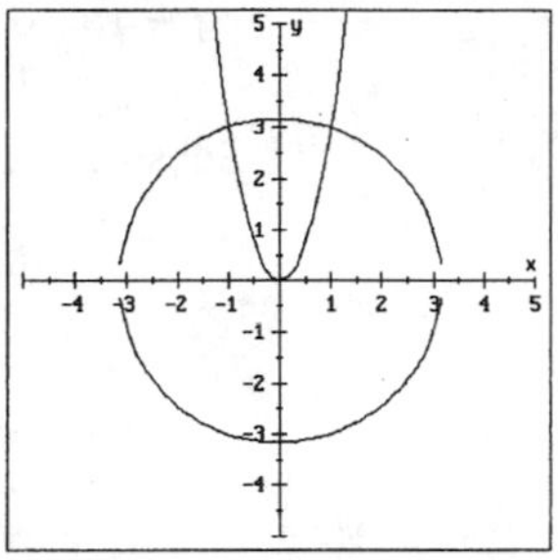

$(-1, 3), (1, 3)$

11. $\begin{cases} x^2 + y^2 = 25 \\ 12x^2 + 64y^2 = 768 \end{cases}$

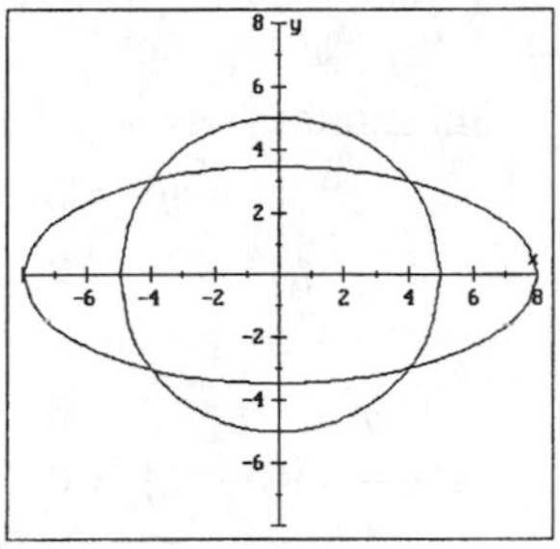

$(-4, 3), (4, 3), (4, -3)$
$(-4, -3)$

13. $\begin{cases} x^2 - 13 = -y^2 \\ y = 2x - 4 \end{cases}$

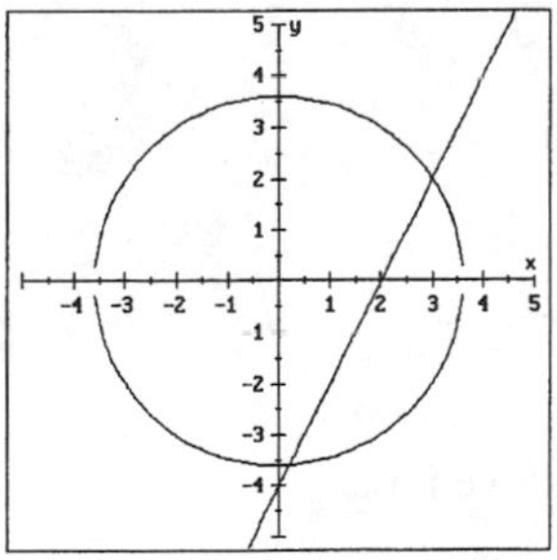

$\left(\frac{1}{5}, -\frac{18}{5}\right), (3, 2)$

15. $\begin{cases} x^2 - 6x - y = -5 \\ x^2 - 6x + y = -5 \end{cases}$
$\begin{cases} y = x^2 - 6x + 5 \\ y = -x^2 + 6x - 5 \end{cases}$

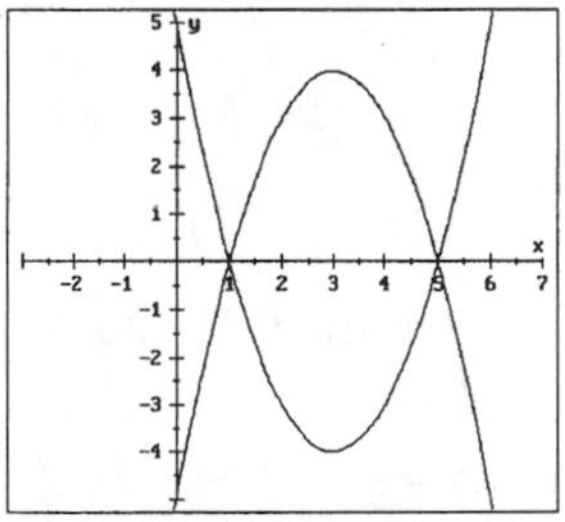

$(1, 0), (5, 0)$

17. $\begin{cases} (1) & 25x^2+9y^2=225 \\ (2) & 5x+3y=15 \end{cases}$

Substitute $x=-\frac{3}{5}y+3$ from (2) into (1):

$$\begin{aligned} 25x^2+9y^2&=225 \\ 25\left(-\tfrac{3}{5}y+3\right)^2+9y^2&=225 \\ 25\left(\tfrac{9}{25}y^2-\tfrac{18}{5}y+9\right)+9y^2&=225 \\ 9y^2-90y+225+9y^2&=225 \\ 18y^2-90y&=0 \\ 18y(y-5)&=0 \end{aligned}$$

$18y=0$ **or** $y-5=0$

$y=0$ $\quad$ $y=5$

Substitute these and solve for x:

$$\begin{aligned} 5x+3y&=15 & 5x+3y&=15 \\ 5x+3(0)&=15 & 5x+3(5)&=15 \\ 5x&=15 & 5x&=0 \\ x&=3 & x&=0 \end{aligned}$$

Solutions: $(3,0),(0,5)$

19. $\begin{cases} (1) & x^2+y^2=2 \\ (2) & x+y=2 \end{cases}$

Substitute $x=2-y$ from (2) into (1):

$$\begin{aligned} x^2+y^2&=2 \\ (2-y)^2+y^2&=2 \\ 4-4y+y^2+y^2&=2 \\ 2y^2-4y+2&=0 \\ 2(y-1)(y-1)&=0 \end{aligned}$$

$y-1=0$ **or** $y-1=0$

$y=1$ $\quad$ $y=1$

Substitute this and solve for x:

$x=2-y$

$x=2-1$

$x=1$

Solution: $(1,1)$

21. $\begin{cases} (1) & x^2+y^2=5 \\ (2) & x+y=3 \end{cases}$

Substitute $x=3-y$ from (2) into (1):

$$\begin{aligned} x^2+y^2&=5 \\ (3-y)^2+y^2&=5 \\ 9-6y+y^2+y^2&=5 \\ 2y^2-6y+4&=0 \\ 2(y-2)(y-1)&=0 \end{aligned}$$

$y-2=0$ **or** $y-1=0$

$y=2$ $\quad$ $y=1$

Substitute these and solve for x:

$$\begin{aligned} x&=3-y & x&=3-y \\ x&=3-2 & x&=3-1 \\ x&=1 & x&=2 \end{aligned}$$

Solutions: $(1,2),(2,1)$

23. $\begin{cases} (1) & x^2+y^2=13 \\ (2) & y=x^2-1 \end{cases}$

Substitute $x^2=13-y^2$ from (1) into (2):

$$\begin{aligned} y&=x^2-1 \\ y&=13-y^2-1 \\ y^2+y-12&=0 \\ (y+4)(y-3)&=0 \end{aligned}$$

$y+4=0$ **or** $y-3=0$

$y=-4$ $\quad$ $y=3$

Substitute these and solve for x:

$$\begin{aligned} x^2&=13-y^2 & x^2&=13-y^2 \\ x^2&=13-16 & x^2&=13-9 \\ x^2&=-4 & x^2&=4 \\ &\text{complex} & x&=\pm 2 \end{aligned}$$

Solutions: $(2,3),(-2,3)$

25. $\begin{cases} (1) & x^2 + y^2 = 30 \\ (2) & y = x^2 \end{cases}$

Substitute $x^2 = y$ from (2) into (1):

$$x^2 + y^2 = 30$$
$$y + y^2 = 30$$
$$y^2 + y - 30 = 0$$
$$(y - 5)(y + 6) = 0$$

$y - 5 = 0$ **or** $y + 6 = 0$

$y = 5 \qquad y = -6$

Substitute these and solve for x:

$x^2 = y \qquad x^2 = y$

$x^2 = 5 \qquad x^2 = -6$

$x = \pm\sqrt{5} \qquad$ complex

Solutions: $\left(\sqrt{5}, 5\right), \left(-\sqrt{5}, 5\right)$

27.

$$\begin{aligned} x^2 + y^2 &= 13 \\ x^2 - y^2 &= 5 \\ \hline 2x^2 &= 18 \\ x^2 &= 9 \\ x &= \pm 3 \end{aligned}$$

Substitute and solve for y:

$x^2 + y^2 = 13 \qquad x^2 + y^2 = 13$

$3^2 + y^2 = 13 \qquad (-3)^2 + y^2 = 13$

$y^2 = 4 \qquad y^2 = 4$

$y = \pm 2 \qquad y = \pm 2$

Solutions: $(3, 2), (3, -2), (-3, 2), (-3, -2)$

29.

$$\begin{aligned} x^2 + y^2 &= 20 \\ x^2 - y^2 &= -12 \\ \hline 2x^2 &= 8 \\ x^2 &= 4 \\ x &= \pm 2 \end{aligned}$$

Substitute and solve for y:

$x^2 + y^2 = 20 \qquad x^2 + y^2 = 20$

$2^2 + y^2 = 20 \qquad (-2)^2 + y^2 = 20$

$y^2 = 16 \qquad y^2 = 16$

$y = \pm 4 \qquad y = \pm 4$

Solutions: $(2, 4), (2, -4), (-2, 4), (-2, -4)$

31. $\begin{cases} (1) & y^2 = 40 - x^2 \\ (2) & y = x^2 - 10 \end{cases}$

Substitute $x^2 = 40 - y^2$ from (1) into (2):

$$y = x^2 - 10$$
$$y = 40 - y^2 - 10$$
$$y^2 + y - 30 = 0$$
$$(y + 6)(y - 5) = 0$$

$y + 6 = 0$ **or** $y - 5 = 0$

$y = -6 \qquad y = 5$

Substitute these and solve for x:

$x^2 = 40 - y^2 \qquad x^2 = 40 - y^2$

$x^2 = 40 - 36 \qquad x^2 = 40 - 25$

$x^2 = 4 \qquad x^2 = 15$

$x = \pm 2 \qquad x = \pm\sqrt{15}$

$(2, -6), (-2, -6), \left(\sqrt{15}, 5\right), \left(-\sqrt{15}, 5\right)$

33. $\begin{cases} (1) & y = x^2 - 4 \\ (2) & x^2 - y^2 = -16 \end{cases}$

Substitute $x^2 = y + 4$ from (1) into (2):

$$x^2 - y^2 = -16$$
$$y + 4 - y^2 = -16$$
$$y^2 - y - 20 = 0$$
$$(y + 4)(y - 5) = 0$$

$y + 4 = 0$ **or** $y - 5 = 0$

$y = -4 \qquad y = 5$

Substitute these and solve for x:

$x^2 = y + 4 \qquad x^2 = y + 4$

$x^2 = -4 + 4 \qquad x^2 = 5 + 4$

$x^2 = 0 \qquad x^2 = 9$

$x = 0 \qquad x = \pm 3$

Solutions: $(0, -4), (3, 5), (-3, 5)$

35.
$$\begin{array}{l} x^2 - y^2 = -5 \Rightarrow (\times 2) \\ 3x^2 + 2y^2 = 30 \Rightarrow \end{array} \qquad \begin{array}{rcl} 2x^2 - 2y^2 &=& -10 \\ 3x^2 + 2y^2 &=& 30 \\ \hline 5x^2 &=& 20 \\ x^2 &=& 4 \\ x &=& \pm 2 \end{array}$$

Substitute and solve for y:

$$\begin{array}{rcl} x^2 - y^2 &=& -5 \\ 2^2 - y^2 &=& -5 \\ -y^2 &=& -9 \\ y^2 &=& 9 \\ y &=& \pm 3 \end{array} \qquad \begin{array}{rcl} x^2 - y^2 &=& -5 \\ (-2)^2 - y^2 &=& -5 \\ -y^2 &=& -9 \\ y^2 &=& 9 \\ y &=& \pm 3 \end{array}$$

Solutions: $(2, 3), (2, -3), (-2, 3), (-2, -3)$

37.
$$\begin{array}{l} \frac{1}{x} + \frac{2}{y} = 1 \Rightarrow \\ \frac{2}{x} - \frac{1}{y} = \frac{1}{3} \Rightarrow (\times 2) \end{array} \qquad \begin{array}{rcl} \frac{1}{x} + \frac{2}{y} &=& 1 \\ \frac{4}{x} - \frac{2}{y} &=& \frac{2}{3} \\ \hline \frac{5}{x} &=& \frac{5}{3} \\ 15 &=& 5x \\ 3 &=& x \end{array}$$

Substitute and solve for y:

$$\begin{array}{rcl} \frac{1}{x} + \frac{2}{y} &=& 1 \\ \frac{1}{3} + \frac{2}{y} &=& 1 \\ \frac{2}{y} &=& \frac{2}{3} \\ 6 &=& 2y \\ 3 &=& y \end{array}$$

Solution: $(3, 3)$

39.
$$\begin{cases} (1) & 3y^2 = xy \\ (2) & 2x^2 + xy - 84 = 0 \end{cases}$$

From (1): $3y^2 - xy = 0$

$y(3y - x) = 0$

$y = 0 \quad \text{or} \quad y = \dfrac{1}{3}x$

Substitute these into (2):

$$\begin{array}{rcl} 2x^2 + xy - 84 &=& 0 \\ 2x^2 + x(0) - 84 &=& 0 \\ 2x^2 &=& 84 \\ x^2 &=& 42 \\ x &=& \pm\sqrt{42} \end{array} \qquad \begin{array}{rcl} 2x^2 + xy - 84 &=& 0 \\ 2x^2 + x\left(\frac{1}{3}x\right) - 84 &=& 0 \\ 2x^2 + \frac{1}{3}x^2 &=& 84 \\ 6x^2 + x^2 &=& 252 \\ 7x^2 &=& 252 \\ x^2 &=& 36 \\ x &=& \pm 6 \end{array}$$

(substitute and solve for y)

Solutions: $\left(\sqrt{42}, 0\right), \left(-\sqrt{42}, 0\right), (6, 2), (-6, -2)$

41.
$$\begin{cases} (1) & xy = \frac{1}{6} \\ (2) & y + x = 5xy \end{cases}$$

Substitute $x = \frac{1}{6y}$ from (1) into (2):

$$\begin{array}{rcl} y + \frac{1}{6y} &=& \frac{5y}{6y} \\ 6y^2 + 1 &=& 5y \\ 6y^2 - 5y + 1 &=& 0 \\ (2y - 1)(3y - 1) &=& 0 \end{array}$$

$$\begin{array}{rcl} 2y - 1 = 0 & \textbf{or} & 3y - 1 = 0 \\ y = \frac{1}{2} & & y = \frac{1}{3} \end{array}$$

Substitute these and solve for x:

$$\begin{array}{rcl} x &=& \frac{1}{6y} \\ x &=& \dfrac{1}{6\left(\frac{1}{2}\right)} \\ x &=& \frac{1}{3} \end{array} \qquad \begin{array}{rcl} x &=& \frac{1}{6y} \\ x &=& \dfrac{1}{6\left(\frac{1}{3}\right)} \\ x &=& \frac{1}{2} \end{array}$$

Solutions: $\left(\frac{1}{3}, \frac{1}{2}\right), \left(\frac{1}{2}, \frac{1}{3}\right)$

43. Let the integers be x and y. Then the equations are

$$\begin{cases} (1) \quad xy = 32 \\ (2) \quad x + y = 12 \end{cases}$$

Substitute $x = \frac{32}{y}$ from (1) into (2):

$$\frac{32}{y} + y = 12$$
$$32 + y^2 = 12y$$
$$y^2 - 12y + 32 = 0$$
$$(y-4)(y-8) = 0$$

$y - 4 = 0$ **or** $y - 8 = 0$

$y = 4$ $\qquad$ $y = 8$

Substitute these and solve for x:

$x = \frac{32}{y} = \frac{32}{4} = 8 \qquad x = \frac{32}{y} = \frac{32}{8} = 4$

The integers are 8 and 4.

45. $\begin{cases} 2x - y > 4 \\ y < -x^2 + 2 \end{cases}$

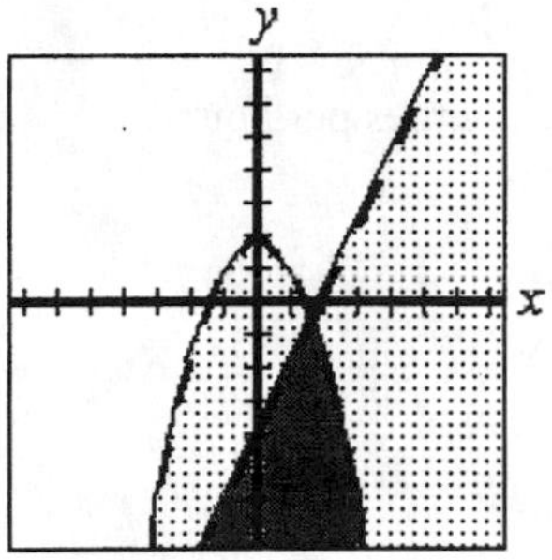

47. $\begin{cases} y > x^2 - 4 \\ y < -x^2 + 4 \end{cases}$

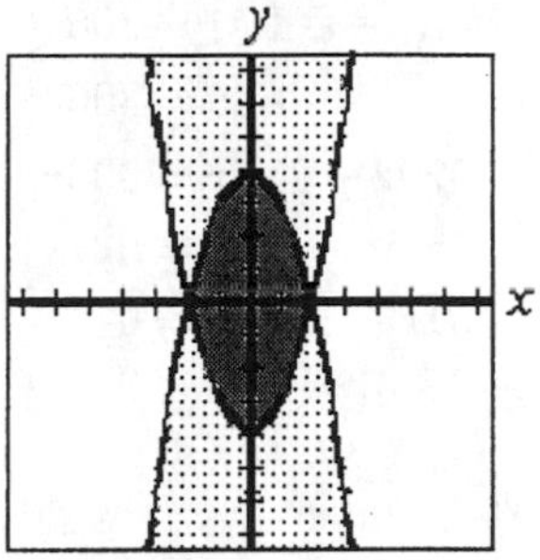

49. Let l = the length of the rectangle, and w = the width of the rectangle. Then the equations are:

$$\begin{cases} (1) \quad lw = 63 \\ (2) \quad 2l + 2w = 32 \end{cases}$$

Substitute $l = \frac{63}{w}$ from (1) into (2):

$$2\left(\frac{63}{w}\right) + 2w = 32$$
$$\frac{126}{w} + 2w = 32$$
$$126 + 2w^2 = 32w$$
$$2w^2 - 32y + 126 = 0$$
$$2(w-7)(w-9) = 0$$

$w - 7 = 0$ **or** $w - 9 = 0$

$w = 7$ $\qquad$ $w = 9$

Substitute these and solve for l:

$l = \frac{63}{w} = \frac{63}{7} = 9 \qquad l = \frac{63}{w} = \frac{63}{9} = 7$

The dimensions are 7 cm by 9 cm.

51. Let r = Carol's rate, and let p = the amount Carol invested.
Then Juan invested $p + 150$ at a rate of $r + 0.015$. The equations are

$$\begin{cases} (1) \; pr = 67.50 \Rightarrow p = \frac{67.5}{r} \\ (2) \; (p + 150)(r + 0.015) = 94.5 \end{cases}$$

Substitute $p = \frac{67.5}{r}$ from (1) into (2):

$$\left(\frac{67.5}{r} + 150\right)(r + 0.015) = 94.5$$
$$67.5 + \frac{1.0125}{r} + 150r + 2.25 = 94.5$$

continued on next page...

51. continued

$$67.5 + \frac{1.0125}{r} + 150r + 2.25 = 94.5$$
$$67.5r + 1.0125 + 150r^2 + 2.25r = 94.5r$$
$$150r^2 - 24.75r - 1.0125 = 0$$
$$12{,}000r^2 - 1980r + 81 = 0$$
$$(100r - 9)(120r - 9) = 0$$

$100r - 9 = 0$ **or** $120r - 9 = 0$

$r = 0.09$ $\qquad$ $r = 0.075$

Substitute and solve for p:

$p = \frac{67.5}{r} = \frac{67.5}{0.09} = 750$ or $p = \frac{67.5}{r} = \frac{67.5}{0.075} = 900$

Carol invested \$750 at 9% or \$900 at 7.5%.

53. Let $r =$ Jim's rate and $t =$ Jim's time. Then his brother's rate was $r - 17$ and his time was $t + 1.5$.

$$\begin{cases} (1)\ rt = 306 \Rightarrow t = \frac{306}{r} \\ (2)\ (r - 17)(t + 1.5) = 306 \end{cases}$$

Substitute $t = \frac{306}{r}$ from (1) into (2):

$$(r - 17)\left(\frac{306}{r} + 1.5\right) = 306$$
$$306 + 1.5r - \frac{5202}{r} - 25.5 = 306$$
$$306r + 1.5r^2 - 5202 - 25.5r = 306r$$
$$1.5r^2 - 25.5r - 5202 = 0$$
$$3r^2 - 51r - 10{,}404 = 0$$
$$(3r + 153)(r - 68) = 0$$

$3r + 153 = 0$ **or** $r - 68 = 0$

$r = -153/3$ $\qquad$ $r = 68$

Substitute and solve for t:

$t = \frac{306}{r} = \frac{306}{68} = 4.5$

Jim drove for 4.5 hours at 68 miles per hour.

55. **Answers may vary.**

57. $0, 1, 2, 3, 4$

Chapter 13 Summary (page 900)

1. $\begin{cases} 2x + y = 11 \\ -x + 2y = 7 \end{cases}$

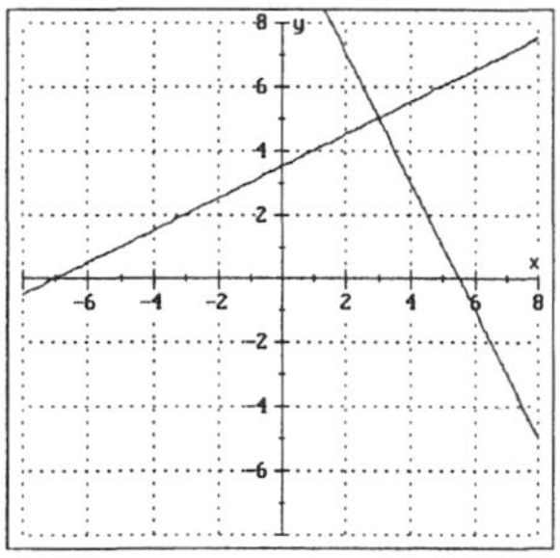

$(3, 5)$ is the solution.

2. $\begin{cases} 3x + 2y = 0 \\ 2x - 3y = -13 \end{cases}$

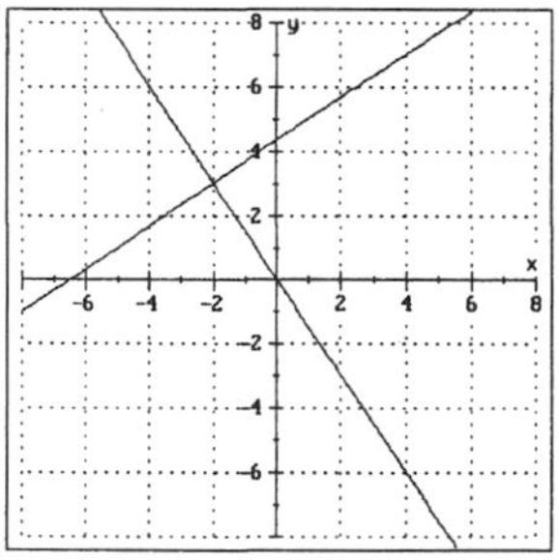

$(-2, 3)$ is the solution.

3. $\begin{cases} \frac{1}{2}x + \frac{1}{3}y = 2 \\ y = 6 - \frac{3}{2}x \end{cases}$

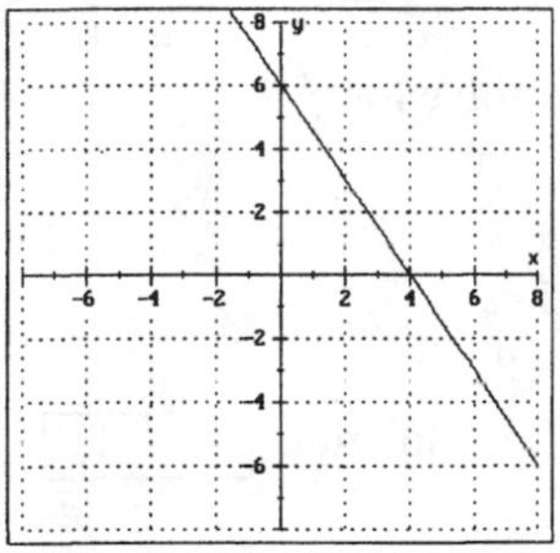

dependent equations

4. $\begin{cases} \frac{1}{3}x - \frac{1}{2}y = 1 \\ 6x - 9y = 2 \end{cases}$

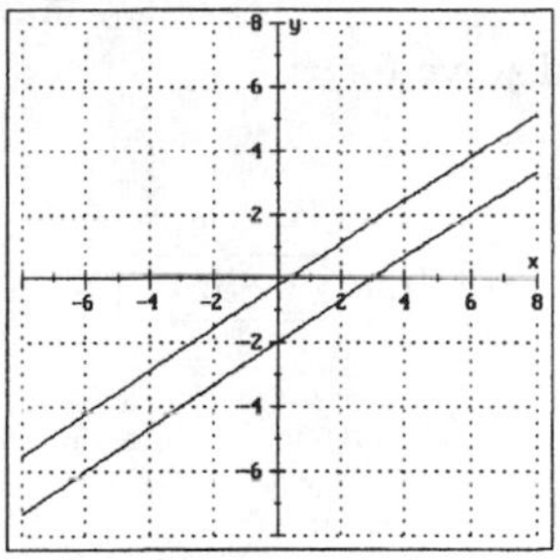

inconsistent system

5. $\begin{cases} (1) \quad y = x + 4 \\ (2) \quad 2x + 3y = 7 \end{cases}$

Substitute $y = x + 4$ from (1) into (2):

$$2x + 3\boldsymbol{y} = 7$$
$$2x + 3(\boldsymbol{x+4}) = 7$$
$$2x + 3x + 12 = 7$$
$$5x = -5$$
$$x = -1$$

Substitute this and solve for y:

$y = x + 4 = -1 + 4 = 3$

Solution: $(-1, 3)$

6. $\begin{cases} (1) \quad y = 2x + 5 \\ (2) \quad 3x - 5y = -4 \end{cases}$

Substitute $y = 2x + 5$ from (1) into (2):

$$3x - 5\boldsymbol{y} = -4$$
$$3x - 5(\boldsymbol{2x+5}) = -4$$
$$3x - 10x - 25 = -4$$
$$-7x = 21$$
$$x = -3$$

Substitute this and solve for y:

$y = 2x + 5 = 2(-3) + 5 = -1$

Solution: $(-3, -1)$

7. $\begin{cases} (1) \quad x + 2y = 11 \\ (2) \quad 2x - y = 2 \end{cases}$

Substitute $x = -2y + 11$ from (1) into (2):

$$2\boldsymbol{x} - y = 2$$
$$2(\boldsymbol{-2y+11}) - y = 2$$
$$-4y + 22 - y = 2$$
$$-5y = -20$$
$$y = 4$$

Substitute this and solve for x:

$x = -2y + 11 = -2(4) + 11 = 3$

Solution: $(3, 4)$

8. $\begin{cases} (1) \quad 2x + 3y = -2 \\ (2) \quad 3x + 5y = -2 \end{cases}$

Substitute $x = \frac{-3y-2}{2}$ from (1) into (2):

$$3\boldsymbol{x} + 5y = -2$$
$$3 \cdot \boldsymbol{\frac{-3y-2}{2}} + 5y = -2$$
$$3(-3y - 2) + 10y = -4$$
$$-9y - 6 + 10y = -4$$
$$y = 2$$

Substitute this and solve for y:

$x = \frac{-3y-2}{2} = \frac{-3(2)-2}{2} = \frac{-8}{2} = -4$

Solution: $(-4, 2)$

9.

$$\begin{array}{rl} x + y = -2 \Rightarrow \times(-2) & -2x - 2y = 4 \\ 2x + 3y = -3 \Rightarrow & 2x + 3y = -3 \\ \hline & y = 1 \end{array}$$

Substitute and solve for x:

$x + y = -2$

$x + 1 = -2$

$x = -3$ Solution: $\boxed{(-3, 1)}$

10.

$$\begin{array}{rl} 3x + 2y = 1 \Rightarrow \times 3 & 9x + 6y = 3 \\ 2x - 3y = 5 \Rightarrow \times 2 & 4x - 6y = 10 \\ \hline & 13x = 13 \\ & x = 1 \end{array}$$

Substitute and solve for y:

$3x + 2y = 1$

$3(1) + 2y = 1$

$3 + 2y = 1$

$2y = -2$

$y = -1$ Solution: $\boxed{(1, -1)}$

11.

$$\begin{array}{rl} x + \frac{1}{2}y = 7 \Rightarrow \times 2 & 2x + y = 14 \Rightarrow \\ -2x = 3y - 6 \Rightarrow & -2x - 3y = -6 \Rightarrow \\ \hline & -2y = 8 \\ & y = -4 \end{array}$$

Solve for x:

$2x + y = 14$

$2x + (-4) = 14$

$2x = 18$

$x = 9$ $\boxed{\text{Solution: } (9, -4)}$

12.

$$\begin{array}{rll} y = \dfrac{x-3}{2} \Rightarrow \times 2 & 2y = x - 3 \Rightarrow & -x + 2y = -3 \\ x = \dfrac{2y+7}{2} \Rightarrow \times 2 & 2x = 2y + 7 \Rightarrow & 2x - 2y = 7 \\ \hline & & x = 4 \end{array}$$

Solve for y:

$y = \dfrac{x-3}{2}$

$y = \dfrac{4-3}{2} = \dfrac{1}{2}$ Solution: $\boxed{(4, \frac{1}{2})}$

13.

$$\begin{array}{ll} (1) & x + y + z = 6 \\ (2) & x - y - z = -4 \\ (3) & -x + y - z = -2 \end{array} \qquad \begin{array}{ll} (1) & x + y + z = 6 \\ (2) & x - y - z = -4 \\ \hline (4) & 2x = 2 \\ & x = 1 \end{array} \qquad \begin{array}{ll} (1) & x + y + z = 6 \\ (3) & -x + y - z = -2 \\ \hline (5) & 2y = 4 \\ & y = 2 \end{array}$$

$x + y + z = 6$

$1 + 2 + z = 6$

$3 + z = 6$

$z = 3$ Solution: $\boxed{(1, 2, 3)}$

14.

$$\begin{array}{ll} (1) & 2x + 3y + z = -5 \\ (2) & -x + 2y - z = -6 \\ (3) & 3x + y + 2z = 4 \end{array} \qquad \begin{array}{ll} (1) & 2x + 3y + z = -5 \\ (2) & -x + 2y - z = -6 \\ \hline (4) & x + 5y = -11 \end{array} \qquad \begin{array}{ll} (3) & 3x + y + 2z = 4 \\ 2 \cdot (2) & -2x + 4y - 2z = -12 \\ \hline (5) & x + 5y = -8 \end{array}$$

$$\begin{array}{rl} x + 5y = -11 \Rightarrow \times(-1) & -x - 5y = 11 \\ x + 5y = -8 \Rightarrow & x + 5y = -8 \\ \hline & 0 = 3 \end{array}$$

Since this is an impossible equation, there is no solution. It is an inconsistent system.

15. $\left[\begin{array}{cc|c} 1 & 2 & 4 \\ 2 & -1 & 3 \end{array}\right] \overset{-2R_1+R_2 \Rightarrow R_2}{\Rightarrow} \left[\begin{array}{cc|c} 1 & 2 & 4 \\ 0 & -5 & -5 \end{array}\right] \overset{-\frac{1}{5}R_2 \Rightarrow R_2}{\Rightarrow} \left[\begin{array}{cc|c} 1 & 2 & 4 \\ 0 & 1 & 1 \end{array}\right]$

From R_2, $y = 1$. From R_1:

$$\begin{aligned} x + 2y &= 4 \\ x + 2(1) &= 4 \\ x &= 2 \end{aligned}$$

Solution: $\boxed{(2,1)}$

16. $\left[\begin{array}{ccc|c} 1 & 1 & 1 & 6 \\ 2 & -1 & 1 & 1 \\ 4 & 1 & -1 & 5 \end{array}\right] \overset{\substack{-2R_1+R_2 \Rightarrow R_2 \\ -4R_1+R_3 \Rightarrow R_3}}{\Rightarrow} \left[\begin{array}{ccc|c} 1 & 1 & 1 & 6 \\ 0 & -3 & -1 & -11 \\ 0 & -3 & -5 & -19 \end{array}\right] \overset{\substack{-R_2 \Rightarrow R_2 \\ -R_3+R_2 \Rightarrow R_3}}{\Rightarrow} \left[\begin{array}{ccc|c} 1 & 1 & 1 & 6 \\ 0 & 3 & 1 & 11 \\ 0 & 0 & 4 & 8 \end{array}\right] \overset{\frac{1}{4}R_3 \Rightarrow R_3}{\Rightarrow} \left[\begin{array}{ccc|c} 1 & 1 & 1 & 6 \\ 0 & 3 & 1 & 11 \\ 0 & 0 & 1 & 2 \end{array}\right]$

From R_3, $z = 2$. From R_2:

$$\begin{aligned} 3y + z &= 11 \\ 3y + 2 &= 11 \\ 3y &= 9 \\ y &= 3 \end{aligned}$$

From R_1:

$$\begin{aligned} x + y + z &= 6 \\ x + 3 + 2 &= 6 \\ x &= 1 \end{aligned}$$

Solution: $\boxed{(1,3,2)}$

17. $\left[\begin{array}{cc|c} 1 & 1 & 3 \\ 1 & -2 & -3 \\ 2 & 1 & 4 \end{array}\right] \overset{\substack{-R_1+R_2 \Rightarrow R_2 \\ -2R_1+R_3 \Rightarrow R_3}}{\Rightarrow} \left[\begin{array}{cc|c} 1 & 1 & 3 \\ 0 & -3 & -6 \\ 0 & -1 & -2 \end{array}\right] \overset{\substack{-\frac{1}{3}R_2 \Rightarrow R_2 \\ -3R_3+R_2 \Rightarrow R_3}}{\Rightarrow} \left[\begin{array}{cc|c} 1 & 1 & 3 \\ 0 & 1 & 2 \\ 0 & 0 & 0 \end{array}\right]$

From R_2, $y = 2$. From R_1:

$$\begin{aligned} x + y &= 3 \\ x + 2 &= 3 \\ x &= 1 \end{aligned}$$

Solution: $\boxed{(1,2)}$

18. $\left[\begin{array}{ccc|c} 1 & 2 & 1 & 2 \\ 2 & 5 & 4 & 5 \end{array}\right] \overset{-2R_1+R_2 \Rightarrow R_2}{\Rightarrow} \left[\begin{array}{ccc|c} 1 & 2 & 1 & 2 \\ 0 & 1 & 2 & 1 \end{array}\right]$

From R_2:

$$\begin{aligned} y + 2z &= 1 \\ y &= 1 - 2z \end{aligned}$$

From R_1:

$$\begin{aligned} x + 2y + z &= 2 \\ x + 2(1 - 2z) + z &= 2 \\ x + 2 - 4z + z &= 2 \\ x - 3z &= 0 \\ x &= 3z \end{aligned}$$

Solution: $\boxed{(3z, 1 - 2z, z)}$

19. $\begin{vmatrix} 2 & 3 \\ -4 & 3 \end{vmatrix} = 2(3) - 3(-4) = 6 + 12 = 18$

20. $\begin{vmatrix} -3 & -4 \\ 5 & -6 \end{vmatrix} = -3(-6) - (-4)(5) = 18 - (-20) = 18 + 20 = 38$

21. $\begin{vmatrix} -1 & 2 & -1 \\ 2 & -1 & 3 \\ 1 & -2 & 2 \end{vmatrix} = -1\begin{vmatrix} -1 & 3 \\ -2 & 2 \end{vmatrix} - 2\begin{vmatrix} 2 & 3 \\ 1 & 2 \end{vmatrix} + (-1)\begin{vmatrix} 2 & -1 \\ 1 & -2 \end{vmatrix} = -1(4) - 2(1) - 1(-3) = -3$

22. $\begin{vmatrix} 3 & -2 & 2 \\ 1 & -2 & -2 \\ 2 & 1 & -1 \end{vmatrix} = 3\begin{vmatrix} -2 & -2 \\ 1 & -1 \end{vmatrix} - (-2)\begin{vmatrix} 1 & -2 \\ 2 & -1 \end{vmatrix} + 2\begin{vmatrix} 1 & -2 \\ 2 & 1 \end{vmatrix} = 3(4) + 2(3) + 2(5) = 28$

23. $x = \dfrac{\begin{vmatrix} 10 & 4 \\ 1 & -3 \end{vmatrix}}{\begin{vmatrix} 3 & 4 \\ 2 & -3 \end{vmatrix}} = \dfrac{-30-4}{-9-8} = \dfrac{-34}{-17} = 2; y = \dfrac{\begin{vmatrix} 3 & 10 \\ 2 & 1 \end{vmatrix}}{\begin{vmatrix} 3 & 4 \\ 2 & -3 \end{vmatrix}} = \dfrac{3-20}{-17} = \dfrac{-17}{-17} = 1$

Solution: $\boxed{(2,1)}$

24. $x = \dfrac{\begin{vmatrix} -17 & -5 \\ 3 & 2 \end{vmatrix}}{\begin{vmatrix} 2 & -5 \\ 3 & 2 \end{vmatrix}} = \dfrac{-34+15}{4+15} = \dfrac{-19}{19} = -1; y = \dfrac{\begin{vmatrix} 2 & -17 \\ 3 & 3 \end{vmatrix}}{\begin{vmatrix} 2 & -5 \\ 3 & 2 \end{vmatrix}} = \dfrac{6+51}{19} = \dfrac{57}{19} = 3$

Solution: $\boxed{(-1,3)}$

25. $D = \begin{vmatrix} 1 & 2 & 1 \\ 2 & 1 & 1 \\ 1 & 1 & 2 \end{vmatrix} = 1\begin{vmatrix} 1 & 1 \\ 1 & 2 \end{vmatrix} - 2\begin{vmatrix} 2 & 1 \\ 1 & 2 \end{vmatrix} + 1\begin{vmatrix} 2 & 1 \\ 1 & 1 \end{vmatrix} = 1(1) - 2(3) + 1(1) = -4$

$N_x = \begin{vmatrix} 0 & 2 & 1 \\ 3 & 1 & 1 \\ 5 & 1 & 2 \end{vmatrix} = 0\begin{vmatrix} 1 & 1 \\ 1 & 2 \end{vmatrix} - 2\begin{vmatrix} 3 & 1 \\ 5 & 2 \end{vmatrix} + 1\begin{vmatrix} 3 & 1 \\ 5 & 1 \end{vmatrix} = 0(1) - 2(1) + 1(-2) = -4$

$N_y = \begin{vmatrix} 1 & 0 & 1 \\ 2 & 3 & 1 \\ 1 & 5 & 2 \end{vmatrix} = 1\begin{vmatrix} 3 & 1 \\ 5 & 2 \end{vmatrix} - 0\begin{vmatrix} 2 & 1 \\ 1 & 2 \end{vmatrix} + 1\begin{vmatrix} 2 & 3 \\ 1 & 5 \end{vmatrix} = 1(1) - 0(3) + 1(7) = 8$

$N_z = \begin{vmatrix} 1 & 2 & 0 \\ 2 & 1 & 3 \\ 1 & 1 & 5 \end{vmatrix} = 1\begin{vmatrix} 1 & 3 \\ 1 & 5 \end{vmatrix} - 2\begin{vmatrix} 2 & 3 \\ 1 & 5 \end{vmatrix} + 0\begin{vmatrix} 2 & 1 \\ 1 & 1 \end{vmatrix} = 1(2) - 2(7) + 0(1) = -12$

$x = \dfrac{N_x}{D} = \dfrac{-4}{-4} = 1; y = \dfrac{N_y}{D} = \dfrac{8}{-4} = -2; z = \dfrac{N_z}{D} = \dfrac{-12}{-4} = 3 \Rightarrow$ solution: $(1,-2,3)$

26. $D = \begin{vmatrix} 2 & 3 & 1 \\ 1 & 3 & 2 \\ 1 & -1 & -1 \end{vmatrix} = 2\begin{vmatrix} 3 & 2 \\ -1 & -1 \end{vmatrix} - 3\begin{vmatrix} 1 & 2 \\ 1 & -1 \end{vmatrix} + 1\begin{vmatrix} 1 & 3 \\ 1 & -1 \end{vmatrix} = 2(-1) - 3(-3) + 1(-4)$

$= 3$

$N_x = \begin{vmatrix} 2 & 3 & 1 \\ 7 & 3 & 2 \\ -7 & -1 & -1 \end{vmatrix} = 2\begin{vmatrix} 3 & 2 \\ -1 & -1 \end{vmatrix} - 3\begin{vmatrix} 7 & 2 \\ -7 & -1 \end{vmatrix} + 1\begin{vmatrix} 7 & 3 \\ -7 & -1 \end{vmatrix}$

$= 2(-1) - 3(7) + 1(14) = -9$

continued on next page...

26. continued

$$N_y = \begin{vmatrix} 2 & 2 & 1 \\ 1 & 7 & 2 \\ 1 & -7 & -1 \end{vmatrix} = 2\begin{vmatrix} 7 & 2 \\ -7 & -1 \end{vmatrix} - 2\begin{vmatrix} 1 & 2 \\ 1 & -1 \end{vmatrix} + 1\begin{vmatrix} 1 & 7 \\ 1 & -7 \end{vmatrix}$$

$$= 2(7) - 2(-3) + 1(-14) = 6$$

$$N_z = \begin{vmatrix} 2 & 3 & 2 \\ 1 & 3 & 7 \\ 1 & -1 & -7 \end{vmatrix} = 2\begin{vmatrix} 3 & 7 \\ -1 & -7 \end{vmatrix} - 3\begin{vmatrix} 1 & 7 \\ 1 & -7 \end{vmatrix} + 2\begin{vmatrix} 1 & 3 \\ 1 & -1 \end{vmatrix}$$

$$= 2(-14) - 3(-14) + 2(-4) = 6$$

$$x = \frac{N_x}{D} = \frac{-9}{3} = -3;\ y = \frac{N_y}{D} = \frac{6}{3} = 2;\ z = \frac{N_z}{D} = \frac{6}{3} = 2 \Rightarrow \text{solution: } (-3, 2, 2)$$

27.

$$\begin{aligned} 3x^2 + y^2 &= 52 \\ x^2 - y^2 &= 12 \\ \hline 4x^2 &= 64 \\ x^2 &= 16 \\ x &= \pm 4 \end{aligned}$$

Substitute and solve for y:

$$\begin{aligned} x^2 - y^2 &= 12 \\ 4^2 - y^2 &= 12 \\ y^2 &= 4 \\ y &= \pm 2 \end{aligned} \qquad \begin{aligned} x^2 - y^2 &= 12 \\ (-4)^2 - y^2 &= 12 \\ y^2 &= 4 \\ y &= \pm 2 \end{aligned}$$

Solutions: $(4, 2), (4, -2), (-4, 2), (-4, -2)$.

28.

$$\begin{aligned} \tfrac{x^2}{16} + \tfrac{y^2}{12} = 1 \Rightarrow \times 48 \qquad 3x^2 + 4y^2 &= 48 \\ x^2 - \tfrac{y^2}{3} = 1 \Rightarrow \times (-3) \quad -3x^2 + y^2 &= -3 \\ \hline 5y^2 &= 45 \\ y^2 &= 9 \\ y &= \pm 3 \end{aligned}$$

Substitute and solve for x:

$$\begin{aligned} 3x^2 + 4y^2 &= 48 \\ 3x^2 + 4(9) &= 48 \\ 3x^2 &= 12 \\ x^2 = 4 \Rightarrow x &= \pm 2 \end{aligned}$$

Solutions: $(2, 3), (2, -3), (-2, 3), (-2, -3)$

29. $\begin{cases} y \geq x^2 - 4 \\ y < x + 3 \end{cases}$

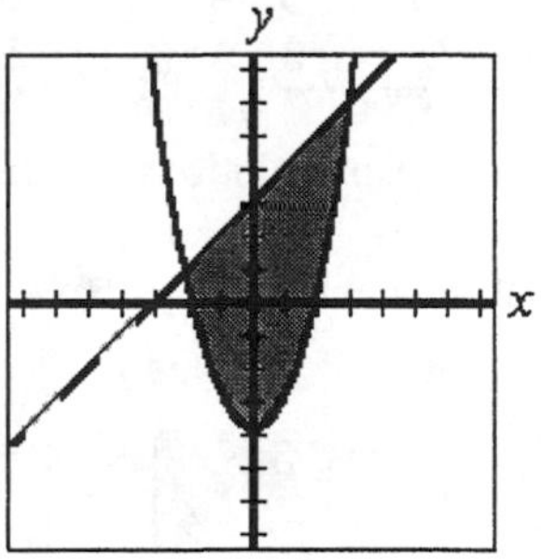

Chapter 13 Test (page 903)

1. $\begin{cases} 2x + y = 5 \\ y = 2x - 3 \end{cases}$

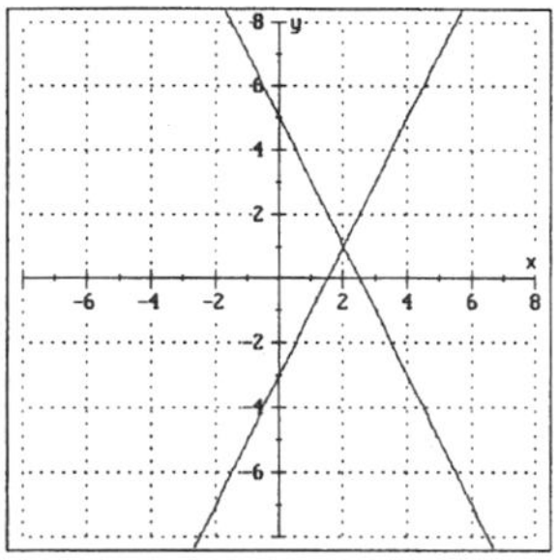

$(2, 1)$ is the solution.

2. $\begin{cases} (1) & 2x - 4y = 14 \\ (2) & x = -2y + 7 \end{cases}$

Substitute $x = -2y + 7$ from (2) into (1):

$$\begin{aligned} 2x - 4y &= 14 \\ 2(-2y + 7) - 4y &= 14 \\ -4y + 14 - 4y &= 14 \\ -8y &= 0 \\ y &= 0 \end{aligned}$$

Substitute this and solve for x:

$$\begin{aligned} x &= -2y + 7 \\ x &= -2(0) + 7 = 7 \end{aligned}$$

Solution: $(7, 0)$

3.
$$\begin{aligned} 2x + 3y &= -5 \Rightarrow \times 2 \quad 4x + 6y = -10 \\ 3x - 2y &= 12 \Rightarrow \times 3 \quad 9x - 6y = 36 \\ & \qquad\qquad 13x = 26 \\ & \qquad\qquad x = 2 \end{aligned}$$

$$\begin{aligned} 2x + 3y &= -5 \\ 2(2) + 3y &= -5 \\ 4 + 3y &= -5 \\ 3y &= -9 \\ y &= -3 \end{aligned}$$

Solution: $\boxed{(2, -3)}$

4.
$$\begin{aligned} \frac{x}{2} - \frac{y}{4} &= -4 \Rightarrow \times 4 \quad 2x - y = -16 \\ x + y &= -2 \Rightarrow \qquad\quad x + y = -2 \\ & \qquad\qquad 3x = -18 \\ & \qquad\qquad x = -6 \end{aligned}$$

$$\begin{aligned} x + y &= -2 \\ -6 + y &= -2 \\ y &= 4 \end{aligned}$$

Solution: $\boxed{(-6, 4)}$

5.
$$\begin{aligned} 3(x + y) = x - 3 &\Rightarrow \quad 2x + 3y = -3 \\ -y = \frac{2x + 3}{3} &\Rightarrow \quad 2x + 3y = -3 \end{aligned}$$

$$\begin{aligned} 2x + 3y = -3 &\Rightarrow \qquad\qquad\quad 2x + 3y = -3 \\ 2x + 3y = -3 &\Rightarrow \times(-1) \quad -2x - 3y = 3 \\ & \qquad\qquad\qquad\qquad 0 = 0 \end{aligned}$$

The equation is an identity, so the system has infinitely many solutions. $\Rightarrow$ dependent equations

6. See #5. Since the system has at least one solution, it is a consistent system.

7. $\begin{bmatrix} 1 & 2 & -1 \\ 2 & -2 & 3 \end{bmatrix} \xrightarrow{-3R_1 + R_2 \Rightarrow R_2} \begin{bmatrix} 1 & 2 & -1 \\ -1 & -8 & \boxed{6} \end{bmatrix}$

8. $\begin{bmatrix} -1 & 3 & 6 \\ 3 & -2 & 4 \end{bmatrix} \xrightarrow{-2R_1 + R_2 \Rightarrow R_2} \begin{bmatrix} -1 & 3 & 6 \\ 5 & -8 & \boxed{-8} \end{bmatrix}$

9. $\left[\begin{array}{ccc|c} 1 & 1 & 1 & 4 \\ 1 & 1 & -1 & 6 \\ 2 & -3 & 1 & -1 \end{array}\right]$

10. $\begin{bmatrix} 1 & 1 & 1 \\ 1 & 1 & -1 \\ 2 & -3 & 1 \end{bmatrix}$

11. $$\left[\begin{array}{cc|c} 1 & 1 & 4 \\ 2 & -1 & 2 \end{array}\right] \xRightarrow{-2R_1+R_2 \Rightarrow R_2} \left[\begin{array}{cc|c} 1 & 1 & 4 \\ 0 & -3 & -6 \end{array}\right] \xRightarrow{-\frac{1}{3}R_2 \Rightarrow R_2} \left[\begin{array}{cc|c} 1 & 1 & 4 \\ 0 & 1 & 2 \end{array}\right]$$

From R_2, $y = 2$. From R_1: $x + y = 4$, $x + 2 = 4$, $x = 2$. Solution: $\boxed{(2,2)}$

12. $$\left[\begin{array}{cc|c} 1 & 1 & 2 \\ 1 & -1 & -4 \\ 2 & 1 & 1 \end{array}\right] \xRightarrow[-2R_1+R_3 \Rightarrow R_3]{-R_2+R_1 \Rightarrow R_2} \left[\begin{array}{cc|c} 1 & 1 & 2 \\ 0 & 2 & 6 \\ 0 & -1 & -3 \end{array}\right] \xRightarrow[2R_3+R_2 \Rightarrow R_3]{\frac{1}{2}R_2 \Rightarrow R_2} \left[\begin{array}{cc|c} 1 & 1 & 2 \\ 0 & 1 & 3 \\ 0 & 0 & 0 \end{array}\right]$$

From R_2, $y = 3$. From R_1: $x + y = 2$, $x + 3 = 2$, $x = -1$. Solution: $\boxed{(-1,3)}$

13. $$\begin{vmatrix} 2 & -3 \\ 4 & 5 \end{vmatrix} = 2(5) - (-3)(4) = 10 - (-12) = 22$$

14. $$\begin{vmatrix} -3 & -4 \\ -2 & 3 \end{vmatrix} = -3(3) - (-4)(-2) = -9 - 8 = -17$$

15. $$\begin{vmatrix} 1 & 2 & 0 \\ 2 & 0 & 3 \\ 1 & -2 & 2 \end{vmatrix} = 1\begin{vmatrix} 0 & 3 \\ -2 & 2 \end{vmatrix} - 2\begin{vmatrix} 2 & 3 \\ 1 & 2 \end{vmatrix} + 0\begin{vmatrix} 2 & 0 \\ 1 & -2 \end{vmatrix} = 1(6) - 2(1) + 0(-4) = 4$$

16. $$\begin{vmatrix} 2 & -1 & 1 \\ 3 & 1 & 0 \\ 0 & 1 & 2 \end{vmatrix} = 2\begin{vmatrix} 1 & 0 \\ 1 & 2 \end{vmatrix} - (-1)\begin{vmatrix} 3 & 0 \\ 0 & 2 \end{vmatrix} + 1\begin{vmatrix} 3 & 1 \\ 0 & 1 \end{vmatrix} = 2(2) + 1(6) + 1(3) = 13$$

17. $$\begin{vmatrix} -6 & -1 \\ -6 & 1 \end{vmatrix}$$

18. $$\begin{vmatrix} 1 & -1 \\ 3 & 1 \end{vmatrix}$$

19. $$x = \frac{\begin{vmatrix} -6 & -1 \\ -6 & 1 \end{vmatrix}}{\begin{vmatrix} 1 & -1 \\ 3 & 1 \end{vmatrix}} = \frac{-6-6}{1-(-3)} = \frac{-12}{4} = -3$$

20. $$y = \frac{\begin{vmatrix} 1 & -6 \\ 3 & -6 \end{vmatrix}}{\begin{vmatrix} 1 & -1 \\ 3 & 1 \end{vmatrix}} = \frac{-6-(-18)}{1-(-3)} = \frac{12}{4} = 3$$

21. $D = \begin{vmatrix} 1 & 1 & 1 \\ 1 & 1 & -1 \\ 2 & -3 & 1 \end{vmatrix} = 1\begin{vmatrix} 1 & -1 \\ -3 & 1 \end{vmatrix} - 1\begin{vmatrix} 1 & -1 \\ 2 & 1 \end{vmatrix} + 1\begin{vmatrix} 1 & 1 \\ 2 & -3 \end{vmatrix} = 1(-2) - 1(3) + 1(-5) = -10$

$N_x = \begin{vmatrix} 4 & 1 & 1 \\ 6 & 1 & -1 \\ -1 & -3 & 1 \end{vmatrix} = 4\begin{vmatrix} 1 & -1 \\ -3 & 1 \end{vmatrix} - 1\begin{vmatrix} 6 & -1 \\ -1 & 1 \end{vmatrix} + 1\begin{vmatrix} 6 & 1 \\ -1 & -3 \end{vmatrix}$ $\boxed{x = \frac{N_x}{D} = \frac{-30}{-10} = 3}$

$= 4(-2) - 1(5) + 1(-17) = -30$

22. See #21. $D = -10$.

$N_z = \begin{vmatrix} 1 & 1 & 4 \\ 1 & 1 & 6 \\ 2 & -3 & -1 \end{vmatrix} = 1\begin{vmatrix} 1 & 6 \\ -3 & -1 \end{vmatrix} - 1\begin{vmatrix} 1 & 6 \\ 2 & -1 \end{vmatrix} + 4\begin{vmatrix} 1 & 1 \\ 2 & -3 \end{vmatrix}$ $\boxed{z = \frac{N_z}{D} = \frac{10}{-10} = -1}$

$= 1(17) - 1(-13) + 4(-5) = 10$

23.
$$\begin{aligned} x^2 + y^2 &= 5 \\ x^2 - y^2 &= 3 \\ \hline 2x^2 &= 8 \\ x^2 &= 4 \\ x &= \pm 2 \end{aligned}$$

Substitute and solve for y:

$$\begin{aligned} x^2 + y^2 &= 5 & x^2 + y^2 &= 5 \\ 2^2 + y^2 &= 5 & (-2)^2 + y^2 &= 5 \\ y^2 &= 1 & y^2 &= 1 \\ y &= \pm 1 & y &= \pm 1 \end{aligned}$$

Solutions: $(2, 1), (2, -1), (-2, 1), (-2, -1)$.

24. $\begin{cases} (1) & x^2 + y^2 = 25 \\ (2) & 4x^2 - 9y = 0 \end{cases}$

Substitute $x^2 = 25 - y^2$ from (1) into (2):

$$\begin{aligned} 4x^2 - 9y &= 0 \\ 4(25 - y^2) - 9y &= 0 \\ 100 - 4y^2 - 9y &= 0 \\ -4y^2 - 9y + 100 &= 0 \\ 4y^2 + 9y - 100 &= 0 \\ (y - 4)(4y + 25) &= 0 \end{aligned}$$

$y - 4 = 0$ **or** $4y + 25 = 0$

$y = 4$ $\quad y = -\frac{25}{4}$

Substitute these and solve for x:

$$\begin{aligned} x^2 &= 25 - y^2 & x^2 &= 25 - y^2 \\ x^2 &= 25 - 4^2 & x^2 &= 25 - \left(-\tfrac{25}{4}\right)^2 \\ x^2 &= 25 - 16 = 9 & x^2 &= 25 - \tfrac{625}{16} = -\tfrac{225}{16} \\ x &= \pm 3 & & x \text{ is nonreal.} \end{aligned}$$

Solutions: $(3, 4), (-3, 4)$

25. $\begin{cases} y \geq x^2 \\ y < x + 3 \end{cases}$

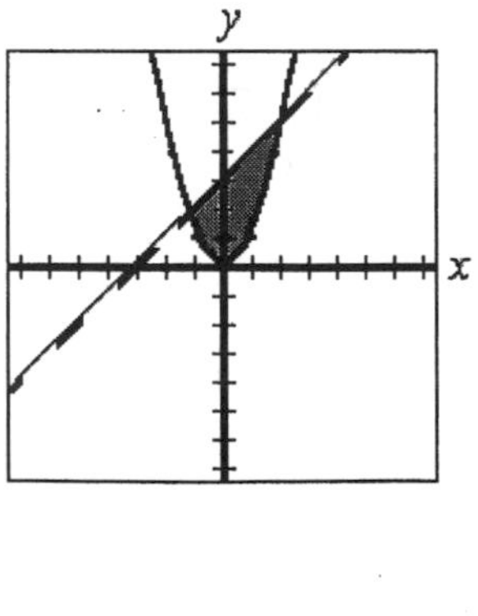

Exercise 14.1 (page 910)

1. $\log_4 16 = x \Rightarrow 4^x = 16 \Rightarrow x = 2$

3. $\log_{25} x = \frac{1}{2} \Rightarrow 25^{1/2} = x \Rightarrow x = 5$

5. one

7. Pascal's

9. 6!

11. 1

13. $3! = 3 \cdot 2 \cdot 1 = 6$

15. $-5! = -1 \cdot 5! = -1 \cdot 5 \cdot 4 \cdot 3 \cdot 2 \cdot 1$
$= -120$

17. $3! + 4! = 3 \cdot 2 \cdot 1 + 4 \cdot 3 \cdot 2 \cdot 1$
$= 6 + 24 = 30$

19. $3!(4!) = 3 \cdot 2 \cdot 1 \cdot 4 \cdot 3 \cdot 2 \cdot 1 = 144$

21. $8(7!) = 8 \cdot 7 \cdot 6 \cdot 5 \cdot 4 \cdot 3 \cdot 2 \cdot 1 = 40{,}320$

23. $\dfrac{9!}{11!} = \dfrac{9!}{11 \cdot 10 \cdot 9!} = \dfrac{1}{11 \cdot 10} = \dfrac{1}{110}$

25. $\dfrac{49!}{47!} = \dfrac{49 \cdot 48 \cdot 47!}{47!} = 49 \cdot 48 = 2352$

27. $\dfrac{9!}{7!\,0!} = \dfrac{9 \cdot 8 \cdot 7!}{7! \cdot 1} = 9 \cdot 8 = 72$

29. $\dfrac{5!}{3!(5-3)!} = \dfrac{5!}{3!2!} = \dfrac{5 \cdot 4 \cdot 3!}{3! \cdot 2 \cdot 1} = \dfrac{5 \cdot 4}{2 \cdot 1} = 10$

31. $\dfrac{7!}{5!(7-5)!} = \dfrac{7!}{5!2!} = \dfrac{7 \cdot 6 \cdot 5!}{5! \cdot 2 \cdot 1} = \dfrac{7 \cdot 6}{2 \cdot 1} = 21$

33. $\dfrac{5!(8-5)!}{4!7!} = \dfrac{5!3!}{4!7!} = \dfrac{5!}{7!} \cdot \dfrac{3!}{4!} = \dfrac{5!}{7 \cdot 6 \cdot 5!} \cdot \dfrac{3!}{4 \cdot 3!} = \dfrac{1}{7 \cdot 6} \cdot \dfrac{1}{4} = \dfrac{1}{168}$

35. $11! = 39{,}916{,}800$

37. $20! = 2.432902008 \times 10^{18}$

39.
$$\begin{aligned}(x+y)^3 &= x^3 + \frac{3!}{1!(3-1)!}x^2y + \frac{3!}{2!(3-2)!}xy^2 + y^3 = x^3 + \frac{3!}{1!2!}x^2y + \frac{3!}{2!1!}xy^2 + y^3 \\ &= x^3 + \frac{3 \cdot 2!}{1!2!}x^2y + \frac{3 \cdot 2!}{2!1!}xy^2 + y^3 \\ &= x^3 + \frac{3}{1}x^2y + \frac{3}{1}xy^2 + y^3 \\ &= x^3 + 3x^2y + 3xy^2 + y^3\end{aligned}$$

41.
$$\begin{aligned}(x-y)^4 &= x^4 + \frac{4!}{1!(4-1)!}x^3(-y) + \frac{4!}{2!(4-2)!}x^2(-y)^2 + \frac{4!}{3!(4-3)!}x(-y)^3 + (-y)^4 \\ &= x^4 + \frac{4!}{1!3!}(-x^3y) + \frac{4!}{2!2!}x^2y^2 + \frac{4!}{3!1!}(-xy^3) + y^4 \\ &= x^4 - \frac{4 \cdot 3!}{1!3!}x^3y + \frac{4 \cdot 3 \cdot 2!}{2! \cdot 2 \cdot 1}x^2y^2 - \frac{4 \cdot 3!}{3!1!}xy^3 + y^4 \\ &= x^4 - \frac{4}{1}x^3y + \frac{12}{2}x^2y^2 - \frac{4}{1}xy^3 + y^4 \\ &= x^4 - 4x^3y + 6x^2y^2 - 4xy^3 + y^4\end{aligned}$$

43. $$\begin{aligned}(2x+y)^3 &= (2x)^3 + \frac{3!}{1!(3-1)!}(2x)^2y + \frac{3!}{2!(3-2)!}2xy^2 + y^3\\ &= 8x^3 + \frac{3!}{1!2!}\cdot 4x^2y + \frac{3!}{2!1!}\cdot 2xy^2 + y^3\\ &= 8x^3 + \frac{3\cdot 2!}{1!2!}\cdot 4x^2y + \frac{3\cdot 2!}{2!1!}\cdot 2xy^2 + y^3\\ &= 8x^3 + \frac{3}{1}\cdot 4x^2y + \frac{3}{1}\cdot 2xy^2 + y^3\\ &= 8x^3 + 12x^2y + 6xy^2 + y^3\end{aligned}$$

45. $$\begin{aligned}(x-2y)^3 &= x^3 + \frac{3!}{1!(3-1)!}x^2(-2y) + \frac{3!}{2!(3-2)!}x(-2y)^2 + (-2y)^3\\ &= x^3 + \frac{3!}{1!2!}\cdot(-2x^2y) + \frac{3!}{2!1!}\cdot 4xy^2 - 8y^3\\ &= x^3 - \frac{3\cdot 2!}{1!2!}\cdot 2x^2y + \frac{3\cdot 2!}{2!1!}\cdot 4xy^2 - 8y^3\\ &= x^3 - \frac{3}{1}\cdot 2x^2y + \frac{3}{1}\cdot 4xy^2 - 8y^3\\ &= x^3 - 6x^2y + 12xy^2 - 8y^3\end{aligned}$$

47. $$\begin{aligned}(2x+3y)^3 &= (2x)^3 + \frac{3!}{1!(3-1)!}(2x)^2(3y) + \frac{3!}{2!(3-2)!}2x(3y)^2 + (3y)^3\\ &= 8x^3 + \frac{3!}{1!2!}\cdot 4x^2(3y) + \frac{3!}{2!1!}\cdot 2x(9y^2) + 27y^3\\ &= 8x^3 + \frac{3\cdot 2!}{1!2!}\cdot 12x^2y + \frac{3\cdot 2!}{2!1!}\cdot 18xy^2 + 27y^3\\ &= 8x^3 + \frac{3}{1}\cdot 12x^2y + \frac{3}{1}\cdot 18xy^2 + 27y^3\\ &= 8x^3 + 36x^2y + 54xy^2 + 27y^3\end{aligned}$$

49. $$\begin{aligned}\left(\frac{x}{2}-\frac{y}{3}\right)^3 &= \left(\frac{x}{2}\right)^3 + \frac{3!}{1!(3-1)!}\left(\frac{x}{2}\right)^2\left(-\frac{y}{3}\right) + \frac{3!}{2!(3-2)!}\left(\frac{x}{2}\right)\left(-\frac{y}{3}\right)^2 + \left(-\frac{y}{3}\right)^3\\ &= \frac{x^3}{8} - \frac{3!}{1!2!}\cdot\frac{x^2}{4}\cdot\frac{y}{3} + \frac{3!}{2!1!}\cdot\frac{x}{2}\cdot\frac{y^2}{9} - \frac{y^3}{27}\\ &= \frac{x^3}{8} - \frac{3\cdot 2!}{1!2!}\cdot\frac{x^2y}{12} + \frac{3\cdot 2!}{2!1!}\cdot\frac{xy^2}{18} - \frac{y^3}{27}\\ &= \frac{x^3}{8} - \frac{3}{1}\cdot\frac{x^2y}{12} + \frac{3}{1}\cdot\frac{xy^2}{18} - \frac{y^3}{27}\\ &= \frac{x^3}{8} - \frac{x^2y}{4} + \frac{xy^2}{6} - \frac{y^3}{27}\end{aligned}$$

51.
$$\begin{aligned}(3+2y)^4 &= 3^4 + \frac{4!}{1!(4-1)!}3^3(2y) + \frac{4!}{2!(4-2)!}3^2(2y)^2 + \frac{4!}{3!(4-3)!}3(2y)^3 + (2y)^4 \\ &= 81 + \frac{4!}{1!3!}\cdot 27(2y) + \frac{4!}{2!2!}\cdot 9(4y^2) + \frac{4!}{3!1!}\cdot 3(8y^3) + 16y^4 \\ &= 81 + \frac{4\cdot 3!}{1!3!}\cdot 54y + \frac{4\cdot 3\cdot 2!}{2!\cdot 2\cdot 1}\cdot 36y^2 + \frac{4\cdot 3!}{3!1!}\cdot 24y^3 + 16y^4 \\ &= 81 + \frac{4}{1}\cdot 54y + \frac{12}{2}\cdot 36y^2 + \frac{4}{1}\cdot 24y^3 + 16y^4 \\ &= 81 + 216y + 216y^2 + 96y^3 + 16y^4\end{aligned}$$

53.
$$\begin{aligned}&\left(\frac{x}{3}-\frac{y}{2}\right)^4 \\ &= \left(\frac{x}{3}\right)^4 + \frac{4!}{1!(4-1)!}\left(\frac{x}{3}\right)^3\left(-\frac{y}{2}\right) + \frac{4!}{2!(4-2)!}\left(\frac{x}{3}\right)^2\left(-\frac{y}{2}\right)^2 + \frac{4!}{3!(4-3)!}\left(\frac{x}{3}\right)\left(-\frac{y}{2}\right)^3 \\ &\qquad + \left(-\frac{y}{2}\right)^4 \\ &= \frac{x^4}{81} - \frac{4!}{1!3!}\cdot\frac{x^3}{27}\cdot\frac{y}{2} + \frac{4!}{2!2!}\cdot\frac{x^2}{9}\cdot\frac{y^2}{4} - \frac{4!}{3!1!}\cdot\frac{x}{3}\cdot\frac{y^3}{8} + \frac{y^4}{16} \\ &= \frac{x^4}{81} - \frac{4\cdot 3!}{1!3!}\cdot\frac{x^3y}{54} + \frac{4\cdot 3\cdot 2!}{2!2!}\cdot\frac{x^2y^2}{36} - \frac{4\cdot 3!}{3!1!}\cdot\frac{xy^3}{24} + \frac{y^4}{16} \\ &= \frac{x^4}{81} - \frac{4}{1}\cdot\frac{x^3y}{54} + \frac{12}{2}\cdot\frac{x^2y^2}{36} - \frac{4}{1}\cdot\frac{xy^3}{24} + \frac{y^4}{16} \\ &= \frac{x^4}{81} - \frac{2x^3y}{27} + \frac{x^2y^2}{6} - \frac{xy^3}{6} + \frac{y^4}{16}\end{aligned}$$

55.

1
1 1
1 2 1
1 3 3 1
1 4 6 4 1
1 5 10 10 5 1
1 6 15 20 15 6 1
1 7 21 35 35 21 7 1
1 8 28 56 70 56 28 8 1
1 9 36 84 126 126 84 36 9 1

57. **Answers may vary.**

59. $\dfrac{n!}{0!(n-0)!} = \dfrac{n!}{0!n!} = \dfrac{n!}{1\cdot n!} = \dfrac{n!}{n!} = 1$

Exercise 14.2 (page 914)

1.
$$\begin{array}{lll} 3x+2y=12 \Rightarrow & 3x+2y=12 & \text{Substitute and solve for } y: \\ 2x-\ \ y=\ 1 \Rightarrow \times 2 & 4x-2y=\ 2 & 2x-y=1 \\ & 7x \qquad = 14 & 2(2)-y=1 \\ & \qquad x=\ 2 & 4-y=1 \\ & & -y=-3 \\ & & y=3 \quad \text{Solution: } (2,3) \end{array}$$

3. $\begin{vmatrix} 2 & -3 \\ 4 & -2 \end{vmatrix} = 2(-2)-(-3)(4) = -4-(-12) = -4+12 = 8$

5. 3

7. 7

9. In the 2nd term, the exponent on b is 1.
Variables: $a^2b^1 = a^2b$
Coef. $= \dfrac{n!}{r!(n-r)!} = \dfrac{3!}{1!2!} = 3$
Term $= 3a^2b$

11. In the 4th term, the exponent on $-y$ is 3.
Variables: $x^1(-y)^3 = -xy^3$
Coef. $= \dfrac{n!}{r!(n-r)!} = \dfrac{4!}{3!1!} = 4$
Term $= 4(-xy^3) = -4xy^3$

13. In the 5th term, the exponent on y is 4.
Variables: x^2y^4
Coef. $= \dfrac{n!}{r!(n-r)!} = \dfrac{6!}{4!2!} = 15$
Term $= 15x^2y^4$

15. In the 3rd term, the exponent on $-y$ is 2.
Variables: $x^6(-y)^2 = x^6y^2$
Coef. $= \dfrac{n!}{r!(n-r)!} = \dfrac{8!}{2!6!} = 28$
Term $= 28x^6y^2$

17. In the 3rd term, the exponent on 3 is 2.
Variables: $x^3(3)^2 = 9x^3$
Coef. $= \dfrac{n!}{r!(n-r)!} = \dfrac{5!}{2!3!} = 10$
Term $= 10(9x^3) = 90x^3$

19. In the 3rd term, the exponent on y is 2.
Variables: $(4x)^3y^2 = 64x^3y^2$
Coef. $= \dfrac{n!}{r!(n-r)!} = \dfrac{5!}{2!3!} = 10$
Term $= 10(64x^3y^2) = 640x^3y^2$

21. In the 2nd term, the exponent on $-3y$ is 1. Variables: $x^3(-3y)^1 = -3x^3y$
Coef. $= \dfrac{n!}{r!(n-r)!} = \dfrac{4!}{1!3!} = 4$; Term $= 4(-3x^3y) = -12x^3y$

23. In the 4th term, the exponent on -5 is 3. Variables: $(2x)^4(-5)^3 = (16x^4)(-125) = -2000x^4$
Coef. $= \dfrac{n!}{r!(n-r)!} = \dfrac{7!}{3!4!} = 35$; Term $= 35(-2000x^4) = -70{,}000x^4$

25. In the 5th term, the exponent on $-3y$ is 4. Variables: $(2x)^1(-3y)^4 = 2x(81y^4) = 162xy^4$
Coef. $= \dfrac{n!}{r!(n-r)!} = \dfrac{5!}{4!1!} = 5$; Term $= 5(162xy^4) = 810xy^4$

27. In the 3rd term, the exponent on $\sqrt{3}y$ is 2. Variables: $(\sqrt{2}x)^4(\sqrt{3}y)^2 = (4x^4)(3y^2) = 12x^4y^2$

Coef. $= \dfrac{n!}{r!(n-r)!} = \dfrac{6!}{2!4!} = 15$; Term $= 15(12x^4y^2) = 180x^4y^2$

29. In the 2nd term, the exponent on $-\dfrac{y}{3}$ is 1. Variables: $\left(\dfrac{x}{2}\right)^3\left(-\dfrac{y}{3}\right)^1 = \left(\dfrac{x^3}{8}\right)\left(-\dfrac{y}{3}\right) = -\dfrac{x^3y}{24}$

Coef. $= \dfrac{n!}{r!(n-r)!} = \dfrac{4!}{1!3!} = 4$; Term $= 4\left(-\dfrac{x^3y}{24}\right) = -\dfrac{x^3y}{6} = -\dfrac{1}{6}x^3y$

31. In the 4th term, the exponent on b is 3.
Variables: $a^{n-3}b^3$

Coef. $= \dfrac{n!}{r!(n-r)!} = \dfrac{n!}{3!(n-3)!}$

Term $= \dfrac{n!}{3!(n-3)!}a^{n-3}b^3$

33. In the 5th term, the exponent on $-b$ is 4.
Variables: $a^{n-4}(-b)^4 = a^{n-4}b^4$

Coef. $= \dfrac{n!}{r!(n-r)!} = \dfrac{n!}{4!(n-4)!}$

Term $= \dfrac{n!}{4!(n-4)!}a^{n-4}b^4$

35. In the rth term, the coefficient on b is $r-1$. Variables: $a^{n-(r-1)}b^{r-1} = a^{n-r+1}b^{r-1}$

Coef. $= \dfrac{n!}{r!(n-r)!} = \dfrac{n!}{(r-1)![n-(r-1)]!} = \dfrac{n!}{(r-1)!(n-r+1)!}$

Term $= \dfrac{n!}{(r-1)!(n-r+1)!}a^{n-r+1}b^{r-1}$

37. **Answers may vary.**

39. $\left(x+\dfrac{1}{x}\right)^{10} = \left(x+x^{-1}\right)^{10}$. The constant term occurs when the exponent is 0.

The $(r+1)$th term of $(x+x^{-1})^{10}$ is $\dfrac{10!}{r!(10-r)!}x^{10-r}\left(x^{-1}\right)^r = \dfrac{10!}{r!(10-r)!}x^{10-r}x^{-r}$.

But $\dfrac{10!}{r!(10-r)!}x^{10-r}x^{-r} = \dfrac{10!}{r!(10-r)!}x^{10-2r}$. If $10-2r=0$, then $r=5$.

The term is $\dfrac{10!}{5!(10-5)!}x^{10-5}x^{-5} = \dfrac{10!}{5!5!} = \dfrac{10\cdot 9\cdot 8\cdot 7\cdot 6\cdot 5!}{5!\cdot 5\cdot 4\cdot 3\cdot 2\cdot 1} = 252$

Exercise 14.3 (page 921)

1. $3(2x^2-4x+7)+4(3x^2+5x-6) = 6x^2-12x+21+12x^2+20x-24 = 18x^2+8x-3$

3. $\dfrac{3a+4}{a-2}+\dfrac{3a-4}{a+2} = \dfrac{(3a+4)(a+2)}{(a-2)(a+2)}+\dfrac{(3a-4)(a-2)}{(a+2)(a-2)}$

$= \dfrac{3a^2+10a+8+3a^2-10a+8}{(a+2)(a-2)} = \dfrac{6a^2+16}{(a+2)(a-2)}$

5. sequence

7. arithmetic; difference

9. arithmetic mean

11. series

13. $1+2+3+4+5$

15. $a_1 = 3(1) - 2 = 1$

17. $a_{25} = 3(25) - 2 = 73$

19. $3, 5, 7, 9, 11$

21. $-5, -8, -11, -14, -17$

23.
$$\begin{aligned} a_n &= a_1 + (n-1)d \\ a_5 &= a_1 + (5-1)d \\ 29 &= 5 + 4d \\ 24 &= 4d \\ 6 &= d \end{aligned}$$
$5, 11, 17, 23, 29$

25.
$$\begin{aligned} a_n &= a_1 + (n-1)d \\ a_6 &= a_1 + (6-1)d \\ -39 &= -4 + 5d \\ -35 &= 5d \\ -7 &= d \end{aligned}$$
$-4, -11, -18, -25, -32$

27.
$$\begin{aligned} a_n &= a_1 + (n-1)d \\ a_6 &= a_1 + (6-1)d \\ -83 &= a_1 + 5(7) \\ -83 &= a_1 + 35 \\ -118 &= a_1 \end{aligned}$$
$-118, -111, -104, -97, -90$

29.
$$\begin{aligned} a_n &= a_1 + (n-1)d \\ a_7 &= a_1 + (7-1)d \\ 16 &= a_1 + 6(-3) \\ 16 &= a_1 - 18 \\ 34 &= a_1 \end{aligned}$$
$34, 31, 28, 25, 22$

31.
$$\begin{aligned} a_n &= a_1 + (n-1)d \\ a_{19} &= a_1 + (19-1)d \\ 131 &= a_1 + 18d \\ a_n &= a_1 + (n-1)d \\ a_{20} &= a_1 + (20-1)d \\ 138 &= a_1 + 19d \end{aligned}$$

$$\begin{aligned} a_1 + 18d = 131 &\Rightarrow \times(-1) & -a_1 - 18d &= -131 \\ a_1 + 19d = 138 &\Rightarrow & a_1 + 19d &= 138 \\ & & d &= 7 \end{aligned}$$

Substitute and solve for a_1:
$$\begin{aligned} a_1 + 18d &= 131 \\ a_1 + 18(7) &= 131 \\ a_1 + 126 &= 131 \\ a_1 &= 5 \Rightarrow 5, 12, 19, 26, 33 \end{aligned}$$

33.
$$\begin{aligned} a_n &= a_1 + (n-1)d \\ a_{30} &= a_1 + (30-1)d \\ &= 7 + 29(12) \\ &= 7 + 348 = 355 \end{aligned}$$

35.
$$\begin{aligned} a_n &= a_1 + (n-1)d \\ a_2 &= a_1 + (2-1)d \\ -4 &= a_1 + d \\ a_n &= a_1 + (n-1)d \\ a_3 &= a_1 + (3-1)d \\ -9 &= a_1 + 2d \end{aligned}$$

$$\begin{aligned} a_1 + d = -4 &\Rightarrow \times(-1) & -a_1 - d &= 4 \\ a_1 + 2d = -9 &\Rightarrow & a_1 + 2d &= -9 \\ & & d &= -5 \end{aligned}$$

Substitute and solve for a_1:
$$\begin{aligned} a_1 + d &= -4 \\ a_1 + (-5) &= -4 \\ a_1 &= 1 \end{aligned}$$

Find the desired term:
$$\begin{aligned} a_n &= a_1 + (n-1)d \\ a_{37} &= a_1 + (37-1)d \\ &= 1 + 36(-5) \\ &= 1 - 180 = \boxed{-179} \end{aligned}$$

37.

$$\begin{aligned} a_n &= a_1 + (n-1)d \\ a_{27} &= a_1 + (27-1)d \\ 263 &= a_1 + 26(11) \\ 263 &= a_1 + 286 \\ -23 &= a_1 \end{aligned}$$

39.

$$\begin{aligned} a_n &= a_1 + (n-1)d \\ a_{44} &= a_1 + (44-1)d \\ 556 &= 40 + 43d \\ 516 &= 43d \\ 12 &= d \end{aligned}$$

41. Form an arithmetic sequence with a 1st term of 2 and a 5th term of 11:

$$\begin{aligned} a_n &= a_1 + (n-1)d \\ a_5 &= a_1 + (5-1)d \\ 11 &= 2 + 4d \\ 9 &= 4d \\ \frac{9}{4} &= d \end{aligned}$$

$$2, \boxed{\tfrac{17}{4}, \tfrac{13}{2}, \tfrac{35}{4}}, 11$$

43. Form an arithmetic sequence with a 1st term of 10 and a 6th term of 20:

$$\begin{aligned} a_n &= a_1 + (n-1)d \\ a_6 &= a_1 + (6-1)d \\ 20 &= 10 + 5d \\ 10 &= 5d \\ 2 &= d \end{aligned}$$

$$10, \boxed{12, 14, 16, 18}, 20$$

45. Form an arithmetic sequence with a 1st term of 10 and a 3rd term of 19:

$$\begin{aligned} a_n &= a_1 + (n-1)d \\ a_3 &= a_1 + (3-1)d \\ 19 &= 10 + 2d \\ 9 &= 2d \\ \frac{9}{2} &= d \end{aligned}$$

$$10, \boxed{\tfrac{29}{2}}, 19$$

47. Form an arithmetic sequence with a 1st term of -4.5 and a 3rd term of 7:

$$\begin{aligned} a_n &= a_1 + (n-1)d \\ a_3 &= a_1 + (3-1)d \\ 7 &= -4.5 + 2d \\ 11.5 &= 2d \\ 5.75 &= d \end{aligned}$$

$$-4.5, \boxed{1.25}, 7$$

49. $a_1 = 1, d = 3, n = 30$

$$a_n = a_1 + (n-1)d = 1 + 29(3) = 88$$

$$S_n = \frac{n(a_1 + a_n)}{2} = \frac{30(1+88)}{2} = 1335$$

51. $a_1 = -5, d = 4, n = 17$

$$a_n = a_1 + (n-1)d = -5 + 16(4) = 59$$

$$S_n = \frac{n(a_1 + a_n)}{2} = \frac{17(-5+59)}{2} = 459$$

53.

$$\begin{aligned} a_n &= a_1 + (n-1)d \\ a_2 &= a_1 + (2-1)d \\ 7 &= a_1 + d \\ a_n &= a_1 + (n-1)d \\ a_3 &= a_1 + (3-1)d \\ 12 &= a_1 + 2d \end{aligned}$$

$$\begin{aligned} a_1 + d &= 7 \Rightarrow \times(-1) & -a_1 - d &= -7 \\ a_1 + 2d &= 12 \Rightarrow & a_1 + 2d &= 12 \\ & & d &= 5 \end{aligned}$$

Substitute and solve for a_1:

$$\begin{aligned} a_1 + d &= 7 \\ a_1 + 5 &= 7 \\ a_1 &= 2, d = 5, n = 12 \\ a_n &= a_1 + (n-1)d = 2 + 11(5) = 57 \end{aligned}$$

$$S_n = \frac{n(a_1 + a_n)}{2} = \frac{12(2+57)}{2} = 354$$

55. $f(n) = 2n + 1 \Rightarrow f(1) = 3$
$f(n) = 2n + 1 = 31$
$2n = 30$
$n = 15$
$S_n = \dfrac{n(a_1 + a_n)}{2} = \dfrac{15(3 + 31)}{2} = 255$

57. $a_1 = 1, d = 1, n = 50$
$a_n = a_1 + (n - 1)d = 1 + 49(1) = 50$
$S_n = \dfrac{n(a_1 + a_n)}{2} = \dfrac{50(1 + 50)}{2} = 1275$

59. $a_1 = 1, d = 2, n = 50$
$a_n = a_1 + (n - 1)d = 1 + 49(2) = 99$
$S_n = \dfrac{n(a_1 + a_n)}{2} = \dfrac{50(1 + 99)}{2} = 2500$

61. $\sum_{k=1}^{4}(3k) = 3(1) + 3(2) + 3(3) + 3(4)$
$= 3 + 6 + 9 + 12$

63. $\sum_{k=4}^{6} k^2 = 4^2 + 5^2 + 6^2 = 16 + 25 + 36$

65. $\sum_{k=1}^{4} 6k = 6(1) + 6(2) + 6(3) + 6(4) = 6 + 12 + 18 + 24 = 60$

67. $\sum_{k=3}^{4}(k^2 + 3) = (3^2 + 3) + (4^2 + 3) = 9 + 3 + 16 + 3 = 31$

69. $\sum_{k=4}^{4}(2k + 4) = 2(4) + 4 = 8 + 4 = 12$

71. $a_1 = 60, d = 50 \Rightarrow 60, 110, 160, 210, 260, 310; n = 121$
$a_n = a_1 + (n - 1)d = 60 + (121 - 1)(50) = 60 + 120(50) = \$6060$

73. $a_1 = 1, d = 1, n = 150, a_n = 150 \Rightarrow S_n = \dfrac{n(a_1 + a_n)}{2} = \dfrac{150(1 + 150)}{2} = 11{,}325$ bricks

75. After 1 sec.: $s = 16(1)^2 = 16$; After 2 sec.: $s = 16(2)^2 = 64$; After 3 sec.: $s = 16(3)^2 = 144$
During 2nd second $\Rightarrow$ falls $64 - 16 = 48$ ft; During 3rd second $\Rightarrow$ falls $144 - 64 = 80$ ft
The sequence of the amounts fallen during each second is $16, 48, 80 \Rightarrow a_1 = 16, d = 32$
$a_n = a_1 + (n - 1)d = 16 + (12 - 1)(32) = 16 + 11(32) = 368$ ft

77. **Answers may vary.**

79. $\sum_{n=1}^{6}\left(\frac{1}{2}n + 1\right)$: $\frac{3}{2}, 2, \frac{5}{2}, 3, \frac{7}{2}, 4$

81. Form an arithmetic sequence with a 1st term of a and a 3rd term of b:

$$a_n = a_1 + (n-1)d$$
$$b = a_1 + (3-1)d$$
$$b = a + 2d$$
$$b - a = 2d$$
$$\frac{b-a}{2} = d \Rightarrow \text{mean} = a_1 + \frac{b-a}{2} = a + \frac{b-a}{2} = \frac{2a}{2} + \frac{b-a}{2} = \frac{a+b}{2}$$

83. $\sum_{k=1}^{5} 5k = 5(1) + 5(2) + 5(3) + 5(4) + 5(5) = 5(1+2+3+4+5) = 5\sum_{k=1}^{5} k.$

85. $\sum_{k=1}^{n} 3 = \sum_{k=1}^{n} 3k^0 = 3(1)^0 + 3(2)^0 + \cdots + 3(n)^0 = 3 + 3 + \cdots + 3 = 3n$

Exercise 14.4 (page 929)

1.

$$x^2 - 5x - 6 \le 0$$
$$(x-6)(x+1) \le 0$$

$x-6$ − − − − − − − − − 0++++
$x+1$ − − − 0+++++++++++

←—— [——————] ——→
-1 6

solution set: $[-1, 6]$

3.

$$\frac{x-4}{x+3} > 0$$

$x-4$ − − − − − − − − 0++++
$x+3$ − − −0+++++++++++

←——)———— (——→
-3 4

solution set: $(-\infty, -3) \cup (4, \infty)$

5. geometric

7. common ratio

9. $S_n = \frac{a_1 - a_1 r^n}{1 - r}$

11. $3, 6, 12, 24, 48$

13. $-5, -1, -\frac{1}{5}, -\frac{1}{25}, -\frac{1}{125}$

15.

$$a_n = a_1 r^{n-1}$$
$$32 = 2r^{3-1}$$
$$32 = 2r^2$$
$$16 = r^2$$
$$\pm 4 = r, \text{ so } r = 4 \ (r > 0)$$

$2, 8, 32, 128, 512$

17.

$$a_n = a_1 r^{n-1}$$
$$-192 = -3r^{4-1}$$
$$-192 = -3r^3$$
$$64 = r^3$$
$$4 = r$$

$-3, -12, -48, -192, -768$

19.

$$a_n = a_1 r^{n-1}$$
$$-4 = -64r^{5-1}$$
$$-4 = -64r^4$$
$$\frac{1}{16} = r^4$$
$$\pm \frac{1}{2} = r, \text{ so } r = -\frac{1}{2} \ (r < 0)$$

$-64, 32, -16, 8, -4$

21.

$$a_n = a_1 r^{n-1}$$
$$-2 = -64r^{6-1}$$
$$-2 = -64r^5$$
$$\frac{1}{32} = r^5$$
$$\frac{1}{2} = r$$

$-64, -32, -16, -8, -4$

23. If the 3rd term is 50 and the 2nd term is 10, then the common ratio $r = 50 \div 10 = 5$.
$a_1 = a_2 \div r = 10 \div 5 = 2$
$2, 10, 50, 250, 1250$

25. $a_n = a_1 r^{n-1} = 7 \cdot 2^{10-1} = 7 \cdot 2^9 = 7 \cdot 512 = 3584$

27.
$$\begin{aligned} a_n &= a_1 r^{n-1} \\ -81 &= a_1(-3)^{8-1} \\ -81 &= a_1(-3)^7 \\ -81 &= -2187 a_1 \\ \tfrac{1}{27} &= a_1 \end{aligned}$$

29.
$$\begin{aligned} a_n &= a_1 r^{n-1} \\ -1944 &= -8r^{6-1} \\ -1944 &= -8r^5 \\ 243 &= r^5 \\ 3 &= r \end{aligned}$$

31. $a_1 = 2,\ a_5 = 162$
$$\begin{aligned} a_n &= a_1 r^{n-1} \\ a_5 &= a_1 r^{5-1} \\ 162 &= 2r^4 \\ 81 &= r^4 \\ \pm 3 &= r \Rightarrow \text{choose } r = 3 \end{aligned}$$
$2, \boxed{6, 18, 54}, 162$

33. $a_1 = -4,\ a_6 = -12500$
$$\begin{aligned} a_n &= a_1 r^{n-1} \\ a_6 &= a_1 r^{6-1} \\ -12500 &= -4r^5 \\ 3125 &= r^5 \\ 5 &= r \end{aligned}$$
$-4, \boxed{-20, -100, -500, -2500}, -12500$

35. $a_1 = 2,\ a_3 = 128$
$$\begin{aligned} a_n &= a_1 r^{n-1} \\ a_3 &= a_1 r^{3-1} \\ 128 &= 2r^2 \\ 64 &= r^2 \\ \pm 8 &= r \Rightarrow \text{choose } r = -2 \end{aligned}$$
$2, \boxed{-16}, 128$

37. $a_1 = 10,\ a_3 = 20$
$$\begin{aligned} a_n &= a_1 r^{n-1} \\ a_3 &= a_1 r^{3-1} \\ 20 &= 10r^2 \\ 2 &= r^2 \\ \pm\sqrt{2} &= r \Rightarrow \text{choose } r = \sqrt{2} \end{aligned}$$
$10, \boxed{10\sqrt{2}}, 20$

39. $a_1 = -50,\ a_3 = 10$
$$\begin{aligned} a_n &= a_1 r^{n-1} \\ a_3 &= a_1 r^{3-1} \\ 10 &= -50r^2 \\ -\tfrac{1}{5} &= r^2 \end{aligned}$$
No such mean exists.

41. $a_1 = 2, r = 3, n = 6;\ S_n = \dfrac{a_1 - a_1 r^n}{1 - r} = \dfrac{2 - 2(3)^6}{1 - 3} = \dfrac{2 - 2(729)}{-2} = \dfrac{-1456}{-2} = 728$

43. $a_1 = 2, r = -3, n = 5;\ S_n = \dfrac{a_1 - a_1 r^n}{1 - r} = \dfrac{2 - 2(-3)^5}{1 - (-3)} = \dfrac{2 - 2(-243)}{4} = \dfrac{488}{4} = 122$

45. $a_1 = 3, r = -2, n = 8;\ S_n = \frac{a_1 - a_1 r^n}{1 - r} = \frac{3 - 3(-2)^8}{1 - (-2)} = \frac{3 - 3(256)}{3} = \frac{-765}{3} = -255$

47. $a_1 = 3, r = 2, n = 7;\ S_n = \frac{a_1 - a_1 r^n}{1 - r} = \frac{3 - 3(2)^7}{1 - 2} = \frac{3 - 3(128)}{-1} = \frac{-381}{-1} = 381$

49. If the 3rd term is $\frac{1}{5}$ and the 2nd term is 1, then the common ratio $r = \frac{1}{5} \div 1 = \frac{1}{5}$.

$a_1 = 1 \div \frac{1}{5} = 5, r = \frac{1}{5}, n = 4;\ S_n = \frac{a_1 - a_1 r^n}{1 - r} = \frac{5 - 5\left(\frac{1}{5}\right)^4}{1 - \frac{1}{5}} = \frac{5 - \frac{1}{125}}{\frac{4}{5}} = \frac{\frac{624}{125}}{\frac{4}{5}} = \frac{156}{25}$

51. If the 4th term is 1 and the 3rd term is -2, then the common ratio $r = 1 \div (-2) = -\frac{1}{2}$.

$a_2 = -2 \div \left(-\frac{1}{2}\right) = 4;\ a_1 = 4 \div \left(-\frac{1}{2}\right) = -8;\ a_1 = -8, r = -\frac{1}{2}, n = 6$

$S_n = \frac{a_1 - a_1 r^n}{1 - r} = \frac{-8 - (-8)\left(-\frac{1}{2}\right)^6}{1 - \left(-\frac{1}{2}\right)} = \frac{-8 + \frac{1}{8}}{\frac{3}{2}} = \frac{-\frac{63}{8}}{\frac{3}{2}} = -\frac{21}{4}$

53. Sequence of population: $500, 500(1.06), 500(1.06)^2, \ldots$

$a_1 = 500, r = 1.06, n = 6 \Rightarrow a_n = a_1 r^{n-1} = 500(1.06)^5 \approx 669$

55. Sequence of amounts: $10000, 10000(0.88), 10000(0.88)^2, \ldots$

$a_1 = 10000, r = 0.88, n = 16 \Rightarrow a_n = a_1 r^{n-1} = 10000(0.88)^{15} \approx \$1{,}469.74$

57. Sequence of values: $70000, 70000(1.06), 70000(1.06)^2, \ldots$

$a_1 = 70000, r = 1.06, n = 13 \Rightarrow a_n = a_1 r^{n-1} = 70000(1.06)^{12} \approx \$140{,}853.75$

59. Sequence of areas: $1, \frac{1}{2}, \frac{1}{4}, \ldots$

$a_1 = 1, r = \frac{1}{2}, n = 12 \Rightarrow a_n = a_1 r^{n-1} 1 = 1\left(\frac{1}{2}\right)^{11} = \left(\frac{1}{2}\right)^{11} \approx 0.0005$

61. Sequence of amounts: $1000(1.03), 1000(1.03)^2, 1000(1.03)^3, \ldots$

$a_1 = 1030, r = 1.03, n = 4 \Rightarrow S_n = \frac{a_1 - a_1 r^n}{1 - r} = \frac{1030 - 1030(1.03)^4}{1 - 1.03} = \frac{-129.2740743}{-0.03}$

$\approx \$4{,}309.14$

63. **Answers may vary.**

65. **Answers may vary.**

67. arithmetic mean

69. **Answers may vary.**

Exercise 14.5 (page 934)

1. $y = 3x^3 - 4$
function

3. $3x = y^2 + 4$
not a function

5. infinite

7. $S_\infty = \frac{a_1}{1 - r}$

9. $a_1 = 8, r = \frac{1}{2}$

$$S_\infty = \frac{a_1}{1-r} = \frac{8}{1-\frac{1}{2}} = \frac{8}{\frac{1}{2}} = 16$$

11. $a_1 = 54, r = \frac{1}{3}$

$$S_\infty = \frac{a_1}{1-r} = \frac{54}{1-\frac{1}{3}} = \frac{54}{\frac{2}{3}} = 81$$

13. $a_1 = 12, r = -\frac{1}{2}$

$$S_\infty = \frac{a_1}{1-r} = \frac{12}{1-\left(-\frac{1}{2}\right)} = \frac{12}{\frac{3}{2}} = 8$$

15. $a_1 = -45, r = -\frac{1}{3}$

$$S_\infty = \frac{a_1}{1-r} = \frac{-45}{1-\left(-\frac{1}{3}\right)} = \frac{-45}{\frac{4}{3}} = -\frac{135}{4}$$

17. $a_1 = \frac{9}{2}, r = \frac{4}{3} \Rightarrow$ no sum $(|r| > 1)$

19. $a_1 = -\frac{27}{2}, r = \frac{2}{3}$

$$S_\infty = \frac{a_1}{1-r} = \frac{-\frac{27}{2}}{1-\frac{2}{3}} = \frac{-\frac{27}{2}}{\frac{1}{3}} = -\frac{81}{2}$$

21. $0.\overline{1} = \frac{1}{10} + \frac{1}{100} + \frac{1}{1000} + \cdots \Rightarrow a_1 = \frac{1}{10}, r = \frac{1}{10} \Rightarrow S_\infty = \frac{a_1}{1-r} = \frac{\frac{1}{10}}{1-\frac{1}{10}} = \frac{\frac{1}{10}}{\frac{9}{10}} = \frac{1}{9}$

23. $-0.\overline{3} = -\frac{3}{10} - \frac{3}{100} - \frac{3}{1000} + \cdots \Rightarrow a_1 = -\frac{3}{10}, r = \frac{1}{10} \Rightarrow S_\infty = \frac{a_1}{1-r} = \frac{-\frac{3}{10}}{1-\frac{1}{10}} = \frac{-\frac{3}{10}}{\frac{9}{10}} = -\frac{1}{3}$

25. $0.\overline{12} = \frac{12}{100} + \frac{12}{10{,}000} + \frac{12}{1{,}000{,}000} + \cdots \Rightarrow a_1 = \frac{12}{100}, r = \frac{1}{100}$

$$S_\infty = \frac{a_1}{1-r} = \frac{\frac{12}{100}}{1-\frac{1}{100}} = \frac{\frac{12}{100}}{\frac{99}{100}} = \frac{12}{99} = \frac{4}{33}$$

27. $0.\overline{75} = \frac{75}{100} + \frac{75}{10{,}000} + \frac{75}{1{,}000{,}000} + \cdots \Rightarrow a_1 = \frac{75}{100}, r = \frac{1}{100}$

$$S_\infty = \frac{a_1}{1-r} = \frac{\frac{75}{100}}{1-\frac{1}{100}} = \frac{\frac{75}{100}}{\frac{99}{100}} = \frac{75}{99} = \frac{25}{33}$$

29. Distance ball travels down $= 10 + 5 + 2.5 + \cdots = \frac{a_1}{1-r} = \frac{10}{1-\frac{1}{2}} = \frac{10}{\frac{1}{2}} = 20$

Distance ball travels up $= 5 + 2.5 + 1.25 + \cdots = \frac{a_1}{1-r} = \frac{5}{1-\frac{1}{2}} = \frac{5}{\frac{1}{2}} = 10$

Total distance $= 20 + 10 = 30$ m

31. $S_\infty = \frac{a_1}{1-r} = \frac{1000}{1-0.8} = \frac{1000}{0.2}$
$= 5{,}000$ moths

33. **Answers may vary.**

35.
$$\begin{aligned} S_\infty &= \frac{a_1}{1-r} \\ 5 &= \frac{1}{1-r} \\ 5(1-r) &= 1 \\ 5-5r &= 1 \\ 4 &= 5r \\ \frac{4}{5} &= r \end{aligned}$$

37. $0.\overline{9} = \frac{9}{10} + \frac{9}{100} + \frac{9}{1000} + \cdots$

$a = \frac{9}{10}, r = \frac{1}{10}$

$$S_\infty = \frac{a_1}{1-r} = \frac{\frac{9}{10}}{1-\frac{1}{10}} = \frac{\frac{9}{10}}{\frac{9}{10}} = \frac{9}{9} = 1$$

39. No. $0.999999 = \frac{999{,}999}{1{,}000{,}000} < 1$

Exercise 14.6 (page 943)

1.
$$\begin{aligned} |2x-3| &= 9 \\ 2x-3 = 9 \quad &\textbf{or} \quad 2x-3 = -9 \\ 2x = 12 \quad &\quad\; 2x = -6 \\ x = 6 \quad &\quad\; x = -3 \end{aligned}$$

3.
$$\begin{aligned} \frac{3}{x-5} &= \frac{8}{x} \\ 3x &= 8(x-5) \\ 3x &= 8x-40 \\ -5x &= -40 \\ x &= 8 \end{aligned}$$

5. $p \cdot q$

7. $P(n,r)$

9. $n!$

11. $\binom{n}{r}$; combinations

13. 1

15. $P(3,3) = \frac{3!}{(3-3)!} = \frac{3!}{0!} = \frac{6}{1} = 6$

17. $P(5,3) = \frac{5!}{(5-3)!} = \frac{5!}{2!} = \frac{120}{2} = 60$

19. $P(2,2) \cdot P(3,3) = \frac{2!}{(2-2)!} \cdot \frac{3!}{(3-3)!} = \frac{2!}{0!} \cdot \frac{3!}{0!} = \frac{2}{1} \cdot \frac{6}{1} = 12$

21. $\frac{P(5,3)}{P(4,2)} = \frac{\frac{5!}{(5-3)!}}{\frac{4!}{(4-2)!}} = \frac{\frac{5!}{2!}}{\frac{4!}{2!}} = \frac{\frac{120}{2}}{\frac{24}{2}} = \frac{60}{12} = 5$

23. $\frac{P(6,2) \cdot P(7,3)}{P(5,1)} = \frac{\frac{6!}{(6-2)!} \cdot \frac{7!}{(7-3)!}}{\frac{5!}{(5-1)!}} = \frac{\frac{6!}{4!} \cdot \frac{7!}{4!}}{\frac{5!}{4!}} = \frac{\frac{720}{24} \cdot \frac{5040}{24}}{\frac{120}{24}} = \frac{30 \cdot 210}{5} = 1{,}260$

25. $C(5,3) = \frac{5!}{3!(5-3)!} = \frac{5!}{3!2!} = \frac{120}{6 \cdot 2} = \frac{120}{12} = 10$

27. $\binom{6}{3} = \frac{6!}{3!(6-3)!} = \frac{6!}{3!3!} = \frac{720}{6 \cdot 6} = \frac{720}{36} = 20$

29. $\binom{5}{4}\binom{5}{3} = \frac{5!}{4!(5-4)!} \cdot \frac{5!}{3!(5-3)!} = \frac{5!}{4!1!} \cdot \frac{5!}{3!2!} = \frac{120}{24 \cdot 1} \cdot \frac{120}{6 \cdot 2} = \frac{120}{24} \cdot \frac{120}{12} = 5 \cdot 10 = 50$

31. $\frac{C(38,37)}{C(19,18)} = \frac{\frac{38!}{37!(38-37)!}}{\frac{19!}{18!(19-18)!}} = \frac{\frac{38 \cdot 37!}{37!1!}}{\frac{19 \cdot 18!}{18!1!}} = \frac{\frac{38}{1}}{\frac{19}{1}} = \frac{38}{19} = 2$

33. $C(12,0)C(12,12) = \frac{12!}{0!(12-0)!} \cdot \frac{12!}{12!(12-12)!} = \frac{12!}{0!12!} \cdot \frac{12!}{12!0!} = \frac{12!}{12!} \cdot \frac{12!}{12!} = 1 \cdot 1 = 1$

35. $C(n,2) = \frac{n!}{2!(n-2)!}$

37. $(x+y)^4 = \binom{4}{0}x^4y^0 + \binom{4}{1}x^3y^1 + \binom{4}{2}x^2y^2 + \binom{4}{3}x^1y^3 + \binom{4}{4}x^0y^4$

$= \frac{4!}{0!4!}x^4 + \frac{4!}{1!3!}x^3y + \frac{4!}{2!2!}x^2y^2 + \frac{4!}{3!1!}xy^3 + \frac{4!}{4!0!}y^4 = x^4 + 4x^3y + 6x^2y^2 + 4xy^3 + y^4$

39. $(2x+y)^3 = \binom{3}{0}(2x)^3y^0 + \binom{3}{1}(2x)^2y^1 + \binom{3}{2}(2x)^1y^2 + \binom{3}{3}(2x)^0y^3$

$= \frac{3!}{0!3!} \cdot 8x^3 + \frac{3!}{1!2!} \cdot 4x^2y + \frac{3!}{2!1!} \cdot 2xy^2 + \frac{3!}{3!0!}y^3$

$= 8x^3 + 12x^2y + 6xy^2 + y^3$

41. $(3x-2)^4$

$= \binom{4}{0}(3x)^4(-2)^0 + \binom{4}{1}(3x)^3(-2)^1 + \binom{4}{2}(3x)^2(-2)^2 + \binom{4}{3}(3x)^1(-2)^3$

$+ \binom{4}{4}(3x)^0(-2)^4$

$= \frac{4!}{0!4!} \cdot 81x^4(1) + \frac{4!}{1!3!} \cdot 27x^3(-2) + \frac{4!}{2!2!} \cdot 9x^2(4) + \frac{4!}{3!1!} \cdot 3x(-8) + \frac{4!}{4!0!} \cdot 1(16)$

$= 81x^4 - 216x^3 + 216x^2 - 96x + 16$

43. $\binom{5}{3}x^2(-5y)^3 = \frac{5!}{3!2!}x^2(-125y^3) = 10x^2(-125y^3) = -1{,}250x^2y^3$

45. $\binom{4}{1}(x^2)^3(-y^3)^1 = \frac{4!}{1!3!}x^6(-y^3) = -4x^6y^3$

47. $7 \cdot 5 = 35$

49. $10 \cdot 10 \cdot 10 \cdot 10 \cdot 10 \cdot 10 = 1{,}000{,}000$

51. $9 \cdot 9 \cdot 8 \cdot 7 \cdot 6 \cdot 5 = 136{,}080$

53. $8 \cdot 10 \cdot 10 \cdot 10 \cdot 10 \cdot 10 \cdot 10 = 8{,}000{,}000$

55. $6! = 720$

57. $4! \cdot 5! = 24 \cdot 120 = 2{,}880$

59. $25 \cdot 24 \cdot 23 = 13{,}800$

61. $P(10,3) = \frac{10!}{(10-3)!} = \frac{10!}{7!} = 720$

63. $9 \cdot 10 \cdot 10 \cdot 1 \cdot 1 = 900$

65. $C(14,3) = \frac{14!}{3!(14-3)!} = \frac{14!}{3!11!} = 364$

67. $C(5,3) = 10 \Rightarrow 5$ persons

69. $C(100,6) = \frac{100!}{6!(100-6)!} = \frac{100!}{6!94!} = \frac{100 \cdot 99 \cdot 98 \cdot 97 \cdot 96 \cdot 95 \cdot 94!}{6!94!} = 1{,}192{,}052{,}400$

71. $C(3,2) \cdot C(4,2) = \frac{3!}{2!1!} \cdot \frac{4!}{2!2!} = 3 \cdot 6 = 18$

73. $C(12,2) \cdot C(10,3) = \frac{12!}{2!10!} \cdot \frac{10!}{3!7!} = 66 \cdot 120 = 7{,}920$

75. **Answers may vary.**

77. Consider the two people who insist on standing together as one person. Then there are a total of 4 "persons" to be arranged. This can be done in $4! = 24$ ways. However, the two people who are standing together can be arranged in 2 different ways, so there are $24 \cdot 2 = 48$ arrangements.

Exercise 14.7 (page 949)

1.
$$\begin{aligned} 5^{4x} &= \frac{1}{125} \\ 5^{4x} &= 5^{-3} \\ 4x &= -3 \\ x &= -\frac{3}{4} \end{aligned}$$

3.
$$\begin{aligned} 2^{x^2-2x} &= 8 \\ 2^{x^2-2x} &= 2^3 \\ x^2 - 2x &= 3 \\ x^2 - 2x - 3 &= 0 \\ (x+1)(x-3) &= 0 \end{aligned}$$
$$x+1=0 \quad \textbf{or} \quad x-3=0$$
$$x=-1 \qquad\qquad x=3$$

5.
$$\begin{aligned} 3^{x^2+4x} &= \frac{1}{81} \\ 3^{x^2+4x} &= 3^{-4} \\ x^2+4x &= -4 \\ x^2+4x+4 &= 0 \\ (x+2)(x+2) &= 0 \end{aligned}$$
$$x+2=0 \quad \textbf{or} \quad x+2=0$$
$$x=-2 \qquad\qquad x=-2$$

7. experiment

9. $\frac{s}{n}$

11. 0

13. **a.** 6 **b.** 52 **c.** $\frac{6}{52}; \frac{3}{26}$

15. $\{(1,H),(2,H),(3,H),(4,H),(5,H),(6,H),(1,T),(2,T),(3,T),(4,T),(5,T),(6,T)\}$

17. $\{A,B,C,D,E,F,G,H,I,J,K,L,M,N,O,P,Q,R,S,T,U,V,W,X,Y,Z\}$

19. $\frac{1}{6}$

21. $\frac{4}{6}=\frac{2}{3}$

23. $\frac{19}{42}$

25. $\frac{13}{42}$

27. $\frac{3}{8}$

29. $\frac{0}{8}=0$

31. rolls of 4: $\{(1,3),(2,2),(3,1)\}$
Probability $=\frac{3}{36}=\frac{1}{12}$

33. $\frac{\#\text{ red}}{\#\text{ eggs}}=\frac{5}{12}$

35. $\frac{\#\text{ ways to get 6 diamonds}}{\#\text{ ways to get 6 cards from the deck of 52}}=\frac{\binom{13}{6}}{\binom{52}{6}}=\frac{1716}{20{,}358{,}520}=\frac{33}{391{,}510}$

37. $\frac{\#\text{ ways to get 5 clubs}}{\#\text{ ways to get 5 cards from the 26 black cards}}=\frac{\binom{13}{5}}{\binom{26}{5}}=\frac{1287}{65780}=\frac{9}{460}$

39. $SSSS, SSSF, SSFS, SSFF, SFSS, SFSF, SFFS, SFFF,$
$FSSS, FSSF, FSFS, FSFF, FFSS, FFSF, FFFS, FFFF$

41. $SFFF, FSFF, FFSF, FFFS \Rightarrow \frac{4}{16}=\frac{1}{4}$

43. $SSSF, SSFS, SFSS, FSSS \Rightarrow \frac{4}{16}=\frac{1}{4}$

45. 1

47. $\frac{32}{119}$

49. $\frac{\binom{8}{4}}{\binom{10}{4}}=\frac{70}{210}=\frac{1}{3}$

51. **Answers may vary.**

53.
$$P(\text{math and art})=P(\text{math})\cdot P(\text{art}|\text{math})$$
$$0.10=0.30\cdot P(\text{art}|\text{math})$$
$$\frac{0.10}{0.30}=P(\text{art}|\text{math})$$
$$\frac{1}{3}=P(\text{art}|\text{math})$$

Chapter 14 Summary (page 952)

1. $(4!)(3!)=4\cdot3\cdot2\cdot1\cdot3\cdot2\cdot1=144$

2. $\frac{5!}{3!}=\frac{5\cdot4\cdot3!}{3!}=5\cdot4=20$

3. $\frac{6!}{2!(6-2)!}=\frac{6!}{2!4!}=\frac{6\cdot5\cdot4!}{2\cdot1\cdot4!}=\frac{30}{2}=15$

4. $\frac{12!}{3!(12-3)!}=\frac{12!}{3!9!}=\frac{12\cdot11\cdot10\cdot9!}{3\cdot2\cdot1\cdot9!}=\frac{1320}{6}=220$

5. $(n-n)! = 0! = 1$

6. $\frac{8!}{7!} = \frac{8 \cdot 7!}{7!} = 8$

7.
$$\begin{aligned}(x+y)^5 &= x^5 + \frac{5!}{1!(5-1)!}x^4y + \frac{5!}{2!(5-2)!}x^3y^2 + \frac{5!}{3!(5-3)!}x^2y^3 + \frac{5!}{4!(5-4)!}xy^4 + y^5 \\ &= x^5 + \frac{5!}{1!4!}x^4y + \frac{5!}{2!3!}x^3y^2 + \frac{5!}{3!2!}x^2y^3 + \frac{5!}{4!1!}xy^4 + y^5 \\ &= x^5 + \frac{5 \cdot 4!}{1!4!}x^4y + \frac{5 \cdot 4 \cdot 3!}{2 \cdot 1 \cdot 3!}x^3y^2 + \frac{5 \cdot 4 \cdot 3!}{3! \cdot 2 \cdot 1}x^2y^3 + \frac{5 \cdot 4!}{4! \cdot 1}xy^4 + y^5 \\ &= x^5 + \frac{5}{1}x^4y + \frac{20}{2}x^3y^2 + \frac{20}{2}x^2y^3 + \frac{5}{1}xy^4 + y^5 \\ &= x^5 + 5x^4y + 10x^3y^2 + 10x^2y^3 + 5xy^4 + y^5\end{aligned}$$

8.
$$\begin{aligned}(x-y)^4 &= x^4 + \frac{4!}{1!(4-1)!}x^3(-y) + \frac{4!}{2!(4-2)!}x^2(-y)^2 + \frac{4!}{3!(4-3)!}x(-y)^3 + (-y)^4 \\ &= x^4 + \frac{4!}{1!3!}(-x^3y) + \frac{4!}{2!2!}x^2y^2 + \frac{4!}{3!1!}(-xy^3) + y^4 \\ &= x^4 - \frac{4 \cdot 3!}{1!3!}x^3y + \frac{4 \cdot 3 \cdot 2!}{2! \cdot 2 \cdot 1}x^2y^2 - \frac{4 \cdot 3!}{3!1!}xy^3 + y^4 \\ &= x^4 - \frac{4}{1}x^3y + \frac{12}{2}x^2y^2 - \frac{4}{1}xy^3 + y^4 = x^4 - 4x^3y + 6x^2y^2 - 4xy^3 + y^4\end{aligned}$$

9.
$$\begin{aligned}(4x-y)^3 &= (4x)^3 + \frac{3!}{1!(3-1)!}(4x)^2(-y) + \frac{3!}{2!(3-2)!}(4x)(-y)^2 + (-y)^3 \\ &= 64x^3 + \frac{3!}{1!2!} \cdot (-16x^2y) + \frac{3!}{2!1!} \cdot 4xy^2 - y^3 \\ &= 64x^3 - \frac{3 \cdot 2!}{1!2!} \cdot 16x^2y + \frac{3 \cdot 2!}{2!1!} \cdot 4xy^2 - y^3 \\ &= 64x^3 - \frac{3}{1} \cdot 16x^2y + \frac{3}{1} \cdot 4xy^2 - y^3 = 64x^3 - 48x^2y + 12xy^2 - y^3\end{aligned}$$

10.
$$\begin{aligned}(x+4y)^3 &= x^3 + \frac{3!}{1!(3-1)!}x^2(4y) + \frac{3!}{2!(3-2)!}x(4y)^2 + (4y)^3 \\ &= x^3 + \frac{3!}{1!2!} \cdot 4x^2y + \frac{3!}{2!1!} \cdot 16xy^2 + 64y^3 \\ &= x^3 + \frac{3 \cdot 2!}{1!2!} \cdot 4x^2y + \frac{3 \cdot 2!}{2!1!} \cdot 16xy^2 + 64y^3 \\ &= x^3 + \frac{3}{1} \cdot 4x^2y + \frac{3}{1} \cdot 16xy^2 + 64y^3 = x^3 + 12x^2y + 48xy^2 + 64y^3\end{aligned}$$

11. In the 3rd term, the exponent on 3 is 2.
Variables: x^2y^2
Coef. $= \frac{n!}{r!(n-r)!} = \frac{4!}{2!2!} = 6$
Term $= 6x^2y^2$

12. In the 4th term, the exponent on $-y$ is 3.
Variables: $x^2(-y)^3 = -x^2y^3$
Coef. $= \frac{n!}{r!(n-r)!} = \frac{5!}{3!2!} = 10$
Term $= 10(-x^2y^3) = -10x^2y^3$

13. 2nd term: The exponent on $-4y$ is 1. Variables: $(3x)^2(-4y)^1 = (9x^2)(-4y) = -36x^2y$

Coef. $= \dfrac{n!}{r!(n-r)!} = \dfrac{3!}{2!1!} = 3$; Term $= 3(-36x^2y) = -108x^2y$

14. 3rd term: The exponent on $3y$ is 2. Variables: $(4x)^2(3y)^2 = (16x^2)(9y^2) = 144x^2y^2$

Coef. $= \dfrac{n!}{r!(n-r)!} = \dfrac{4!}{2!2!} = 6$; Term $= 6(144x^2y^2) = 864x^2y^2$

15. $a_n = a_1 + (n-1)d = 7 + (8-1)5 = 7 + 7 \cdot 5 = 42$

16.
$$\begin{aligned} a_n &= a_1 + (n-1)d \\ 242 &= a_1 + (9-1)d \\ 242 &= a_1 + 8d \\ a_n &= a_1 + (n-1)d \\ 212 &= a_1 + (7-1)d \\ 212 &= a_1 + 6d \end{aligned}$$

$$\begin{array}{ll} a_1 + 8d = 242 \Rightarrow \times(-1) & -a_1 - 8d = -242 \\ \underline{a_1 \ + 6d = 212} \Rightarrow & \underline{a_1 + 6d = \ \ 212} \\ & -2d = -30 \\ & d = \ \ 15 \end{array}$$

Substitute and solve for a_1:

$$\begin{aligned} a_1 + 6d &= 212 \\ a_1 + 6(15) &= 212 \\ a_1 + 90 &= 212 \\ a_1 &= 122 \Rightarrow 122, 137, 152, 167, 182 \end{aligned}$$

17. Form an arithmetic sequence with a 1st term of 8 and a 4th term of 25:

$$\begin{aligned} a_n &= a_1 + (n-1)d \\ 25 &= 8 + (4-1)d \\ 17 &= 3d \\ \tfrac{17}{3} &= d \Rightarrow 8, \boxed{\tfrac{41}{3}, \tfrac{58}{3}}, 25 \end{aligned}$$

18. $a_1 = 11, d = 7, n = 20$

$a_n = a_1 + (n-1)d = 11 + 19(7) = 144$

$S_n = \dfrac{n(a_1 + a_n)}{2} = \dfrac{20(11 + 144)}{2} = 1{,}550$

19. $a_1 = 9, d = -\frac{5}{2}, n = 10$

$a_n = a_1 + (n-1)d = 9 + 9\left(-\frac{5}{2}\right) = -\frac{27}{2}$

$S_n = \dfrac{n(a_1 + a_n)}{2} = \dfrac{10\left(9 - \frac{27}{2}\right)}{2} = -\dfrac{45}{2}$

20. $\sum_{k=4}^{6} \frac{1}{2}k = \frac{1}{2}(4) + \frac{1}{2}(5) + \frac{1}{2}(6) = 2 + \frac{5}{2} + 3 = \frac{15}{2}$

21. $\sum_{k=2}^{5} 7k^2 = 7(2)^2 + 7(3)^2 + 7(4)^2 + 7(5)^2 = 28 + 63 + 112 + 175 = 378$

22. $\sum_{k=1}^{4} (3k-4) = (3(1)-4) + (3(2)-4) + (3(3)-4) + (3(4)-4) = -1 + 2 + 5 + 8 = 14$

23. $\sum_{k=10}^{10} 36k = 36(10) = 360$

24. If the 5th term is $\frac{3}{2}$ and the 4th term is 3, then the common ratio $r = \frac{3}{2} \div 3 = \frac{1}{2}$.

$a_3 = a_4 \div r = 3 \div \frac{1}{2} = 6$

$a_2 = a_3 \div r = 6 \div \frac{1}{2} = 12$

$a_1 = a_2 \div r = 12 \div \frac{1}{2} = 24$

$24, 12, 6, 3, \frac{3}{2}$

25.
$$\begin{aligned} a_n &= a_1 r^{n-1} \\ &= \frac{1}{8}(2)^{6-1} \\ &= \frac{1}{8}(32) = 4 \end{aligned}$$

26. $a_1 = -6$, $a_4 = 384$
$$\begin{aligned} a_n &= a_1 r^{n-1} \\ 384 &= -6r^{4-1} \\ -64 &= r^3 \\ -4 &= r \Rightarrow -6, \boxed{24, -96}, 384 \end{aligned}$$

27. $a_1 = 162, r = \frac{1}{3}, n = 7;\ S_n = \dfrac{a_1 - a_1 r^n}{1 - r} = \dfrac{162 - 162\left(\frac{1}{3}\right)^7}{1 - \frac{1}{3}} = \dfrac{162 - 162\left(\frac{1}{2187}\right)}{-\frac{2}{3}} = \dfrac{\frac{4372}{27}}{-\frac{2}{3}} = \dfrac{2186}{9}$

28. $a_1 = \frac{1}{8}, r = -2, n = 8;\ S_n = \dfrac{a_1 - a_1 r^n}{1 - r} = \dfrac{\frac{1}{8} - \frac{1}{8}(-2)^8}{1 - (-2)} = \dfrac{\frac{1}{8} - \frac{1}{8}(256)}{3} = \dfrac{-\frac{255}{8}}{3} = -\dfrac{85}{8}$

29. $a_1 = 25, r = \frac{4}{5};\ S_\infty = \dfrac{a_1}{1 - r} = \dfrac{25}{1 - \frac{4}{5}} = \dfrac{25}{\frac{1}{5}} = 125$

30. $0.\overline{05} = \dfrac{5}{100} + \dfrac{5}{10{,}000} + \dfrac{5}{1{,}000{,}000} + \cdots \Rightarrow a_1 = \dfrac{5}{100}, r = \dfrac{1}{100}$

$S_\infty = \dfrac{a_1}{1 - r} = \dfrac{\frac{5}{100}}{1 - \frac{1}{100}} = \dfrac{\frac{5}{100}}{\frac{99}{100}} = \dfrac{5}{99}$

31. $17 \cdot 8 = 136$

32. $P(7,7) = \dfrac{7!}{(7-7)!} = \dfrac{7!}{0!} = 7! = 5{,}040$

33. $P(7,0) = \dfrac{7!}{(7-0)!} = \dfrac{7!}{7!} = 1$

34. $P(8,6) = \dfrac{8!}{(8-6)!} = \dfrac{8!}{2!} = \dfrac{40{,}320}{2} = 20{,}160$

35. $\dfrac{P(9,6)}{P(10,7)} = \dfrac{\frac{9!}{(9-6)!}}{\frac{10!}{(10-7)!}} = \dfrac{\frac{9!}{3!}}{\frac{10!}{3!}} = \dfrac{9!}{3!} \cdot \dfrac{3!}{10!} = \dfrac{9!}{10!} = \dfrac{9!}{10 \cdot 9!} = \dfrac{1}{10}$

36. $C(7,7) = \dfrac{7!}{7!(7-7)!} = \dfrac{7!}{7!0!} = \dfrac{7!}{7!} = 1$

37. $C(7,0) = \dfrac{7!}{0!(7-0)!} = \dfrac{7!}{0!7!} = \dfrac{7!}{7!} = 1$

38. $\dbinom{8}{6} = \dfrac{8!}{6!(8-6)!} = \dfrac{8 \cdot 7 \cdot 6!}{6!2!} = \dfrac{56}{2} = 28$

39. $\dbinom{9}{6} = \dfrac{9!}{6!(9-6)!} = \dfrac{9 \cdot 8 \cdot 7 \cdot 6!}{6!3!} = \dfrac{504}{6} = 84$

40. $C(6,3) \cdot C(7,3) = \dfrac{6!}{3!(6-3)!} \cdot \dfrac{7!}{3!(7-3)!} = \dfrac{6!}{3!3!} \cdot \dfrac{7!}{3!4!} = 20 \cdot 35 = 700$

41. $\frac{C(7,3)}{C(6,3)} = \frac{\frac{7!}{3!(7-3)!}}{\frac{6!}{3!(6-3)!}} = \frac{\frac{7!}{3!4!}}{\frac{6!}{3!3!}} = \frac{35}{20} = \frac{7}{4}$

42. Sequence of amounts: $5000, 5000(0.80), 5000(0.80)^2, \ldots$
$a_1 = 5000, r = 0.80, n = 6 \Rightarrow a_n = a_1 r^{n-1} = 5000(0.80)^5 \approx \$1{,}638.40$

43. Sequence of amounts: $25700, 25700(1.18), 25700(1.18)^2, \ldots$
$a_1 = 25700, r = 1.18, n = 11 \Rightarrow a_n = a_1 r^{n-1} = 25700(1.18)^{10} \approx \$134{,}509.57$

44. $a_1 = 300, d = 75;$
$$\begin{aligned} a_n &= a_1 + (n-1)d \\ 1200 &= 300 + (n-1)(75) \\ 900 &= 75n - 75 \\ 975 &= 75n \Rightarrow n = 13 \Rightarrow \text{in 12 years} \end{aligned}$$

45. $a_1 = 16, d = 32.$
$a_n = a_1 + (n-1)d = 16 + (10-1)(32) = 16 + 9(32) = 304$
$S_n = \frac{n(a_1 + a_n)}{2} = \frac{10(16 + 304)}{2} = 1600 \text{ ft}$

46. $5! = 120$

47. $5! \cdot 3! = 720$

48. $\binom{10}{3} = \frac{10!}{3!7!} = 120$

49. $\binom{5}{2}\binom{6}{2} = 10 \cdot 15 = 150$

50. $\frac{6}{16} = \frac{3}{8}$

51. $\frac{8}{16} = \frac{1}{2}$

52. $\frac{14}{16} = \frac{7}{8}$

53. roll of 11 $\Rightarrow \{(5, 6), (6, 5)\}$
$\frac{\text{\# roll of 11}}{\text{total \# rolls}} = \frac{2}{36} = \frac{1}{18}$

54. impossible $\Rightarrow 0$

55. $\frac{\text{\# tens}}{\text{total \# cards}} = \frac{4}{52} = \frac{1}{13}$

56. $\frac{\text{\# ways to get 3 aces} \cdot \text{\# ways to get 2 other cards}}{\text{\# ways to draw 5 cards}} = \frac{\binom{4}{3}\binom{48}{2}}{\binom{52}{5}} = \frac{4 \cdot 1128}{2{,}598{,}960} = \frac{4512}{2{,}598{,}960} = \frac{94}{54145}$

57. $\frac{\text{\# ways to get 5 spades}}{\text{\# ways to get 5 cards}} = \frac{\binom{13}{5}}{\binom{52}{5}} = \frac{1287}{2{,}598{,}960} = \frac{33}{66{,}640}$

Chapter 14 Test (page 956)

1. $\frac{7!}{4!} = \frac{7 \cdot 6 \cdot 5 \cdot 4!}{4!} = 7 \cdot 6 \cdot 5 = 210$

2. $0! = 1$

3. In the 2nd term, the exponent on $-y$ is 1.
Variables: $x^4(-y)^1 = -x^4y$
Coef. $= \dfrac{n!}{r!(n-r)!} = \dfrac{5!}{1!4!} = 5$
Term $= 5(-x^4y) = -5x^4y$

4. In the 3rd term, the exponent on $2y$ is 2.
Variables: $(x)^2(2y)^2 = 4x^2y^2$
Coef. $= \dfrac{n!}{r!(n-r)!} = \dfrac{4!}{2!2!} = 6$
Term $= 6(4x^2y^2) = 24x^2y^2$

5. $a_1 = 3, d = 7, n = 10;\ a_n = a_1 + (n-1)d = 3 + (10-1)(7) = 3 + 9(7) = 66$

6. $a_1 = -2, d = 5, n = 12;\ a_n = a_1 + (n-1)d = -2 + (12-1)(5) = -2 + 11(5) = 53$
$S_n = \dfrac{n(a_1 + a_n)}{2} = \dfrac{12(-2+53)}{2} = \dfrac{12(51)}{2} = 306$

7. Form an arithmetic sequence with a 1st term of 2 and a 4th term of 98:
$a_n = a_1 + (n-1)d$
$98 = 2 + (4-1)d$
$96 = 3d$
$32 = d \Rightarrow 2, \boxed{34, 66}, 98$

8. $\sum_{k=1}^{3}(2k-3) = (2(1)-3) + (2(2)-3) + (2(3)-3) = -1 + 1 + 3 = 3$

9. $a_1 = -\frac{1}{9}, r = 3, n = 7;\ a_n = a_1r^{n-1} = -\frac{1}{9}(3)^{7-1} = -\frac{1}{9}(3)^6 = -\frac{1}{9}(729) = -81$

10. $a_1 = \frac{1}{27}, r = 3, n = 6;\ S_n = \dfrac{a_1 - a_1r^n}{1-r} = \dfrac{\frac{1}{27} - \frac{1}{27}(3)^6}{1-3} = \dfrac{\frac{1}{27} - \frac{1}{27}(729)}{-2} = \dfrac{-\frac{728}{27}}{-2} = \dfrac{364}{27}$

11. $a_1 = 3, a_4 = 648$
$a_n = a_1r^{n-1}$
$648 = 3r^{4-1}$
$216 = r^3$
$6 = r \Rightarrow 3, \boxed{18, 108}, 648$

12. $a_1 = 9, r = \dfrac{1}{3}$
$S_\infty = \dfrac{a_1}{1-r} = \dfrac{9}{1-\frac{1}{3}} = \dfrac{9}{\frac{2}{3}} = \dfrac{27}{2}$

13. $P(5,4) = \dfrac{5!}{(5-4)!} = \dfrac{5!}{1!} = 5! = 120$

14. $P(8,8) = \dfrac{8!}{(8-8)!} = \dfrac{8!}{0!} = 8! = 40{,}320$

15. $C(6,4) = \dfrac{6!}{4!2!} = \dfrac{6 \cdot 5 \cdot 4!}{4! \cdot 2 \cdot 1} = \dfrac{30}{2} = 15$

16. $C(8,3) = \dfrac{8!}{3!5!} = \dfrac{8 \cdot 7 \cdot 6 \cdot 5!}{3 \cdot 2 \cdot 1 \cdot 5!} = \dfrac{56 \cdot 6}{6} = 56$

17. $C(6,0) \cdot P(6,5) = \dfrac{6!}{0!6!} \cdot \dfrac{6!}{1!} = 1 \cdot 6!$
$= 720$

18. $P(8,7) \cdot C(8,7) = \dfrac{8!}{1!} \cdot \dfrac{8!}{7!1!} = 8! \cdot 8$
$= 322{,}560$

19. $\dfrac{P(6,4)}{C(6,4)} = \dfrac{\frac{6!}{2!}}{\frac{6!}{4!2!}} = \dfrac{6!}{2!} \cdot \dfrac{4!2!}{6!} = 4! = 24$

20. $\dfrac{C(9,6)}{P(6,4)} = \dfrac{\frac{9!}{6!3!}}{\frac{6!}{2!}} = \dfrac{84}{360} = \dfrac{7}{30}$

21. $\binom{7}{3} = \dfrac{7!}{3!4!} = 35$

22. $\binom{5}{1}\binom{4}{2} = 5 \cdot 6 = 30$

23. $\dfrac{1}{6}$

24. $\dfrac{8}{52} = \dfrac{2}{13}$

25. $\dfrac{\text{\# ways to get 5 hearts}}{\text{\# ways to get 5 cards}} = \dfrac{\binom{13}{5}}{\binom{52}{5}} = \dfrac{1287}{2{,}598{,}960} = \dfrac{33}{66{,}640}$

26. $\dfrac{\text{\# ways to get 2 heads}}{\text{\# ways to toss 5 times}} = \dfrac{\binom{5}{2}}{2^5} = \dfrac{20}{32} = \dfrac{5}{16}$

Cumulative Review Exercises (page 956)

1. $\begin{cases} 2x + y = 5 \\ x - 2y = 0 \end{cases}$

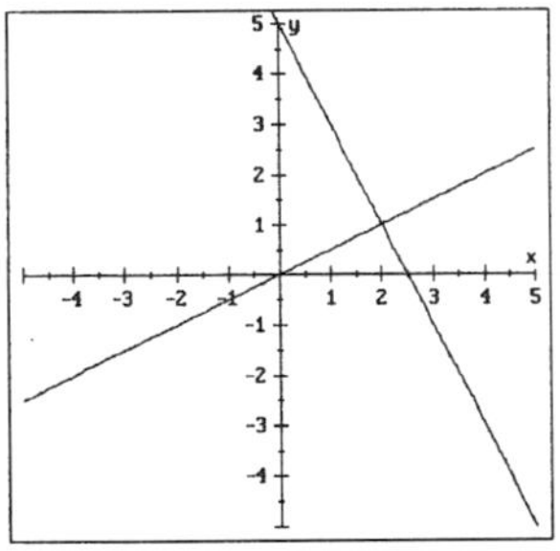

Solution: $(2, 1)$

2. $\begin{cases} (1) & 3x + y = 4 \\ (2) & 2x - 3y = -1 \end{cases}$

Substitute $y = 4 - 3x$ from (1) into (2):

$$\begin{aligned} 2x - 3y &= -1 \\ 2x - 3(4 - 3x) &= -1 \\ 2x - 12 + 9x &= -1 \\ 11x &= 11 \\ x &= 1 \end{aligned}$$

Substitute this and solve for y:

$y = 4 - 3x = 4 - 3(1) = 1$

Solution: $(1, 1)$

3.

$$\begin{array}{l} x + 2y = -2 \\ 2x - y = 6 \Rightarrow \times (2) \end{array} \quad \begin{array}{rcl} x + 2y &=& -2 \\ 4x - 2y &=& 12 \\ \hline 5x &=& 10 \\ x &=& 2 \end{array} \quad \begin{aligned} x + 2y &= -2 \\ 2 + 2y &= -2 \\ 2y &= -4 \\ y &= -2 \end{aligned}$$

Solution: $\boxed{(2, -2)}$

4.

$$\begin{array}{rcll} \dfrac{x}{10} + \dfrac{y}{5} &=& \dfrac{1}{2} \Rightarrow \times 10 & x + 2y = 5 \\ \dfrac{x}{2} - \dfrac{y}{5} &=& \dfrac{13}{10} \Rightarrow \times 10 & 5x - 2y = 13 \\ \hline & & & 6x = 18 \\ & & & x = 3 \end{array} \quad \begin{aligned} x + 2y &= 5 \\ 3 + 2y &= 5 \\ 2y &= 2 \\ y &= 1 \end{aligned}$$

$\Rightarrow$ Solution: $\boxed{(3, 1)}$

5. $\begin{vmatrix} 3 & -2 \\ 1 & -1 \end{vmatrix} = 3(-1) - (-2)(1) = -3 + 2 = -1$

6. $y = \dfrac{\begin{vmatrix} 4 & -1 \\ 3 & -7 \end{vmatrix}}{\begin{vmatrix} 4 & -3 \\ 3 & 4 \end{vmatrix}} = \dfrac{4(-7) - (-1)(3)}{4(4) - (-3)(3)} = \dfrac{-25}{25} = -1$

7.

$$\begin{array}{lr} (1) & x+y+z=1 \\ (2) & 2x-y-z=-4 \\ (3) & x-2y+z=4 \end{array} \qquad \begin{array}{lrcr} (1) & x+y+z & = & 1 \\ (2) & 2x-y-z & = & -4 \\ \hline (4) & 3x & = & -3 \\ & x & = & -1 \end{array} \qquad \begin{array}{lrcr} (2) & 2x-\ y-z & = & -4 \\ (3) & x-2y+z & = & 4 \\ \hline (5) & 3x-3y & = & 0 \end{array}$$

$$\begin{aligned} 3x-3y&=0 \\ 3(-1)-3y&=0 \\ -3-3y&=0 \\ -3y&=3 \\ y&=-1 \end{aligned} \qquad \begin{aligned} x+y+z&=1 \\ -1+(-1)+z&=1 \\ -2+z&=1 \\ z&=3 \end{aligned}$$

The solution is $(-1,-1,3)$.

8.

$$z=\frac{\begin{vmatrix}1&2&6\\3&2&6\\2&3&6\end{vmatrix}}{\begin{vmatrix}1&2&3\\3&2&1\\2&3&1\end{vmatrix}}=\frac{1\begin{vmatrix}2&6\\3&6\end{vmatrix}-2\begin{vmatrix}3&6\\2&6\end{vmatrix}+6\begin{vmatrix}3&2\\2&3\end{vmatrix}}{1\begin{vmatrix}2&1\\3&1\end{vmatrix}-2\begin{vmatrix}3&1\\2&1\end{vmatrix}+3\begin{vmatrix}3&2\\2&3\end{vmatrix}}=\frac{1(-6)-2(6)+6(5)}{1(-1)-2(1)+3(5)}=\frac{12}{12}=1$$

9. $\begin{cases}3x-2y<6\\y<-x+2\end{cases}$

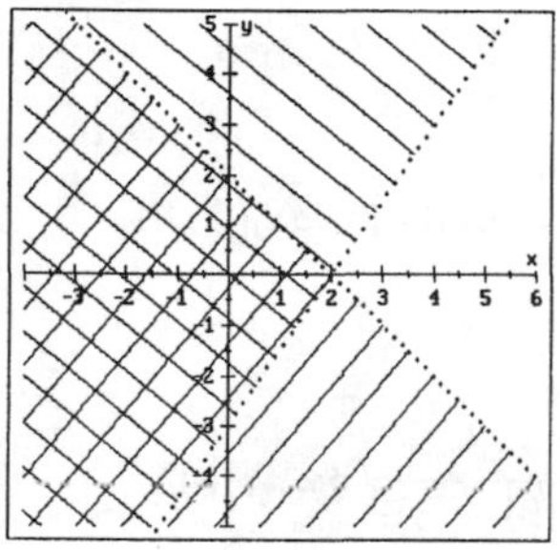

10. $\begin{cases}y<x+2\\3x+y\le 6\end{cases}$

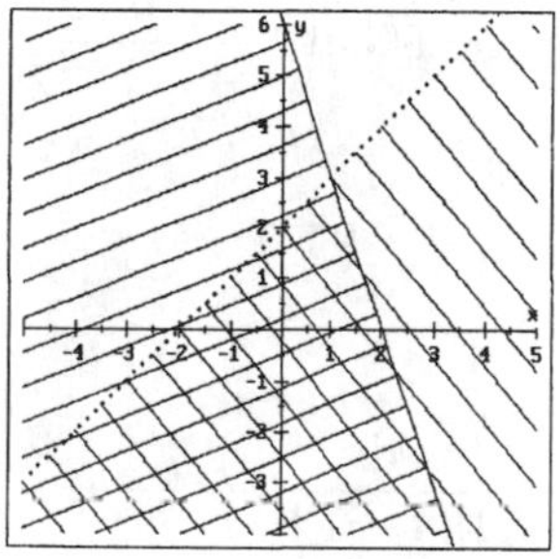

11. $y=\left(\frac{1}{2}\right)^x$

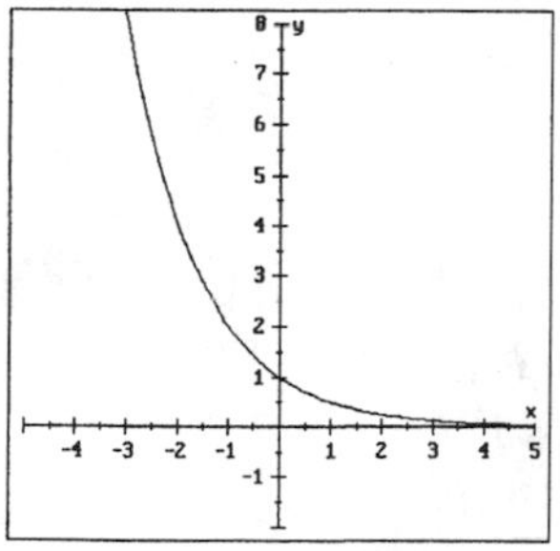

12. $y=\log_2 x \Rightarrow 2^y=x$

13. $\log_x 25=2 \Rightarrow x^2=25 \Rightarrow x=5$

14. $\log_5 125=x \Rightarrow 5^x=125 \Rightarrow x=3$

15. $\log_3 x=-3 \Rightarrow 3^{-3}=x \Rightarrow x=\frac{1}{27}$

16. $\log_5 x=0 \Rightarrow 5^0=x \Rightarrow x=1$

17. $y = 2^x$

18. x

19. $\log 98 = \log(14 \cdot 7) = \log 14 + \log 7 = 1.1461 + 0.8451 = 1.9912$

20. $\log 2 = \log \frac{14}{7} = \log 14 - \log 7 = 1.1461 - 0.8451 = 0.3010$

21. $\log 49 = \log 7^2 = 2 \log 7 = 2(0.8451) = 1.6902$

22. $\log \frac{7}{5} = \log \frac{14}{10} = \log 14 - \log 10 = 1.1461 - 1 = 0.1461$

23.
$$\begin{aligned} 2^{x+5} &= 3^x \\ \log 2^{x+2} &= \log 3^x \\ (x+5)\log 2 &= x \log 3 \\ x\log 2 + 5\log 2 &= x\log 3 \\ 5\log 2 &= x\log 3 - x\log 2 \\ 5\log 2 &= x(\log 3 - \log 2) \\ \frac{5\log 2}{\log 3 - \log 2} &= x \end{aligned}$$

24.
$$\begin{aligned} \log 5 + \log x - \log 4 &= 1 \\ \log \frac{5x}{4} &= 1 \\ 10^1 &= \frac{5x}{4} \\ 40 &= 5x \\ 8 &= x \end{aligned}$$

25. $A = P\left(1 + \frac{r}{k}\right)^{kt} = 9000\left(1 + \frac{-0.12}{1}\right)^{1(9)} \approx \$2{,}848.31$

26. $\log_6 8 = \dfrac{\log 8}{\log 6} \approx 1.16056$

27. $\dfrac{6!7!}{5!} = \dfrac{6 \cdot 5! \cdot 7!}{5!} = 6 \cdot 7! = 30{,}240$

28. $(3a - b)^4$
$$\begin{aligned} &= (3a)^4 + \frac{4!}{1!(4-1)!}(3a)^3(-b) + \frac{4!}{2!(4-2)!}(3a)^2(-b)^2 + \frac{4!}{3!(4-3)!}(3a)(-b)^3 + (-b)^4 \\ &= 81a^4 + \frac{4!}{1!3!}(-27a^3b) + \frac{4!}{2!2!}(9a^2b^2) + \frac{4!}{3!1!}(-3ab^3) + b^4 \\ &= 81a^4 - \frac{4 \cdot 3!}{1!3!}(27a^3b) + \frac{4 \cdot 3 \cdot 2!}{2! \cdot 2 \cdot 1}(9a^2b^2) - \frac{4 \cdot 3!}{3!1!}(3ab^3) + b^4 \\ &= 81a^4 - \frac{4}{1}(27a^3b) + \frac{12}{2}(9a^2b^2) - \frac{4}{1}(3ab^3) + b^4 \\ &= 81a^4 - 108a^3b + 54a^2b^2 - 12ab^3 + b^4 \end{aligned}$$

29. In the 7th term, the exponent on $-y$ is 6.
Variables: $(2x)^2(-y)^6 = 4x^2y^6$
Coef. $= \dfrac{n!}{r!(n-r)!} = \dfrac{8!}{6!2!} = 28$
Term $= 28(4x^2y^6) = 112x^2y^6$

30.
$$\begin{aligned} a_1 &= -11, d = 6, n = 20 \\ a_n &= a_1 + (n-1)d \\ &= -11 + (19)(6) \\ &= -11 + 114 = 103 \end{aligned}$$

CUMULATIVE REVIEW EXERCISES

31. $a_1 = 6, d = 3, n = 20;\ a_n = a_1 + (n-1)d = 6 + (20-1)(3) = 6 + 19(3) = 63$

$$S_n = \frac{n(a_1 + a_n)}{2} = \frac{20(6+63)}{2} = \frac{20(69)}{2} = 690$$

32. $a_1 = -3;\ a_4 = 30:$

$a_n = a_1 + (n-1)d$

$30 = -3 + (4-1)d$

$33 = 3d$

$11 = d \Rightarrow -3, \boxed{8, 19}, 30$

33. $\sum_{k=1}^{3} 3k^2 = 3(1)^2 + 3(2)^2 + 3(3)^2$

$= 3 + 12 + 27 = 42$

34. $\sum_{k=3}^{5} (2k+1) = (2(3)+1) + (2(4)+1) + (2(5)+1) = 7 + 9 + 11 = 27$

35. $a_1 = \frac{1}{27}, r = 3, n = 7;\ a_n = a_1 r^{n-1} = \frac{1}{27}(3)^{7-1} = \frac{1}{27}(3)^6 = \frac{1}{27}(729) = 27$

36. $a_1 = \frac{1}{64}, r = 2, n = 10;\ S_n = \frac{a_1 - a_1 r^n}{1-r} = \frac{\frac{1}{64} - \frac{1}{64}(2)^{10}}{1-2} = \frac{\frac{1}{64} - \frac{1}{64}(1024)}{-1} = \frac{-\frac{1023}{64}}{-1} = \frac{1023}{64}$

37. $a_1 = -3,\ a_4 = 192$

$a_n = a_1 r^{n-1}$

$192 = -3r^{4-1}$

$-64 = r^3$

$-4 = r \Rightarrow -3, \boxed{12, -48}, 192$

38. $a_1 = 9, r = \frac{1}{3}$

$$S_\infty = \frac{a_1}{1-r} = \frac{9}{1-\frac{1}{3}} = \frac{9}{\frac{2}{3}} = \frac{27}{2}$$

39. $P(9,3) = \frac{9!}{(9-3)!} = \frac{9!}{6!} = \frac{9 \cdot 8 \cdot 7 \cdot 6!}{6!} = 9 \cdot 8 \cdot 7 = 504$

40. $C(7,4) = \frac{7!}{4!(7-4)!} = \frac{7!}{4!3!} = \frac{7 \cdot 6 \cdot 5 \cdot 4!}{4! \cdot 3 \cdot 2 \cdot 1} = \frac{210}{6} = 35$

41. $\frac{C(8,4)C(8,0)}{P(6,2)} = \frac{\frac{8!}{4!4!} \cdot \frac{8!}{0!8!}}{\frac{6!}{4!}} = \frac{70 \cdot 1}{30} = \frac{7}{3}$

42. $C(n,n) = 1$ is smaller than $P(n,n) = n!$.

43. $7! = 5{,}040$

44. $\binom{9}{3} = \frac{9!}{3!6!} = 84$

45. $\frac{12}{52} = \frac{3}{13}$

Appendix 1 (page A-4)

1. $y = x^2 - 1$

x-axis	y-axis	origin
$-y = x^2 - 1$	$y = (-x)^2 - 1$	$-y = (-x)^2 - 1$
not equivalent: no symmetry	$y = x^2 - 1$	$-y = x^2 - 1$
	equivalent: $\boxed{\text{symmetry}}$	not equivalent: no symmetry

3. $y = x^5$

x-axis	y-axis	origin
$-y = x^5$	$y = (-x)^3$	$-y = (-x)^3$
not equivalent: no symmetry	$y = -x^5$	$-y = -x^5$
	not equivalent: no symmetry	$y = x^5$
		equivalent: $\boxed{\text{symmetry}}$

5. $y = -x^2 + 2$

x-axis	y-axis	origin
$-y = -x^2 + 2$	$y = (-x)^2 + 2$	$-y = (-x)^2 + 2$
not equivalent: no symmetry	$y = -x^2 + 2$	$-y = x^2 + 2$
	equivalent: $\boxed{\text{symmetry}}$	not equivalent: no symmetry

7. $y = x^2 - x$

x-axis	y-axis	origin
$-y = x^2 - x$	$y = (-x)^2 - (-x)$	$-y = (-x)^2 - (-x)$
not equivalent: no symmetry	$y = x^2 + x$	$-y = x^2 + x$
	not equivalent: no symmetry	not equivalent: no symmetry

9. $y = -\lvert x + 2\rvert$

x-axis	y-axis	origin
$-y = -\lvert x + 2\rvert$	$y = -\lvert -x + 2\rvert$	$-y = -\lvert -x + 2\rvert$
not equivalent: no symmetry	not equivalent: no symmetry	not equivalent: no symmetry

11. $\lvert y\rvert = x$

x-axis	y-axis	origin
$\lvert -y\rvert = x$	$\lvert y\rvert = -x$	$\lvert -y\rvert = -x$
$\lvert -1\rvert\lvert y\rvert = x$	not equivalent: no symmetry	$\lvert -1\rvert\lvert y\rvert = -x$
$\lvert y\rvert = x$		$\lvert y\rvert = -x$
equivalent: $\boxed{\text{symmetry}}$		not equivalent: no symmetry

13. $y = x^4 - 4$

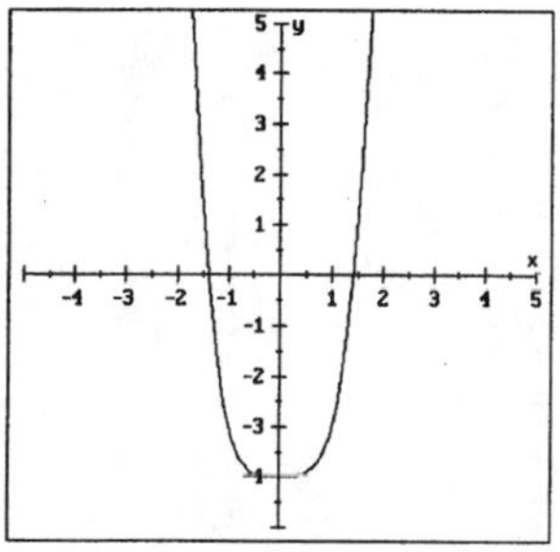

D $(-\infty, \infty)$; R $[-4, \infty)$

15. $y = -x^3$

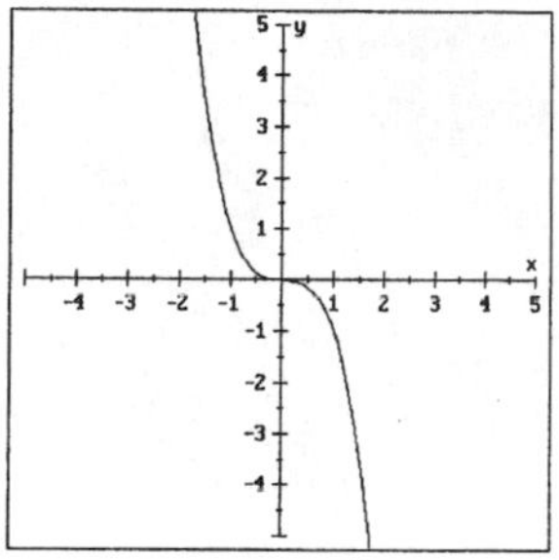

D $(-\infty, \infty)$; R $(-\infty, \infty)$

17. $y = x^4 + x^2$

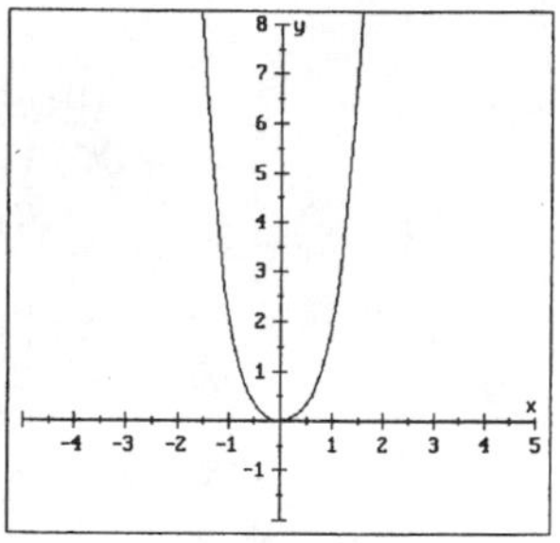

D $(-\infty, \infty)$; R $[0, \infty)$

19. $y = x^3 - x$

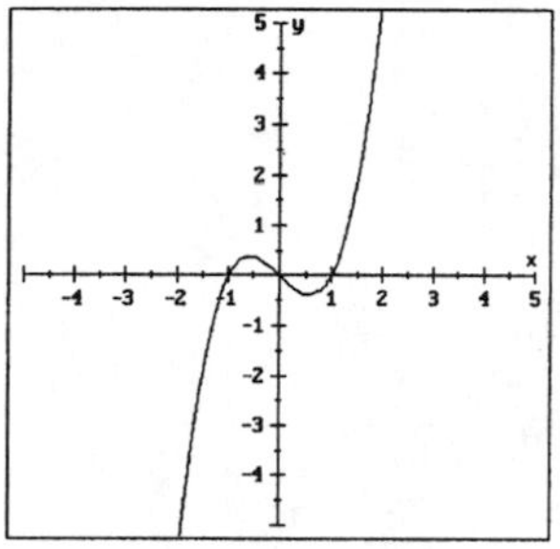

D $(-\infty, \infty)$; R $(-\infty, \infty)$

21. $y = \dfrac{1}{2}|x| - 1$

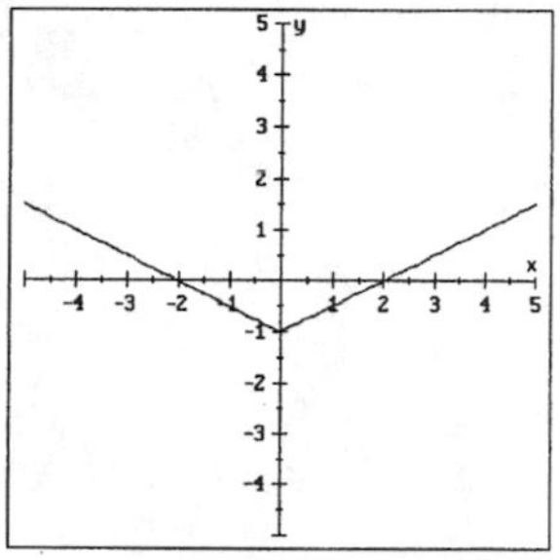

D $(-\infty, \infty)$; R $[-1, \infty)$

23. $y = -|x + 2|$

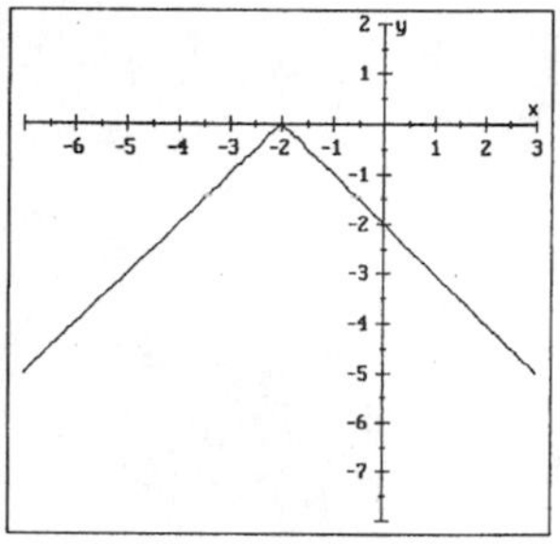

D $(-\infty, \infty)$; R $(-\infty, 0]$